NONLINEAR SYSTEMS

Second Edition

Hassan K. Khalil
Michigan State University

Prentice Hall, Upper Saddle River, NJ 07458

Library of Congress Cataloging-in-Publication Data

Khalil, Hassan K.
    Nonlinear systems / Hassan K. Khalil.--[2nd ed.]
        p.   cm.
    Includes index
    ISBN 0-13-228024-8
    1. Nonlinear theories.   I . Title.
QA427.K48   1996
003' . 75--dc20   95-45804
                                                        CIP

Acquisitions editor: ALICE DWORKIN
Editorial/production supervision
  and interior design: SHARYN VITRANO
Cover designer: BRUCE KENSELAAR
Manufacturing buyer: DONNA SULLIVAN
Supplements editor: BARBARA MURRAY

©1996 by Prentice-Hall, Inc.
Simon & Schuster/ A Viacom Company
Upper Saddle River, New Jersey 07458

The author and publisher of this book have used their best efforts in preparing this book. These efforts include the development, research, and testing of the theories and programs to determine their effectiveness. The author and publisher make no warranty of any kind, expressed or implied, with regard to these programs or the documentation contained in this book. The author and publisher shall not be liable in any event for incidental or consequential dam in connection with, or arising out of, the furnishing, performance, or use of these programs.

Printed in the United States of America

10 9 8 7 6 5 4 3 2 1

0-13-228024-8

Prentice-Hall International (UK) Limited, London
Prentice-Hall of Australia Pty. Limited, Sydney
Prentice-Hall Canada Inc., Toronto
Prentice-Hall Hispanoamericana, S.A., Mexico
Prentice-Hall of India Private Limited, New Delhi
Prentice-Hall of Japan, Inc., Tokyo
Simon & Schuster Asia Pte. Ltd., Singapore
Editora Prentice-Hall do Brasil, Ltda., Rio de Janeiro

To
Petar V. Kokotović

Preface

This text is intended for a first-year graduate-level course on nonlinear systems. It may also be used for self-study or reference by engineers and applied mathematicians. It is an outgrowth of my experience in teaching the nonlinear systems course at Michigan State University, East Lansing, several times since 1985. Students taking this course have had background in electrical engineering, mechanical engineering, or applied mathematics. The prerequisite for the course is a graduate-level course in linear systems, taught at the level of the texts by Chen [29], DeCarlo [41], Kailath [86], or Rugh [142]. In writing this text, I have assumed the linear systems prerequisite which allowed me not to worry about introducing the concept of a "state" and to refer freely to "transfer functions," "state transition matrices," and other linear system concepts. As for the mathematics background, I have assumed the usual level of calculus, differential equations, and matrix theory that any graduate student in engineering or mathematics would have. In one section, I have collected a few mathematical facts which are used throughout the book. I do not assume that the student has previous knowledge of this material but I expect, with the level of maturity of graduate students, that with the brief account of these facts provided here they should be able to read through the text without difficulty.

I have written the text in such a way that the level of mathematical sophistication builds up as we advance from chapter to chapter. This is why the first chapter is written in an elementary context. Actually this chapter could be taught at senior, or even junior, level courses without difficulty. This is also the reason I have split the treatment of Lyapunov stability into two parts. In Sections 3.1 to 3.3, I introduce the essence of Lyapunov stability for autonomous systems where I do not have to worry about technicalities such as uniformity, class $\mathcal{K}$ functions, etc. In Sections 3.4 to 3.6, I present Lyapunov stability in a more general setup that accommodates nonautonomous systems and allows for a deeper look into advanced aspects of stability theory. The level of mathematical sophistication at the end of Chapter 3 is the level that I like to bring the students to, so that they can comfortably read chapters 4 to 13.

There is yet a higher level of mathematical sophistication that is assumed in writing the proofs given in Appendix A. The proofs collected in this appendix are not intended for classroom use. They are included to make the text self contained on one hand, and to respond to the need or desire of some students to read these proofs, like students continuing on to conduct Ph.D. research into nonlinear systems or control theory. Those students can continue to read Appendix A in a self-study manner. By the higher level of sophistication in Appendix A, I do not mean more background knowledge; rather I refer to the writing style. A derivation step that would usually take three or four lines to describe in the main text, would probably be done in one line in the appendix. Having said that, let me add that many of the proofs in Appendix A are, in my opinion, simpler to read than their write-ups in the original references.

The second edition of the book is organized into thirteen chapters. The first nine chapters deal with nonlinear systems analysis. Chapter 10 is also an analysis chapter, but deals with feedback systems. The last three chapter deal with the design of feedback control. Chapter 1 starts with introductory remarks on nonlinear phenomena which define the scope of the text. These remarks are followed by two sections. Section 1.1 introduces examples of nonlinear state equations which are used later in the book to illustrate the various results and tools covered. In teaching this material, I have found it beneficial to get the students to appreciate the versatility of the nonlinear state equation before we overwhelm them with the technical results for that equation. Section 1.2 deals with classical material on second-order systems. I have used this material to introduce the reader to the two nonlinear phenomena which are studied rigorously later in the book; namely, multiple equilibria and their stability properties, and limit cycles.

Chapter 2 starts with a section of mathematical preliminaries. The contraction mapping theorem of Section 2.1.4 is used a few times in Appendix A, but only once in the main text (in the proof of Theorem 2.2). If the instructor chooses to skip the proof of Theorem 2.2, then the contraction mapping theorem might be skipped. Notice, however, that normed linear spaces, introduced in Section 2.1.4, are used in Chapter 6. Sections 2.2–2.5 contain the fundamental theory of ordinary differential equations, which should be treated, at least briefly without proofs.

Chapter 3 presents the Lyapunov theory. It is the core chapter of the book and should be thoroughly covered. Section 3.6 contains converse Lyapunov theorems which are referenced several times later on. It is important to mention these theorems in the classroom. Notice, however, that both Sections 3.5 and 3.6 can be taught very briefly since their proofs can be left as reading assignments.

Beyond Chapter 3, all chapters are designed to be basically independent of each other, to give the instructor flexibility in shaping his/her course. I list here the exceptions to this general rule. The invariance theorems of Section 4.3 are used in Section 13.4 on adaptive control. Sections 5.1 to 5.4 are used in Sections 6.2, 6.3, 6.4, 12.3, 13.1, 13.2, and 13.3, and in a few examples at the end of Section 13.4. I strongly recommend coverage of Sections 5.1 to 5.4. They deal with the important topic of stability in the presence of perturbations. Despite the importance of this topic, its coverage in introductory books on Lyapunov stability has been limited, if at all touched. To be introduced to this important aspect of Lyapunov theory, students had to consult advanced stability monographs, such as Hahn's [62] and Krasovskii's [96]. Section 5.7 on slowly varying systems is beneficial for Chapter 9 on singular perturbations and Section 11.3 on gain scheduling, but it is not required. The notion of $\mathcal{L}$-stability, introduced in Section 6.1, is used in Sections 10.2 and 10.3. Familiarity with the singular perturbation theory of Chapter 9 is needed in Section 11.3, but no technical results from Chapter 9 are explicitly used. Sections 11.1 and 11.2 should be included in any coverage of the last three chapters. Feedback linearizable systems, introduced in Sections 12.1 and 12.2, are used in several examples in Sections 13.1 to 13.3, but they are not needed for the main ideas of these sections.

Every chapter from 1 to 13 ends with a section of exercises which are designed to help the reader understand and apply the results and tools covered. Many of these exercises have been used in homework assignments and tests in my class over the years. A solution manual is available from the publisher.

It is my pleasure to acknowledge all those who helped me in the writing of this book. I would like, first of all, to thank my wife Amina for her encouragement and complete support throughout this project. I would like also to thank my sons Mohammad and Omar for their understanding and moral support. I am especially appreciative of the discussions I had with Petar Kokotović at different stages of this project. His remarks and suggestions were very valuable. Thanks are also due to Eyad Abed, Nazir Atassi, Martin Corless, Ahmed Dabroom, Farzad Esfandiari, Jessy Grizzle, Suhada Jayasuriya, Rolf Johansson, Ioannis Kanellakopoulos, Mustafa Khammash, P.S. Krishnaprasad, Miroslav Krstić, Zongli Lin, Semyon Meerkov, Jack Rugh, Fathi Salam, Eduardo Sontag, Mark Spong, and Steve Yurkovich, for their suggestions, constructive comments, feedback on the first edition, and/or reviewing parts of the new material of the second edition. I would like, in particular, to acknowledge Semyon Meerkov for his comprehensive feedback on the first edition and his suggestion to reorganize the book into its new form. The students in my Nonlinear Systems class at Michigan State University have provided me with excellent feedback over the years. Finally, I would like to thank Ioannis Kanellakopoulos and Semyon Meerkov for reviewing the full second edition manuscript for Prentice Hall.

I am grateful to Michigan State University for an environment that allowed me to write this book, and to the National Science Foundation for supporting my research on singular perturbation and nonlinear feedback control; from this research endeavor I gained the experience that is necessary for writing such a book.

I am indebted to Jiansheng Hou, Sharyn Vitrano, and Elias Strangas for their help in producing the book. Jiansheng generated all the figures of the first edition, many of which remain in the second one, Sharyn did an excellent job as the copy editor of the second edition, and Elias applied his LaTeX expertise to help me at critical times.

The book was typeset using LATEX. All computations, including numerical solution of differential equations, were done using MATLAB. The figures were generated using FrameMaker, MATLAB, or the graphics tool of LATEX.

Hassan Khalil
East Lansing, Michigan

Contents

Chapter 1

Introduction

When engineers analyze and design nonlinear dynamical systems in electrical circuits, mechanical systems, control systems, and other engineering disciplines, they need to absorb and digest a wide range of nonlinear analysis tools. In this book, we introduce some of the these tools. In particular, we present tools for stability analysis of nonlinear systems, with an emphasis on Lyapunov's method. We give special attention to the stability of feedback systems. We present tools for the detection and analysis of "free" oscillations, including the describing function method. We introduce the asymptotic tools of perturbation theory, including averaging and singular perturbations. Finally, we introduce nonlinear feedback control tools, including linearization, gain scheduling, exact feedback linearization, Lyapunov redesign, backstepping, sliding mode control, and adaptive control.

We shall deal with dynamical systems which are modeled by a finite number of coupled first-order ordinary differential equations

$$
\begin{aligned}
\dot{x}_1 &= f_1(t, x_1, \ldots, x_n, u_1, \ldots, u_p) \\
\dot{x}_2 &= f_2(t, x_1, \ldots, x_n, u_1, \ldots, u_p) \\
&\vdots \qquad \vdots \\
\dot{x}_n &= f_n(t, x_1, \ldots, x_n, u_1, \ldots, u_p)
\end{aligned}
$$

where $\dot{x}_i$ denotes the derivative of x_i with respect to the time variable t, and u_1, u_2, ..., u_p are specified input variables. We call the variables x_1, x_2, ..., x_n the state variables. They represent the memory that the dynamical system has of its past. We usually use vector notation to write the above equations in a compact

1

form. Define

$$
x = \begin{bmatrix} x_1 \\ x_2 \\ \vdots \\ \vdots \\ x_n \end{bmatrix}, \quad u = \begin{bmatrix} u_1 \\ u_2 \\ \vdots \\ u_p \end{bmatrix}, \quad f(t, x, u) = \begin{bmatrix} f_1(t, x, u) \\ f_2(t, x, u) \\ \vdots \\ \vdots \\ f_n(t, x, u) \end{bmatrix}
$$

and rewrite the n first-order differential equations as one n-dimensional first-order vector differential equation

$$
\dot{x} = f(t, x, u) \tag{1.1}
$$

We call (1.1) the state equation, and refer to x as the *state* and u as the *input*. Sometimes, we associate with (1.1) another equation

$$
y = h(t, x, u) \tag{1.2}
$$

which defines a q-dimensional *output* vector that comprises variables of particular interest in the analysis of the dynamical system, like variables which can be physically measured or variables which are required to behave in a specified manner. We call (1.2) the output equation and refer to equations (1.1) and (1.2) together as the state-space model, or simply the state model. Mathematical models of finite-dimensional physical systems do not always come in the form of a state-space model. However, more often than not, we can model physical systems in this form by carefully choosing the state variables. Examples and exercises that will appear later in this chapter will demonstrate the versatility of the state-space model.

Most of our analysis in this book will deal with the state equation, and many times without explicit presence of an input u, that is, the so-called unforced state equation

$$
\dot{x} = f(t, x) \tag{1.3}
$$

Working with an unforced state equation does not necessarily mean that the input to the system is zero. It could be that the input has been specified as a given function of time, $u = \gamma(t)$, a given feedback function of the state, $u = \gamma(x)$, or both, $u = \gamma(t, x)$. Substitution of $u = \gamma$ in (1.1) eliminates u and yields an unforced state equation.

A special case of (1.3) arises when the function f does not depend explicitly on t; that is,

$$
\dot{x} = f(x) \tag{1.4}
$$

in which case the system is said to be *autonomous* or *time invariant*. The behavior of an autonomous system is invariant to shifts in the time origin, since changing the time variable from t to $\tau = t - a$ does not change the right-hand side of the state equation. If the system is not autonomous, then it is called *nonautonomous* or *time varying*.

An important concept in dealing with the state equation is the concept of an equilibrium point. A point $x = x^*$ in the state space is said to be an equilibrium point of (1.3) if it has the property that whenever the state of the system starts at x^* it will remain at x^* for all future time. For the autonomous system (1.4), the equilibrium points are the real roots of the equation

$$f(x) = 0$$

An equilibrium point could be isolated; that is, there are no other equilibrium points in its vicinity, or there could be a continuum of equilibrium points.

For linear systems, the state-space model (1.1)–(1.2) takes the special form

$$\dot{x} = A(t)x + B(t)u$$
$$y = C(t)x + D(t)u$$

We assume that the reader is familiar with the powerful analysis tools for linear systems, founded on the basis of the *superposition principle*. As we move from linear to nonlinear systems, we are faced with a more difficult situation. The superposition principle does not hold any longer, and analysis tools involve more advanced mathematics. Because of the powerful tools we know for linear systems, the first step in analyzing a nonlinear system is usually to linearize it, if possible, about some nominal operating point and analyze the resulting linear model. This is a common practice in engineering, and it is a useful one. There is no question that, whenever possible, we should make use of linearization to learn as much as we can about the behavior of a nonlinear system. However, linearization alone will not be sufficient; we must develop tools for the analysis of nonlinear systems. There are two basic limitations of linearization. First, since linearization is an approximation in the neighborhood of an operating point, it can only predict the "local" behavior of the nonlinear system in the vicinity of that point. It cannot predict the "nonlocal" behavior far from the operating point, and certainly not the "global" behavior throughout the state space. Second, the dynamics of a nonlinear system are much richer than the dynamics of a linear system. There are "essentially nonlinear phenomena" that can take place only in the presence of nonlinearity; hence, they cannot be described or predicted by linear models. Examples of essentially nonlinear phenomena are:

- *Finite escape time*: The state of an unstable linear system goes to infinity as time approaches infinity; a nonlinear system's state, however, can go to infinity in finite time.

- *Multiple isolated equilibria*: A linear system can have only one isolated equilibrium point; hence, it can have only one steady-state operating point which attracts the state of the system irrespective of the initial state. A nonlinear system can have more than one isolated equilibrium point. The state may converge to one of several steady-state operating points, depending on the initial state of the system.

- *Limit cycles*: For a linear time-invariant system to oscillate, it must have a pair of eigenvalues on the imaginary axis, which is a nonrobust condition that is almost impossible to maintain in the presence of perturbations. Even if we do, the amplitude of oscillation will be dependent on the initial state. In real life, stable oscillation must be produced by nonlinear systems. There are nonlinear systems which can go into an oscillation of fixed amplitude and frequency, irrespective of the initial state. This type of oscillation is known as a limit cycle.

- *Subharmonic, harmonic, or almost-periodic oscillations*: A stable linear system under a periodic input produces an output of the same frequency. A nonlinear system under periodic excitation can oscillate with frequencies which are submultiples or multiples of the input frequency. It may even generate an almost-periodic oscillation, an example of which is the sum of periodic oscillations with frequencies which are not multiples of each other.

- *Chaos*: A nonlinear system can have a more complicated steady-state behavior that is not equilibrium, periodic oscillation, or almost-periodic oscillation. Such behavior is usually referred to as chaos. Some of these chaotic motions exhibit randomness, despite the deterministic nature of the system.

- *Multiple modes of behavior*: It is not unusual for two or more modes of behavior to be exhibited by the same nonlinear system. An unforced system may have more than one limit cycle. A forced system with periodic excitation may exhibit harmonic, subharmonic, or more complicated steady-state behavior, depending upon the amplitude and frequency of the input. It may even exhibit a discontinuous jump in the mode of behavior as the amplitude or frequency of the excitation is smoothly changed.

In this book, we shall encounter only the first three of these phenomena.[1] Multiple equilibria and limit cycles will be introduced later in this chapter, as we examine second-order autonomous systems, while the phenomenon of finite escape time will be introduced in Chapter 2.

---

[1] To read about forced oscillation, chaos, bifurcation, and other important topics, the reader may consult [59], [64], and [183].

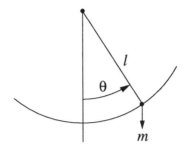

Figure 1.1: Pendulum.

1.1 Examples

1.1.1 Pendulum Equation

Consider the simple pendulum shown in Figure 1.1, where l denotes the length of
the rod and m denotes the mass of the bob. Assume the rod is rigid and has zero
mass. Let θ denote the angle subtended by the rod and the vertical axis through
the pivot point. The pendulum is free to swing in the vertical plane. The bob of
the pendulum moves in a circle of radius l. To write the equation of motion of
the pendulum, let us identify the forces acting on the bob. There is a downward
gravitational force equal to mg, where g is the acceleration due to gravity. There
is also a frictional force resisting the motion, which we assume to be proportional
to the speed of the bob with a coefficient of friction k. Using Newton's second law
of motion, we can write the equation of motion in the tangential direction as

$$ml\ddot{\theta} = -mg\sin\theta - kl\dot{\theta}$$

Writing the equation of motion in the tangential direction has the advantage that
the rod tension, which is in the normal direction, does not appear in the equation.
Note that we could have arrived at the same equation by writing the moment
equation about the pivot point. To obtain a state-space model of the pendulum, let
us take the state variables as $x_1 = \theta$ and $x_2 = \dot{\theta}$. Then, the state equation is

$$\dot{x}_1 = x_2 \tag{1.5}$$

$$\dot{x}_2 = -\frac{g}{l}\sin x_1 - \frac{k}{m}x_2 \tag{1.6}$$

To find the equilibrium points, we set $\dot{x}_1 = \dot{x}_2 = 0$ and solve for x_1 and x_2

$$0 = x_2$$

$$0 = -\frac{g}{l}\sin x_1 - \frac{k}{m}x_2$$

The equilibrium points are located at $(n\pi, 0)$, for $n = 0, \pm 1, \pm 2, \ldots$. From the physical description of the pendulum, it is clear that the pendulum has only two equilibrium positions corresponding to the equilibrium points $(0,0)$ and $(\pi, 0)$. Other equilibrium points are repetitions of these two positions which correspond to the number of full swings the pendulum would make before it rests at one of the two equilibrium positions. For example, if the pendulum makes m complete 360° revolutions before it rests at the downward vertical position, then mathematically we say that the pendulum approaches the equilibrium point $(2m\pi, 0)$. In our investigation of the pendulum, we will limit our attention to the two "nontrivial" equilibrium points at $(0,0)$ and $(\pi, 0)$. Physically, we can see that these two equilibrium positions are quite distinct from each other. While the pendulum can indeed rest at the $(0,0)$ equilibrium point, it can hardly maintain rest at the $(\pi, 0)$ equilibrium point because infinitesimally small disturbance from that equilibrium will take the pendulum away. The difference between the two equilibrium points is in their stability properties, a topic we shall study in some depth.

Sometimes it is instructive to consider a version of the pendulum equation where the frictional resistance is neglected by setting $k = 0$. The resulting equation

$$\dot{x}_1 \;=\; x_2 \tag{1.7}$$

$$\dot{x}_2 \;=\; -\frac{g}{l}\sin x_1 \tag{1.8}$$

is conservative in the sense that if the pendulum is given an initial push, it will keep oscillating forever with a nondissipative energy exchange between kinetic and potential energies. This, of course, is not realistic but gives insight into the behavior of the pendulum. It may also help in finding approximate solutions of the pendulum equation when the friction coefficient k is small. Another version of the pendulum equation arises if we can apply a torque T to the pendulum. This torque may be viewed as a control input in the equation

$$\dot{x}_1 \;=\; x_2 \tag{1.9}$$

$$\dot{x}_2 \;=\; -\frac{g}{l}\sin x_1 - \frac{k}{m}x_2 + \frac{1}{ml^2}T \tag{1.10}$$

Interestingly enough, several unrelated physical systems are modeled by equations similar to the pendulum equation. Such examples are the model of a synchronous generator connected to an infinite bus (Exercise 1.7), the model of a Josephson junction circuit (Exercise 1.8), and the model of a phase-locked loop (Exercise 1.10). Consequently, the pendulum equation is of great practical importance.

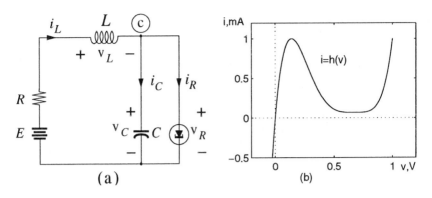

Figure 1.2: (a) Tunnel diode circuit; (b) Tunnel diode $v_R - i_R$ characteristic.

1.1.2 Tunnel Diode Circuit

Consider the tunnel diode circuit shown in Figure 1.2,[2] where the tunnel diode is characterized by $i_R = h(v_R)$. The energy-storing elements in this circuit are the capacitor C and the inductor L. Assuming they are linear and time-invariant, we can model them by the equations

$$i_C = C\frac{dv_C}{dt}$$
$$v_L = L\frac{di_L}{dt}$$

where i and v are the current through and the voltage across an element, with the subscript specifying the element. To write a state-space model for the system, let us take $x_1 = v_C$ and $x_2 = i_L$ as the state variables, and $u = E$ as a constant input. To write the state equation for x_1, we need to express i_C as a function of the state variables x_1, x_2 and the input u. Using Kirchhoff's current law, we can write an equation that the algebraic sum of all currents leaving node ⓒ is equal to zero:

$$i_C + i_R - i_L = 0$$

Hence,

$$i_C = -h(x_1) + x_2$$

Similarly, we need to express v_L as a function of the state variables x_1, x_2 and the input u. Using Kirchhoff's voltage law, we can write an equation that the algebraic sum of all voltages across elements in the left loop is equal to zero:

$$v_C - E + Ri_L + v_L = 0$$

---

[2] This figure, as well as Figures 1.3 and 1.6, are taken from [33].

Hence,

$$v_L = -x_1 - Rx_2 + u$$

We can now write the state-space model for the circuit as

$$\dot{x}_1 = \frac{1}{C}[-h(x_1) + x_2] \tag{1.11}$$

$$\dot{x}_2 = \frac{1}{L}[-x_1 - Rx_2 + u] \tag{1.12}$$

The equilibrium points of the system are determined by setting $\dot{x}_1 = \dot{x}_2 = 0$ and solving for x_1 and x_2

$$0 = -h(x_1) + x_2$$
$$0 = -x_1 - Rx_2 + u$$

Therefore, the equilibrium points correspond to the roots of the equation

$$h(x_1) = \frac{E}{R} - \frac{1}{R}x_1$$

Figure 1.3 shows graphically that for certain values of E and R this equation has three isolated roots which correspond to three isolated equilibrium points of the system. The number of equilibrium points might change as the values of E and R change. For example, if we increase E for the same value of R we will reach a point beyond which only the point Q_3 will exist. On the other hand, if we decrease E for the same value of R, we will end up with the point Q_1 as the only equilibrium. Suppose we are in the multiple equilibria situation: Which of these equilibrium points can we observe in an experimental setup of this circuit? The answer depends on the stability properties of the equilibrium points. We will come back to this example later in the chapter and answer this question.

1.1.3   Mass-Spring System

In the mass-spring mechanical system, shown in Figure 1.4, we consider a mass m sliding on a horizontal surface and attached to a vertical surface through a spring. The mass is subjected to an external force F. We define y as the displacement from a reference position and write Newton's law of motion

$$m\ddot{y} + F_f + F_{sp} = F$$

where F_f is a resistive force due to friction and F_{sp} is the restoring force of the spring. We assume that F_{sp} is a function only of the displacement y and write it as $F_{sp} = g(y)$. We assume also that the reference position has been chosen such that

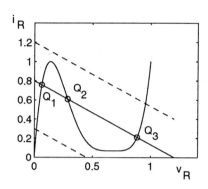

Figure 1.3: Equilibrium points of the tunnel diode circuit.

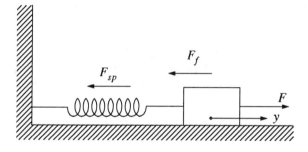

Figure 1.4: Mass-spring mechanical system.

$g(0) = 0$. The external force F is at our disposal. Depending upon F, F_f, and g, several interesting autonomous and nonautonomous second-order models arise.

For a relatively small displacement, the restoring force of the spring can be modeled as a linear function $g(y) = ky$, where k is the spring constant. For a large displacement, however, the restoring force may depend nonlinearly on y. For example, the function

$$g(y) = k(1 - a^2y^2)y, \quad |ay| < 1$$

models the so-called *softening spring* where, beyond a certain displacement, a large displacement increment produces a small force increment. On the other hand, the function

$$g(y) = k(1 + a^2y^2)y$$

models the so-called *hardening spring* where, beyond a certain displacement, a small displacement increment produces a large force increment.

As the mass moves in the air (a viscous medium) there will be a frictional force due to viscosity. This force is usually modeled as a nonlinear function of the velocity; that is, $F_v = h(\dot{y})$ where $h(0) = 0$. For small velocity, we can assume that $F_v = c\dot{y}$. The combination of a hardening spring, linear viscous damping, and a periodic external force $F = A\cos\omega t$ results in the Duffing's equation

$$m\ddot{y} + c\dot{y} + ky + ka^2y^3 = A\cos\omega t \qquad (1.13)$$

which is a classical example in the study of periodic excitation of nonlinear systems.

As another example of a frictional force, we consider the so-called *dry* or *Coulomb friction*. This type of damping arises when the mass slides on a dry surface. When the mass is at rest, there is a static friction force F_s that acts parallel to the surface and is limited to $\pm\mu_s mg$, where $0 < \mu_s < 1$ is the static friction coefficient. This force takes whatever value, between its limits, to keep the mass at rest. For motion to begin, there must be a force acting on the mass to overcome the resistance to motion caused by static friction. In the absence of an external force, $F = 0$, the static friction force will balance the restoring force of the spring and maintain equilibrium for $|g(y)| \leq \mu_s mg$. Once motion has started, there is a slipping friction force of magnitude $\mu_k mg$ that resists the motion of the mass, where μ_k is the kinetic friction coefficient, which we assume to be constant. As an idealization of the dry friction force, we may define it as

$$F_d = \left\{ \begin{array}{lll} -\mu_k mg, & \text{for} & \dot{y} < 0 \\ F_s, & \text{for} & \dot{y} = 0 \\ \mu_k mg, & \text{for} & \dot{y} > 0 \end{array} \right.$$

The combination of a linear spring, linear viscous damping, dry friction, and zero external force results in

$$m\ddot{y} + ky + c\dot{y} + \eta(y, \dot{y}) = 0$$

where

$$\eta(y, \dot{y}) = \left\{ \begin{array}{lll} \mu_k mg\, \text{sign}(\dot{y}), & \text{for} & |\dot{y}| > 0 \\ -ky, & \text{for} & \dot{y} = 0 \ \text{ and } \ |y| \leq \mu_s mg/k \\ -\mu_s mg\, \text{sign(y)}, & \text{for} & \dot{y} = 0 \ \text{ and } \ |y| > \mu_s mg/k \end{array} \right.$$

The value of $\eta(y, \dot{y})$ for $\dot{y} = 0$ and $|y| \leq \mu_s mg/k$ is obtained from the equilibrium condition $\ddot{y} = \dot{y} = 0$. With $x_1 = y$ and $x_2 = \dot{y}$, the state equation is

$$\dot{x}_1 = x_2 \qquad (1.14)$$

$$\dot{x}_2 = -\frac{k}{m}x_1 - \frac{c}{m}x_2 - \frac{1}{m}\eta(x_1, x_2) \qquad (1.15)$$

Let us note two features of this state equation. First, it has an equilibrium set rather than isolated equilibrium points. Second, the right-hand side function is a discontinuous function of the state. The discontinuity is a consequence of the idealization we adopted in modeling the dry friction. One would expect the physical dry friction to change from its static friction mode into its slipping friction mode in a smooth way, not abruptly as our idealization suggests. The advantage of the discontinuous idealization is that it allows us to simplify the analysis. For example, when $x_2 > 0$, we can model the system by the linear model

$$
\begin{aligned}
\dot{x}_1 &= x_2 \\
\dot{x}_2 &= -\frac{k}{m}x_1 - \frac{c}{m}x_2 - \mu_k g
\end{aligned}
$$

Similarly, when $x_2 < 0$, we can model it by the linear model

$$
\begin{aligned}
\dot{x}_1 &= x_2 \\
\dot{x}_2 &= -\frac{k}{m}x_1 - \frac{c}{m}x_2 + \mu_k g
\end{aligned}
$$

Thus, in each region we can predict the behavior of the system via linear analysis. This is an example of the so-called *piecewise linear analysis*, where a system is represented by linear models in various regions of the state space, certain coefficients changing from region to region.

1.1.4   Negative-Resistance Oscillator

Figure 1.5 shows the basic circuit structure of an important class of electronic oscillators. The inductor and capacitor are assumed to be linear, time-invariant and passive, that is, $L > 0$ and $C > 0$. The resistive element is an active circuit characterized by the *voltage-controlled* $i - v$ characteristic $i = h(v)$, shown in the figure. The function $h(\cdot)$ satisfies the conditions

$$
h(0) = 0, \quad h'(0) < 0
$$

$$
h(v) \to \infty \text{ as } v \to \infty, \quad \text{and} \quad h(v) \to -\infty \text{ as } v \to -\infty
$$

where $h'(v)$ is the first derivative of $h(v)$ with respect to v. Such $i - v$ characteristic can be realized, for example, by the twin-tunnel-diode circuit of Figure 1.6. Using Kirchhoff's current law, we can write the equation

$$
i_C + i_L + i = 0
$$

Hence,

$$
C\frac{dv}{dt} + \frac{1}{L}\int_{-\infty}^{t} v(s)\ ds + h(v) = 0
$$

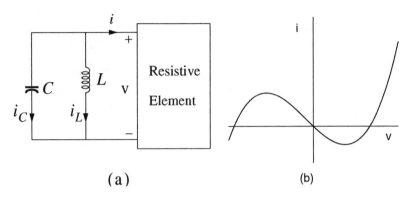

Figure 1.5: (a) Basic oscillator circuit; (b) Typical nonlinear driving-point characteristic.

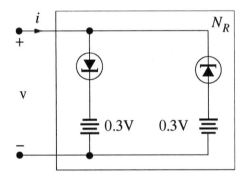

Figure 1.6: A negative-resistance twin-tunnel-diode circuit.

Differentiating once with respect to t and multiplying through by L, we obtain

$$CL\frac{d^2v}{dt^2} + v + Lh'(v)\frac{dv}{dt} = 0$$

This equation can be written in a form that coincides with some well-known equations in nonlinear systems theory. To do that, let us change the time variable from t to $\tau = t/\sqrt{CL}$. The derivatives of v with respect to t and τ are related by

$$\frac{dv}{d\tau} = \sqrt{CL}\frac{dv}{dt}$$
$$\frac{d^2v}{d\tau^2} = CL\frac{d^2v}{dt^2}$$

Denoting the derivative of v with respect to τ by $\dot{v}$, we can rewrite the circuit equation as

$$\ddot{v} + \epsilon h'(v)\dot{v} + v = 0$$

where $\epsilon = \sqrt{L/C}$. This equation is a special case of *Liénard's equation*

$$\ddot{v} + f(v)\dot{v} + g(v) = 0 \tag{1.16}$$

When

$$h(v) = -v + \tfrac{1}{3}v^3$$

the circuit equation takes the form

$$\ddot{v} - \epsilon(1 - v^2)\dot{v} + v = 0 \tag{1.17}$$

which is known as the *Van der Pol equation*. This equation, which was used by Van der Pol to study oscillations in vacuum tube circuits, is a fundamental example in nonlinear oscillation theory. It possesses a periodic solution that attracts every other solution except the trivial one at the unique equilibrium point $v = \dot{v} = 0$. To write a state-space model for the circuit, let us take $x_1 = v$ and $x_2 = \dot{v}$ to obtain

$$\dot{x}_1 = x_2 \tag{1.18}$$
$$\dot{x}_2 = -x_1 - \epsilon h'(x_1)x_2 \tag{1.19}$$

Note that an alternate state-space model could have been obtained by choosing the state variables as the voltage across the capacitor and the current through the inductor, as we did in the previous example. Denoting the state variables by $z_1 = i_L$ and $z_2 = v_C$, the state-space model is given by

$$\frac{dz_1}{dt} = \frac{1}{L}z_2$$
$$\frac{dz_2}{dt} = -\frac{1}{C}[z_1 + h(z_2)]$$

Since the first state-space model has been written with respect to the time variable $\tau = t/\sqrt{CL}$, let us write this model with respect to τ.

$$\dot{z}_1 = \frac{1}{\epsilon}z_2 \tag{1.20}$$
$$\dot{z}_2 = -\epsilon[z_1 + h(z_2)] \tag{1.21}$$

The state-space models in x and z look different, but they are equivalent representations of the system. This equivalence can be seen by noting that these models can be obtained from each other by a change of coordinates

$$z = T(x)$$

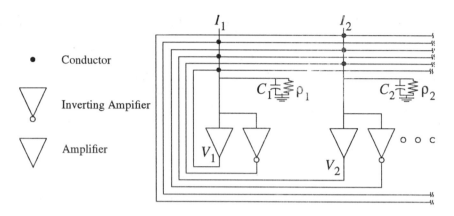

Figure 1.7: Hopfield neural network model.

Since we have chosen both x and z in terms of the physical variables of the circuit, it is not hard to find the map $T(\cdot)$. We have

$$x_1 = v = z_2$$
$$x_2 = \frac{dv}{d\tau} = \sqrt{CL}\frac{dv}{dt} = \sqrt{\frac{L}{C}}[-i_L - h(v_C)] = \epsilon[-z_1 - h(z_2)]$$

Thus,

$$z = T(x) = \begin{bmatrix} -h(x_1) - \frac{1}{\epsilon}x_2 \\ x_1 \end{bmatrix}$$

and the inverse mapping is

$$x = T^{-1}(z) = \begin{bmatrix} z_2 \\ -\epsilon z_1 - \epsilon h(z_2) \end{bmatrix}$$

1.1.5   Artificial Neural Network

Artificial neural networks, in analogy to biological structures, take advantage of distributed information processing and their inherent potential for parallel computation. Figure 1.7 shows an electric circuit that implements one model of neural networks, known as the *Hopfield model*. The circuit is based on an RC network connecting amplifiers. The input-output characteristics of the amplifiers are given by $V_i = g_i(u_i)$, where u_i and V_i are the input and output voltages of the ith amplifier. The function $g_i(\cdot) : R \to (-V_M, V_M)$ is a sigmoid function with asymptotes $-V_M$ and V_M, as shown in Figure 1.8 for a case with $V_M = 1$. It is continuously differentiable, odd, monotonically increasing, and $g_i(u_i) = 0$ if and only if $u_i = 0$.

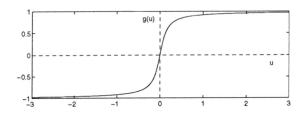

Figure 1.8: A typical input-output characteristic for the amplifiers in Hopfield network.

Examples of possible $g_i(\cdot)$ are

$$g_i(u_i) = \frac{2V_M}{\pi} \tan^{-1} \frac{\lambda \pi u_i}{2V_M}, \quad \lambda > 0$$

and

$$g_i(u_i) = V_M \frac{e^{\lambda u_i} - e^{-\lambda u_i}}{e^{\lambda u_i} + e^{-\lambda u_i}}, \quad \lambda > 0$$

where λ is the slope of $g_i(u_i)$ at $u_i = 0$. Such sigmoid input-output characteristics can be realized using operational amplifiers. For each amplifier, the circuit contains an inverting amplifier whose output is $-V_i$, which permits a choice of the sign of the amplifier output that is connected to a given input line. The outputs V_i and $-V_i$ are usually provided by two output terminals of the same operational amplifier circuit. The pair of nonlinear amplifiers is referred to as a "neuron." The circuit also contains an RC section at the input of each amplifier. The capacitance $C_i > 0$ and the resistance $\rho_i > 0$ represent the total shunt capacitance and shunt resistance at the ith amplifier input. Writing Kirchhoff's current law at the input node of the ith amplifier, we obtain

$$C_i \frac{du_i}{dt} = \sum_j T_{ij} V_j - \frac{1}{R_i} u_i + I_i$$

where

$$\frac{1}{R_i} = \frac{1}{\rho_i} + \sum_j \frac{1}{R_{ij}}$$

R_{ij} is the resistor connecting the output of the jth amplifier to the ith input line, T_{ij} is a signed conductance whose magnitude is $1/R_{ij}$ and whose sign is determined by the choice of the positive or negative output of the jth amplifier, and I_i represent any other constant input current (not shown in the figure). For a circuit containing n amplifiers, the motion is described by n first-order differential equations. To write a state-space model for the circuit, let us choose the state variables as $x_i = V_i$ for

$i = 1, 2, \ldots, n$. Then

$$\dot{x}_i = \frac{dg_i}{du_i}(u_i) \times \dot{u}_i = \frac{dg_i}{du_i}(u_i) \times \frac{1}{C_i} \left(\sum_j T_{ij} x_j - \frac{1}{R_i} u_i + I_i \right)$$

Defining

$$h_i(x_i) = \frac{dg_i}{du_i}(u_i) \Bigg|_{u_i = g_i^{-1}(x_i)}$$

we can write the state equation as

$$\dot{x}_i = \frac{1}{C_i} h_i(x_i) \left[\sum_j T_{ij} x_j - \frac{1}{R_i} g_i^{-1}(x_i) + I_i \right] \tag{1.22}$$

for $i = 1, 2, \ldots, n$. Note that, due to the sigmoid characteristic of $g_i(\cdot)$, the function $h_i(\cdot)$ satisfies

$$h_i(x_i) > 0, \quad \forall \; x_i \in (-V_M, V_M)$$

The equilibrium points of the system are the roots of the n simultaneous equations

$$0 = \sum_j T_{ij} x_j - \frac{1}{R_i} g_i^{-1}(x_i) + I_i$$

They are determined by the sigmoid characteristics, the linear resistive connection, and the input currents. We can obtain an equivalent state-space model by choosing the state variables as the voltage at the amplifier inputs u_i.

Stability analysis of this neural network depends critically on whether the symmetry condition $T_{ij} = T_{ji}$ is satisfied. An example of the analysis when $T_{ij} = T_{ji}$ is given in Section 3.2, while an example when $T_{ij} \neq T_{ji}$ is given in Section 5.6.

1.1.6   Adaptive Control

Consider a first-order linear system described by the model

$$\dot{y}_p = a_p y_p + k_p u$$

where u is the control input and y_p is the measured output. We shall refer to this system as the plant. Suppose it is desirable to obtain a closed-loop system whose input-output behavior is described by the reference model

$$\dot{y}_m = a_m y_m + k_m r$$

where r is the reference input and the model has been chosen such that $y_m(t)$ represents the desired output of the closed-loop system. This goal can be achieved by the linear feedback control

$$u(t) = \theta_1^* r(t) + \theta_2^* y_p(t)$$

provided that the plant parameters a_p and k_p are known, $k_p \neq 0$, and the controller parameters θ_1^* and θ_2^* are chosen as

$$\theta_1^* = \frac{k_m}{k_p}, \qquad \theta_2^* = \frac{a_m - a_p}{k_p}$$

When a_p and k_p are unknown, we may consider the controller

$$u(t) = \theta_1(t) r(t) + \theta_2(t) y_p(t)$$

where the time-varying gains $\theta_1(t)$ and $\theta_2(t)$ are adjusted on-line using the available data, namely, $r(\tau)$, $y_m(\tau)$, $y_p(\tau)$, and $u(\tau)$ for $\tau < t$. The adaptation should be such that $\theta_1(t)$ and $\theta_2(t)$ evolve to their nominal values θ_1^* and θ_2^*. The adaptation rule is chosen based on stability considerations. One such rule, known as the gradient algorithm,[3] is to use

$$\begin{aligned}
\dot{\theta}_1 &= -\gamma(y_p - y_m) r \\
\dot{\theta}_2 &= -\gamma(y_p - y_m) y_p
\end{aligned}$$

where γ is a positive constant that determines the speed of adaptation. This adaptive control law assumes that the sign of k_p is known and, without loss of generality, takes it to be positive. To write a state-space model that describes the closed-loop system under the adaptive control law, it is more convenient to define the output error e_o and the parameter errors ϕ_1 and ϕ_2 as

$$e_o = y_p - y_m, \quad \phi_1 = \theta_1 - \theta_1^*, \quad \phi_2 = \theta_2 - \theta_2^*$$

Using the definition of θ_1^* and θ_2^*, the reference model can be rewritten as

$$\dot{y}_m = a_p y_m + k_p(\theta_1^* r + \theta_2^* y_m)$$

On the other hand, the plant output y_p satisfies the equation

$$\dot{y}_p = a_p y_p + k_p(\theta_1 r + \theta_2 y_p)$$

Subtracting the above two equations, we obtain the error equation

$$\begin{aligned}
\dot{e}_o &= a_p e_o + k_p(\theta_1 - \theta_1^*) r + k_p(\theta_2 y_p - \theta_2^* y_m) \\
&= a_p e_o + k_p(\theta_1 - \theta_1^*) r + k_p(\theta_2 y_p - \theta_2^* y_m + \theta_2^* y_p - \theta_2^* y_p) \\
&= (a_p + k_p \theta_2^*) e_o + k_p(\theta_1 - \theta_1^*) r + k_p(\theta_2 - \theta_2^*) y_p
\end{aligned}$$

---

[3] This adaptation rule will be derived in Section 13.4.

Thus, the closed-loop system is described by the nonlinear, nonautonomous, third-order state equation

$$\dot{e}_o = a_m e_o + k_p \phi_1 r(t) + k_p \phi_2 (e_o + y_m(t)) \qquad (1.23)$$

$$\dot{\phi}_1 = -\gamma e_o r(t) \qquad (1.24)$$

$$\dot{\phi}_2 = -\gamma e_o (e_o + y_m(t)) \qquad (1.25)$$

where we have used $\dot{\phi}_i(t) = \dot{\theta}_i(t)$ and wrote $r(t)$ and $y_m(t)$ as explicit functions of time to emphasize the nonautonomous nature of the system. The signals $r(t)$ and $y_m(t)$ are the external driving inputs of the closed-loop system.

A simpler version of this model arises if we know k_p. In this case, we can take $\theta_1 = \theta_1^*$ and only θ_2 needs to be adjusted on-line. The closed-loop model reduces to

$$\dot{e}_o = a_m e_o + k_p \phi (e_o + y_m(t)) \qquad (1.26)$$

$$\dot{\phi} = -\gamma e_o (e_o + y_m(t)) \qquad (1.27)$$

where we have dropped the subscript from ϕ_2. If the goal of the control design is to regulate the plant output y_p to zero, we take $r(t) \equiv 0$ (hence, $y_m(t) \equiv 0$) and the closed-loop model simplifies to the autonomous second-order model

$$\dot{e}_o = (a_m + k_p \phi) e_o$$

$$\dot{\phi} = -\gamma e_o^2$$

The equilibrium points of this system are determined by setting $\dot{e}_o = \dot{\phi} = 0$ to obtain the algebraic equations

$$0 = (a_m + k_p \phi) e_o$$

$$0 = -\gamma e_o^2$$

The system has equilibrium at $e_o = 0$ for all values of ϕ; that is, it has an equilibrium set $e_o = 0$. There are no isolated equilibrium points.

The particular adaptive control scheme described here is called *direct model reference adaptive control*. The term "model reference" stems from the fact that the controller's task is to match a given closed-loop reference model, while the term "direct" is used to indicate that the controller parameters are adapted directly as opposed, for example, to an adaptive control scheme that would estimate the plant parameters a_p and k_p on line and use their estimates to calculate the controller parameters. We shall not describe the various possible adaptive control schemes.[4] However, the analysis of direct model reference adaptive control that we shall present is typical for other adaptive controls. The adaptive control problem

---

[4] For a comprehensive treatment of adaptive control, the reader may consult [4], [11], [79], [122], or [151].

will generate some interesting nonlinear models like the second- and third-order models we have just derived, which will be used to illustrate some of the stability and perturbation techniques of this book. A more general description of model reference adaptive control and its stability analysis will be pursued in Section 13.4.

1.2 Second-Order Systems

Second-order autonomous systems occupy an important place in the study of nonlinear systems because solution trajectories can be represented by curves in the plane. This allows easy visualization of the qualitative behavior of the system. The purpose of this section is to use second-order systems to introduce, in an elementary context, some of the basic ideas of nonlinear systems. In particular, we shall look at the behavior of a nonlinear system near equilibrium points and the phenomenon of nonlinear oscillation.

A second-order autonomous system is represented by two scalar differential equations

$$\dot{x}_1 = f_1(x_1, x_2) \tag{1.28}$$
$$\dot{x}_2 = f_2(x_1, x_2) \tag{1.29}$$

Let $x(t) = (x_1(t), x_2(t))$ be the solution[5] of (1.28)–(1.29) that starts at a certain initial state $x_0 = (x_{10}, x_{20})$; that is, $x(0) = x_0$. The locus in the x_1–x_2 plane of the solution $x(t)$ for all $t \geq 0$ is a curve that passes through the point x_0. This curve is called a *trajectory* or *orbit* of (1.28)–(1.29) from x_0. The x_1–x_2 plane is usually called the *state plane* or *phase plane*. The right-hand side of (1.28)–(1.29) expresses the tangent vector $\dot{x}(t) = (\dot{x}_1(t), \dot{x}_2(t))$ to the curve. Using the vector notation

$$\dot{x} = f(x)$$

where $f(x)$ is the vector $(f_1(x), f_2(x))$, we consider $f(x)$ as a *vector field* on the state plane. This means that to each point x in the plane we assign a vector $f(x)$. For easy visualization, we represent $f(x)$ as a vector based at x; that is, we assign to x the directed line segment from x to $x + f(x)$. For example, if $f(x) = (2x_1^2, x_2)$, then at $x = (1, 1)$ we draw an arrow pointing from $(1, 1)$ to $(1, 1) + (2, 1) = (3, 2)$; see Figure 1.9. Repeating this at every point in the plane, we obtain the vector field diagram in Figure 1.10. In this figure, the length of the arrow at a given point x is proportional to $\sqrt{f_1^2(x) + f_2^2(x)}$. Since the vector field at a point is tangent to the trajectory through that point, we can, in essence, construct the trajectory starting at a given point x_0 from the vector field diagram.

The family of all trajectories or solution curves is called the *phase portrait* of (1.28)–(1.29). An (approximate) picture of the phase portrait can be constructed

---

[5] It is assumed that there is a unique solution.

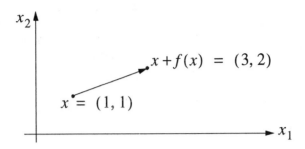

Figure 1.9: Vector field representation.

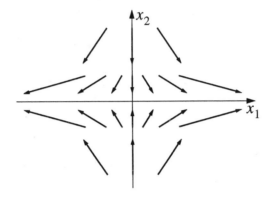

Figure 1.10: Vector field diagram.

by plotting trajectories from a large number of initial states spread all over the x_1–x_2 plane. Since numerical subroutines for solving general nonlinear differential equations are widely available, we can easily construct the phase portrait using computer simulations.[6] There are, however, several other methods for constructing the phase portrait. While we do not need to dwell on describing these methods, it might be useful to know one of them. The method we chose to describe here is known as the *isocline method*. To understand the idea behind the method, note that the slope of a trajectory at any given point x, denoted by $s(x)$, is given by

$$s(x) = \frac{f_2(x_1, x_2)}{f_1(x_1, x_2)}$$

---

[6]Some hints for the construction of phase portraits using computer simulation are given in Section 1.2.5.

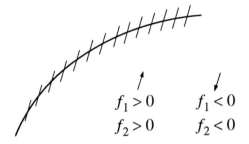

$$f_1 > 0 \qquad f_1 < 0$$
$$f_2 > 0 \qquad f_2 < 0$$

Figure 1.11: Isocline with positive slope.

Therefore, the equation

$$s(x) = c$$

defines a curve in the x_1-x_2 plane along which the trajectories of (1.28)–(1.29) have slope c. Thus, whenever a trajectory of (1.28)–(1.29) crosses the curve $s(x) = c$, the slope of the trajectory at the intersection point must be c. The procedure is to plot the curve $s(x) = c$ in the x_1-x_2 plane and along this curve draw short line segments having the slope c. These line segments are parallel and their directions are determined by the signs of $f_1(x)$ and $f_2(x)$ at x; see Figure 1.11. The curve $s(x) = c$ is known as an isocline. The procedure is repeated for sufficiently many values of the constant c, until the plane is filled with isoclines. Then, starting from a given initial point x_0, one can construct the trajectory from x_0 by moving, in the direction of the short line segments, from one isocline to the next.

Example 1.1 Consider the pendulum equation without friction:

$$\dot{x}_1 = x_2$$
$$\dot{x}_2 = -\sin x_1$$

The slope function $s(x)$ is given by

$$s(x) = \frac{-\sin x_1}{x_2}$$

Hence, the isoclines are defined by

$$x_2 = -\frac{1}{c}\sin x_1$$

Figure 1.12 shows the isoclines for several values of c. One can easily sketch the trajectory starting at any point, as shown in the figure for the trajectory starting at

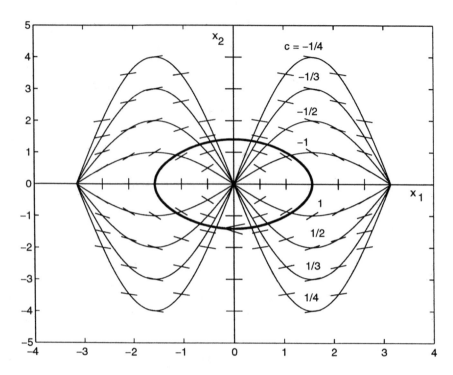

Figure 1.12: Graphical construction of the phase portrait of the pendulum equation (without friction) by the isocline method.

$(\pi/2, 0)$. If the sketch is done carefully, one might even detect that the trajectory is a closed curve. Consider next the pendulum equation with friction

$$\dot{x}_1 = x_2$$
$$\dot{x}_2 = -0.5x_2 - \sin x_1$$

This time the isoclines are defined by

$$x_2 = -\frac{1}{0.5 + c}\sin x_1$$

The isoclines shown in Figure 1.13 are identical to those of Figure 1.12, except for the slopes of the line segments. A sketch of the trajectory starting at $(\pi/2, 0)$ shows that the trajectory is a shrinking spiral that moves toward the origin. $\triangle$

Note that since the time t is suppressed in a trajectory, it is not possible to recover the solution $(x_1(t), x_2(t))$ associated with a given trajectory. Hence, a trajectory gives only the *qualitative* but not *quantitative* behavior of the associated

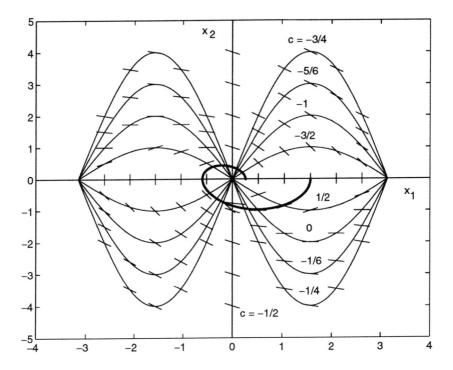

Figure 1.13: Graphical construction of the phase portrait of the pendulum equation (with friction) by the isocline method.

solution. For example, a closed trajectory shows that there is a periodic solution; that is, the system has a sustained oscillation, whereas a shrinking spiral shows a decaying oscillation. In the rest of this section, we will qualitatively analyze the behavior of second-order systems using their phase portraits.

1.2.1 Qualitative Behavior of Linear Systems

Consider the linear time-invariant system

$$\dot{x} = Ax \qquad (1.30)$$

where A is a 2×2 real matrix. The solution of (1.30) for a given initial state x_0 is given by

$$x(t) = M \exp(J_r t) M^{-1} x_0$$

where J_r is the real Jordan form of A and M is a real nonsingular matrix such that $M^{-1}AM = J_r$. Depending on the eigenvalues of A, the real Jordan form may take one of the following three forms:

$$\begin{bmatrix} \lambda_1 & 0 \\ 0 & \lambda_2 \end{bmatrix}, \quad \begin{bmatrix} \lambda & k \\ 0 & \lambda \end{bmatrix}, \quad \begin{bmatrix} \alpha & -\beta \\ \beta & \alpha \end{bmatrix}$$

where k is either 0 or 1. The first form corresponds to the case when the eigenvalues λ_1 and λ_2 are real and distinct, the second form corresponds to the case when the eigenvalues are real and equal, and the third form corresponds to the case of complex eigenvalues $\lambda_{1,2} = \alpha \pm j\beta$. In our analysis, we have to distinguish between these three cases. Moreover, in the case of real eigenvalues we have to isolate the case when at least one of the eigenvalues is zero. In that case, the origin is not an isolated equilibrium point and the qualitative behavior of the system is quite different from the other cases.

Case 1. Both eigenvalues are real: $\lambda_1 \neq \lambda_2 \neq 0$.

In this case, $M = [v_1, v_2]$ where v_1 and v_2 are the real eigenvectors associated with λ_1 and λ_2. The change of coordinates $z = M^{-1}x$ transforms the system into two decoupled first-order differential equations

$$\begin{aligned} \dot{z}_1 &= \lambda_1 z_1 \\ \dot{z}_2 &= \lambda_2 z_2 \end{aligned}$$

whose solution, for a given initial state (z_{10}, z_{20}), is given by

$$z_1(t) = z_{10}e^{\lambda_1 t}, \qquad z_2(t) = z_{20}e^{\lambda_2 t}$$

Eliminating t between the two equations, we obtain

$$z_2 = c z_1^{\lambda_2/\lambda_1} \tag{1.31}$$

where $c = z_{20}/(z_{10})^{\lambda_2/\lambda_1}$. The phase portrait of the system is given by the family of curves generated from (1.31) by allowing the constant c to take arbitrary values in R. The shape of the phase portrait depends on the signs of λ_1 and λ_2.

Consider first the case when both eigenvalues are negative. Without loss of generality, let $\lambda_2 < \lambda_1 < 0$. In this case, both exponential terms $e^{\lambda_1 t}$ and $e^{\lambda_2 t}$ tend to zero as $t \to \infty$. Moreover, since $\lambda_2 < \lambda_1 < 0$, the term $e^{\lambda_2 t}$ will tend to zero faster than the term $e^{\lambda_1 t}$. Hence, we call λ_2 the fast eigenvalue and λ_1 the slow eigenvalue. For later reference, we call v_2 the fast eigenvector and v_1 the slow eigenvector. The trajectory tends to the origin of the z_1–z_2 plane along the curve (1.31), which now has a ratio λ_2/λ_1 that is greater than one. The slope of the curve is given by

$$\frac{dz_2}{dz_1} = c\frac{\lambda_2}{\lambda_1}z_1^{[(\lambda_2/\lambda_1)-1]}$$

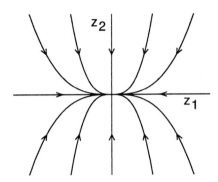

Figure 1.14: Phase portrait of a stable node in modal coordinates.

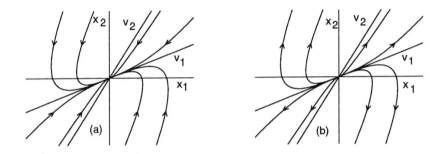

Figure 1.15: Phase portraits for (a) a stable node; (b) an unstable node.

Since $[(\lambda_2/\lambda_1) - 1]$ is positive, the slope of the curve approaches zero as $|z_1| \to 0$ and approaches ∞ as $|z_1| \to \infty$. Therefore, as the trajectory approaches the origin, it becomes tangent to the z_1-axis; as it approaches ∞, it becomes parallel to the z_2-axis. These observations allow us to sketch the typical family of trajectories shown in Figure 1.14. When transformed back into the x-coordinates, the family of trajectories will have the typical portrait shown in Figure 1.15(a). Note that in the x_1-x_2 plane, the trajectories become tangent to the slow eigenvector v_1 as they approach the origin and parallel to the fast eigenvector v_2 far from the origin. In this case, the equilibrium point $x = 0$ is called a *stable node*.

When λ_1 and λ_2 are positive, the phase portrait will retain the character of Figure 1.15(a) but with the trajectory directions reversed, since the exponential terms $e^{\lambda_1 t}$ and $e^{\lambda_2 t}$ grow exponentially as t increases. Figure 1.15(b) shows the phase portrait for the case $\lambda_2 > \lambda_1 > 0$. The equilibrium point $x = 0$ is referred to in this

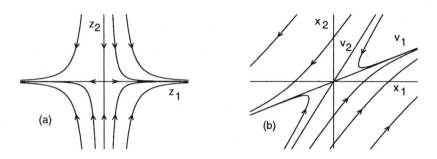

Figure 1.16: Phase portrait of a saddle point (a) in modal coordinates; (b) in original coordinates.

case as an *unstable node*.

Suppose now that the eigenvalues have opposite signs. In particular, let $\lambda_2 < 0 < \lambda_1$. In this case, $e^{\lambda_1 t} \to \infty$ while $e^{\lambda_2 t} \to 0$ as $t \to \infty$. Hence, we call λ_2 the stable eigenvalue and λ_1 the unstable eigenvalue. Correspondingly, v_2 and v_1 are called the stable and unstable eigenvectors, respectively. The trajectory equation (1.31) will have a negative exponent (λ_2/λ_1). Thus, the family of trajectories in the z_1-z_2 plane will take the typical form shown in Figure 1.16(a). Trajectories have hyperbolic shapes. They become tangent to the z_1-axis as $|z_1| \to \infty$ and tangent to the z_2-axis as $|z_1| \to 0$. The only exception to these hyperbolic shapes are the four trajectories along the axes. The two trajectories along the z_2-axis are called the stable trajectories since they approach the origin as $t \to \infty$, while the two trajectories along the z_1-axis are called the unstable trajectories since they approach infinity as $t \to \infty$. The phase portrait in the x_1-x_2 plane is shown in Figure 1.16(b). Here the stable trajectories are along the stable eigenvector v_2 and the unstable trajectories are along the unstable eigenvector v_1. In this case, the equilibrium point is called a *saddle*.

Case 2. Complex eigenvalues: $\lambda_{1,2} = \alpha \pm j\beta$.

The change of coordinates $z = M^{-1}x$ transforms the system (1.30) into the form

$$\begin{aligned} \dot{z}_1 &= \alpha z_1 - \beta z_2 \\ \dot{z}_2 &= \beta z_1 + \alpha z_2 \end{aligned}$$

The solution of this equation is oscillatory and can be expressed more conveniently

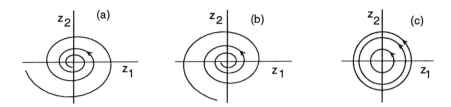

Figure 1.17: Typical trajectories in the case of complex eigenvalues.
(a) $\alpha < 0$; (b) $\alpha > 0$; (c) $\alpha = 0$.

Figure 1.18: Phase portraits for (a) a stable focus; (b) an unstable focus; (c) a center.

in the polar coordinates

$$r = \sqrt{z_1^2 + z_2^2}, \quad \theta = \tan^{-1}\left(\frac{z_2}{z_1}\right)$$

where we have two uncoupled first-order differential equations:

$$\dot{r} = \alpha r$$
$$\dot{\theta} = \beta$$

The solution for a given initial state (r_0, θ_0) is given by

$$r(t) = r_0 e^{\alpha t}, \quad \theta(t) = \theta_0 + \beta t$$

which defines a logarithmic spiral in the z_1–z_2 plane. Depending on the value of α, the trajectory will take one of the three forms shown in Figure 1.17. When $\alpha < 0$, the spiral converges to the origin; when $\alpha > 0$, it diverges away from the origin. When $\alpha = 0$, the trajectory is a circle of radius r_0. Figure 1.18 shows the trajectories in the x_1–x_2 plane. The equilibrium point $x = 0$ is referred to as a *stable focus* if $\alpha < 0$, *unstable focus* if $\alpha > 0$, and *center* if $\alpha = 0$.

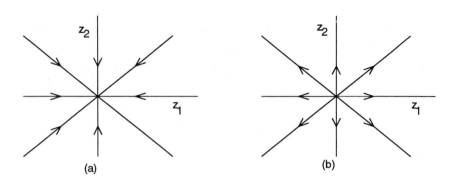

Figure 1.19: Phase portraits for the case of nonzero multiple eigenvalues when $k = 0$: (a) $\lambda < 0$; (b) $\lambda > 0$.

Case 3. Nonzero multiple eigenvalues: $\lambda_1 = \lambda_2 = \lambda \neq 0$.

The change of coordinates $z = M^{-1}x$ transforms the system (1.30) into the form

$$\dot{z}_1 = \lambda z_1 + k z_2$$
$$\dot{z}_2 = \lambda z_2$$

whose solution, for a given initial state (z_{10}, z_{20}), is given by

$$z_1(t) = e^{\lambda t}(z_{10} + k z_{20} t), \qquad z_2(t) = e^{\lambda t} z_{20}$$

Eliminating t, we obtain the trajectory equation

$$z_1 = z_2 \left[\frac{z_{10}}{z_{20}} + \frac{k}{\lambda} \ln \left(\frac{z_2}{z_{20}} \right) \right]$$

Figure 1.19 shows the form of the trajectories when $k = 0$, while Figure 1.20 shows their form when $k = 1$. The phase portrait has some similarity with the portrait of a node. Therefore, the equilibrium point $x = 0$ is usually referred to as a stable node if $\lambda < 0$ and unstable node if $\lambda > 0$. Note, however, that the phase portraits of Figures 1.19 and 1.20 do not have the asymptotic slow-fast behavior that we saw in Figures 1.14 and 1.15.

Before we discuss the degenerate case when one or both of the eigenvalues are zero, let us summarize our findings about the qualitative behavior of the system when the equilibrium point $x = 0$ is isolated. We have seen that the system can display six qualitatively different phase portraits, which are associated with different types of equilibria: stable node, unstable node, saddle point, stable focus, unstable focus, and center. The type of equilibrium point is completely specified by the location of the eigenvalues of A. Note that the global (throughout the phase plane)

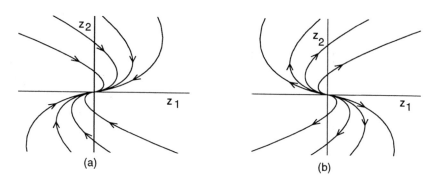

Figure 1.20: Phase portraits for the case of nonzero multiple eigenvalues when $k = 1$: (a) $\lambda < 0$; (b) $\lambda > 0$.

qualitative behavior of the system is determined by the type of equilibrium point. This is a characteristic of linear systems. When we study the qualitative behavior of nonlinear systems in the next section, we shall see that the type of equilibrium point can only determine the qualitative behavior of the trajectories in the vicinity of that point.

Case 4. One or both eigenvalues are zero.

When one or both eigenvalues of A are zero, the phase portrait is in some sense degenerate. In this case, the matrix A has a nontrivial null space. Any vector in the null space of A is an equilibrium point for the system; that is, the system has an equilibrium subspace rather than an equilibrium point. The dimension of the null space could be one or two; if it is two, the matrix A will be the zero matrix. This is a trivial case where every point in the plane is an equilibrium point. When the dimension of the null space is one, the shape of the Jordan form of A will depend on the multiplicity of the zero eigenvalue. When $\lambda_1 = 0$ and $\lambda_2 \neq 0$, the matrix M is given by $M = [v_1, v_2]$ where v_1 and v_2 are the associated eigenvectors. Note that v_1 spans the null space of A. The change of variables $z = M^{-1}x$ results in

$$
\begin{aligned}
\dot{z}_1 &= 0 \\
\dot{z}_2 &= \lambda_2 z_2
\end{aligned}
$$

whose solution is

$$z_1(t) = z_{10}, \quad z_2(t) = z_{20}e^{\lambda_2 t}$$

The exponential term will grow or decay, depending on the sign of λ_2. Figure 1.21 shows the phase portrait in the x_1–x_2 plane. All trajectories converge to the equilibrium subspace when $\lambda_2 < 0$, and diverge away from it when $\lambda_2 > 0$.

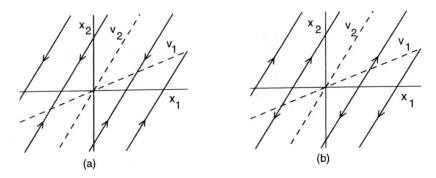

Figure 1.21: Phase portraits for (a) $\lambda_1 = 0$, $\lambda_2 < 0$; (b) $\lambda_1 = 0$, $\lambda_2 > 0$.

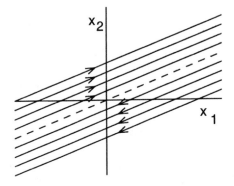

Figure 1.22: Phase portrait when $\lambda_1 = \lambda_2 = 0$.

When both eigenvalues are at the origin, the change of variables $z = M^{-1}x$ results in

$$\begin{aligned} \dot{z}_1 &= z_2 \\ \dot{z}_2 &= 0 \end{aligned}$$

whose solution is

$$z_1(t) = z_{10} + z_{20}t, \quad z_2(t) = z_{20}$$

The term $z_{20}t$ will increase or decrease, depending on the sign of z_{20}. The z_1-axis is the equilibrium subspace. Figure 1.22 shows the phase portrait in the x_1–x_2 plane; the dashed line is the equilibrium subspace. The phase portrait in Figure 1.22 is quite different from that in Figure 1.21. Trajectories starting off the equilibrium subspace move parallel to it.

The study of the behavior of linear systems about the equilibrium point $x = 0$ is important because, in many cases, the local behavior of a nonlinear system near an equilibrium point can be deduced by linearizing the system about that point and studying the behavior of the resultant linear system. How conclusive this approach is depends to a great extent on how the various qualitative phase portraits of a linear system persist under perturbations. We can gain insight into the behavior of a linear system under perturbations by examining the special case of linear perturbations. Suppose A has distinct eigenvalues and consider $A + \Delta A$, where ΔA is a 2×2 real matrix whose elements have arbitrarily small magnitudes. From the perturbation theory of matrices,[7] we know that the eigenvalues of a matrix depend continuously on its parameters. This means that, given any positive number ϵ, there is a corresponding positive number δ such that if the magnitude of the perturbation in each element of A is less than δ, the eigenvalues of the perturbed matrix $A + \Delta A$ will lie in open balls of radius ϵ centered at the eigenvalues of A. Consequently, any eigenvalue of A which lies in the open right-half plane (positive real part) or in the open left-half plane (negative real part) will remain in its respective half of the plane after arbitrarily small perturbations. On the other hand, eigenvalues on the imaginary axis, when perturbed, might go into either the right-half or the left-half of the plane since a ball centered on the imaginary axis will extend in both halves no matter how small ϵ is. Consequently, we can conclude that if the equilibrium point $x = 0$ of $\dot{x} = Ax$ is a node, focus, or saddle point, then the equilibrium point $x = 0$ of $\dot{x} = (A + \Delta A)x$ will be of the same type for sufficiently small perturbations. The situation is quite different if the equilibrium point is a center. Consider the following perturbation of the real Jordan form in the case of a center

$$\begin{bmatrix} \mu & 1 \\ -1 & \mu \end{bmatrix}$$

where μ is a perturbation parameter. When μ is positive, the equilibrium point of the perturbed system is an unstable focus; when μ is negative it is a stable focus. This is true no matter how small μ is, as long as it is different from zero. Because the phase portraits of a stable focus and unstable focus are qualitatively different from the phase portrait of a center, we see that a center equilibrium point will not persist under perturbations. The node, focus, and saddle equilibrium points are said to be *structurally stable* because they maintain their qualitative behavior under infinitesimally small perturbations,[8] while the center equilibrium point is not structurally stable. The distinction between the two cases is due to the location of the eigenvalues of A, with the eigenvalues on the imaginary axis being vulnerable to perturbations. This brings in the definition of a *hyperbolic equilibrium point*. The origin $x = 0$ is said to be a hyperbolic equilibrium point of $\dot{x} = Ax$ if A has no

---

[7] See [57, Chapter 7].

[8] See [70, Chapter 16] for a rigorous and more general definition of structural stability.

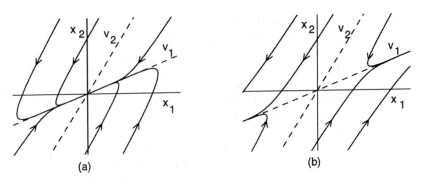

Figure 1.23: Phase portraits of a perturbed system when $\lambda_1 = 0$ and $\lambda_2 < 0$: (a) $\mu < 0$; (b) $\mu > 0$.

eigenvalues with zero real part.[9]

When A has multiple nonzero real eigenvalues, infinitesimally small perturbations could result in a pair of complex eigenvalues. Hence, a stable (respectively, unstable) node would either remain a stable (respectively, unstable) node or become a stable (respectively, unstable) focus.

When A has eigenvalues at zero, one would expect perturbations to move these eigenvalues away from zero, resulting in a major change in the phase portrait. It turns out, however, that there is an important difference between the case when there is only one eigenvalue at zero and the case when both eigenvalues are at zero ($A \neq 0$). In the first case, perturbation of the zero eigenvalue results in a real eigenvalue $\lambda_1 = \mu$ where μ could be positive or negative. Since the other eigenvalue λ_2 is different from zero, its perturbation will keep it away from zero. Moreover, since we are talking about arbitrarily small perturbations, $|\lambda_1| = |\mu|$ will be much smaller than $|\lambda_2|$. Thus, we end up with two real distinct eigenvalues, which means that the equilibrium point of the perturbed system will be a node or a saddle point, depending on the signs of λ_2 and μ. This is already an important change in the phase portrait. However, a careful examination of the phase portrait gives more insight into the qualitative behavior of the system. Since $|\lambda_1| \ll |\lambda_2|$, the exponential term $e^{\lambda_2 t}$ will change with t much faster than the exponential term $e^{\lambda_1 t}$. This results in the typical phase portraits of a node and a saddle shown in Figure 1.23, for the case $\lambda_2 < 0$. Comparison of these phase portraits with Figure 1.21(a) shows some similarity. In particular, similar to Figure 1.21, trajectories starting off the eigenvector v_1 converge to that vector along lines (almost) parallel to the eigenvector v_2. As they approach the vector v_1, they become tangent to it and move along it.

---

[9] This definition of a hyperbolic equilibrium point extends to higher-dimensional systems. It also carries over to equilibria of nonlinear systems by applying it to the eigenvalues of the linearized system.

When $\mu < 0$, the motion along v_1 converges to the origin (stable node) while when $\mu > 0$ the motion along v_1 tends to infinity (saddle point). This qualitative behavior is characteristic of singularly perturbed systems, which will be studied in Chapter 9.

When both eigenvalues of A are zeros, the effect of perturbations is more dramatic. Consider the following four possible perturbations of the Jordan form:

$$\begin{bmatrix} 0 & 1 \\ -\mu^2 & 0 \end{bmatrix}, \quad \begin{bmatrix} \mu & 1 \\ -\mu^2 & \mu \end{bmatrix}, \quad \begin{bmatrix} \mu & 1 \\ 0 & \mu \end{bmatrix}, \quad \begin{bmatrix} \mu & 1 \\ 0 & -\mu \end{bmatrix}$$

where μ is a perturbation parameter which could be positive or negative. It can be easily seen that the equilibrium points in these four cases are a center, a focus, a node, and a saddle point, respectively. In other words, all the possible phase portraits of an isolated equilibrium could result from perturbations.

1.2.2   Multiple Equilibria

The linear system (1.30) has an isolated equilibrium point at $x = 0$ if A has no zero eigenvalues, that is, if $\det A \neq 0$. When $\det A = 0$, the system has a continuum of equilibrium points. These are the only possible equilibria patterns that a linear system may have. A nonlinear system can have multiple isolated equilibrium points. In the following two examples, we explore the qualitative behavior of the tunnel diode circuit of Section 1.1.2 and the pendulum equation of Section 1.1.1. Both systems exhibit multiple isolated equilibria.

Example 1.2 The state-space model of a tunnel diode circuit is given by

$$\begin{aligned} \dot{x}_1 &= \frac{1}{C}[-h(x_1) + x_2] \\ \dot{x}_2 &= \frac{1}{L}[-x_1 - Rx_2 + u] \end{aligned}$$

Assume the circuit parameters are[10] $u = 1.2\ V$, $R = 1.5\ k\Omega = 1.5 \times 10^3\ \Omega$, $C = 2\ pF = 2 \times 10^{-12}\ F$, and $L = 5\ \mu H = 5 \times 10^{-6}\ H$. Measuring time in nanoseconds and the currents x_2 and $h(x_1)$ in mA, the state-space model is given by

$$\begin{aligned} \dot{x}_1 &= 0.5[-h(x_1) + x_2] \\ \dot{x}_2 &= 0.2(-x_1 - 1.5x_2 + 1.2) \end{aligned}$$

Suppose $h(\cdot)$ is given by

$$h(x_1) = 17.76x_1 - 103.79x_1^2 + 229.62x_1^3 - 226.31x_1^4 + 83.72x_1^5$$

---

[10] The numerical data are taken from [33].

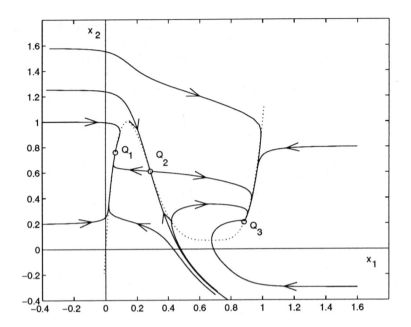

Figure 1.24: Phase portrait of the tunnel diode circuit of Example 1.2.

Setting $\dot{x}_1 = \dot{x}_2 = 0$ and solving for the equilibrium points, it can be verified that there are three equilibrium points at $(0.063, 0.758)$, $(0.285, 0.61)$, and $(0.884, 0.21)$. The phase portrait of the system, generated by a computer program, is shown in Figure 1.24. The three equilibrium points are denoted in the portrait by Q_1, Q_2, and Q_3, respectively. Examination of the phase portrait shows that, except for two special trajectories which tend to the point Q_2, all trajectories eventually tend to either the point Q_1 or the point Q_3. The two special trajectories converging to Q_2 form a curve that divides the plane into two halves. All trajectories originating from the left side of the curve will tend to the point Q_1, while all trajectories originating from the right side will tend to point Q_3. This special curve is called a *separatrix* because it partitions the plane into two regions of different qualitative behavior.[11] In an experimental setup, we shall observe one of the two steady-state operating points Q_1 or Q_3, depending on the initial capacitor voltage and inductor current. The equilibrium point at Q_2 is never observed in practice because the ever-present physical noise would cause the trajectory to diverge from Q_2 even if it were possible to set up the exact initial conditions corresponding to Q_2. This tunnel

---

[11] In general, the state plane decomposes into a number of regions, within each of which the trajectories may show a different type of behavior. The curves separating these regions are called separatrices.

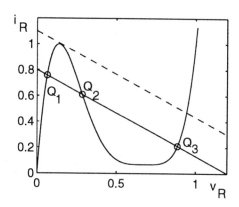

Figure 1.25: Adjustment of the load line of the tunnel diode circuit during triggering.

diode circuit with multiple equilibria is referred to as a *bistable* circuit, because it has two steady-state operating points. It has been used as a computer memory, where the equilibrium point Q_1 is associated with the binary state "0" and the equilibrium point Q_3 is associated with the binary state "1." Triggering from Q_1 to Q_3 or vice versa is achieved by a triggering signal of sufficient amplitude and duration which allows the trajectory to move to the other side of the separatrix. For example, if the circuit is initially at Q_1, then a positive pulse added to the supply voltage u will carry the trajectory to the right side of the separatrix. The pulse must be adequate in amplitude to raise the load line beyond the dashed line in Figure 1.25 and long enough to allow the trajectory to reach the right side of the separatrix.

The phase portrait in Figure 1.24 tells us the global qualitative behavior of the tunnel diode circuit. The range of x_1 and x_2 was chosen so that all essential qualitative features are displayed. The portrait outside this range does not contain any new qualitative features.   △

Example 1.3 Consider the pendulum equation with friction:

$$
\begin{aligned}
\dot{x}_1 &= x_2 \\
\dot{x}_2 &= -\frac{g}{l}\sin x_1 - \frac{k}{m}x_2
\end{aligned}
$$

A computer-generated phase portrait for $g/l = 1$ and $k/m = 0.5$ is shown in Figure 1.26. The phase portrait is periodic with a period equal to 2π. Consequently, all distinct features of the system's qualitative behavior can be captured by drawing the portrait in the vertical strip $-\pi \leq x_1 \leq \pi$, as shown in Figure 1.27. To follow a trajectory starting from an arbitrary initial state, we may have to switch back

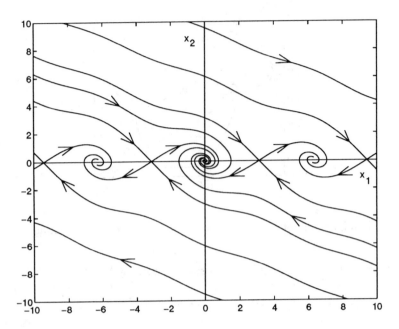

Figure 1.26: Phase portrait of the pendulum equation of Example 1.3.

and forth between the vertical boundaries of this strip. For example, consider the trajectory starting from point a in Figure 1.27. Follow this trajectory until it hits the right boundary at point b; then identify point c on the left boundary, which has the same value of the x_2 coordinate as point b. We then continue following the trajectory from c. This process is equivalent to following the trajectory on a cylinder formed by cutting out the phase portrait in Figure 1.27 along the vertical lines $x_1 = \pm\pi$ and pasting them together. When represented in this manner, the pendulum equation is said to be represented in a cylindrical phase space rather than a phase plane. From the phase portrait we see that, except for special trajectories which terminate at the "unstable" equilibrium position $(\pi, 0)$, all other trajectories tend to the "stable" equilibrium position $(0, 0)$. Once again, the "unstable" equilibrium position $(\pi, 0)$ cannot be maintained in practice, because noise would cause the trajectory to diverge away from that position.   △

1.2.3 Qualitative Behavior Near Equilibrium Points

Examination of the phase portraits in Examples 1.2 and 1.3 shows that the qualitative behavior in the vicinity of each equilibrium point looks just like those we saw in Section 1.2.1 for linear systems. In particular, in Figure 1.24 the trajectories near

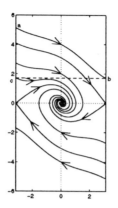

Figure 1.27: Phase portrait of the pendulum equation over the region $-\pi \leq x_1 \leq \pi$.

Q_1, Q_2, and Q_3 are similar to those associated with a stable node, saddle point, and stable node, respectively. Similarly, in Figure 1.26 the trajectories near $(0,0)$ and $(\pi, 0)$ are similar to those associated with a stable focus and a saddle point, respectively. These observations are actually quite general. Except for some special cases, the qualitative behavior of a nonlinear system near an equilibrium point can be determined via *linearization* with respect to that point.

Let $p = (p_1, p_2)$ be an equilibrium point of the nonlinear system (1.28)–(1.29), and suppose that the functions f_1 and f_2 are continuously differentiable. Expanding the right-hand side of (1.28)–(1.29) into its Taylor series about the point (p_1, p_2), we obtain

$$\dot{x}_1 = f_1(p_1, p_2) + a_{11}(x_1 - p_1) + a_{12}(x_2 - p_2) + \text{H.O.T.}$$
$$\dot{x}_2 = f_2(p_1, p_2) + a_{21}(x_1 - p_1) + a_{22}(x_2 - p_2) + \text{H.O.T.}$$

where

$$a_{11} = \left.\frac{\partial f_1(x_1, x_2)}{\partial x_1}\right|_{x_1=p_1, x_2=p_2}, \qquad a_{12} = \left.\frac{\partial f_1(x_1, x_2)}{\partial x_2}\right|_{x_1=p_1, x_2=p_2}$$

$$a_{21} = \left.\frac{\partial f_2(x_1, x_2)}{\partial x_1}\right|_{x_1=p_1, x_2=p_2}, \qquad a_{22} = \left.\frac{\partial f_2(x_1, x_2)}{\partial x_2}\right|_{x_1=p_1, x_2=p_2}$$

and H.O.T. denotes higher-order terms of the expansion, that is, terms of the form $(x_1 - p_1)^2$, $(x_2 - p_2)^2$, $(x_1 - p_1) \times (x_2 - p_2)$, and so on. Since (p_1, p_2) is an equilibrium point, we have

$$f_1(p_1, p_2) = f_2(p_1, p_2) = 0$$

Moreover, since we are interested in the trajectories near (p_1, p_2), we define

$$y_1 = x_1 - p_1, \quad y_2 = x_2 - p_2$$

and rewrite the state equation as

$$\dot{y}_1 = \dot{x}_1 = a_{11}y_1 + a_{12}y_2 + \text{H.O.T.}$$
$$\dot{y}_2 = \dot{x}_2 = a_{21}y_1 + a_{22}y_2 + \text{H.O.T.}$$

If we restrict attention to a sufficiently small neighborhood of the equilibrium point such that the higher-order terms are negligible, then we may drop these terms and approximate the nonlinear state equation by the linear state equation

$$\dot{y}_1 = a_{11}y_1 + a_{12}y_2$$
$$\dot{y}_2 = a_{21}y_1 + a_{22}y_2$$

Rewriting this equation in a vector form, we obtain

$$\dot{y} = Ay$$

where

$$A = \begin{bmatrix} a_{11} & a_{12} \\ a_{21} & a_{22} \end{bmatrix} = \begin{bmatrix} \frac{\partial f_1}{\partial x_1} & \frac{\partial f_1}{\partial x_2} \\ \frac{\partial f_2}{\partial x_1} & \frac{\partial f_2}{\partial x_2} \end{bmatrix}\Bigg|_{x=p} = \frac{\partial f}{\partial x}\Bigg|_{x=p}$$

The matrix $[\partial f / \partial x]$ is called the Jacobian matrix of $f(x)$ and A is the Jacobian matrix evaluated at the equilibrium point p.

It is reasonable to expect that the trajectories of the nonlinear system in a small neighborhood of an equilibrium point are "close" to the trajectories of its linearization about that point. Indeed it is true that[12] *if the origin of the linearized state equation is a stable (respectively, unstable) node with distinct eigenvalues, a stable (respectively, unstable) focus, or a saddle point, then, in a small neighborhood of the equilibrium point, the trajectories of the nonlinear state equation will behave like a stable (respectively, unstable) node, a stable (respectively, unstable) focus, or a saddle point.* Consequently, we call an equilibrium point of the nonlinear state equation (1.28)–(1.29) a stable (respectively, unstable) node, a stable (respectively, unstable) focus, or a saddle point if its linearized state equation about the equilibrium point has the same behavior. The type of equilibrium points in Examples 1.2 and 1.3 could have been determined by linearization without the need to construct the global phase portrait of the system.

---

[12] The proof of this linearization property can be found in [66]. It is valid under the assumption that $f_1(x_1, x_2)$ and $f_2(x_1, x_2)$ have continuous first partial derivatives in a neighborhood of the equilibrium point (p_1, p_2). A related, but different, linearization result will be proved in Chapter 3 for higher-dimensional systems; see Theorem 3.7.

Example 1.4 The Jacobian matrix of the function $f(x)$ of the tunnel diode circuit in Example 1.2 is given by

$$\frac{\partial f}{\partial x} = \begin{bmatrix} -0.5h'(x_1) & 0.5 \\ -0.2 & -0.3 \end{bmatrix}$$

where

$$h'(x_1) = \frac{dh}{dx_1} = 17.76 - 207.58x_1 + 688.86x_1^2 - 905.24x_1^3 + 418.6x_1^4$$

Evaluating the Jacobian matrix at the equilibrium points $Q_1 = (0.063, 0.758)$, $Q_2 = (0.285, 0.61)$, and $Q_3 = (0.884, 0.21)$, respectively, yields the three matrices

$$A_1 = \begin{bmatrix} -3.598 & 0.5 \\ -0.2 & -0.3 \end{bmatrix}, \quad \text{Eigenvalues}: \ -3.57, \ -0.33$$

$$A_2 = \begin{bmatrix} 1.82 & 0.5 \\ -0.2 & -0.3 \end{bmatrix}, \quad \text{Eigenvalues}: 1.77, \ -0.25$$

$$A_3 = \begin{bmatrix} -1.427 & 0.5 \\ -0.2 & -0.3 \end{bmatrix}, \quad \text{Eigenvalues}: \ -1.33, \ -0.4$$

Thus, Q_1 is a stable node, Q_2 is a saddle point, and Q_3 is a stable node.   △

Example 1.5 The Jacobian matrix of the function $f(x)$ of the pendulum equation in Example 1.3 is given by

$$\frac{\partial f}{\partial x} = \begin{bmatrix} 0 & 1 \\ -\cos x_1 & -0.5 \end{bmatrix}$$

Evaluating the Jacobian matrix at the equilibrium points $(0,0)$ and $(\pi, 0)$ yields, respectively, the two matrices

$$A_1 = \begin{bmatrix} 0 & 1 \\ -1 & -0.5 \end{bmatrix}, \quad \text{Eigenvalues}: \ -0.25 \pm j0.97$$

$$A_2 = \begin{bmatrix} 0 & 1 \\ 1 & -0.5 \end{bmatrix}, \quad \text{Eigenvalues}: \ -1.28, \ 0.78$$

Thus, the equilibrium point $(0,0)$ is a stable focus and the equilibrium point $(\pi, 0)$ is a saddle point.   △

Note that the foregoing linearization property dealt only with cases when the linearized state equation has no eigenvalues on the imaginary axis, that is, when the origin is a hyperbolic equilibrium point of the linear system. We extend this definition to nonlinear systems and say that an equilibrium point is hyperbolic if the Jacobian matrix, evaluated at that point, has no eigenvalues on the imaginary axis. If the Jacobian matrix has eigenvalues on the imaginary axis, then the qualitative behavior of the nonlinear state equation near the equilibrium point could be quite distinct from that of the linearized state equation. This should come as no surprise in view of our earlier discussion on the effect of linear perturbations on the qualitative behavior of a linear system when the origin is not a hyperbolic equilibrium point. The following example considers a case when the origin of the linearized state equation is a center.

Example 1.6 The system

$$\dot{x}_1 = -x_2 - \mu x_1(x_1^2 + x_2^2)$$
$$\dot{x}_2 = x_1 - \mu x_2(x_1^2 + x_2^2)$$

has an equilibrium point at the origin. The linearized state equation at the origin has eigenvalues $\pm j$. Thus, the origin is a center equilibrium point for the linearized system. In this particular example, we can determine the qualitative behavior of the nonlinear system by representing the system in polar coordinates:

$$x_1 = r\cos\theta, \quad x_2 = r\sin\theta$$

which yields

$$\dot{r} = -\mu r^3, \quad \dot{\theta} = 1$$

From these equations, it can be easily seen that the trajectories of the nonlinear system will resemble a stable focus when $\mu > 0$ and an unstable focus when $\mu < 0$.
$$\triangle$$

This example shows that the qualitative behavior describing a center in the linearized state equation is not preserved in the nonlinear state equation.

The foregoing discussion excludes the case when the linearized state equation has a node with multiple eigenvalues. Exercise 1.15 shows a case where the linearization has a stable node, while the trajectories of the nonlinear state equation behave like a stable focus. It should be mentioned, however, that a smoother function $f(x)$ will not allow this to happen. In particular, if $f_1(x_1, x_2)$ and $f_2(x_1, x_2)$ are analytic functions[13] in a neighborhood of the equilibrium point, then it is true that[14] *if the origin of the linearized state equation is a stable (respectively, unstable) node, then, in a small neighborhood of the equilibrium point, the trajectories of the nonlinear state equation will behave like a stable (respectively, unstable) node whether or not the eigenvalues of the linearization are distinct.*

---

[13] That is, f_1 and f_2 have convergent Taylor series representations.
[14] See [102, Theorem 3.4, page 188].

Determining the type of equilibrium points via linearization provides useful information that should be used when we construct a global phase portrait of a second-order system, whether we do that graphically or numerically. In fact, the first step in constructing a phase portrait should be the calculation of all equilibrium points and determining the type of isolated ones via linearization. This will give us a clear idea about the expected portrait in the neighborhood of the equilibrium points.

1.2.4 Limit Cycles

Oscillation is one of the most important phenomena that occur in dynamical systems. A system oscillates when it has a *nontrivial periodic solution*

$$x(t + T) = x(t), \quad \forall \, t \geq 0$$

for some $T > 0$. The word "nontrivial" is used to exclude constant solutions corresponding to equilibrium points. A constant solution satisfies the above equation but it is not what we have in mind when we talk of oscillation or periodic solutions. Unless otherwise specified, from this point on whenever we refer to a periodic solution we shall mean a nontrivial one. The image of a periodic solution in the phase portrait is a closed trajectory, that is usually called a *periodic orbit* or a *closed orbit*.

We have already seen an example of oscillation in Section 1.2.1: the second-order linear system with eigenvalues $\pm j\beta$. The origin of that system is a center, and the trajectories are closed orbits. When the system is transformed into its real Jordan form, the solution is given by

$$z_1(t) = r_0 \cos(\beta t + \theta_0), \quad z_2(t) = r_0 \sin(\beta t + \theta_0)$$

where

$$r_0 = \sqrt{z_1^2(0) + z_2^2(0)}, \quad \theta_0 = \tan^{-1} \left[\frac{z_2(0)}{z_1(0)} \right]$$

Therefore, the system has a sustained oscillation of amplitude r_0. It is usually referred to as the *harmonic oscillator*. If we think of the harmonic oscillator as a model for the linear LC circuit of Figure 1.28, then we can see that the physical mechanism leading to these oscillations is a periodic exchange (without dissipation) of the energy stored in the capacitor electric field with the energy stored in the inductor magnetic field. There are, however, two fundamental problems with this linear oscillator. The first problem is one of robustness. We have seen that infinitesimally small right-hand side (linear or nonlinear) perturbations will destroy the oscillation. That is, *the linear oscillator is not structurally stable.* In fact, it is impossible to build an LC circuit that realizes the harmonic oscillator, for the resistance in the electric wires alone will eventually consume whatever energy was

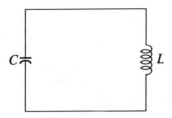

Figure 1.28: A linear LC circuit for the harmonic oscillator.

initially stored in the capacitor and inductor. Even if we succeeded in building this linear oscillator, we would face the second problem: *the amplitude of oscillation is dependent on the initial conditions.*

The two fundamental problems of the linear oscillator can be eliminated in nonlinear oscillators. It is possible to build physical nonlinear oscillators such that

- The nonlinear oscillator is structurally stable.

- The amplitude of oscillation (at steady state) is independent of initial conditions.

The negative-resistance oscillator of Section 1.1.4 is an example of such nonlinear oscillators. The state equation of the system is given by

$$\dot{x}_1 = x_2$$
$$\dot{x}_2 = -x_1 - \epsilon h'(x_1)x_2$$

where the function h satisfies certain properties, stated in Section 1.1.4. The system has only one equilibrium point at $x_1 = x_2 = 0$. The Jacobian matrix at this point is given by

$$A = \frac{\partial f}{\partial x}\bigg|_{x=0} = \begin{bmatrix} 0 & 1 \\ -1 & -\epsilon h'(0) \end{bmatrix}$$

Since $h'(0) < 0$, the origin is either an unstable node or unstable focus depending on the value of $\epsilon h'(0)$. In either case, all trajectories starting near the origin would diverge away from it and head toward infinity. This repelling feature of the origin is due to the negative resistance of the resistive element near the origin, which means that the resistive element is "active" and supplies energy. This point can be seen analytically by writing an expression for the rate of change of energy. The total energy stored in the capacitor and inductor at any time t is given by

$$E = \tfrac{1}{2}Cv_C^2 + \tfrac{1}{2}Li_L^2$$

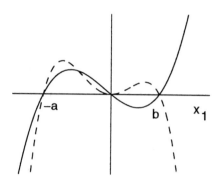

Figure 1.29: A sketch of $h(x_1)$ (solid) and $-x_1 h(x_1)$ (dashed) which shows that $\dot{E}$ is positive for $-a \leq x_1 \leq b$.

We have seen in Section 1.1.4 that

$$v_C = x_1, \quad i_L = -h(x_1) - \frac{1}{\epsilon} x_2$$

Thus, recalling that $\epsilon = \sqrt{L/C}$, we can rewrite the energy expression as

$$E = \tfrac{1}{2} C \{x_1^2 + [\epsilon h(x_1) + x_2]^2\}$$

The rate of change of energy is given by

$$
\begin{aligned}
\dot{E} &= C\{x_1 \dot{x}_1 + [\epsilon h(x_1) + x_2][\epsilon h'(x_1)\dot{x}_1 + \dot{x}_2]\} \\
&= C\{x_1 x_2 + [\epsilon h(x_1) + x_2][\epsilon h'(x_1)x_2 - x_1 - \epsilon h'(x_1)x_2]\} \\
&= C[x_1 x_2 - \epsilon x_1 h(x_1) - x_1 x_2] \\
&= -\epsilon C x_1 h(x_1)
\end{aligned}
$$

This expression confirms that, near the origin, the trajectory gains energy since for small $|x_1|$ the term $x_1 h(x_1)$ is negative. It also shows that there is a strip $-a \leq x_1 \leq b$ such that the trajectory gains energy within the strip and loses energy outside the strip. The strip boundaries $-a$ and b are roots of $h(x_1) = 0$, as shown in Figure 1.29. As a trajectory moves in and out of the strip, there is an exchange of energy with the trajectory gaining energy inside the strip and losing it outside. A stationary oscillation will occur if, along a trajectory, the net exchange of energy over one cycle is zero. Such a trajectory will be a closed orbit. It turns out that the negative-resistance oscillator has an isolated closed orbit, which is illustrated in the next example for the Van der Pol oscillator.

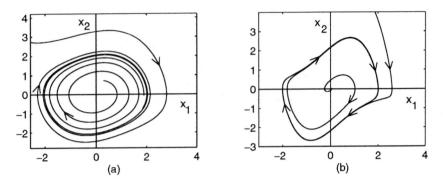

Figure 1.30: Phase portraits of the Van der Pol oscillator: (a) $\epsilon = 0.2$; (b) $\epsilon = 1.0$.

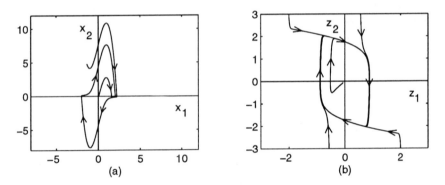

Figure 1.31: Phase portrait of the Van der Pol oscillator with $\epsilon = 5.0$: (a) in x_1–x_2 plane; (b) in z_1–z_2 plane.

Example 1.7 Figures 1.30(a), 1.30(b), and 1.31(a) show the phase portraits of the Van der Pol equation

$$\dot{x}_1 = x_2 \qquad\qquad (1.32)$$
$$\dot{x}_2 = -x_1 + \epsilon(1 - x_1^2)x_2 \qquad\qquad (1.33)$$

for three different values of the parameter ϵ: a small value of 0.2, a medium value of 1.0, and a large value of 5.0. In all three cases, the phase portraits show that there is a unique closed orbit that attracts all trajectories starting off the orbit. For $\epsilon = 0.2$, the closed orbit is a smooth orbit that is close to a circle of radius 2. This is typical for small ϵ (say, $\epsilon < 0.3$). For the medium value of $\epsilon = 1.0$, the circular shape of the closed orbit is distorted as shown in Figure 1.30(b). For the large value of $\epsilon = 5.0$, the closed orbit is severely distorted as shown in Figure 1.31(a). A more

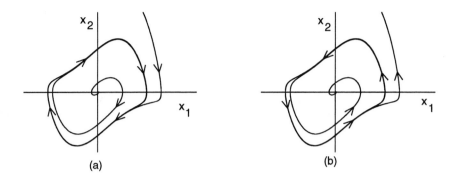

Figure 1.32: (a) A stable limit cycle; (b) an unstable limit cycle.

revealing phase portrait in this case can be obtained when the state variables are chosen as $z_1 = i_L$ and $z_2 = v_C$, resulting in the state equation

$$\dot{z}_1 = \frac{1}{\epsilon} z_2$$
$$\dot{z}_2 = -\epsilon(z_1 - z_2 + \tfrac{1}{3} z_2^3)$$

The phase portrait in the z_1–z_2 plane for $\epsilon = 5.0$ is shown in Figure 1.31(b). The closed orbit is very close to the curve $z_1 = z_2 - \tfrac{1}{3} z_2^3$ except at the corners, where it becomes nearly vertical. This vertical portion of the closed orbit can be viewed as if the closed orbit jumps from one branch of the curve to the other as it reaches the corner. Oscillations where this *jump phenomenon* takes place are usually referred to as *relaxation oscillations*. This phase portrait is typical for large values of ϵ (say, $\epsilon > 3.0$).   △

The closed orbit we have seen in Example 1.7 is different from what we have seen in the harmonic oscillator. In the case of the harmonic oscillator, there is a continuum of closed orbits, while in the Van der Pol example there is only one isolated periodic orbit. An isolated periodic orbit is called a *limit cycle*.[15] The limit cycle of the Van der Pol oscillator has the property that all trajectories in the vicinity of the limit cycle ultimately tend toward the limit cycle as $t \to \infty$. A limit cycle with this property is classically known as a stable limit cycle. We shall also encounter *unstable limit cycles*, which have the property that all trajectories starting from points arbitrarily close to the limit cycle will tend away from it as $t \to \infty$; see Figure 1.32. To see an example of an unstable limit cycle, consider the Van der Pol equation in reverse time; that is,

$$\dot{x}_1 = -x_2$$

---

[15] A formal definition of limit cycles will be given in Chapter 6.

$$\dot{x}_2 \;=\; x_1 - \epsilon(1 - x_1^2)x_2$$

The phase portrait of this system is identical to that of the Van der Pol oscillator, except that the arrowheads are reversed. Consequently, the limit cycle is unstable.

The limit cycle of the Van der Pol oscillator of Example 1.7 takes special forms in the limiting cases when ϵ is very small and very large. These special forms can be predicted analytically using asymptotic methods. In Chapter 8, we shall use the *averaging method* to derive the special form of the limit cycle as $\epsilon \to 0$, while in Chapter 9 we shall use the *singular perturbation method* to derive the special form of the limit cycle as $\epsilon \to \infty$.

1.2.5   Numerical Construction of Phase Portraits

Computer programs for numerical solution of ordinary differential equations are widely available. They can be effectively used to construct phase portraits for second-order systems. In this section, we give some hints[16] that might be useful for beginners.

The first step in constructing the phase portrait is to find all equilibrium points and determine the type of isolated ones via linearization.

Drawing trajectories involves three tasks:[17]

- Selection of a bounding box in the state plane where trajectories are to be drawn. The box takes the form

$$x_{1min} \le x_1 \le x_{1max}, \quad x_{2min} \le x_2 \le x_{2max}$$

- Selection of initial points (conditions) inside the bounding box.

- Calculation of trajectories.

Let us talk first about calculating trajectories. To find the trajectory passing through a point x_0, solve the equation

$$\dot{x} = f(x), \quad x(0) = x_0$$

in forward time (with positive t) and in reverse time (with negative t). Solution in reverse time is equivalent to solution in forward time of the equation

$$\dot{x} = -f(x), \quad x(0) = x_0$$

---

[16]These hints are taken from [132, Chapter 10], which contains more instructions on how to generate informative phase portraits.

[17]A fourth task that we left out is placing arrowheads on the trajectory. For the purpose of this textbook, this can be conveniently done manually.

since the change of time variable $\tau = -t$ reverses the sign of the right-hand side. The arrowhead on the forward trajectory is placed heading away from x_0, while the one on the reverse trajectory is placed heading into x_0. Note that solution in reverse time is the only way we can get a good portrait in the neighborhood of unstable focus, unstable node, or unstable limit cycle. Trajectories are continued until they get out of the bounding box. If processing time is a concern, you may want to add a stopping criterion when trajectories converge to an equilibrium point.

The bounding box should be selected so that all essential qualitative features are displayed. Since some of these features will not be known *a priori*, we may have to adjust the bounding box interactively. However, our initial choice should make use of all prior information. For example, the box should include all equilibrium points. Care should be exercised when a trajectory travels out of bounds, for such a trajectory is either unbounded or is attracted to a stable limit cycle.

The simplest approach to select initial points is to place them uniformly on a grid throughout the bounding box. However, an evenly spaced set of initial conditions rarely yields an evenly spaced set of trajectories. A better approach is to select the initial points interactively after plotting the already calculated trajectories. Since most computer programs have sophisticated plotting tools, this approach should be quite feasible.

For a saddle point, we can use linearization to generate the stable and unstable trajectories. This is useful because, as we saw in Examples 1.2 and 1.3, the stable trajectories of a saddle define a separatrix. Let the eigenvalues of the linearization be $\lambda_1 > 0 > \lambda_2$, and the corresponding eigenvectors be v_1 and v_2. The stable and unstable trajectories of the nonlinear saddle will be tangent to the stable eigenvector v_2 and the unstable eigenvector v_1, respectively, as they approach the equilibrium point p. Therefore, the two unstable trajectories can be generated from the initial points $x_0 = p \pm \alpha v_1$, where α is a small positive number. Similarly, the two stable trajectories can be generated from the initial points $x_0 = p \pm \alpha v_2$. The major parts of the unstable trajectories will be generated by solution in forward time, while the major parts of the stable ones will be generated by solution in reverse time.

1.3 Exercises

Exercise 1.1 A mathematical model that describes a wide variety of physical non-linear systems is the nth-order differential equation

$$\frac{d^n y}{dt^n} = g\left(t, y, \dot{y}, \ldots, \frac{d^{n-1}y}{dt^{n-1}}, u\right)$$

where u and y are scalar variables. With u as input and y as output, find a state-space model.

Exercise 1.2 Consider a single-input–single-output system described by the nth-order differential equation

$$\frac{d^n y}{dt^n} = g_1\left(t, y, \dot{y}, \ldots, \frac{d^{n-1}y}{dt^{n-1}}, u\right) + g_2\left(t, y, \dot{y}, \ldots, \frac{d^{n-2}y}{dt^{n-2}}\right)\dot{u}$$

where $g_2(\cdot)$ is a differentiable function of its arguments. With u as input and y as output, find a state-space model.

Hint: Take $x_n = \frac{d^{n-1}y}{dt^{n-1}} - g_2\left(t, y, \dot{y}, \ldots, \frac{d^{n-2}y}{dt^{n-2}}\right)u$.

Exercise 1.3 The nonlinear dynamic equations for an m-link robot [166] take the form

$$M(q)\ddot{q} + C(q, \dot{q})\dot{q} + g(q) = u$$

where q is an m-dimensional vector of generalized coordinates representing joint positions, u is an m-dimensional control (torque) input, and $M(q)$ is a symmetric inertia matrix which is positive definite $\forall\ q \in R^m$. The term $C(q, \dot{q})\dot{q}$ accounts for centrifugal and Coriolis forces. The matrix C has the property that $\dot{M} - 2C$ is a skew-symmetric matrix $\forall\ q, \dot{q} \in R^m$, where $\dot{M}$ is the total derivative of $M(q)$ with respect to t. The term $g(q)$, which accounts for gravity forces, is given by $g(q) = [\partial P(q)/\partial q]^T$, where $P(q)$ is the total potential energy of the links due to gravity. We assume that $P(q)$ is a positive definite function of q, and $g(q) = 0$ has an isolated root at $q = 0$.[18] Choose state variables for this system and write down the state equation.

Exercise 1.4 The nonlinear dynamic equations for a single-link manipulator with flexible joints [166], damping ignored, is given by

$$\begin{aligned} I\ddot{q}_1 + MgL\sin q_1 + k(q_1 - q_2) &= 0 \\ J\ddot{q}_2 - k(q_1 - q_2) &= u \end{aligned}$$

where q_1 and q_2 are angular positions, I and J are moments of inertia, k is a spring constant, M is the total mass, L is a distance, and u is a torque input. Choose state variables for this system and write down the state equation.

Exercise 1.5 The nonlinear dynamic equations for an m-link robot with flexible joints [166] take the form

$$\begin{aligned} M(q_1)\ddot{q}_1 + h(q_1, \dot{q}_1) + K(q_1 - q_2) &= 0 \\ J\ddot{q}_2 - K(q_1 - q_2) &= u \end{aligned}$$

---

[18] the properties of C, M, and P are not needed in this exercise. They will be used in later exercises.

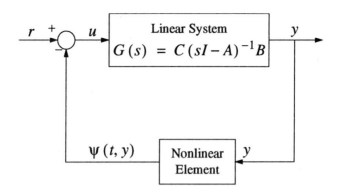

Figure 1.33: Exercise 1.6.

where q_1 and q_2 are m-dimensional vectors of generalized coordinates, $M(q_1)$ and J are symmetric nonsingular inertia matrices, and u is an m-dimensional control input. The term $h(q, \dot{q})$ accounts for centrifugal, Coriolis, and gravity forces, and K is a diagonal matrix of joint spring constants. Choose state variables for this system and write down the state equation.

Exercise 1.6 Figure 1.33 shows a feedback connection of a linear time- invariant system and a nonlinear time-varying element. The variables r, u, and y are vectors of the same dimension, and $\psi(t, y)$ is a vector-valued function. With r as input and y as output, find a state-space model.

Exercise 1.7 A synchronous generator connected to an infinite bus can be represented [131] by

$$M\ddot{\delta} = P - D\dot{\delta} - \eta_1 E_q \sin\delta$$
$$\tau\dot{E}_q = -\eta_2 E_q + \eta_3 \cos\delta + E_{FD}$$

where δ is an angle in radians, E_q is voltage, P is mechanical input power, E_{FD} is field voltage (input), D is damping coefficient, M is inertial coefficient, τ is a time constant, and η_1, η_2, and η_3 are constant parameters.

(a) Using δ, $\dot{\delta}$, and E_q as state variables, find the state equation.

(b) Let $P = 0.815$, $E_{FD} = 1.22$, $\eta_1 = 2.0$, $\eta_2 = 2.7$, $\eta_3 = 1.7$, $\tau = 6.6$, $M = 0.0147$, and $D/M = 4$. Find all equilibrium points.

(c) Suppose that τ is relatively large so that $\dot{E}_q \approx 0$. Show that assuming E_q to be constant reduces the model to a pendulum equation.

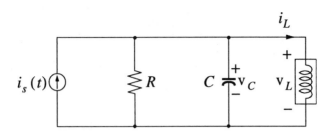

Figure 1.34: Exercises 1.8 and 1.9.

Exercise 1.8 The circuit shown in Figure 1.34 contains a nonlinear inductor and is driven by a time-dependent current source. Suppose that the nonlinear inductor is a Josephson junction [33], described by $i_L = I_0 \sin k\phi_L$ where ϕ_L is the magnetic flux of the inductor and I_0 and k are constants.

(a) Using ϕ_L and v_C as state variables, find the state equation.

(b) Is it easier to choose i_L and v_C as state variables?

Exercise 1.9 The circuit shown in Figure 1.34 contains a nonlinear inductor and is driven by a time-dependent current source. Suppose that the nonlinear inductor is described by $i_L = L\phi_L + \mu\phi_L^3$, where ϕ_L is the magnetic flux of the inductor and L and μ are positive constants.

(a) Using ϕ_L and v_C as state variables, find the state equation.

(b) Let $i_s = 0$. Find all equilibrium points and determine the type of each point.

Exercise 1.10 A phase-locked loop [54] can be represented by the block diagram of Figure 1.35. Let $\{A, B, C\}$ be a minimal realization of the scalar, strictly proper, mth-order transfer function $G(s)$. Assume that all eigenvalues of A have negative real parts, $G(0) \neq 0$, and $\theta_i = $ constant. Let z be the state of the realization $\{A, B, C\}$.

(a) Show that the closed-loop system can be represented by the state equation

$$
\begin{aligned}
\dot{z} &= Az + B \sin e \\
\dot{e} &= -Cz
\end{aligned}
$$

(b) Find all equilibrium points of the system.

(c) Show that when $G(s) = 1/(\tau s + 1)$, the closed-loop model coincides with the model of a pendulum equation.

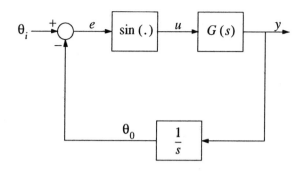

Figure 1.35: Exercise 1.10.

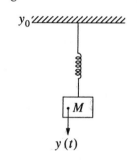

Figure 1.36: Exercise 1.11.

Exercise 1.11 Consider the mass-spring system shown in Figure 1.36. Assuming a linear spring and nonlinear viscous damping described by $c_1\dot{y} + c_2\dot{y}|\dot{y}|$, find a state equation that describes the motion of the system.

Exercise 1.12 A field-controlled DC motor [159] can be described by

$$v_f = R_f i_f + L_f \frac{di_f}{dt}$$

$$v_a = c_1 i_f \omega + L_a \frac{di_a}{dt} + R_a i_a$$

$$J \frac{d\omega}{dt} = c_2 i_f i_a - c_3 \omega$$

The first equation is for the field circuit with v_f, i_f, R_f, and L_f being its voltage, current, resistance, and inductance. The variables v_a, i_a, R_a, and L_a are the corresponding variables for the armature circuit described by the second equation. The third equation is a torque equation for the shaft, with J as the rotor inertia and c_3 as damping coefficient. The term $c_1 i_f \omega$ is the back e.m.f. induced in the armature circuit, and $c_2 i_f i_a$ is the torque produced by the interaction of the armature current with the field circuit flux. The voltage v_a is held constant, and the voltage v_f is the control input. Choose appropriate state variables and write down the state equation.

Exercise 1.13 For each of the following systems, find all equilibrium points and determine the type of each isolated equilibrium.

$$(1) \qquad \dot{x}_1 = x_2$$
$$\dot{x}_2 = -x_1 + \frac{x_1^3}{6} - x_2$$

$$(2) \qquad \dot{x}_1 = -x_1 + x_2$$
$$\dot{x}_2 = 0.1x_1 - 2x_2 - x_1^2 - 0.1x_1^3$$

$$(3) \qquad \dot{x}_1 = (1 - x_1)x_1 - \frac{2x_1 x_2}{1 + x_1}$$
$$\dot{x}_2 = \left(2 - \frac{x_2}{1 + x_1}\right) x_2$$

$$(4) \qquad \dot{x}_1 = x_2$$
$$\dot{x}_2 = -x_1 + x_2(1 - 3x_1^2 - 2x_2^2)$$

$$(5) \qquad \dot{x}_1 = -x_1 + x_2(1 + x_1)$$
$$\dot{x}_2 = -x_1(1 + x_1)$$

$$(6) \qquad \dot{x}_1 = (x_1 - x_2)\left(x_1^2 + x_2^2 - 1\right)$$
$$\dot{x}_2 = (x_1 + x_2)\left(x_1^2 + x_2^2 - 1\right)$$

$$(7) \qquad \dot{x}_1 = -x_1^3 + x_2$$
$$\dot{x}_2 = x_1 - x_2^3$$

Exercise 1.14 Find all equilibrium points of the system

$$\dot{x}_1 = ax_1 - x_1 x_2$$
$$\dot{x}_2 = bx_1^2 - cx_2$$

for all positive real values of a, b, and c, and determine the type of each equilibrium.

Exercise 1.15 The system

$$\dot{x}_1 = -x_1 - \frac{x_2}{\ln \sqrt{x_1^2 + x_2^2}}$$

$$\dot{x}_2 = -x_2 + \frac{x_1}{\ln \sqrt{x_1^2 + x_2^2}}$$

has an equilibrium point at the origin.

(a) Linearize the system about the origin and find the type of the origin as an equilibrium point of the linear system.

(b) Find the phase portrait of the nonlinear system near the origin, and show that the portrait resembles a stable focus.
   Hint: Transform the equations into polar coordinates.

(c) Explain the discrepancy between the results of parts (a) and (b).

Exercise 1.16 For each of the following systems, construct the phase portrait using the isocline method and discuss the qualitative behavior of the system. You can use information about the equilibrium points or the vector field, but do not use a computer program to generate the phase portrait.

$$(1) \qquad \dot{x}_1 = x_2 \cos x_1$$
$$\dot{x}_2 = \sin x_1$$

$$(2) \qquad \dot{x}_1 = (x_1 - x_2)\left(x_1^2 + x_2^2 - 1\right)$$
$$\dot{x}_2 = (x_1 + x_2)\left(x_1^2 + x_2^2 - 1\right)$$

$$(3) \qquad \dot{x}_1 = \left(1 - \frac{x_1}{x_2}\right)$$
$$\dot{x}_2 = -\frac{x_1}{x_2}\left(1 - \frac{x_1}{x_2}\right)$$

$$(4) \qquad \dot{x}_1 = x_2$$
$$\dot{x}_2 = -x_1 - \tfrac{1}{3}x_1^3 - x_2$$

$$(5) \qquad \dot{x}_1 = -x_1^3 + x_2$$
$$\dot{x}_2 = -x_1^3 - 4x_2$$

$$(6) \qquad \dot{x}_1 = x_2$$
$$\dot{x}_2 = -x_1 - \tfrac{1}{6}x_1^3$$

Exercise 1.17 For each of the following systems, construct the phase portrait (preferably using a computer program) and discuss the qualitative behavior of the system.

$$
\begin{aligned}
\textbf{(1)} \qquad \dot{x}_1 &= x_2 \\
\dot{x}_2 &= x_1 - 2\tan^{-1}(x_1 + x_2)
\end{aligned}
$$

$$
\begin{aligned}
\textbf{(2)} \qquad \dot{x}_1 &= x_2 \\
\dot{x}_2 &= -x_1 + x_2\left(1 - 3x_1^2 - 2x_2^2\right)
\end{aligned}
$$

$$
\begin{aligned}
\textbf{(3)} \qquad \dot{x}_1 &= 2x_1 - x_1 x_2 \\
\dot{x}_2 &= 2x_1^2 - x_2
\end{aligned}
$$

$$
\begin{aligned}
\textbf{(4)} \qquad \dot{x}_1 &= x_1 + x_2 - x_1(|x_1| + |x_2|) \\
\dot{x}_2 &= -2x_1 + x_2 - x_2(|x_1| + |x_2|)
\end{aligned}
$$

Exercise 1.18 Consider the system

$$
\begin{aligned}
\dot{x}_1 &= -(x_1 - x_1^2) + 1 - x_1 - x_2 \\
\dot{x}_2 &= -(x_2 - x_2^2) + 1 - x_1 - x_2
\end{aligned}
$$

(a) Find all equilibrium points and determine the type of each point.

(b) Construct the phase portrait, preferably using a computer program, and discuss the qualitative behavior of the system.

Exercise 1.19 Consider the tunnel-diode circuit of Section 1.1.2 with the numerical data used in Example 1.2, except for R and E which are taken as $E = 0.2V$ and $R = 0.2k\Omega$.

(a) Find all equilibrium points and determine the type of each point.

(b) Construct the phase portrait, preferably using a computer program, and discuss the qualitative behavior of the circuit.

Exercise 1.20 Repeat the previous problem with $E = 0.4V$ and $R = 0.2k\Omega$.

Exercise 1.21 Consider the Hopfield neural network model of Section 1.1.5 with $n = 2$, $V_M = 1$, and $T_{21} = T_{12} = 1$. For $i = 1, 2$, take $I_i = 0$, $C_i = 1$, $\rho_i = 1$, $T_{ii} = 0$, and $g_i(u) = (2/\pi)\tan^{-1}(\lambda\pi u/2)$.

(a) Find all equilibrium points and determine the type of each point.

(b) Take $\lambda = 5$. Construct the phase portrait, preferably using a computer program, and discuss the qualitative behavior of the system.

Exercise 1.22 The phase portraits of the following four systems are shown in Figures 1.37: parts (a), (b), (c), and (d), respectively. Mark the arrowheads and discuss the qualitative behavior of each system.

$$
\begin{aligned}
\textbf{(1)} \qquad \dot{x}_1 &= -x_2 \\
\dot{x}_2 &= x_1 - x_2\left(1 - x_1^2 + 0.1x_1^4\right)
\end{aligned}
$$

$$
\begin{aligned}
\textbf{(2)} \qquad \dot{x}_1 &= x_2 \\
\dot{x}_2 &= x_1 + x_2 - 3\tan^{-1}(x_1 + x_2)
\end{aligned}
$$

$$
\begin{aligned}
\textbf{(3)} \qquad \dot{x}_1 &= x_2 \\
\dot{x}_2 &= -(0.5x_1 + x_1^3)
\end{aligned}
$$

$$
\begin{aligned}
\textbf{(4)} \qquad \dot{x}_1 &= x_2 \\
\dot{x}_2 &= -x_2 - \psi(x_1 - x_2)
\end{aligned}
$$

where $\psi(y) = y^3 + 0.5y$ if $|y| \leq 1$ and $\psi(y) = 2y - 0.5$ if $|y| > 1$.

Exercise 1.23 An equivalent circuit of the Wien-Bridge oscillator is shown in Figure 1.38 [34], where $g(v)$ is a nonlinear voltage-controlled voltage source.

(a) With $x_1 = v_{C1}$ and $x_2 = v_{C2} = v$ as the state variables, show that the state equation is

$$
\begin{aligned}
\dot{x}_1 &= \frac{1}{C_1 R_1}[-x_1 + x_2 - g(x_2)] \\
\dot{x}_2 &= -\frac{1}{C_2 R_1}[-x_1 + x_2 - g(x_2)] - \frac{1}{C_2 R_2}x_2
\end{aligned}
$$

(b) Let $C_1 = C_2 = R_1 = R_2 = 1$ and $g(v) = 3.234v - 2.195v^3 + 0.666v^5$. Construct the phase portrait, preferably using a computer program, and discuss the qualitative behavior of the system.

Exercise 1.24 Consider the mass-spring system with dry friction

$$
\ddot{y} + ky + c\dot{y} + \eta(y, \dot{y}) = 0
$$

where η is defined in Section 1.1.3. Use piecewise linear analysis to construct the phase portrait qualitatively (without numerical data), and discuss the qualitative behavior of the system.

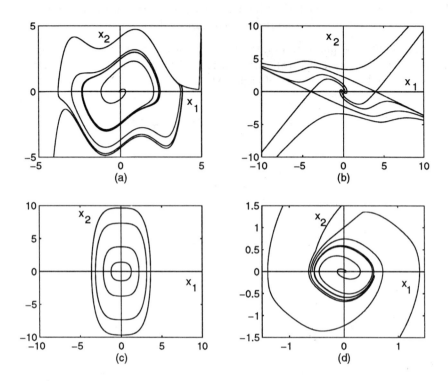

Figure 1.37: Exercise 1.22.

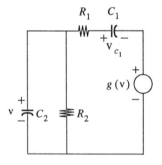

Figure 1.38: Exercise 1.23.

Chapter 2

Fundamental Properties

This chapter starts by recalling a few elements of mathematical analysis which will be used throughout the book. We then consider some fundamental properties of the solutions of ordinary differential equations, like existence, uniqueness, continuous dependence on initial conditions, and continuous dependence on parameters. These properties are essential for the state equation $\dot{x} = f(t, x)$ to be a useful mathematical model of a physical system. In experimenting with a physical system such as the pendulum, we expect that starting the experiment from a given initial state at time t_0, the system will move and its state will be defined in the (at least immediate) future time $t > t_0$. Moreover, with a deterministic system, we expect that if we could repeat the experiment exactly, we would get exactly the same motion and the same state at $t > t_0$. For the mathematical model to predict the future state of the system from its current state at t_0, the initial value problem

$$\dot{x} = f(t, x), \qquad x(t_0) = x_0 \tag{2.1}$$

must have a unique solution. This is the question of existence and uniqueness that is addressed in Section 2.2. It is shown that existence and uniqueness can be ensured by imposing some constraints on the right-hand side function $f(t, x)$. The key constraint used in Section 2.2 is the Lipschitz condition, whereby $f(t, x)$ satisfies the inequality

$$\|f(t, x) - f(t, y)\| \ \leq \ L\|x - y\| \tag{2.2}$$

for all (t, x) and (t, y) in some neighborhood of (t_0, x_0).

An essential factor in the validity of any mathematical model is the continuous dependence of its solutions on the data of the problem. The least we should expect from a mathematical model is that arbitrarily small errors in the data will not result in large errors in the solutions obtained by the model. The data of the initial value problem (2.1) are the initial state x_0, the initial time t_0, and the right-hand side function $f(t, x)$. Continuous dependence on the initial conditions (t_0, x_0) and on the parameters of f are studied in Section 2.3. If f is differentiable with respect to its parameters, then the solution will be differentiable with respect to these parameters. This is shown in Section 2.4 and is used to derive sensitivity equations which describe the effect of small parameter variations on the performance of the system. The continuity and differentiability results of Sections 2.3 and 2.4 are valid only on finite (compact) time intervals. Continuity results on the infinite-time interval will be given later, after stability concepts have been introduced.[1]

The chapter ends with a brief statement of a comparison principle that bounds the solution of a scalar differential inequality $\dot{v} \leq f(t, v)$ by the solution of the differential equation $\dot{u} = f(t, u)$.

2.1   Mathematical Preliminaries

2.1.1   Euclidean Space

The set of all n-dimensional vectors $x = (x_1, \ldots, x_n)^T$, where $x_1, \ldots, x_n$ are real numbers, defines the n-dimensional Euclidean space denoted by R^n. The one-dimensional Euclidean space consists of all real numbers and is denoted by R. Vectors in R^n can be added by adding their corresponding components. They can be multiplied by a scalar by multiplying each component by the scalar. The inner product of two vectors x and y is $x^T y = \sum_{i=1}^n x_i y_i$.

Vector and Matrix Norms

The norm $\|x\|$ of a vector x is a real-valued function with the properties

- $\|x\| \geq 0$ for all $x \in R^n$, with $\|x\| = 0$ if and only if $x = 0$.

- $\|x + y\| \leq \|x\| + \|y\|$, for all x, $y \in R^n$.

- $\|\alpha x\| = |\alpha| \, \|x\|$, for all $\alpha \in R$ and $x \in R^n$.

The second property is the triangle inequality. We shall consider the class of p-norms, defined by

$$\|x\|_p = (|x_1|^p + \cdots + |x_n|^p)^{1/p}, \quad 1 \leq p < \infty$$

and

$$\|x\|_\infty = \max_i |x_i|$$

---

[1] See, in particular, Section 5.5.

The three most commonly used norms are $\|x\|_1$, $\|x\|_\infty$, and the Euclidean norm

$$\|x\|_2 = \left(|x_1|^2 + \cdots + |x_n|^2\right)^{1/2} = \left(x^T x\right)^{1/2}$$

All p-norms are equivalent in the sense that if $\|\cdot\|_\alpha$ and $\|\cdot\|_\beta$ are two different p-norms, then there exist positive constants c_1 and c_2 such that

$$c_1 \|x\|_\alpha \leq \|x\|_\beta \leq c_2 \|x\|_\alpha$$

for all $x \in R^n$. Exercise 2.1 gives these constants for the most popular norms. A classic result concerning p-norms is the *Hölder inequality*

$$|x^T y| \leq \|x\|_p \|y\|_q, \quad \frac{1}{p} + \frac{1}{q} = 1$$

for all $x \in R^n$, $y \in R^n$. Quite often when we use norms, we only use properties deduced from the three basic properties satisfied by any norm. In those cases, the subscript p is dropped, indicating that the norm can be any p-norm.

An $m \times n$ matrix A of real elements defines a linear mapping $y = Ax$ from R^n into R^m. The induced p-norm of A is defined by[2]

$$\|A\|_p = \sup_{x \neq 0} \frac{\|Ax\|_p}{\|x\|_p} = \max_{\|x\|_p = 1} \|Ax\|_p$$

which for $p = 1, 2, \infty$ is given by

$$\|A\|_1 = \max_j \sum_{i=1}^m |a_{ij}|, \quad \|A\|_2 = \left[\lambda_{max}(A^T A)\right]^{1/2}, \quad \|A\|_\infty = \max_i \sum_{j=1}^n |a_{ij}|$$

$\lambda_{max}(A^T A)$ is the maximum eigenvalue of $A^T A$. Some useful properties of induced matrix norms are summarized in Exercise 2.2.

Topological Concepts in R^n

Convergence of Sequences: A sequence of vectors x_0, x_1, ..., x_k, ... in R^n, denoted by $\{x_k\}$, is said to converge to a limit vector x if

$$\|x_k - x\| \to 0 \text{ as } k \to \infty$$

which is equivalent to saying that, given any $\epsilon > 0$, there is N such that

$$\|x_k - x\| < \epsilon, \quad \forall \, k \geq N$$

---

[2] sup denotes supremum, the least upper bound; inf denotes infimum, the greatest lower bound.

The symbol "∀" reads "for all." A vector x is an accumulation point of a sequence $\{x_k\}$ if there is a subsequence of $\{x_k\}$ that converges to x; that is, if there is an infinite subset K of the nonnegative integers such that $\{x_k\}_{k \in K}$ converges to x. A bounded sequence $\{x_k\}$ in R^n has at least one accumulation point in R^n. A sequence of real numbers $\{r_k\}$ is said to be increasing (monotonically increasing or nondecreasing) if $r_k \leq r_{k+1} \; \forall \; k$. If $r_k < r_{k+1}$, it is said to be strictly increasing. Decreasing (monotonically decreasing or nonincreasing) and strictly decreasing sequences are defined similarly with $r_k \geq r_{k+1}$. An increasing sequence of real numbers that is bounded from above converges to a real number. Similarly, a decreasing sequence of real numbers that is bounded from below converges to a real number.

Sets: A subset $S \subset R^n$ is said to be *open* if, for every vector $x \in S$, one can find an ϵ-neighborhood of x

$$N(x, \epsilon) = \{ z \in R^n \mid \|z - x\| < \epsilon \}$$

such that $N(x, \epsilon) \subset S$. A set S is *closed* if and only if its complement in R^n is open. Equivalently, S is closed if and only if every convergent sequence $\{x_k\}$ with elements in S converges to a point in S. A set S is *bounded* if there is $r > 0$ such that $\|x\| \leq r$ for all $x \in S$. A set S is *compact* if it is closed and bounded. A point p is a *boundary point* of a set S if every neighborhood of p contains at least one point of S and one point not belonging to S. The set of all boundary points of S, denoted by ∂S, is called the boundary of S. A closed set contains all its boundary points. An open set contains none of its boundary points. The *interior* of a set S is $S - \partial S$. An open set is equal to its interior. The *closure* of a set S, denoted by $\bar{S}$, is the union of S and its boundary. A closed set is equal to its closure. An open set S is connected if every pair of points in S can be joined by an arc lying in S. A set S is called a *region* if it is the union of an open connected set with some, none, or all of its boundary points. If none of the boundary points are included, the region is called an open region or *domain*. A set S is *convex* if, for every $x, y \in S$ and every real number θ, $0 < \theta < 1$, the point $\theta x + (1 - \theta)y \in S$. If $x \in X \subset R^n$ and $y \in Y \subset R^m$, we say that (x, y) belongs to the product set $X \times Y \subset R^n \times R^m$.

Continuous Functions: A function f mapping a set S_1 into a set S_2 is denoted by $f : S_1 \to S_2$. A function $f : R^n \to R^m$ is said to be *continuous* at a point x if $f(x_k) \to f(x)$ whenever $x_k \to x$. Equivalently, f is continuous at x if, given $\epsilon > 0$, there is $\delta > 0$ such that

$$\|x - y\| < \delta \Rightarrow \|f(x) - f(y)\| < \epsilon$$

The symbol "⇒" reads "implies." A function f is continuous on a set S if it is continuous at every point of S, and it is *uniformly continuous* on S if, given $\epsilon > 0$ there is $\delta > 0$ (dependent only on ϵ) such that the inequality holds for all $x, \; y \in S$.

Note that uniform continuity is defined on a set, while continuity is defined at a point. For uniform continuity, the same constant δ works for all points in the set. Clearly, if f is uniformly continuous on a set S, then it is continuous on S. The opposite statement is not true in general. However, if S is a compact set, then continuity and uniform continuity on S are equivalent. The function

$$(a_1 f_1 + a_2 f_2)(\cdot) = a_1 f_1(\cdot) + a_2 f_2(\cdot)$$

is continuous for any two scalars a_1 and a_2 and any two continuous functions f_1 and f_2. If S_1, S_2, and S_3 are any sets and $f_1 : S_1 \to S_2$ and $f_2 : S_2 \to S_3$ are functions, then the function $f_2 \circ f_1 : S_1 \to S_3$, defined by

$$(f_2 \circ f_1)(\cdot) = f_2(f_1(\cdot))$$

is called the *composition* of f_1 and f_2. The composition of two continuous functions is continuous. If $S \subset R^n$ and $f : S \to R^m$, then the set of $f(x)$ such that $x \in S$ is called the image of S under f and is denoted by $f(S)$. If f is a continuous function defined on a compact set S, then $f(S)$ is compact; hence, continuous functions on compact sets are bounded. Moreover, if f is real-valued, that is, $f : S \to R$, then there are points p and q in the compact set S such that $f(x) \leq f(p)$ and $f(x) \geq f(q)$ for all $x \in S$. If f is a continuous function defined on a connected set S, then $f(S)$ is connected. A function f defined on a set S is said to be *one-to-one* on S if whenever x, $y \in S$ and $x \neq y$, then $f(x) \neq f(y)$. If $f : S \to R^m$ is a continuous, one-to-one function on a compact set $S \subset R^n$, then f has a continuous inverse f^{-1} on $f(S)$. The composition of f and f^{-1} is identity; that is, $f^{-1}(f(x)) = x$. A function $f : R \to R^n$ is said to be *piecewise continuous* on an interval $J \subset R$ if for every bounded subinterval $J_0 \subset J$, f is continuous for all $x \in J_0$ except, possibly, at a finite number of points where f may have discontinuities. Moreover, at each point of discontinuity x_0, the right-side limit $\lim_{h \to 0} f(x_0 + h)$ and the left-side limit $\lim_{h \to 0} f(x_0 - h)$ exist; that is, the function has a finite jump at x_0. Examples of piecewise continuous functions are given in Exercise 2.4.

Differentiable functions: A function $f : R \to R$ is said to be *differentiable* at x if the limit

$$f'(x) = \lim_{h \to 0} \frac{f(x + h) - f(x)}{h}$$

exists. The limit $f'(x)$ is called the derivative of f at x. A function $f : R^n \to R^m$ is said to be *continuously differentiable* at a point x_0 if the partial derivatives $\partial f_i / \partial x_j$ exist and are continuous at x_0 for $1 \leq i \leq m$, $1 \leq j \leq n$. A function f is continuously differentiable on a set S if it is continuously differentiable at every point of S. For a continuously differentiable function $f : R^n \to R$, the row vector $\partial f / \partial x$ is defined by

$$\frac{\partial f}{\partial x} = \left[\frac{\partial f}{\partial x_1}, \ \ldots, \ \frac{\partial f}{\partial x_n} \right]$$

The *gradient vector*, denoted by $\nabla f(x)$, is

$$\nabla f(x) = \left[\frac{\partial f}{\partial x}\right]^T$$

For a continuously differentiable function $f : R^n \rightarrow R^m$, the *Jacobian matrix* $[\partial f/\partial x]$ is an $m \times n$ matrix whose element in the ith row and jth column is $\partial f_i/\partial x_j$. Suppose $S \subset R^n$ is open, f maps S into R^m, f is continuously differentiable at $x_0 \in S$, g maps an open set containing $f(S)$ into R^k, and g is continuously differentiable at $f(x_0)$. Then the mapping h of S into R^k, defined by $h(x) = g(f(x))$, is continuously differentiable at x_0 and its Jacobian matrix is given by the *chain rule*

$$\left.\frac{\partial h}{\partial x}\right|_{x=x_0} = \left.\frac{\partial g}{\partial f}\right|_{f=f(x_0)} \left.\frac{\partial f}{\partial x}\right|_{x=x_0}$$

2.1.2   Mean Value and Implicit Function Theorems

If x and y are two distinct points in R^n, then the *line segment* $L(x,y)$ joining x and y is

$$L(x,y) = \{z \mid z = \theta x + (1 - \theta)y, \ 0 < \theta < 1\}$$

Mean Value Theorem

Assume that $f : R^n \rightarrow R$ is continuously differentiable at each point x of an open set $S \subset R^n$. Let x and y be two points of S such that the line segment $L(x,y) \subset S$. Then there exists a point z of $L(x,y)$ such that

$$f(y) - f(x) = \left.\frac{\partial f}{\partial x}\right|_{x=z} (y - x)$$

Implicit Function Theorem

Assume that $f : R^n \times R^m \rightarrow R^n$ is continuously differentiable at each point (x,y) of an open set $S \subset R^n \times R^m$. Let (x_0, y_0) be a point in S for which $f(x_0, y_0) = 0$ and for which the Jacobian matrix $[\partial f/\partial x](x_0, y_0)$ is nonsingular. Then there exist neighborhoods $U \subset R^n$ of x_0 and $V \subset R^m$ of y_0 such that for each $y \in V$ the equation $f(x, y) = 0$ has a unique solution $x \in U$. Moreover, this solution can be given as $x = g(y)$ where g is continuously differentiable at $y = y_0$.

The proof of these two theorems, as well as the other facts stated earlier in this section, can be found in any textbook on advanced calculus or mathematical analysis.[3]

---

[3] See, for example, [7].

2.1.3 Gronwall-Bellman Inequality

Lemma 2.1 *Let $\lambda : [a, b] \to R$ be continuous and $\mu : [a, b] \to R$ be continuous and nonnegative. If a continuous function $y : [a, b] \to R$ satisfies*

$$y(t) \leq \lambda(t) + \int_a^t \mu(s)y(s) \; ds$$

for $a \leq t \leq b$, then on the same interval

$$y(t) \leq \lambda(t) + \int_a^t \lambda(s)\mu(s) \exp\left[\int_s^t \mu(\tau) \; d\tau\right] \; ds$$

In particular, if $\lambda(t) \equiv \lambda$ is a constant, then

$$y(t) \leq \lambda \exp\left[\int_a^t \mu(\tau) \; d\tau\right]$$

If, in addition, $\mu(t) \equiv \mu \geq 0$ is a constant, then

$$y(t) \leq \lambda \exp[\mu(t - a)]$$

$\diamondsuit$

Proof: Let

$$z(t) = \int_a^t \mu(s)y(s) \; ds$$

and

$$v(t) = z(t) + \lambda(t) - y(t) \geq 0$$

Then, z is differentiable and

$$\dot{z} = \mu(t)y(t) = \mu(t)z(t) + \mu(t)\lambda(t) - \mu(t)v(t)$$

This is a scalar linear state equation with the state transition function

$$\phi(t, s) = \exp\left[\int_s^t \mu(\tau) \; d\tau\right]$$

Since $z(a) = 0$, we have

$$z(t) = \int_a^t \phi(t, s)[\mu(s)\lambda(s) - \mu(s)v(s)] \; ds$$

The term

$$\int_a^t \phi(t, s)\mu(s)v(s) \; ds$$

is nonnegative. Therefore,

$$z(t) \leq \int_a^t \exp\left[\int_s^t \mu(\tau)\,d\tau\right]\mu(s)\lambda(s)\,ds$$

Since $y(t) \leq \lambda(t) + z(t)$, this completes the proof in the general case. In the special case when $\lambda(t) \equiv \lambda$, we have

$$\int_a^t \mu(s)\exp\left[\int_s^t \mu(\tau)\,d\tau\right]\,ds = -\int_a^t \frac{d}{ds}\left\{\exp\left[\int_s^t \mu(\tau)\,d\tau\right]\right\}\,ds$$

$$= -\left\{\exp\left[\int_s^t \mu(\tau)\,d\tau\right]\right\}\Bigg|_{s=a}^{s=t}$$

$$= -1 + \exp\left[\int_a^t \mu(\tau)\,d\tau\right]$$

which proves the lemma when λ is a constant. The proof when both λ and μ are constants follows by integration.   □

2.1.4   Contraction Mapping

Consider an equation of the form $x = T(x)$. A solution x^* to this equation is said to be a *fixed point* of the mapping T since T leaves x^* invariant. A classical idea for finding a fixed point is the successive approximation method. We begin with an initial trial vector x_1 and compute $x_2 = T(x_1)$. Continuing in this manner iteratively, we compute successive vectors $x_{k+1} = T(x_k)$. The contraction mapping theorem gives sufficient conditions under which there is a fixed point x^* of $x = T(x)$ and the sequence $\{x_k\}$ converges to x^*. It is a powerful analysis tool for proving the existence of a solution of an equation of the form $x = T(x)$. The theorem is valid, not only when T is a mapping from one Euclidean space into another, but also when T is a mapping between *Banach spaces*. We shall use the contraction mapping theorem in that general setting. We start by introducing Banach spaces.[4]

Vector Spaces: A linear vector space $\mathcal{X}$ over the field R is a set of elements x, y, z, ... called vectors such that for any two vectors x, $y \in \mathcal{X}$, the sum $x + y$ is defined, $x + y \in \mathcal{X}$, $x + y = y + x$, $(x + y) + z = x + (y + z)$, and there is a zero vector $0 \in \mathcal{X}$ such that $x + 0 = x$ for all $x \in \mathcal{X}$. Also for any numbers α, $\beta \in R$, the scalar multiplication αx is defined, $\alpha x \in \mathcal{X}$, $1 \cdot x = x$, $0 \cdot x = 0$, $(\alpha\beta)x = \alpha(\beta x)$, $\alpha(x + y) = \alpha x + \alpha y$, and $(\alpha + \beta)x = \alpha x + \beta x$, for all x, $y \in \mathcal{X}$.

---

[4]For a complete treatment of Banach spaces, consult any textbook on functional analysis. A lucid treatment can be found in [106, Chapter 2].

Normed Linear Space: A linear space $\mathcal{X}$ is a normed linear space if, to each vector $x \in \mathcal{X}$, there is a real-valued norm $\|x\|$ which satisfies

- $\|x\| \geq 0$ for all $x \in \mathcal{X}$, with $\|x\| = 0$ if and only if $x = 0$.

- $\|x + y\| \leq \|x\| + \|y\|$ for all $x,\ y \in \mathcal{X}$.

- $\|\alpha x\| = |\alpha|\ \|x\|$ for all $\alpha \in R$ and $x \in \mathcal{X}$.

If it is not clear from the context whether $\|\cdot\|$ is a norm on $\mathcal{X}$ or a norm on R^n, we shall write $\|\cdot\|_{\mathcal{X}}$ for the norm on $\mathcal{X}$.

Convergence: A sequence $\{x_k\} \in \mathcal{X}$, a normed linear space, converges to $x \in \mathcal{X}$ if

$$\|x_k - x\| \to 0 \quad \text{as} \quad k \to \infty$$

Closed Set: A set $S \subset \mathcal{X}$ is closed if and only if every convergent sequence with elements in S has its limit in S.

Cauchy Sequence: A sequence $\{x_k\} \in \mathcal{X}$ is said to be a *Cauchy sequence* if

$$\|x_k - x_m\| \to 0 \quad \text{as} \quad k,\ m \to \infty$$

Every convergent sequence is Cauchy, but not vice versa.

Banach Space: A normed linear space $\mathcal{X}$ is complete if every Cauchy sequence in $\mathcal{X}$ converges to a vector in $\mathcal{X}$. A complete normed linear space is a Banach space.

Example 2.1 Consider the set of all continuous functions $f : [a, b] \to R^n$, denoted by $C[a, b]$. This set forms a vector space on R. The sum $x + y$ is defined by $(x + y)(t) = x(t) + y(t)$. The scalar multiplication is defined by $(\alpha x)(t) = \alpha x(t)$. The zero vector is the function identically zero on $[a, b]$. We define a norm by

$$\|x\|_C = \max_{t \in [a,b]} \|x(t)\|$$

where the right-hand side norm is any p-norm on R^n. Clearly $\|x\|_C \geq 0$ and is zero only for the function which is identically zero. The triangle inequality follows from

$$\max \|x(t) + y(t)\| \leq \max[\|x(t)\| + \|y(t)\|] \leq \max \|x(t)\| + \max \|y(t)\|$$

Also

$$\max \|\alpha x(t)\| = \max |\alpha|\ \|x(t)\| = |\alpha| \max \|x(t)\|$$

where all maxima are taken over $[a, b]$. Hence, $C[a, b]$ together with the norm $\|\cdot\|_C$ is a normed linear space. It is also a Banach space. To prove this claim, we need to show that every Cauchy sequence in $C[a, b]$ converges to a vector in $C[a, b]$. Suppose that $\{x_k\}$ is a Cauchy sequence in $C[a, b]$. For each fixed $t \in [a, b]$,

$$\|x_k(t) - x_m(t)\| \leq \|x_k - x_m\|_C \to 0 \quad \text{as} \quad k,\ m \to \infty$$

So $\{x_k(t)\}$ is a Cauchy sequence in R^n. But R^n with any p-norm is complete because convergence implies componentwise convergence and R is complete. Therefore, there is a real vector $x(t)$ to which the sequence converges: $x_k(t) \rightarrow x(t)$. This proves pointwise convergence. We prove next that the convergence is uniform in $t \in [a, b]$. Given $\epsilon > 0$, choose N such that $\|x_k - x_m\|_C < \epsilon/2$ for k, $m > N$. Then for $k > N$

$$\begin{aligned} \|x_k(t) - x(t)\| &\leq \|x_k(t) - x_m(t)\| + \|x_m(t) - x(t)\| \\ &\leq \|x_k - x_m\|_C + \|x_m(t) - x(t)\| \end{aligned}$$

By choosing m sufficiently large (which may depend on t), each term on the right-hand side can be made smaller than $\epsilon/2$; so $\|x_k(t) - x(t)\| < \epsilon$ for $k > N$. Hence $\{x_k\}$ converges to x, uniformly in $t \in [a, b]$. To complete the proof, we need to show that $x(t)$ is continuous and $\{x_k\}$ converges to x in the norm of $C[a, b]$. To prove continuity, consider

$$\begin{aligned} \|x(t + \delta) - x(t)\| &\leq \|x(t + \delta) - x_k(t + \delta)\| \\ &\quad + \|x_k(t + \delta) - x_k(t)\| + \|x_k(t) - x(t)\| \end{aligned}$$

Since $\{x_k\}$ converges uniformly to x, given any $\epsilon > 0$, we can choose k large enough to make both the first and third terms on the right-hand side less than $\epsilon/3$. Since $x_k(t)$ is continuous, we can choose δ small enough to make the second term less than $\epsilon/3$. Therefore $x(t)$ is continuous. The convergence of x_k to x in $\| \cdot \|_C$ is a direct consequence of the uniform convergence. $\triangle$

Other examples of Banach spaces are given in exercises.

Theorem 2.1 (Contraction Mapping) *Let S be a closed subset of a Banach space $\mathcal{X}$ and let T be a mapping that maps S into S. Suppose that*

$$\|T(x) - T(y)\| \leq \rho\|x - y\|, \quad \forall \ x, \ y \in S, \quad 0 \leq \rho < 1$$

then

- *There exists a unique vector $x^* \in S$ satisfying $x^* = T(x^*)$.*

- *x^* can be obtained by the method of successive approximation starting from any arbitrary initial vector in S.* $\diamond$

Proof: Select an arbitrary $x_1 \in S$ and define the sequence $\{x_k\}$ by the formula $x_{k+1} = T(x_k)$. Since T maps S into S, $x_k \in S$ for all $k \geq 1$. The first step of the proof is to show that $\{x_k\}$ is Cauchy. We have

$$\begin{aligned} \|x_{k+1} - x_k\| &= \|T(x_k) - T(x_{k-1})\| \\ &\leq \rho\|x_k - x_{k-1}\| \leq \rho^2\|x_{k-1} - x_{k-2}\| \leq \cdots \leq \rho^{k-1}\|x_2 - x_1\| \end{aligned}$$

It follows that

$$
\begin{aligned}
\|x_{k+r} - x_k\| &\leq \|x_{k+r} - x_{k+r-1}\| + \|x_{k+r-1} - x_{k+r-2}\| + \cdots + \|x_{k+1} - x_k\| \\
&\leq \left[\rho^{k+r-2} + \rho^{k+r-3} + \cdots + \rho^{k-1}\right] \|x_2 - x_1\| \\
&\leq \rho^{k-1} \sum_{i=0}^{\infty} \rho^i \|x_2 - x_1\| = \frac{\rho^{k-1}}{1 - \rho}\|x_2 - x_1\|
\end{aligned}
$$

The right-hand side tends to zero as $k \to \infty$. Hence, the sequence is Cauchy. Since $\mathcal{X}$ is a Banach space,

$$x_k \to x^* \in \mathcal{X} \quad \text{as} \quad k \to \infty$$

Moreover, since S is closed, $x^* \in S$. We now show that $x^* = T(x^*)$. For any $x_k = T(x_{k-1})$, we have

$$\|x^* - T(x^*)\| \leq \|x^* - x_k\| + \|x_k - T(x^*)\| \leq \|x^* - x_k\| + \rho\|x_{k-1} - x^*\|$$

By choosing k large enough, the right-hand side of the inequality can be made arbitrarily small. Thus, $\|x^* - T(x^*)\| = 0$; that is, $x^* = T(x^*)$. It remains to show that x^* is the unique fixed point of T in S. Suppose that x^* and y^* are fixed points. Then,

$$\|x^* - y^*\| = \|T(x^*) - T(y^*)\| \leq \rho\|x^* - y^*\|$$

Since $\rho < 1$, we have $x^* = y^*$.   □

2.2   Existence and Uniqueness

In this section, we derive sufficient conditions for the existence and uniqueness of the solution of the initial value problem (2.1). By a solution of (2.1) over an interval $[t_0, t_1]$ we mean a continuous function $x : [t_0, t_1] \to R^n$ such that $\dot{x}(t)$ is defined and $\dot{x}(t) = f(t, x(t))$ for all $t \in [t_0, t_1]$. If $f(t, x)$ is continuous in t and x, then the solution $x(t)$ will be continuously differentiable. We will assume that $f(t, x)$ is continuous in x but only piecewise continuous in t, in which case a solution $x(t)$ could only be piecewise continuously differentiable. The assumption that $f(t, x)$ be piecewise continuous in t allows us to include the case when $f(t, x)$ depends on a time-varying input that may experience step changes with time.

A differential equation with a given initial condition might have several solutions. For example, the scalar equation

$$\dot{x} = x^{1/3}, \quad \text{with} \quad x(0) = 0 \tag{2.3}$$

has a solution $x(t) = (2t/3)^{3/2}$. This solution is not unique, since $x(t) \equiv 0$ is another solution. Noting that the right-hand side of (2.3) is continuous in x, it is clear that continuity of $f(t, x)$ in its arguments is not sufficient to ensure uniqueness of the solution. Extra conditions must be imposed on the function f. The question of existence of a solution is less stringent. In fact, continuity of $f(t, x)$ in its arguments ensures that there is at least one solution. We shall not prove this fact here.[5] Instead, we prove an easier theorem that employs the Lipschitz condition to show existence and uniqueness.

Theorem 2.2 (Local Existence and Uniqueness) *Let $f(t, x)$ be piecewise continuous in t and satisfy the Lipschitz condition*

$$\|f(t, x) - f(t, y)\| \leq L\|x - y\|$$

$\forall\ x, y \in B = \{x \in R^n \mid \|x - x_0\| \leq r\}$, $\forall\ t \in [t_0, t_1]$. *Then, there exists some $\delta > 0$ such that the state equation*

$$\dot{x} = f(t, x), \quad \text{with}\ \ x(t_0) = x_0$$

has a unique solution over $[t_0, t_0 + \delta]$. $\diamond$

Proof: We start by noting that if $x(t)$ is a solution of

$$\dot{x} = f(t, x), \quad x(t_0) = x_0 \tag{2.4}$$

then, by integration, we have

$$x(t) = x_0 + \int_{t_0}^{t} f(s, x(s))\ ds \tag{2.5}$$

Conversely, if $x(t)$ satisfies (2.5), then $x(t)$ satisfies (2.4). Thus, the study of existence and uniqueness of the solution of the differential equation (2.4) is equivalent to the study of existence and uniqueness of the solution of the integral equation (2.5). We proceed with (2.5). Viewing the right-hand side of (2.5) as a mapping of the continuous function $x : [t_0, t_1]\ \rightarrow\ R^n$, and denoting it by $(Px)(t)$, we can rewrite (2.5) as

$$x(t) = (Px)(t) \tag{2.6}$$

Note that $(Px)(t)$ is continuous in t. A solution of (2.6) is a fixed point of the mapping P that maps x into Px. Existence of a fixed point of (2.6) can be established using the contraction mapping theorem. This requires defining a Banach space $\mathcal{X}$ and a closed set $S \subset \mathcal{X}$ such that P maps S into S and is a contraction over S. Let

$$\mathcal{X} = C[t_0, t_0 + \delta], \quad \text{with norm}\ \ \|x\|_C = \max_{t \in [t_0, t_0 + \delta]} \|x(t)\|$$

---

[5] See [118, Theorem 2.3] for a proof.

and

$$S = \{x \in \mathcal{X} \mid \|x - x_0\|_C \le r\}$$

where r is the radius of the ball B and δ is a positive constant to be chosen. We shall restrict the choice of δ to satisfy $\delta \le t_1 - t_0$, so that $[t_0, t_0 + \delta] \subset [t_0, t_1]$. Notice that $\|x(t)\|$ denotes a norm on R^n, while $\|x\|_C$ denotes a norm on $\mathcal{X}$. Also, B is a ball in R^n while S is a ball in $\mathcal{X}$. By definition, P maps $\mathcal{X}$ into $\mathcal{X}$. To show that it maps S into S, write

$$(Px)(t) - x_0 = \int_{t_0}^{t} f(s, x(s)) \, ds = \int_{t_0}^{t} [f(s, x(s)) - f(s, x_0) + f(s, x_0)] \, ds$$

By piecewise continuity of f, we know that $f(t, x_0)$ is bounded on $[t_0, t_1]$. Let

$$h = \max_{t \in [t_0, t_1]} \|(f(t, x_0))\|$$

Using the Lipschitz condition (2.2) and the fact that for each $x \in S$,

$$\|x(t) - x_0\| \le r, \ \forall \ t \in [t_0, t_0 + \delta]$$

we obtain

$$
\begin{aligned}
\|(Px)(t) - x_0\| &\le \int_{t_0}^{t} [\|f(s, x(s)) - f(s, x_0)\| + \|f(s, x_0)\|] \, ds \\
&\le \int_{t_0}^{t} [L\|x(s) - x_0\| + h] \, ds \ \le \ \int_{t_0}^{t} (Lr + h) \, ds \\
&= (t - t_0)(Lr + h) \ \le \ \delta(Lr + h)
\end{aligned}
$$

and

$$\|Px - x_0\|_C = \max_{t \in [t_0, t_0 + \delta]} \|(Px)(t) - x_0\| \le \delta(Lr + h)$$

Hence, choosing $\delta \le r/(Lr + h)$ ensures that P maps S into S. To show that P is a contraction mapping over S, let x and $y \in S$ and consider

$$
\begin{aligned}
\|(Px)(t) - (Py)(t)\| &= \left\| \int_{t_0}^{t} [f(s, x(s)) - f(s, y(s))] \, ds \right\| \\
&\le \int_{t_0}^{t} \|f(s, x(s)) - f(s, y(s))\| \, ds \\
&\le \int_{t_0}^{t} L\|x(s) - y(s)\| \, ds \ \le \ \int_{t_0}^{t} ds \, L\|x - y\|_C
\end{aligned}
$$

Therefore,

$$\|Px - Py\|_C \le L\delta\|x - y\|_C \le \rho\|x - y\|_C \quad \text{for } \delta \le \frac{\rho}{L}$$

Thus, choosing $\rho < 1$ and $\delta \leq \rho/L$ ensures that P is a contraction mapping over S. By the contraction mapping theorem, we can conclude that if δ is chosen to satisfy

$$\delta \leq \min\left\{t_1 - t_0, \frac{r}{Lr + h}, \frac{\rho}{L}\right\}, \quad \text{for} \;\; \rho < 1 \tag{2.7}$$

then (2.5) will have a unique solution in S. This is not the end of the proof, though, since we are interested in establishing uniqueness of the solution among all continuous functions $x(t)$, that is, uniqueness in $\mathcal{X}$. It turns out that any solution of (2.5) in $\mathcal{X}$ will lie in S. To see this, note that since $x(t_0) = x_0$ is inside the ball B, any continuous solution $x(t)$ must lie inside B for some interval of time. Suppose that $x(t)$ leaves the ball B, and let $t_0 + \mu$ be the first time $x(t)$ intersects the boundary of B. Then,

$$\|x(t_0 + \mu) - x_0\| = r$$

On the other hand, for all $t \leq t_0 + \mu$,

$$\|x(t) - x_0\| \leq \int_{t_0}^{t} [\|f(s, x(s)) - f(s, x_0)\| + \|f(s, x_0)\|] \, ds$$

$$\leq \int_{t_0}^{t} [L\|x(s) - x_0\| + h] \, ds \leq \int_{t_0}^{t} (Lr + h) \, ds$$

Therefore,

$$r = \|x(t_0 + \mu) - x_0\| \leq (Lr + h)\mu \;\Rightarrow\; \mu \geq \frac{r}{Lr + h} \geq \delta$$

Hence, the solution $x(t)$ cannot leave the set B within the time interval $[t_0, t_0 + \delta]$, which implies that any solution in $\mathcal{X}$ lies in S. Consequently, uniqueness of the solution in S implies uniqueness in $\mathcal{X}$. $\qquad\qquad\qquad\qquad\qquad\qquad\qquad\qquad\qquad\qquad\quad\Box$

The key assumption in Theorem 2.2 is the Lipschitz condition (2.2). A function satisfying (2.2) is said to be *Lipschitz* in x, and the positive constant L is called a *Lipschitz constant*. We also use the words *locally Lipschitz* and *globally Lipschitz* to indicate the domain over which the Lipschitz condition holds. Let us introduce the terminology first for the case when f depends only on x. A function $f(x)$ is said to be *locally Lipschitz* on a domain (open and connected set) $D \subset R^n$ if each point of D has a neighborhood D_0 such that f satisfies the Lipschitz condition (2.2) for all points in D_0 with some Lipschitz constant L_0. We say that f is Lipschitz on a set W if it satisfies (2.2) for all points in W, with the same Lipschitz constant L. A locally Lipschitz function on a domain D is not necessarily Lipschitz on D, since the Lipschitz condition may not hold uniformly (with the same constant L) for all points in D. However, a locally Lipschitz function on a domain D is Lipschitz on every compact (closed and bounded) subset of D (Exercise 2.16). A function $f(x)$

is said to be *globally Lipschitz* if it is Lipschitz on R^n. The same terminology is extended to a function $f(t, x)$, provided the Lipschitz condition holds uniformly in t for all t in a given interval of time. For example, $f(t, x)$ is locally Lipschitz in x on $[a, b] \times D \subset R \times R^n$ if each point $x \in D$ has a neighborhood D_0 such that f satisfies (2.2) on $[a, b] \times D_0$ with some Lipschitz constant L_0. We shall say that $f(t, x)$ is locally Lipschitz in x on $[t_0, \infty) \times D$ if it is locally Lipschitz in x on $[a, b] \times D$ for every compact interval $[a, b] \subset [t_0, \infty)$. A function $f(t, x)$ is Lipschitz in x on $[a, b] \times W$ if it satisfies (2.2) for all $t \in [a, b]$ and all points in W, with the same Lipschitz constant L.

When $f : R \to R$, the Lipschitz condition can be written as

$$\frac{|f(y) - f(x)|}{|y - x|} \leq L$$

which implies that on a plot of $f(x)$ versus x, a straight line joining any two points of $f(x)$ cannot have a slope whose absolute value is greater than L. Therefore, any function $f(x)$ that has infinite slope at some point is not locally Lipschitz at that point. For example, any discontinuous function is not locally Lipschitz at the point of discontinuity. As another example, the function $f(x) = x^{1/3}$, which was used in (2.3), is not locally Lipschitz at $x = 0$ since $f'(x) = \frac{1}{3} x^{-2/3} \to \infty$ as $x \to 0$. On the other hand, if $|f'(x)|$ is bounded by a constant k over the interval of interest, then $f(x)$ is Lipschitz on the same interval with Lipschitz constant $L = k$. This observation extends to vector-valued functions, as demonstrated by the following lemma.

Lemma 2.2 *Let $f : [a, b] \times D \to R^m$ be continuous for some domain $D \subset R^n$. Suppose $[\partial f / \partial x]$ exists and is continuous on $[a, b] \times D$. If, for a convex subset $W \subset D$, there is a constant $L \geq 0$ such that*

$$\left\| \frac{\partial f}{\partial x}(t, x) \right\| \leq L$$

on $[a, b] \times W$, then

$$\|f(t, x) - f(t, y)\| \leq L\|x - y\|$$

for all $t \in [a, b]$, $x \in W$, and $y \in W$. ◇

Proof: Let $\|\cdot\|_p$ be the underlying norm for any $p \in [1, \infty]$, and determine $q \in [1, \infty]$ from the relationship $1/p + 1/q = 1$. Fix $t \in [a, b]$, $x \in W$, and $y \in W$. Define $\gamma(s) = (1 - s)x + sy$ for all $s \in R$ such that $\gamma(s) \in D$. Since $W \subset D$ is convex, $\gamma(s) \in W$ for $0 \leq s \leq 1$. Take $z \in R^m$ such that[6]

$$\|z\|_q = 1 \quad \text{and} \quad z^T [f(t, y) - f(t, x)] = \|f(t, y) - f(t, x)\|_p$$

---

[6] Such z always exists; see Exercise 2.19.

Set $g(s) = z^T f(t, \gamma(s))$. Since $g(s)$ is a real-valued function which is continuously differentiable in an open interval that includes $[0, 1]$, we conclude by the mean value theorem that there is $s_1 \in (0, 1)$ such that

$$g(1) - g(0) = g'(s_1)$$

Evaluating g at $s = 0$, $s = 1$, and calculating $g'(s)$ using the chain rule, we obtain

$$z^T [f(t, y) - f(t, x)] = z^T \frac{\partial f}{\partial x}(t, \gamma(s_1))(y - x)$$

$$\|f(t, y) - f(t, x)\|_p \leq \|z\|_q \left\| \frac{\partial f}{\partial x}(t, \gamma(s_1)) \right\|_p \|y - x\|_p \leq L\|y - x\|_p$$

where we used the Hölder inequality $|z^T w| \leq \|z\|_q \|w\|_p$.   □

The proof of the lemma shows how a Lipschitz constant can be calculated using knowledge of $[\partial f/\partial x]$.

The Lipschitz property of a function is stronger than continuity. It can be easily seen that if $f(x)$ is Lipschitz on W, then it is uniformly continuous on W (Exercise 2.17). The converse is not true, as seen from the function $f(x) = x^{1/3}$, which is continuous but not locally Lipschitz at $x = 0$. The Lipschitz property is weaker than continuous differentiability, as stated in the following lemma.

Lemma 2.3 *Let $f(t, x)$ be continuous on $[a, b] \times D$, for some domain $D \subset R^n$. If $[\partial f/\partial x]$ exists and is continuous on $[a, b] \times D$, then f is locally Lipschitz in x on $[a, b] \times D$.*   ◇

Proof: For $x_0 \in D$, let r be so small that the ball $D_0 = \{x \in R^n \mid \|x - x_0\| \leq r\}$ is contained in D. The set D_0 is convex and compact. By continuity, $[\partial f/\partial x]$ is bounded on $[a, b] \times D$. Let L_0 be a bound for $\|\partial f/\partial x\|$ on $[a, b] \times D_0$. By Lemma 2.2, f is Lipschitz on $[a, b] \times D_0$ with Lipschitz constant L_0.   □

It is left as an exercise to the reader (Exercise 2.20) to extend the proof of Lemma 2.2 to prove the following lemma.

Lemma 2.4 *Let $f(t, x)$ be continuous on $[a, b] \times R^n$. If $[\partial f/\partial x]$ exists and is continuous on $[a, b] \times R^n$, then f is globally Lipschitz in x on $[a, b] \times R^n$ if and only if $[\partial f/\partial x]$ is uniformly bounded on $[a, b] \times R^n$.*   ◇

Example 2.2 The function

$$f(x) = \begin{bmatrix} -x_1 + x_1 x_2 \\ x_2 - x_1 x_2 \end{bmatrix}$$

is continuously differentiable on R^2. Hence, it is locally Lipschitz on R^2. It is not globally Lipschitz since $[\partial f/\partial x]$ is not uniformly bounded on R^2. On any compact subset of R^2, f is Lipschitz. Suppose we are interested in calculating a Lipschitz constant over the convex set

$$W = \{x \in R^2 \mid |x_1| \le a_1, \ |x_2| \le a_2\}$$

The Jacobian matrix $[\partial f/\partial x]$ is given by

$$\left[\frac{\partial f}{\partial x}\right] = \left[\begin{array}{cc} -1 + x_2 & x_1 \\ -x_2 & 1 - x_1 \end{array}\right]$$

Using $\|.\|_\infty$ for vectors in R^2 and the induced matrix norm for matrices, we have

$$\left\|\frac{\partial f}{\partial x}\right\|_\infty = \max\{|-1 + x_2| + |x_1|, \ |x_2| + |1 - x_1|\}$$

All points in W satisfy

$$\begin{aligned} |-1 + x_2| + |x_1| &\le& 1 + a_2 + a_1 \\ |x_2| + |1 - x_1| &\le& a_2 + 1 + a_1 \end{aligned}$$

Hence,

$$\left\|\frac{\partial f}{\partial x}\right\|_\infty \le 1 + a_1 + a_2$$

and a Lipschitz constant can be taken as $L = 1 + a_1 + a_2$. $\triangle$

Example 2.3 Consider the function

$$f(x) = \left[\begin{array}{c} x_2 \\ -\mathrm{sat}(x_1 + x_2) \end{array}\right]$$

where the saturation function $\mathrm{sat}(\cdot)$ is defined by

$$\mathrm{sat}(y) = \left\{\begin{array}{ll} -1, & \text{for } y < -1 \\ y, & \text{for } |y| \le 1 \\ 1, & \text{for } y > 1 \end{array}\right.$$

The function f is not continuously differentiable on R^2. Let us check its Lipschitz property by examining $f(x) - f(y)$. Using $\|.\|_2$ for vectors in R^2 and the fact that the saturation function $\mathrm{sat}(\cdot)$ satisfies

$$|\mathrm{sat}(\eta) - \mathrm{sat}(\xi)| \le |\eta - \xi|$$

we obtain

$$\|f(x) - f(y)\|_2^2 \leq (x_2 - y_2)^2 + (x_1 + x_2 - y_1 - y_2)^2$$
$$\leq (x_1 - y_1)^2 + 2(x_1 - y_1)(x_2 - y_2) + 2(x_2 - y_2)^2$$

Using the inequality

$$a^2 + 2ab + 2b^2 = \begin{bmatrix} a \\ b \end{bmatrix}^T \begin{bmatrix} 1 & 1 \\ 1 & 2 \end{bmatrix} \begin{bmatrix} a \\ b \end{bmatrix} \leq \lambda_{\max} \left\{ \begin{bmatrix} 1 & 1 \\ 1 & 2 \end{bmatrix} \right\} \times \left\| \begin{bmatrix} a \\ b \end{bmatrix} \right\|_2^2$$

we conclude that

$$\|f(x) - f(y)\|_2 \leq 1.618\|x - y\|_2, \quad \forall\, x, y \in R^2$$

Here we have used a property of positive semidefinite symmetric matrices; that is, $x^T P x \leq \lambda_{\max}(P)\, x^T x$, for all $x \in R^n$, where $\lambda_{\max}(\cdot)$ is the maximum eigenvalue of the matrix. A more conservative (larger) Lipschitz constant will be obtained if we use the more conservative inequality

$$a^2 + 2ab + 2b^2 \leq 2a^2 + 3b^2 \leq 3(a^2 + b^2)$$

resulting in a Lipschitz constant $L = \sqrt{3}$. $\triangle$

In these two examples, we have used $\|\cdot\|_\infty$ in one case and $\|\cdot\|_2$ in the other. Due to equivalence of norms, the choice of a norm on R^n does not affect the Lipschitz property of a function. It only affects the value of the Lipschitz constant (Exercise 2.21). Example 2.3 illustrates the fact that the Lipschitz condition (2.2) does not uniquely define the Lipschitz constant L. If (2.2) is satisfied with some positive constant L, it is satisfied with any positive constant larger than L. This nonuniqueness can be removed by defining L to be the smallest constant for which (2.2) is satisfied, but we seldom need to do that.

Theorem 2.2 is a local theorem since it guarantees existence and uniqueness only over an interval $[t_0, t_0 + \delta]$, where δ may be very small. In other words, we have no control on δ; hence, we cannot ensure existence and uniqueness over a given time interval $[t_0, t_1]$. However, one may try to extend the interval of existence by repeated applications of the local theorem. Starting at a time t_0 with an initial state $x(t_0) = x_0$, Theorem 2.2 shows that there is a positive constant δ (dependent on x_0) such that the state equation(2.1) has a unique solution over the time interval $[t_0, t_0 + \delta]$. Now, taking $t_0 + \delta$ as a new initial time and $x(t_0 + \delta)$ as a new initial state, one may try to apply Theorem 2.2 to establish existence of the solution beyond $t_0 + \delta$. If all the conditions of the theorem are satisfied at $[t_0 + \delta, x(t_0 + \delta)]$, then there exists $\delta_2 > 0$ such that the equation has a unique solution over $[t_0 + \delta, t_0 + \delta + \delta_2]$ that passes through the point $[t_0 + \delta, x(t_0 + \delta)]$. We piece together the solutions over

$[t_0, t_0 + \delta]$ and $[t_0 + \delta, t_0 + \delta + \delta_2]$ to establish the existence of a unique solution over $[t_0, t_0 + \delta + \delta_2]$. This idea can be repeated to keep extending the solution. However in general, the interval of existence of the solution cannot be extended indefinitely because the conditions of Theorem 2.2 may cease to hold. There is a maximum interval $[t_0, T)$ where the unique solution starting at (t_0, x_0) exists.[7] In general, T may be less than t_1, in which case as $t \to T$, the solution leaves any compact set over which f is locally Lipschitz in x (Exercise 2.33).

Example 2.4 Consider the scalar system

$$\dot{x} = -x^2, \quad \text{with} \ x(0) = -1$$

The function $f(x) = -x^2$ is locally Lipschitz for all $x \in R$. Hence, it is Lipschitz on any compact subset of R. The unique solution

$$x(t) = \frac{1}{t - 1}$$

exists over $[0, 1)$. As $t \to 1$, $x(t)$ leaves any compact set. $\triangle$

The phrase "finite escape time" is used to describe the phenomenon that a trajectory escapes to infinity at a finite time. In Example 2.4, we say that the trajectory has a finite escape time at $t = 1$.

In view of the discussion preceding Example 2.4, one may pose the question: When is it guaranteed that the solution can be extended indefinitely? One way to answer this question is to require additional conditions which ensure that the solution $x(t)$ will always be in a set where $f(t, x)$ is uniformly Lipschitz in x. This is done in the next theorem by requiring f to satisfy a global Lipschitz condition. The theorem establishes the existence of a unique solution over $[t_0, t_1]$ where t_1 may be arbitrarily large.

Theorem 2.3 (Global Existence and Uniqueness) *Suppose $f(t, x)$ is piecewise continuous in t and satisfies*

$$\|f(t, x) - f(t, y)\| \leq L\|x - y\|$$
$$\|f(t, x_0)\| \leq h$$

$\forall \ x, y \in R^n, \ \forall \ t \in [t_0, t_1]$. *Then, the state equation*

$$\dot{x} = f(t, x), \quad \text{with} \ x(t_0) = x_0$$

has a unique solution over $[t_0, t_1]$. $\diamond$

---

[7]For a proof of this statement, see [70, Section 8.5] or [118, Section 2.3].

Proof: The key point of the proof is to show that the constant δ of Theorem 2.2 can be made independent of the initial state x_0. From (2.7), we see that the dependence of δ on the initial state comes through the constant h in the term $r/(Lr+h)$. Since in the current case the Lipschitz condition holds globally, we can choose r arbitrarily large. Therefore, for any finite h, we can choose r large enough so that $r/(Lr+h) > \rho/L$. This reduces (2.7) to the requirement

$$\delta \leq \min\left\{t_1 - t_0, \frac{\rho}{L}\right\}, \quad \text{for } \rho < 1$$

If $t_1 - t_0 \leq \rho/L$, we could choose $\delta = t_1 - t_0$ and be done. Otherwise, we choose δ to satisfy $\delta \leq \rho/L$. Now, divide $[t_0, t_1]$ into a finite number of subintervals of length $\delta \leq \rho/L$, and apply Theorem 2.2 repeatedly.[8] $\square$

Example 2.5 Consider the linear system

$$\dot{x} = A(t)x + g(t) = f(t, x)$$

where $A(\cdot)$ and $g(\cdot)$ are piecewise continuous functions of t. Over any finite interval of time $[t_0, t_1]$, the elements of $A(t)$ and $g(t)$ are bounded. Hence, $\|A(t)\| \leq a$ and $\|g(t)\| \leq b$, where $\|g\|$ can be any norm on R^n and $\|A\|$ is the induced matrix norm. The conditions of Theorem 2.3 are satisfied since

$$\begin{aligned}
\|f(t, x) - f(t, y)\| &= \|A(t)(x - y)\| \leq \|A(t)\|\,\|x - y\| \\
&\leq a\|x - y\|, \quad \forall\, x, y \in R^n,\ \forall\, t \in [t_0, t_1]
\end{aligned}$$

and

$$\begin{aligned}
\|f(t, x_0)\| &= \|A(t)x_0 + g(t)\| \leq a\|x_0\| + b \\
&\leq h, \quad \text{for each finite } x_0,\ \forall\, t \in [t_0, t_1]
\end{aligned}$$

Therefore, Theorem 2.3 shows that the linear system has a unique solution over $[t_0, t_1]$. Since t_1 can be arbitrarily large, we can also conclude that if $A(t)$ and $g(t)$ are piecewise continuous $\forall\, t \geq t_0$, then the system has a unique solution $\forall\, t \geq t_0$. Therefore, the system cannot have a finite escape time. $\triangle$

For the linear system of Example 2.5, the global Lipschitz condition of Theorem 2.3 is a reasonable requirement. This may not be the case for nonlinear systems, in general. We should distinguish between the local Lipschitz requirement of Theorem 2.2 and the global Lipschitz requirement of Theorem 2.3. Local Lipschitz property of a function is basically a smoothness requirement. It is implied by continuous

---

[8]Note that the initial state of each subinterval x_1, say, will satisfy $\|f(t, x_1)\| \leq h_1$ for some finite h_1.

differentiability. Except for discontinuous nonlinearities, which are idealizations of physical phenomena, it is reasonable to expect models of physical systems to have locally Lipschitz right-hand side functions. Examples of continuous functions which are not locally Lipschitz are quite exceptional and rarely arise in practice. The global Lipschitz property, on the other hand, is restrictive. Models of many physical systems fail to satisfy it. One can easily construct smooth meaningful examples which do not have the global Lipschitz property, but do have unique global solutions, which is an indication of the conservative nature of Theorem 2.3.

Example 2.6 Consider the scalar system

$$\dot{x} = -x^3 = f(x)$$

The function $f(x)$ does not satisfy a global Lipschitz condition since the Jacobian $\partial f/\partial x = -3x^2$ is not globally bounded. Nevertheless, for any initial state $x(t_0) = x_0$, the equation has the unique solution

$$x(t) = \text{sign}(x_0)\sqrt{\frac{x_0^2}{1 + 2x_0^2(t - t_0)}}$$

which is well defined for all $t \geq t_0$.   △

In view of the conservative nature of the global Lipschitz condition, it would be useful to have a global existence and uniqueness theorem that requires the function f to be only locally Lipschitz. The following theorem achieves that at the expense of having to know more about the solution of the system.

Theorem 2.4 *Let $f(t, x)$ be piecewise continuous in t and locally Lipschitz in x for all $t \geq t_0$ and all x in a domain $D \subset R^n$. Let W be a compact subset of D, $x_0 \in W$, and suppose it is known that every solution of*

$$\dot{x} = f(t, x), \quad x(t_0) = x_0$$

lies entirely in W. Then, there is a unique solution that is defined for all $t \geq t_0$. ◇

Proof: Recall the discussion on extending solutions, preceding Example 2.4. By Theorem 2.2, there is a unique local solution over $[t_0, t_0 + \delta]$. Let $[t_0, T)$ be its maximal interval of existence. We want to show that $T = \infty$. Recall (Exercise 2.33) the fact that if T is finite, then the solution must leave any compact subset of D. Since the solution never leaves the compact set W, we conclude that $T = \infty$.   □

The trick in applying Theorem 2.4 is in checking the assumption that every solution lies in a compact set without actually solving the differential equation. We shall see in Chapter 3 that Lyapunov's method for studying stability is very valuable in that regard. For now, let us illustrate the application of the theorem by a simple example.

Example 2.7 Consider again the system

$$\dot{x} = -x^3 = f(x)$$

of Example 2.6. The function $f(x)$ is locally Lipschitz on R. If, at any instant of time, $x(t)$ is positive, the derivative $\dot{x}(t)$ will be negative. Similarly, if $x(t)$ is negative, the derivative $\dot{x}(t)$ will be positive. Therefore, starting from any initial condition $x(0) = a$, the solution cannot leave the compact set $\{x \in R \mid |x| \leq |a|\}$. Thus, without calculating the solution, we conclude by Theorem 2.4 that the equation has a unique solution for all $t \geq 0$. $\triangle$

2.3   Continuous Dependence on Initial Conditions and Parameters

For the solution of the state equation (2.1) to be of any interest, it must depend continuously on the initial state x_0, the initial time t_0, and the right-hand side function $f(t, x)$. Continuous dependence on the initial time is obvious from the integral expression (2.5). Therefore, we leave it as an exercise (Exercise 2.44) and concentrate our attention on continuous dependence on the initial state x_0 and the function f. Let $y(t)$ be a solution of (2.1) that starts at $y(t_0) = y_0$ and is defined on the compact time interval $[t_0, t_1]$. The solution depends continuously on y_0 if solutions starting at nearby points are defined on the same time interval and remain close to each other in this interval. This statement can be made precise with the ϵ–δ argument: given $\epsilon > 0$, there is $\delta > 0$ such that for all z_0 in the ball $\{x \in R^n \mid \|x - y_0\| < \delta\}$, the equation $\dot{x} = f(t, x)$ has a unique solution $z(t)$ defined on $[t_0, t_1]$, with $z(t_0) = z_0$, and satisfies $\|z(t) - y(t)\| < \epsilon$ for all $t \in [t_0, t_1]$. Continuous dependence on the right-hand side function f is defined similarly, but to state the definition precisely we need a mathematical representation of the perturbation of f. One possible representation is to replace f by a sequence of functions f_m which converge uniformly to f as $m \to \infty$. For each function f_m, the solution of $\dot{x} = f_m(t, x)$ with $x(t_0) = x_0$ is denoted by $x_m(t)$. The solution is said to depend continuously on the right-hand side function if $x_m(t) \to x(t)$ as $m \to \infty$. This approach is a little bit involved, and will not be pursued here.[9] A more restrictive, but simpler, mathematical representation is to assume that f depends continuously on a set of constant parameters; that is, $f = f(t, x, \lambda)$ where $\lambda \in R^p$. The constant parameters could represent physical parameters of the system, and the study of perturbation of these parameters accounts for modeling errors or changes in the parameter values due to aging. Let $x(t, \lambda_0)$ be a solution of $\dot{x} = f(t, x, \lambda_0)$ defined on $[t_0, t_1]$, with $x(t_0, \lambda_0) = x_0$. The solution is said to

---

[9] See [36, Section 1.3], [65, Section 1.3], or [118, Section 2.5] for results on continuous dependence on parameters using this approach.

depend continuously on λ if for any $\epsilon > 0$, there is $\delta > 0$ such that for all λ in the ball $\{\lambda \in R^p \mid \|\lambda - \lambda_0\| < \delta\}$, the equation $\dot{x} = f(t, x, \lambda)$ has a unique solution $x(t, \lambda)$ defined on $[t_0, t_1]$ with $x(t_0, \lambda) = x_0$, and satisfies $\|x(t, \lambda) - x(t, \lambda_0)\| < \epsilon$ for all $t \in [t_0, t_1]$.

Continuous dependence on initial states and continuous dependence on parameters can be studied simultaneously. We start with a simple result which bypasses the issue of existence and uniqueness and concentrates on the closeness of solutions.

Theorem 2.5 *Let $f(t, x)$ be piecewise continuous in t and Lipschitz in x on $[t_0, t_1] \times W$ with a Lipschitz constant L, where $W \subset R^n$ is an open connected set. Let $y(t)$ and $z(t)$ be solutions of*

$$\dot{y} = f(t, y), \quad y(t_0) = y_0$$

and

$$\dot{z} = f(t, z) + g(t, z), \quad z(t_0) = z_0$$

such that $y(t)$, $z(t) \in W$ for all $t \in [t_0, t_1]$. Suppose that

$$\|g(t, x)\| \leq \mu, \quad \forall \ (t, x) \in [t_0, t_1] \times W$$

for some $\mu > 0$, and

$$\|y_0 - z_0\| \leq \gamma$$

Then,

$$\|y(t) - z(t)\| \leq \gamma \exp[L(t - t_0)] + \frac{\mu}{L} \left\{ \exp[L(t - t_0)] - 1 \right\}$$

$\forall \ t \in [t_0, t_1]$. $\diamond$

Proof: The solutions $y(t)$ and $z(t)$ are given by

$$y(t) = y_0 + \int_{t_0}^{t} f(s, y(s)) \ ds$$

$$z(t) = z_0 + \int_{t_0}^{t} [f(s, z(s)) + g(s, z(s))] \ ds$$

Subtracting the two equations and taking norms yield

$$
\begin{aligned}
\|y(t) - z(t)\| &\leq \|y_0 - z_0\| + \int_{t_0}^{t} \|f(s, y(s)) - f(s, z(s))\| \ ds \\
&\quad + \int_{t_0}^{t} \|g(s, z(s))\| \ ds \\
&\leq \gamma + \mu(t - t_0) + \int_{t_0}^{t} L\|y(s) - z(s)\| \ ds
\end{aligned}
$$

Application of the Gronwall-Bellman inequality to the function $\|y(t) - z(t)\|$ results in

$$\|y(t) - z(t)\| \leq \gamma + \mu(t - t_0) + \int_{t_0}^{t} L[\gamma + \mu(s - t_0)] \exp[L(t - s)] \, ds$$

Integrating the right-hand side by parts, we obtain

$$
\begin{aligned}
\|y(t) - z(t)\| \quad \leq \quad & \gamma + \mu(t - t_0) - \gamma - \mu(t - t_0) + \gamma \exp[L(t - t_0)] \\
& + \int_{t_0}^{t} \mu \exp[L(t - s)] \, ds \\
= \quad & \gamma \exp[L(t - t_0)] + \frac{\mu}{L} \{\exp[L(t - t_0)] - 1\}
\end{aligned}
$$

which completes the proof of the theorem.   □

With Theorem 2.5 in hand, we can prove the following theorem on the continuity of solutions in terms of initial states and parameters.

Theorem 2.6 *Let $f(t, x, \lambda)$ be continuous in (t, x, λ) and locally Lipschitz in x (uniformly in t and λ) on $[t_0, t_1] \times D \times \{\|\lambda - \lambda_0\| \leq c\}$, where $D \subset R^n$ is an open connected set. Let $y(t, \lambda_0)$ be a solution of $\dot{x} = f(t, x, \lambda_0)$ with $y(t_0, \lambda_0) = y_0 \in D$. Suppose $y(t, \lambda_0)$ is defined and belongs to D for all $t \in [t_0, t_1]$. Then, given $\epsilon > 0$, there is $\delta > 0$, such that if*

$$\|z_0 - y_0\| < \delta \quad \text{and} \quad \|\lambda - \lambda_0\| < \delta$$

then there is a unique solution $z(t, \lambda)$ of $\dot{x} = f(t, x, \lambda)$ defined on $[t_0, t_1]$, with $z(t_0, \lambda) = z_0$, and $z(t, \lambda)$ satisfies

$$\|z(t, \lambda) - y(t, \lambda_0)\| < \epsilon, \quad \forall \, t \in [t_0, t_1]$$

◇

Proof: By continuity of $y(t, \lambda_0)$ in t and the compactness of $[t_0, t_1]$, we know that $y(t, \lambda_0)$ is uniformly bounded on $[t_0, t_1]$. Define a "tube" U around the solution $y(t, \lambda_0)$ (see Figure 2.1) by

$$U = \{(t, x) \in [t_0, t_1] \times R^n \mid \|x - y(t, \lambda_0)\| \leq \epsilon\}$$

Suppose $U \subset [t_0, t_1] \times D$; if not, replace ϵ by an $\epsilon_1 < \epsilon$ that is small enough to ensure that $U \subset [t_0, t_1] \times D$ and continue the proof with ϵ_1. The set U is compact; hence, $f(t, x, \lambda)$ is Lipschitz in x on U with a Lipschitz constant, say, L. By continuity of f in λ, for any $\alpha > 0$ there is $\beta > 0$ (with $\beta < c$) such that

$$\|f(t, x, \lambda) - f(t, x, \lambda_0)\| < \alpha, \quad \forall \, (t, x) \in U, \, \forall \, \|\lambda - \lambda_0\| < \beta$$

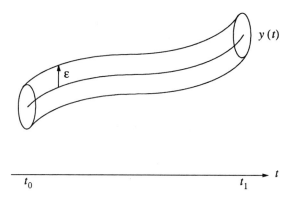

Figure 2.1: A tube constructed around the solution $y(t)$.

Take $\alpha < \epsilon$ and $\|z_0 - y_0\| < \alpha$. By the local existence and uniqueness theorem, there is a unique solution $z(t, \lambda)$ on some time interval $[t_0, t_0 + \Delta]$. The solution starts inside the tube U, and as long as it remains in the tube it can be extended. We will show that, by choosing a small enough α, the solution remains in U for all $t \in [t_0, t_1]$. In particular, we let τ be the first time the solution leaves the tube and show that we can make $\tau > t_1$. On the time interval $[t_0, \tau]$, all the conditions of Theorem 2.5 are satisfied with $\gamma = \mu = \alpha$. Hence,

$$
\begin{aligned}
\|z(t, \lambda) - y(t, \lambda_0)\| &\leq \alpha \exp[L(t - t_0)] + \frac{\alpha}{L}\{\exp[L(t - t_0)] - 1\} \\
&< \alpha \left(1 + \frac{1}{L}\right) \exp[L(t - t_0)]
\end{aligned}
$$

Choosing $\alpha \leq \epsilon L \exp[-L(t_1 - t_0)]/(1 + L)$ ensures that the solution $z(t, \lambda)$ cannot leave the tube during the interval $[t_0, t_1]$. Therefore, $z(t, \lambda)$ is defined on $[t_0, t_1]$ and satisfies $\|z(t, \lambda) - y(t, \lambda_0)\| < \epsilon$. Taking $\delta = \min\{\alpha, \beta\}$ completes the proof of the theorem. $\qquad\square$

2.4 Differentiability of Solutions and Sensitivity Equations

Suppose $f(t, x, \lambda)$ is continuous in (t, x, λ) and has continuous first partial derivatives with respect to x and λ for all $(t, x, \lambda) \in [t_0, t_1] \times R^n \times R^p$. Let λ_0 be a nominal value of λ, and suppose the nominal state equation

$$
\dot{x} = f(t, x, \lambda_0), \quad \text{with} \quad x(t_0) = x_0
$$

has a unique solution $x(t, \lambda_0)$ over $[t_0, t_1]$. From Theorem 2.6, we know that for all λ sufficiently close to λ_0, that is, $\|\lambda - \lambda_0\|$ sufficiently small, the state equation

$$\dot{x} = f(t, x, \lambda), \quad \text{with} \quad x(t_0) = x_0$$

has a unique solution $x(t, \lambda)$ over $[t_0, t_1]$ that is close to the nominal solution $x(t, \lambda_0)$. The continuous differentiability of f with respect to x and λ implies the additional property that the solution $x(t, \lambda)$ is differentiable with respect to λ near λ_0. To see this, write

$$x(t, \lambda) = x_0 + \int_{t_0}^{t} f(s, x(s, \lambda), \lambda) \, ds$$

Taking partial derivatives with respect to λ yields

$$x_\lambda(t, \lambda) = \int_{t_0}^{t} \left[\frac{\partial f}{\partial x}(s, x(s, \lambda), \lambda) \, x_\lambda(s, \lambda) + \frac{\partial f}{\partial \lambda}(s, x(s, \lambda), \lambda) \right] ds$$

where $x_\lambda(t, \lambda) = [\partial x(t, \lambda)/\partial \lambda]$ and $[\partial x_0/\partial \lambda] = 0$, since x_0 is independent of λ. Differentiating with respect to t, it can be seen that $x_\lambda(t, \lambda)$ satisfies the differential equation

$$\frac{\partial}{\partial t} x_\lambda(t, \lambda) = A(t, \lambda) x_\lambda(t, \lambda) + B(t, \lambda), \quad x_\lambda(t_0, \lambda) = 0 \qquad (2.8)$$

where

$$A(t, \lambda) = \left. \frac{\partial f(t, x, \lambda)}{\partial x} \right|_{x = x(t, \lambda)}, \quad B(t, \lambda) = \left. \frac{\partial f(t, x, \lambda)}{\partial \lambda} \right|_{x = x(t, \lambda)}$$

For λ sufficiently close to λ_0, the matrices $A(t, \lambda)$ and $B(t, \lambda)$ are defined on $[t_0, t_1]$. Hence, $x_\lambda(t, \lambda)$ is defined on the same interval. At $\lambda = \lambda_0$, the right-hand side of (2.8) depends only on the nominal solution $x(t, \lambda_0)$. Let $S(t) = x_\lambda(t, \lambda_0)$; then $S(t)$ is the unique solution of the equation

$$\dot{S}(t) = A(t, \lambda_0) S(t) + B(t, \lambda_0), \quad S(t_0) = 0 \qquad (2.9)$$

The function $S(t)$ is called the *sensitivity function*, and (2.9) is called the *sensitivity equation*. Sensitivity functions provide first-order estimates of the effect of parameter variations on solutions. They can also be used to approximate the solution when λ is sufficiently close to its nominal value λ_0. For small $\|\lambda - \lambda_0\|$, $x(t, \lambda)$ can be expanded in a Taylor series about the nominal solution $x(t, \lambda_0)$ to obtain

$$x(t, \lambda) = x(t, \lambda_0) + S(t)(\lambda - \lambda_0) + \text{ higher-order terms}$$

Neglecting the higher-order terms, the solution $x(t, \lambda)$ can be approximated by

$$x(t, \lambda) \approx x(t, \lambda_0) + S(t)(\lambda - \lambda_0) \qquad (2.10)$$

We shall not justify this approximation here. It will be justified in Chapter 7 when we study the perturbation theory. The significance of (2.10) is in the fact that knowledge of the nominal solution and the sensitivity function suffices to approximate the solution for all values of λ in a (small) ball centered at λ_0.

The procedure for calculating the sensitivity function $S(t)$ is summarized by the following steps:

- Solve the nominal state equation for the nominal solution $x(t, \lambda_0)$.

- Evaluate the Jacobian matrices

$$A(t, \lambda_0) = \left. \frac{\partial f(t, x, \lambda)}{\partial x} \right|_{x=x(t,\lambda_0),\ \lambda=\lambda_0}$$

$$B(t, \lambda_0) = \left. \frac{\partial f(t, x, \lambda)}{\partial \lambda} \right|_{x=x(t,\lambda_0),\ \lambda=\lambda_0}$$

- Solve the sensitivity equation (2.10) for $S(t)$.

In this procedure, we need to solve a nonlinear nominal state equation and a linear time-varying sensitivity equation. Except for some trivial cases, we will be forced to solve these equations numerically. An alternative approach for calculating $S(t)$ is to solve for the nominal solution and the sensitivity function simultaneously. A procedure for calculating $x_\lambda(t, \lambda)$ is to append the variational equation (2.8) with the original state equation to obtain the $(n + np)$ augmented equation

$$
\begin{aligned}
\dot{x} &= f(t, x, \lambda), & x(t_0) &= x_0 \\
\dot{x}_\lambda &= \left[\tfrac{\partial f(t,x,\lambda)}{\partial x} \right] x_\lambda + \left[\tfrac{\partial f(t,x,\lambda)}{\partial \lambda} \right], & x_\lambda(t_0) &= 0
\end{aligned}
\tag{2.11}
$$

which is solved numerically. The sensitivity function $S(t)$ can be calculated by solving (2.11) at the nominal value $\lambda = \lambda_0$. Notice that if the original state equation is autonomous, that is, $f(t, x, \lambda) = f(x, \lambda)$, then the augmented equation (2.11) will be autonomous as well. We illustrate the latter procedure by the following example.

Example 2.8 Consider the phase-locked loop equation

$$
\begin{aligned}
\dot{x}_1 &= x_2 & &= f_1(x_1, x_2) \\
\dot{x}_2 &= -c \sin x_1 - (a + b \cos x_1)x_2 & &= f_2(x_1, x_2)
\end{aligned}
$$

and suppose the parameters a, b, and c have the nominal values $a_0 = 1$, $b_0 = 0$, and $c_0 = 1$. The nominal system is given by

$$
\begin{aligned}
\dot{x}_1 &= x_2 \\
\dot{x}_2 &= -\sin x_1 - x_2
\end{aligned}
$$

The Jacobian matrices $[\partial f/\partial x]$ and $[\partial f/\partial \lambda]$ are given by

$$\frac{\partial f}{\partial x} = \begin{bmatrix} 0 & 1 \\ -c \cos x_1 + b x_2 \sin x_1 & -(a + b \cos x_1) \end{bmatrix}$$

$$\frac{\partial f}{\partial \lambda} = \begin{bmatrix} \frac{\partial f}{\partial a} & \frac{\partial f}{\partial b} & \frac{\partial f}{\partial c} \end{bmatrix} = \begin{bmatrix} 0 & 0 & 0 \\ -x_2 & -x_2 \cos x_1 & -\sin x_1 \end{bmatrix}$$

Evaluate these Jacobian matrices at the nominal parameters $a = 1$, $b = 0$, and $c = 1$ to obtain

$$\left. \frac{\partial f}{\partial x} \right|_{\text{nominal}} = \begin{bmatrix} 0 & 1 \\ -\cos x_1 & -1 \end{bmatrix}$$

$$\left. \frac{\partial f}{\partial \lambda} \right|_{\text{nominal}} = \begin{bmatrix} 0 & 0 & 0 \\ -x_2 & -x_2 \cos x_1 & -\sin x_1 \end{bmatrix}$$

To solve for $S(t)$ numerically, we solve (2.11) at the nominal values of the parameters. Let

$$S = \begin{bmatrix} x_3 & x_5 & x_7 \\ x_4 & x_6 & x_8 \end{bmatrix} = \left. \begin{bmatrix} \frac{\partial x_1}{\partial a} & \frac{\partial x_1}{\partial b} & \frac{\partial x_1}{\partial c} \\ \frac{\partial x_2}{\partial a} & \frac{\partial x_2}{\partial b} & \frac{\partial x_2}{\partial c} \end{bmatrix} \right|_{\text{nominal}}$$

Then (2.11) is given by

$$
\begin{aligned}
\dot{x}_1 &= x_2, & x_1(0) &= x_{10} \\
\dot{x}_2 &= -\sin x_1 - x_2, & x_2(0) &= x_{20} \\
\dot{x}_3 &= x_4, & x_3(0) &= 0 \\
\dot{x}_4 &= -x_3 \cos x_1 - x_4 - x_2, & x_4(0) &= 0 \\
\dot{x}_5 &= x_6, & x_5(0) &= 0 \\
\dot{x}_6 &= -x_5 \cos x_1 - x_6 - x_2 \cos x_1, & x_6(0) &= 0 \\
\dot{x}_7 &= x_8, & x_7(0) &= 0 \\
\dot{x}_8 &= -x_7 \cos x_1 - x_8 - \sin x_1, & x_8(0) &= 0
\end{aligned}
$$

The solution of this equation was computed for the initial state $x_{10} = x_{20} = 1$. Figure 2.2(a) shows x_3, x_5, and x_7, which are the sensitivities of x_1 with respect to a, b, and c, respectively. Figure 2.2(b) shows the corresponding quantities for x_2. Inspection of these figures shows that the solution is more sensitive to variations in the parameter c than to variations in the parameters a and b. This pattern is consistent when we solve for other initial states. $\triangle$

2.5 Comparison Principle

Quite often when we study the state equation $\dot{x} = f(t, x)$ we need to compute bounds on the solution $x(t)$ without computing the solution itself. The Gronwall-Bellman inequality (Lemma 2.1) is one tool that can be used toward that goal. Another tool is the comparison lemma. It applies to a situation where the derivative of

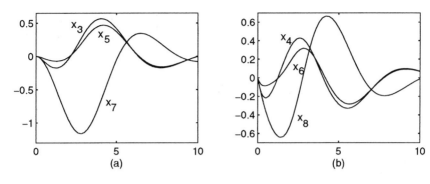

Figure 2.2: Sensitivity functions for Example 2.8.

a scalar differentiable function $v(t)$ satisfies inequality of the form $\dot{v}(t) \leq f(t, v(t))$ for all t in a certain time interval. Such inequality is called a *differential inequality* and a function $v(t)$ satisfying the inequality is called a solution of the differential inequality. The comparison lemma compares the solution of the differential inequality $\dot{v}(t) \leq f(t, v(t))$ to the solution of the differential equation $\dot{u} = f(t, u)$. The lemma applies even when $v(t)$ is not differentiable, but has an upper right-hand derivative $D^+ v(t)$ which satisfies a differential inequality. The upper right-hand derivative $D^+ v(t)$ is defined in Appendix A.1. For our purposes, it is enough to know two facts:

- if $v(t)$ is differentiable at t, then $D^+ v(t) = \dot{v}(t)$;

- if

$$\frac{1}{h} |v(t + h) - v(t)| \leq g(t, h), \quad \forall \, h \in (0, b]$$

and

$$\lim_{h \to 0^+} g(t, h) = g_0(t)$$

then $D^+ v(t) \leq g_0(t)$.

The limit $h \to 0^+$ means that h approaches zero from above.

Lemma 2.5 (Comparison Lemma) *Consider the scalar differential equation*

$$\dot{u} = f(t, u), \quad u(t_0) = u_0$$

where $f(t, u)$ is continuous in t and locally Lipschitz in u, for all $t \geq 0$ and all $u \in J \subset R$. Let $[t_0, T)$ (T could be infinity) be the maximal interval of existence *of the solution $u(t)$, and suppose $u(t) \in J$ for all $t \in [t_0, T)$. Let $v(t)$ be a continuous function whose upper right-hand derivative $D^+ v(t)$ satisfies the differential inequality*

$$D^+ v(t) \le f(t, v(t)), \qquad v(t_0) \le u_0$$

with $v(t) \in J$ for all $t \in [t_0, T)$. Then, $v(t) \le u(t)$ for all $t \in [t_0, T)$. $\diamond$

Proof: Appendix A.1.

Example 2.9 The scalar differential equation

$$\dot{x} = f(x) = -(1 + x^2)x, \qquad x(0) = a$$

has a unique solution on $[0, t_1)$ for some $t_1 > 0$, because $f(x)$ is locally Lipschitz. Let $v(t) = x^2(t)$. The function $v(t)$ is differentiable and its derivative is given by

$$\dot{v}(t) = 2x(t)\dot{x}(t) = -2x^2(t) - 2x^4(t) \le -2x^2(t)$$

Hence, $v(t)$ satisfies the differential inequality

$$\dot{v}(t) \le -2v(t), \qquad v(0) = a^2$$

Let $u(t)$ be the solution of the differential equation

$$\dot{u} = -2u, \quad u(0) = a^2 \quad \Rightarrow \quad u(t) = a^2 e^{-2t}$$

Then, by the comparison lemma, the solution $x(t)$ is defined for all $t \ge 0$ and satisfies

$$|x(t)| = \sqrt{v(t)} \le e^{-t}|a|, \quad \forall \, t \ge 0$$

$\triangle$

Example 2.10 The scalar differential equation

$$\dot{x} = f(t, x) = -(1 + x^2)x + e^t, \qquad x(0) = a$$

has a unique solution on $[0, t_1)$ for some $t_1 > 0$, because $f(t, x)$ is locally Lipschitz in x. We want to find an upper bound on $|x(t)|$ similar to the one we obtained in the previous example. Let us start with $v(t) = x^2(t)$ as in the previous example. The derivative of v is given by

$$\dot{v}(t) = 2x(t)\dot{x}(t) = -2x^2(t) - 2x^4(t) + 2x(t)e^t \le -2v(t) + 2\sqrt{v(t)}e^t$$

We can apply the comparison lemma to this differential inequality, but the resulting differential equation will not be easy to solve. Instead, we consider a different choice of $v(t)$. Let $v(t) = |x(t)|$. For $x(t) \neq 0$, the function $v(t)$ is differentiable and its derivative is given by

$$\dot{v}(t) = \frac{d}{dt}\sqrt{x^2(t)} = \frac{x(t)\dot{x}(t)}{|x(t)|} = -|x(t)|[1 + x^2(t)] + \frac{x(t)}{|x(t)|}e^t$$

Since $1 + x^2(t) \geq 1$, we have $-|x(t)|[1 + x^2(t)] \leq -|x(t)|$ and $\dot{v}(t) \leq -v(t) + e^t$. On the other hand, when $x(t) = 0$ we have

$$\frac{|v(t+h) - v(t)|}{h} = \frac{|x(t+h)|}{h} = \frac{1}{h}\left| \int_t^{t+h} f(\tau, x(\tau)) \, d\tau \right|$$

$$= \left| f(t, 0) + \frac{1}{h}\int_t^{t+h} [f(\tau, x(\tau)) - f(t, x(t))] \, d\tau \right|$$

$$\leq |f(t, 0)| + \frac{1}{h}\int_t^{t+h} |f(\tau, x(\tau)) - f(t, x(t))| \, d\tau$$

Since $f(t, x(t))$ is a continuous function of t, given any $\epsilon > 0$ there is $\delta > 0$ such that for all $|\tau - t| < \delta$, $|f(\tau, x(\tau)) - f(t, x(t))| < \epsilon$. Hence, for all $h < \delta$,

$$\frac{1}{h}\int_t^{t+h} |f(\tau, x(\tau)) - f(t, x(t))| \, d\tau < \epsilon$$

which shows that

$$\lim_{h \to 0^+} \frac{1}{h}\int_t^{t+h} |f(\tau, x(\tau)) - f(t, x(t))| \, d\tau = 0$$

Thus, $D^+ v(t) \leq |f(t, 0)| = e^t$ whenever $x(t) = 0$. Thus, for all $t \in [0, t_1)$, we have

$$D^+ v(t) \leq -v(t) + e^t, \quad v(0) = |a|$$

Letting $u(t)$ be the solution of the linear differential equation

$$\dot{u} = -u + e^t, \quad u(0) = |a|$$

we conclude by the comparison lemma that

$$v(t) \leq u(t) = e^{-t}|a| + \frac{1}{2}\left[e^t - e^{-t}\right], \quad \forall \, t \in [0, t_1)$$

The upper bound on $v(t)$ is finite for every finite t_1 and approaches infinity only as $t_1 \to \infty$. Thus, the solution $x(t)$ is defined for all $t \geq 0$ and satisfies

$$|x(t)| \leq e^{-t}|a| + \frac{1}{2}\left[e^t - e^{-t}\right], \quad \forall \, t \geq 0$$

$\triangle$

2.6   Exercises

Exercise 2.1 Show that, for any $x \in R^n$, we have

$$\|x\|_2 \leq \|x\|_1 \leq \sqrt{n}\, \|x\|_2$$

$$\|x\|_\infty \leq \|x\|_2 \leq \sqrt{n}\, \|x\|_\infty$$

$$\|x\|_\infty \leq \|x\|_1 \leq n\, \|x\|_\infty$$

Exercise 2.2 Show that, for any $m \times n$ real matrix A and any $n \times q$ real matrix B, we have

$$\|A\|_2 \leq \sqrt{\|A\|_1\, \|A\|_\infty}$$

$$\frac{1}{\sqrt{n}}\, \|A\|_\infty \leq \|A\|_2 \leq \sqrt{m}\, \|A\|_\infty$$

$$\frac{1}{\sqrt{m}}\, \|A\|_1 \leq \|A\|_2 \leq \sqrt{n}\, \|A\|_1$$

$$\|AB\|_p \leq \|A\|_p\, \|B\|_p$$

Exercise 2.3 Consider the set $S = \{x \in R^2 \mid -1 < x_i \leq 1,\ i = 1, 2\}$. Is S open? Is it closed? Find the closure, interior, and boundary of S.

Exercise 2.4 Let $u_T(t)$ be the unit step function, defined by $u_T(t) = 0$ for $t < T$ and $u_T(t) = 1$ for $t \geq T$.

(a) Show that $u_T(t)$ is piecewise continuous.

(b) Show that $f(t) = g(t)u_T(t)$, for any continuous function $g(t)$, is piecewise continuous.

(c) Show that the periodic square waveform is piecewise continuous.

Exercise 2.5 Let $f(x)$ be a continuously differentiable function that maps a convex domain $D \subset R^n$ into R^n. Suppose D contains the origin $x = 0$ and $f(0) = 0$. Show that

$$f(x) = \int_0^1 \frac{\partial f}{\partial x}(\sigma x)\, d\sigma\ x, \quad \forall\, x \in D$$

Hint: Set $g(\sigma) = f(\sigma x)$ for $0 \leq \sigma \leq 1$ and use the fact that $g(1) - g(0) = \int_0^1 g'(\sigma)\, d\sigma$.

Exercise 2.6 Let $f(x)$ be continuously differentiable. Show that an equilibrium point x^* of $\dot{x} = f(x)$ is isolated if the Jacobian matrix $[\partial f/\partial x](x^*)$ is nonsingular. **Hint:** Use the implicit function theorem.

Exercise 2.7 Let $y(t)$ be a nonnegative scalar function that satisfies the inequality

$$y(t) \leq k_1 e^{-\alpha(t-t_0)} + \int_{t_0}^{t} e^{-\alpha(t-\tau)} [k_2 y(\tau) + k_3] \, d\tau$$

where k_1, k_2, k_3 are nonnegative constants and α is a positive constant that satisfies $\alpha > k_2$. Using the Gronwall-Bellman inequality, show that

$$y(t) \leq k_1 e^{-(\alpha-k_2)(t-t_0)} + \frac{k_3}{\alpha - k_2} \left[1 - e^{-(\alpha-k_2)(t-t_0)} \right]$$

Hint: Take $z(t) = y(t)e^{\alpha(t-t_0)}$ and find the inequality satisfied by z.

Exercise 2.8 Let $\mathcal{L}_2$ be the set of all piecewise continuous functions $u : [0, \infty) \rightarrow R^n$ with the property that each component is square integrable on $[0, \infty)$; that is, $\int_0^\infty |u_i(t)|^2 \, dt < \infty$. Define $\|u\|_{\mathcal{L}_2} = \sqrt{\int_0^\infty u^T(t)u(t) \, dt}$. Show that $\|u\|_{\mathcal{L}_2}$ is a well-defined norm.
Hint: Use the Cauchy-Schwartz inequality

$$\int_a^b v(t)u(t) \, dt \leq \sqrt{\int_a^b v^2(t) \, dt \int_a^b u^2(t) \, dt}$$

for all nonnegative scalar functions $u(t)$ and $v(t)$.

Exercise 2.9 Let l_2 be the set of all sequences of scalars $\{\eta_1, \eta_2, \ldots\}$ for which $\sum_{i=1}^{\infty} |\eta_i|^2 < \infty$ and define

$$\|x\|_l = \left[\sum_{i=1}^{\infty} |\eta_i|^2 \right]^{1/2}$$

(a) Show that $\|x\|_l$ is a well-defined norm.

(b) Show that l_2 with the norm $\|x\|_l$ is a Banach space.

Exercise 2.10 Let $\mathcal{S}$ be the set of all half-wave symmetric periodic signals of fundamental frequency ω, which have finite energy on any finite interval. A signal $y \in \mathcal{S}$ can be represented by its Fourier series

$$y(t) = \sum_{k \text{ odd}} a_k \exp(jk\omega t), \quad \sum_{k \text{ odd}} |a_k|^2 < \infty$$

Define

$$\|y\|_{\mathcal{S}} = \left[\frac{\omega}{\pi} \int_0^{2\pi/\omega} y^2(t) \, dt \right]^{1/2}$$

(a) Show that $\|y\|_S$ is a well-defined norm.

(b) Show that S with the norm $\|y\|_S$ is a Banach space.

Exercise 2.11 Let $\mathcal{L}_\infty$ be the set of all piecewise continuous functions $u : [0, \infty) \to R^n$ with the property that each component $u_i(t)$ is uniformly bounded for all $t \geq 0$. Let $\|u\|_{\mathcal{L}_\infty} = \sup_{t \geq 0} \|u(t)\|$, where $\| \cdot \|$ is any p-norm.

(a) Show that $\|u\|_{\mathcal{L}_\infty}$ is a well-defined norm.

(b) Show that $\mathcal{L}_\infty$ with the norm $\|u\|_{\mathcal{L}_\infty}$ is a Banach space.

Exercise 2.12 Consider the linear algebraic equation $Ax = b$, where the matrix A is $n \times n$. Let a_{ij} denote the element of A in the ith row and jth column. Assume that $a_{ii} \neq 0$ for all i and the matrix A has the strict diagonal dominance property; that is,

$$|a_{ii}| > \sum_{j \neq i} |a_{ij}|$$

Show that $Ax = b$ has a unique solution which can be computed via the successive approximations:

$$Dx^{(k+1)} = (D - A)x^{(k)} + b, \quad k \geq 1$$

where $D = \text{diag}[a_{11}, a_{22}, \ldots, a_{nn}]$ and $x^{(1)}$ is any vector in R^n.

Exercise 2.13 Consider the nonlinear algebraic equation

$$c - Dx + \epsilon x(a - b^T x) = 0$$

where D is an $n \times n$ nonsingular matrix, b and c are n-vectors, and a and $\epsilon > 0$ are scalars. Show that, for sufficiently small ϵ, the equation has a unique real root in the set $\{x \in R^n \mid \|x - D^{-1}c\|_2 \leq 1\}$.

Exercise 2.14 Consider the discrete-time dynamical system $x(k + 1) = Ax(k)$.

(a) Show that $x(k) \to 0$ as $k \to \infty$ if and only if all eigenvalues of A satisfy $|\lambda_i| < 1$.

(b) Can you conclude from part (a) that the eigenvalue condition $|\lambda_i| < 1$ implies that the mapping Ax is a contraction mapping in any p-norm on R^n?

Exercise 2.15 Consider the nonlinear equations

$$\left(1 + \frac{\epsilon x_1}{1 + x_1^2}\right) x_1 + \epsilon x_2 = 1$$

$$\epsilon x_1 + \left(1 + \frac{\epsilon x_2}{1 + x_2^2}\right) x_2 = 2$$

where $|\epsilon| < \frac{1}{2}$. Using the contraction mapping principle, show that the equations have a unique real solution.

Exercise 2.16 Let $f : R^n \rightarrow R^n$ be locally Lipschitz in a domain $D \subset R^n$. Let $S \subset D$ be a compact set. Show that there is a positive constant L such that for all $x, y \in S$,

$$\|f(x) - f(y)\| \leq L\|x - y\|$$

Hint: The set S can be covered by a finite number of neighborhoods; that is,

$$S \subset N(a_1, r_1) \cup N(a_2, r_2) \cup \cdots \cup N(a_k, r_k)$$

Consider the following two cases separately:

- $x, y \in S \cap N(a_i, r_i)$ for some i.

- $x, y \notin S \cap N(a_i, r_i)$ for any i; in this case, $\|x - y\| \geq \min_i r_i$.

In the second case, use the fact that $f(x)$ is uniformly bounded on S.

Exercise 2.17 Show that if $f : R^n \rightarrow R^n$ is Lipschitz on $W \subset R^n$, then $f(x)$ is uniformly continuous on W.

Exercise 2.18 Show that if $f : R^n \rightarrow R^n$ is Lipschitz on a domain $D = \{x \in R^n \mid \|x\| < r\}$ and $f(0) = 0$, then there is a positive constant k such that $\|f(x)\| \leq k\|x\|$ for all $x \in D$.

Exercise 2.19 For any $x \in R^n$ and any $p \in [1, \infty)$, define the vector $y \in R^n$ by

$$y_i = \frac{x_i^{p-1}}{\|x\|_p^{p-1}} \, \text{sign}(x_i^p)$$

Show that $y^T x = \|x\|_p$ and $\|y\|_q = 1$, where $q \in (1, \infty]$ is determined from $1/p + 1/q = 1$. For $p = \infty$, find a vector y such that $y^T x = \|x\|_\infty$ and $\|y\|_1 = 1$.

Exercise 2.20 Prove Lemma 2.4.

Exercise 2.21 Let $\|\cdot\|_\alpha$ and $\|\cdot\|_\beta$ be two different norms of the class of p-norms on R^n. Show that $f : R^n \rightarrow R^n$ is Lipschitz in $\|\cdot\|_\alpha$ if and only if it is Lipschitz in $\|\cdot\|_\beta$.

Exercise 2.22 Consider the initial value problem (2.1) and let $D \subset R^n$ be a domain that contains $x = 0$. Suppose $f(t, 0) = 0$, $f(t, x)$ is Lipschitz in x on $[t_0, \infty) \times D$ with a Lipschitz constant L in $\|\cdot\|_2$, and $x(t)$, the solution of (2.1), is defined for all $t \geq t_0$ and belongs to D.

(a) Show that

$$\left| \frac{d}{dt} \left[x^T(t) x(t) \right] \right| \leq 2L \|x(t)\|_2^2$$

(b) Show that

$$\|x_0\|_2 \exp[-L(t - t_0)] \le \|x(t)\|_2 \le \|x_0\|_2 \exp[L(t - t_0)]$$

Exercise 2.23 Let $x : R \to R^n$ be a differentiable function that satisfies

$$\|\dot{x}(t)\| \le g(t), \quad \forall\, t \ge t_0$$

Show that

$$\|x(t)\| \le \|x(t_0)\| + \int_{t_0}^{t} g(s)\, ds$$

Exercise 2.24 ([36]) Give another proof of Theorem 2.2 by direct application of the contraction principle in the space of continuous functions with the norm

$$\|x\|_{\mathcal{X}} = \max_{t_0 \le t \le t_1} \left\{ e^{-L(t - t_0)/\rho} \|x(t)\| \right\}, \quad 0 < \rho < 1, \quad L > 0$$

Exercise 2.25 ([7]) Let $f : R \to R$. Suppose f satisfies the inequality

$$|f(x) - f(x_0)| \le M|x - x_0|^{\alpha}, \quad M \ge 0, \ \alpha \ge 0$$

in some neighborhood of x_0. Show that f is continuous at x_0 if $\alpha > 0$, and differentiable at x_0 if $\alpha > 1$.

Exercise 2.26 For each of the following functions $f : R \to R$, find whether (a) f is continuously differentiable at $x = 0$; (b) f is locally Lipschitz at $x = 0$; (c) f is continuous at $x = 0$; (d) f is globally Lipschitz; (e) f is uniformly continuous on R; (f) f is Lipschitz on $(-1, 1)$.

(1) $f(x) = \begin{cases} x^2 \sin(1/x), & \text{for } x \ne 0 \\ \\ 0, & \text{for } x = 0 \end{cases}$

(2) $f(x) = \begin{cases} x^3 \sin(1/x), & \text{for } x \ne 0 \\ \\ 0, & \text{for } x = 0 \end{cases}$

(3) $f(x) = \tan(\pi x/2)$

Exercise 2.27 For each of the following functions $f : R^n \to R^n$, find whether (a) f is continuously differentiable; (b) f is locally Lipschitz; (c) f is continuous; (d) f is globally Lipschitz; (e) f is uniformly continuous on R^n.

(1) $f(x) = \begin{bmatrix} x_1 + \text{sgn}\,(x_2) \\ x_2 \end{bmatrix}$ **(2)** $f(x) = \begin{bmatrix} x_1 + \text{sat}\,(x_2) \\ x_1 + \sin x_2 \end{bmatrix}$

(3) $f(x) = \begin{bmatrix} x_3 \text{sat}\,(x_1 + x_2) \\ x_2^2 \\ x_1 \end{bmatrix}$

Exercise 2.28 Let $D_r = \{x \in R^n \mid \|x\| < r\}$. For each of the following systems, represented as $\dot{x} = f(t, x)$, find whether (a) f is locally Lipschitz in x on D_r, for sufficiently small r; (b) f is locally Lipschitz in x on D_r, for any finite $r > 0$; (c) f is globally Lipschitz in x.

(1) The pendulum equation with friction and constant input torque (Section 1.1.1).

(2) The tunnel diode circuit (Example 1.2).

(3) The mass-spring equation with linear spring, linear viscous damping, dry friction, and zero external force (Section 1.1.3).

(4) The Van der Pol oscillator (Example 1.7).

(5) The closed-loop equation of a third-order adaptive control system (Section 1.1.6).

(6) The system $\dot{x} = Ax - B\psi(Cx)$, where A, B, and C are $n \times n$, $n \times 1$, and $1 \times n$ matrices, respectively, and $\psi(\cdot)$ is a dead-zone nonlinearity, defined by

$$\psi(y) = \begin{cases} y + d, & \text{for } y < -d \\ 0, & \text{for } -d \le y \le d \\ y - d, & \text{for } y > d \end{cases}$$

Exercise 2.29 Show that if $f_1 : R \to R$ and $f_2 : R \to R$ are locally Lipschitz, then $f_1 + f_2$, $f_1 f_2$ and $f_2 \circ f_1$ are locally Lipschitz.

Exercise 2.30 Let $f : R^n \to R^n$ be defined by

$$f(x) = \begin{cases} \frac{1}{\|Kx\|} Kx, & \text{if } g(x)\|Kx\| \ge \mu > 0 \\[2mm] \frac{g(x)}{\mu} Kx, & \text{if } g(x)\|Kx\| < \mu \end{cases}$$

where $g : R^n \to R$ is locally Lipschitz and nonnegative, and K is a constant matrix. Show that $f(x)$ is Lipschitz on any compact subset of R^n.

Exercise 2.31 Let $V : R \times R^n \to R$ be continuously differentiable. Suppose $V(t, 0) = 0$ for all $t \ge 0$ and

$$V(t, x) \ge c_1 \|x\|^2; \quad \left\| \frac{\partial V}{\partial x}(t, x) \right\| \le c_4 \|x\|, \quad \forall\, (t, x) \in [0, \infty) \times D$$

where c_1 and c_4 are positive constants and $D \subset R^n$ is a convex domain that contains the origin $x = 0$.

(a) Show that $V(t, x) \leq \frac{1}{2}c_4\|x\|^2$ for all $x \in D$.

   Hint: Use the representation $V(t, x) = \int_0^1 \frac{\partial V}{\partial x}(t, \sigma x)\, d\sigma\, x$.

(b) Show that the constants c_1 and c_4 must satisfy $2c_1 \leq c_4$.

(c) Show that $W(t, x) = \sqrt{V(t, x)}$ satisfies the Lipschitz condition

$$|W(t, x_2) - W(t, x_1)| \leq \frac{c_4}{2\sqrt{c_1}}\|x_2 - x_1\|, \quad \forall\, t \geq 0,\ \forall\, x_1, x_2 \in D$$

Exercise 2.32 Let $f(t, x)$ be piecewise continuous in t and locally Lipschitz in x on $[t_0, t_1] \times D$, for some domain $D \subset R^n$. Let W be a compact subset of D. Let $x(t)$ be the solution of $\dot{x} = f(t, x)$ starting at $x(t_0) = x_0 \in W$. Suppose $x(t)$ is defined and $x(t) \in W$ for all $t \in [t_0, T)$, $T < t_1$.

(a) Show that $x(t)$ is uniformly continuous on $[t_0, T)$.

(b) Show that $x(T)$ is defined and belongs to W, and $x(t)$ is a solution on $[t_0, T]$.

(c) Show that there is $\delta > 0$ such that the solution can be extended to $[t_0, T + \delta]$.

Exercise 2.33 Let $f(t, x)$ be piecewise continuous in t and locally Lipschitz in x on $[t_0, t_1] \times D$, for some domain $D \subset R^n$. Let $y(t)$ be a solution of (2.1) on a maximal open interval $[t_0, T) \subset [t_0, t_1]$ with $T < \infty$. Let W be any compact subset of D. Show that there is some $t \in [t_0, T)$ with $y(t) \notin W$.
Hint: Use the previous exercise.

Exercise 2.34 Let $f(t, x)$ be piecewise continuous in t, locally Lipschitz in x, and

$$\|f(t, x)\| \leq k_1 + k_2\|x\|, \quad \forall\ (t, x) \in [t_0, \infty) \times R^n$$

(a) Show that the solution of (2.1) satisfies

$$\|x(t)\| \leq \|x_0\|\exp[k_2(t - t_0)] + \frac{k_1}{k_2}\{\exp[k_2(t - t_0)] - 1\}$$

   for all $t \geq t_0$ for which the solution exists.

(b) Can the solution have a finite escape time?

Exercise 2.35 Let $g : R^n \to R^n$ be continuously differentiable for all $x \in R^n$, and define $f(x)$ by

$$f(x) = \frac{1}{1 + g^T(x)g(x)}g(x)$$

Show that $\dot{x} = f(x)$, $x(0) = x_0$, has a unique solution defined for all $t \geq 0$.

Exercise 2.36 Show that the state equation

$$\dot{x}_1 = -x_1 + \frac{2x_2}{1 + x_2^2}, \quad x_1(0) = a$$

$$\dot{x}_2 = -x_2 + \frac{2x_1}{1 + x_1^2}, \quad x_2(0) = b$$

has a unique solution defined for all $t \geq 0$.

Exercise 2.37 Suppose the second-order system $\dot{x} = f(x)$, with a locally Lipschitz $f(x)$, has a limit cycle. Show that any solution that starts in the region enclosed by the limit cycle cannot have a finite escape time.

Exercise 2.38 ([36]) Let $x_1 : R \to R^n$ and $x_2 : R \to R^n$ be differentiable functions such that

$$\|x_1(a) - x_2(a)\| \leq \gamma, \quad \|\dot{x}_i(t) - f((t, x_i(t))\| \leq \mu_i, \text{ for } i = 1, 2$$

for $a \leq t \leq b$. If f satisfies the Lipschitz condition (2.2), show that

$$\|x_1(t) - x_2(t)\| \leq \gamma e^{L(t-a)} + (\mu_1 + \mu_2) \left[\frac{e^{L(t-a)} - 1}{L} \right], \quad \text{for } a \leq t \leq b$$

Exercise 2.39 Derive the sensitivity equations for the tunnel diode circuit of Example 1.2 as L and C vary from their nominal values.

Exercise 2.40 Derive the sensitivity equations for the Van der Pol oscillator of Example 1.7 as ϵ varies from its nominal value. Use the state equation in the x-coordinates.

Exercise 2.41 Repeat the previous exercise using the state equation in the z-coordinates.

Exercise 2.42 Derive the sensitivity equations for the system

$$\dot{x}_1 = \tan^{-1}(ax_1) - x_1 x_2$$

$$\dot{x}_2 = bx_1^2 - cx_2$$

as the parameters a, b, c vary from their nominal values $a_0 = 1$, $b_0 = 0$, and $c_0 = 1$.

Exercise 2.43 Let $f(x)$ be continuously differentiable, $f(0) = 0$, and

$$\left\| \frac{\partial f_i}{\partial x}(x) - \frac{\partial f_i}{\partial x}(0) \right\|_2 \leq L_i \|x\|_2, \text{ for } 1 \leq i \leq n$$

Show that

$$\left\| f(x) - \frac{\partial f}{\partial x}(0)x \right\|_2 \leq L \|x\|_2^2, \text{ where } L = \sqrt{\sum_{i=1}^{n} L_i^2}$$

Exercise 2.44 Show, under the assumptions of Theorem 2.6, that the solution of (2.1) depends continuously on the initial time t_0.

Exercise 2.45 Let $f(t, x)$ and its partial derivative with respect to x be continuous in (t, x) for all $(t, x) \in [t_0, t_1] \times R^n$. Let $x(t, \eta)$ be the solution of (2.1) that starts at $x(t_0) = \eta$ and suppose $x(t, \eta)$ is defined on $[t_0, t_1]$. Show that $x(t, \eta)$ is continuously differentiable with respect to η and find the variational equation satisfied by $[\partial x / \partial \eta]$. **Hint:** Put $y = x - \eta$ to transform (2.1) into

$$\dot{y} = f(t, y + \eta), \quad y(t_0) = 0$$

with η as a parameter.

Exercise 2.46 Let $f(t, x)$ and its partial derivative with respect to x be continuous in (t, x) for all $(t, x) \in R \times R^n$. Let $x(t, a, \eta)$ be the solution of (2.1) that starts at $x(a) = \eta$ and suppose that $x(t, a, \eta)$ is defined on $[a, t_1]$. Show that $x(t, a, \eta)$ is continuously differentiable with respect to a and η and let $x_a(t)$ and $x_\eta(t)$ denote $[\partial x / \partial a]$ and $[\partial x / \partial \eta]$, respectively. Show that $x_a(t)$ and $x_\eta(t)$ satisfy the identity

$$x_a(t) + x_\eta(t) f(a, \eta) \equiv 0, \quad \forall \, t \in [a, t_1]$$

Exercise 2.47 ([36]) Let $f : R \times R \to R$ be a continuous function. Suppose $f(t, x)$ is locally Lipschitz and nondecreasing in x for each fixed value of t. Let $x(t)$ be a solution of $\dot{x} = f(t, x)$ on an interval $[a, b]$. If the continuous function $y(t)$ satisfies the integral inequality

$$y(t) \leq x(a) + \int_a^t f(s, y(s)) \, ds$$

for $a \leq t \leq b$, show that $y(t) \leq x(t)$ throughout this interval.

Chapter 3

Lyapunov Stability

Stability theory plays a central role in systems theory and engineering. There are different kinds of stability problems that arise in the study of dynamical systems. This chapter is concerned with stability of equilibrium points. In later chapters we shall see other kinds of stability, such as stability of periodic orbits and input-output stability. Stability of equilibrium points is usually characterized in the sense of Lyapunov, a Russian mathematician and engineer who laid the foundation of the theory which now carries his name. An equilibrium point is stable if all solutions starting at nearby points stay nearby; otherwise, it is unstable. It is asymptotically stable if all solutions starting at nearby points not only stay nearby, but also tend to the equilibrium point as time approaches infinity. These notions are made precise in Section 3.1, where the basic theorems of Lyapunov's method for autonomous systems are given. An extension of the basic theory, due to LaSalle, is given in Section 3.2. For a linear time-invariant system $\dot{x}(t) = Ax(t)$, the stability of the equilibrium point $x = 0$ can be completely characterized by the location of the eigenvalues of A. This is discussed in Section 3.3. In the same section, it is shown when and how the stability of an equilibrium point can be determined by linearization about that point. In Sections 3.4 and 3.5, we extend Lyapunov's method to nonautonomous systems. In Section 3.4, we define the concepts of uniform stability, uniform asymptotic stability, and exponential stability of the equilibrium of a nonautonomous system. We give Lyapunov's method for testing uniform asymptotic stability, as well as exponential stability. In Section 3.5, we study linear time-varying systems and linearization.

Lyapunov stability theorems give sufficient conditions for stability, asymptotic stability, and so on. They do not say whether the given conditions are also necessary. There are theorems which establish, at least conceptually, that for many of Lyapunov stability theorems the given conditions are indeed necessary. Such theorems are usually called converse theorems. We present two converse theorems in Section 3.6. Moreover, we use the converse theorem for exponential stability to show that an equilibrium point of a nonlinear system is exponentially stable if and only if the linearization of the system about that point has an exponentially stable equilibrium at the origin.

3.1 Autonomous Systems

Consider the autonomous system

$$\dot{x} = f(x) \tag{3.1}$$

where $f : D \rightarrow R^n$ is a locally Lipschitz map from a domain $D \subset R^n$ into R^n. Suppose $\bar{x} \in D$ is an equilibrium point of (3.1); that is,

$$f(\bar{x}) = 0$$

Our goal is to characterize and study stability of $\bar{x}$. For convenience, we state all definitions and theorems for the case when the equilibrium point is at the origin of R^n; that is, $\bar{x} = 0$. There is no loss of generality in doing so because any equilibrium point can be shifted to the origin via a change of variables. Suppose $\bar{x} \neq 0$, and consider the change of variables $y = x - \bar{x}$. The derivative of y is given by

$$\dot{y} = \dot{x} \ = f(x) = f(y + \bar{x}) \ \stackrel{\text{def}}{=} \ g(y), \quad \text{where } g(0) = 0$$

In the new variable y, the system has equilibrium at the origin. Therefore, without loss of generality, we shall always assume that $f(x)$ satisfies $f(0) = 0$, and study stability of the origin $x = 0$.

Definition 3.1 *The equilibrium point $x = 0$ of* (3.1) *is*

- *stable if, for each $\epsilon > 0$, there is $\delta = \delta(\epsilon) > 0$ such that*

$$\|x(0)\| < \delta \Rightarrow \|x(t)\| < \epsilon, \quad \forall \ t \geq 0$$

- *unstable if not stable.*

- *asymptotically stable if it is stable and δ can be chosen such that*

$$\|x(0)\| < \delta \Rightarrow \lim_{t \to \infty} x(t) = 0$$

The ϵ-δ requirement for stability takes a challenge-answer form. To demonstrate that the origin is stable, then, for any value of ϵ that a challenger may care to designate, we must produce a value of δ, possibly dependent on ϵ, such that a trajectory starting in a δ neighborhood of the origin will never leave the ϵ neighborhood. The three types of stability properties can be illustrated by the pendulum example of Section 1.1.1. The pendulum equation

$$\dot{x}_1 = x_2$$
$$\dot{x}_2 = -\left(\frac{g}{l}\right)\sin x_1 - \left(\frac{k}{m}\right)x_2$$

has two equilibrium points at $(x_1 = 0, \ x_2 = 0)$ and $(x_1 = \pi, \ x_2 = 0)$. Neglecting friction, by setting $k = 0$, we have seen in Example 1.1 that trajectories in the neighborhood of the first equilibrium are closed orbits. Therefore, by starting sufficiently close to the equilibrium point, trajectories can be guaranteed to stay within any specified ball centered at the equilibrium point. Therefore, the ϵ-δ requirement for stability is satisfied. The equilibrium point, however, is not asymptotically stable since trajectories starting off the equilibrium point do not tend to it eventually. Instead, they remain in their closed orbits. When friction is taken into consideration $(k > 0)$, the equilibrium point at the origin becomes a stable focus. Inspection of the phase portrait of a stable focus shows that the ϵ-δ requirement for stability is satisfied. In addition, trajectories starting close to the equilibrium point tend to it as t tends to ∞. The second equilibrium point at $x_1 = \pi$ is a saddle point. Clearly the ϵ-δ requirement cannot be satisfied since, for any $\epsilon > 0$, there is always a trajectory that will leave the ball $\{x \in R^n \mid \|x - \bar{x}\| \leq \epsilon\}$ even when $x(0)$ is arbitrarily close to the equilibrium point.

Implicit in Definition 3.1 is a requirement that solutions of (3.1) be defined for all $t \geq 0$.[1] Such global existence of the solution is not guaranteed by the local Lipschitz property of f. It will be shown, however, that the additional conditions needed in Lyapunov's theorem will ensure global existence of the solution. This will come as an application of Theorem 2.4.

Having defined stability and asymptotic stability of equilibrium points, our task now is to find ways to determine stability. The approach we used in the pendulum example relied on our knowledge of the phase portrait of the pendulum equation. Trying to generalize that approach amounts to actually finding all solutions of (3.1), which may be difficult or even impossible. However, the conclusions we reached about the stable equilibrium of the pendulum can also be reached by using energy concepts. Let us define the energy of the pendulum $E(x)$ as the sum of its potential and kinetic energies, with the reference of the potential energy chosen such that $E(0) = 0$; that is,

$$E(x) = \int_0^{x_1} \left(\frac{g}{l}\right)\sin y \ dy + \tfrac{1}{2}x_2^2 = \left(\frac{g}{l}\right)(1 - \cos x_1) + \tfrac{1}{2}x_2^2$$

---

[1] It is possible to change the definition to alleviate the implication of global existence of the solution. In [137], stability is defined on the maximal interval of existence $[0, t_1)$, without assuming that $t_1 = \infty$.

When friction is neglected ($k = 0$), the system is conservative; that is, there is no dissipation of energy. Hence, $E =$ constant during the motion of the system or, in other words, $dE/dt = 0$ along the trajectories of the system. Since $E(x) = c$ forms a closed contour around $x = 0$ for small c, we can again arrive at the conclusion that $x = 0$ is a stable equilibrium point. When friction is accounted for ($k > 0$), energy will dissipate during the motion of the system, that is, $dE/dt \leq 0$ along the trajectories of the system. Due to friction, E cannot remain constant indefinitely while the system is in motion. Hence, it keeps decreasing until it eventually reaches zero, showing that the trajectory tends to $x = 0$ as t tends to ∞. Thus, by examining the derivative of E along the trajectories of the system, it is possible to determine the stability of the equilibrium point. In 1892, Lyapunov showed that certain other functions could be used instead of energy to determine stability of an equilibrium point. Let $V : D \to R$ be a continuously differentiable function defined in a domain $D \subset R^n$ that contains the origin. The derivative of V along the trajectories of (3.1), denoted by $\dot{V}(x)$, is given by

$$\dot{V}(x) \;=\; \sum_{i=1}^{n} \frac{\partial V}{\partial x_i} \dot{x}_i \;=\; \sum_{i=1}^{n} \frac{\partial V}{\partial x_i} f_i(x)$$

$$=\; \left[\; \frac{\partial V}{\partial x_1}, \;\; \frac{\partial V}{\partial x_2}, \;\; \cdots \;, \;\; \frac{\partial V}{\partial x_n} \;\right] \begin{bmatrix} f_1(x) \\ f_2(x) \\ \vdots \\ f_n(x) \end{bmatrix} \;=\; \frac{\partial V}{\partial x} f(x)$$

The derivative of V along the trajectories of a system is dependent on the system's equation. Hence, $\dot{V}(x)$ will be different for different systems. If $\phi(t; x)$ is the solution of (3.1) that starts at initial state x at time $t = 0$, then

$$\dot{V}(x) = \frac{d}{dt} V(\phi(t; x)) \Big|_{t=0}$$

Therefore, if $\dot{V}(x)$ is negative, V will decrease along the solution of (3.1). We are now ready to state Lyapunov's stability theorem.

Theorem 3.1 *Let $x = 0$ be an equilibrium point for (3.1) and $D \subset R^n$ be a domain containing $x = 0$. Let $V : D \to R$ be a continuously differentiable function, such that*

$$V(0) = 0 \quad \text{and} \quad V(x) > 0 \text{ in } D - \{0\} \tag{3.2}$$

$$\dot{V}(x) \leq 0 \text{ in } D \tag{3.3}$$

Then, $x = 0$ is stable. Moreover, if

$$\dot{V}(x) < 0 \text{ in } D - \{0\} \tag{3.4}$$

then $x = 0$ is asymptotically stable.   ◇

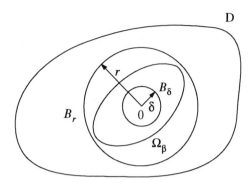

Figure 3.1: Geometric representation of sets in the proof of Theorem 3.1.

Proof: Given $\epsilon > 0$, choose $r \in (0, \epsilon]$ such that

$$B_r = \{x \in R^n \mid \|x\| \leq r\} \subset D$$

Let $\alpha = \min_{\|x\|=r} V(x)$. Then, $\alpha > 0$ by (3.2). Take $\beta \in (0, \alpha)$, and let

$$\Omega_\beta = \{x \in B_r \mid V(x) \leq \beta\}$$

Then, Ω_β is in the interior of B_r;[2] see Figure 3.1. The set Ω_β has the property that any trajectory starting in Ω_β at $t = 0$ stays in Ω_β for all $t \geq 0$. This follows from (3.3) since

$$\dot{V}(x(t)) \leq 0 \;\Rightarrow\; V(x(t)) \leq V(x(0)) \leq \beta, \; \forall\, t \geq 0$$

Since Ω_β is a compact set,[3] we conclude from Theorem 2.4 that (3.1) has a unique solution defined for all $t \geq 0$ whenever $x(0) \in \Omega_\beta$. Since $V(x)$ is continuous and $V(0) = 0$, there is $\delta > 0$ such that

$$\|x\| \leq \delta \;\Rightarrow\; V(x) < \beta$$

Then,

$$B_\delta \subset \Omega_\beta \subset B_r$$

and

$$x(0) \in B_\delta \Rightarrow x(0) \in \Omega_\beta \Rightarrow x(t) \in \Omega_\beta \Rightarrow x(t) \in B_r$$

Therefore,

$$\|x(0)\| < \delta \Rightarrow \|x(t)\| < r \leq \epsilon, \quad \forall\, t \geq 0$$

---

[2] This fact can be shown by contradiction. Suppose Ω_β is not in the interior of B_r, then there is a point $p \in \Omega_\beta$ that lies on the boundary of B_r. At this point, $V(p) \geq \alpha > \beta$ but, for all $x \in \Omega_\beta$, $V(x) \leq \beta$, which is a contradiction.

[3] Ω_β is closed by definition, and bounded since it is contained in B_r.

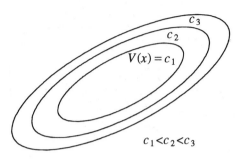

Figure 3.2: Level surfaces of a Lyapunov function.

which shows that the equilibrium point $x = 0$ is stable. Now, assume that (3.4) holds as well. To show asymptotic stability, we need to show that $x(t) \to 0$ as $t \to \infty$; that is, for every $a > 0$, there is $T > 0$ such that $\|x(t)\| < a$, for all $t > T$. By repetition of previous arguments, we know that for every $a > 0$, we can choose $b > 0$ such that $\Omega_b \subset B_a$. Therefore, it is sufficient to show that $V(x(t)) \to 0$ as $t \to \infty$. Since $V(x(t))$ is monotonically decreasing and bounded from below by zero,

$$V(x(t)) \to c \geq 0 \quad \text{as} \quad t \to \infty$$

To show that $c = 0$, we use a contradiction argument. Suppose $c > 0$. By continuity of $V(x)$, there is $d > 0$ such that $B_d \subset \Omega_c$. The limit $V(x(t)) \to c > 0$ implies that the trajectory $x(t)$ lies outside the ball B_d for all $t \geq 0$. Let $-\gamma = \max_{d \leq \|x\| \leq r} \dot{V}(x)$, which exists because the continuous function $\dot{V}(x)$ has a maximum over the compact set $\{d \leq \|x\| \leq r\}$.[4] By (3.4), $-\gamma < 0$. It follows that

$$V(x(t)) = V(x(0)) + \int_0^t \dot{V}(x(\tau)) \, d\tau \leq V(x(0)) - \gamma t$$

Since the right-hand side will eventually become negative, the inequality contradicts the assumption that $c > 0$.   □

A continuously differentiable function $V(x)$ satisfying (3.2) and (3.3) is called a *Lyapunov function* . The surface $V(x) = c$, for some $c > 0$, is called a *Lyapunov surface* or a *level surface*. Using Lyapunov surfaces, Figure 3.2 makes the theorem intuitively clear. It shows Lyapunov surfaces for decreasing values of c. The condition $\dot{V} \leq 0$ implies that when a trajectory crosses a Lyapunov surface $V(x) = c$, it moves inside the set $\Omega_c = \{x \in R^n \mid V(x) \leq c\}$ and can never come out again. When $\dot{V} < 0$, the trajectory moves from one Lyapunov surface to an inner Lyapunov

---

[4] See [7, Theorem 4-20].

surface with a smaller c. As c decreases, the Lyapunov surface $V(x) = c$ shrinks to the origin, showing that the trajectory approaches the origin as time progresses. If we only know that $\dot{V} \leq 0$, we cannot be sure that the trajectory will approach the origin,[5] but we can conclude that the origin is stable since the trajectory can be contained inside any ball B_ϵ by requiring the initial state $x(0)$ to lie inside a Lyapunov surface contained in that ball.

A function $V(x)$ satisfying condition (3.2), that is, $V(0) = 0$ and $V(x) > 0$ for $x \neq 0$, is said to be *positive definite*. If it satisfies the weaker condition $V(x) \geq 0$ for $x \neq 0$, it is said to be *positive semidefinite*. A function $V(x)$ is said to be *negative definite* or *negative semidefinite* if $-V(x)$ is positive definite or positive semidefinite, respectively. If $V(x)$ does not have a definite sign as per one of these four cases, it is said to be *indefinite*. With this terminology, we can rephrase Lyapunov's theorem to say that *the origin is stable if there is a continuously differentiable positive definite function $V(x)$ so that $\dot{V}(x)$ is negative semidefinite, and it is asymptotically stable if $\dot{V}(x)$ is negative definite.*

A class of scalar functions $V(x)$ for which sign definiteness can be easily checked is the class of functions of the quadratic form

$$V(x) = x^T P x = \sum_{i=1}^{n} \sum_{j=1}^{n} p_{ij} x_i x_j$$

where P is a real symmetric matrix. In this case, $V(x)$ is positive definite (positive semidefinite) if and only if all the eigenvalues of P are positive (nonnegative), which is true if and only if all the leading principal minors of P are positive (all principal minors of P are nonnegative).[6] If $V(x) = x^T P x$ is positive definite (positive semidefinite), we say that the matrix P is positive definite (positive semidefinite) and write $P > 0$ ($P \geq 0$).

Example 3.1 Consider

$$
\begin{aligned}
V(x) &= ax_1^2 + 2x_1x_3 + ax_2^2 + 4x_2x_3 + ax_3^2 \\
&= [x_1\ x_2\ x_3] \begin{bmatrix} a & 0 & 1 \\ 0 & a & 2 \\ 1 & 2 & a \end{bmatrix} \begin{bmatrix} x_1 \\ x_2 \\ x_3 \end{bmatrix} = x^T P x
\end{aligned}
$$

The leading principal minors of P are a, a^2, and $a(a^2-5)$. Therefore, $V(x)$ is positive definite if $a > \sqrt{5}$. For negative definiteness, the leading principal minors of $-P$ should be positive; that is, the leading principal minors of P should have alternating signs, with the odd-numbered minors being negative and the even-numbered minors being positive. Therefore, $V(x)$ is negative definite if $a < -\sqrt{5}$. By calculating all

---

[5] See, however, LaSalle's theorem in Section 3.2

[6] This is a well-known fact in matrix theory. Its proof can be found in [16] or [53].

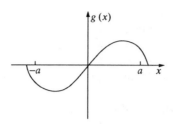

Figure 3.3: A possible nonlinearity in Example 3.2.

principal minors, it can be seen that $V(x)$ is positive semidefinite if $a \geq \sqrt{5}$ and negative semidefinite if $a \leq -\sqrt{5}$. For $a \in (-\sqrt{5}, \sqrt{5})$, $V(x)$ is indefinite. $\triangle$

Lyapunov's theorem can be applied without solving the differential equation (3.1). On the other hand, there is no systematic method for finding Lyapunov functions. In some cases, there are natural Lyapunov function candidates like energy functions in electrical or mechanical systems. In other cases, it is basically a matter of trial and error. The situation, however, is not as bad as it might seem. As we go over various examples and applications throughout the book, some ideas and approaches for searching for Lyapunov functions will be delineated.

Example 3.2 Consider the first-order differential equation

$$\dot{x} = -g(x)$$

where $g(x)$ is locally Lipschitz on $(-a, a)$ and satisfies

$$g(0) = 0; \quad xg(x) > 0, \quad \forall\, x \neq 0, \; x \in (-a, a)$$

A sketch of a possible $g(x)$ is shown in Figure 3.3. The system has an isolated equilibrium point at the origin. It is not difficult in this simple example to see that the origin is asymptotically stable, because solutions starting on either side of the origin will have to move toward the origin due to the sign of the derivative $\dot{x}$. To arrive at the same conclusion using Lyapunov's theorem, consider the function

$$V(x) = \int_0^x g(y)\, dy$$

Over the domain $D = (-a, a)$, $V(x)$ is continuously differentiable, $V(0) = 0$, and $V(x) > 0$ for all $x \neq 0$. Thus, $V(x)$ is a valid Lyapunov function candidate. To see whether or not $V(x)$ is indeed a Lyapunov function, we calculate its derivative along the trajectories of the system.

$$\dot{V}(x) = \frac{\partial V}{\partial x}[-g(x)] = -g^2(x) < 0, \; \forall\, x \in D - \{0\}$$

Thus, by Theorem 3.1 we conclude that the origin is asymptotically stable. $\triangle$

Example 3.3 Consider the pendulum equation without friction:

$$\begin{aligned}
\dot{x}_1 &= x_2 \\
\dot{x}_2 &= -\left(\frac{g}{l}\right)\sin x_1
\end{aligned}$$

and let us study the stability of the equilibrium point at the origin. A natural Lyapunov function candidate is the energy function

$$V(x) = \left(\frac{g}{l}\right)(1 - \cos x_1) + \tfrac{1}{2}x_2^2$$

Clearly, $V(0) = 0$ and $V(x)$ is positive definite over the domain $-2\pi < x_1 < 2\pi$. The derivative of $V(x)$ along the trajectories of the system is given by

$$\dot{V}(x) = \left(\frac{g}{l}\right)\dot{x}_1 \sin x_1 + x_2\dot{x}_2 = \left(\frac{g}{l}\right)x_2 \sin x_1 - \left(\frac{g}{l}\right)x_2 \sin x_1 = 0$$

Thus, conditions (3.2) and (3.3) of Theorem 3.1 are satisfied and we conclude that the origin is stable. Since $\dot{V}(x) \equiv 0$, we can also conclude that the origin is not asymptotically stable; for trajectories starting on a Lyapunov surface $V(x) = c$ remain on the same surface for all future time. $\triangle$

Example 3.4 Consider again the pendulum equation, but this time with friction:

$$\begin{aligned}
\dot{x}_1 &= x_2 \\
\dot{x}_2 &= -\left(\frac{g}{l}\right)\sin x_1 - \left(\frac{k}{m}\right)x_2
\end{aligned}$$

Let us try again $V(x) = (g/l)(1 - \cos x_1) + \tfrac{1}{2}x_2^2$ as a Lyapunov function candidate.

$$\dot{V}(x) = \left(\frac{g}{l}\right)\dot{x}_1 \sin x_1 + x_2\dot{x}_2 = -\left(\frac{k}{m}\right)x_2^2$$

$\dot{V}(x)$ is negative semidefinite. It is not negative definite because $\dot{V}(x) = 0$ for $x_2 = 0$ irrespective of the value of x_1; that is, $\dot{V}(x) = 0$ along the x_1-axis. Therefore, we can only conclude that the origin is stable. However, using the phase portrait of the pendulum equation, we have seen that when $k > 0$, the origin is asymptotically stable. The energy Lyapunov function fails to show this fact. We shall see later in Section 3.2 that a theorem due to LaSalle will enable us to arrive at a different conclusion using the energy Lyapunov function. For now, let us look for a Lyapunov function $V(x)$ that would have a negative definite $\dot{V}(x)$. Starting from the energy

Lyapunov function, let us replace the term $\frac{1}{2}x_2^2$ by the more general quadratic form $\frac{1}{2}x^T P x$ for some 2×2 positive definite matrix P.

$$
\begin{aligned}
V(x) &= \tfrac{1}{2}x^T P x + \left(\frac{g}{l}\right)(1 - \cos x_1) \\
&= \tfrac{1}{2}[x_1\ x_2]\begin{bmatrix} p_{11} & p_{12} \\ p_{12} & p_{22} \end{bmatrix}\begin{bmatrix} x_1 \\ x_2 \end{bmatrix} + \left(\frac{g}{l}\right)(1 - \cos x_1)
\end{aligned}
$$

For the quadratic form $\frac{1}{2}x^T P x$ to be positive definite, the elements of the matrix P must satisfy

$$p_{11} > 0;\ p_{22} > 0;\ p_{11}p_{22} - p_{12}^2 > 0$$

The derivative $\dot{V}(x)$ is given by

$$
\begin{aligned}
\dot{V}(x) &= \left[p_{11}x_1 + p_{12}x_2 + \left(\frac{g}{l}\right)\sin x_1\right]x_2 \\
&\quad + (p_{12}x_1 + p_{22}x_2)\left[-\left(\frac{g}{l}\right)\sin x_1 - \left(\frac{k}{m}\right)x_2\right] \\
&= \left(\frac{g}{l}\right)(1 - p_{22})x_2 \sin x_1 - \left(\frac{g}{l}\right)p_{12}x_1 \sin x_1 \\
&\quad + \left[p_{11} - p_{12}\left(\frac{k}{m}\right)\right]x_1 x_2 + \left[p_{12} - p_{22}\left(\frac{k}{m}\right)\right]x_2^2
\end{aligned}
$$

Now we want to choose p_{11}, p_{12}, and p_{22} such that $\dot{V}(x)$ is negative definite. Since the cross product terms $x_2 \sin x_1$ and $x_1 x_2$ are sign indefinite, we will cancel them by taking $p_{22} = 1$ and $p_{11} = (k/m)p_{12}$. With these choices, p_{12} must satisfy $0 < p_{12} < (k/m)$ for $V(x)$ to be positive definite. Let us take $p_{12} = 0.5(k/m)$. Then, $\dot{V}(x)$ is given by

$$\dot{V}(x) = -\tfrac{1}{2}\left(\frac{g}{l}\right)\left(\frac{k}{m}\right)x_1 \sin x_1 - \tfrac{1}{2}\left(\frac{k}{m}\right)x_2^2$$

The term $x_1 \sin x_1 > 0$ for all $0 < |x_1| < \pi$. Taking $D = \{x \in R^2 \mid |x_1| < \pi\}$, we see that $V(x)$ is positive definite and $\dot{V}(x)$ is negative definite over D. Thus, by Theorem 3.1 we conclude that the origin is asymptotically stable. $\triangle$

This example emphasizes an important feature of Lyapunov's stability theorem; namely, *the theorem's conditions are only sufficient*. Failure of a Lyapunov function candidate to satisfy the conditions for stability or asymptotic stability does not mean that the equilibrium is not stable or asymptotically stable. It only means that such stability property cannot be established by using this Lyapunov function candidate. Whether the equilibrium point is stable (asymptotically stable) or not can be determined only by further investigation.

In searching for a Lyapunov function in Example 3.4, we approached the problem in a backward manner. We investigated an expression for $\dot{V}(x)$ and went back to choose the parameters of $V(x)$ so as to make $\dot{V}(x)$ negative definite. This is a useful idea in searching for a Lyapunov function. A procedure that exploits this idea is known as the *variable gradient method*. To describe this procedure, let $V(x)$ be a scalar function of x and $g(x) = \nabla V = (\partial V/\partial x)^T$. The derivative $\dot{V}(x)$ along the trajectories of (3.1) is given by

$$\dot{V}(x) = \frac{\partial V}{\partial x} f(x) = g^T(x) f(x)$$

The idea now is to try to choose $g(x)$ such that it would be the gradient of a positive definite function $V(x)$ and, at the same time, $\dot{V}(x)$ would be negative definite. It is not difficult (Exercise 3.5) to verify that $g(x)$ is the gradient of a scalar function if and only if the Jacobian matrix $[\partial g/\partial x]$ is symmetric; that is,

$$\frac{\partial g_i}{\partial x_j} = \frac{\partial g_j}{\partial x_i}, \quad \forall\, i, j = 1, \ldots, n$$

Under this constraint, we start by choosing $g(x)$ such that $g^T(x) f(x)$ is negative definite. The function $V(x)$ is then computed from the integral

$$V(x) = \int_0^x g^T(y)\, dy = \int_0^x \sum_{i=1}^n g_i(y)\, dy_i$$

The integration is taken over any path joining the origin to x.[7] Usually, this is done along the axes; that is,

$$\begin{aligned}
V(x) &= \int_0^{x_1} g_1(y_1, 0, \ldots, 0)\, dy_1 + \int_0^{x_2} g_2(x_1, y_2, 0, \ldots, 0)\, dy_2 \\
&\quad + \cdots + \int_0^{x_n} g_n(x_1, x_2, \ldots, x_{n-1}, y_n)\, dy_n
\end{aligned}$$

By leaving some parameters of $g(x)$ undetermined, one would try to choose them to ensure that $V(x)$ is positive definite. The variable gradient method can be used to arrive at the Lyapunov function of Example 3.4. Instead of repeating the example, we illustrate the method on a slightly more general system.

Example 3.5 Consider the second-order system

$$\begin{aligned}
\dot{x}_1 &= x_2 \\
\dot{x}_2 &= -h(x_1) - ax_2
\end{aligned}$$

---

[7] The line integral of a gradient vector is independent of the path; see [7, Theorem 10-37].

where $a > 0$, $h(\cdot)$ is locally Lipschitz, $h(0) = 0$ and $yh(y) > 0$ for all $y \neq 0$, $y \in (-b, c)$ for some positive constants b and c. The pendulum equation is a special case of this system. To apply the variable gradient method, we want to choose a second-order vector $g(x)$ that satisfies

$$\frac{\partial g_1}{\partial x_2} = \frac{\partial g_2}{\partial x_1}$$

$$\dot{V}(x) = g_1(x)x_2 - g_2(x)[h(x_1) + ax_2] < 0, \quad \text{for } x \neq 0$$

and

$$V(x) = \int_0^x g^T(y) \, dy > 0, \quad \text{for } x \neq 0$$

Let us try

$$g(x) = \left[\begin{array}{c} \alpha(x)x_1 + \beta(x)x_2 \\ \gamma(x)x_1 + \delta(x)x_2 \end{array} \right]$$

where the scalar functions $\alpha(\cdot)$, $\beta(\cdot)$, $\gamma(\cdot)$, and $\delta(\cdot)$ are to be determined. To satisfy the symmetry requirement, we must have

$$\beta(x) + \frac{\partial \alpha}{\partial x_2}x_1 + \frac{\partial \beta}{\partial x_2}x_2 = \gamma(x) + \frac{\partial \gamma}{\partial x_1}x_1 + \frac{\partial \delta}{\partial x_1}x_2$$

The derivative $\dot{V}(x)$ is given by

$$\begin{aligned} \dot{V}(x) = & \; \alpha(x)x_1x_2 + \beta(x)x_2^2 - a\gamma(x)x_1x_2 \\ & -a\delta(x)x_2^2 - \delta(x)x_2h(x_1) - \gamma(x)x_1h(x_1) \end{aligned}$$

To cancel the cross product terms, we choose

$$\alpha(x)x_1 - a\gamma(x)x_1 - \delta(x)h(x_1) = 0$$

so that

$$\dot{V}(x) = -[a\delta(x) - \beta(x)]x_2^2 - \gamma(x)x_1h(x_1)$$

To simplify our choices, let us take $\delta(x) = \delta = $ constant, $\gamma(x) = \gamma = $ constant, and $\beta(x) = \beta = $ constant. Then, $\alpha(x)$ depends only on x_1, and the symmetry requirement is satisfied by choosing $\beta = \gamma$. The expression for $g(x)$ reduces to

$$g(x) = \left[\begin{array}{c} a\gamma x_1 + \delta h(x_1) + \gamma x_2 \\ \gamma x_1 + \delta x_2 \end{array} \right]$$

By integration, we obtain

$$\begin{aligned} V(x) = & \int_0^{x_1} [a\gamma y_1 + \delta h(y_1)] \, dy_1 + \int_0^{x_2} (\gamma x_1 + \delta y_2) \, dy_2 \\ = & \; \tfrac{1}{2}a\gamma x_1^2 + \delta \int_0^{x_1} h(y) \, dy + \gamma x_1 x_2 + \tfrac{1}{2}\delta x_2^2 \; = \; \tfrac{1}{2}x^T P x + \delta \int_0^{x_1} h(y) \, dy \end{aligned}$$

where

$$P = \begin{bmatrix} a\gamma & \gamma \\ \gamma & \delta \end{bmatrix}$$

Choosing $\delta > 0$ and $0 < \gamma < a\delta$ ensures that $V(x)$ is positive definite and $\dot{V}(x)$ is negative definite. For example, taking $\gamma = ak\delta$ for $0 < k < 1$ yields the Lyapunov function

$$V(x) = \frac{\delta}{2} x^T \begin{bmatrix} ka^2 & ka \\ ka & 1 \end{bmatrix} x + \delta \int_0^{x_1} h(y) \, dy$$

which satisfies conditions (3.2) and (3.4) of Theorem 3.1 over the domain $D = \{x \in R^2 \mid -b < x_1 < c\}$. Therefore, the origin is asymptotically stable. △

When the origin $x = 0$ is asymptotically stable, we are often interested in determining how far from the origin the trajectory can be and still converge to the origin as t approaches ∞. This gives rise to the definition of the *region of attraction* (also called *region of asymptotic stability, domain of attraction,* or *basin*). Let $\phi(t; x)$ be the solution of (3.1) that starts at initial state x at time $t = 0$. Then, the region of attraction is defined as the set of all points x such that $\lim_{t \to \infty} \phi(t; x) = 0$. Finding the exact region of attraction analytically might be difficult or even impossible. However, Lyapunov functions can be used to estimate the region of attraction, that is, to find sets contained in the region of attraction. From the proof of Theorem 3.1 we see that if there is a Lyapunov function that satisfies the conditions of asymptotic stability over a domain D, and if $\Omega_c = \{x \in R^n \mid V(x) \leq c\}$ is bounded and contained in D, then every trajectory starting in Ω_c remains in Ω_c and approaches the origin as $t \to \infty$. Thus, Ω_c is an estimate of the region of attraction. This estimate, however, may be conservative; that is, it may be much smaller than the actual region of attraction. In Section 4.2, we shall solve examples on estimating the region of attraction and see some ideas to enlarge the estimates. Here, we want to pursue another question: under what conditions will the region of attraction be the whole space R^n? This will be the case if we can show that for any initial state x, the trajectory $\phi(t; x)$ approaches the origin as $t \to \infty$, no matter how large $\|x\|$ is. If an asymptotically stable equilibrium point at the origin has this property, it is said to be *globally asymptotically stable*. Recalling again the proof of Theorem 3.1, we can see that global asymptotic stability can be established if any point $x \in R^n$ can be included in the interior of a bounded set Ω_c. It is obvious that for this to hold, the conditions of the theorem must hold globally, that is, $D = R^n$; but, is this enough? It turns out that we need more conditions to ensure that any point in R^n can be included in a bounded set Ω_c. The problem is that for large c, the set Ω_c need not be bounded. Consider, for example, the function

$$V(x) = \frac{x_1^2}{1 + x_1^2} + x_2^2$$

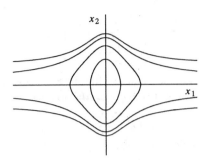

Figure 3.4: Lyapunov surfaces for $V(x) = x_1^2/(1 + x_1^2) + x_2^2$.

Figure 3.4 shows the surfaces $V(x) = c$ for various positive values of c. For small c, the surface $V(x) = c$ is closed; hence, Ω_c is bounded since it is contained in a closed ball B_r for some $r > 0$. This is a consequence of the continuity and positive definiteness of $V(x)$. As c increases, a value is reached after which the surface $V(x) = c$ is open and Ω_c is unbounded. For Ω_c to be in the interior of a ball B_r, c must satisfy $c < \inf_{\|x\| \geq r} V(x)$. If

$$l = \lim_{r \to \infty} \inf_{\|x\| \geq r} V(x) < \infty$$

then Ω_c will be bounded if $c < l$. In the preceding example,

$$l = \lim_{r \to \infty} \min_{\|x\| = r} \left[\frac{x_1^2}{1 + x_1^2} + x_2^2 \right] = \lim_{|x_1| \to \infty} \frac{x_1^2}{1 + x_1^2} = 1$$

Hence, Ω_c is bounded only for $c < 1$. An extra condition that ensures that Ω_c is bounded for all values of $c > 0$ is

$$V(x) \to \infty \quad \text{as} \quad \|x\| \to \infty$$

A function satisfying this condition is said to be *radially unbounded*.

Theorem 3.2 *Let $x = 0$ be an equilibrium point for* (3.1). *Let $V : R^n \to R$ be a continuously differentiable function such that*

$$V(0) = 0 \quad \text{and} \quad V(x) > 0, \quad \forall\, x \neq 0 \tag{3.5}$$

$$\|x\| \to \infty \quad \Rightarrow \quad V(x) \to \infty \tag{3.6}$$

$$\dot{V}(x) < 0, \quad \forall\, x \neq 0 \tag{3.7}$$

then $x = 0$ is globally asymptotically stable. $\diamond$

Proof: Given any point $p \in R^n$, let $c = V(p)$. Condition (3.6) implies that for any $c > 0$, there is $r > 0$ such that $V(x) > c$ whenever $\|x\| > r$. Thus $\Omega_c \subset B_r$, which implies that Ω_c is bounded. The rest of the proof is similar to that of Theorem 3.1.
□

Theorem 3.2 is known as Barbashin-Krasovskii theorem. Exercise 3.7 gives a counterexample to show that the radial unboundedness condition of the theorem is indeed needed.

Example 3.6 Consider again the second-order system of Example 3.5, but this time assume that the condition $yh(y) > 0$ holds for all $y \neq 0$. The Lyapunov function

$$V(x) = \frac{\delta}{2} x^T \begin{bmatrix} ka^2 & ka \\ ka & 1 \end{bmatrix} x + \delta \int_0^{x_1} h(y) \, dy$$

is positive definite for all $x \in R^2$ and radially unbounded. The derivative

$$\dot{V}(x) = -a\delta(1-k)x_2^2 - a\delta k x_1 h(x_1)$$

is negative definite for all $x \in R^2$ since $0 < k < 1$. Therefore, the origin is globally asymptotically stable.   △

If the origin $x = 0$ is a globally asymptotically stable equilibrium point of a system, then it must be the unique equilibrium point of the system. For if there was another equilibrium point $\bar{x}$, the trajectory starting at $\bar{x}$ would remain at $\bar{x}$ for all $t \geq 0$; hence, it would not approach the origin which contradicts the claim that the origin is globally asymptotically stable. Therefore, global asymptotic stability is not studied for multiple equilibria systems like the pendulum equation.

Theorems 3.1 and 3.2 are concerned with establishing the stability or asymptotic stability of an equilibrium point. There are also instability theorems for establishing that an equilibrium point is unstable. The most powerful of these theorems is Chetaev's theorem, which will be stated as Theorem 3.3. Before we state the theorem, let us introduce some terminology that will be used in the theorem's statement. Let $V : D \to R$ be a continuously differentiable function on a domain $D \subset R^n$ that contains the origin $x = 0$. Suppose $V(0) = 0$ and there is a point x_0 arbitrarily close to the origin such that $V(x_0) > 0$. Choose $r > 0$ such that the ball $B_r = \{x \in R^n \mid \|x\| \leq r\}$ is contained in D, and let

$$U = \{x \in B_r \mid V(x) > 0\} \tag{3.8}$$

The set U is a nonempty set contained in B_r. Its boundary is given by the surface $V(x) = 0$ and the sphere $\|x\| = r$. Since $V(0) = 0$, the origin lies on the boundary of U inside B_r. Notice that U may contain more than one component. For example, Figure 3.5 shows the set U for $V(x) = \frac{1}{2}(x_1^2 - x_2^2)$. The set U can be always constructed provided that $V(0) = 0$ and $V(x_0) > 0$ for some x_0 arbitrarily close to the origin.

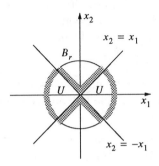

Figure 3.5: The set U for $V(x) = \frac{1}{2}(x_1^2 - x_2^2)$.

Theorem 3.3 *Let $x = 0$ be an equilibrium point for (3.1). Let $V : D \rightarrow R$ be a continuously differentiable function such that $V(0) = 0$ and $V(x_0) > 0$ for some x_0 with arbitrarily small $\|x_0\|$. Define a set U as in (3.8) and suppose that $\dot{V}(x) > 0$ in U. Then, $x = 0$ is unstable.* $\diamond$

Proof: The point x_0 is in the interior of U and $V(x_0) = a > 0$. The trajectory $x(t)$ starting at $x(0) = x_0$ must leave the set U. To see this point, notice that as long as $x(t)$ is inside U, $V(x(t)) \geq a$ since $\dot{V}(x) > 0$ in U. Let

$$\gamma = \min\{\dot{V}(x) \mid x \in U \text{ and } V(x) \geq a\}$$

which exists since the continuous function $\dot{V}(x)$ has a minimum over the compact set $\{x \in U \text{ and } V(x) \geq a\} = \{x \in B_r \text{ and } V(x) \geq a\}$.[8] Then, $\gamma > 0$ and

$$V(x(t)) = V(x_0) + \int_0^t \dot{V}(x(s)) \, ds \geq a + \int_0^t \gamma \, ds = a + \gamma t$$

This inequality shows that $x(t)$ cannot stay forever in U because $V(x)$ is bounded on U. Now, $x(t)$ cannot leave U through the surface $V(x) = 0$ since $V(x(t)) \geq a$. Hence, it must leave U through the sphere $\|x\| = r$. Since this can happen for an arbitrarily small $\|x_0\|$, the origin is unstable. $\square$

There are other instability theorems which were proved before Chetaev's theorem, but they are corollaries of the theorem; see Exercises 3.10 and 3.11.

Example 3.7 Consider the second-order system

$$\begin{aligned} \dot{x}_1 &= x_1 + g_1(x) \\ \dot{x}_2 &= -x_2 + g_2(x) \end{aligned}$$

---

[8] See [7, Theorem 4-20].

where $g_1(\cdot)$ and $g_2(\cdot)$ satisfy

$$|g_i(x)| \leq k\|x\|_2^2$$

in a neighborhood D of the origin. This inequality implies that $g_i(0) = 0$. Hence, the origin is an equilibrium point. Consider the function

$$V(x) = \tfrac{1}{2}(x_1^2 - x_2^2)$$

On the line $x_2 = 0$, $V(x) > 0$ at points arbitrarily close to the origin. The set U is shown in Figure 3.5. The derivative of $V(x)$ along the trajectories of the system is given by

$$\dot{V}(x) = x_1^2 + x_2^2 + x_1 g_1(x) - x_2 g_2(x)$$

The magnitude of the term $x_1 g_1(x) - x_2 g_2(x)$ satisfies the inequality

$$|x_1 g_1(x) - x_2 g_2(x)| \leq \sum_{i=1}^{2} |x_i| \cdot |g_i(x)| \leq 2k\|x\|_2^3$$

Hence,

$$\dot{V}(x) \geq \|x\|_2^2 - 2k\|x\|_2^3 = \|x\|_2^2(1 - 2k\|x\|_2)$$

Choosing r such that $B_r \subset D$ and $r < 1/2k$, it is seen that all the conditions of Theorem 3.3 are satisfied. Hence, the origin is unstable. $\triangle$

3.2 The Invariance Principle

In our study of the pendulum equation with friction (Example 3.4), we saw that the energy Lyapunov function fails to satisfy the asymptotic stability condition of Theorem 3.1 because $\dot{V}(x) = -(k/m)x_2^2$ is only negative semidefinite. Notice, however, that $\dot{V}(x)$ is negative everywhere except on the line $x_2 = 0$, where $\dot{V}(x) = 0$. For the system to maintain the $\dot{V}(x) = 0$ condition, the trajectory of the system must be confined to the line $x_2 = 0$. Unless $x_1 = 0$, this is impossible because from the pendulum equation

$$x_2(t) \equiv 0 \;\Rightarrow\; \dot{x}_2(t) \equiv 0 \;\Rightarrow\; \sin x_1(t) \equiv 0$$

Hence, on the segment $-\pi < x_1 < \pi$ of the $x_2 = 0$ line, the system can maintain the $\dot{V}(x) = 0$ condition only at the origin $x = 0$. Therefore, $V(x(t))$ must decrease toward 0, and consequently $x(t) \to 0$ as $t \to \infty$. This is consistent with the physical understanding that, due to friction, energy cannot remain constant while the system is in motion.

The above argument shows, formally, that if in a domain about the origin we can find a Lyapunov function whose derivative along the trajectories of the system is negative semidefinite, and if we can establish that no trajectory can stay identically at points where $\dot{V}(x) = 0$ except at the origin, then the origin is asymptotically stable. This idea follows from LaSalle's *invariance principle*, which is the subject of this section. To state and prove LaSalle's invariance theorem, we need to introduce a few definitions. Let $x(t)$ be a solution of (3.1). A point p is said to be a *positive limit point* of $x(t)$ if there is a sequence $\{t_n\}$, with $t_n \to \infty$ as $n \to \infty$, such that $x(t_n) \to p$ as $n \to \infty$. The set of all positive limit points of $x(t)$ is called the *positive limit set* of $x(t)$. A set M is said to be an *invariant set* with respect to (3.1) if

$$x(0) \in M \Rightarrow x(t) \in M, \quad \forall\, t \in R$$

That is, if a solution belongs to M at some time instant, then it belongs to M for all future and past time. A set M is said to be a *positively invariant set* if

$$x(0) \in M \Rightarrow x(t) \in M, \quad \forall\, t \geq 0$$

We also say that $x(t)$ approaches a set M as t approaches infinity, if for each $\epsilon > 0$ there is $T > 0$ such that

$$\text{dist}(x(t), M) < \epsilon, \quad \forall\, t > T$$

where $\text{dist}(p, M)$ denotes the distance from a point p to a set M, that is, the smallest distance from p to any point in M. More precisely,

$$\text{dist}(p, M) = \inf_{x \in M} \|p - x\|$$

These few concepts can be illustrated by examining an asymptotically stable equilibrium point and a stable limit cycle in the plane. The asymptotically stable equilibrium is the positive limit set of every solution starting sufficiently near the equilibrium point. The stable limit cycle is the positive limit set of every solution starting sufficiently near the limit cycle. The solution approaches the limit cycle as $t \to \infty$. Notice, however, that the solution does not approach any specific point on the limit cycle. In other words, the statement $x(t) \to M$ as $t \to \infty$ does not imply that the limit $\lim_{t \to \infty} x(t)$ exists. The equilibrium point and the limit cycle are invariant sets, since any solution starting in either set remains in the set for all $t \in R$. The set $\Omega_c = \{x \in R^n \mid V(x) \leq c\}$ with $\dot{V}(x) \leq 0$ for all $x \in \Omega_c$ is a positively invariant set since, as we saw in the proof of Theorem 3.1, a solution starting in Ω_c remains in Ω_c for all $t \geq 0$.

A fundamental property of limit sets is stated in the following lemma, whose proof is given in Appendix A.2.

Lemma 3.1 *If a solution $x(t)$ of (3.1) is bounded and belongs to D for $t \geq 0$, then its positive limit set L^+ is a nonempty, compact, invariant set. Moreover,*

$$x(t) \to L^+ \text{ as } t \to \infty$$

$\diamond$

We are now ready to state LaSalle's theorem.

Theorem 3.4 *Let $\Omega \subset D$ be a compact set that is positively invariant with respect to (3.1). Let $V : D \to R$ be a continuously differentiable function such that $\dot{V}(x) \leq 0$ in Ω. Let E be the set of all points in Ω where $\dot{V}(x) = 0$. Let M be the largest invariant set in E. Then every solution starting in Ω approaches M as $t \to \infty$.* $\diamond$

Proof: Let $x(t)$ be a solution of (3.1) starting in Ω. Since $\dot{V}(x) \leq 0$ in Ω, $V(x(t))$ is a decreasing function of t. Since $V(x)$ is continuous on the compact set Ω, it is bounded from below on Ω. Therefore, $V(x(t))$ has a limit a as $t \to \infty$. Note also that the positive limit set L^+ is in Ω because Ω is a closed set. For any $p \in L^+$, there is a sequence t_n with $t_n \to \infty$ and $x(t_n) \to p$ as $n \to \infty$. By continuity of $V(x)$, $V(p) = \lim_{n \to \infty} V(x(t_n)) = a$. Hence, $V(x) = a$ on L^+. Since L^+ is an invariant set (by Lemma 3.1), $\dot{V}(x) = 0$ on L^+. Thus,

$$L^+ \subset M \subset E \subset \Omega$$

Since $x(t)$ is bounded, $x(t)$ approaches L^+ as $t \to \infty$ (by Lemma 3.1). Hence, $x(t)$ approaches M as $t \to \infty$. $\square$

Unlike Lyapunov's theorem, Theorem 3.4 does not require the function $V(x)$ to be positive definite. Note also that the construction of the set Ω does not have to be tied in with the construction of the function $V(x)$. However, in many applications the construction of $V(x)$ will itself guarantee the existence of a set Ω. In particular, if $\Omega_c = \{x \in R^n \mid V(x) \leq c\}$ is bounded and $\dot{V}(x) \leq 0$ in Ω_c, then we can take $\Omega = \Omega_c$. When $V(x)$ is positive definite, Ω_c is bounded for sufficiently small $c > 0$. This is not necessarily true when $V(x)$ is not positive definite. For example, if $V(x) = (x_1 - x_2)^2$, the set Ω_c is not bounded no matter how small c is. If $V(x)$ is radially unbounded, that is, $V(x) \to \infty$ as $\|x\| \to \infty$, the set Ω_c is bounded for all values of c. This is true whether or not $V(x)$ is positive definite. However, checking radial unboundedness is easier for positive definite functions since it is enough to let x approach ∞ along the principal axes. This may not be sufficient if the function is not positive definite, as can be seen from $V(x) = (x_1 - x_2)^2$. Here, $V(x) \to \infty$ as $\|x\| \to \infty$ along the lines $x_1 = 0$ and $x_2 = 0$, but not when $\|x\| \to \infty$ along the line $x_1 = x_2$.

When our interest is in showing that $x(t) \to 0$ as $t \to \infty$, we need to establish that the largest invariant set in E is the origin. This is done by showing that no solution can stay identically in E, other than the trivial solution $x(t) = 0$. Specializing Theorem 3.4 to this case and taking $V(x)$ to be positive definite, we obtain the following two corollaries which extend Theorems 3.1 and 3.2.[9]

---

[9] Corollaries 3.1 and 3.2 are known as the theorems of Barbashin and Krasovskii, who proved them before the introduction of LaSalle's invariance principle.

Corollary 3.1 *Let $x = 0$ be an equilibrium point for* (3.1). *Let $V : D \to R$ be a continuously differentiable positive definite function on a domain D containing the origin $x = 0$, such that $\dot{V}(x) \leq 0$ in D. Let $S = \{x \in D \mid \dot{V}(x) = 0\}$ and suppose that no solution can stay identically in S, other than the trivial solution. Then, the origin is asymptotically stable.* $\diamond$

Corollary 3.2 *Let $x = 0$ be an equilibrium point for* (3.1). *Let $V : R^n \to R$ be a continuously differentiable, radially unbounded, positive definite function such that $\dot{V}(x) \leq 0$ for all $x \in R^n$. Let $S = \{x \in R^n \mid \dot{V}(x) = 0\}$ and suppose that no solution can stay identically in S, other than the trivial solution. Then, the origin is globally asymptotically stable.* $\diamond$

When $\dot{V}(x)$ is negative definite, $S = \{0\}$. Then, Corollaries 3.1 and 3.2 coincide with Theorems 3.1 and 3.2, respectively.

Example 3.8 Consider the system

$$\begin{aligned} \dot{x}_1 &= x_2 \\ \dot{x}_2 &= -g(x_1) - h(x_2) \end{aligned}$$

where $g(\cdot)$ and $h(\cdot)$ are locally Lipschitz and satisfy

$$g(0) = 0, \quad yg(y) > 0, \quad \forall\, y \neq 0, \; y \in (-a, a)$$

$$h(0) = 0, \quad yh(y) > 0, \quad \forall\, y \neq 0, \; y \in (-a, a)$$

The system has an isolated equilibrium point at the origin. Depending upon the functions $g(\cdot)$ and $h(\cdot)$, it might have other equilibrium points. The equation of this system can be viewed as a generalized pendulum equation with $h(x_2)$ as the friction term. Therefore, a Lyapunov function candidate may be taken as the energy-like function

$$V(x) = \int_0^{x_1} g(y)\, dy \; + \; \tfrac{1}{2} x_2^2$$

Let $D = \{x \in R^2 \mid -a < x_i < a\}$. $V(x)$ is positive definite in D. The derivative of $V(x)$ along the trajectories of the system is given by

$$\dot{V}(x) = g(x_1) x_2 + x_2[-g(x_1) - h(x_2)] = -x_2 h(x_2) \leq 0$$

Thus, $\dot{V}(x)$ is negative semidefinite. To characterize the set $S = \{x \in D \mid \dot{V}(x) = 0\}$, note that

$$\dot{V}(x) = 0 \;\Rightarrow\; x_2 h(x_2) = 0 \;\Rightarrow\; x_2 = 0, \quad \text{since } -a < x_2 < a$$

Hence, $S = \{x \in D \mid x_2 = 0\}$. Suppose $x(t)$ is a trajectory that belongs identically to S.

$$x_2(t) \equiv 0 \Rightarrow \dot{x}_2(t) \equiv 0 \Rightarrow g(x_1(t)) \equiv 0 \Rightarrow x_1(t) \equiv 0$$

Therefore, the only solution that can stay identically in S is the trivial solution $x(t) = 0$. Thus, the origin is asymptotically stable.   △

Example 3.9 Consider again the system of Example 3.8, but this time let $a = \infty$ and assume that $g(\cdot)$ satisfies the additional condition:

$$\int_0^y g(z)\,dz \;\rightarrow\; \infty \;\text{ as }\; |y| \rightarrow \infty$$

The Lyapunov function

$$V(x) = \int_0^{x_1} g(y)\,dy \;+\; \tfrac{1}{2}x_2^2$$

is radially unbounded. Similar to the previous example, it can be shown that $\dot{V}(x) \leq 0$ in R^2, and the set

$$S = \{x \in R^2 \mid \dot{V}(x) = 0\} = \{x \in R^2 \mid x_2 = 0\}$$

contains no solutions other than the trivial solution. Hence, the origin is globally asymptotically stable.   △

Not only does LaSalle's theorem relax the negative definiteness requirement of Lyapunov's theorem, but it also extends Lyapunov's theorem in three different directions. First, it gives an estimate of the region of attraction which is not necessarily of the form $\Omega_c = \{x \in R^n \mid V(x) \leq c\}$. The set Ω of Theorem 3.4 can be any compact positively invariant set. We shall use this feature in Section 4.2 to obtain less conservative estimates of the region of attraction. Second, LaSalle's theorem can be used in cases where the system has an equilibrium set rather than an isolated equilibrium point. This will be illustrated by an application to a simple adaptive control example from Section 1.1.6. Third, the function $V(x)$ does not have to be positive definite. The utility of this feature will be illustrated by an application to the neural network example of Section 1.1.5.

Example 3.10 Consider the first-order system

$$\dot{y} = ay + u$$

together with the adaptive control law

$$u = -ky, \qquad \dot{k} = \gamma y^2, \quad \gamma > 0$$

Taking $x_1 = y$ and $x_2 = k$, the closed-loop system is represented by

$$\begin{aligned} \dot{x}_1 &= -(x_2 - a)x_1 \\ \dot{x}_2 &= \gamma x_1^2 \end{aligned}$$

The line $x_1 = 0$ is an equilibrium set for this system. We want to show that the trajectory of the system approaches this equilibrium set as $t \to \infty$, which means that the adaptive controller succeeds in regulating y to zero. Consider the Lyapunov function candidate

$$V(x) = \tfrac{1}{2}x_1^2 + \frac{1}{2\gamma}(x_2 - b)^2$$

where $b > a$. The derivative of V along the trajectories of the system is given by

$$\begin{aligned} \dot{V}(x) &= x_1\dot{x}_1 + \frac{1}{\gamma}(x_2 - b)\dot{x}_2 \\ &= -x_1^2(x_2 - a) + x_1^2(x_2 - b) = -x_1^2(b - a) \leq 0 \end{aligned}$$

Hence, $\dot{V}(x) \leq 0$. Since $V(x)$ is radially unbounded, the set $\Omega_c = \{x \in R^2 \mid V(x) \leq c\}$ is a compact, positively invariant set. Thus, taking $\Omega = \Omega_c$, all the conditions of Theorem 3.4 are satisfied. The set E is given by

$$E = \{x \in \Omega_c \mid x_1 = 0\}$$

Since any point on the line $x_1 = 0$ is an equilibrium point, E is an invariant set. Therefore, in this example $M = E$. From Theorem 3.4, we conclude that every trajectory starting in Ω_c approaches E as $t \to \infty$; that is, $x_1(t) \to 0$ as $t \to \infty$. Moreover, since $V(x)$ is radially unbounded, this conclusion is global; that is, it holds for all initial conditions $x(0)$ because for any $x(0)$ the constant c can be chosen large enough that $x(0) \in \Omega_c$.   △

Note that the Lyapunov function in Example 3.10 is dependent on a constant b which is required to satisfy $b > a$. Since in this adaptive control problem the constant a is not known, we may not know the constant b explicitly but we know that it always exists. This highlights another feature of Lyapunov's method, which we have not seen before; namely, in some situations we may be able to assert the existence of a Lyapunov function that satisfies the conditions of a certain theorem even though we may not explicitly know that function. In Example 3.10 we can determine the Lyapunov function explicitly if we know some bound on a. For example, if we know that $|a| \leq \alpha$, where the bound α is known, we can choose $b > \alpha$.

Example 3.11 The neural network of Section 1.1.5 is represented by

$$\dot{x}_i = \frac{1}{C_i}h_i(x_i)\left[\sum_j T_{ij}x_j - \frac{1}{R_i}g_i^{-1}(x_i) + I_i\right]$$

for $i = 1, 2, \ldots, n$, where the state variables x_i are the voltages at the amplifier outputs. They can only take values in the set

$$H = \{x \in R^n \mid -V_M < x_i < V_M\}$$

The functions $g_i : R \rightarrow (-V_M, V_M)$ are sigmoid functions,

$$h_i(x_i) = \left. \frac{dg_i}{du_i} \right|_{u_i = g_i^{-1}(x_i)} > 0, \quad \forall \, x_i \in (-V_M, V_M)$$

I_i are constant current inputs, $R_i > 0$, and $C_i > 0$. Assume that the symmetry condition $T_{ij} = T_{ji}$ is satisfied. The system may have several equilibrium points in H. We assume that all equilibrium points in H are isolated. Due to the symmetry property $T_{ij} = T_{ji}$, the vector whose ith component is

$$- \left[\sum_j T_{ij} x_j - \frac{1}{R_i} g_i^{-1}(x_i) + I_i \right]$$

is a gradient vector of a scalar function. By integration, similar to what we have done in the variable gradient method, it can be shown that this scalar function is given by

$$V(x) = -\tfrac{1}{2} \sum_i \sum_j T_{ij} x_i x_j + \sum_i \frac{1}{R_i} \int_0^{x_i} g_i^{-1}(y) \, dy - \sum_i I_i x_i$$

This function is continuously differentiable, but (typically) not positive definite. We rewrite the state equations as

$$\dot{x}_i = -\frac{1}{C_i} h_i(x_i) \frac{\partial V}{\partial x_i}$$

Let us now apply Theorem 3.4 with $V(x)$ as a candidate function. The derivative of $V(x)$ along the trajectories of the system is given by

$$\dot{V}(x) = \sum_{i=1}^n \frac{\partial V}{\partial x_i} \dot{x}_i = -\sum_{i=1}^n \frac{1}{C_i} h_i(x_i) \left(\frac{\partial V}{\partial x_i} \right)^2 \leq 0$$

Moreover,

$$\dot{V}(x) = 0 \Rightarrow \frac{\partial V}{\partial x_i} = 0 \Rightarrow \dot{x}_i = 0, \quad \forall \, i$$

Hence, $\dot{V}(x) = 0$ only at equilibrium points. To apply Theorem 3.4, we need to construct a set Ω. Let

$$\Omega(\epsilon) = \{ x \in R^n \mid -(V_M - \epsilon) \leq x_i \leq (V_M - \epsilon) \}$$

where $\epsilon > 0$ is arbitrarily small. This set $\Omega(\epsilon)$ is closed and bounded, and $\dot{V}(x) \leq 0$ in $\Omega(\epsilon)$. It remains to show that $\Omega(\epsilon)$ is a positively invariant set; that is, every trajectory starting in $\Omega(\epsilon)$ stays for all future time in $\Omega(\epsilon)$. To simplify this task, we assume a specific form for the sigmoid function $g_i(\cdot)$. Let

$$g_i(u_i) = \frac{2V_M}{\pi} \tan^{-1} \frac{\lambda \pi u_i}{2V_M}, \quad \lambda > 0$$

Then

$$\dot{x}_i = \frac{1}{C_i} h_i(x_i) \left[\sum_j T_{ij} x_j - \frac{2V_M}{\lambda \pi R_i} \tan \frac{\pi x_i}{2V_M} + I_i \right]$$

For $|x_i| \geq V_M - \epsilon$,

$$\left| \tan \frac{\pi x_i}{2V_M} \right| \geq \tan \frac{\pi (V_M - \epsilon)}{2V_M} \to \infty \text{ as } \epsilon \to 0$$

Since x_i and I_i are bounded, ϵ can be chosen small enough to ensure that

$$x_i \sum_j T_{ij} x_j - \frac{2V_M x_i}{\lambda \pi R_i} \tan \frac{\pi x_i}{2V_M} + x_i I_i < 0, \quad \text{for } V_M - \epsilon \leq |x_i| < V_M$$

Hence,

$$\frac{d}{dt} \left(x_i^2 \right) = 2 x_i \dot{x}_i < 0, \quad \text{for } V_M - \epsilon \leq |x_i| < V_M, \ \forall \ i$$

Consequently, trajectories starting in $\Omega(\epsilon)$ will stay in $\Omega(\epsilon)$ for all future time. In fact, trajectories starting in $H - \Omega(\epsilon)$ will converge to $\Omega(\epsilon)$. This implies that all the equilibrium points of the system lie in the compact set $\Omega(\epsilon)$. Hence, there can be only a finite number of isolated equilibrium points. In $\Omega(\epsilon)$, $E = M = $ the set of equilibrium points inside $\Omega(\epsilon)$. By Theorem 3.4, we know that every trajectory starting inside $\Omega(\epsilon)$ approaches M as t approaches infinity. Since M consists of isolated equilibrium points, it can be shown (Exercise 3.22) that a trajectory approaching M must approach one of these equilibria. Thus, for all possible initial conditions the trajectory of the system will always converge to one of the equilibrium points. This ensures that the system will not oscillate. $\triangle$

3.3   Linear Systems and Linearization

The linear time-invariant system

$$\dot{x} = Ax \tag{3.9}$$

has an equilibrium point at the origin. The equilibrium point is isolated if and only if $\det(A) \neq 0$. If $\det(A) = 0$, the matrix A has a nontrivial null space. Every point in the null space of A is an equilibrium point for the system (3.9). In other words, if $\det(A) = 0$, the system has an equilibrium subspace. Notice that a linear system cannot have multiple isolated equilibrium points. For, if $\bar{x}_1$ and $\bar{x}_2$ are two equilibrium points for (3.9), then by linearity every point on the line connecting $\bar{x}_1$ and $\bar{x}_2$ is an equilibrium point for the system. Stability properties of the origin can be characterized by the locations of the eigenvalues of the matrix A. Recall from linear system theory[10] that the solution of (3.9) for a given initial state $x(0)$ is given by

$$x(t) = \exp(At)x(0) \tag{3.10}$$

and that for any matrix A there is a nonsingular matrix P (possibly complex) that transforms A into its Jordan form; that is,

$$P^{-1}AP = J = \text{block diag}[J_1, J_2, \ldots, J_r]$$

where J_i is the Jordan block associated with the eigenvalue λ_i of A. A Jordan block of order m takes the form

$$J_i = \begin{bmatrix} \lambda_i & 1 & 0 & \ldots & \ldots & 0 \\ 0 & \lambda_i & 1 & 0 & \ldots & 0 \\ \vdots & & \ddots & & & \vdots \\ \vdots & & & \ddots & & 0 \\ \vdots & & & & \ddots & 1 \\ 0 & \ldots & \ldots & \ldots & 0 & \lambda_i \end{bmatrix}_{m \times m}$$

Therefore,

$$\exp(At) = P \exp(Jt) P^{-1} = \sum_{i=1}^{r} \sum_{k=1}^{m_i} t^{k-1} \exp(\lambda_i t) R_{ik} \tag{3.11}$$

where m_i is the order of the Jordan block associated with the eigenvalue λ_i.[11] The following theorem characterizes the stability properties of the origin.

Theorem 3.5 *The equilibrium point $x = 0$ of (3.9) is stable if and only if all eigenvalues of A satisfy $\text{Re}\lambda_i \leq 0$ and every eigenvalue with $\text{Re}\lambda_i = 0$ has an associated Jordan block of order one. The equilibrium point $x = 0$ is (globally) asymptotically stable if and only if all eigenvalues of A satisfy $\text{Re}\lambda_i < 0$.* ◇

Proof: From (3.10) we can see that the origin is stable if and only if $\exp(At)$ is a bounded function of t for all $t \geq 0$. If one of the eigenvalues of A is in the open right-half complex plane, the corresponding exponential term $\exp(\lambda_i t)$ in (3.11) will grow unbounded as $t \to \infty$. Therefore, we must restrict the eigenvalues to be in the closed left-half complex plane. However, those eigenvalues on the imaginary axis

---

[10] See, for example, [29], [41], [70], [86], or [142].

[11] Equivalently, m_i is the multiplicity of λ_i as a zero of the minimal polynomial of A.

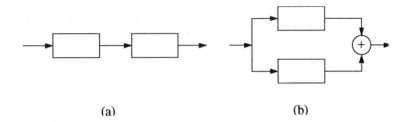

(a)   (b)

Figure 3.6: (a) Series connection; (b) parallel connection.

(if any) could give rise to unbounded terms if the order of the associated Jordan block is higher than one, due to the term t^{k-1} in (3.11). Therefore, we must restrict eigenvalues on the imaginary axis to have Jordan blocks of order one. Thus, we conclude that the condition for stability is a necessary one. It is clear that the condition is also sufficient to ensure that $\exp(At)$ is bounded. For asymptotic stability of the origin, $\exp(At)$ must approach 0 as $t \to \infty$. From (3.11), this is the case if and only if $Re\lambda_i < 0$, $\forall\ i$. Since $x(t)$ depends linearly on the initial state $x(0)$, asymptotic stability of the origin is global. $\square$

The proof shows, mathematically, why eigenvalues on the imaginary axis must have Jordan blocks of order one to ensure stability. The following example may shed some light on the physical meaning of this requirement.

Example 3.12 Figure 3.6 shows a series connection and a parallel connection of two identical systems. Each system is represented by the state-space model

$$\dot{x} = \begin{bmatrix} 0 & 1 \\ -1 & 0 \end{bmatrix} x + \begin{bmatrix} 0 \\ 1 \end{bmatrix} u$$

$$y = \begin{bmatrix} 1 & 0 \end{bmatrix} x$$

where u and y are the input and output of the system. Let A_s and A_p be the matrices of the series and parallel connections, when modeled in the form (3.9) (no driving inputs). Let J_s and J_p be the corresponding Jordan blocks. Then

$$A_p = \begin{bmatrix} 0 & 1 & 0 & 0 \\ -1 & 0 & 0 & 0 \\ 0 & 0 & 0 & 1 \\ 0 & 0 & -1 & 0 \end{bmatrix}, \quad J_p = \begin{bmatrix} j & 0 & 0 & 0 \\ 0 & -j & 0 & 0 \\ 0 & 0 & j & 0 \\ 0 & 0 & 0 & -j \end{bmatrix}$$

and

$$A_s = \begin{bmatrix} 0 & 1 & 0 & 0 \\ -1 & 0 & 0 & 0 \\ 0 & 0 & 0 & 1 \\ 1 & 0 & -1 & 0 \end{bmatrix}, \quad J_s = \begin{bmatrix} j & 1 & 0 & 0 \\ 0 & j & 0 & 0 \\ 0 & 0 & -j & 1 \\ 0 & 0 & 0 & -j \end{bmatrix}$$

where $j = \sqrt{-1}$. The matrices A_p and A_s have the same eigenvalues on the imaginary axis, but the associated Jordan blocks for the parallel connection are of order one while those of the series connection are of order two. Thus, by Theorem 3.5, the origin of the parallel connection is stable while the origin of the series connection is unstable. To physically see the difference between the two connections, notice that in the parallel connection nonzero initial conditions produce sinusoidal oscillations of frequency 1 rad/sec, which are bounded functions of time. The sum of these sinusoidal signals remains bounded. On the other hand, nonzero initial conditions in the first component of the series connection produce a sinusoidal oscillation of frequency 1 rad/sec, which acts as a driving force for the second component. Since the second component has an undamped natural frequency of 1 rad/sec, the driving force causes "resonance" and the response grows unbounded. $\qquad \triangle$

When all eigenvalues of A satisfy $Re\lambda_i < 0$, A is called a *stability matrix* or a *Hurwitz matrix*. The origin of (3.9) is asymptotically stable if and only if A is a stability matrix. Asymptotic stability of the origin can be also investigated using Lyapunov's method. Consider a quadratic Lyapunov function candidate

$$V(x) = x^T P x$$

where P is a real symmetric positive definite matrix. The derivative of V along the trajectories of the linear system (3.9) is given by

$$\dot{V}(x) = x^T P \dot{x} + \dot{x}^T P x = x^T (PA + A^T P)x = -x^T Q x$$

where Q is a symmetric matrix defined by

$$PA + A^T P = -Q \qquad (3.12)$$

If Q is positive definite, we can conclude by Theorem 3.1 that the origin is asymptotically stable; that is, $Re\lambda_i < 0$ for all eigenvalues of A. Here we follow the usual procedure of Lyapunov's method, where we choose $V(x)$ to be positive definite and then check the negative definiteness of $\dot{V}(x)$. In the case of linear systems, we can reverse the order of these two steps. Suppose we start by choosing Q as a real symmetric positive definite matrix, and then solve (3.12) for P. If (3.12) has a positive definite solution, then we can again conclude that the origin is asymptotically stable. Equation (3.12) is called the *Lyapunov equation*. The following theorem characterizes asymptotic stability of the origin in terms of the solution of the Lyapunov equation.

Theorem 3.6 *A matrix A is a stability matrix; that is, $Re\lambda_i < 0$ for all eigenvalues of A, if and only if for any given positive definite symmetric matrix Q there exists a positive definite symmetric matrix P that satisfies the Lyapunov equation* (3.12). *Moreover, if A is a stability matrix, then P is the unique solution of* (3.12). $\qquad \diamond$

Proof: Sufficiency follows from Theorem 3.1 with the Lyapunov function $V(x) = x^T P x$, as we have already shown. To prove necessity, assume that all eigenvalues of A satisfy $Re\lambda_i < 0$ and consider the matrix P, defined by

$$P = \int_0^\infty \exp(A^T t) Q \exp(At) \, dt \tag{3.13}$$

The integrand is a sum of terms of the form $t^{k-1} \exp(\lambda_i t)$, where $Re\lambda_i < 0$. There-fore, the integral exists. The matrix P is symmetric and positive definite. The fact that it is positive definite can be shown as follows. Supposing it is not so, there is a vector $x \neq 0$ such that $x^T P x = 0$. However,

$$x^T P x = 0 \quad \Rightarrow \quad \int_0^\infty x^T \exp(A^T t) Q \exp(At) x \, dt = 0$$
$$\Rightarrow \quad \exp(At) x \equiv 0, \ \forall \ t \geq 0 \quad \Rightarrow \quad x = 0$$

since $\exp(At)$ is nonsingular for all t. This contradiction shows that P is positive definite. Now, substitution of (3.13) in the left-hand side of (3.12) yields

$$PA + A^T P = \int_0^\infty \exp(A^T t) Q \exp(At) A \, dt + \int_0^\infty A^T \exp(A^T t) Q \exp(At) \, dt$$
$$= \int_0^\infty \frac{d}{dt} \exp(A^T t) Q \exp(At) \, dt = \exp(A^T t) Q \exp(At) \Big|_0^\infty = -Q$$

which shows that P is indeed a solution of (3.12). To show that it is the unique solution, suppose there is another solution $\tilde{P} \neq P$. Then,

$$(P - \tilde{P})A + A^T(P - \tilde{P}) = 0$$

Premultiplying by $\exp(A^T t)$ and postmultiplying by $\exp(At)$, we obtain

$$0 = \exp(A^T t)[(P - \tilde{P})A + A^T(P - \tilde{P})] \exp(At) = \frac{d}{dt} \exp(A^T t)(P - \tilde{P}) \exp(At)$$

Hence,

$$\exp(A^T t)(P - \tilde{P}) \exp(At) \equiv \text{a constant} \ \forall \ t$$

In particular, since $\exp(A0) = I$, we have

$$(P - \tilde{P}) = \exp(A^T t)(P - \tilde{P}) \exp(At) \to 0 \text{ as } t \to \infty$$

Therefore, $\tilde{P} = P$.   □

The positive definiteness requirement on Q can be relaxed. It is left as an exercise to the reader (Exercise 3.24) to verify that Q can be taken as a positive semidefinite matrix of the form $Q = C^T C$, where the pair (A, C) is observable.

Equation (3.12) is a linear algebraic equation which can be solved by rearranging it in the form $Mx = y$ where x and y are defined by stacking the elements of P and Q in vectors, as will be illustrated in the next example. There are numerically efficient methods for solving such equations.[12]

Example 3.13 Let

$$A = \begin{bmatrix} 0 & -1 \\ 1 & -1 \end{bmatrix}, \quad Q = \begin{bmatrix} 1 & 0 \\ 0 & 1 \end{bmatrix}, \text{ and } P = \begin{bmatrix} p_{11} & p_{12} \\ p_{12} & p_{22} \end{bmatrix}$$

where, due to symmetry, $p_{12} = p_{21}$. The Lyapunov equation (3.12) can be rewritten as

$$\begin{bmatrix} 0 & 2 & 0 \\ -1 & -1 & 1 \\ 0 & -2 & -2 \end{bmatrix} \begin{bmatrix} p_{11} \\ p_{12} \\ p_{22} \end{bmatrix} = \begin{bmatrix} -1 \\ 0 \\ -1 \end{bmatrix}$$

The unique solution of this equation is given by

$$\begin{bmatrix} p_{11} \\ p_{12} \\ p_{22} \end{bmatrix} = \begin{bmatrix} 1.5 \\ -0.5 \\ 1.0 \end{bmatrix} \quad \Rightarrow \quad P = \begin{bmatrix} 1.5 & -0.5 \\ -0.5 & 1.0 \end{bmatrix}$$

The matrix P is positive definite since its leading principal minors (1.5 and 1.25) are positive. Hence, all eigenvalues of A are in the open left-half complex plane.
$\triangle$

The Lyapunov equation can be used to test whether or not a matrix A is a stability matrix, as an alternative to calculating the eigenvalues of A. One starts by choosing a positive definite matrix Q (for example, $Q = I$) and solves the Lyapunov equation (3.12) for P. If the equation has a positive definite solution, we conclude that A is a stability matrix; otherwise, it is not so. However, there is no computational advantage in solving the Lyapunov equation over calculating the eigenvalues of A.[13] Besides, the eigenvalues provide more direct information about the response of the linear system. The interest in the Lyapunov equation is not in its use as a stability test for linear systems;[14] rather, it is in the fact that it provides

---

[12] Consult [57] on numerical methods for solving linear algebraic equations. The Lyapunov equation can also be solved by viewing it as a special case of the Sylvester equation $PA + BP + C = 0$, which is treated in [57]. Almost all commercial software programs for control systems include commands for solving the Lyapunov equation.

[13] A typical procedure for solving the Lyapunov equation, the Bartels-Stewart algorithm [57], starts by transforming A into its real Schur form which gives the eigenvalues of A. Hence, the computational effort for solving the Lyapunov equation is more than calculating the eigenvalues of A. Other algorithms for solving the Lyapunov equation take an amount of computations comparable to the Bartels-Stewart algorithm.

[14] It might be of interest, however, to know that one can use the Lyapunov equation to derive the classical Routh-Hurwitz criterion; see [29, pp. 417–419].

a procedure for finding a Lyapunov function for any linear system (3.9) when A is a stability matrix. The mere existence of a Lyapunov function will allow us to draw conclusions about the system when the right-hand side Ax is perturbed, whether such perturbation is a linear perturbation in the coefficients of A or a nonlinear perturbation. This advantage will unfold as we continue our study of Lyapunov's method.

Let us go back to the nonlinear system (3.1)

$$\dot{x} = f(x)$$

where $f : D \to R^n$ is a continuously differentiable map from a domain $D \subset R^n$ into R^n. Suppose that the origin $x = 0$ is in the interior of D and is an equilibrium point for the system; that is, $f(0) = 0$. By the mean value theorem,

$$f_i(x) = f_i(0) + \frac{\partial f_i}{\partial x}(z_i)\, x$$

where z_i is a point on the line segment connecting x to the origin. The above equality is valid for any point $x \in D$ such that the line segment connecting x to the origin lies entirely in D. Since $f(0) = 0$, we can write $f_i(x)$ as

$$f_i(x) = \frac{\partial f_i}{\partial x}(z_i)x = \frac{\partial f_i}{\partial x}(0)x + \left[\frac{\partial f_i}{\partial x}(z_i) - \frac{\partial f_i}{\partial x}(0) \right] x$$

Hence,

$$f(x) = Ax + g(x)$$

where

$$A = \frac{\partial f}{\partial x}(0), \quad \text{and} \quad g_i(x) = \left[\frac{\partial f_i}{\partial x}(z_i) - \frac{\partial f_i}{\partial x}(0) \right] x$$

The function $g_i(x)$ satisfies

$$|g_i(x)| \leq \left\| \frac{\partial f_i}{\partial x}(z_i) - \frac{\partial f_i}{\partial x}(0) \right\| \|x\|$$

By continuity of $[\partial f/\partial x]$, we see that

$$\frac{\|g(x)\|}{\|x\|} \to 0 \quad \text{as} \quad \|x\| \to 0$$

This suggests that in a small neighborhood of the origin we can approximate the nonlinear system (3.1) by its linearization about the origin

$$\dot{x} = Ax, \quad \text{where} \quad A = \frac{\partial f}{\partial x}(0)$$

The following theorem spells out conditions under which we can draw conclusions about stability of the origin as an equilibrium point for the nonlinear system by investigating its stability as an equilibrium point for the linear system. The theorem is known as *Lyapunov's indirect method.*

Theorem 3.7 *Let $x = 0$ be an equilibrium point for the nonlinear system*

$$\dot{x} = f(x)$$

where $f : D \rightarrow R^n$ is continuously differentiable and D is a neighborhood of the origin. Let

$$A = \left.\frac{\partial f}{\partial x}(x)\right|_{x=0}$$

Then,

   1. *The origin is asymptotically stable if $Re\lambda_i < 0$ for all eigenvalues of A.*

   2. *The origin is unstable if $Re\lambda_i > 0$ for one or more of the eigenvalues of A.*

$\diamond$

Proof: To prove the first part, let A be a stability matrix. Then, by Theorem 3.6 we know that for any positive definite symmetric matrix Q, the solution P of the Lyapunov equation (3.12) is positive definite. We use

$$V(x) = x^T P x$$

as a Lyapunov function candidate for the nonlinear system. The derivative of $V(x)$ along the trajectories of the system is given by

$$
\begin{aligned}
\dot{V}(x) &= x^T P f(x) + f^T(x) P x \\
&= x^T P[Ax + g(x)] + [x^T A^T + g^T(x)] P x \\
&= x^T (PA + A^T P)x + 2x^T P g(x) \\
&= -x^T Q x + 2x^T P g(x)
\end{aligned}
$$

The first term on the right-hand side is negative definite, while the second term is (in general) indefinite. The function $g(x)$ satisfies

$$\frac{\|g(x)\|_2}{\|x\|_2} \rightarrow 0 \quad \text{as} \quad \|x\|_2 \rightarrow 0$$

Therefore, for any $\gamma > 0$ there exists $r > 0$ such that

$$\|g(x)\|_2 < \gamma\|x\|_2, \quad \forall \ \|x\|_2 < r$$

Hence,

$$\dot{V}(x) < -x^T Q x + 2\gamma \|P\|_2 \|x\|_2^2, \quad \forall \ \|x\|_2 < r$$

but

$$x^T Q x \geq \lambda_{\min}(Q) \|x\|_2^2$$

where $\lambda_{\min}(\cdot)$ denotes the minimum eigenvalue of a matrix. Note that $\lambda_{\min}(Q)$ is real and positive since Q is symmetric and positive definite. Thus

$$\dot{V}(x) < -[\lambda_{\min}(Q) - 2\gamma \|P\|_2] \|x\|_2^2, \quad \forall \ \|x\|_2 < r$$

Choosing $\gamma < \lambda_{\min}(Q)/2\|P\|_2$ ensures that $\dot{V}(x)$ is negative definite. By Theorem 3.1, we conclude that the origin is asymptotically stable. To prove the second part of the theorem, let us consider first the special case when A has no eigenvalues on the imaginary axis. If the eigenvalues of A cluster into a group of eigenvalues in the open right-half plane and a group of eigenvalues in the open left-half plane, then there is a nonsingular matrix T such that[15]

$$TAT^{-1} = \left[\begin{array}{cc} -A_1 & 0 \\ 0 & A_2 \end{array} \right]$$

where A_1 and A_2 are stability matrices. Let

$$z = Tx = \left[\begin{array}{c} z_1 \\ z_2 \end{array} \right]$$

where the partition of z is compatible with the dimensions of A_1 and A_2. The change of variables $z = Tx$ transforms the system

$$\dot{x} = Ax + g(x)$$

into the form

$$\begin{array}{rcl} \dot{z}_1 & = & -A_1 z_1 + g_1(z) \\ \dot{z}_2 & = & A_2 z_2 + g_2(z) \end{array}$$

where the functions $g_i(z)$ have the property that for any $\gamma > 0$, there exists $r > 0$ such that

$$\|g_i(z)\|_2 < \gamma \|z\|_2, \quad \forall \ \|z\|_2 \leq r, \ i = 1, 2$$

The origin $z = 0$ is an equilibrium point for the system in the z-coordinates. Clearly, any conclusion we arrive at concerning stability properties of $z = 0$ carries over to the equilibrium point $x = 0$ in the x-coordinates, since T is nonsingular.[16] To show

---

[15] There are several methods for finding the matrix T, one of which is to transform the matrix A into its real Jordan form [57].

[16] See Exercise 3.27 for a general discussion of stability-preserving maps.

that the origin is unstable, we apply Theorem 3.3. The construction of a function $V(z)$ will be done basically as in Example 3.7, except for working with vectors instead of scalars. Let Q_1 and Q_2 be positive definite symmetric matrices of the dimensions of A_1 and A_2, respectively. Since A_1 and A_2 are stability matrices, we know from Theorem 3.6 that the Lyapunov equations

$$P_i A_i + A_i^T P_i = -Q_i, \ i = 1, 2$$

have unique positive definite solutions P_1 and P_2. Let

$$V(z) = z_1^T P_1 z_1 - z_2^T P_2 z_2 = z^T \begin{bmatrix} P_1 & 0 \\ 0 & -P_2 \end{bmatrix} z$$

In the subspace $z_2 = 0$, $V(z) > 0$ at points arbitrarily close to the origin. Let

$$U = \{z \in R^n \mid \|z\|_2 \leq r \text{ and } V(z) > 0\}$$

In U,

$$
\begin{aligned}
\dot{V}(z) &= -z_1^T (P_1 A_1 + A_1^T P_1)z_1 + 2z_1^T P_1 g_1(z) \\
&\quad -z_2^T (P_2 A_2 + A_2^T P_2)z_2 - 2z_2^T P_2 g_2(z) \\
&= z_1^T Q_1 z_1 + z_2^T Q_2 z_2 + 2z^T \begin{bmatrix} P_1 g_1(z) \\ -P_2 g_2(z) \end{bmatrix} \\
&\geq \lambda_{\min}(Q_1)\|z_1\|_2^2 + \lambda_{\min}(Q_2)\|z_2\|_2^2 \\
&\quad -2\|z\|_2 \sqrt{\|P_1\|_2^2 \|g_1(z)\|_2^2 + \|P_2\|_2^2 \|g_2(z)\|_2^2} \\
&> (\alpha - 2\sqrt{2}\beta\gamma)\|z\|_2^2
\end{aligned}
$$

where

$$\alpha = \min\{\lambda_{\min}(Q_1), \lambda_{\min}(Q_2)\}, \quad \beta = \max\{\|P_1\|_2, \|P_2\|_2\}$$

Thus, choosing $\gamma < \alpha/2\sqrt{2}\beta$ ensures that $\dot{V}(z) > 0$ in U. Therefore, by Theorem 3.3, the origin is unstable. Notice that we could have applied Theorem 3.3 in the original coordinates by defining the matrices

$$P = T^T \begin{bmatrix} P_1 & 0 \\ 0 & -P_2 \end{bmatrix} T; \quad Q = T^T \begin{bmatrix} Q_1 & 0 \\ 0 & Q_2 \end{bmatrix} T$$

which satisfy the equation

$$PA + A^T P = Q$$

The matrix Q is positive definite, and $V(x) = x^T P x$ is positive for points arbitrarily close to the origin $x = 0$. Let us consider now the general case when A may have eigenvalues on the imaginary axis, in addition to eigenvalues in the open right-half complex plane. We can reduce this case to the special case we have just studied by a simple trick of shifting the imaginary axis. Suppose A has m eigenvalues with $Re\lambda_i > \delta > 0$. Then, the matrix $[A - (\delta/2)I]$ has m eigenvalues in the open right-half plane, but no eigenvalues on the imaginary axis. By previous arguments, there exist matrices $P = P^T$ and $Q = Q^T > 0$ such that

$$P\left[A - \frac{\delta}{2}I\right] + \left[A - \frac{\delta}{2}I\right]^T P = Q$$

where $V(x) = x^T P x$ is positive for points arbitrarily close to the origin. The derivative of $V(x)$ along the trajectories of the system is given by

$$
\begin{aligned}
\dot{V}(x) &= x^T (PA + A^T P)x + 2x^T Pg(x) \\
&= x^T \left[P\left(A - \frac{\delta}{2}I\right) + \left(A - \frac{\delta}{2}I\right)^T P\right] x + \delta x^T Px + 2x^T Pg(x) \\
&= x^T Qx + \delta V(x) + 2x^T Pg(x)
\end{aligned}
$$

In the set

$$\{x \in R^n \mid \|x\|_2 \leq r \text{ and } V(x) > 0\}$$

where r is chosen such that $\|g(x)\|_2 \leq \gamma \|x\|_2$ for $\|x\|_2 < r$, $\dot{V}(x)$ satisfies

$$\dot{V}(x) \geq \lambda_{\min}(Q)\|x\|_2^2 - 2\|P\|_2\|x\|_2\|g(x)\|_2 \geq (\lambda_{\min}(Q) - 2\gamma\|P\|_2)\|x\|_2^2$$

which is positive for $\gamma < \lambda_{\min}(Q)/2\|P\|_2$. Application of Theorem 3.3 concludes the proof.   □

Theorem 3.7 provides us with a simple procedure for determining stability of an equilibrium point at the origin. We calculate the *Jacobian matrix*

$$A = \left.\frac{\partial f}{\partial x}\right|_{x=0}$$

and test its eigenvalues. If $Re\lambda_i < 0$ for all i or $Re\lambda_i > 0$ for some i, we conclude that the origin is asymptotically stable or unstable, respectively. Moreover, the proof of the theorem shows that when $Re\lambda_i < 0$ for all i, we can also find a Lyapunov function for the system that will work locally in some neighborhood of the origin. The Lyapunov function is the quadratic form $V(x) = x^T P x$, where P is the solution of the Lyapunov equation (3.12) for any positive definite symmetric matrix Q. Note that Theorem 3.7 does not say anything about the case when $Re\lambda_i \leq 0$ for all i, with $Re\lambda_i = 0$ for some i. In this case, linearization fails to determine stability of the equilibrium point.[17]

---

[17] See Section 4.1 for further investigation of the critical case when linearization fails.

Example 3.14 Consider the scalar system

$$\dot{x} = ax^3$$

Linearization of the system about the origin $x = 0$ yields

$$A = \left. \frac{\partial f}{\partial x} \right|_{x=0} = 3ax^2 \big|_{x=0} = 0$$

There is one eigenvalue which lies on the imaginary axis. Hence, linearization fails to determine stability of the origin. This failure is genuine in the sense that the origin could be asymptotically stable, stable, or unstable, depending on the value of the parameter a. If $a < 0$, the origin is asymptotically stable as can be seen from the Lyapunov function $V(x) = x^4$ whose derivative along the trajectories of the system is given by $\dot{V}(x) = 4ax^6 < 0$ when $a < 0$. If $a = 0$, the system is linear and the origin is stable according to Theorem 3.5. If $a > 0$, the origin is unstable as can be seen from Theorem 3.3 and the function $V(x) = x^4$, whose derivative along the trajectories of the system is given by $\dot{V}(x) = 4ax^6 > 0$ when $a > 0$. △

Example 3.15 The pendulum equation

$$\begin{aligned} \dot{x}_1 &= x_2 \\ \dot{x}_2 &= -\left(\frac{g}{l}\right)\sin x_1 - \left(\frac{k}{m}\right)x_2 \end{aligned}$$

has two equilibrium points at $(x_1 = 0, \ x_2 = 0)$ and $(x_1 = \pi, \ x_2 = 0)$. Let us investigate the stability of both points using linearization. The Jacobian matrix is given by

$$\frac{\partial f}{\partial x} = \begin{bmatrix} \frac{\partial f_1}{\partial x_1} & \frac{\partial f_1}{\partial x_2} \\ \frac{\partial f_2}{\partial x_1} & \frac{\partial f_2}{\partial x_2} \end{bmatrix} = \begin{bmatrix} 0 & 1 \\ -\left(\frac{g}{l}\right)\cos x_1 & -\left(\frac{k}{m}\right) \end{bmatrix}$$

To determine stability of the origin, we evaluate the Jacobian at $x = 0$.

$$A = \left. \frac{\partial f}{\partial x} \right|_{x=0} = \begin{bmatrix} 0 & 1 \\ -\left(\frac{g}{l}\right) & -\left(\frac{k}{m}\right) \end{bmatrix}$$

The eigenvalues of A are

$$\lambda_{1,2} = -\frac{k}{2m} \pm \frac{1}{2}\sqrt{\left(\frac{k}{m}\right)^2 - \frac{4g}{l}}$$

For all g, l, m, $k > 0$, the eigenvalues satisfy $Re\lambda_i < 0$. Hence, the equilibrium point at the origin is asymptotically stable. In the absence of friction ($k = 0$), both

eigenvalues are on the imaginary axis. In this case, we cannot determine stability of
the origin through linearization. We have seen in Example 3.3 that, in this case, the
origin is a stable equilibrium point as determined by an energy Lyapunov function.
To determine stability of the equilibrium point at $(x_1 = \pi, \ x_2 = 0)$, we evaluate
the Jacobian at this point. This is equivalent to performing a change of variables
$z_1 = x_1 - \pi$, $z_2 = x_2$ to shift the equilibrium point to the origin, and evaluating
the Jacobian $[\partial f/\partial z]$ at $z = 0$.

$$\tilde{A} = \left.\frac{\partial f}{\partial x}\right|_{x_1=\pi, x_2=0} = \left[\begin{array}{cc} 0 & 1 \\ \left(\frac{g}{l}\right) & -\left(\frac{k}{m}\right) \end{array}\right]$$

The eigenvalues of $\tilde{A}$ are

$$\lambda_{1,2} = -\frac{k}{2m} \pm \frac{1}{2}\sqrt{\left(\frac{k}{m}\right)^2 + \frac{4g}{l}}$$

For all g, l, $m > 0$ and all $k \geq 0$, there is one eigenvalue in the open right-half
plane. Hence, the equilibrium point at $(x_1 = \pi, \ x_2 = 0)$ is unstable. $\triangle$

3.4   Nonautonomous Systems

Consider the nonautonomous system

$$\dot{x} = f(t, x) \tag{3.14}$$

where $f : [0, \infty) \times D \to R^n$ is piecewise continuous in t and locally Lipschitz in x
on $[0, \infty) \times D$, and $D \subset R^n$ is a domain that contains the origin $x = 0$. The origin
is an equilibrium point for (3.14) at $t = 0$ if

$$f(t, 0) = 0, \quad \forall\, t \geq 0$$

An equilibrium at the origin could be a translation of a nonzero equilibrium point
or, more generally, a translation of a nonzero solution of the system. To see the
latter point, suppose that $\bar{y}(\tau)$ is a solution of the system

$$\frac{dy}{d\tau} = g(\tau, y)$$

defined for all $\tau \geq a$. The change of variables

$$x = y - \bar{y}(\tau); \quad t = \tau - a$$

transforms the system into the form

$$\dot{x} = g(\tau, y) - \dot{\bar{y}}(\tau) = g(t + a, x + \bar{y}(t + a)) - \dot{\bar{y}}(t + a) \stackrel{\text{def}}{=} f(t, x)$$

Since
$$\dot{\bar{y}}(t+a) = g(t+a, \bar{y}(t+a)), \quad \forall \, t \geq 0$$

the origin $x = 0$ is an equilibrium point of the transformed system at $t = 0$. Therefore, by examining the stability behavior of the origin as an equilibrium point for the transformed system, we determine the stability behavior of the solution $\bar{y}(\tau)$ of the original system. Notice that if $\bar{y}(\tau)$ is not constant, the transformed system will be nonautonomous even when the original system is autonomous, that is, even when $g(\tau, y) = g(y)$. That is why studying the stability behavior of solutions in the sense of Lyapunov can only be done in the context of studying the stability behavior of the equilibria of nonautonomous systems.

The notions of stability and asymptotic stability of the equilibrium of a nonautonomous system are basically the same as we introduced in Definition 3.1 for autonomous systems. The new element here is that, while the solution of an autonomous system depends only on $(t - t_0)$, the solution of a nonautonomous system may depend on both t and t_0. Therefore, the stability behavior of the equilibrium point will, in general, be dependent on t_0. The origin $x = 0$ is a stable equilibrium point for (3.14) if for each $\epsilon > 0$ and any $t_0 \geq 0$ there is $\delta = \delta(\epsilon, t_0) > 0$ such that

$$\|x(t_0)\| < \delta \Rightarrow \|x(t)\| < \epsilon, \quad \forall \, t \geq t_0$$

The constant δ is, in general, dependent on the initial time t_0. The existence of δ for every t_0 does not necessarily guarantee that there is one constant δ, dependent only on ϵ, that would work for all t_0, as illustrated by the following example.

Example 3.16 The linear first-order system

$$\dot{x} = (6t \sin t - 2t)x$$

has the closed-form solution

$$
\begin{aligned}
x(t) &= x(t_0) \exp\left[\int_{t_0}^{t} (6\tau \sin \tau - 2\tau) \, d\tau \right] \\
&= x(t_0) \exp\left[6 \sin t - 6t \cos t - t^2 - 6 \sin t_0 + 6t_0 \cos t_0 + t_0^2 \right]
\end{aligned}
$$

For any t_0, the term $-t^2$ will eventually dominate, which shows that the exponential term is bounded for all $t \geq t_0$ by a constant $c(t_0)$ dependent on t_0. Hence,

$$|x(t)| < |x(t_0)| c(t_0), \quad \forall \, t \geq t_0$$

For any $\epsilon > 0$, the choice $\delta = \epsilon / c(t_0)$ shows that the origin is stable. Now, suppose t_0 takes on the successive values $t_0 = 2n\pi$ for $n = 0, 1, 2, \ldots$, and suppose that $x(t)$ is evaluated π seconds later in each case. Then,

$$x(t_0 + \pi) = x(t_0) \exp\left[(4n+1)(6 - \pi)\pi \right]$$

This implies that, for $x(t_0) \neq 0$,

$$\frac{x(t_0 + \pi)}{x(t_0)} \to \infty \quad \text{as} \quad n \to \infty$$

Thus, given $\epsilon > 0$, there is no δ independent of t_0 that would satisfy the stability requirement uniformly in t_0. $\triangle$

Nonuniformity with respect to t_0 could also appear in studying asymptotic stability of the origin, as the following example shows.

Example 3.17 The linear first-order system

$$\dot{x} = -\frac{x}{1+t}$$

has the closed-form solution

$$x(t) = x(t_0) \exp\left(\int_{t_0}^{t} \frac{-1}{1+\tau} \, d\tau\right) = x(t_0)\frac{1+t_0}{1+t}$$

Since $|x(t)| \leq |x(t_0)|$, $\forall\, t \geq t_0$, the origin is clearly stable. Actually, given any $\epsilon > 0$, we can choose δ independent of t_0. It is also clear that

$$x(t) \to 0 \quad \text{as} \quad t \to \infty$$

Hence, according to Definition 3.1, the origin is asymptotically stable. Notice, however, that the convergence of $x(t)$ to the origin is not uniform with respect to the initial time t_0. Recall that convergence of $x(t)$ to the origin is equivalent to saying that, given any $\epsilon > 0$, there is $T = T(\epsilon, t_0) > 0$ such that $|x(t)| < \epsilon$ for all $t \geq t_0 + T$. Although this is true for every t_0, the constant T cannot be chosen independent of t_0. $\triangle$

As a consequence, we need to refine Definition 3.1 to emphasize the dependence of the stability behavior of the origin on the initial time t_0. We are interested in a refinement that defines stability and asymptotic stability of the origin as uniform properties with respect to the initial time.[18]

Definition 3.2 *The equilibrium point $x = 0$ of (3.14) is*

- *stable if, for each $\epsilon > 0$, there is $\delta = \delta(\epsilon, t_0) > 0$ such that*

$$\|x(t_0)\| < \delta \Rightarrow \|x(t)\| < \epsilon, \quad \forall\, t \geq t_0 \geq 0 \tag{3.15}$$

---

[18] See [62] or [87] for other refinements of Definition 3.1.

- *uniformly stable if, for each $\epsilon > 0$, there is $\delta = \delta(\epsilon) > 0$, independent of t_0, such that (3.15) is satisfied.*

- *unstable if not stable.*

- *asymptotically stable if it is stable and there is $c = c(t_0) > 0$ such that $x(t) \rightarrow 0$ as $t \rightarrow \infty$, for all $\|x(t_0)\| < c$.*

- *uniformly asymptotically stable if it is uniformly stable and there is $c > 0$, independent of t_0, such that for all $\|x(t_0)\| < c$, $x(t) \rightarrow 0$ as $t \rightarrow \infty$, uniformly in t_0; that is, for each $\epsilon > 0$, there is $T = T(\epsilon) > 0$ such that*

$$\|x(t)\| < \epsilon, \quad \forall \, t \geq t_0 + T(\epsilon), \ \forall \, \|x(t_0)\| < c$$

- *globally uniformly asymptotically stable if it is uniformly stable and, for each pair of positive numbers ϵ and c, there is $T = T(\epsilon, c) > 0$ such that*

$$\|x(t)\| < \epsilon, \quad \forall \, t \geq t_0 + T(\epsilon, c), \ \forall \, \|x(t_0)\| < c$$

Uniform stability and asymptotic stability can be characterized in terms of special scalar functions, known as class $\mathcal{K}$ and class $\mathcal{KL}$ functions.

Definition 3.3 *A continuous function $\alpha : [0, a) \rightarrow [0, \infty)$ is said to belong to class $\mathcal{K}$ if it is strictly increasing and $\alpha(0) = 0$. It is said to belong to class $\mathcal{K}_\infty$ if $a = \infty$ and $\alpha(r) \rightarrow \infty$ as $r \rightarrow \infty$.*

Definition 3.4 *A continuous function $\beta : [0, a) \times [0, \infty) \rightarrow [0, \infty)$ is said to belong to class $\mathcal{KL}$ if, for each fixed s, the mapping $\beta(r, s)$ belongs to class $\mathcal{K}$ with respect to r and, for each fixed r, the mapping $\beta(r, s)$ is decreasing with respect to s and $\beta(r, s) \rightarrow 0$ as $s \rightarrow \infty$.*

Example 3.18

- $\alpha(r) = \tan^{-1} r$ is strictly increasing since $\alpha'(r) = 1/(1 + r^2) > 0$. It belongs to class $\mathcal{K}$, but not to class $\mathcal{K}_\infty$ since $\lim_{r \to \infty} \alpha(r) = \pi/2 < \infty$.

- $\alpha(r) = r^c$, for any positive real number c, is strictly increasing since $\alpha'(r) = cr^{c-1} > 0$. Moreover, $\lim_{r \to \infty} \alpha(r) = \infty$; thus, it belongs to class $\mathcal{K}_\infty$.

- $\beta(r, s) = r/(ksr + 1)$, for any positive real number k, is strictly increasing in r since

$$\frac{\partial \beta}{\partial r} = \frac{1}{(ksr + 1)^2} > 0$$

and strictly decreasing in s since

$$\frac{\partial \beta}{\partial s} = \frac{-kr^2}{(ksr + 1)^2} < 0$$

Moreover, $\beta(r, s) \rightarrow 0$ as $s \rightarrow \infty$. Hence, it belongs to class $\mathcal{KL}$.

- $\beta(r, s) = r^c e^{-s}$, for any positive real number c, belongs to class $\mathcal{KL}$.   △

The following lemma states some obvious properties of class $\mathcal{K}$ and class $\mathcal{KL}$ functions, which will be needed later on. The proof of the lemma is left as an exercise for the reader (Exercise 3.32).

Lemma 3.2 *Let $\alpha_1(\cdot)$ and $\alpha_2(\cdot)$ be class $\mathcal{K}$ functions on $[0, a)$, $\alpha_3(\cdot)$ and $\alpha_4(\cdot)$ be class $\mathcal{K}_\infty$ functions, and $\beta(\cdot, \cdot)$ be a class $\mathcal{KL}$ function. Denote the inverse of $\alpha_i(\cdot)$ by $\alpha_i^{-1}(\cdot)$. Then,*

- α_1^{-1} *is defined on $[0, \alpha_1(a))$ and belongs to class $\mathcal{K}$.*

- α_3^{-1} *is defined on $[0, \infty)$ and belongs to class $\mathcal{K}_\infty$.*

- $\alpha_1 \circ \alpha_2$ *belongs to class $\mathcal{K}$.*

- $\alpha_3 \circ \alpha_4$ *belongs to class $\mathcal{K}_\infty$.*

- $\sigma(r, s) = \alpha_1(\beta(\alpha_2(r), s))$ *belongs to class $\mathcal{KL}$.*   ◇

The following lemma gives equivalent definitions of uniform stability and uniform asymptotic stability using class $\mathcal{K}$ and class $\mathcal{KL}$ functions.

Lemma 3.3 *The equilibrium point $x = 0$ of (3.14) is*

- *uniformly stable if and only if there exist a class $\mathcal{K}$ function $\alpha(\cdot)$ and a positive constant c, independent of t_0, such that*

$$\|x(t)\| \le \alpha(\|x(t_0)\|), \quad \forall\, t \ge t_0 \ge 0, \ \forall\, \|x(t_0)\| < c \qquad (3.16)$$

- *uniformly asymptotically stable if and only if there exist a class $\mathcal{KL}$ function $\beta(\cdot, \cdot)$ and a positive constant c, independent of t_0, such that*

$$\|x(t)\| \le \beta(\|x(t_0)\|, t - t_0), \quad \forall\, t \ge t_0 \ge 0, \ \forall\, \|x(t_0)\| < c \qquad (3.17)$$

- *globally uniformly asymptotically stable if and only if inequality (3.17) is satisfied for any initial state $x(t_0)$.*   ◇

Proof: Appendix A.3.

As a consequence of Lemma 3.3, we see that in the case of autonomous systems stability and asymptotic stability per Definition 3.1 imply the existence of class $\mathcal{K}$ and class $\mathcal{KL}$ functions that satisfy inequalities (3.16) and (3.17). This is the case because, for autonomous systems, stability and asymptotic stability of the origin are uniform with respect to the initial time t_0.

A special case of uniform asymptotic stability arises when the class $\mathcal{KL}$ function β in (3.17) takes the form $\beta(r, s) = kr e^{-\gamma s}$. This case is very important and will be designated as a distinct stability property of equilibrium points.

Definition 3.5 *The equilibrium point* $x = 0$ *of* (3.14) *is exponentially stable if inequality* (3.17) *is satisfied with*

$$\beta(r, s) = kre^{-\gamma s}, \quad k > 0, \quad \gamma > 0$$

and is globally exponentially stable if this condition is satisfied for any initial state.

Lyapunov theory for autonomous systems can be extended to nonautonomous systems. For each of Theorems 3.1–3.4, one can state various extensions to nonautonomous systems. We shall not document all these extensions here.[19] Instead, we concentrate on uniform asymptotic stability.[20] This is the case we encounter in most nonautonomous applications of Lyapunov's method. To establish uniform asymptotic stability of the origin, we need to verify inequality (3.17). The class $\mathcal{KL}$ function in this inequality will appear in our analysis through the solution of an autonomous scalar differential equation. We start with two preliminary lemmas; the first one states properties of the solution of this special equation and the second one gives upper and lower bounds on a positive definite function in terms of class $\mathcal{K}$ functions.

Lemma 3.4 *Consider the scalar autonomous differential equation*

$$\dot{y} = -\alpha(y), \quad y(t_0) = y_0$$

where $\alpha(\cdot)$ *is a locally Lipschitz class* $\mathcal{K}$ *function defined on* $[0, a)$. *For all* $0 \le y_0 < a$, *this equation has a unique solution* $y(t)$ *defined for all* $t \ge t_0$. *Moreover,*

$$y(t) = \sigma(y_0, t - t_0)$$

where $\sigma(r, s)$ *is a class* $\mathcal{KL}$ *function defined on* $[0, a) \times [0, \infty)$. $\diamond$

Proof: Appendix A.4.

We can see that the claim of this lemma is true by examining specific examples where a closed-form solution of the scalar equation can be found. For example, if $\dot{y} = -ky$, $k > 0$, then the solution is

$$y(t) = y_0 \exp[-k(t - t_0)] \quad \Rightarrow \quad \sigma(r, s) = r \exp(-ks)$$

As another example, if $\dot{y} = -ky^2$, $k > 0$, then the solution is

$$y(t) = \frac{y_0}{ky_0(t - t_0) + 1} \quad \Rightarrow \quad \sigma(r, s) = \frac{r}{krs + 1}$$

---

[19]Lyapunov theory for nonautonomous systems is well documented in the literature. Good references on the subject include [62] and [137], while good introductions can be found in [181] and [118].

[20]See Exercise 3.35 for a uniform stability theorem.

Lemma 3.5 *Let $V(x) : D \to R$ be a continuous positive definite function defined on a domain $D \subset R^n$ that contains the origin. Let $B_r \subset D$ for some $r > 0$. Then, there exist class $\mathcal{K}$ functions α_1 and α_2, defined on $[0, r]$, such that*

$$\alpha_1(\|x\|) \leq V(x) \leq \alpha_2(\|x\|)$$

for all $x \in B_r$. Moreover, if $D = R^n$ and $V(x)$ is radially unbounded, then α_1 and α_2 can be chosen to belong to class $\mathcal{K}_\infty$ and the foregoing inequality holds for all $x \in R^n$.   ◇

Proof: Appendix A.5.

For a quadratic positive definite function $V(x) = x^T P x$, Lemma 3.5 follows from the inequalities

$$\lambda_{\min}(P)\|x\|_2^2 \leq x^T P x \leq \lambda_{\max}(P)\|x\|_2^2$$

We are now ready to state and prove the main result of this section.

Theorem 3.8 *Let $x = 0$ be an equilibrium point for (3.14) and $D \subset R^n$ be a domain containing $x = 0$. Let $V : [0, \infty) \times D \to R$ be a continuously differentiable function such that*

$$W_1(x) \leq V(t, x) \leq W_2(x) \tag{3.18}$$

$$\frac{\partial V}{\partial t} + \frac{\partial V}{\partial x} f(t, x) \leq -W_3(x) \tag{3.19}$$

$\forall\, t \geq 0,\ \forall\, x \in D$ where $W_1(x)$, $W_2(x)$, and $W_3(x)$ are continuous positive definite functions on D. Then, $x = 0$ is uniformly asymptotically stable.   ◇

Proof: The derivative of V along the trajectories of (3.14) is given by

$$\dot{V}(t, x) = \frac{\partial V}{\partial t} + \frac{\partial V}{\partial x} f(t, x) \leq -W_3(x)$$

Choose $r > 0$ and $\rho > 0$ such that $B_r \subset D$ and $\rho < \min_{\|x\|=r} W_1(x)$. Then, $\{x \in B_r \mid W_1(x) \leq \rho\}$ is in the interior of B_r. Define a time-dependent set $\Omega_{t,\rho}$ by

$$\Omega_{t,\rho} = \{x \in B_r \mid V(t, x) \leq \rho\}$$

The set $\Omega_{t,\rho}$ contains $\{x \in B_r \mid W_2(x) \leq \rho\}$ since

$$W_2(x) \leq \rho \Rightarrow V(t, x) \leq \rho$$

On the other hand, $\Omega_{t,\rho}$ is a subset of $\{x \in B_r \mid W_1(x) \leq \rho\}$ since

$$V(t, x) \leq \rho \Rightarrow W_1(x) \leq \rho$$

Thus,
$$\{x \in B_r \mid W_2(x) \leq \rho\} \subset \Omega_{t,\rho} \subset \{x \in B_r \mid W_1(x) \leq \rho\} \subset B_r$$

for all $t \geq 0$. For any $t_0 \geq 0$ and any $x_0 \in \Omega_{t_0,\rho}$, the solution starting at (t_0, x_0) stays in $\Omega_{t,\rho}$ for all $t \geq t_0$. This follows from the fact that $\dot{V}(t, x)$ is negative on $D - \{0\}$; hence, $V(t, x)$ is decreasing. Therefore, the solution starting at (t_0, x_0) is defined for all $t \geq t_0$ and $x(t) \in B_r$. For the rest of the proof, we will assume that $x_0 \in \{x \in B_r \mid W_2(x) \leq \rho\}$. By Lemma 3.5, there exist class $\mathcal{K}$ functions α_1, α_2, and α_3, defined on $[0, r]$, such that

$$W_1(x) \geq \alpha_1(\|x\|), \quad W_2(x) \leq \alpha_2(\|x\|), \quad W_3(x) \geq \alpha_3(\|x\|)$$

Hence, V and $\dot{V}$ satisfy the inequalities

$$\alpha_1(\|x\|) \leq V(t, x) \leq \alpha_2(\|x\|)$$

$$\dot{V}(t, x) \leq -\alpha_3(\|x\|)$$

Consequently,
$$\dot{V} \leq -\alpha_3(\|x\|) \leq -\alpha_3\left(\alpha_2^{-1}(V)\right) \overset{\text{def}}{=} -\alpha(V)$$

The function $\alpha(\cdot)$ is a class $\mathcal{K}$ function defined on $[0, r]$ (see Lemma 3.2). Assume, without loss of generality,[21] that $\alpha(\cdot)$ is locally Lipschitz. Let $y(t)$ satisfy the autonomous first-order differential equation

$$\dot{y} = -\alpha(y), \quad y(t_0) = V(t_0, x(t_0)) \geq 0$$

By the comparison lemma (Lemma 2.5),

$$V(t, x(t)) \leq y(t), \quad \forall\, t \geq t_0$$

By Lemma 3.4, there exists a class $\mathcal{KL}$ function $\sigma(r, s)$ defined on $[0, r] \times [0, \infty)$ such that

$$V(t, x(t)) \leq \sigma(V(t_0, x(t_0)), t - t_0), \quad \forall\, V(t_0, x(t_0)) \in [0, \rho]$$

Therefore, any solution starting in $\Omega_{t_0,\rho}$ satisfies the inequality

$$\begin{aligned}
\|x(t)\| &\leq \alpha_1^{-1}(V(t, x(t))) \leq \alpha_1^{-1}(\sigma(V(t_0, x(t_0)), t - t_0)) \\
&\leq \alpha_1^{-1}(\sigma(\alpha_2(\|x(t_0)\|), t - t_0)) \overset{\text{def}}{=} \beta(\|x(t_0)\|, t - t_0)
\end{aligned}$$

---

[21] If α is not locally Lipschitz, we can choose a locally Lipschitz class $\mathcal{K}$ function β such that $\alpha(r) \geq \beta(r)$ over the domain of interest. Then, $\dot{V} \leq -\beta(V)$ and we can continue the proof with β instead of α. For example, suppose $\alpha(r) = \sqrt{r}$. The function $\sqrt{r}$ is a class $\mathcal{K}$ function, but not locally Lipschitz at $r = 0$. Define β as $\beta(r) = r$ for $r < 1$ and $\beta(r) = \sqrt{r}$ for $r \geq 1$. The function β is class $\mathcal{K}$ and locally Lipschitz. Moreover, $\alpha(r) \geq \beta(r)$ for all $r \geq 0$.

Lemma 3.2 shows that $\beta(\cdot, \cdot)$ is a class $\mathcal{KL}$ function. Thus, inequality (3.17) is satisfied for all $x(t_0) \in \{x \in B_r \mid W_2(x) \leq \rho\}$, which implies that $x = 0$ is uniformly asymptotically stable. $\Box$

A function $V(t, x)$ satisfying the left inequality of (3.18) is said to be *positive definite*. A function satisfying the right inequality of (3.18) is said to be *decrescent*. A function $V(t, x)$ is said to be *negative definite* if $-V(t, x)$ is positive definite. Therefore, Theorem 3.8 says that the origin is uniformly asymptotically stable if there is a continuously differentiable, positive definite, decrescent function $V(t, x)$ whose derivative along the trajectories of the system is negative definite. In this case, $V(t, x)$ is called a Lyapunov function.

The proof of Theorem 3.8 estimates the region of attraction of the origin by the set

$$\{x \in B_r \mid W_2(x) \leq \rho\}$$

This estimate allows us to obtain a global version of Theorem 3.8.

Corollary 3.3 *Suppose that all the assumptions of Theorem 3.8 are satisfied globally (for all $x \in R^n$) and $W_1(x)$ is radially unbounded. Then, $x = 0$ is globally uniformly asymptotically stable.* $\Diamond$

Proof: Since $W_1(x)$ is radially unbounded, so is $W_2(x)$. Therefore, the set $\{x \in R^n \mid W_2(x) \leq \rho\}$ is bounded for any $\rho > 0$. For any $x_0 \in R^n$, we can choose ρ large enough so that $x_0 \in \{x \in R^n \mid W_2(x) \leq \rho\}$. The rest of the proof is the same as that of Theorem 3.8. $\Box$

A Lyapunov function $V(t, x)$ satisfying (3.18) with radially unbounded $W_1(x)$ is said to be *radially unbounded*.

The manipulation of class $\mathcal{K}$ functions in the proof of Theorem 3.8 simplifies when the class $\mathcal{K}$ functions take the special form $\alpha_i(r) = k_i r^c$. In this case, we can actually show that the origin is exponentially stable.

Corollary 3.4 *Suppose all the assumptions of Theorem 3.8 are satisfied with*

$$W_1(x) \geq k_1 \|x\|^c, \quad W_2(x) \leq k_2 \|x\|^c, \quad W_3(x) \geq k_3 \|x\|^c$$

for some positive constants k_1, k_2, k_3, and c. Then, $x = 0$ is exponentially stable. Moreover, if the assumptions hold globally, then $x = 0$ is globally exponentially stable. $\Diamond$

Proof: V and $\dot{V}$ satisfy the inequalities

$$k_1 \|x\|^c \leq V(t, x) \leq k_2 \|x\|^c$$

$$\dot{V}(t, x) \leq -k_3 \|x\|^c \leq -\frac{k_3}{k_2} V(t, x)$$

By the comparison lemma (Lemma 2.5),

$$V(t, x(t)) \leq V(t_0, x(t_0))e^{-(k_3/k_2)(t-t_0)}$$

Hence,

$$
\|x(t)\| \leq \left[\frac{V(t, x(t))}{k_1}\right]^{1/c} \leq \left[\frac{V(t_0, x(t_0))e^{-(k_3/k_2)(t-t_0)}}{k_1}\right]^{1/c}
$$

$$
\leq \left[\frac{k_2\|x(t_0)\|^c e^{-(k_3/k_2)(t-t_0)}}{k_1}\right]^{1/c} = \left(\frac{k_2}{k_1}\right)^{1/c} \|x(t_0)\|e^{-(k_3/k_2 c)(t-t_0)}
$$

Hence, the origin is exponentially stable. If all the assumptions hold globally, the foregoing inequality holds for all $x(t_0) \in R^n$.   □

Example 3.19 Consider the scalar system

$$\dot{x} = -[1 + g(t)]x^3$$

where $g(t)$ is continuous and $g(t) \geq 0$ for all $t \geq 0$. Using the Lyapunov function candidate $V(x) = \frac{1}{2}x^2$, we obtain

$$\dot{V} = -[1 + g(t)]x^4 \leq -x^4, \quad \forall\, x \in R, \,\forall\, t \geq 0$$

The assumptions of Theorem 3.8 are satisfied globally with $W_1(x) = W_2(x) = V(x)$ and $W_3(x) = x^4$. Hence, the origin is globally uniformly asymptotically stable.   △

Example 3.20 Consider the system

$$
\begin{aligned}
\dot{x}_1 &= -x_1 - g(t)x_2 \\
\dot{x}_2 &= x_1 - x_2
\end{aligned}
$$

where $g(t)$ is continuously differentiable and satisfies

$$0 \leq g(t) \leq k \quad \text{and} \quad \dot{g}(t) \leq g(t), \quad \forall\, t \geq 0$$

Taking $V(t, x) = x_1^2 + [1 + g(t)]x_2^2$ as a Lyapunov function candidate, it can be easily seen that

$$x_1^2 + x_2^2 \leq V(t, x) \leq x_1^2 + (1 + k)x_2^2, \quad \forall\, x \in R^2$$

Hence, $V(t, x)$ is positive definite, decrescent, and radially unbounded. The derivative of V along the trajectories of the system is given by

$$\dot{V}(t, x) = -2x_1^2 + 2x_1x_2 - [2 + 2g(t) - \dot{g}(t)]x_2^2$$

Using the inequality

$$2 + 2g(t) - \dot{g}(t) \geq 2 + 2g(t) - g(t) \geq 2$$

we obtain

$$\dot{V}(t, x) \leq -2x_1^2 + 2x_1 x_2 - 2x_2^2 = - \begin{bmatrix} x_1 \\ x_2 \end{bmatrix}^T \begin{bmatrix} 2 & -1 \\ -1 & 2 \end{bmatrix} \begin{bmatrix} x_1 \\ x_2 \end{bmatrix} \stackrel{\text{def}}{=} -x^T Q x$$

where Q is positive definite; hence, $\dot{V}(t, x)$ is negative definite. Thus, all the assumptions of Theorem 3.8 are satisfied globally with quadratic positive definite functions W_1, W_2, and W_3. From Corollary 3.4, we conclude that the origin is globally exponentially stable.   △

Example 3.21 The linear time-varying system

$$\dot{x} = A(t)x \tag{3.20}$$

has an equilibrium point at $x = 0$. Let $A(t)$ be continuous for all $t \geq 0$. Suppose there is a continuously differentiable, symmetric, bounded, positive definite matrix $P(t)$, that is,

$$0 < c_1 I \leq P(t) \leq c_2 I, \quad \forall\, t \geq 0$$

which satisfies the matrix differential equation

$$-\dot{P}(t) = P(t)A(t) + A^T(t)P(t) + Q(t) \tag{3.21}$$

where $Q(t)$ is continuous, symmetric, and positive definite; that is,

$$Q(t) \geq c_3 I > 0, \quad \forall\, t \geq 0$$

Consider a Lyapunov function candidate

$$V(t, x) = x^T P(t)x$$

The function $V(t, x)$ is positive definite, decrescent, and radially unbounded since

$$c_1 \|x\|_2^2 \leq V(t, x) \leq c_2 \|x\|_2^2$$

The derivative of $V(t, x)$ along the trajectories of the system (3.20) is given by

$$\begin{aligned} \dot{V}(t, x) &= x^T \dot{P}(t)x + x^T P(t)\dot{x} + \dot{x}^T P(t)x \\ &= x^T[\dot{P}(t) + P(t)A(t) + A^T(t)P(t)]x = -x^T Q(t)x \leq -c_3 \|x\|_2^2 \end{aligned}$$

Hence, $\dot{V}(t, x)$ is negative definite. All the assumptions of Corollary 3.4 are satisfied globally with $c = 2$. Therefore, the origin is globally exponentially stable.   △

3.5 Linear Time-Varying Systems and Linearization

The stability behavior of the origin as an equilibrium point for the linear time-varying system (3.20):

$$\dot{x}(t) = A(t)x$$

can be completely characterized in terms of the state transition matrix of the system. From linear system theory,[22] we know that the solution of (3.20) is given by

$$x(t) = \Phi(t, t_0)x(t_0)$$

where $\Phi(t, t_0)$ is the state transition matrix. The following theorem characterizes uniform asymptotic stability in terms of $\Phi(t, t_0)$.

Theorem 3.9 *The equilibrium point $x = 0$ of (3.20) is (globally) uniformly asymptotically stable if and only if the state transition matrix satisfies the inequality*

$$\|\Phi(t, t_0)\| \le ke^{-\gamma(t-t_0)}, \quad \forall\, t \ge t_0 \ge 0 \tag{3.22}$$

for some positive constants k and γ.   ◇

Proof: Due to the linear dependence of $x(t)$ on $x(t_0)$, if the origin is uniformly asymptotically stable it is globally so. Sufficiency of (3.22) is obvious since

$$\|x(t)\| \le \|\Phi(t, t_0)\|\, \|x(t_0)\| \le k\|x(t_0)\|e^{-\gamma(t-t_0)}$$

To prove necessity, suppose the origin is uniformly asymptotically stable. Then, there is a class $\mathcal{KL}$ function $\beta(\cdot, \cdot)$ such that

$$\|x(t)\| \le \beta(\|x(t_0)\|, t - t_0), \quad \forall\, t \ge t_0, \ \forall\, x(t_0) \in R^n$$

From the definition of an induced matrix norm (Section 2.1), we have

$$\|\Phi(t, t_0)\| = \max_{\|x\|=1} \|\Phi(t, t_0)x\| \le \max_{\|x\|=1} \beta(\|x\|, t - t_0) = \beta(1, t - t_0)$$

Since

$$\beta(1, s) \to 0 \quad \text{as} \quad s \to \infty$$

there exists $T > 0$ such that $\beta(1, T) \le 1/e$. For any $t \ge t_0$, let N be the smallest positive integer such that $t \le t_0 + NT$. Divide the interval $[t_0, t_0 + (N-1)T]$ into $(N-1)$ equal subintervals of width T each. Using the transition property of $\Phi(t, t_0)$, we can write

$$\Phi(t, t_0) = \Phi(t, t_0 + (N-1)T)\Phi(t_0 + (N-1)T, t_0 + (N-2)T)\cdots\Phi(t_0 + T, t_0)$$

---

[22] See, for example, [29], [41], [86], or [142].

Hence,

$$\|\Phi(t,t_0)\| \leq \|\Phi(t,t_0+(N-1)T)\| \prod_{k=1}^{k=N-1} \|\Phi(t_0+kT,t_0+(k-1)T)\|$$

$$\leq \beta(1,0) \prod_{k=1}^{k=N-1} \frac{1}{e} = e\beta(1,0)e^{-N}$$

$$\leq e\beta(1,0)e^{-(t-t_0)/T} = ke^{-\gamma(t-t_0)}$$

where $k = e\beta(1,0)$ and $\gamma = 1/T$. $\square$

Theorem 3.9 shows that, for linear systems, uniform asymptotic stability of the origin is equivalent to exponential stability. Although inequality (3.22) characterizes uniform asymptotic stability of the origin without the need to search for a Lyapunov function, it is not as useful as the eigenvalue criterion we have for linear time-invariant systems because knowledge of the state transition matrix $\Phi(t,t_0)$ requires solving the state equation (3.20). Note that, for linear time-varying systems, uniform asymptotic stability cannot be characterized by the location of the eigenvalues of the matrix A [23] as the following example shows.

Example 3.22 Consider a second-order linear system with

$$A(t) = \begin{bmatrix} -1+1.5\cos^2 t & 1-1.5\sin t\cos t \\ -1-1.5\sin t\cos t & -1+1.5\sin^2 t \end{bmatrix}$$

For each t, the eigenvalues of $A(t)$ are given by $-0.25 \pm 0.25\sqrt{7}j$. Thus, the eigenvalues are independent of t and lie in the open left-half plane. Yet, the origin is unstable. It can be verified that

$$\Phi(t,0) = \begin{bmatrix} e^{0.5t}\cos t & e^{-t}\sin t \\ -e^{0.5t}\sin t & e^{-t}\cos t \end{bmatrix}$$

which shows that there are initial states $x(0)$, arbitrarily close to the origin, for which the solution is unbounded and escapes to infinity. $\triangle$

Although Theorem 3.9 may not be very helpful as a stability test, we shall see that it guarantees the existence of a Lyapunov function for the linear system (3.20). We saw in Example 3.21 that if we can find a positive definite, bounded matrix $P(t)$ that satisfies the differential equation (3.21) for some positive definite $Q(t)$,

---

[23] There are special cases where uniform asymptotic stability of the origin as equilibrium for (3.20) is equivalent to an eigenvalue condition. One case is periodic systems; see Exercise 3.40 and Example 8.8. Another case is slowly-varying systems; see Example 5.13.

then $V(t, x) = x^T P(t)x$ is a Lyapunov function for the system. If the matrix $Q(t)$ is chosen to be bounded in addition to being positive definite, that is,

$$0 < c_3 I \leq Q(t) \leq c_4 I, \quad \forall\, t \geq 0$$

and if $A(t)$ is continuous and bounded, then it can be shown that when the origin is uniformly asymptotically stable, there is a solution of (3.21) that possesses the desired properties.

Theorem 3.10 *Let $x = 0$ be the uniformly asymptotically stable equilibrium of (3.20). Suppose $A(t)$ is continuous and bounded. Let $Q(t)$ be a continuous, bounded, positive definite, symmetric matrix. Then, there is a continuously differentiable, bounded, positive definite, symmetric matrix $P(t)$ which satisfies (3.21). Hence, $V(t, x) = x^T P(t)x$ is a Lyapunov function for the system that satisfies the conditions of Theorem 3.8.* ◇

Proof: Let

$$P(t) = \int_t^\infty \Phi^T(\tau, t)Q(\tau)\Phi(\tau, t)\, d\tau$$

and $\phi(\tau, t, x)$ be the solution of (3.20) that starts at (t, x). Due to linearity, $\phi(\tau, t, x) = \Phi(\tau, t)x$. In view of the definition of $P(t)$, we have

$$x^T P(t)x = \int_t^\infty \phi^T(\tau, t, x)Q(\tau)\phi(\tau, t, x)\, d\tau$$

Use of (3.22) yields

$$
\begin{aligned}
x^T P(t)x &\leq \int_t^\infty c_4 \|\Phi(\tau, t)\|_2^2 \, \|x\|_2^2 d\tau \\
&\leq \int_t^\infty k^2 e^{-2\gamma(\tau - t)}\, d\tau\, c_4 \|x\|_2^2 \;=\; \frac{k^2 c_4}{2\gamma} \|x\|_2^2 \stackrel{\text{def}}{=} c_2 \|x\|_2^2
\end{aligned}
$$

On the other hand, since

$$\|A(t)\|_2 \leq L, \quad \forall\, t \geq 0$$

the linear system (3.20) satisfies a global Lipschitz condition with a Lipschitz constant L. Therefore,[24] the solution $\phi(\tau, t, x)$ satisfies the lower bound

$$\|\phi(\tau, t, x)\|_2^2 \geq \|x\|_2^2 e^{-2L(\tau - t)}$$

Hence,

$$
\begin{aligned}
x^T P(t)x &\geq \int_t^\infty c_3 \|\phi(\tau, t, x)\|_2^2\, d\tau \\
&\geq \int_t^\infty e^{-2L(\tau - t)}\, d\tau\, c_3 \|x\|_2^2 \;=\; \frac{c_3}{2L} \|x\|_2^2 \stackrel{\text{def}}{=} c_1 \|x\|_2^2
\end{aligned}
$$

---

[24] See Exercise 2.22.

Thus,

$$c_1\|x\|_2^2 \le x^T P(t)x \le c_2\|x\|_2^2$$

which shows that $P(t)$ is positive definite and bounded. The definition of $P(t)$ shows that it is symmetric and continuously differentiable. The fact that $P(t)$ satisfies (3.21) can be shown by differentiation of $P(t)$ using the property

$$\frac{\partial}{\partial t}\Phi(\tau,t) = -\Phi(\tau,t)A(t)$$

In particular,

$$
\begin{aligned}
\dot{P}(t) &= \int_t^\infty \Phi^T(\tau,t)Q(\tau)\frac{\partial}{\partial t}\Phi(\tau,t)\,d\tau \\
&\quad + \int_t^\infty \frac{\partial}{\partial t}\Phi^T(\tau,t)Q(\tau)\Phi(\tau,t)\,d\tau \;-\; Q(t) \\
&= -\int_t^\infty \Phi^T(\tau,t)Q(\tau)\Phi(\tau,t)\,d\tau\, A(t) \\
&\quad - A^T(t)\int_t^\infty \Phi^T(\tau,t)Q(\tau)\Phi(\tau,t)\,d\tau \;-\; Q(t) \\
&= -P(t)A(t) - A^T(t)P(t) - Q(t)
\end{aligned}
$$

The fact that $V(t,x) = x^T P(t)x$ is a Lyapunov function is shown in Example 3.21.

$\square$

When the linear system (3.20) is time invariant, that is, when A is constant, the Lyapunov function $V(t,x)$ of Theorem 3.10 can be chosen to be independent of t. Recall that, for linear time-invariant systems,

$$\Phi(\tau,t) = \exp[(\tau - t)A]$$

which satisfies (3.22) when A is a stability matrix. Choosing Q to be a positive definite, symmetric (constant) matrix, the matrix $P(t)$ is given by

$$P = \int_t^\infty \exp[(\tau - t)A^T]Q\exp[(\tau - t)A]\,d\tau = \int_0^\infty \exp[A^T s]Q\exp[As]\,ds$$

which is independent of t. Comparison of this expression for P with (3.13) shows that P is the unique solution of the Lyapunov equation (3.12). Thus, the Lyapunov function of Theorem 3.10 reduces to the one we used in Section 3.3.

The existence of Lyapunov functions for linear systems per Theorem 3.10 will now be used to prove a linearization result that extends Theorem 3.7 to the nonautonomous case. Consider the nonlinear nonautonomous system (3.14)

$$\dot{x} = f(t,x)$$

where $f : [0, \infty) \times D \to R^n$ is continuously differentiable and $D = \{x \in R^n \mid \|x\|_2 < r\}$. Suppose the origin $x = 0$ is an equilibrium point for the system at $t = 0$; that is, $f(t, 0) = 0$ for all $t \geq 0$. Furthermore, suppose the Jacobian matrix $[\partial f / \partial x]$ is bounded and Lipschitz on D, uniformly in t; thus,

$$\left\| \frac{\partial f_i}{\partial x}(t, x_1) - \frac{\partial f_i}{\partial x}(t, x_2) \right\|_2 \leq L_1 \|x_1 - x_2\|_2, \quad \forall \, x_1, x_2 \in D, \ \forall \, t \geq 0$$

for all $1 \leq i \leq n$. By the mean value theorem,

$$f_i(t, x) = f_i(t, 0) + \frac{\partial f_i}{\partial x}(t, z_i) \, x$$

where z_i is a point on the line segment connecting x to the origin. Since $f(t, 0) = 0$, we can write $f_i(t, x)$ as

$$f_i(t, x) = \frac{\partial f_i}{\partial x}(t, z_i) \, x = \frac{\partial f_i}{\partial x}(t, 0) \, x + \left[\frac{\partial f_i}{\partial x}(t, z_i) - \frac{\partial f_i}{\partial x}(t, 0) \right] x$$

Hence,

$$f(t, x) = A(t)x + g(t, x)$$

where

$$A(t) = \frac{\partial f}{\partial x}(t, 0) \quad \text{and} \quad g_i(t, x) = \left[\frac{\partial f_i}{\partial x}(t, z_i) - \frac{\partial f_i}{\partial x}(t, 0) \right] x$$

The function $g(t, x)$ satisfies

$$\|g(t, x)\|_2 \leq \left(\sum_{i=1}^{n} \left\| \frac{\partial f_i}{\partial x}(t, z_i) - \frac{\partial f_i}{\partial x}(t, 0) \right\|_2^2 \right)^{1/2} \|x\|_2 \leq L \|x\|_2^2$$

where $L = \sqrt{n} L_1$. Therefore, in a small neighborhood of the origin, we may approximate the nonlinear system (3.14) by its linearization about the origin. The following theorem states Lyapunov's indirect method for showing uniform asymptotic stability of the origin in the nonautonomous case.

Theorem 3.11 *Let $x = 0$ be an equilibrium point for the nonlinear system*

$$\dot{x} = f(t, x)$$

where $f : [0, \infty) \times D \to R^n$ is continuously differentiable, $D = \{x \in R^n \mid \|x\|_2 < r\}$, and the Jacobian matrix $[\partial f / \partial x]$ is bounded and Lipschitz on D, uniformly in t. Let

$$A(t) = \left. \frac{\partial f}{\partial x}(t, x) \right|_{x=0}$$

Then, the origin is an exponentially stable equilibrium point for the nonlinear system
if it is an exponentially stable equilibrium point for the linear system

$$\dot{x} = A(t)x$$

$$\diamond$$

Proof: Since the linear system has an exponentially stable equilibrium point at
the origin and $A(t)$ is continuous and bounded, Theorem 3.10 ensures the existence
of a continuously differentiable, bounded, positive definite symmetric matrix $P(t)$
that satisfies (3.21) where $Q(t)$ is continuous, positive definite, and symmetric. We
use $V(t, x) = x^T P(t)x$ as a Lyapunov function candidate for the nonlinear system.
The derivative of $V(t, x)$ along the trajectories of the system is given by

$$
\begin{aligned}
\dot{V}(t, x) &= x^T P(t)f(t, x) + f^T(t, x)P(t)x + x^T \dot{P}(t)x \\
&= x^T[P(t)A(t) + A^T(t)P(t) + \dot{P}(t)]x + 2x^T P(t)g(t, x) \\
&= -x^T Q(t)x + 2x^T P(t)g(t, x) \\
&\leq -c_3\|x\|_2^2 + 2c_2 L\|x\|_2^3 \\
&\leq -(c_3 - 2c_2 L\rho)\|x\|_2^2, \quad \forall \ \|x\|_2 < \rho
\end{aligned}
$$

Choosing $\rho < \min\{r, c_3/2c_2 L\}$ ensures that $\dot{V}(t, x)$ is negative definite in $\|x\|_2 < \rho$.
Hence, all the conditions of Corollary 3.4 are satisfied in $\|x\|_2 < \rho$, and we conclude
that the origin is exponentially stable.   □

3.6   Converse Theorems

Theorem 3.8 and Corollaries 3.3 and 3.4 establish uniform asymptotic stability (or
exponential stability) of the origin by requiring the existence of a Lyapunov function $V(t, x)$ that satisfies certain conditions. Requiring the existence of an auxiliary
function $V(t, x)$ that satisfies certain conditions is typical in many theorems of Lyapunov's method. The conditions of these theorems cannot be checked directly on
the data of the problem. Instead, one has to search for the auxiliary function. Faced
with this searching problem, two questions come to mind. First, is there a function
that satisfies the conditions of the theorem? Second, how can we search for such
a function? In many cases, Lyapunov theory provides an affirmative answer to the
first question. The answer takes the form of a converse Lyapunov theorem, which
is the inverse of one of Lyapunov's theorems. For example, a converse theorem for
uniform asymptotic stability would confirm that if the origin is uniformly asymptotically stable, then there is a Lyapunov function that satisfies the conditions of
Theorem 3.8. Most of these converse theorems are proven by actually constructing
auxiliary functions that satisfy the conditions of the respective theorems. Unfor-
tunately, this construction almost always assumes the knowledge of the solutions

of the differential equation. Therefore, these theorems do not help in the practical search for an auxiliary function. The mere knowledge that a function exists is, however, better than nothing. At least we know that our search is not hopeless. The theorems are also useful in using Lyapunov theory to draw conceptual conclusions about the behavior of dynamical systems. Theorem 3.13 is an example of such use. Other examples will appear in the following chapters. In this section, we give two converse Lyapunov theorems.[25] The first one is a converse Lyapunov theorem when the origin is an exponentially stable equilibrium, and the second one is when the origin is uniformly asymptotically stable.

The idea of constructing a converse Lyapunov function is not new to us. We have done it for linear systems in the proof of Theorem 3.10. A careful reading of that proof shows that linearity of the system does not play a crucial role in the proof, except for showing that $V(t, x)$ is quadratic in x. This observation leads to the first of our two converse theorems, whose proof is a simple extension of the proof of Theorem 3.10.

Theorem 3.12 *Let $x = 0$ be an equilibrium point for the nonlinear system*

$$\dot{x} = f(t, x)$$

where $f : [0, \infty) \times D \to R^n$ is continuously differentiable, $D = \{x \in R^n \mid \|x\| < r\}$, and the Jacobian matrix $[\partial f / \partial x]$ is bounded on D, uniformly in t. Let k, γ, and r_0 be positive constants with $r_0 < r/k$. Let $D_0 = \{x \in R^n \mid \|x\| < r_0\}$. Assume that the trajectories of the system satisfy

$$\|x(t)\| \le k \|x(t_0)\| e^{-\gamma(t - t_0)}, \quad \forall \, x(t_0) \in D_0, \, \forall \, t \ge t_0 \ge 0$$

Then, there is a function $V : [0, \infty) \times D_0 \to R$ that satisfies the inequalities

$$c_1 \|x\|^2 \le V(t, x) \le c_2 \|x\|^2$$

$$\frac{\partial V}{\partial t} + \frac{\partial V}{\partial x} f(t, x) \le -c_3 \|x\|^2$$

$$\left\| \frac{\partial V}{\partial x} \right\| \le c_4 \|x\|$$

for some positive constants c_1, c_2, c_3, and c_4. Moreover, if $r = \infty$ and the origin is globally exponentially stable, then $V(t, x)$ is defined and satisfies the above inequalities on R^n. Furthermore, if the system is autonomous, V can be chosen independent of t. ◇

---

[25] See [62] or [96] for a comprehensive treatment of converse Lyapunov theorems.

Proof: Due to the equivalence of norms, it is sufficient to prove the theorem for the 2-norm. Let $\phi(\tau, t, x)$ denote the solution of the system that starts at (t, x); that is, $\phi(t, t, x) = x$. For all $x \in D_0$, $\phi(\tau, t, x) \in D$ for all $\tau \geq t$. Let

$$V(t, x) = \int_t^{t+T} \phi^T(\tau, t, x) \phi(\tau, t, x) \, d\tau$$

where T is a positive constant to be chosen later. Due to the exponentially decaying bound on the trajectories, we have

$$
\begin{aligned}
V(t, x) &= \int_t^{t+T} \|\phi(\tau, t, x)\|_2^2 \, d\tau \\
&\leq \int_t^{t+T} k^2 e^{-2\gamma(\tau - t)} \, d\tau \, \|x\|_2^2 = \frac{k^2}{2\gamma}(1 - e^{-2\gamma T})\|x\|_2^2
\end{aligned}
$$

On the other hand, the Jacobian matrix $[\partial f / \partial x]$ is bounded on D. Let

$$\left\| \frac{\partial f}{\partial x}(t, x) \right\|_2 \leq L, \quad \forall \, x \in D$$

The function $f(t, x)$ is Lipschitz on D with a Lipschitz constant L. Therefore, the solution $\phi(\tau, t, x)$ satisfies the lower bound[26]

$$\|\phi(\tau, t, x)\|_2^2 \geq \|x\|_2^2 e^{-2L(\tau - t)}$$

Hence,

$$V(t, x) \geq \int_t^{t+T} e^{-2L(\tau - t)} \, d\tau \, \|x\|_2^2 = \frac{1}{2L}(1 - e^{-2LT})\|x\|_2^2$$

Thus, $V(t, x)$ satisfies the first inequality of the theorem with

$$c_1 = \frac{(1 - e^{-2LT})}{2L} \quad \text{and} \quad c_2 = \frac{k^2(1 - e^{-2\gamma T})}{2\gamma}$$

To calculate the derivative of V along the trajectories of the system, define the sensitivity functions

$$\phi_t(\tau, t, x) = \frac{\partial}{\partial t}\phi(\tau, t, x); \quad \phi_x(\tau, t, x) = \frac{\partial}{\partial x}\phi(\tau, t, x)$$

Then,

$$\frac{\partial V}{\partial t} + \frac{\partial V}{\partial x} f(t, x) = \phi^T(t+T, t, x)\phi(t+T, t, x) - \phi^T(t, t, x)\phi(t, t, x)$$

---

[26] See Exercise 2.22.

$$+ \int_t^{t+T} 2\phi^T(\tau, t, x)\phi_t(\tau, t, x) \, d\tau$$

$$+ \int_t^{t+T} 2\phi^T(\tau, t, x)\phi_x(\tau, t, x) \, d\tau f(t, x)$$

$$= \phi^T(t+T, t, x)\phi(t+T, t, x) - \|x\|_2^2$$

$$+ \int_t^{t+T} 2\phi^T(\tau, t, x)[\phi_t(\tau, t, x) + \phi_x(\tau, t, x)f(t, x)] \, d\tau$$

It is not difficult to show that[27]

$$\phi_t(\tau, t, x) + \phi_x(\tau, t, x)f(t, x) \equiv 0, \quad \forall \, \tau \geq t$$

Therefore,

$$\frac{\partial V}{\partial t} + \frac{\partial V}{\partial x} f(t, x) = \phi^T(t+T, t, x)\phi(t+T, t, x) - \|x\|_2^2$$

$$\leq -(1 - k^2 e^{-2\gamma T})\|x\|_2^2$$

Choosing $T = \ln(2k^2)/2\gamma$, the second inequality of the theorem is satisfied with $c_3 = 1/2$. To show the last inequality, let us note that $\phi_x(\tau, t, x)$ satisfies the sensitivity equation

$$\frac{\partial}{\partial \tau} \phi_x = \frac{\partial f}{\partial x}(\tau, \phi(\tau, t, x)) \, \phi_x, \quad \phi_x(t, t, x) = I$$

Since

$$\left\| \frac{\partial f}{\partial x}(t, x) \right\|_2 \leq L$$

on D, ϕ_x satisfies the bound[28]

$$\|\phi_x(\tau, t, x)\|_2 \leq e^{L(\tau - t)}$$

Hence,

$$\left\| \frac{\partial V}{\partial x} \right\|_2 = \left\| \int_t^{t+T} 2\phi^T(\tau, t, x)\phi_x(\tau, t, x) \, d\tau \right\|_2$$

$$\leq \int_t^{t+T} 2\|\phi(\tau, t, x)\|_2 \, \|\phi_x(\tau, t, x)\|_2 \, d\tau$$

$$\leq \int_t^{t+T} 2k e^{-\gamma(\tau - t)} \, e^{L(\tau - t)} \, d\tau \, \|x\|_2$$

$$= \frac{2k}{(\gamma - L)} [1 - e^{-(\gamma - L)T}]\|x\|_2$$

---

[27] See Exercise 2.46.

[28] See Exercise 2.22.

Thus, the last inequality of the theorem is satisfied with

$$c_4 = \frac{2k}{(\gamma - L)}[1 - e^{-(\gamma - L)T}]$$

If all the assumptions hold globally, then clearly r_0 can be chosen arbitrarily large. If the system is autonomous, then $\phi(\tau, t, x)$ depends only on $(\tau - t)$; that is,

$$\phi(\tau, t, x) = \psi(\tau - t, x)$$

Then,

$$V(t, x) = \int_t^{t+T} \psi^T(\tau - t, x)\psi(\tau - t, x) \, d\tau = \int_0^T \psi^T(s, x)\psi(s, x) \, ds$$

which is independent of t.   □

In Theorem 3.11, we saw that if the linearization of a nonlinear system about the origin has an exponentially stable equilibrium, then the origin is an exponentially stable equilibrium for the nonlinear system. We will use Theorem 3.12 to prove that exponential stability of the linearization is a necessary and sufficient condition for exponential stability of the origin.

Theorem 3.13 *Let $x = 0$ be an equilibrium point for the nonlinear system*

$$\dot{x} = f(t, x)$$

where $f : [0, \infty) \times D \to R^n$ is continuously differentiable, $D = \{x \in R^n \mid \|x\|_2 < r\}$, and the Jacobian matrix $[\partial f/\partial x]$ is bounded and Lipschitz on D, uniformly in t. Let

$$A(t) = \left. \frac{\partial f}{\partial x}(t, x) \right|_{x=0}$$

Then, the origin is an exponentially stable equilibrium point for the nonlinear system if and only if it is an exponentially stable equilibrium point for the linear system

$$\dot{x} = A(t)x$$

◇

Proof: The "if" part follows from Theorem 3.11. To prove the "only if" part, write the linear system as

$$\dot{x} = f(t, x) - [f(t, x) - A(t)x] = f(t, x) - g(t, x)$$

Recalling the argument preceding Theorem 3.11, we know that

$$\|g(t,x)\|_2 \leq L\|x\|_2^2, \quad \forall\, x \in D, \; \forall\, t \geq 0$$

Since the origin is an exponentially stable equilibrium of the nonlinear system, there are positive constants k, γ, and c such that

$$\|x(t)\|_2 \leq k\|x(t_0)\|_2 e^{-\gamma(t-t_0)}, \quad \forall\, t \geq t_0 \geq 0, \; \forall\, \|x(t_0)\|_2 < c$$

Choosing $r_0 < \min\{c, r/k\}$, all the conditions of Theorem 3.12 are satisfied. Let $V(t,x)$ be the function provided by Theorem 3.12, and use it as a Lyapunov function candidate for the linear system. Then,

$$\begin{aligned}
\frac{\partial V}{\partial t} + \frac{\partial V}{\partial x}A(t)x &= \frac{\partial V}{\partial t} + \frac{\partial V}{\partial x}f(t,x) - \frac{\partial V}{\partial x}g(t,x) \\
&\leq -c_3\|x\|_2^2 + c_4 L\|x\|_2^3 \\
&\leq -(c_3 - c_4 L\rho)\|x\|_2^2, \quad \forall\, \|x\|_2 < \rho
\end{aligned}$$

The choice $\rho < \min\{r_0, c_3/c_4 L\}$ ensures that $\dot{V}(t,x)$ is negative definite in $\|x\|_2 < \rho$. Hence, all the conditions of Corollary 3.4 are satisfied in $\|x\|_2 < \rho$, and we conclude that the origin is an exponentially stable equilibrium point for the linear system. $\qquad\square$

Example 3.23 Consider the first-order system

$$\dot{x} = -x^3$$

We saw in Example 3.14 that the origin is asymptotically stable, but linearization about the origin results in the linear system

$$\dot{x} = 0$$

whose A matrix is not Hurwitz. Using Theorem 3.13, we conclude that the origin is not exponentially stable. $\qquad\triangle$

We conclude this section by stating another converse theorem which applies to the more general case of uniformly asymptotically stable equilibria. The proof of the theorem is more involved than that of Theorem 3.13.

Theorem 3.14 Let $x = 0$ be an equilibrium point for the nonlinear system

$$\dot{x} = f(t,x)$$

where $f : [0, \infty) \times D \to R^n$ is continuously differentiable, $D = \{x \in R^n \mid \|x\| < r\}$, and the Jacobian matrix $[\partial f/\partial x]$ is bounded on D, uniformly in t. Let $\beta(\cdot, \cdot)$ be a class $\mathcal{KL}$ function and r_0 be a positive constant such that $\beta(r_0, 0) < r$. Let $D_0 = \{x \in R^n \mid \|x\| < r_0\}$. Assume that the trajectory of the system satisfies

$$\|x(t)\| \leq \beta(\|x(t_0)\|, t - t_0), \quad \forall \; x(t_0) \in D_0, \; \forall \; t \geq t_0 \geq 0$$

Then, there is a continuously differentiable function $V : [0, \infty) \times D_0 \to R$ that satisfies the inequalities

$$\alpha_1(\|x\|) \leq V(t, x) \leq \alpha_2(\|x\|)$$

$$\frac{\partial V}{\partial t} + \frac{\partial V}{\partial x} f(t, x) \leq -\alpha_3(\|x\|)$$

$$\left\| \frac{\partial V}{\partial x} \right\| \leq \alpha_4(\|x\|)$$

where $\alpha_1(\cdot)$, $\alpha_2(\cdot)$, $\alpha_3(\cdot)$, and $\alpha_4(\cdot)$ are class $\mathcal{K}$ functions defined on $[0, r_0]$. If the system is autonomous, V can be chosen independent of t. $\diamond$

Proof: Appendix A.6.

3.7   Exercises

Exercise 3.1 Consider a second-order autonomous system $\dot{x} = f(x)$. For each of the following types of equilibrium points, classify whether the equilibrium point is stable, unstable, or asymptotically stable. Justify your answer using phase portraits.
   (1) stable node   **(2)** unstable node   **(3)** stable focus
   (4) unstable focus   **(5)** center   **(6)** saddle

Exercise 3.2 Consider the scalar system $\dot{x} = ax^p + g(x)$, where p is a positive integer and $g(x)$ satisfies $|g(x)| \leq k|x|^{p+1}$ in some neighborhood of the origin $x = 0$. Show that the origin is asymptotically stable if p is odd and $a < 0$. Show that it is unstable if p is odd and $a > 0$ or p is even and $a \neq 0$.

Exercise 3.3 For each of the following systems, use a quadratic Lyapunov function candidate to show that the origin is asymptotically stable. Then, investigate whether the origin is globally asymptotically stable.

(1) $\begin{aligned} \dot{x}_1 &= -x_1 + x_2^2 \\ \dot{x}_2 &= -x_2 \end{aligned}$

(2) $\begin{aligned} \dot{x}_1 &= (x_1 - x_2)(x_1^2 + x_2^2 - 1) \\ \dot{x}_2 &= (x_1 + x_2)(x_1^2 + x_2^2 - 1) \end{aligned}$

(3) $\begin{aligned} \dot{x}_1 &= -x_1 + x_1^2 x_2 \\ \dot{x}_2 &= -x_2 + x_1 \end{aligned}$

(4) $\begin{aligned} \dot{x}_1 &= -x_1 - x_2 \\ \dot{x}_2 &= x_1 - x_2^3 \end{aligned}$

Exercise 3.4 Using $V(x) = x_1^2 + x_2^2$, study stability of the origin of the system

$$\dot{x}_1 = x_1(k^2 - x_1^2 - x_2^2) + x_2(x_1^2 + x_2^2 + k^2)$$
$$\dot{x}_2 = -x_1(k^2 + x_1^2 + x_2^2) + x_2(k^2 - x_1^2 - x_2^2)$$

when   (a) $k = 0$ and   (b) $k \neq 0$.

Exercise 3.5 Let $g(x)$ be a map from R^n into R^n. Show that $g(x)$ is the gradient vector of a scalar function $V : R^n \to R$ if and only if

$$\frac{\partial g_i}{\partial x_j} = \frac{\partial g_j}{\partial x_i}, \quad \forall \, i, j = 1, 2, \ldots, n$$

Exercise 3.6 Using the variable gradient method, find a Lyapunov function $V(x)$ that shows asymptotic stability of the origin of the system

$$\dot{x}_1 = x_2$$
$$\dot{x}_2 = -(x_1 + x_2) - \sin(x_1 + x_2)$$

Exercise 3.7 ([62]) Consider the second-order system

$$\dot{x}_1 = \frac{-6x_1}{u^2} + 2x_2$$
$$\dot{x}_2 = \frac{-2(x_1 + x_2)}{u^2}$$

where $u = 1 + x_1^2$. Let $V(x) = x_1^2/(1 + x_1^2) + x_2^2$.

(a) Show that $V(x) > 0$ and $\dot{V}(x) < 0$ for all $x \in R^2 - \{0\}$.

(b) Consider the hyperbola $x_2 = 2/(x_1 - \sqrt{2})$. Show, by investigating the vector field on the boundary of this hyperbola, that trajectories to the right of the branch in the first quadrant cannot cross that branch.

(c) Show that the origin is not globally asymptotically stable.

Hint: In part (b), show that $\dot{x}_2/\dot{x}_1 = -1/(1 + 2\sqrt{2}x_1 + 2x_1^2)$ on the hyperbola, and compare with the slope of the tangents to the hyperbola.

Exercise 3.8 Consider the system

$$\dot{x}_1 = x_2$$
$$\dot{x}_2 = x_1 - \text{sat}(2x_1 + x_2)$$

(a) Show that the origin is asymptotically stable.

(b) Show that all trajectories starting in the first quadrant to the right of the curve $x_1 x_2 = c$ (with sufficiently large $c > 0$) cannot reach the origin.

(c) Show that the origin is not globally asymptotically stable.

Hint: In part (b), consider $V(x) = x_1 x_2$; calculate $\dot{V}(x)$ and show that on the curve $V(x) = c$ the derivative $\dot{V}(x) > 0$ when c is large enough.

Exercise 3.9 (Krasovskii's Method) Consider the system $\dot{x} = f(x)$ with $f(0) = 0$. Assume that $f(x)$ is continuously differentiable and its Jacobian $[\partial f/\partial x]$ satisfies

$$P\left[\frac{\partial f}{\partial x}(x)\right] + \left[\frac{\partial f}{\partial x}(x)\right]^T P \leq -I, \quad \forall\, x \in R^n, \quad \text{where } P = P^T > 0$$

(a) Using the representation $f(x) = \int_0^1 \frac{\partial f}{\partial x}(\sigma x) x\, d\sigma$, show that

$$x^T P f(x) + f^T(x) P x \leq -x^T x, \quad \forall\, x \in R^n$$

(b) Show that $V(x) = f^T(x) P f(x)$ is positive definite for all $x \in R^n$.

(c) Show that $V(x)$ is radially unbounded.

(d) Using $V(x)$ as a Lyapunov function candidate, show that the origin is globally asymptotically stable.

Exercise 3.10 Using Theorem 3.3, prove Lyapunov's first instability theorem: For the system (3.1), if a continuous function $V_1(x)$ with continuous first partial derivatives can be found in a neighborhood of the origin such that $V_1(0) = 0$, and $\dot{V}_1$ along the trajectories of the system is positive definite, but V_1 itself is not negative definite or negative semidefinite arbitrarily near the origin, then the origin is unstable.

Exercise 3.11 Using Theorem 3.3, prove Lyapunov's second instability theorem: For the system (3.1), if in a neighborhood D of the origin, a continuously differentiable function $V_1(x)$ exists such that $V_1(0) = 0$ and $\dot{V}_1$ along the trajectories of the system is of the form $\dot{V}_1 = \lambda V_1 + W(x)$ where $\lambda > 0$ and $W(x) \geq 0$ in D, and if $V_1(x)$ is not negative definite or negative semidefinite arbitrarily near the origin, then the origin is unstable.

Exercise 3.12 Show that the origin of the following system is unstable.

$$\begin{aligned} \dot{x}_1 &= -x_1 + x_2^6 \\ \dot{x}_2 &= x_2^3 + x_1^6 \end{aligned}$$

Exercise 3.13 Show that the origin of the following system is unstable.

$$\dot{x}_1 = -x_1^3 + x_2$$
$$\dot{x}_2 = x_1^6 - x_2^3$$

Hint: Show that the set $\Gamma = \{0 \le x_1 \le 1\} \cap \{x_2 \ge x_1^3\} \cap \{x_2 \le x_1^2\}$ is a nonempty positively invariant set, and investigate the behavior of the trajectories inside Γ.

Exercise 3.14 ([23]) Consider the system

$$\dot{z} = -\sum_{i=1}^{m} a_i y_i$$
$$\dot{y}_i = -h(z, y)y_i + b_i g(z), \quad i = 1, 2, \ldots, m$$

where z is a scalar, $y^T = (y_1, \ldots, y_m)$. The functions $h(\cdot, \cdot)$ and $g(\cdot)$ are continuously differentiable for all (z, y) and satisfy $zg(z) > 0$, $\forall z \ne 0$, $h(z, y) > 0$, $\forall (z, y) \ne 0$, and $\int_0^z g(\sigma) \, d\sigma \to \infty$ as $|z| \to \infty$. The constants a_i and b_i satisfy $b_i \ne 0$ and $a_i/b_i > 0$, $\forall i = 1, 2, \ldots, m$. Show that the origin is an equilibrium point, and investigate its stability using a Lyapunov function candidate of the form

$$V(z, y) = \alpha \int_0^z g(\sigma) \, d\sigma + \sum_{i=1}^{m} \beta_i y_i^2$$

Exercise 3.15 ([143]) Consider the system

$$\dot{x}_1 = x_2$$
$$\dot{x}_2 = -x_1 - x_2 \, \mathrm{sat}(x_2^2 - x_3^2)$$
$$\dot{x}_3 = x_3 \, \mathrm{sat}(x_2^2 - x_3^2)$$

where $\mathrm{sat}(\cdot)$ is the saturation function. Show that the origin is the unique equilibrium point, and use $V(x) = x^T x$ to show that it is globally asymptotically stable.

Exercise 3.16 The origin $x = 0$ is an equilibrium point of the system

$$\dot{x}_1 = -kh(x)x_1 + x_2$$
$$\dot{x}_2 = -h(x)x_2 - x_1^3$$

Let $D = \{x \in R^2 \mid \|x\|_2 < 1\}$. Using $V(x) = \frac{1}{4}x_1^4 + \frac{1}{2}x_2^2$, investigate stability of the origin in each of the following cases.
(1) $k > 0$, $h(x) > 0$, $\forall x \in D$;   (2) $k > 0$, $h(x) > 0$, $\forall x \in R^2$;
(3) $k > 0$, $h(x) < 0$, $\forall x \in D$;   (4) $k > 0$, $h(x) = 0$, $\forall x \in D$;
(5) $k = 0$, $h(x) > 0$, $\forall x \in D$;   (6) $k = 0$, $h(x) > 0$, $\forall x \in R^2$.

Exercise 3.17 Consider the system

$$
\begin{aligned}
\dot{x}_1 &= -x_1 + g(x_3) \\
\dot{x}_2 &= -g(x_3) \\
\dot{x}_3 &= -ax_1 + bx_2 - cg(x_3)
\end{aligned}
$$

where a, b, and c are positive constants and $g(\cdot)$ satisfies

$$
g(0) = 0 \quad \text{and} \quad yg(y) > 0, \quad \forall\, 0 < |y| < k, \quad k > 0
$$

(a) Show that the origin is an isolated equilibrium point.

(b) With $V(x) = \frac{1}{2}ax_1^2 + \frac{1}{2}bx_2^2 + \int_0^{x_3} g(y)\, dy$ as a Lyapunov function candidate, show that the origin is asymptotically stable.

(c) Suppose $yg(y) > 0 \ \forall\, y \in R - \{0\}$. Is the origin globally asymptotically stable?

Exercise 3.18 ([67]) Consider Liénard's equation

$$
\ddot{y} + h(y)\dot{y} + g(y) = 0
$$

where g and h are continuously differentiable.

(a) Using $x_1 = y$ and $x_2 = \dot{y}$, write the state equation and find conditions on g and h to ensure that the origin is an isolated equilibrium point.

(b) Using $V(x) = \int_0^{x_1} g(y)\, dy + \frac{1}{2}x_2^2$ as a Lyapunov function candidate, find conditions on g and h to ensure that the origin is asymptotically stable.

(c) Repeat (b) using $V(x) = \frac{1}{2}\left[x_2 + \int_0^{x_1} h(y)\, dy\right]^2 + \int_0^{x_1} g(y)\, dy$.

Exercise 3.19 Consider the system

$$
\begin{aligned}
\dot{x}_1 &= x_2 \\
\dot{x}_2 &= -a\sin x_1 - kx_1 - dx_2 - cx_3 \\
\dot{x}_3 &= -x_3 + x_2
\end{aligned}
$$

where all coefficients are positive and $k > a$. Using

$$
V(x) = 2a \int_0^{x_1} \sin y\, dy + kx_1^2 + x_2^2 + px_3^2
$$

with some $p > 0$, show that the origin is globally asymptotically stable.

Exercise 3.20 The mass-spring system of Exercise 1.11 is modeled by

$$M\ddot{y} = Mg - ky - c_1\dot{y} - c_2\dot{y}|\dot{y}|$$

Show that the system has a globally asymptotically stable equilibrium point.

Exercise 3.21 Consider the equations of motion of an m-link robot, described in Exercise 1.3.

(a) With $u = 0$, use the total energy $V(q, \dot{q}) = \frac{1}{2}\dot{q}^T M(q)\dot{q} + P(q)$ as a Lyapunov function candidate to show that the origin $(q = 0, \dot{q} = 0)$ is stable.

(b) With $u = -K_d\dot{q}$, where K_d is a positive diagonal matrix, show that the origin is asymptotically stable.

(c) With $u = g(q) - K_p(q - q^*) - K_d\dot{q}$, where K_p and K_d are positive diagonal matrices and q^* is a desired robot position in R^m, show that the point $(q = q^*, \dot{q} = 0)$ is an asymptotically stable equilibrium point.

Exercise 3.22 Suppose the set M in LaSalle's theorem consists of a finite number of isolated points. Show that $\lim_{t\to\infty} x(t)$ exists and equals one of these points.

Exercise 3.23 ([70]) A gradient system is a dynamical system of the form $\dot{x} = -\nabla V(x)$, where $\nabla V(x) = [\partial V/\partial x]^T$ and $V : D \subset R^n \to R$ is twice continuously differentiable.

(a) Show that $\dot{V}(x) \leq 0$ for all $x \in D$, and $\dot{V}(x) = 0$ if and only if x is an equilibrium point.

(b) Suppose the set $\Omega_c = \{x \in R^n \mid V(x) \leq c\}$ is compact for every $c \in R$. Show that every solution of the system is defined for all $t \geq 0$.

(c) Continuing with part (b), suppose $\nabla V(x) \neq 0$ except for a finite number of points $p_1, \ldots, p_r$. Show that for every solution $x(t)$, the limit $\lim_{t\to\infty} x(t)$ exists and equals one of the points $p_1, \ldots, p_r$.

Exercise 3.24 Consider the Lyapunov equation $PA + A^T P = -C^T C$, where the pair (A, C) is observable. Show that A is Hurwitz if and only if there exists $P = P^T > 0$ that satisfies the equation. Furthermore, show that if A is Hurwitz, the Lyapunov equation will have a unique solution.
Hint: Apply LaSalle's theorem and recall that for an observable pair (A, C), the vector $C\exp(At)x \equiv 0 \; \forall \; t$ if and only if $x = 0$.

Exercise 3.25 Consider the linear system $\dot{x} = (A - BR^{-1}B^T P)x$, where (A, B) is controllable, $P = P^T > 0$ satisfies the Riccati equation

$$PA + A^T P + Q - PBR^{-1}B^T P = 0$$

$R = R^T > 0$, and $Q = Q^T \geq 0$. Using $V(x) = x^T Px$ as a Lyapunov function candidate, show that the origin is globally asymptotically stable when

(1) $Q > 0$

(2) $Q = C^T C$ and (A, C) is observable; see the hint of the previous exercise.

Exercise 3.26 Consider the linear system $\dot{x} = Ax + Bu$, where (A, B) is controllable. Let $W = \int_0^T e^{-At} BB^T e^{-A^T} \, dt$ for some $T > 0$. Let $K = -B^T W^{-1}$.

(a) Show that W^{-1} is positive definite.

(b) Using $V(x) = x^T W^{-1} x$ as a Lyapunov function candidate for the system $\dot{x} = (A + BK)x$, show that $(A + BK)$ is Hurwitz.

Exercise 3.27 Let $\dot{x} = f(x)$, where $f : R^n \to R^n$. Consider the change of variables $z = T(x)$, where $T(0) = 0$ and $T : R^n \to R^n$ is a diffeomorphism in the neighborhood of the origin; that is, the inverse map $T^{-1}(\cdot)$ exists, and both $T(\cdot)$ and $T^{-1}(\cdot)$ are continuously differentiable. The transformed system is

$$\dot{z} = \hat{f}(z), \quad \text{where } \hat{f}(z) = \left. \frac{\partial T}{\partial x} f(x) \right|_{x=T^{-1}(z)}$$

(a) Show that $x = 0$ is an isolated equilibrium point of $\dot{x} = f(x)$ if and only if $z = 0$ is an isolated equilibrium point of $\dot{z} = \hat{f}(z)$.

(b) Show that $x = 0$ is stable (asymptotically stable/unstable) if and only if $z = 0$ is stable (asymptotically stable/unstable).

Exercise 3.28 Show that the system

$$
\begin{aligned}
\dot{x}_1 &= \frac{1}{1 + x_3} - x_1 \\
\dot{x}_2 &= x_1 - 2x_2 \\
\dot{x}_3 &= x_2 - 3x_3
\end{aligned}
$$

has a unique equilibrium point in the region $x_i \geq 0$, $i = 1, 2, 3$, and investigate stability of this point using linearization.

Exercise 3.29 Consider the system

$$\dot{x}_1 = (x_1 x_2 - 1)x_1^3 + (x_1 x_2 - 1 + x_2^2)x_1$$
$$\dot{x}_2 = -x_2$$

(a) Show that $x = 0$ is the unique equilibrium point.

(b) Show, using linearization, that $x = 0$ is asymptotically stable.

(c) Show that $\Gamma = \{x \in R^2 \mid x_1 x_2 \geq 2\}$ is a positively invariant set.

(d) Is $x = 0$ globally asymptotically stable? Justify your answer.

Exercise 3.30 For each of the following systems, use linearization to show that the origin is asymptotically stable. Then, show that the origin is globally asymptotically stable.

(1)
$$\dot{x}_1 = -x_1 + x_2$$
$$\dot{x}_2 = (x_1 + x_2)\sin x_1 - 3x_2$$

(2)
$$\dot{x}_1 = -x_1^3 + x_2$$
$$\dot{x}_2 = -ax_1 - bx_2, \quad a, b > 0$$

Exercise 3.31 For each for the following systems, investigate stability of the origin.

(1)
$$\dot{x}_1 = -x_1 + x_1^2$$
$$\dot{x}_2 = -x_2 + x_3^2$$
$$\dot{x}_3 = x_3 - x_1^2$$

(2)
$$\dot{x}_1 = x_2$$
$$\dot{x}_2 = -\sin x_3 + x_1[-2x_3 - \mathrm{sat}(y)]^2$$
$$\dot{x}_3 = -2x_3 - \mathrm{sat}(y)$$
where $y = -2x_1 - 5x_2 + 2x_3$

(3)
$$\dot{x}_1 = -2x_1 + x_1^3$$
$$\dot{x}_2 = -x_2 + x_1^2$$
$$\dot{x}_3 = -x_3$$

(4)
$$\dot{x}_1 = -x_1$$
$$\dot{x}_2 = -x_1 - x_2 - x_3 - x_1 x_3$$
$$\dot{x}_3 = (x_1 + 1)x_2$$

Exercise 3.32 Prove Lemma 3.2.

Exercise 3.33 Let α be a class $\mathcal{K}$ function on $[0, a)$. Show that

$$\alpha(r_1 + r_2) \leq \alpha(2r_1) + \alpha(2r_2), \quad \forall \, r_1, r_2 \in [0, a)$$

Exercise 3.34 Is the origin of the scalar system $\dot{x} = -x/t, \; t \geq 1$, uniformly asymptotically stable?

Exercise 3.35 Suppose the conditions of Theorem 3.8 are satisfied except that $\dot{V}(t, x) \leq -W_3(x)$ where $W_3(x)$ is positive semidefinite. Show that the origin is uniformly stable.

Exercise 3.36 Consider the linear time-varying system

$$\dot{x} = \begin{bmatrix} -1 & \alpha(t) \\ -\alpha(t) & -1 \end{bmatrix} x$$

where $\alpha(t)$ is continuous for all $t \geq 0$. Show that the origin is exponentially stable.

Exercise 3.37 ([87]) An RLC circuit with time-varying elements is represented by

$$\dot{x}_1 = \frac{1}{L(t)} x_2$$

$$\dot{x}_2 = -\frac{1}{C(t)} x_1 - \frac{R(t)}{L(t)} x_2$$

Suppose that $L(t)$, $C(t)$, and $R(t)$ are continuously differentiable and satisfy the inequalities $k_1 \leq L(t) \leq k_2$, $k_3 \leq C(t) \leq k_4$, and $k_5 \leq R(t) \leq k_6$ for all $t \geq 0$, where k_1, k_2, and k_3 are positive. Consider a Lyapunov function candidate

$$V(t, x) = \left[R(t) + \frac{2L(t)}{R(t)C(t)} \right] x_1^2 + 2x_1 x_2 + \frac{2}{R(t)} x_2^2$$

(a) Show that $V(t, x)$ is positive definite and decrescent.

(b) Find conditions on $\dot{L}(t)$, $\dot{C}(t)$, and $\dot{R}(t)$ that will ensure exponential stability of the origin.

Exercise 3.38 Consider the system

$$\dot{x}_1 = -x_1^3 + \alpha(t)x_2$$

$$\dot{x}_2 = -\alpha(t)x_1 - x_2^3$$

where $\alpha(t)$ is a continuous function. Show that the origin is globally uniformly asymptotically stable. Is it exponentially stable?

Exercise 3.39 ([137]) A pendulum with time-varying friction is represented by

$$\dot{x}_1 = x_2$$

$$\dot{x}_2 = -\sin x_1 - g(t)x_2$$

Suppose that $g(t)$ is continuously differentiable and satisfies

$$0 < a < \alpha \leq g(t) \leq \beta < \infty; \quad \dot{g}(t) \leq \gamma < 2$$

for all $t \geq 0$. Consider the Lyapunov function candidate

$$V(t, x) = \tfrac{1}{2}(a \sin x_1 + x_2)^2 + [1 + ag(t) - a^2](1 - \cos x_1)$$

(a) Show that $V(t, x)$ is positive definite and decrescent.

(b) Show that $\dot{V} \leq -(\alpha - a)x_2^2 - a(2 - \gamma)(1 - \cos x_1) + O(\|x\|^3)$, where $O(\|x\|^3)$ is a term bounded by $k\|x\|^3$ in some neighborhood of the origin.

(c) Show that the origin is uniformly asymptotically stable.

Exercise 3.40 (Floquet theory) [29] Consider the linear system $\dot{x} = A(t)x$, where $A(t) = A(t+T)$. Let $\Phi(\cdot, \cdot)$ be the state transition matrix. Define a constant matrix B via the equation $\exp(BT) = \Phi(T, 0)$, and let $P(t) = \exp(Bt)\Phi(0, t)$. Show that

(a) $P(t + T) = P(t)$.

(b) $\Phi(t, \tau) = P^{-1}(t) \exp[(t - \tau)B]P(\tau)$.

(c) the origin of $\dot{x} = A(t)x$ is exponentially stable if and only if B is Hurwitz.

Exercise 3.41 Consider the system

$$\begin{aligned}
\dot{x}_1 &= x_2 \\
\dot{x}_2 &= 2x_1 x_2 + 3t + 2 - 3x_1 - 2(t + 1)x_2
\end{aligned}$$

(a) Verify that $x_1(t) = t$, $x_2(t) = 1$ is a solution.

(b) Show that if $x(0)$ is sufficiently close to $\begin{bmatrix} 0 \\ 1 \end{bmatrix}$ then $x(t)$ approaches $\begin{bmatrix} t \\ 1 \end{bmatrix}$ as $t \to \infty$.

Exercise 3.42 Consider the system

$$\begin{aligned}
\dot{x}_1 &= -2x_1 + g(t)x_2 \\
\dot{x}_2 &= g(t)x_1 - 2x_2
\end{aligned}$$

where $g(t)$ is continuously differentiable and $|g(t)| \leq 1$ for all $t \geq 0$. Show that the origin is uniformly asymptotically stable.

Exercise 3.43 Consider the system

$$\begin{aligned}
\dot{x}_1 &= x_2 \\
\dot{x}_2 &= -x_1 - (1 + b \cos t)x_2
\end{aligned}$$

Find $b^* > 0$ such that the origin is exponentially stable for all $|b| < b^*$.

---

[29] See [142] for a comprehensive treatment of Floquet theory.

Exercise 3.44 Consider the system

$$\dot{x}_1 = x_2 - g(t)x_1(x_1^2 + x_2^2)$$
$$\dot{x}_2 = -x_1 - g(t)x_2(x_1^2 + x_2^2)$$

where $g(t)$ is continuously differentiable and $g(t) \geq k > 0$ for all $t \geq 0$. Is the origin uniformly asymptotically stable? Is it exponentially stable?

Exercise 3.45 Consider the system

$$\dot{x}_1 = -x_1 + (x_1^2 + x_2^2)\sin t$$
$$\dot{x}_2 = -x_2 + (x_1^2 + x_2^2)\cos t$$

Show that the origin is exponentially stable and estimate the region of attraction.

Exercise 3.46 Consider two systems represented by

$$\dot{x} = f(x) \tag{3.23}$$
$$\dot{x} = h(x)f(x) \tag{3.24}$$

where $f : R^n \to R^n$ and $h : R^n \to R$ are continuously differentiable, $f(0) = 0$, and $h(0) > 0$. Show that the origin of (3.23) is exponentially stable if and only if the origin of (3.24) is exponentially stable.

Exercise 3.47 Show that the system

$$\dot{x}_1 = -ax_1 + b$$
$$\dot{x}_2 = -cx_2 + x_1(\alpha - \beta x_1 x_2)$$

where all coefficients are positive, has a globally exponentially stable equilibrium point.
Hint: Shift the equilibrium point to the origin and use V of the form $V = k_1 y_1^2 + k_2 y_2^2 + k_3 y_1^4$, where (y_1, y_2) are the new coordinates.

Exercise 3.48 Consider the system

$$\dot{x} = f(t, x); \quad f(t, 0) = 0$$

where $[\partial f / \partial x]$ is bounded and Lipschitz in x in a neighborhood of the origin, uniformly in t for all $t \geq t_0 \geq 0$. Suppose that the origin of the linearization at $x = 0$ is exponentially stable, and the solutions of the system satisfy

$$\|x(t)\| \leq \beta(\|x(t_0)\|, t - t_0), \quad \forall\, t \geq t_0 \geq 0, \quad \forall\, \|x(t_0)\| < c \tag{3.25}$$

for some class $\mathcal{KL}$ function β and some positive constant c.

(a) Show that there is a class $\mathcal{K}$ function $\alpha(\cdot)$ and a positive constant γ such that

$$\|x(t)\| \leq \alpha(\|x(t_0)\|)\exp[-\gamma(t - t_0)], \quad \forall\, t \geq t_0, \quad \forall\, \|x(t_0)\| < c$$

(b) Show that there is a positive constant M, possibly dependent on c, such that

$$\|x(t)\| \leq M\|x(t_0)\|\exp[-\gamma(t - t_0)], \quad \forall\, t \geq t_0, \quad \forall\, \|x(t_0)\| < c \qquad (3.26)$$

(c) If inequality (3.25) holds globally, can you state inequality (3.26) globally?

In the next few exercises, we deal with the discrete-time dynamical system[30]

$$x(k + 1) = f(x(k)), \quad f(0) = 0 \qquad (3.27)$$

The rate of change of a scalar function $V(x)$ along the motion of (3.27) is defined by

$$\Delta V(x) = V(f(x)) - V(x)$$

Exercise 3.49 Restate Definition 3.1 for the origin of the discrete-time system (3.27).

Exercise 3.50 Show that the origin of (3.27) is stable if, in a neighborhood of the origin, there is a continuous positive definite function $V(x)$ so that $\Delta V(x)$ is negative semidefinite. Show that it is asymptotically stable if, in addition, $\Delta V(x)$ is negative definite. Finally, show that the origin is globally asymptotically stable if the conditions for asymptotic stability hold globally and $V(x)$ is radially unbounded.

Exercise 3.51 Show that the origin of (3.27) is asymptotically stable if, in a neighborhood of the origin, there is a continuous positive definite function $V(x)$ so that $\Delta V(x)$ is negative semidefinite and $\Delta V(x)$ does not vanish identically for any $x \neq 0$.

Exercise 3.52 Consider the linear system $x(k + 1) = Ax(k)$. Show that the following statements are equivalent:

(1) $x = 0$ is asymptotically stable.

(2) $|\lambda_i| < 1$ for all eigenvalues of A.

(3) Given any $Q = Q^T > 0$ there exists $P = P^T > 0$, which is the unique solution of the linear equation

$$A^T PA - P = -Q$$

---

[30] See [87] for a detailed treatment of Lyapunov stability for discrete-time dynamical systems.

Exercise 3.53 Let A be the linearization of (3.27) at the origin; that is,

$$A = \frac{\partial f}{\partial x}(0)$$

Show that the origin is asymptotically stable if all the eigenvalues of A have magnitudes less than one.

Exercise 3.54 Let $x = 0$ be an equilibrium point for the nonlinear discrete-time system

$$x(k + 1) = f(x(k))$$

where $f : D \rightarrow R^n$ is continuously differentiable and $D = \{x \in R^n \mid \|x\| < r\}$. Let C, $\gamma < 1$, and r_0 be positive constants with $r_0 < r/C$. Let $D_0 = \{x \in R^n \mid \|x\| < r_0\}$. Assume that the solutions of the system satisfy

$$\|x(k)\| \leq C\|x(0)\|\gamma^k, \quad \forall\, x(0) \in D_0, \,\forall\, k \geq 0$$

Show that there is a function $V : D_0 \rightarrow R$ that satisfies

$$c_1\|x\|^2 \leq V(x) \leq c_2\|x\|^2$$

$$\Delta V(x) = V(f(x)) - V(x) \leq -c_3\|x\|^2$$

$$|V(x) - V(y)| \leq c_4\|x - y\|\,(\|x\| + \|y\|)$$

for all $x, y \in D_0$ for some positive constants c_1, c_2, c_3, and c_4.

Chapter 4

Advanced Stability Analysis

In Chapter 3, we gave the basic concepts and tools of Lyapunov stability. In this chapter, we examine some of these concepts more closely and present a number of extensions and refinements.

We saw in Chapter 3 how to use linearization to study stability of equilibrium points of an autonomous system. We saw also that linearization fails when the Jacobian matrix, evaluated at the equilibrium point, has some eigenvalues with zero real parts and no eigenvalues with positive real parts. In Section 4.1, we introduce the center manifold theorem and use it to study stability of the origin of an autonomous system in the critical case when linearization fails.

The concept of the region of attraction of an asymptotically stable equilibrium point was introduced in Section 3.1. In Section 4.2, we elaborate further on this concept and present some ideas for providing estimates of this region.

LaSalle's invariance principle for autonomous systems is very useful in applications. For a general nonautonomous system, there is no invariance principle in the same form that was presented in Theorem 3.4. There are, however, theorems which capture some features of the invariance principle. Two such theorems are given in Section 3.6. The first theorem shows convergence of the trajectory to a set, while the second one shows uniform asymptotic stability of the origin.

4.1   The Center Manifold Theorem

Consider the autonomous system

$$\dot{x} = f(x) \tag{4.1}$$

where $f : D \to R^n$ is continuously differentiable and D is a neighborhood of the origin $x = 0$. Suppose that the origin is an equilibrium point of (4.1). Theorem 3.7

states that if the linearization of f at the origin, that is, the matrix

$$A = \left.\frac{\partial f}{\partial x}(x)\right|_{x=0}$$

has all eigenvalues with negative real parts, then the origin is asymptotically stable; if it has some eigenvalues with positive real parts, then the origin is unstable. If A has some eigenvalues with zero real parts with the rest of the eigenvalues having negative real parts, then linearization fails to determine the stability properties of the origin. In this section, we take a closer look into the case when linearization fails. Failure of linearization leaves us with the task of analyzing the nth-order nonlinear system (4.1) in order to determine stability of the origin. The interesting finding that we shall present in the next few pages is that stability properties of the origin can actually be determined by analyzing a lower-order nonlinear system, whose order is exactly equal to the number of eigenvalues of A with zero real parts. This will follow as an application of the *center manifold theory*.[1]

A k-dimensional manifold in R^n $(1 \leq k < n)$ has a rigorous mathematical definition.[2] For our purpose here, it is sufficient to think of a k-dimensional manifold as the solution of the equation

$$\eta(x) = 0$$

where $\eta : R^n \to R^{n-k}$ is sufficiently smooth (that is, sufficiently many times continuously differentiable). For example, the unit circle

$$\{x \in R^2 \mid x_1^2 + x_2^2 = 1\}$$

is a one-dimensional manifold in R^2. Similarly, the unit sphere

$$\{x \in R^n \mid \sum_{i=1}^{n} x_i^2 = 1\}$$

is an $(n-1)$-dimensional manifold in R^n. A manifold $\{\eta(x) = 0\}$ is said to be an invariant manifold for (4.1) if

$$\eta(x(0)) = 0 \Rightarrow \eta(x(t)) \equiv 0, \quad \forall\, t \in [0, t_1) \subset R$$

where $[0, t_1)$ is any time interval over which the solution $x(t)$ is defined.

Suppose now that $f(x)$ is twice continuously differentiable. Equation (4.1) can be represented as

$$\dot{x} = Ax + \left[f(x) - \frac{\partial f}{\partial x}(0)\, x \right] = Ax + \tilde{f}(x)$$

---

[1] The center manifold theory has several applications to dynamical systems. It is presented here only insofar as it relates to determining the stability of the origin. For a broader viewpoint of the theory, the reader may consult [28].

[2] See, for example, [60].

where

$$\tilde{f}(x) = f(x) - \frac{\partial f}{\partial x}(0)\, x$$

is twice continuously differentiable and

$$\tilde{f}(0) = 0; \quad \frac{\partial \tilde{f}}{\partial x}(0) = 0$$

Since our interest is in the case when linearization fails, assume that A has k eigenvalues with zero real parts and $m = n - k$ eigenvalues with negative real parts. We can always find a similarity transformation T that transforms A into a block diagonal matrix, that is,

$$TAT^{-1} = \begin{bmatrix} A_1 & 0 \\ 0 & A_2 \end{bmatrix}$$

where all eigenvalues of A_1 have zero real parts and all eigenvalues of A_2 have negative real parts. Clearly, A_1 is $k \times k$ and A_2 is $m \times m$. The change of variables

$$\begin{bmatrix} y \\ z \end{bmatrix} = Tx; \quad y \in R^k; \ z \in R^m$$

transforms (4.1) into the form

$$\dot{y} = A_1 y + g_1(y, z) \tag{4.2}$$
$$\dot{z} = A_2 z + g_2(y, z) \tag{4.3}$$

where g_1 and g_2 inherit properties of $\tilde{f}$. In particular, they are twice continuously differentiable and

$$g_i(0,0) = 0; \quad \frac{\partial g_i}{\partial y}(0,0) = 0; \quad \frac{\partial g_i}{\partial z}(0,0) = 0 \tag{4.4}$$

for $i = 1, 2$. If $z = h(y)$ is an invariant manifold for (4.2)–(4.3) and h is smooth, then it is called a *center manifold* if

$$h(0) = 0; \quad \frac{\partial h}{\partial y}(0) = 0$$

Theorem 4.1 *If g_1 and g_2 are twice continuously differentiable and satisfy (4.4), all eigenvalues of A_1 have zero real parts, and all eigenvalues of A_2 have negative real parts, then there exist $\delta > 0$ and a continuously differentiable function $h(y)$, defined for all $\|y\| < \delta$, such that $z = h(y)$ is a center manifold for (4.2)–(4.3).* $\diamond$

Proof: Appendix A.7.

If the initial state of the system (4.2)–(4.3) lies in the center manifold, that is, $z(0) = h(y(0))$, then the solution $(y(t), z(t))$ will lie in the manifold for all $t \geq 0$; that is, $z(t) = h(y(t))$. In this case, the motion of the system in the center manifold is described by the kth-order differential equation

$$\dot{y} = A_1 y + g_1(y, h(y)) \tag{4.5}$$

which we shall refer to as the *reduced system*. If $z(0) \neq h(y(0))$, then the difference $z(t) - h(y(t))$ represents the deviation of the trajectory from the center manifold at any time t. The change of variables

$$\left[\begin{array}{c} y \\ w \end{array} \right] = \left[\begin{array}{c} y \\ z - h(y) \end{array} \right]$$

transforms (4.2)–(4.3) into

$$\dot{y} = A_1 y + g_1(y, w + h(y)) \tag{4.6}$$

$$\dot{w} = A_2[w + h(y)] + g_2(y, w + h(y)) - \frac{\partial h}{\partial y}(y) \left[A_1 y + g_1(y, w + h(y)) \right] \tag{4.7}$$

In the new coordinates, the center manifold is $w = 0$. The motion in the manifold is characterized by

$$w(t) \equiv 0 \ \Rightarrow \ \dot{w}(t) \equiv 0$$

Substitution of these identities in (4.7) results in

$$0 = A_2 h(y) + g_2(y, h(y)) - \frac{\partial h}{\partial y}(y) \left[A_1 y + g_1(y, h(y)) \right] \tag{4.8}$$

Since this equation must be satisfied by any solution which lies in the center manifold, we conclude that the function $h(y)$ must satisfy the partial differential equation (4.8). Adding and subtracting $g_1(y, h(y))$ to the right-hand side of (4.6), and subtracting (4.8) from (4.7), we can rewrite the equation in the transformed coordinates as

$$\dot{y} = A_1 y + g_1(y, h(y)) + N_1(y, w) \tag{4.9}$$

$$\dot{w} = A_2 w + N_2(y, w) \tag{4.10}$$

where

$$N_1(y, w) = g_1(y, w + h(y)) - g_1(y, h(y))$$

and

$$N_2(y, w) = g_2(y, w + h(y)) - g_2(y, h(y)) - \frac{\partial h}{\partial y}(y) \, N_1(y, w)$$

It is not difficult to verify that N_1 and N_2 are twice continuously differentiable, and

$$N_i(y, 0) = 0; \quad \frac{\partial N_i}{\partial w}(0, 0) = 0$$

for $i = 1, 2$. Consequently, in the domain

$$\left\| \begin{array}{c} y \\ w \end{array} \right\|_2 < \rho$$

N_1 and N_2 satisfy

$$\|N_i(y, w)\|_2 \leq k_i \|w\|, \quad i = 1, 2$$

where the positive constants k_1 and k_2 can be made arbitrarily small by choosing ρ small enough. These inequalities, together with the fact that A_2 is Hurwitz, suggest that the stability properties of the origin are determined by the reduced system (4.5). The following theorem, known as the *reduction principle*, confirms this conjecture.

Theorem 4.2 *Under the assumptions of Theorem 4.1, if the origin $y = 0$ of the reduced system* (4.5) *is asymptotically stable (respectively, unstable) then the origin of the full system* (4.2)–(4.3) *is also asymptotically stable (respectively, unstable).*
$\diamond$

Proof: The change of coordinates from (y, z) to (y, w) does not change the stability properties of the origin (Exercise 3.27); therefore, we can work with the system (4.9)–(4.10). If the origin of the reduced system (4.5) is unstable, then by invariance, the origin of (4.9)–(4.10) is unstable. In particular, for any solution $y(t)$ of (4.5) there is a corresponding solution $(y(t), 0)$ of (4.9)–(4.10). Therefore, stability of the origin of (4.9)–(4.10) implies stability of the origin of (4.5). This is equivalent to saying that instability of the origin of (4.5) implies instability of the origin of (4.9)–(4.10). Suppose now that the origin of the reduced system (4.5) is asymptotically stable. By (the converse Lyapunov) Theorem 3.14, there is a continuously differentiable function $V(y)$ which is positive definite and satisfies the following inequalities in a neighborhood of the origin

$$\frac{\partial V}{\partial y}[A_1 y + g_1(y, h(y))] \leq -\alpha_3(\|y\|_2)$$

$$\left\| \frac{\partial V}{\partial y} \right\|_2 \leq \alpha_4(\|y\|_2) \leq k$$

where α_3 and α_4 are class $\mathcal{K}$ functions. On the other hand, since A_2 is Hurwitz, the Lyapunov equation

$$PA_2 + A_2^T P = -I$$

has a unique positive definite solution P. Consider

$$\nu(y, w) = V(y) + \sqrt{w^T P w}$$

as a Lyapunov function candidate[3] for the full system (4.9)–(4.10). The derivative of ν along the trajectories of the system is given by

$$
\begin{aligned}
\dot{\nu}(y, w) &= \frac{\partial V}{\partial y} \left[A_1 y + g_1(y, h(y)) + N_1(y, w) \right] \\
&\quad + \frac{1}{2\sqrt{w^T P w}} \left[w^T (P A_2 + A_2^T P) w + 2 w^T P N_2(y, w) \right] \\
&\leq -\alpha_3(\|y\|_2) + k k_1 \|w\|_2 - \frac{\|w\|_2}{2\sqrt{\lambda_{\max}(P)}} + \frac{k_2 \lambda_{\max}(P)}{\sqrt{\lambda_{\min}(P)}} \|w\|_2 \\
&= -\alpha_3(\|y\|_2) - \frac{1}{4\sqrt{\lambda_{\max}(P)}} \|w\|_2 \\
&\quad - \left[\frac{1}{4\sqrt{\lambda_{\max}(P)}} - k k_1 - k_2 \frac{\lambda_{\max}(P)}{\sqrt{\lambda_{\min}(P)}} \right] \|w\|_2
\end{aligned}
$$

Since k_1 and k_2 can be made arbitrarily small by restricting the domain around the origin to be sufficiently small, we can choose them small enough to ensure that

$$\frac{1}{4\sqrt{\lambda_{\max}(P)}} - k k_1 - k_2 \frac{\lambda_{\max}(P)}{\sqrt{\lambda_{\min}(P)}} > 0$$

Hence,

$$\dot{\nu}(y, w) \leq -\alpha_3(\|y\|_2) - \frac{1}{4\sqrt{\lambda_{\max}(P)}} \|w\|_2$$

which shows that $\dot{\nu}(y, w)$ is negative definite. Consequently, the origin of the full system (4.9)–(4.10) is asymptotically stable.   □

We leave it to the reader (Exercises 4.1 and 4.2) to extend the proof of Theorem 4.2 to prove the following two corollaries.

Corollary 4.1 *Under the assumptions of Theorem 4.1, if the origin $y = 0$ of the reduced system (4.5) is stable and there is a continuously differentiable Lyapunov*

---

[3]The function $\nu(y, w)$ is continuously differentiable everywhere around the origin, except on the manifold $w = 0$. Both $\nu(y, w)$ and $\dot{\nu}(y, w)$ are defined and continuous around the origin. It can be easily seen that the statement of Theorem 3.1 is still valid.

function $V(y)$ such that [4]

$$\frac{\partial V}{\partial y} \left[A_1 y + g_1(y, h(y)) \right] \leq 0$$

in some neighborhood of $y = 0$, then the origin of the full system (4.2)–(4.3) is stable. ◇

Corollary 4.2 *Under the assumptions of Theorem 4.1, the origin of the reduced system (4.5) is asymptotically stable if and only if the origin of the full system (4.2)–(4.3) is asymptotically stable.* ◇

To use Theorem 4.2, we need to find the center manifold $z = h(y)$. The function h is a solution of the partial differential equation

$$\mathcal{N}(h(y)) \stackrel{\text{def}}{=} \frac{\partial h}{\partial y}(y) \left[A_1 y + g_1(y, h(y)) \right] - A_2 h(y) - g_2(y, h(y)) = 0 \qquad (4.11)$$

with boundary conditions

$$h(0) = 0; \quad \frac{\partial h}{\partial t}(0) = 0 \qquad (4.12)$$

This equation for h cannot be solved exactly in most cases (to do so would imply that a solution of the full system (4.2)–(4.3) has been found), but its solution can be approximated arbitrarily closely as a Taylor series in y.

Theorem 4.3 *If a continuously differentiable function $\phi(y)$ with $\phi(0) = 0$ and $[\partial\phi/\partial y](0) = 0$ can be found such that $\mathcal{N}(\phi(y)) = O(\|y\|^p)$ for some $p > 1$, then for sufficiently small $\|y\|$*

$$h(y) - \phi(y) = O(\|y\|^p)$$

and the reduced system can be represented as

$$\dot{y} = A_1 y + g_1(y, \phi(y)) + O(\|y\|^{p+1})$$

◇

Proof: Appendix A.7.

---

[4] The existence of the Lyapunov function $V(y)$ cannot be inferred from a converse Lyapunov theorem. The converse Lyapunov theorem for stability [62, 96] guarantees the existence of a Lyapunov function $V(t, y)$ whose derivative satisfies $\dot{V}(t, y) \leq 0$. In general, this function cannot be made independent of t [62, page 228]. Even though we can choose $V(t, y)$ to be continuously differentiable in its arguments, it cannot be guaranteed that the partial derivatives $\partial V/\partial y_i$, $\partial V/\partial t$ will be uniformly bounded in a neighborhood of the origin for all $t \geq 0$ [96, page 53].

The order of magnitude notation $O(\cdot)$ will be formally introduced in Chapter 8 (Definition 8.1). For our purpose here, it is enough to think of $f(y) = O(\|y\|^p)$ as a shorthand notation for $\|f(y)\| \le k\|y\|^p$ for sufficiently small $\|y\|$. Let us now illustrate the application of the center manifold theorem by examples. In the first two examples, we shall make use of the observation that for a scalar state equation of the form

$$\dot{y} = ay^p + O\left(|y|^{p+1}\right)$$

where p is a positive integer, the origin is asymptotically stable if p is odd and $a < 0$. It is unstable if p is odd and $a > 0$, or p is even and $a \ne 0$.[5]

Example 4.1 Consider the system

$$\begin{aligned}
\dot{x}_1 &= x_2 \\
\dot{x}_2 &= -x_2 + ax_1^2 + bx_1x_2
\end{aligned}$$

where $a \ne 0$. The system has a unique equilibrium point at the origin. The linearization at the origin results in the matrix

$$A = \begin{bmatrix} 0 & 1 \\ 0 & -1 \end{bmatrix}$$

which has eigenvalues at 0 and -1. Let M be a matrix whose columns are the eigenvectors of A; that is,

$$M = \begin{bmatrix} 1 & 1 \\ 0 & -1 \end{bmatrix}$$

and take $T = M^{-1}$. Then,

$$TAT^{-1} = \begin{bmatrix} 0 & 0 \\ 0 & -1 \end{bmatrix}$$

The change of variables

$$\begin{bmatrix} y \\ z \end{bmatrix} = T \begin{bmatrix} x_1 \\ x_2 \end{bmatrix} = \begin{bmatrix} x_1 + x_2 \\ -x_2 \end{bmatrix}$$

puts the system into the form

$$\begin{aligned}
\dot{y} &= a(y+z)^2 - b(yz+z^2) \\
\dot{z} &= -z - a(y+z)^2 + b(yz+z^2)
\end{aligned}$$

The center manifold equation (4.11)–(4.12) becomes

$$\begin{aligned}
\mathcal{N}(h(y)) &= h'(y)[a(y+h(y))^2 - b(yh(y) + h^2(y))] + h(y) \\
&\quad + a(y+h(y))^2 - b(yh(y) + h^2(y)) = 0, \quad h(0) = h'(0) = 0
\end{aligned}$$

---

[5] See Exercise 3.2.

We set $h(y) = h_2 y^2 + h_3 y^3 + \cdots$ and substitute this series in the center manifold equation to find the unknown coefficients $h_2, h_3, \ldots$ by matching coefficients of like powers in y (since the equation holds as an identity in y). We do not know in advance how many terms of the series we need. We start with the simplest approximation $h(y) \approx 0$. We substitute $h(y) = O(|y|^2)$ in the reduced system and study stability of its origin. If the stability properties of the origin can be determined, we are done. Otherwise, we calculate the coefficient h_2, substitute $h(y) = h_2 y^2 + O(|y|^3)$, and study stability of the origin. If it cannot be resolved, we proceed to the approximation $h(y) \approx h_2 y^2 + h_3 y^3$, and so on. Let us start with the approximation $h(y) \approx 0$. The reduced system is

$$\dot{y} = ay^2 + O(|y|^3)$$

Notice that an $O(|y|^2)$ error in $h(y)$ results in an $O(|y|^3)$ error in the right-hand side of the reduced system. This is a consequence of the fact that the function $g_1(y, z)$ (which appears on the right-hand side of the reduced system (4.5) as $g_1(y, h(y))$) has a partial derivative with respect to z which vanishes at the origin. Clearly, this observation is also valid for higher-order approximations; that is, an error of order $O(|y|^k)$ in $h(y)$ results in an error of order $O(|y|^{k+1})$ in $g_1(y, h(y))$, for $k \geq 2$. The term ay^2 is the dominant term on the right-hand side of the reduced system. For $a \neq 0$, the origin of the reduced system is unstable. Consequently, by Theorem 4.2, the origin of the full system is unstable.   △

Example 4.2 Consider the system

$$\begin{aligned} \dot{y} &= yz \\ \dot{z} &= -z + ay^2 \end{aligned}$$

which is already represented in the (y, z) coordinates. The center manifold equation (4.11)–(4.12) is

$$h'(y)[yh(y)] + h(y) - ay^2 = 0, \quad h(0) = h'(0) = 0$$

We start by trying $\phi(y) = 0$. The reduced system is

$$\dot{y} = O(|y|^3)$$

Clearly, we cannot reach any conclusion about the stability of the origin. Therefore, we substitute $h(y) = h_2 y^2 + O(|y|^3)$ in the center manifold equation and calculate h_2, by matching coefficients of y^2, to obtain $h_2 = a$. The reduced system is[6]

$$\dot{y} = ay^3 + O(|y|^4)$$

---

[6] The error on the right-hand side of the reduced system is actually $O(|y|^5)$ since if we write $h(y) = h_2 y^2 + h_3 y^3 + \cdots$, we will find that $h_3 = 0$.

Therefore, the origin is asymptotically stable if $a < 0$ and unstable if $a > 0$. Consequently, by Theorem 4.2, we conclude that the origin of the full system is asymptotically stable if $a < 0$ and unstable if $a > 0$. $\triangle$

Example 4.3 Consider the system (4.2)–(4.3) with

$$A_1 = \begin{bmatrix} 0 & 1 \\ -1 & 0 \end{bmatrix}, \ g_1 = \begin{bmatrix} -y_1^3 \\ -y_2^3 + z^2 \end{bmatrix}, \ A_2 = -1, \ g_2 = y_1^3 - 3y_1^5 + 3y_1^2 y_2$$

It can be verified that $\phi(y) = 0$ results in $\mathcal{N}(\phi(y)) = O\left(\|y\|_2^3\right)$ and

$$\dot{y} = \begin{bmatrix} -y_1^3 + y_2 \\ -y_1 - y_2^3 \end{bmatrix} + O\left(\|y\|_2^4\right)$$

Using $V(y) = \frac{1}{2}(y_1^2 + y_2^2)$ as a Lyapunov function candidate, we obtain

$$\dot{V} = -y_1^4 - y_2^4 + y^T O\left(\|y\|_2^4\right) \le -\|y\|_2^4 + k\|y\|_2^5$$

in some neighborhood of the origin where $k > 0$. Hence,

$$\dot{V} \le -\tfrac{1}{2}\|y\|_2^4, \quad \text{for } \|y\|_2 < \frac{1}{2k}$$

which shows that the origin of the reduced system is asymptotically stable. Consequently, the origin of the full system is asymptotically stable. $\triangle$

Notice that in the preceding example it is not enough to study the system

$$\dot{y} = \begin{bmatrix} -y_1^3 + y_2 \\ -y_1 - y_2^3 \end{bmatrix}$$

We have to find a Lyapunov function that confirms asymptotic stability of the origin for all perturbations of the order $O\left(\|y\|_2^4\right)$. The importance of this observation is illustrated by the following example.

Example 4.4 Consider the previous example, but change A_1 to

$$A_1 = \begin{bmatrix} 0 & 1 \\ 0 & 0 \end{bmatrix}$$

With $\phi(y) = 0$, the reduced system can be represented as

$$\dot{y} = \begin{bmatrix} -y_1^3 + y_2 \\ -y_2^3 \end{bmatrix} + O\left(\|y\|_2^4\right)$$

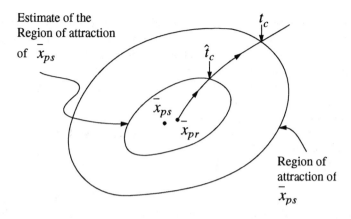

Figure 4.1: Critical clearance time.

Without the perturbation term $O\left(\|y\|_2^4\right)$, the origin of this system is asymptotically stable.[7] If you try to find a Lyapunov function $V(y)$ to show asymptotic stability in the presence of the perturbation term, you will not succeed. In fact, it can be verified that the center manifold equation (4.11)–(4.12) has the exact solution $h(y) = y_1^3$, so that the reduced system is given by the equation

$$\dot{y} = \begin{bmatrix} -y_1^3 + y_2 \\ y_1^6 - y_2^3 \end{bmatrix}$$

whose origin is unstable.[8]   △

4.2 Region of Attraction

Quite often, it is not sufficient to determine that a given system has an asymptotically stable equilibrium point. Rather, it is important to find the region of attraction of that point, or at least an estimate of it. To appreciate the importance of determining the region of attraction, let us run a scenario of events that could happen in the operation of a nonlinear system. Suppose that a nonlinear system has an asymptotically stable equilibrium point, which is denoted by $\bar{x}_{pr}$ in Figure 4.1. Suppose the system is operating at steady state at $\bar{x}_{pr}$. Then, at time t_0 a fault that changes the structure of the system takes place, for example, a short circuit in an electrical network. Suppose the faulted system does not have equilibrium at $\bar{x}_{pr}$ or in its neighborhood. The trajectory of the system will be driven away from

---

[7] See Exercise 5.24.

[8] See Exercise 3.13.

$\bar{x}_{pr}$. Suppose further that the fault is cleared at time t_1 and the postfault system has an asymptotically stable equilibrium point at $\bar{x}_{ps}$, where either $\bar{x}_{ps} = \bar{x}_{pr}$ or $\bar{x}_{ps}$ is sufficiently close to $\bar{x}_{pr}$ so that steady-state operation at $\bar{x}_{ps}$ is still acceptable. At time t_1 the state of the system, say, $x(t_1)$, could be far from the postfault equilibrium $\bar{x}_{ps}$. Whether or not the system will return to steady-state operation at $\bar{x}_{ps}$ depends on whether $x(t_1)$ belongs to the region of attraction of $\bar{x}_{ps}$, as determined by the post-fault system equation. A crucial factor in determining how far $x(t_1)$ could be from $\bar{x}_{ps}$ is the time it takes the operators of the system to remove the fault, that is, the time difference $(t_1 - t_0)$. If $(t_1 - t_0)$ is very short, then, by continuity of the solution with respect to t, it is very likely that $x(t_1)$ will be in the region of attraction of $\bar{x}_{ps}$. However, operators need time to detect the fault and fix it. How much time they have is a critical question. In planning such a system, it is valuable to give operators a "critical clearance time," say t_c, such that they have to clear the fault within this time; that is, $(t_1 - t_0)$ must be less than t_c. If we know the region of attraction of $\bar{x}_{ps}$, we can find t_c by integrating the faulted system equation starting from the prefault equilibrium $\bar{x}_{pr}$ until it hits the boundary of the region of attraction. The time it takes the trajectory to reach the boundary can be taken as the critical clearance time because if the fault is cleared before that time the state $x(t_1)$ will be within the region of attraction. Of course, we are assuming that $\bar{x}_{pr}$ belongs to the region of attraction of $\bar{x}_{ps}$, which is reasonable. If the actual region of attraction is not known, and an estimate $\hat{t}_c$ of t_c is obtained using an estimate of the region of attraction, then $\hat{t}_c < t_c$ since the boundary of the estimate of the region of attraction will be inside the actual boundary of the region; see Figure 4.1. This scenario shows an example where finding the region of attraction is needed in planning the operation of a nonlinear system. It also shows the importance of finding estimates of the region of attraction which are not too conservative. A very conservative estimate of the region of attraction would result in $\hat{t}_c$ that is too small to be useful. Let us conclude this motivating discussion by saying that the scenario of events described here is not hypothetical. It is the essence of the transient stability problem in power systems.[9]

Let the origin $x = 0$ be an asymptotically stable equilibrium point for the nonlinear system

$$\dot{x} = f(x) \tag{4.13}$$

where $f : D \rightarrow R^n$ is locally Lipschitz and $D \subset R^n$ is a domain containing the origin. Let $\phi(t; x)$ be the solution of (4.13) that starts at initial state x at time $t = 0$. The region of attraction of the origin, denoted by R_A, is defined by

$$R_A = \{ x \in D \mid \phi(t; x) \rightarrow 0 \text{ as } t \rightarrow \infty \}$$

Some properties of the region of attraction are stated in the following lemma, whose proof is given in Appendix A.8.

---

[9] See [131] for an introduction to the transient stability problem in power systems.

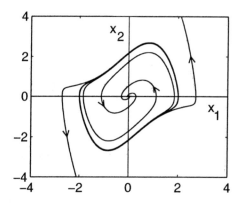

Figure 4.2: Phase portrait for Example 4.5.

Lemma 4.1 *If $x = 0$ is an asymptotically stable equilibrium point for* (4.13), *then its region of attraction R_A is an open, connected, invariant set. Moreover, the boundary of R_A is formed by trajectories.*   ◇

This lemma suggests that one way to determine the region of attraction is to characterize those trajectories which lie on the boundary of R_A. There are some methods which approach the problem from this viewpoint, but they use geometric notions from the theory of dynamical systems which are not introduced in this book. Therefore, we will not describe this class of methods.[10] We may, however, get a flavor of these geometric methods in the case of second-order systems ($n = 2$) by employing phase portraits. Examples 4.5 and 4.6 show typical cases in the state plane. In the first example, the boundary of the region of attraction is a limit cycle, while in the second one the boundary is formed of stable trajectories of saddle points. Example 4.7 shows a rather pathological case where the boundary is a closed curve of equilibrium points.

Example 4.5 The second-order system

$$\begin{aligned} \dot{x}_1 &= -x_2 \\ \dot{x}_2 &= x_1 + (x_1^2 - 1)x_2 \end{aligned}$$

is a Van der Pol equation in reverse time, that is, with t replaced by $-t$. The system has one equilibrium point at the origin and one unstable limit cycle, as determined from the phase portrait shown in Figure 4.2. The phase portrait shows that the origin is a stable focus; hence, it is asymptotically stable. This can be confirmed by

---

[10]Examples of these methods can be found in [30] and [191].

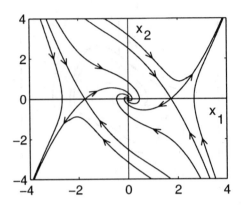

Figure 4.3: Phase portrait for Example 4.6.

linearization, since

$$A = \left.\frac{\partial f}{\partial x}\right|_{x=0} = \begin{bmatrix} 0 & -1 \\ 1 & -1 \end{bmatrix}$$

has eigenvalues at $-1/2 \pm j\sqrt{3}/2$. Clearly, the region of attraction is bounded because trajectories starting outside the limit cycle cannot cross it to reach the origin. Since there are no other equilibrium points, the boundary of R_A must be the limit cycle. Inspection of the phase portrait shows that indeed all trajectories starting inside the limit cycle spiral toward the origin.   △

Example 4.6 Consider the second-order system

$$\begin{aligned} \dot{x}_1 &= x_2 \\ \dot{x}_2 &= -x_1 + \tfrac{1}{3}x_1^3 - x_2 \end{aligned}$$

This system has three isolated equilibrium points at $(0,0)$, $(\sqrt{3},0)$, and $(-\sqrt{3},0)$. The phase portrait of the system is shown in Figure 4.3. The phase portrait shows that the origin is a stable focus, and the other two equilibria are saddle points. Hence, the origin is asymptotically stable and the other equilibria are unstable; a fact which can be confirmed by linearization. From the phase portrait, we can also see that the stable trajectories of the saddle points form two separatrices which are the boundaries of the region of attraction. The region is unbounded.   △

Example 4.7 The system

$$\begin{aligned} \dot{x}_1 &= -x_1(1 - x_1^2 - x_2^2) \\ \dot{x}_2 &= -x_2(1 - x_1^2 - x_2^2) \end{aligned}$$

has an isolated equilibrium point at the origin and a continuum of equilibrium points on the unit circle; that is, every point on the unit circle is an equilibrium point. Clearly, R_A must be confined to the interior of the unit circle. The trajectories of this system are the radii of the unit circle. This can be seen by transforming the system into polar coordinates. The change of variables

$$x_1 = \rho \cos \theta, \quad x_2 = \rho \sin \theta$$

yields

$$\dot{\rho} = -\rho(1 - \rho^2), \quad \dot{\theta} = 0$$

All trajectories starting with $\rho < 1$ approach the origin as $t \to \infty$. Therefore, R_A is the interior of the unit circle. $\triangle$

Lyapunov's method can be used to find the region of attraction R_A or an estimate of it. The basic tool for finding the boundary of R_A is Zubov's theorem, which is given in Exercise 4.11. The theorem, however, has the character of an existence theorem and requires the solution of a partial differential equation. Via much simpler procedures, we can find estimates of R_A using Lyapunov's method. By an estimate of R_A, we mean a set $\Omega \subset R_A$ such that every trajectory starting in Ω approaches the origin as $t \to \infty$. For the rest of this section, we shall discuss some aspects of estimating R_A. Let us start by showing that the domain D of Theorem 3.1 is not an estimate of R_A. We have seen in Theorem 3.1 that if D is a domain that contains the origin, and if we can find a Lyapunov function $V(x)$ that is positive definite in D and $\dot{V}(x)$ is negative definite in D, then the origin is asymptotically stable. One may jump to the conclusion that D is an estimate of R_A. This conjecture is not true, as illustrated by the following example.

Example 4.8 Consider again the system of Example 4.6.

$$\begin{aligned} \dot{x}_1 &= x_2 \\ \dot{x}_2 &= -x_1 + \tfrac{1}{3}x_1^3 - x_2 \end{aligned}$$

This system is a special case of that of Example 3.5 with

$$h(x_1) = x_1 - \tfrac{1}{3}x_1^3 \quad \text{and} \quad a = 1$$

Therefore, a Lyapunov function is given by

$$\begin{aligned} V(x) &= \tfrac{1}{2}x^T \begin{bmatrix} \tfrac{1}{2} & \tfrac{1}{2} \\ \tfrac{1}{2} & 1 \end{bmatrix} x + \int_0^{x_1} (y - \tfrac{1}{3}y^3) \, dy \\ &= \tfrac{3}{4}x_1^2 - \tfrac{1}{12}x_1^4 + \tfrac{1}{2}x_1 x_2 + \tfrac{1}{2}x_2^2 \end{aligned}$$

and

$$\dot{V}(x) = -\tfrac{1}{2}x_1^2(1 - \tfrac{1}{3}x_1^2) - \tfrac{1}{2}x_2^2$$

Defining a domain D by

$$D = \{x \in R^2 \mid -\sqrt{3} < x_1 < \sqrt{3}\}$$

it can be easily seen that $V(x) > 0$ and $\dot{V}(x) < 0$ in $D - \{0\}$. Inspection of the phase portrait in Figure 4.3 shows that D is not a subset of R_A. $\triangle$

In view of this example, it is not difficult to see why D of Theorem 3.1 is not an estimate of R_A. Even though a trajectory starting in D will move from one Lyapunov surface $V(x) = c_1$ to an inner Lyapunov surface $V(x) = c_2$, with $c_2 < c_1$, there is no guarantee that the trajectory will remain forever in D. Once the trajectory leaves D, there is no guarantee that $\dot{V}(x)$ will be negative. Hence, the whole argument about $V(x)$ decreasing to zero falls apart. This problem does not arise in the estimates of R_A, which we are going to present, since they are positively invariant sets; that is, a trajectory starting in the set remains for all future time in the set. The simplest estimate is provided by the set

$$\Omega_c = \{x \in R^n \mid V(x) \le c\}$$

when Ω_c is bounded and contained in D. This follows as a corollary of Theorem 3.4. The simplicity of obtaining Ω_c has increased significance in view of the linearization results of Section 3.3. There, we saw that if the Jacobian matrix

$$A = \left. \frac{\partial f}{\partial x} \right|_{x=0}$$

is a stability matrix, then we can always find a quadratic Lyapunov function $V(x) = x^T P x$ by solving the Lyapunov equation (3.12) for any positive definite matrix Q. Putting the pieces together, we see that *whenever A is a stability matrix, we can estimate the region of attraction of the origin.* This is illustrated by the following two examples.

Example 4.9 The second-order system

$$\begin{aligned}
\dot{x}_1 &= -x_2 \\
\dot{x}_2 &= x_1 + (x_1^2 - 1)x_2
\end{aligned}$$

was treated in Example 4.5. There, we saw that the origin is asymptotically stable since

$$A = \left. \frac{\partial f}{\partial x} \right|_{x=0} = \begin{bmatrix} 0 & -1 \\ 1 & -1 \end{bmatrix}$$

is a stability matrix. A Lyapunov function for the system can be found by taking $Q = I$ and solving the Lyapunov equation

$$PA + A^T P = -I$$

for P. The unique solution is the positive definite matrix

$$P = \begin{bmatrix} 1.5 & -0.5 \\ -0.5 & 1 \end{bmatrix}$$

The quadratic function $V(x) = x^T P x$ is a Lyapunov function for the system in a certain neighborhood of the origin. Since our interest here is in estimating the region of attraction, we need to determine a domain D about the origin where $\dot{V}(x)$ is negative definite and a set $\Omega_c \subset D$, which is bounded. We are also interested in the largest set Ω_c that we can determine, that is, the largest value for the constant c, because Ω_c will be our estimate of R_A. Notice that we do not have to worry about checking positive definiteness of $V(x)$ in D because $V(x)$ is positive definite for all x. The derivative of $V(x)$ along the trajectories of the system is given by

$$\dot{V}(x) = -(x_1^2 + x_2^2) - (x_1^3 x_2 - 2x_1^2 x_2^2)$$

The right-hand side of $\dot{V}(x)$ is written as the sum of two terms. The first term, $-\|x\|_2^2$, is the contribution of the linear part Ax while the second term is the contribution of the nonlinear term $g(x) = f(x) - Ax$, which we may refer to as the perturbation term. Since

$$\frac{\|g(x)\|_2}{\|x\|_2} \to 0 \quad \text{as} \quad \|x\|_2 \to 0$$

we know that there is an open ball $D = \{x \in R^2 \mid \|x\|_2 < r\}$ such that $\dot{V}(x)$ is negative definite in D. Once we find such a ball, we can find $\Omega_c \subset D$ by choosing

$$c < \min_{\|x\|_2 = r} V(x) = \lambda_{\min}(P) r^2$$

Thus, to enlarge the estimate of the region of attraction we need to find the largest ball on which $\dot{V}(x)$ is negative definite. We have

$$\dot{V}(x) \le -\|x\|_2^2 + |x_1| \, |x_1 x_2| \, |x_1 - 2x_2| \le -\|x\|_2^2 + \frac{\sqrt{5}}{2}\|x\|_2^4$$

where we used $|x_1| \le \|x\|_2$, $|x_1 x_2| \le \frac{1}{2}\|x\|_2^2$, and $|x_1 - 2x_2| \le \sqrt{5}\|x\|_2$. Thus, $\dot{V}(x)$ is negative definite on a ball D of radius given by $r^2 = 2/\sqrt{5} = 0.8944$. In this second-order example, a less conservative estimate of Ω_c can be found by searching for the ball D in polar coordinates. Taking

$$x_1 = \rho \cos \theta, \quad x_2 = \rho \sin \theta$$

we have

$$
\begin{aligned}
\dot{V} &= -\rho^2 + \rho^4 \cos^2\theta \sin\theta(2\sin\theta - \cos\theta) \\
&\leq -\rho^2 + \rho^4 |\cos^2\theta \sin\theta| \cdot |2\sin\theta - \cos\theta| \\
&\leq -\rho^2 + \rho^4 \times 0.3849 \times 2.2361 \\
&\leq -\rho^2 + 0.861\rho^4 \ < 0, \quad \text{for} \ \ \rho^2 < \frac{1}{0.861}
\end{aligned}
$$

Using this, together with $\lambda_{\min}(P) \geq 0.69$, we choose

$$
c = 0.8 < \frac{0.69}{0.861} = 0.801
$$

The set Ω_c with $c = 0.8$ is an estimate of the region of attraction. $\triangle$

Example 4.10 Consider the second-order system

$$
\begin{aligned}
\dot{x}_1 &= -2x_1 + x_1 x_2 \\
\dot{x}_2 &= -x_2 + x_1 x_2
\end{aligned}
$$

There are two equilibrium points at $(0,0)$ and $(1,2)$. The equilibrium point at $(1,2)$ is unstable since

$$
\left. \frac{\partial f}{\partial x} \right|_{x_1=1, x_2=2} = \begin{bmatrix} 0 & 1 \\ 2 & 0 \end{bmatrix}
$$

has eigenvalues at $\pm\sqrt{2}$; it is a saddle point. The origin is asymptotically stable since

$$
A = \left. \frac{\partial f}{\partial x} \right|_{x=0} = \begin{bmatrix} -2 & 0 \\ 0 & -1 \end{bmatrix}
$$

is a stability matrix. A Lyapunov function for the system can be found by taking $Q = I$ and solving the Lyapunov equation

$$
PA + A^T P = -I
$$

for P. The unique solution is the positive definite matrix

$$
P = \begin{bmatrix} \frac{1}{4} & 0 \\ 0 & \frac{1}{2} \end{bmatrix}
$$

The derivative of $V(x) = x^T P x$ along the trajectories of the system is given by

$$
\begin{aligned}
\dot{V}(x) &= -(x_1^2 + x_2^2) + (\tfrac{1}{2}x_1^2 x_2 + x_1 x_2^2) \\
&\leq -\|x\|_2^2 + \tfrac{1}{2}|x_1^2 x_2| \, |x_1 + 2x_2| \ \leq \ -\|x\|_2^2 + \frac{\sqrt{5}}{4}\|x\|_2^3
\end{aligned}
$$

Thus, $\dot{V}(x)$ is negative definite in a ball of radius $r = 4/\sqrt{5}$. Since $\lambda_{\min}(P) = 1/4$, we choose

$$c = 0.79 < \frac{1}{4} \times \left(\frac{4}{\sqrt{5}} \right)^2 = 0.8$$

The set Ω_c with $c = 0.79$ is an estimate of the region of attraction. $\triangle$

Estimating the region of attraction by means of the sets Ω_c is simple, but is usually conservative. We shall present two ideas to obtain better estimates of the region of attraction. The first idea is based on LaSalle's theorem (Theorem 3.4) while the second one is the *trajectory-reversing method*, which uses computer simulation. LaSalle's theorem provides an estimate of R_A by the set Ω, which is a compact positively invariant set. Working with the set Ω_c, whose boundary is a Lyapunov surface, is a special case where the invariance of the set follows from the negative definiteness of $\dot{V}(x)$. We can work with more general sets, but then we need to establish that the set is a positively invariant one. This typically requires investigating the vector field at the boundary of the set to ensure that trajectories starting in the set cannot leave it. The following example illustrates this idea.

Example 4.11 Consider the system

$$\begin{aligned}
\dot{x}_1 &= x_2 \\
\dot{x}_2 &= -4(x_1 + x_2) - h(x_1 + x_2)
\end{aligned}$$

where $h : R \to R$ satisfies

$$h(0) = 0; \quad uh(u) \geq 0, \ \forall \ |u| \leq 1$$

Consider the quadratic function

$$V(x) = x^T \begin{bmatrix} 2 & 1 \\ 1 & 1 \end{bmatrix} x = 2x_1^2 + 2x_1x_2 + x_2^2$$

as a Lyapunov function candidate.[11] The derivative $\dot{V}(x)$ is given by

$$\begin{aligned}
\dot{V}(x) &= (4x_1 + 2x_2)\dot{x}_1 + 2(x_1 + x_2)\dot{x}_2 \\
&= -2x_1^2 - 6(x_1 + x_2)^2 - 2(x_1 + x_2)h(x_1 + x_2) \\
&\leq -2x_1^2 - 6(x_1 + x_2)^2, \ \forall \ |x_1 + x_2| \leq 1 \\
&= -x^T \begin{bmatrix} 8 & 6 \\ 6 & 6 \end{bmatrix} x
\end{aligned}$$

---

[11] This Lyapunov function candidate can be derived using the variable gradient method. See Exercise 4.14.

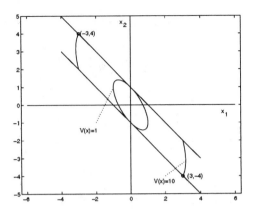

Figure 4.4: Estimates of the region of attraction for Example 4.11.

Therefore, $\dot{V}(x)$ is negative definite in the set

$$G = \{x \in R^2 \mid |x_1 + x_2| \le 1\}$$

and we can conclude that the origin is asymptotically stable. To estimate R_A, let us start by an estimate of the form Ω_c. We need to find the largest $c > 0$ such that $\Omega_c \subset G$. It can be easily seen that this c is given by

$$c = \min_{|x_1+x_2|=1} V(x) = \min\{\min_{x_1+x_2=1} V(x),\ \min_{x_1+x_2=-1} V(x)\}$$

The first minimization yields

$$\min_{x_1+x_2=1} V(x) = \min_{x_1}\{2x_1^2 + 2x_1(1 - x_1) + (1 - x_1)^2\} = 1$$

Similarly,

$$\min_{x_1+x_2=-1} V(x) = 1$$

Hence, Ω_c with $c = 1$ is an estimate of R_A; see Figure 4.4. In this example, we can obtain a better estimate of R_A by not restricting ourselves to estimates of the form Ω_c. A key point in the following development is to observe that trajectories inside G cannot leave through certain segments of the boundary $|x_1 + x_2| = 1$. This can be seen by plotting the phase portrait of the system, or by the following analysis. Let

$$\sigma = x_1 + x_2$$

such that the boundary of G is given by $\sigma = 1$ and $\sigma = -1$. The derivative of σ^2 along the trajectories of the system is given by

$$\frac{d}{dt}\sigma^2 = 2\sigma(\dot{x}_1 + \dot{x}_2) = 2\sigma x_2 - 8\sigma^2 - 2\sigma h(\sigma) \le 2\sigma x_2 - 8\sigma^2, \quad \forall\ |\sigma| \le 1$$

On the boundary $\sigma = 1$,

$$\frac{d}{dt}\sigma^2 \leq 2x_2 - 8 \leq 0, \quad \forall \; x_2 \leq 4$$

This implies that when the trajectory is at any point on the segment of the boundary $\sigma = 1$ for which $x_2 \leq 4$, it cannot move outside the set G because at such point σ^2 is nonincreasing. Similarly, on the boundary $\sigma = -1$,

$$\frac{d}{dt}\sigma^2 \leq -2x_2 - 8 \leq 0, \quad \forall \; x_2 \geq -4$$

Hence, the trajectory cannot leave the set G through the segment of the boundary $\sigma = -1$ for which $x_2 \geq -4$. This information can be used to form a closed, bounded, positively invariant set Ω that satisfies the conditions of Theorem 3.4. Using the two segments of the boundary of G identified above to define the boundary of Ω, we now need two other segments to close the set. These segments should have the property that trajectories cannot leave the set through them. We can take them as segments of a Lyapunov surface. Let c_1 be such that the Lyapunov surface $V(x) = c_1$ intersects the boundary $x_1 + x_2 = 1$ at $x_2 = 4$, that is, at the point $(-3, 4)$; see Figure 4.4. Let c_2 be such that the Lyapunov surface $V(x) = c_2$ intersects the boundary $x_1 + x_2 = -1$ at $x_2 = -4$, that is, at the point $(3, -4)$. The required Lyapunov surface is defined by $V(x) = \min\{c_1, c_2\}$. The constant c_1 is given by

$$c_1 = V(x)|_{x_1 = -3, x_2 = 4} = 10$$

Similarly, $c_2 = 10$. Therefore, we take $c = 10$ and define the set Ω by

$$\Omega = \{x \in R^2 \mid V(x) \leq 10 \text{ and } |x_1 + x_2| \leq 1\}$$

This set is closed, bounded, and positively invariant. Moreover, $\dot{V}(x)$ is negative definite in Ω since $\Omega \subset G$. Thus, all the conditions of Theorem 3.4 are satisfied and we can conclude that all trajectories starting in Ω approach the origin as $t \to \infty$; that is, $\Omega \subset R_A$. $\triangle$

The trajectory-reversing method uses computer simulation to enlarge an initial estimate of the region of attraction. It draws its name from reversing the direction of the trajectories of (4.13) via backward integration. This is equivalent to forward integration of the system

$$\dot{x} = -f(x) \tag{4.14}$$

which is obtained from (4.13) via replacing t by $-t$. System (4.14) has the same trajectory configuration in the state-space as system (4.13), but with reversed arrow heads on the trajectories. Let $W_0 \subset D$ be a compact set that contains the origin in its interior. Suppose W_0 is positively invariant with respect to (4.13) and there is a positive definite, continuously differentiable function $V : W_0 \to R$ such that

$$\dot{V}(x) < 0 \ \ \forall \ x \in W_0 - \{0\}$$

According to Theorems 3.1 and 3.4, the origin is asymptotically stable equilibrium for (4.13) and $W_0 \subset R_A$. Consequently, W_0 is connected.[12] For $x \in W_0$, let $\psi(t, x)$ be the solution of (4.14) that starts at x at time $t = 0$. This is the backward integration of (4.13) starting at x. Since $f(x)$ is locally Lipschitz and W_0 is compact, there is $T > 0$ such that $\psi(t, x)$ is defined on $[0, T)$ for all $x \in W_0$. Notice that backward integration starting at $x \in W_0$ may have a finite escape time even though forward integration is defined for all t since $x \in R_A$. A simple example that illustrates this point is the scalar system

$$\dot{x} = -x^3, \ \ x(0) = x_0$$

which, in reverse time, becomes

$$\dot{x} = x^3, \ \ x(0) = x_0$$

Forward integration is defined for all t and $x(t)$ approaches the origin, while backward integration results in

$$x(t) = \frac{x_0}{\sqrt{1 - 2tx_0^2}}$$

which has a finite escape time at $t = 1/2x_0^2$. Let $t_1 \in (0, T)$, and consider the backward mapping

$$F(x) = \psi(t_1, x), \ \ x \in W_0$$

Due to uniqueness and continuity of solutions in both forward and backward integration, this mapping is continuous and one-to-one on W_0, and its inverse F^{-1} is continuous on $F(W_0)$. Let

$$W_1 = F(W_0)$$

Then, W_1 is compact and connected.[13] Moreover,

$$W_0 \subset W_1 \subset R_A$$

The fact that W_0 is a subset of W_1 follows from the fact that W_0 is a positively invariant set in forward integration. Suppose there is a point $\bar{x} \in W_0$ such that $\bar{x} \neq \psi(t_1, x)$ for any $x \in W_0$. Then, the forward trajectory starting at $\bar{x}$ must leave

---

[12] See Exercise 4.15.

[13] This follows from the fact that compactness and connectedness of a set remain invariant under a topological mapping or a homeomorphism [7]. A mapping $F(\cdot)$ is a topological mapping on a set S if it is continuous, one-to-one, and its inverse F^{-1} is continuous on $F(S)$. The backward mapping $F(x) = \psi(t_1, x)$ is a topological mapping on W_0.

the set W_0, which contradicts the forward invariance of W_0. Actually, W_0 is a proper subset[14] of W_1. To see this, note that no point on the boundary of W_0 may be taken by backward integration into the interior of W_0 since, again, this would contradict the fact that W_0 is a positively invariant set in forward integration. Moreover, since every forward trajectory starting on the boundary of W_0 must eventually approach the origin, there must be some points on the boundary where the vector fields of forward trajectories point into the set W_0. Backward integration must take these points outside W_0. The fact that $W_1 \subset R_A$ follows from the fact that forward integration of any point in W_1 would take it into W_0, which is a subset of R_A. Thus, W_1 provides a better estimate of the region of attraction than W_0. Now, let us increase t_1 to $t_2, t_3, \ldots, t_i, \ldots$, where $t_i < t_{i+1} < T$ for all i. Repetition of the argument shows that the backward mappings $\psi(t_i, x)$ define nested compact sets W_i such that

$$W_i = F_i(W_0), \quad W_i \subset W_{i+1} \subset R_A$$

where $F_i(x) = \psi(t_i, x)$. Therefore, the sequence $W_1, W_2, \ldots$ provides increasingly better estimates of R_A. The utility of the trajectory-reversing method is not in generating the sets W_i analytically, but in using computer simulations to approximate these sets.[15] Let Γ_i be the boundary of W_i. It is not difficult to see that the backward mapping $F_i(x) = \psi(t_i, x)$ transforms Γ_0 into Γ_i. Discretize Γ_0 using an adequate number of points and carry out backward integration of these points over the interval $[0, t_1]$ to obtain a discretized image of Γ_1. To obtain a better estimate, choose $t_2 > t_1$ and repeat the backward integration. It is enough, however, to integrate the points on Γ_1 over the interval $[t_1, t_2]$ since, due to uniqueness of solution, this would be equivalent to integrating the points on Γ_0 over the interval $[t_0, t_2]$. The process can be repeated to obtain $\Gamma_3, \Gamma_4, \ldots$. The method will function particularly well when R_A is bounded since backward integration cannot escape to infinity.

Example 4.12 Consider the second-order system

$$
\begin{aligned}
\dot{x}_1 &= -x_2 \\
\dot{x}_2 &= x_1 + (x_1^2 - 1)x_2
\end{aligned}
$$

We found in Example 4.5 that the region of attraction R_A is the interior of the limit cycle shown in Figure 4.2. In Example 4.9, we used Lyapunov's method to estimate R_A by the set

$$W_0 = \{x \in R^2 \mid x^T P x \leq 0.8\}, \quad \text{where } P = \begin{bmatrix} 1.5 & -0.5 \\ -0.5 & 1.0 \end{bmatrix}$$

---

[14] A is a proper subset of B if $A \subset B$ but $A \neq B$.

[15] The approximation described here can be applied, in essence, to any n-dimensional system. When $n = 2$, topological considerations can be combined with the trajectory-reversing idea to come up with a set of guidelines that could be very effective in determining the region of attraction. For further details, see [55, 127].

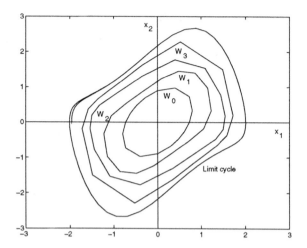

Figure 4.5: Estimates of the region of attraction for Example 4.12.

Let us use the trajectory-reversing method to improve the estimate W_0. We have discretized the surface $x^T P x = 0.8$ using 16 points which are equally-spaced angle-wise, that is, the points are given in polar coordinates by (ρ_k, θ_k), where

$$\theta_k = \frac{2\pi k}{16}, \quad \rho_k^2 = 0.8 \left/ \left[\begin{array}{c} \cos\theta_k \\ \sin\theta_k \end{array} \right]^T P \left[\begin{array}{c} \cos\theta_k \\ \sin\theta_k \end{array} \right] \right.$$

for $k = 0, 1, 2, \ldots, 15$. Figure 4.5 shows the surfaces obtained by backward integration of these points over the interval $[0, t_i]$ for $t_i = 1, 2, 3$. It is clear that if we continue this process, the estimate W_i will approach R_A.   △

If R_A is unbounded, the method will still enlarge the initial estimate of the region of attraction but we might not be able to asymptotically recover R_A. One problem is that backward integration may have a finite escape time. Even without facing this problem, we may still experience numerical difficulties due to the fact that backward trajectories approaching infinity may have different speeds. While the backward integration of some points on Γ_0 might still be close to Γ_0, the backward integration of other points might produce solutions large enough to induce numerical difficulties in the computer simulation routine. The following example illustrates this point.

Example 4.13 Consider the second-order system

$$\dot{x}_1 = -2x_1 + x_1 x_2$$
$$\dot{x}_2 = -x_2 + x_1 x_2$$

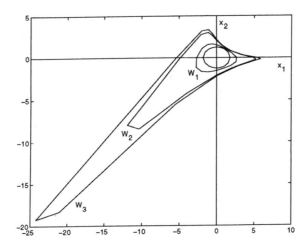

Figure 4.6: Estimates of the region of attraction for Example 4.13.

In Example 4.10, we estimated R_A by the set

$$W_0 = \{x \in R^2 \mid x^T P x \le 0.79\}, \quad \text{where} \quad P = \begin{bmatrix} 0.25 & 0 \\ 0 & 0.5 \end{bmatrix}$$

Discretizing the surface $x^T P x = 0.79$ using the scheme from Example 4.12, we obtain the surfaces Γ_i for $t_i = 0.2$, 0.5, and 0.55, which are shown in Figure 4.6. The wedge-like shape of the surfaces is due to backward integration of two points on Γ_0 which approach infinity rapidly. At $t_i = 0.6$, the simulation program suffers from singularities and the calculations are no longer reliable.[16] $\triangle$

4.3 Invariance Theorems

In the case of autonomous systems, LaSalle's invariance theorem (Theorem 3.4) shows that the trajectory of the system approaches the largest invariant set in E, where E is the set of all points in Ω where $\dot{V}(x) = 0$. In the case of nonautonomous systems, it may not even be clear how to define a set E since $\dot{V}(t, x)$ is a function of both t and x. The situation will be simpler if it can be shown that

$$\dot{V}(t, x) \le -W(x) \le 0$$

---

[16] The integration is carried out using a classical fourth-order Runge-Kutta algorithm with automatic step adjustment. Of course, using another algorithm we might be able to exceed the 0.6 limit, but the problem will happen again at a higher value of t_i.

for, then, a set E may be defined as the set of points where $W(x) = 0$. We may expect that the trajectory of the system approaches E as t tends to ∞. This is, basically, the statement of the next theorem. Before we state the theorem, we state a lemma that will be used in the proof of the theorem. The lemma is interesting in its own sake and is known as *Barbalat's lemma*.

Lemma 4.2 *Let* $\phi : R \to R$ *be a uniformly continuous function on* $[0, \infty)$. *Suppose that* $\lim_{t \to \infty} \int_0^t \phi(\tau) \, d\tau$ *exists and is finite. Then,*

$$\phi(t) \to 0 \quad \text{as} \quad t \to \infty$$

$\diamond$

Proof: If it is not true, then there is a positive constant k_1 such that for every $T > 0$ we can find $T_1 \geq T$ with $|\phi(T_1)| \geq k_1$. Since $\phi(t)$ is uniformly continuous, there is a positive constant k_2 such that $|\phi(t + \tau) - \phi(t)| < k_1/2$ for all $t \geq 0$ and all $0 \leq \tau \leq k_2$. Hence,

$$\begin{aligned}
|\phi(t)| &= |\phi(t) - \phi(T_1) + \phi(T_1)| \\
&\geq |\phi(T_1)| - |\phi(t) - \phi(T_1)| \\
&> k_1 - \tfrac{1}{2}k_1 = \tfrac{1}{2}k_1, \quad \forall \, t \in [T_1, T_1 + k_2]
\end{aligned}$$

Therefore,

$$\left| \int_{T_1}^{T_1 + k_2} \phi(t) \, dt \right| = \int_{T_1}^{T_1 + k_2} |\phi(t)| \, dt > \tfrac{1}{2}k_1 k_2$$

where the equality holds since $\phi(t)$ retains the same sign for $T_1 \leq t \leq T_1 + k_2$. Thus, the integral $\int_0^t \phi(\tau) \, d\tau$ cannot converge to a finite limit as $t \to \infty$, a contradiction.

$\square$

Theorem 4.4 *Let* $D = \{x \in R^n \mid \|x\| < r\}$ *and suppose that* $f(t, x)$ *is piecewise continuous in* t *and locally Lipschitz in* x, *uniformly in* t, *on* $[0, \infty) \times D$. *Let* $V : [0, \infty) \times D \to R$ *be a continuously differentiable function such that*

$$W_1(x) \leq V(t, x) \leq W_2(x)$$

$$\dot{V}(t, x) = \frac{\partial V}{\partial t} + \frac{\partial V}{\partial x} f(t, x) \leq -W(x)$$

$\forall \, t \geq 0, \, \forall \, x \in D$, *where* $W_1(x)$ *and* $W_2(x)$ *are continuous positive definite functions and* $W(x)$ *is a continuous positive semidefinite function on* D. *Let* $\rho < \min_{\|x\|=r} W_1(x)$. *Then, all solutions of* $\dot{x} = f(t, x)$ *with* $x(t_0) \in \{x \in B_r \mid W_2(x) \leq \rho\}$ *are bounded and satisfy*

$$W(x(t)) \to 0 \quad \text{as} \quad t \to \infty$$

Moreover, if all the assumptions hold globally and $W_1(x)$ is radially unbounded, the statement is true for all $x(t_0) \in R^n$. ◇

Proof: Similar to the proof of Theorem 3.8, it can be shown that

$$x(t_0) \in \{x \in B_r \mid W_2(x) \leq \rho\} \Rightarrow x(t) \in \Omega_{t,\rho}, \quad \forall \, t \geq t_0$$

since $\dot{V}(t, x) \leq 0$. Hence, $\|x(t)\| < r$ for all $t \geq t_0$. Since $V(t, x(t))$ is monotonically nonincreasing and bounded from below by zero, it converges as $t \to \infty$. Now,

$$\int_{t_0}^t W(x(\tau)) \, d\tau \leq - \int_{t_0}^t \dot{V}(\tau, x(\tau)) \, d\tau = V(t_0, x(t_0)) - V(t, x(t))$$

Therefore, $\lim_{t \to \infty} \int_{t_0}^t W(x(\tau)) \, d\tau$ exists and is finite. Since $\|x(t)\| < r$ for all $t \geq t_0$ and $f(t, x)$ is locally Lipschitz in x, uniformly in t, we conclude that $x(t)$ is uniformly continuous in t on $[t_0, \infty)$. Consequently, $W(x(t))$ is uniformly continuous in t on $[t_0, \infty)$ since $W(x)$ is uniformly continuous in x on the compact set B_r. Hence, by Lemma 4.2, we conclude that $W(x(t)) \to 0$ as $t \to \infty$. If all the assumptions hold globally and $W_1(x)$ is radially unbounded, then for any $x(t_0)$ we can choose ρ so large that $x(t_0) \in \{x \in R^n \mid W_2(x) \leq \rho\}$. □

The limit $W(x(t)) \to 0$ implies that $x(t)$ approaches E as $t \to \infty$, where

$$E = \{x \in D \mid W(x) = 0\}$$

Therefore, the positive limit set of $x(t)$ is a subset of E. The mere knowledge that $x(t)$ approaches E is much weaker than the invariance principle for autonomous systems which states that $x(t)$ approaches the largest invariant set in E. The stronger conclusion in the case of autonomous systems is a consequence of the property of autonomous systems stated in Lemma 3.1, namely the positive limit set is an invariant set. There are some special classes of nonautonomous systems where positive limit sets have some sort of an invariance property.[17] However, for a general nonautonomous system, the positive limit sets are not invariant. The fact that, in the case of autonomous systems, $x(t)$ approaches the largest invariant set in E allowed us to arrive at Corollary 3.1, where asymptotic stability of the origin is established by showing that the set E does not contain an entire trajectory of the system, other than the trivial solution. For a general nonautonomous system, there is no extension of Corollary 3.1 that would show uniform asymptotic stability. However, the following theorem shows that it is possible to conclude uniform asymptotic stability if, in addition to $\dot{V}(t, x) \leq 0$, the integral of $\dot{V}(t, x)$ satisfies a certain inequality.

---

[17]Examples are periodic systems, almost-periodic systems, and asymptotically autonomous systems. See [137, Chapter 8] for invariance theorems for these classes of systems. See, also, [119] for a different generalization of the invariance principle.

Theorem 4.5 *Let* $D = \{x \in R^n \mid \|x\| < r\}$ *and suppose that* $f(t, x)$ *is piecewise continuous in* t *and locally Lipschitz in* x *on* $[0, \infty) \times D$. *Let* $x = 0$ *be an equilibrium point for* $\dot{x} = f(t, x)$ *at* $t = 0$. *Let* $V : [0, \infty) \times D \to R$ *be a continuously differentiable function such that*

$$W_1(x) \leq V(t, x) \leq W_2(x)$$

$$\dot{V}(t, x) = \frac{\partial V}{\partial t} + \frac{\partial V}{\partial x} f(t, x) \leq 0$$

$$\int_t^{t+\delta} \dot{V}(\tau, \phi(\tau, t, x)) \, d\tau \leq -\lambda V(t, x), \quad 0 < \lambda < 1^{18}$$

$\forall \, t \geq 0, \, \forall \, x \in D$, *for some* $\delta > 0$, *where* $W_1(x)$ *and* $W_2(x)$ *are continuous positive definite functions on* D *and* $\phi(\tau, t, x)$ *is the solution of the system that starts at* (t, x). *Then, the origin is uniformly asymptotically stable. If all the assumptions hold globally and* $W_1(x)$ *is radially unbounded, then the origin is globally uniformly asymptotically stable. If*

$$W_1(x) \geq k_1 \|x\|^c, \quad W_2(x) \leq k_2 \|x\|^c, \quad k_1 > 0, \quad k_2 > 0, \quad c > 0$$

then the origin is exponentially stable. $\diamond$

Proof: Similar to the proof of Theorem 3.8, it can be shown that

$$x(t_0) \in \{x \in B_r \mid W_2(x) \leq \rho\} \Rightarrow x(t) \in \Omega_{t,\rho}, \quad \forall \, t \geq t_0$$

where $\rho < \min_{\|x\|=r} W_1(x)$ because $\dot{V}(t, x) \leq 0$. Now, for all $t \geq t_0$, we have

$$
\begin{aligned}
V(t + \delta, x(t + \delta)) &= V(t, x(t)) + \int_t^{t+\delta} \dot{V}(\tau, \phi(\tau, t, x(t))) \, d\tau \\
&\leq V(t, x(t)) - \lambda V(t, x(t)) = (1 - \lambda)V(t, x(t))
\end{aligned}
$$

Moreover, since $\dot{V}(t, x) \leq 0$,

$$V(\tau, x(\tau)) \leq V(t, x(t)), \quad \forall \tau \in [t, t + \delta]$$

For any $t \geq t_0$, let N be the smallest positive integer such that $t \leq t_0 + N\delta$. Divide the interval $[t_0, t_0 + (N-1)\delta]$ into $(N-1)$ equal subintervals of length δ each. Then,

$$
\begin{aligned}
V(t, x(t)) &\leq V(t_0 + (N-1)\delta, x(t_0 + (N-1)\delta)) \\
&\leq (1 - \lambda)V(t_0 + (N-2)\delta, x(t_0 + (N-2)\delta))
\end{aligned}
$$

---

[18] There is no loss of generality in assuming that $\lambda < 1$, for if the inequality is satisfied with $\lambda_1 \geq 1$, then it is satisfied for any positive $\lambda < 1$ since $-\lambda_1 V \leq -\lambda V$. Notice, however, that this inequality could not be satisfied with $\lambda > 1$ since $V(t, x) > 0$, $\forall \, x \neq 0$.

$$\vdots$$

$$\leq \ (1 - \lambda)^{(N-1)} V(t_0, x(t_0))$$

$$\leq \ \frac{1}{(1 - \lambda)} (1 - \lambda)^{(t-t_0)/\delta} V(t_0, x(t_0))$$

$$= \ \frac{1}{(1 - \lambda)} e^{-b(t-t_0)} V(t_0, x(t_0))$$

where

$$b = \frac{1}{\delta} \ln \frac{1}{(1 - \lambda)}$$

Taking

$$\sigma(r, s) = \frac{r}{(1 - \lambda)} e^{-bs}$$

it can be easily seen that $\sigma(r, s)$ is a class $\mathcal{KL}$ function and $V(t, x(t))$ satisfies

$$V(t, x(t)) \leq \sigma(V(t_0, x(t_0)), t - t_0), \quad \forall \ V(t_0, x(t_0)) \in [0, \rho]$$

From this point on, the rest of the proof is identical to that of Theorem 3.8. The proof of the statements on global uniform asymptotic stability and exponential stability are the same as the proof of Corollaries 3.3 and 3.4. $\qquad \square$

Example 4.14 Consider the linear time-varying system

$$\dot{x} = A(t)x$$

where $A(t)$ is continuous for all $t \geq 0$. Suppose there is a continuously differentiable, symmetric matrix $P(t)$ which satisfies

$$0 < c_1 I \leq P(t) \leq c_2 I, \quad \forall \ t \geq 0$$

as well as the matrix differential equation

$$-\dot{P}(t) = P(t)A(t) + A^T(t)P(t) + C^T(t)C(t)$$

where $C(t)$ is continuous in t. The derivative of the quadratic function

$$V(t, x) = x^T P(t)x$$

along the trajectories of the system is

$$\dot{V}(t, x) = -x^T C^T(t)C(t)x \leq 0$$

The solution of the linear system is given by

$$\phi(\tau, t, x) = \Phi(\tau, t)x$$

where $\Phi(\tau, t)$ is the state transition matrix. Therefore,

$$
\begin{aligned}
\int_t^{t+\delta} \dot{V}(\tau, \phi(\tau, t, x))\, d\tau &= -x^T \int_t^{t+\delta} \Phi^T(\tau, t) C^T(\tau) C(\tau) \Phi(\tau, t)\, d\tau\, x \\
&= -x^T W(t, t+\delta) x
\end{aligned}
$$

where

$$
W(t, t+\delta) = \int_t^{t+\delta} \Phi^T(\tau, t) C^T(\tau) C(\tau) \Phi(\tau, t)\, d\tau
$$

Suppose there is a positive constant $k < c_2$ such that

$$
W(t, t+\delta) \geq kI, \quad \forall\, t \geq 0
$$

then

$$
\int_t^{t+\delta} \dot{V}(\tau, \phi(\tau, t, x))\, d\tau \leq -k\|x\|_2^2 \leq -\frac{k}{c_2} V(t, x)
$$

Thus, all the assumptions of Theorem 4.5 are satisfied globally with

$$
W_i(x) = c_i \|x\|_2^2, \quad i = 1, 2, \quad \lambda = \frac{k}{c_2} < 1
$$

and we conclude that the origin is globally exponentially stable. Readers familiar with linear system theory will recognize that the matrix $W(t, t+\delta)$ is the observability Gramian of the pair $(A(t), C(t))$ and that the inequality $W(t, t+\delta) \geq kI$ is implied by uniform observability of $(A(t), C(t))$. Comparison of this example with Example 3.21 shows that Theorem 4.5 allows us to replace the positive definiteness requirement on the matrix $Q(t)$ of (3.21) by the weaker requirement $Q(t) = C^T(t) C(t)$, where the pair $(A(t), C(t))$ is uniformly observable. $\triangle$

4.4   Exercises

Exercise 4.1 Prove Corollary 4.1.

Exercise 4.2 Prove Corollary 4.2.

Exercise 4.3 Suppose the conditions of Theorem 4.1 are satisfied in a case where $g_1(y, 0) = 0$, $g_2(y, 0) = 0$, and $A_1 = 0$. Show that the origin of the full system is stable.

Exercise 4.4 Reconsider Example 4.1 with $a = 0$, and Example 4.2 with $a = 0$. In each case, apply Corollary 4.1 to show that the origin is stable.

Exercise 4.5 ([80]) Consider the system

$$\dot{x}_a = f_a(x_a, x_b)$$
$$\dot{x}_b = A_b x_b + f_b(x_a, x_b)$$

where $\dim(x_a) = n_1$, $\dim(x_b) = n_2$, A_b is a Hurwitz matrix, f_a and f_b are continuously differentiable, $[\partial f_b / \partial x_b](0,0) = 0$, and $f_b(x_a, 0) = 0$ in a neighborhood of $x_a = 0$.

(a) Show that if the origin $x_a = 0$ is an exponentially stable equilibrium point of $\dot{x}_a = f_a(x_a, 0)$, then the origin $(x_a, x_b) = (0,0)$ is an exponentially stable equilibrium point of the full system.

(b) Using the center manifold theorem, show that if the origin $x_a = 0$ is an asymptotically (but not exponentially) stable equilibrium point of $\dot{x}_a = f_a(x_a, 0)$, then the origin $(x_a, x_b) = (0,0)$ is an asymptotically stable equilibrium point of the full system.

Exercise 4.6 ([59]) For each of the following systems, investigate the stability of the origin using the center manifold theorem.

(1)
$$\dot{x}_1 = -x_2^2$$
$$\dot{x}_2 = -x_2 + x_1^2 + x_1 x_2$$

(2)
$$\dot{x}_1 = a x_1^2 - x_2^2, \quad a \neq 0$$
$$\dot{x}_2 = -x_2 + x_1^2 + x_1 x_2$$

(3)
$$\dot{x}_1 = -x_2 + x_1 x_3$$
$$\dot{x}_2 = x_1 + x_2 x_3$$
$$\dot{x}_3 = -x_3 - (x_1^2 + x_2^2) + x_3^2$$

Exercise 4.7 For each of the following systems, investigate the stability of the origin using the center manifold theorem.

(1)
$$\dot{x}_1 = x_1 x_2^3$$
$$\dot{x}_2 = -x_2 - x_1^2 + 2x_1^8$$

(2)
$$\dot{x}_1 = -x_1 + x_2^3(x_1 + x_2 - 1)$$
$$\dot{x}_2 = x_2^3(x_1 + x_2 - 1)$$

(3)
$$\dot{x}_1 = x_2$$
$$\dot{x}_2 = -x_2 + a x_1^3/(1 + x_1^2)$$
$$a \neq 0$$

(4)
$$\dot{x}_1 = -2x_1 - 3x_2 + x_3 + x_3^2$$
$$\dot{x}_2 = x_1 + x_1^2 + x_2$$
$$\dot{x}_3 = x_1^2$$

Exercise 4.8 ([28]) Consider the system

$$\dot{x}_1 = x_1 x_2 + a x_1^3 + b x_1 x_2^2$$
$$\dot{x}_2 = -x_2 + c x_1^2 + d x_1^2 x_2$$

Investigate the stability of the origin using the center manifold theorem for each of the following cases.

(1) $a + c > 0$ **(2)** $a + c < 0$
(3) $a + c = 0$ and $cd + bc^2 < 0$ **(4)** $a + c = 0$ and $cd + bc^2 > 0$
(5) $a + c = cd + bc^2 = 0$

Exercise 4.9 ([28]) Consider the system

$$\dot{x}_1 = ax_1^3 + x_1^2 x_2$$
$$\dot{x}_2 = -x_2 + x_2^2 + x_1 x_2 - x_1^3$$

Investigate the stability of the origin using the center manifold theorem for all possible values of the real parameter a.

Exercise 4.10 ([80]) Consider the system

$$\dot{x}_1 = ax_1 x_2 - x_1^3$$
$$\dot{x}_2 = -x_2 + bx_1 x_2 + cx_1^2$$

Investigate the stability of the origin using the center manifold theorem for all possible values of the real constants a, b, c.

Exercise 4.11 (Zubov's Theorem) Consider the system (3.1) and let $G \subset R^n$ be a domain containing the origin. Suppose there exist two functions $V : G \to R$ and $h : R^n \to R$ with the following properties:

- V is continuously differentiable and positive definite in G and satisfies the inequality
$$0 < V(x) < 1, \quad \forall \, x \in G - \{0\}$$

- As x approaches the boundary of G, or in case of unbounded G as $\|x\| \to \infty$, $\lim V(x) = 1$.

- h is continuous and positive definite on R^n.

- For $x \in G$, $V(x)$ satisfies the partial differential equation
$$\frac{\partial V}{\partial x} f(x) = -h(x)[1 - V(x)] \qquad (4.15)$$

Show that $x = 0$ is asymptotically stable and G is the region of attraction.

Exercise 4.12 ([62]) Consider the second-order system

$$\dot{x}_1 = -h(x_1) + g_2(x_2)$$
$$\dot{x}_2 = -g_1(x_1)$$

where

$$h(0) = 0, \quad zh(z) > 0 \ \forall \ -a_1 < z < b_1$$

$$g_i(0) = 0, \quad zg_i(z) > 0 \ \forall \ -a_i < z < b_i$$

$$\int_0^z g_i(\sigma) \, d\sigma \to \infty \text{ as } z \to -a_i \text{ or } z \to b_i$$

for some positive constants a_i, b_i ($a_i = \infty$ or $b_i = \infty$ is allowed). Apply Zubov's theorem to show that the region of attraction is $\{x \in R^2 \mid -a_i < x_i < b_i\}$.

Hint: Take $h(x) = g_1(x_1)h(x_1)$ and seek a solution of the partial differential equation (4.15) in the form $V(x) = 1 - W_1(x_1)W_2(x_2)$. Note that, with this choice of h, $\dot{V}(x)$ is only negative semidefinite; apply LaSalle's invariance principle.

Exercise 4.13 Find the region of attraction of the system

$$\dot{x}_1 = -x_1 + x_2$$
$$\dot{x}_2 = -\tan(x_1)$$

Hint: Use the previous exercise.

Exercise 4.14 Consider the system of Example 4.11. Derive the Lyapunov function $V(x) = 2x_1^2 + 2x_1x_2 + x_2^2$ using the variable gradient method.

Exercise 4.15 Let Ω be an open, positively invariant set containing the origin. Suppose every trajectory in Ω approaches the origin as $t \to \infty$. Show that Ω is connected.

Exercise 4.16 Consider a second-order system $\dot{x} = f(x)$ with asymptotically stable origin. Let $V(x) = x_1^2 + x_2^2$, and $D = \{x \in R^2 \mid |x_2| < 1, |x_1 - x_2| < 1\}$. Suppose that $[\partial V/\partial x] f(x)$ is negative definite in D. Estimate the region of attraction.

Exercise 4.17 Consider the system

$$\dot{x}_1 = x_2$$
$$\dot{x}_2 = -x_1 - x_2 - (2x_2 + x_1)(1 - x_2^2)$$

(a) Using $V(x) = 5x_1^2 + 2x_1x_2 + 2x_2^2$, show that the origin is asymptotically stable.

(b) Let

$$S = \{x \in R^2 \mid V(x) \leq 5\} \cap \{x \in R^2 \mid |x_2| \leq 1\}$$

Show that S is an estimate of the region of attraction.

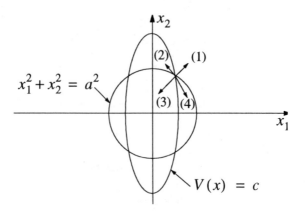

Figure 4.7: Exercise 4.19.

Exercise 4.18 Show that the origin of

$$\dot{x}_1 = x_2$$
$$\dot{x}_2 = -x_2 - x_1 + x_1^3$$

is asymptotically stable and estimate the region of attraction.

Exercise 4.19 Consider a second-order system $\dot{x} = f(x)$, together with a Lyapunov function $V(x)$. Suppose that $\dot{V}(x) < 0$ for all $x_1^2 + x_2^2 \geq a^2$. The sketch, given in Figure 4.7, shows four different directions of the vector field at a point on the circle $x_1^2 + x_2^2 = a^2$. Which of these directions are possible and which are not? Justify your answer.

Exercise 4.20 Consider the system

$$\dot{x}_1 = x_2$$
$$\dot{x}_2 = -x_2 - \sin x_1 - 2 \operatorname{sat}(x_1 + x_2)$$

where $\operatorname{sat}(\cdot)$ is the saturation function.

(a) Show that the origin is the unique equilibrium point.

(b) Show, using linearization, that the origin is asymptotically stable.

(c) Let $\sigma = x_1 + x_2$. Show that

$$\sigma \dot{\sigma} \leq -|\sigma|, \quad \text{for } |\sigma| \geq 1$$

(d) Let $V(x) = x_1^2 + 0.5x_2^2 + 1 - \cos x_1$, and M_c be the set defined, for any $c > 0$, by

$$M_c = \{x \in R^2 \mid V(x) \le c\} \cap \{x \in R^2 \mid |\sigma| \le 1\}$$

Show that M_c is positively invariant and that every trajectory inside M_c must approach the origin as $t \to \infty$.

(e) Show that the origin is globally asymptotically stable.

Exercise 4.21 Consider the tunnel diode circuit of Example 1.2. The circuit has three equilibrium points.

(a) Investigate the stability of each point using linearization.

(b) For each asymptotically stable equilibrium point, find a quadratic Lyapunov function $V(x) = x^T P x$ and use it to estimate the region of attraction.

(c) Continuing part (b), extend the region of attraction using the trajectory reversing method.

Exercise 4.22 Consider the third-order synchronous generator model described in Exercise 1.7. Take the state variables and parameters as in parts (a) and (b) of the exercise. Moreover, take $\tau = 6.6$ sec, $M = 0.0147$ (per unit power) $\times$ sec^2/rad, and $D/M = 4$ sec^{-1}.

(a) Find all equilibrium points in the region $-\pi \le x_1 \le \pi$, and determine the stability properties of each equilibrium using linearization.

(b) For each asymptotically stable equilibrium point, estimate the region of attraction using a quadratic Lyapunov function.

Exercise 4.23 ([100]) Consider the system

$$\begin{aligned}
\dot{x}_1 &= x_2 \\
\dot{x}_2 &= -x_1 - g(t)x_2
\end{aligned}$$

where $g(t)$ is continuously differentiable and $0 < k_1 \le g(t) \le k_2$ for all $t \ge 0$.

(a) Show that the origin is exponentially stable.

(b) Would (a) be true if $g(t)$ were not bounded? Consider $g(t) = 2 + \exp(t)$ in your answer.

Hint: Part (a) uses observability properties of linear systems. In particular, for any bounded matrix $K(t)$, uniform observability of the pair $(A(t), C(t))$ is equivalent to uniform observability of the pair $(A(t) - K(t)C(t), C(t))$.

Exercise 4.24 Consider the system

$$\dot{x}_1 = x_2$$
$$\dot{x}_2 = -\sin x_1 - g(t)x_2$$

where $g(t)$ is continuously differentiable and $0 < k_1 \leq g(t) \leq k_2$ for all $t \geq 0$. Show that the origin is exponentially stable.
Hint: Use the previous exercise.

Exercise 4.25 Consider the system

$$\dot{x}_1 = -x_1 - x_2 - \alpha(t)x_3$$
$$\dot{x}_2 = x_1$$
$$\dot{x}_3 = \alpha(t)x_1$$

where $\alpha(t)$ is bounded and $\dot{\alpha}(t) \not\equiv 0$. Show that the origin is exponentially stable.
Hint: Try $V(x) = x^T x$ and apply the result of Example 4.14.

Chapter 5

Stability of Perturbed Systems

Consider the system

$$\dot{x} = f(t, x) + g(t, x) \tag{5.1}$$

where $f : [0, \infty) \times D \to R^n$ and $g : [0, \infty) \times D \to R^n$ are piecewise continuous in t and locally Lipschitz in x on $[0, \infty) \times D$, and $D \subset R^n$ is a domain that contains the origin $x = 0$. We think of this system as a perturbation of the nominal system

$$\dot{x} = f(t, x) \tag{5.2}$$

The perturbation term $g(t, x)$ could result from modeling errors, aging, or uncertainties and disturbances which exist in any realistic problem. In a typical situation, we do not know $g(t, x)$ but we know some information about it, like knowing an upper bound on $\|g(t, x)\|$. Here, we represent the perturbation as an additive term on the right-hand side of the state equation. Uncertainties which do not change the system order can always be represented in this form. For if the perturbed right-hand side is some function $\tilde{f}(t, x)$, then by adding and subtracting $f(t, x)$ we can rewrite the right-hand side as

$$\tilde{f}(t, x) = f(t, x) + [\tilde{f}(t, x) - f(t, x)]$$

and define

$$g(t, x) = \tilde{f}(t, x) - f(t, x)$$

Suppose the nominal system (5.2) has a uniformly asymptotically stable equilibrium point at the origin, what can we say about the stability behavior of the perturbed system (5.1)? A natural approach to address this question is to use a Lyapunov function for the nominal system as a Lyapunov function candidate for the perturbed system. This is what we have done in the analysis of the linearization approach in Sections 3.3 and 3.5. The new element here is that the perturbation term could be more general than the perturbation term in the case of linearization. The conclusions we can arrive at depend critically on whether the perturbation term vanishes at the origin. If $g(t, 0) = 0$, the perturbed system (5.2) has an equilibrium point at the origin. In this case, we analyze the stability behavior of the origin as an equilibrium point of the perturbed system. If $g(t, 0) \neq 0$, the origin will not be an equilibrium point of the perturbed system. Therefore, we can no longer study the problem as a question of stability of equilibria. Some new concepts will have to be introduced.

The cases of vanishing and nonvanishing perturbations are treated in Sections 5.1 and 5.2, respectively. In Section 5.3, we use the results on nonvanishing perturbations to introduce the concept of input-to-state stability. In Section 5.4, we restrict our attention to the case when the nominal system has an exponentially stable equilibrium point at the origin, and use the comparison lemma to derive some sharper results on the asymptotic behavior of the solution of the perturbed system. In Section 5.5, we give a result that establishes continuity of the solution of the state equation on the infinite-time interval.

The last two sections deal with interconnected systems and slowly varying systems, respectively. In both cases, stability analysis is simplified by viewing the system as a perturbation of a simpler system. In the case of interconnected systems, the analysis is simplified by decomposing the system into smaller isolated subsystems, while in the case of slowly varying systems, a nonautonomous system with slowly varying inputs is approximated by an autonomous system where the slowly varying inputs are treated as constant parameters.

5.1   Vanishing Perturbation

Let us start with the case $g(t, 0) = 0$. Suppose $x = 0$ is an exponentially stable equilibrium point of the nominal system (5.2), and let $V(t, x)$ be a Lyapunov function that satisfies

$$c_1 \|x\|^2 \leq V(t, x) \leq c_2 \|x\|^2 \tag{5.3}$$

$$\frac{\partial V}{\partial t} + \frac{\partial V}{\partial x} f(t, x) \leq -c_3 \|x\|^2 \tag{5.4}$$

$$\left\| \frac{\partial V}{\partial x} \right\| \leq c_4 \|x\| \tag{5.5}$$

for all $(t, x) \in [0, \infty) \times D$ for some positive constants c_1, c_2, c_3, and c_4. The existence of a Lyapunov function satisfying (5.3)–(5.5) is guaranteed by Theorem 3.12, under some additional assumptions. Suppose the perturbation term $g(t, x)$ satisfies the linear growth bound

$$\|g(t, x)\| \leq \gamma \|x\|, \quad \forall \, t \geq 0, \, \forall \, x \in D \tag{5.6}$$

where γ is a nonnegative constant. This bound is natural in view of the assumptions on $g(t, x)$. In fact, any function $g(t, x)$ that vanishes at the origin and is locally Lipschitz in x, uniformly in t for all $t \geq 0$, in a bounded neighborhood of the origin satisfies (5.6) over that neighborhood.[1] We use V as a Lyapunov function candidate to investigate the stability of the origin as an equilibrium point for the perturbed system (5.1). The derivative of V along the trajectories of (5.1) is given by

$$\dot{V}(t, x) = \frac{\partial V}{\partial t} + \frac{\partial V}{\partial x} f(t, x) + \frac{\partial V}{\partial x} g(t, x)$$

The first two terms on the right-hand side constitute the derivative of $V(t, x)$ along the trajectories of the nominal system, which is negative definite and satisfies (5.4). The third term, $[\partial V / \partial x] g$, is the effect of the perturbation. Since we do not have complete knowledge of g, we cannot judge whether this term helps or hurts the cause of making $\dot{V}(t, x)$ negative definite. With the growth bound (5.6) as our only information on g, the best we can do is worst-case analysis where $[\partial V / \partial x] g$ is majorized by a nonnegative term. Using (5.4)–(5.6), we obtain

$$\dot{V}(t, x) \leq -c_3 \|x\|^2 + \left\| \frac{\partial V}{\partial x} \right\| \|g(t, x)\| \leq -c_3 \|x\|^2 + c_4 \gamma \|x\|^2$$

If γ is small enough to satisfy the bound

$$\gamma < \frac{c_3}{c_4} \tag{5.7}$$

then

$$\dot{V}(t, x) \leq -(c_3 - \gamma c_4) \|x\|^2, \quad (c_3 - \gamma c_4) > 0$$

Therefore, by Corollary 3.4, we conclude the following lemma.

Lemma 5.1 *Let $x = 0$ be an exponentially stable equilibrium point of the nominal system (5.2). Let $V(t, x)$ be a Lyapunov function of the nominal system that satisfies (5.3)–(5.5) in $[0, \infty) \times D$. Suppose the perturbation term $g(t, x)$ satisfies (5.6)–(5.7). Then, the origin is an exponentially stable equilibrium point of the perturbed system (5.1). Moreover, if all the assumptions hold globally, then the origin is globally exponentially stable.* $\diamond$

This lemma is conceptually important because it shows that exponential stability of the origin is robust with respect to a class of perturbations that satisfy (5.6)–(5.7). To assert this robustness property, we do not have to know $V(t, x)$ explicitly. It is just enough to know that the origin is an exponentially stable equilibrium of the nominal system. Sometimes, we may be able to show that the origin is

---

[1] Note, however, that the linear growth bound (5.6) becomes restrictive when required to hold globally, because that would require g to be globally Lipschitz in x.

exponentially stable without actually finding a Lyapunov function that satisfies (5.3)–(5.5).[2] Irrespective of the method we use to show exponential stability of the origin, we can assert the existence of $V(t, x)$ satisfying (5.3)–(5.5) by application of Theorem 3.12 (provided the Jacobian matrix $[\partial f / \partial x]$ is bounded). However, if we do not know the Lyapunov function $V(t, x)$ we cannot calculate the bound (5.7). Consequently, our robustness conclusion becomes a qualitative one where we say that the origin is exponentially stable for all perturbations satisfying

$$\|g(t, x)\| \leq \gamma \|x\|$$

with sufficiently small γ. On the other hand, if we know $V(t, x)$ we can calculate the bound (5.7), which is an additional piece of information. We should be careful not to overemphasize such bounds because they could be conservative for a given perturbation $g(t, x)$. The conservatism is a consequence of the worst-case analysis we have adopted from the beginning.

Example 5.1 Consider the system

$$\dot{x} = Ax + g(t, x)$$

where A is Hurwitz and $\|g(t, x)\|_2 \leq \gamma \|x\|_2$ for all $t \geq 0$ and all $x \in R^n$. Let $Q = Q^T > 0$ and solve the Lyapunov equation

$$PA + A^T P = -Q$$

for P. From Theorem 3.6, we know that there is a unique solution $P = P^T > 0$. The quadratic Lyapunov function $V(x) = x^T P x$ satisfies (5.3)–(5.5). In particular,

$$\lambda_{\min}(P)\|x\|_2^2 \leq V(x) \leq \lambda_{\max}(P)\|x\|_2^2$$

$$\frac{\partial V}{\partial x} Ax = -x^T Q x \leq -\lambda_{\min}(Q)\|x\|_2^2$$

$$\left\|\frac{\partial V}{\partial x}\right\|_2 = \|2x^T P\|_2 \leq 2\|P\|_2\|x\|_2 = 2\lambda_{\max}(P)\|x\|_2$$

The derivative of $V(x)$ along the trajectories of the perturbed system satisfies

$$\dot{V}(x) \leq -\lambda_{\min}(Q)\|x\|_2^2 + 2\lambda_{\max}(P)\gamma \|x\|_2^2$$

Hence, the origin is globally exponentially stable if $\gamma < \lambda_{\min}(Q)/2\lambda_{\max}(P)$. Since this bound depends on the choice of Q, one may wonder how to choose Q to maximize the ratio $\lambda_{\min}(Q)/\lambda_{\max}(P)$. It turns out that this ratio is maximized with the choice $Q = I$ (Exercise 5.1). $\triangle$

---

[2] This is the case, for example, when exponential stability of the origin is shown using Theorem 4.5.

Example 5.2 Consider the second-order system

$$\dot{x}_1 = x_2$$
$$\dot{x}_2 = -4x_1 - 2x_2 + \beta x_2^3$$

where the constant $\beta \geq 0$ is unknown. We view the system as a perturbed system of the form (5.1) with

$$f(x) = Ax = \begin{bmatrix} 0 & 1 \\ -4 & -2 \end{bmatrix} \begin{bmatrix} x_1 \\ x_2 \end{bmatrix} \quad \text{and} \quad g(x) = \begin{bmatrix} 0 \\ \beta x_2^3 \end{bmatrix}$$

The eigenvalues of A are $-1 \pm j\sqrt{3}$. Hence, A is Hurwitz. The solution of the Lyapunov equation

$$PA + A^T P = -I$$

is given by

$$P = \begin{bmatrix} \frac{3}{2} & \frac{1}{8} \\ \frac{1}{8} & \frac{5}{16} \end{bmatrix}$$

As we saw in Example 5.1, the Lyapunov function $V(x) = x^T P x$ satisfies inequalities (5.3)–(5.5) with $c_3 = 1$ and

$$c_4 = 2\lambda_{\max}(P) = 2 \times 1.513 = 3.026$$

The perturbation term $g(x)$ satisfies

$$\|g(x)\|_2 = \beta |x_2|^3 \leq \beta k_2^2 |x_2| \leq \beta k_2^2 \|x\|_2$$

for all $|x_2| \leq k_2$. At this point in the analysis, we do not know a bound on $x_2(t)$, although we know that $x_2(t)$ will be bounded whenever the trajectory $x(t)$ is confined to a compact set. We keep k_2 undetermined and proceed with the analysis. Using $V(x)$ as a Lyapunov function candidate for the perturbed system, we obtain

$$\dot{V}(x) \leq -\|x\|_2^2 + 3.026\beta k_2^2 \|x\|_2^2$$

Hence, $\dot{V}(x)$ will be negative definite if

$$\beta < \frac{1}{3.026k_2^2}$$

To estimate the bound k_2, let $\Omega_c = \{x \in R^2 \mid V(x) \leq c\}$. For any positive constant c, the set Ω_c is closed and bounded. The boundary of Ω_c is the Lyapunov surface

$$V(x) = \frac{3}{2}x_1^2 + \frac{1}{4}x_1 x_2 + \frac{5}{16}x_2^2 = c$$

The largest value of $|x_2|$ on the surface $V(x) = c$ can be determined by differentiating the surface equation partially with respect to x_1. This results in

$$3x_1 + \tfrac{1}{4}x_2 = 0$$

Therefore, the extreme values of x_2 are obtained at the intersection of the line $x_1 = -x_2/12$ with the Lyapunov surface. Simple calculations show that the largest value of x_2^2 on the Lyapunov surface is $96c/29$. Thus, all points inside Ω_c satisfy the bound

$$|x_2| \leq k_2, \quad \text{where} \quad k_2^2 = \frac{96c}{29}$$

Therefore, if

$$\beta < \frac{29}{3.026 \times 96c} \approx \frac{0.1}{c}$$

$\dot{V}(x)$ will be negative definite in Ω_c and we can conclude that the origin $x = 0$ is exponentially stable with Ω_c as an estimate of the region of attraction. The inequality $\beta < 0.1/c$ shows a tradeoff between the estimate of the region of attraction and the estimate of the upper bound on β. The smaller the upper bound on β, the larger the estimate of the region of attraction. This tradeoff is not artificial; it does exist in this example. The change of variables

$$z_1 = \sqrt{\frac{3\beta}{2}} x_2$$

$$z_2 = \sqrt{\frac{3\beta}{8}} (4x_1 + 2x_2 - \beta x_2^3) = -\sqrt{\frac{3\beta}{8}} \dot{x}_2$$

$$\tau = 2t$$

transforms the state equation into

$$\frac{dz_1}{d\tau} = -z_2$$

$$\frac{dz_2}{d\tau} = z_1 + (z_1^2 - 1)z_2$$

which was shown in Example 4.5 to have a bounded region of attraction surrounded by an unstable limit cycle. When transformed into the x-coordinates, the region of attraction will expand with decreasing β and shrink with increasing β. Finally, let us use this example to illustrate our remarks on the conservative nature of the bound (5.7). Using this bound, we came up with the inequality $\beta < 1/3.026k_2^2$. This inequality allows the perturbation term $g(t, x)$ to be any second-order vector that satisfies $\|g(t, x)\|_2 \leq \beta k_2^2 \|x\|_2$. This class of perturbations is more general than the perturbation we have in this specific problem. We have a *structured perturbation* in the sense that the first component of g is always zero, while our analysis allowed for an *unstructured perturbation* where the vector g could change in all directions. Such disregard of the structure of the perturbation will, in general, lead to conservative bounds. Suppose we repeat the analysis, this time taking into consideration the structure of the perturbation. Instead of using the general bound (5.7), we calculate the derivative of $V(t, x)$ along the trajectories of the perturbed system to obtain

$$
\begin{aligned}
\dot{V}(x) &= -\|x\|_2^2 + 2x^T P g(x) \\
&= -\|x\|_2^2 + 2\beta x_2^2 \left(\tfrac{1}{8}x_1 x_2 + \tfrac{5}{16}x_2^2\right) \\
&\leq -\|x\|_2^2 + 2\beta x_2^2 \left(\tfrac{1}{16}\|x\|_2^2 + \tfrac{5}{16}\|x\|_2^2\right) \\
&\leq -\|x\|_2^2 + \tfrac{3}{4}\beta k_2^2 \|x\|_2^2
\end{aligned}
$$

Hence, $\dot{V}(x)$ is negative definite for $\beta < 4/3k_2^2$. Using, again, the fact that for all $x \in \Omega_c$, $|x_2|^2 \leq k_2^2 = 96c/29$, we arrive at the bound $\beta < 0.4/c$, which is four times the bound we obtained using (5.7). $\triangle$

When the origin of the nominal system (5.2) is uniformly asymptotically stable but not exponentially stable, the stability analysis of the perturbed system is more involved. Suppose the nominal system has a positive definite, decrescent Lyapunov function $V(t, x)$ that satisfies

$$
\frac{\partial V}{\partial t} + \frac{\partial V}{\partial x} f(t, x) \leq -W_3(x)
$$

for all $(t, x) \in [0, \infty) \times D$, where $W_3(x)$ is positive definite and continuous. The derivative of V along the trajectories of (5.1) is given by

$$
\dot{V}(t, x) = \frac{\partial V}{\partial t} + \frac{\partial V}{\partial x} f(t, x) + \frac{\partial V}{\partial x} g(t, x) \leq -W_3(x) + \left\|\frac{\partial V}{\partial x} g(t, x)\right\|
$$

Our task now is to show that

$$
\left\|\frac{\partial V}{\partial x} g(t, x)\right\| < W_3(x)
$$

for all $(t, x) \in [0, \infty) \times D$, a task which cannot be done by a putting a simple order of magnitude bound on $\|g(t, x)\|$ as we have done in the exponentially stable case. The growth bound on $\|g(t, x)\|$ will depend on the nature of the Lyapunov function of the nominal system. One class of Lyapunov functions for which the analysis is almost as simple as in exponential stability is the case when $V(t, x)$ is positive definite, decrescent, and satisfies

$$
\frac{\partial V}{\partial t} + \frac{\partial V}{\partial x} f(t, x) \leq -c_3 \phi^2(x) \tag{5.8}
$$

$$\left\|\frac{\partial V}{\partial x}\right\| \le c_4\phi(x) \tag{5.9}$$

for all $(t, x) \in [0, \infty) \times D$ for some positive constants c_3 and c_4, where $\phi : R^n \to R$ is positive definite and continuous. A Lyapunov function satisfying (5.8)–(5.9) is usually called a *quadratic-type* Lyapunov function. It is clear that a Lyapunov function satisfying (5.3)–(5.5) is quadratic type, but a quadratic-type Lyapunov function may exist even when the origin is not exponentially stable. We shall illustrate this point shortly by an example. If the nominal system (5.2) has a quadratic-type Lyapunov function $V(t, x)$, then its derivative along the trajectories of (5.1) satisfies

$$\dot{V}(t, x) \le -c_3\phi^2(x) + c_4\phi(x)\|g(t, x)\|$$

Suppose now that the perturbation term satisfies the bound

$$\|g(t, x)\| \le \gamma\phi(x), \quad \gamma < \frac{c_3}{c_4}$$

Then,

$$\dot{V}(t, x) \le -(c_3 - c_4\gamma)\phi^2(x)$$

which shows that $\dot{V}(t, x)$ is negative definite.

Example 5.3 Consider the scalar system

$$\dot{x} = -x^3 + g(t, x)$$

The nominal system

$$\dot{x} = -x^3$$

has a globally asymptotically stable equilibrium point at the origin but, as we saw in Example 3.23, the origin is not exponentially stable. Thus, there is no Lyapunov function that satisfies (5.3)–(5.5). The Lyapunov function $V(x) = x^4$ satisfies (5.8)–(5.9), with $\phi(x) = |x|^3$, $c_3 = 4$, and $c_4 = 4$. Suppose the perturbation term $g(t, x)$ satisfies the bound $|g(t, x)| \le \gamma|x|^3$ for all x, with $\gamma < 1$. Then, the derivative of V along the trajectories of the perturbed system satisfies

$$\dot{V}(t, x) \le -4(1 - \gamma)\phi^2(x)$$

Hence, the origin is a globally uniformly asymptotically stable equilibrium point of the perturbed system. $\triangle$

In contrast to the case of exponential stability, it is important to notice that a nominal system with uniformly asymptotically stable, but not exponentially stable, origin is not robust to smooth perturbations with arbitrarily small linear growth bounds of the form (5.6). This point is illustrated by the following example.[3]

---

[3] See, also, Exercise 5.8.

Example 5.4 Consider the scalar system of the previous example with perturbation $g = \gamma x$ where $\gamma > 0$; that is,

$$\dot{x} = -x^3 + \gamma x$$

It can be easily seen, via linearization, that for any $\gamma > 0$, no matter how small γ, the origin is unstable. $\triangle$

5.2   Nonvanishing Perturbation

Let us turn now to the more general case when we do not know that $g(t,0) = 0$. The origin $x = 0$ may not be an equilibrium point of the perturbed system (5.1). We can no longer study stability of the origin as an equilibrium point, nor should we expect the solution of the perturbed system to approach the origin as $t \to \infty$. The best we can hope for is that if the perturbation term $g(t, x)$ is small in some sense, then $x(t)$ will be ultimately bounded by a small bound; that is, $\|x(t)\|$ will be small for sufficiently large t. This brings in the concept of ultimate boundedness.

Definition 5.1 *The solutions of $\dot{x} = f(t, x)$ are said to be uniformly ultimately bounded if there exist positive constants b and c, and for every $\alpha \in (0, c)$ there is a positive constant $T = T(\alpha)$ such that*

$$\|x(t_0)\| < \alpha \Rightarrow \|x(t)\| \leq b, \quad \forall \, t \geq t_0 + T \tag{5.10}$$

They are said to be globally uniformly ultimately bounded if (5.10) holds for arbitrarily large α.

We shall refer to the constant b in (5.10) as the ultimate bound. In the case of autonomous systems, we may drop the word "uniform" since the solution depends only on $t - t_0$. The following Lyapunov-like theorem is very useful in showing uniform ultimate boundedness.

Theorem 5.1 *Let $D \subset R^n$ be a domain containing the origin and $f : [0, \infty) \times D \to R^n$ be piecewise continuous in t and locally Lipschitz in x. Let $V : [0, \infty) \times D \to R$ be a continuously differentiable function such that*

$$W_1(x) \leq V(t, x) \leq W_2(x) \tag{5.11}$$

$$\frac{\partial V}{\partial t} + \frac{\partial V}{\partial x} f(t, x) \leq -W_3(x), \quad \forall \, \|x\| \geq \mu > 0 \tag{5.12}$$

$\forall \, t \geq 0, \, \forall \, x \in D$ where $W_1(x)$, $W_2(x)$, and $W_3(x)$ are continuous positive definite functions on D. Take $r > 0$ such that $B_r \subset D$ and suppose that μ is small enough that

$$\max_{\|x\| \leq \mu} W_2(x) < \min_{\|x\| = r} W_1(x) \tag{5.13}$$

Let $\eta = \max_{\|x\| \leq \mu} W_2(x)$ and take ρ such that $\eta < \rho < \min_{\|x\|=r} W_1(x)$. Then, there exist a finite time t_1 (dependent on $x(t_0)$ and μ) and a class $\mathcal{KL}$ function $\beta(\cdot, \cdot)$ such that $\forall\, x(t_0) \in \{x \in B_r \mid W_2(x) \leq \rho\}$, the solutions of $\dot{x} = f(t, x)$ satisfy

$$\|x(t)\| \leq \beta(\|x(t_0)\|, t - t_0), \quad \forall\, t_0 \leq t < t_1 \tag{5.14}$$

$$x(t) \in \{x \in B_r \mid W_1(x) \leq \eta\}, \quad \forall\, t \geq t_1 \tag{5.15}$$

Moreover, if $D = R^n$ and $W_1(x)$ is radially unbounded, then (5.14)–(5.15) hold for any initial state $x(t_0)$ and any μ. $\diamond$

Proof: The current proof shares many points with the proof of Theorem 3.8. To shorten the current proof, we shall use terminology and ideas from the proof of Theorem 3.8 as needed. Define $\Omega_{t,\eta} = \{x \in B_r \mid V(t, x) \leq \eta\}$. Then,

$$B_\mu \subset \{x \in B_r \mid W_2(x) \leq \eta\} \subset \Omega_{t,\eta} \subset \{x \in B_r \mid W_1(x) \leq \eta\} \subset \{x \in B_r \mid W_1(x) \leq \rho\}$$

and

$$\Omega_{t,\eta} \subset \Omega_{t,\rho} \subset \{x \in B_r \mid W_1(x) \leq \rho\} \subset B_r \subset D$$

The sets $\Omega_{t,\rho}$ and $\Omega_{t,\eta}$ have the property that a solution starting inside either set cannot leave it because $\dot{V}(t, x)$ is negative on the boundary. Therefore, if $x(t_0) \in \{x \in B_r \mid W_2(x) \leq \rho\}$, the solution $x(t)$ will belong to $\Omega_{t,\rho}$ for all $t \geq t_0$. For a solution starting inside $\Omega_{t,\eta}$, the relation (5.15) is satisfied for all $t \geq t_0$. For a solution starting inside $\Omega_{t,\rho}$ but outside $\Omega_{t,\eta}$, let t_1 be the first time it enters $\Omega_{t,\eta}$. This time t_1 could be ∞ if the solution never enters $\Omega_{t,\eta}$. For all $t \in [t_0, t_1)$, the inequalities (5.11)–(5.12) are satisfied. Hence, similar to the proof of Theorem 3.8, there is a class $\mathcal{K}$ function $\alpha(\cdot)$ such that $\dot{V} \leq -\alpha(V)$. Consequently, there is a class $\mathcal{KL}$ function $\beta(\cdot, \cdot)$ that satisfies (5.14). Since $\beta(\|x(t_0)\|, t - t_0) \to 0$ as $t \to \infty$, there is a finite time after which $\beta(\|x(t_0)\|, t - t_0) < \mu$ for all t. Therefore, the time t_1 must be finite; that is, the solution must enter the set $\Omega_{t,\eta}$ in finite time. Once inside the set, the solution remains inside for all $t \geq t_1$. Therefore, $x(t) \in \{x \in B_r \mid W_1(x) \leq \eta\}$ for all $t \geq t_1$. If $D = R^n$ and $W_1(x)$ is radially unbounded, then for any initial state $x(t_0)$ and any μ, we can choose r and ρ large enough that (5.13) is satisfied and $x(t_0) \in \{x \in B_r \mid W_2(x) \leq \rho\}$. $\square$

We leave it to the reader (Exercises 5.10, 5.11, and 5.12) to verify the following corollaries.

Corollary 5.1 *Under the assumption of Theorem 5.1, let $\alpha_1(\cdot)$ and $\alpha_2(\cdot)$ be class $\mathcal{K}$ functions defined on $[0, r]$ such that*

$$\alpha_1(\|x\|) \leq W_1(x) \quad \text{and} \quad W_2(x) \leq \alpha_2(\|x\|), \quad \forall\, x \in D$$

Suppose $\mu < \alpha_2^{-1}(\alpha_1(r))$ and $\|x(t_0)\| < \alpha_2^{-1}(\alpha_1(r))$. Then, the solutions of $\dot{x} = f(t, x)$ are uniformly ultimately bounded with an ultimate bound $\alpha_1^{-1}(\alpha_2(\mu))$.

Corollary 5.2 *Under the assumptions of Corollary 5.1, the solutions of* $\dot{x} = f(t, x)$ *satisfy*

$$\|x(t)\| \leq \beta(\|x(t_0)\|, t - t_0) + \alpha_1^{-1}(\alpha_2(\mu)), \quad \forall\, t \geq t_0 \tag{5.16}$$

Corollary 5.3 *Suppose the assumptions of Theorem 5.1 are satisfied with*

$$W_1(x) \geq k_1\|x\|^c, \quad W_2(x) \leq k_2\|x\|^c, \quad W_3(x) \geq k_3\|x\|^c$$

for some positive constants k_i *and* c. *Suppose* $\mu < r(k_1/k_2)^{1/c}$ *and* $\|x(t_0)\| < r(k_1/k_2)^{1/c}$. *Then,* (5.14)–(5.15) *take the form*

$$\|x(t)\| \leq k\|x(t_0)\|e^{-\gamma(t-t_0)}, \quad \forall\, t_0 \leq t < t_1 \tag{5.17}$$

$$\|x(t)\| \leq \mu \left(\frac{k_2}{k_1}\right)^{1/c}, \quad \forall\, t \geq t_1 \tag{5.18}$$

where $k = (k_2/k_1)^{1/c}$ *and* $\gamma = (k_3/k_2 c)$.

It is significant that the ultimate bound shown in Corollary 5.1 is a class $\mathcal{K}$ function of μ, because the smaller the value of μ the smaller the ultimate bound. As $\mu \to 0$, the ultimate bound approaches zero.

Let us illustrate how Theorem 5.1 is used in the analysis of the perturbed system (5.1) when the origin of the nominal system is exponentially stable.

Lemma 5.2 *Let* $x = 0$ *be an exponentially stable equilibrium point of the nominal system* (5.2). *Let* $V(t, x)$ *be a Lyapunov function of the nominal system that satisfies* (5.3)–(5.5) *in* $[0, \infty) \times D$, *where* $D = \{x \in R^n \mid \|x\| < r\}$. *Suppose the perturbation term* $g(t, x)$ *satisfies*

$$\|g(t, x)\| \leq \delta < \frac{c_3}{c_4}\sqrt{\frac{c_1}{c_2}}\theta r \tag{5.19}$$

for all $t \geq 0$, *all* $x \in D$, *and some positive constant* $\theta < 1$. *Then, for all* $\|x(t_0)\| < \sqrt{c_1/c_2}\,r$, *the solution* $x(t)$ *of the perturbed system* (5.1) *satisfies*

$$\|x(t)\| \leq k \exp[-\gamma(t - t_0)]\|x(t_0)\|, \quad \forall\, t_0 \leq t < t_1$$

and

$$\|x(t)\| \leq b, \quad \forall\, t \geq t_1$$

for some finite time t_1, *where*

$$k = \sqrt{\frac{c_2}{c_1}}, \quad \gamma = \frac{(1-\theta)c_3}{2c_2}, \quad b = \frac{c_4}{c_3}\sqrt{\frac{c_2}{c_1}}\frac{\delta}{\theta}$$

Proof: We use $V(t, x)$ as a Lyapunov function candidate for the perturbed system (5.1). The derivative of $V(t, x)$ along the trajectories of (5.1) satisfies

$$
\begin{aligned}
\dot{V}(t, x) &\leq -c_3\|x\|^2 + \left\| \frac{\partial V}{\partial x} \right\| \|g(t, x)\| \\
&\leq -c_3\|x\|^2 + c_4\delta\|x\| \\
&= -(1 - \theta)c_3\|x\|^2 - \theta c_3\|x\|^2 + c_4\delta\|x\|, \quad 0 < \theta < 1 \\
&\leq -(1 - \theta)c_3\|x\|^2, \quad \forall \ \|x\| \geq \delta c_4/\theta c_3
\end{aligned}
$$

Application of Corollary 5.3 completes the proof. $\square$

Note that the ultimate bound b in Lemma 5.2 is proportional to the upper bound on the perturbation δ. Once again, this result can be viewed as a robustness property of nominal systems having exponentially stable equilibria at the origin because it shows that arbitrarily small (uniformly bounded) perturbations will not result in large steady-state deviations from the origin.

Example 5.5 Consider the second-order system

$$
\begin{aligned}
\dot{x}_1 &= x_2 \\
\dot{x}_2 &= -4x_1 - 2x_2 + \beta x_2^3 + d(t)
\end{aligned}
$$

where $\beta \geq 0$ is unknown and $d(t)$ is a uniformly bounded disturbance that satisfies $|d(t)| \leq \delta$ for all $t \geq 0$. This is the same system we studied in Example 5.2 except for the additional perturbation term $d(t)$. Again, the system can be viewed as a perturbation of a nominal linear system that has a Lyapunov function $V(x) = x^T P x$ where

$$
P = \begin{bmatrix} \frac{3}{2} & \frac{1}{8} \\ \frac{1}{8} & \frac{5}{16} \end{bmatrix}
$$

We use $V(x)$ as a Lyapunov function candidate for the perturbed system, but we treat the two perturbation terms βx_2^3 and $d(t)$ differently since the first term vanishes at the origin while the second one does not. Calculating the derivative of $V(x)$ along the trajectories of the perturbed system, we obtain

$$
\begin{aligned}
\dot{V}(t, x) &= -\|x\|_2^2 + 2\beta x_2^2 \left(\tfrac{1}{8}x_1x_2 + \tfrac{5}{16}x_2^2 \right) + 2d(t)\left(\tfrac{1}{8}x_1 + \tfrac{5}{16}x_2 \right) \\
&\leq -\|x\|_2^2 + \tfrac{3}{4}\beta k_2^2\|x\|_2^2 + \frac{\sqrt{29}\delta}{8}\|x\|_2
\end{aligned}
$$

where we have used the inequality

$$
|2x_1 + 5x_2| \leq \|x\|_2\sqrt{4 + 25}
$$

and k_2 is an upper bound on $|x_2|$. Suppose $\beta \leq 4(1 - \zeta)/3k_2^2$, where $0 < \zeta < 1$. Then,

$$\dot{V}(t,x) \leq -\zeta \|x\|_2^2 + \frac{\sqrt{29}\delta}{8}\|x\|_2$$

$$\leq -(1-\theta)\zeta\|x\|_2^2, \ \forall \ \|x\|_2 \geq \mu = \frac{\sqrt{29}\delta}{8\zeta\theta}$$

where $0 < \theta < 1$. As we saw in Example 5.2, $|x_2|^2$ is bounded on Ω_c by $96c/29$. Thus, if $\beta \leq 0.4(1 - \zeta)/c$ and δ is so small that $\mu^2 \lambda_{\max}(P) < c$, then $B_\mu \subset \Omega_c$ and all trajectories starting inside Ω_c remain for all future time in Ω_c. Furthermore, the conditions of Theorem 5.1 are satisfied in Ω_c. Therefore, the solutions of the perturbed system are uniformly ultimately bounded with an ultimate bound

$$b = \frac{\sqrt{29}\delta}{8\zeta\theta}\sqrt{\frac{\lambda_{\max}(P)}{\lambda_{\min}(P)}}$$

$$\triangle$$

In the more general case when the origin $x = 0$ is a uniformly asymptotically stable equilibrium point of the nominal system (5.2), rather than exponentially stable, the analysis of the perturbed system proceeds in a similar manner.

Lemma 5.3 *Let $x = 0$ be a uniformly asymptotically stable equilibrium point of the nominal system (5.2). Let $V(t, x)$ be a Lyapunov function of the nominal system that satisfies the inequalities*[4]

$$\alpha_1(\|x\|) \leq V(t,x) \leq \alpha_2(\|x\|) \tag{5.20}$$

$$\frac{\partial V}{\partial t} + \frac{\partial V}{\partial x}f(t,x) \leq -\alpha_3(\|x\|) \tag{5.21}$$

$$\left\|\frac{\partial V}{\partial x}\right\| \leq \alpha_4(\|x\|) \tag{5.22}$$

in $[0, \infty) \times D$, where $D = \{x \in R^n \mid \|x\| < r\}$ and $\alpha_i(\cdot)$, $i = 1, 2, 3, 4$, are class $\mathcal{K}$ functions. Suppose the perturbation term $g(t, x)$ satisfies the uniform bound

$$\|g(t,x)\| \leq \delta < \frac{\theta\alpha_3(\alpha_2^{-1}(\alpha_1(r)))}{\alpha_4(r)} \tag{5.23}$$

---

[4]The existence of a Lyapunov function satisfying these inequalities (on a bounded domain) is guaranteed by Theorem 3.14 under some additional assumptions.

for all $t \geq 0$, all $x \in D$, and some positive constant $\theta < 1$. Then, for all $\|x(t_0)\| < \alpha_2^{-1}(\alpha_1(r))$, the solution $x(t)$ of the perturbed system (5.1) satisfies

$$\|x(t)\| \leq \beta(\|x(t_0)\|, t - t_0), \quad \forall\, t_0 \leq t < t_1$$

and

$$\|x(t)\| \leq \rho(\delta), \quad \forall\, t \geq t_1$$

for some class $\mathcal{KL}$ function $\beta(\cdot, \cdot)$ and some finite time t_1, where $\rho(\delta)$ is a class $\mathcal{K}$ function of δ defined by

$$\rho(\delta) = \alpha_1^{-1}\left(\alpha_2\left(\alpha_3^{-1}\left(\frac{\delta \alpha_4(r)}{\theta}\right)\right)\right)$$

$\diamondsuit$

Proof: We use $V(t, x)$ as a Lyapunov function candidate for the perturbed system (5.1). The derivative of $V(t, x)$ along the trajectories of (5.1) satisfies

$$
\begin{aligned}
\dot{V}(t, x) &\leq -\alpha_3(\|x\|) + \left\|\frac{\partial V}{\partial x}\right\| \|g(t, x)\| \\
&\leq -\alpha_3(\|x\|) + \delta \alpha_4(\|x\|) \\
&\leq -(1 - \theta)\alpha_3(\|x\|) - \theta \alpha_3(\|x\|) + \delta \alpha_4(r), \quad 0 < \theta < 1 \\
&\leq -(1 - \theta)\alpha_3(\|x\|), \quad \forall\, \|x\| \geq \alpha_3^{-1}\left(\frac{\delta \alpha_4(r)}{\theta}\right)
\end{aligned}
$$

Application of Theorem 5.1 and Corollary 5.1 completes the proof. $\qquad\square$

This lemma is similar to the one we arrived at in the special case of exponential stability. However, there is an important feature of our analysis in the case of exponential stability which has no counterpart in the more general case of uniform asymptotic stability. In the case of exponential stability, δ is required to satisfy (5.19). The right-hand side of (5.19) approaches ∞ as $r \to \infty$. Therefore, if all the assumptions hold globally, we can conclude that *for all uniformly bounded disturbances, the solution of the perturbed system will be uniformly bounded.* This is the case because, for any δ, we can choose r large enough to satisfy (5.19). In the case of uniform asymptotic stability, δ is required to satisfy (5.23). Inspection of the right-hand side of (5.23) shows that, without further information about the class $\mathcal{K}$ functions, we cannot say anything about the limit of the right-hand side as $r \to \infty$. Thus, we cannot conclude that uniformly bounded perturbations of a nominal system with a uniformly asymptotically stable equilibrium at the origin will have bounded solutions irrespective of the size of the perturbation. Of course the fact that we cannot show it does not mean it is not true. It turns out, however, that such a statement is not true. It is possible to construct examples (Exercise 5.19) where the origin is globally uniformly asymptotically stable, but a bounded perturbation could drive the solution of the perturbed system to infinity.

5.3   Input-to-State Stability

Consider the system

$$\dot{x} = f(t, x, u) \qquad\qquad (5.24)$$

where $f : [0, \infty) \times D \times D_u \to R^n$ is piecewise continuous in t and locally Lipschitz in x and u, $D \subset R^n$ is a domain that contains $x = 0$, and $D_u \subset R^m$ is a domain that contains $u = 0$. The input $u(t)$ is a piecewise continuous, bounded function of t for all $t \geq 0$. Suppose that the unforced system

$$\dot{x} = f(t, x, 0) \qquad\qquad (5.25)$$

has a uniformly asymptotically stable equilibrium point at the origin $x = 0$. By viewing the system (5.24) as a perturbation of the unforced system (5.25), we can apply the techniques of the preceding section to analyze the input-to-state behavior of (5.24). For example, if the unforced system satisfies the assumptions of Lemma 5.3 and the perturbation term satisfies the bound

$$\|f(t, x, u) - f(t, x, 0)\| \leq L\|u\|, \quad L \geq 0 \qquad\qquad (5.26)$$

for all $t \geq 0$ and all (x, u) in some bounded neighborhood of $(x = 0, u = 0)$, then the conclusion of Lemma 5.3 shows that for sufficiently small $\|x(t_0)\|$ and $\sup_{t \geq t_0} \|u(t)\|$, the solution of (5.24) satisfies

$$\|x(t)\| \leq \beta(\|x(t_0)\|, t - t_0) + \rho\left(L \sup_{\tau \geq t_0} \|u(\tau)\|\right), \quad \forall\, t \geq t_0$$

This inequality motivates the following definition of *input-to-state stability*.

Definition 5.2   *The system* (5.24) *is said to be locally input-to-state stable if there exist a class $\mathcal{KL}$ function β, a class $\mathcal{K}$ function γ, and positive constants k_1 and k_2 such that for any initial state $x(t_0)$ with $\|x(t_0)\| < k_1$ and any input $u(t)$ with $\sup_{t \geq t_0} \|u(t)\| < k_2$, the solution $x(t)$ exists and satisfies*

$$\|x(t)\| \leq \beta(\|x(t_0)\|, t - t_0) + \gamma\left(\sup_{t_0 \leq \tau \leq t} \|u(\tau)\|\right) \qquad\qquad (5.27)$$

for all $t \geq t_0 \geq 0$. It is said to be input-to-state stable if $D = R^n$, $D_u = R^m$, and inequality (5.27) *is satisfied for any initial state $x(t_0)$ and any bounded input $u(t)$.*

Inequality (5.27) guarantees that for a bounded input $u(t)$, the state $x(t)$ will be bounded. Furthermore, as t increases, the state $x(t)$ will be ultimately bounded with an ultimate bound that is a class $\mathcal{K}$ function of $\sup_{t \geq t_0} \|u(t)\|$. We leave it to the reader (Exercise 5.12) to use inequality (5.27) to show that if $u(t)$ converges to zero as $t \to \infty$, so does $x(t)$.[5] Since, with $u(t) \equiv 0$, (5.27) reduces to

$$\|x(t)\| \leq \beta(\|x(t_0)\|, t - t_0)$$

local input-to-state stability implies that the origin of the unforced system (5.25) is uniformly asymptotically stable, while input-to-state stability implies that it is globally uniformly asymptotically stable.

The following Lyapunov-like theorem gives a sufficient condition for input-to-state stability.[6]

Theorem 5.2 *Let* $D = \{x \in R^n \mid \|x\| < r\}$, $D_u = \{u \in R^m \mid \|u\| < r_u\}$, *and* $f : [0, \infty) \times D \times D_u \to R^n$ *be piecewise continuous in* t *and locally Lipschitz in* x *and* u. *Let* $V : [0, \infty) \times D \to R$ *be a continuously differentiable function such that*

$$\alpha_1(\|x\|) \leq V(t, x) \leq \alpha_2(\|x\|) \tag{5.28}$$

$$\frac{\partial V}{\partial t} + \frac{\partial V}{\partial x} f(t, x, u) \leq -\alpha_3(\|x\|), \quad \forall \ \|x\| \geq \rho(\|u\|) > 0 \tag{5.29}$$

$\forall \ (t, x, u) \in [0, \infty) \times D \times D_u$ *where* α_1, α_2, α_3 *and* ρ *are class* $\mathcal{K}$ *functions. Then, the system* (5.24) *is locally input-to-state stable with* $\gamma = \alpha_1^{-1} \circ \alpha_2 \circ \rho$, $k_1 = \alpha_2^{-1}(\alpha_1(r))$, *and* $k_2 = \rho^{-1} (\min\{k_1, \rho(r_u)\})$. *Moreover, if* $D = R^n$, $D_u = R^m$, *and* α_1 *is a class* $\mathcal{K}_\infty$ *function, then the system* (5.24) *is input-to-state stable with* $\gamma = \alpha_1^{-1} \circ \alpha_2 \circ \rho$.
$$\diamond$$

Proof: Application of Theorem 5.1 (with Corollaries 5.1 and 5.2) shows that for any $x(t_0)$ and any input $u(t)$ such that

$$\|x(t_0)\| < \alpha_2^{-1}(\alpha_1(r)), \quad \rho \left(\sup_{t \geq t_0} \|u(t)\| \right) < \min\{\alpha_2^{-1}(\alpha_1(r)), \rho(r_u)\} \tag{5.30}$$

the solution $x(t)$ exists and satisfies

$$\|x(t)\| \leq \beta(\|x(t_0)\|, t - t_0) + \gamma \left(\sup_{\tau \geq t_0} \|u(\tau)\| \right), \quad \forall \ t \geq t_0 \tag{5.31}$$

Since the solution $x(t)$ depends only on $u(\tau)$ for $t_0 \leq \tau \leq t$, the supremum on the right-hand side of (5.31) can be taken over $[t_0, t]$ which yields (5.27).[7] In the

---

[5] Another interesting use of inequality (5.27) will be given shortly in Lemma 5.6.

[6] For autonomous systems, it has been shown that the conditions of Theorem 5.2 are also necessary; see [164].

[7] In particular, repeat the above argument over the period $[0, T]$ to show that

$$\|x(\sigma)\| \leq \beta(\|x(t_0)\|, \sigma - t_0) + \gamma \left(\sup_{t_0 \leq \tau \leq T} \|u(\tau)\| \right), \quad \forall \ t_0 \leq \sigma \leq T$$

Then, set $\sigma = T = t$.

global case, the function $\alpha_2^{-1} \circ \alpha_1$ belongs to class $\mathcal{K}_\infty$. Hence, for any initial state $x(t_0)$ and any bounded input $u(t)$, we can choose r and r_u large enough that the inequalities (5.30) are satisfied.   □

The next two lemmas are immediate consequences of converse Lyapunov theorems.

Lemma 5.4 *Suppose that, in some neighborhood of* $(x = 0, u = 0)$*, the function* $f(t, x, u)$ *is continuously differentiable and the Jacobian matrices* $[\partial f/\partial x]$ *and* $[\partial f/\partial u]$ *are bounded, uniformly in* t*. If the unforced system (5.25) has a uniformly asymptotically stable equilibrium point at the origin* $x = 0$*, then the system (5.24) is locally input-to-state stable.*   ◇

Proof: (The converse Lyapunov) Theorem 3.14 shows that the unforced system (5.25) has a Lyapunov function $V(t, x)$ that satisfies (5.20)–(5.22) over some bounded neighborhood of $x = 0$. Since $[\partial f/\partial u]$ is bounded, the perturbation term satisfies (5.26) for all $t \geq t_0$ and all (x, u) in some bounded neighborhood of $(x = 0, u = 0)$. It can be verified that $V(t, x)$ satisfies the conditions of Theorem 5.2 in some neighborhood of $(x = 0, u = 0)$.   □

In the case of autonomous systems, the boundedness assumptions of Lemma 5.4 follow from continuous differentiability of $f(x, u)$. Therefore, for autonomous systems the lemma says that *if* $f(x, u)$ *is continuously differentiable and the origin of (5.25) is asymptotically stable, then (5.24) is locally input-to-state stable.*

Lemma 5.5 *Suppose that* $f(t, x, u)$ *is continuously differentiable and globally Lipschitz in* (x, u)*, uniformly in* t*. If the unforced system (5.25) has a globally exponentially stable equilibrium point at the origin* $x = 0$*, then the system (5.24) is input-to-state stable.*   ◇

Proof: (The converse Lyapunov) Theorem 3.12 shows that the unforced system (5.25) has a Lyapunov function $V(t, x)$ that satisfies (5.3)–(5.5) globally. Due to the uniform global Lipschitz property of f, the perturbation term satisfies (5.26) for all $t \geq t_0$ and all (x, u). It can be verified that $V(t, x)$ satisfies the conditions of Theorem 5.2 globally.   □

Recalling the last paragraph of the preceding section and Exercise 5.19, it is clear that if the origin of the unforced system (5.25) is globally uniformly asymptotically stable but not globally exponentially stable, then the system (5.24) is not necessarily input-to-state stable even when f is globally Lipschitz in (x, u). We illustrate the use of Theorem 5.2 by three first-order examples.

Example 5.6 The system

$$\dot{x} = -x^3 + u$$

has a globally asymptotically stable origin when $u = 0$. Taking $V = \frac{1}{2}x^2$, the derivative of V along the trajectories of the system is given by

$$\dot{V} = -x^4 + xu = -(1 - \theta)x^4 - \theta x^4 + xu \leq -(1 - \theta)x^4, \quad \forall \, |x| \geq \left(\frac{|u|}{\theta}\right)^{1/3}$$

where θ is any constant such that $0 < \theta < 1$. Hence, the system is input-to-state stable with $\gamma(a) = (a/\theta)^{1/3}$. $\triangle$

Example 5.7 The system

$$\dot{x} = f(x, u) = -x - 2x^3 + (1 + x^2)u^2$$

has a globally exponentially stable origin when $u = 0$, but Lemma 5.5 does not apply since f is not globally Lipschitz. Taking $V = \frac{1}{2}x^2$, we obtain

$$\dot{V} = -x^2 - 2x^4 + x(1 + x^2)u^2 \leq -x^4, \quad \forall \, |x| \geq u^2$$

Hence, the system is input-to-state stable with $\gamma(a) = a^2$. $\triangle$

Example 5.8 Consider the system

$$\dot{x} = f(x, u) = -x + (1 + x^2)u$$

Similar to the previous example, when $u = 0$ the origin is globally exponentially stable but Lemma 5.5 does not apply since f is not globally Lipschitz in x. This time, however, the system is not input-to-state stable as can be seen by taking $u(t) \equiv 1$. The solution of the resulting system

$$\dot{x} = x^2 - x + 1$$

that starts at $x(0) = 0$ diverges to ∞; note that $\dot{x} \geq \frac{3}{4}$. According to Lemma 5.4, the system is locally input-to-state stable. Theorem 5.2 can be used to estimate the bounds on the initial state and input (the constants k_1 and k_2 in Definition 5.2). Let $D = \{|x| < r\}$ and $D_u = R$. With $V(x) = \frac{1}{2}x^2$, we have

$$
\begin{aligned}
\dot{V} &= -x^2 + x(1 + x^2)u \leq -(1 - \theta)x^2 - \theta x^2 + |x|(1 + r^2)|u| \\
&\leq -(1 - \theta)x^2, \quad \forall \, \frac{(1 + r^2)|u|}{\theta} \leq |x| < r
\end{aligned}
$$

where $0 < \theta < 1$. Thus, the system is locally input-to-state stable with $k_1 = r$, $k_2 = r\theta/(1 + r^2)$, and $\gamma(a) = a(1 + r^2)/\theta$. $\triangle$

An interesting application of the concept of input-to-state stability arises in the stability analysis of the interconnected system

$$\dot{x}_1 = f_1(t, x_1, x_2) \tag{5.32}$$

$$\dot{x}_2 = f_2(t, x_2) \tag{5.33}$$

where $f_1 : [0, \infty) \times D_1 \times D_2 \to R^{n_1}$ and $f_2 : [0, \infty) \times D_2 \to R^{n_2}$ are piecewise continuous in t and locally Lipschitz in x;

$$x \stackrel{\text{def}}{=} \left[\begin{array}{c} x_1 \\ x_2 \end{array} \right]$$

The set D_i is a domain in R^{n_i} that contains the origin $x_i = 0$; in the global case, we take $D_i = R^{n_i}$. Suppose that

$$\dot{x}_1 = f_1(t, x_1, 0)$$

has a uniformly asymptotically stable equilibrium point at $x_1 = 0$, and (5.33) has a uniformly asymptotically stable equilibrium point at $x_2 = 0$. It is intuitively clear that if the input x_2 into the $\dot{x}_1$–equation is "well behaved" in some sense, then the interconnected system (5.32)–(5.33) will have a uniformly asymptotically stable equilibrium point at $x = 0$. The following lemma shows that this will be the case if (5.32), with x_2 viewed as input, is input-to-state stable.

Lemma 5.6 *Under the stated assumptions,*

- *If the system (5.32), with x_2 as input, is locally input-to-state stable and the origin of (5.33) is uniformly asymptotically stable, then the origin of the interconnected system (5.32)–(5.33) is uniformly asymptotically stable.*

- *If the system (5.32), with x_2 as input, is input-to-state stable and the origin of (5.33) is globally uniformly asymptotically stable, then the origin of the interconnected system (5.32)–(5.33) is globally uniformly asymptotically stable.*

$$\diamond$$

Proof: We prove it only for the global case; the local case is similar except for restricting the domain of analysis. Let $t_0 \geq 0$ be the initial time. The solutions of (5.32) and (5.33) satisfy

$$\|x_1(t)\| \leq \beta_1(\|x_1(s)\|, t - s) + \gamma_1 \left(\sup_{s \leq \tau \leq t} \|x_2(\tau)\| \right) \tag{5.34}$$

$$\|x_2(t)\| \leq \beta_2(\|x_2(s)\|, t - s) \tag{5.35}$$

globally, where $t \geq s \geq t_0$, β_1, β_2 are class $\mathcal{KL}$ functions and γ_1 is a class $\mathcal{K}$ function. Apply (5.34) with $s = (t + t_0)/2$ to obtain

$$\|x_1(t)\| \leq \beta_1 \left(\left\| x_1 \left(\frac{t + t_0}{2} \right) \right\|, \frac{t - t_0}{2} \right) + \gamma_1 \left(\sup_{\frac{t+t_0}{2} \leq \tau \leq t} \|x_2(\tau)\| \right) \qquad (5.36)$$

To estimate $x_1 \left(\frac{t+t_0}{2} \right)$, apply (5.34) with $s = t_0$ and t replaced by $(t+t_0)/2$ to obtain

$$\left\| x_1 \left(\frac{t + t_0}{2} \right) \right\| \leq \beta_1 \left(\|x_1(t_0)\|, \frac{t - t_0}{2} \right) + \gamma_1 \left(\sup_{t_0 \leq \tau \leq \frac{t+t_0}{2}} \|x_2(\tau)\| \right) \qquad (5.37)$$

Using (5.35), we obtain

$$\sup_{t_0 \leq \tau \leq \frac{t+t_0}{2}} \|x_2(\tau)\| \quad \leq \quad \beta_2(\|x_2(t_0)\|, 0) \qquad (5.38)$$

$$\sup_{\frac{t+t_0}{2} \leq \tau \leq t} \|x_2(\tau)\| \quad \leq \quad \beta_2 \left(\|x_2(t_0)\|, \frac{t - t_0}{2} \right) \qquad (5.39)$$

Substituting (5.37)–(5.39) into (5.36) and using the inequalities

$$\|x_1(t_0)\| \leq \|x(t_0)\|, \quad \|x_2(t_0)\| \leq \|x(t_0)\|, \quad \|x(t)\| \leq \|x_1(t)\| + \|x_2(t)\|$$

yield

$$\|x(t)\| \leq \beta(\|x(t_0)\|, t - t_0)$$

where

$$\beta(r, s) = \beta_1 \left(\beta_1 \left(r, \frac{s}{2} \right) + \gamma_1 \left(\beta_2 \left(r, 0 \right) \right), \frac{s}{2} \right) + \gamma_1 \left(\beta_2 \left(r, \frac{s}{2} \right) \right) + \beta_2(r, s)$$

It can be easily verified that β is a class $\mathcal{KL}$ function for all $r \geq 0$. Hence, the origin of (5.32)–(5.33) is globally uniformly asymptotically stable. $\qquad \Box$

5.4   Comparison Method

Consider the perturbed system (5.1). Let $V(t, x)$ be a Lyapunov function for the nominal system (5.2) and suppose the derivative of V along the trajectories of (5.1) satisfies the differential inequality

$$\dot{V} \leq h(t, V)$$

By the comparison lemma (Lemma 2.5),

$$V(t, x(t)) \leq y(t)$$

where $y(t)$ is the solution of the differential equation

$$\dot{y} = h(t, y), \quad y(t_0) = V(t_0, x(t_0))$$

This approach is particularly useful when the differential inequality is linear, that is, when $h(t, V) = a(t)V + b(t)$, for then we can write down a closed-form expression for the solution of the first-order linear differential equation of y. Arriving at a linear differential inequality is possible when the origin of the nominal system (5.2) is exponentially stable.

Let $V(t, x)$ be a Lyapunov function of the nominal system (5.2) that satisfies (5.3)–(5.5) for all $(t, x) \in [0, \infty) \times D$, where $D = \{x \in R^n \mid \|x\| < r\}$. Suppose the perturbation term $g(t, x)$ satisfies the bound

$$\|g(t, x)\| \leq \gamma(t)\|x\| + \delta(t), \quad \forall \, t \geq 0, \, \forall \, x \in D \tag{5.40}$$

where $\gamma : R \to R$ is nonnegative and continuous for all $t \geq 0$, and $\delta : R \to R$ is nonnegative, continuous, and bounded for all $t \geq 0$. The derivative of V along the trajectories of (5.1) satisfies

$$
\begin{aligned}
\dot{V}(t, x) &= \frac{\partial V}{\partial t} + \frac{\partial V}{\partial x}f(t, x) + \frac{\partial V}{\partial x}g(t, x) \\
&\leq -c_3\|x\|^2 + \left\| \frac{\partial V}{\partial x} \right\| \, \|g(t, x)\| \\
&\leq -c_3\|x\|^2 + c_4\gamma(t)\|x\|^2 + c_4\delta(t)\|x\| \tag{5.41}
\end{aligned}
$$

Using (5.3), we can majorize $\dot{V}$ as

$$\dot{V} \leq -\left[\frac{c_3}{c_2} - \frac{c_4}{c_1}\gamma(t) \right] V + c_4\delta(t)\sqrt{\frac{V}{c_1}}$$

To obtain a linear differential inequality, we take $W(t) = \sqrt{V(t, x(t))}$ and use the fact $\dot{W} = \dot{V}/2\sqrt{V}$, when $V(t) \neq 0$, to obtain

$$\dot{W} \leq -\frac{1}{2}\left[\frac{c_3}{c_2} - \frac{c_4}{c_1}\gamma(t) \right] W + \frac{c_4}{2\sqrt{c_1}}\delta(t) \tag{5.42}$$

When $V(t) = 0$, it can be shown[8] that $D^+W(t) \leq c_4\delta(t)/2\sqrt{c_1}$. Hence, $D^+W(t)$ satisfies (5.42) for all values of $V(t)$. By the comparison lemma, $W(t)$ satisfies the inequality

$$W(t) \leq \phi(t, t_0)W(t_0) + \frac{c_4}{2\sqrt{c_1}} \int_{t_0}^{t} \phi(t, \tau)\delta(\tau) \, d\tau \tag{5.43}$$

---

[8] See Exercise 5.27.

where the transition function $\phi(t, \tau)$ is given by

$$\phi(t, \tau) = \exp\left[-\frac{c_3}{2c_2}(t - t_0) + \frac{c_4}{2c_1}\int_{t_0}^{t}\gamma(\tau)\,d\tau\right]$$

Using (5.3) in (5.43), we obtain

$$\|x(t)\| \leq \sqrt{\frac{c_2}{c_1}}\phi(t, t_0)\|x(t_0)\| + \frac{c_4}{2c_1}\int_{t_0}^{t}\phi(t, \tau)\delta(\tau)\,d\tau \qquad (5.44)$$

Suppose now that $\gamma(t)$ satisfies the condition

$$\int_{t_0}^{t}\gamma(\tau)\,d\tau \leq \epsilon(t - t_0) + \eta \qquad (5.45)$$

for some nonnegative constants ϵ and η, where

$$\epsilon < \frac{c_1 c_3}{c_2 c_4} \qquad (5.46)$$

Defining the constants α and ρ by

$$\alpha = \frac{1}{2}\left[\frac{c_3}{c_2} - \epsilon\frac{c_4}{c_1}\right] > 0, \quad \rho = \exp\left(\frac{c_4\eta}{2c_1}\right) \geq 1 \qquad (5.47)$$

and using (5.45)–(5.46) in (5.44), we obtain

$$\|x(t)\| \leq \sqrt{\frac{c_2}{c_1}}\rho\|x(t_0)\|e^{-\alpha(t-t_0)} + \frac{c_4\rho}{2c_1}\int_{t_0}^{t}e^{-\alpha(t-\tau)}\delta(\tau)\,d\tau \qquad (5.48)$$

For this bound to be valid, we must ensure that $\|x(t)\| < r$ for all $t \geq t_0$. Noting that[9]

$$
\begin{aligned}
\|x(t)\| &\leq \sqrt{\frac{c_2}{c_1}}\rho\|x(t_0)\|e^{-\alpha(t-t_0)} + \frac{c_4\rho}{2\alpha c_1}\left[1 - e^{-\alpha(t-t_0)}\right]\sup_{t \geq t_0}\delta(t) \\
&\leq \max\left\{\sqrt{\frac{c_2}{c_1}}\rho\|x(t_0)\|, \frac{c_4\rho}{2\alpha c_1}\sup_{t \geq t_0}\delta(t)\right\}
\end{aligned}
$$

we see that the condition $\|x(t)\| < r$ will be satisfied if

$$\|x(t_0)\| < \frac{r}{\rho}\sqrt{\frac{c_1}{c_2}} \qquad (5.49)$$

---

[9] We use the fact that the function $ae^{-\alpha t} + b(1 - e^{-\alpha t})$, with positive a, b, and α, relaxes monotonically from its initial value a to its steady-state value b, Hence, it is bounded by the maximum of the two numbers.

and

$$\sup_{t \geq t_0} \delta(t) < \frac{2c_1 \alpha r}{c_4 \rho} \qquad (5.50)$$

For easy reference, we summarize our findings in the following lemma.

Lemma 5.7 *Let $x = 0$ be an exponentially stable equilibrium point of the nominal system (5.2). Let $V(t, x)$ be a Lyapunov function of the nominal system that satisfies (5.3)–(5.5) in $[0, \infty) \times D$, where $D = \{x \in R^n \mid \|x\|_2 < r\}$. Suppose the perturbation term $g(t, x)$ satisfies (5.40), where $\gamma(t)$ satisfies (5.45)–(5.46). Then, provided $x(t_0)$ satisfies (5.49) and $\sup_{t \geq t_0} \delta(t)$ satisfies (5.50), the solution of the perturbed system (5.1) satisfies (5.48). Furthermore, if all the assumptions hold globally, then (5.48) is satisfied for any $x(t_0)$ and any bounded $\delta(t)$.* ◇

Specializing the foregoing lemma to the case of vanishing perturbations, that is, when $\delta(t) \equiv 0$, we obtain the following result.

Corollary 5.4 *Let $x = 0$ be an exponentially stable equilibrium point of the nominal system (5.2). Let $V(t, x)$ be a Lyapunov function of the nominal system that satisfies (5.3)–(5.5) in $[0, \infty) \times D$. Suppose the perturbation term $g(t, x)$ satisfies*

$$\|g(t, x)\| \leq \gamma(t)\|x\|$$

where $\gamma(t)$ satisfies (5.45)–(5.46). Then, the origin is an exponentially stable equilibrium point of the perturbed system (5.1). Moreover, if all the assumptions hold globally, then the origin is globally exponentially stable. ◇

If $\gamma(t) \equiv \gamma = \text{constant}$, then Corollary 5.4 requires γ to satisfy the bound

$$\gamma < \frac{c_1 c_3}{c_2 c_4}$$

which has no advantage over the bound (5.7)

$$\gamma < \frac{c_3}{c_4}$$

required by Lemma 5.1, since $(c_1/c_2) \leq 1$. In fact, whenever $(c_1/c_2) < 1$, the current bound will be more conservative (that is, smaller) than (5.7). The advantage of Corollary 5.4 is seen in the case when the integral of $\gamma(t)$ satisfies the conditions (5.45)–(5.46), even when $\sup_{t \geq t_0} \gamma(t)$ is not small enough to satisfy (5.7). Three such cases are given in the following lemma.

Lemma 5.8

1. If

$$\int_{t_0}^{\infty} \gamma(\tau) \ d\tau \leq k$$

then (5.45) is satisfied with $\epsilon = 0$ and $\eta = k$.

2. If

$$\gamma(t) \rightarrow 0 \ \ \text{as} \ \ t \rightarrow \infty$$

then for any $\epsilon > 0$ there is $\eta = \eta(\epsilon) > 0$ such that (5.45) is satisfied.

3. If there are constants $\Delta > 0$, $T \geq 0$, and $\epsilon_1 > 0$ such that

$$\frac{1}{\Delta} \int_{t}^{t+\Delta} \gamma(\tau) \ d\tau \leq \epsilon_1, \ \ \forall \ t \geq t_0 + T$$

then (5.45) is satisfied with $\epsilon = \epsilon_1$ and $\eta = \epsilon_1 \Delta + T \max_{[t_0, t_0+T]} \gamma(t)$. $\diamond$

Proof: The first case is obvious. To prove the second case, note that for any $\epsilon > 0$, there is $T = T(\epsilon)$ such that $\gamma(t) < \epsilon$ for all $t \geq t_0 + T$. For $t \geq t_0 + T$,

$$\int_{t_0}^{t} \gamma(\tau) \ d\tau \leq T \max_{[t_0, t_0+T]} \gamma(\tau) + \epsilon(t - t_0 - T) \leq T \max_{[t_0, t_0+T]} \gamma(\tau) + \epsilon(t - t_0)$$

Thus, (5.45) is satisfied with the given ϵ and $\eta = T \max_{[t_0, t_0+T]} \gamma(\tau)$.
To prove the last case, consider $t \geq t_0 + T$ and let N be the integer for which $(N-1)\Delta \leq t - t_0 - T \leq N\Delta$. Let $k_1 = \max_{[t_0, t_0+T]} \gamma(\tau)$. Now,

$$
\begin{aligned}
\int_{t_0}^{t} \gamma(\tau) \ d\tau &= \int_{t_0}^{t_0+T} \gamma(\tau) \ d\tau + \int_{t_0+T}^{t} \gamma(\tau) \ d\tau \\
&= \int_{t_0}^{t_0+T} \gamma(\tau) \ d\tau + \sum_{i=0}^{i=N-2} \int_{t_0+T+i\Delta}^{t_0+T+(i+1)\Delta} \gamma(\tau) \ d\tau \\
&\quad + \int_{t_0+T+(N-1)\Delta}^{t} \gamma(\tau) \ d\tau \\
&\leq k_1 T + \sum_{i=0}^{i=N-2} \epsilon_1 \Delta + \epsilon_1 \Delta \ \leq \ k_1 T + \epsilon_1(t - t_0 - T) + \epsilon_1 \Delta \\
&\leq \epsilon_1(t - t_0) + \eta
\end{aligned}
$$

$\square$

It should be noted that in the first two cases of the foregoing lemma, the condition (5.45) is satisfied with $\epsilon = 0$ in the first case and ϵ can be made arbitrarily small in the second case. Therefore, the condition (5.46) is always satisfied in these two cases. In the third case of the lemma, the condition (5.46) requires smallness of a *moving average* of $\gamma(t)$, asymptotically as $t \to \infty$.

Example 5.9 Consider the linear system

$$\dot{x} = [A(t) + B(t)]x$$

where $A(t)$ and $B(t)$ are continuous and $A(t)$ is bounded on $[0, \infty)$. Suppose the origin is an exponentially stable equilibrium point of the nominal system

$$\dot{x} = A(t)x$$

and

$$B(t) \to 0 \text{ as } t \to \infty$$

From Theorem 3.10, we know that there is a quadratic Lyapunov function $V(t, x) = x^T P(t)x$ that satisfies (5.3)–(5.5) globally. The perturbation term $B(t)x$ satisfies the inequality

$$\|B(t)x\| \leq \|B(t)\| \, \|x\|$$

Since $\|B(t)\| \to 0$ as $t \to \infty$, we conclude from Corollary 5.4 and the second case of Lemma 5.8 that the origin is a globally exponentially stable equilibrium of the perturbed system. $\triangle$

Similar conclusions can be drawn when $\int_0^\infty \|B(t)\| \, dt < \infty$ (Exercise 5.28) and $\int_0^\infty \|B(t)\|^2 \, dt < \infty$ (Exercise 5.29).

In the case of nonvanishing perturbations, that is, when $\delta(t) \not\equiv 0$, the following lemma states a number of conclusions concerning the asymptotic behavior of $x(t)$ as $t \to \infty$.

Lemma 5.9 *Suppose the conditions of Lemma 5.7 are satisfied, and let $x(t)$ denote the solution of the perturbed system (5.1).*

1. If

$$\int_{t_0}^t e^{-\alpha(t-\tau)}\delta(\tau) \, d\tau \leq \beta, \quad \forall \, t \geq t_0$$

for some positive constant β, then $x(t)$ is uniformly ultimately bounded with the ultimate bound

$$b = \frac{c_4 \rho \beta}{2c_1 \theta}$$

where $\theta \in (0, 1)$ is an arbitrary positive constant.

2. If

$$\lim_{t \to \infty} \delta(t) = \delta_\infty$$

then $x(t)$ is uniformly ultimately bounded with the ultimate bound

$$b = \frac{c_4 \rho \delta_\infty}{2\alpha c_1 \theta}$$

where $\theta \in (0, 1)$ is an arbitrary positive constant.

3. If

$$\lim_{t \to \infty} \delta(t) = 0, \quad \text{then} \quad \lim_{t \to \infty} x(t) = 0$$

If the conditions of Lemma 5.7 are satisfied globally, then the foregoing statements hold for any initial state $x(t_0)$. $\diamond$

Proof: All three cases follow easily from inequality (5.48). $\square$

5.5   Continuity of Solutions on the Infinite Interval

In Section 2.3, we studied continuous dependence of the solution of the state equation on initial states and parameters. In particular, in Theorem 2.5 we examined the nominal system

$$\dot{x} = f(t, x) \tag{5.51}$$

and the perturbed system

$$\dot{x} = f(t, x) + g(t, x) \tag{5.52}$$

under the assumption that $\|g(t, x)\| \le \delta$ in the domain of interest. Using the Gronwall-Bellman inequality, we found that if $y(t)$ and $z(t)$ are well-defined solutions of the nominal and perturbed systems, respectively, then

$$\|y(t) - z(t)\| \le \|y(t_0) - z(t_0)\| \exp[L(t - t_0)] + \frac{\delta}{L}\{\exp[L(t - t_0)] - 1\} \tag{5.53}$$

where L is a Lipschitz constant for f. This bound is valid only on compact time intervals since the exponential term $\exp[L(t - t_0)]$ grows unbounded as $t \to \infty$. In fact, the bound is useful only on an interval $[t_0, t_1]$ where t_1 is reasonably small, for if t_1 is large, the bound will be too large to be of any use. This is not surprising, since in Section 2.3 we did not impose any stability conditions on the system. In this section, we use Theorem 5.1 to calculate a bound on the error between the solutions of (5.51) and (5.52) that is valid uniformly in t for all $t \ge t_0$.

Theorem 5.3 *Let $D = \{x \in R^n \mid \|x\| < r\}$ and suppose the following assumptions are satisfied for all $(t, x) \in [0, \infty) \times D$:*

- *$f(t, x)$ is continuously differentiable and the Jacobian matrix $\partial f / \partial x$ is bounded and Lipschitz in x, uniformly in t.*

- *The origin $x = 0$ is an exponentially stable equilibrium point of the nominal system (5.51).*

- *The perturbation term $g(t, x)$ is piecewise continuous in t and locally Lipschitz in x, and satisfies the bound*

$$\|g(t, x)\| \leq \delta, \quad \forall\, t \geq t_0 \geq 0, \ \forall\, x \in D \tag{5.54}$$

Let $y(t)$ and $z(t)$ denote solutions of the nominal system (5.51) and the perturbed system (5.52), respectively. Then, there exist positive constants β, γ, η, μ, λ, and k, independent of δ, such that if $\delta < \eta$, $\|y(t_0)\| < \lambda$, and $\|z(t_0) - y(t_0)\| < \mu$, then the solutions $y(t)$ and $z(t)$ will be uniformly bounded for all $t \geq t_0 \geq 0$ and

$$\|z(t) - y(t)\| \leq k e^{-\gamma(t - t_0)} \|z(t_0) - y(t_0)\| + \beta \delta \tag{5.55}$$

$\diamond$

Proof: By exponential stability of the origin, there are positive constants $\rho < r$ and $\lambda \leq \rho$ such that

$$\|y(t_0)\| < \lambda \ \Rightarrow\ \|y(t)\| < \rho, \quad \forall\, t \geq t_0$$

The error $e(t) = z(t) - y(t)$ satisfies the equation

$$\begin{aligned} \dot{e} &= \dot{z} - \dot{y} \\ &= f(t, z) + g(t, z) - f(t, y) \\ &= f(t, e) + \Delta F(t, e) + g(t, z) \end{aligned} \tag{5.56}$$

where

$$\Delta F(t, e) = f(t, y(t) + e) - f(t, y(t)) - f(t, e)$$

Noting that

$$\|e\| < r - \rho \ \Rightarrow\ \|e + y\| < r$$

we analyze the error equation (5.56) over the ball $\{\|e\| < r - \rho\}$. Equation (5.56) can be viewed as a perturbation of the system

$$\dot{e} = f(t, e)$$

whose origin is exponentially stable. By Theorem 3.12, there exists a Lyapunov function $V(t, e)$ which satisfies (5.3)–(5.5) for $\|e\| < r_0 < r - \rho$. By the mean value theorem, the error term ΔF_i can be written as

$$\Delta F_i(t, e) = \left[\frac{\partial f_i}{\partial x}(t, \lambda_1 e + y) - \frac{\partial f_i}{\partial x}(t, \lambda_2 e)\right] e$$

where $0 < \lambda_i < 1$. Since the Jacobian matrix $[\partial f / \partial x]$ is Lipschitz in x, uniformly in t, we have

$$\|\Delta F\| \le L_1 \|e\|_2^2 + L_2 \|e\| \|y(t)\|$$

Thus, the perturbation term $(\Delta F + g)$ satisfies

$$\|\Delta F(t, e) + g(t, z)\| \le L_1 \|e\|^2 + L_2 \|e\| k_1 \exp[-\gamma_1(t - t_0)] + \delta$$

where we have used the fact that the solution of the nominal system decays to zero exponentially fast. Now,

$$\|\Delta F(t, e) + g(t, z)\| \le \{L_1 r_1 + L_2 k_1 \exp[-\gamma_1(t - t_0)]\} \|e\| + \delta$$

for all $\|e\| \le r_1 < r_0$. This inequality takes the form (5.40) with

$$\gamma(t) = \{L_1 r_1 + L_2 k_1 \exp[-\gamma_1(t - t_0)]\}, \quad \text{and} \quad \delta(t) \equiv \delta$$

The condition (5.45) is satisfied since

$$\int_{t_0}^{t} \gamma(\tau)\, d\tau \le L_1 r_1 (t - t_0) + \frac{L_2 k_1}{\gamma_1}$$

By taking r_1 small enough, we can satisfy the condition (5.46). Thus, all the assumptions of Lemma 5.7 are satisfied, and inequality (5.55) follows from (5.48).
□

5.6 Interconnected Systems

When we analyze the stability of a nonlinear dynamical system, the complexity of the analysis grows rapidly as the order of the system increases. This situation motivates us to look for ways to simplify the analysis. If the system can be modeled as an interconnection of lower-order subsystems, then we may pursue the stability analysis in two steps. In the first step, we decompose the system into smaller isolated subsystems by ignoring interconnections, and analyze the stability of each subsystem. In the second step, we combine our conclusions from the first step with information about the interconnections to draw conclusions about the stability of the interconnected system. In this section, we shall illustrate how this idea can be utilized in searching for Lyapunov functions for interconnected systems.

Consider the interconnected system

$$\dot{x}_i = f_i(t, x_i) + g_i(t, x), \quad i = 1, 2, \ldots, m \tag{5.57}$$

where $x_i(\cdot) \in R^{n_i}$, $n_1 + \cdots + n_m = n$, and $x = \left(x_1^T, \ldots, x_m^T\right)^T$. Suppose that $f_i(\cdot, \cdot)$ and $g_i(\cdot, \cdot)$ are smooth enough to ensure local existence and uniqueness of the solution for all initial conditions in a domain of interest, and that

$$f_i(t, 0) = 0, \quad g_i(t, 0) = 0, \quad \forall\, i$$

so that the origin $x = 0$ is an equilibrium point of the system. We want to study stability of the equilibrium $x = 0$ using Lyapunov's method. Ignoring the interconnection terms $g_i(\cdot, \cdot)$, the system decomposes into m isolated subsystems

$$\dot{x}_i = f_i(t, x_i) \tag{5.58}$$

with each one having equilibrium at its origin $x_i = 0$. We start by searching for Lyapunov functions that establish uniform asymptotic stability of the origin for each isolated subsystem. Suppose this search has been successful and that, for each subsystem, we have a positive definite decrescent Lyapunov function $V_i(t, x_i)$ whose derivative along the trajectories of the isolated subsystem (5.58) is negative definite. The function

$$V(t, x) = \sum_{i=1}^{m} d_i V_i(t, x_i), \quad d_i > 0$$

is a *composite Lyapunov function* for the collection of the m isolated subsystems for all values of the positive constants d_i. Viewing the interconnected system (5.57) as a perturbation of the isolated subsystems (5.58), it is reasonable to try $V(t, x)$ as a Lyapunov function candidate for (5.57). The derivative of $V(t, x)$ along the trajectories of (5.57) is given by

$$\dot{V}(t, x) = \sum_{i=1}^{m} d_i \left[\frac{\partial V_i}{\partial t} + \frac{\partial V_i}{\partial x_i} f_i(t, x_i) + \frac{\partial V_i}{\partial x_i} g_i(t, x)\right]$$

The term

$$\frac{\partial V_i}{\partial t} + \frac{\partial V_i}{\partial x_i} f_i(t, x_i)$$

is negative definite by virtue of the fact that V_i is a Lyapunov function for the ith isolated subsystem, but the term

$$\frac{\partial V_i}{\partial x_i} g_i(t, x)$$

is, in general, indefinite. The situation is similar to our earlier investigation of perturbed systems in Section 5.1. Therefore, we may approach the problem by performing worst-case analysis where the term $[\partial V_i / \partial x_i] g_i$ is majorized by a non-negative upper bound. Let us illustrate the idea using the quadratic-type Lyapunov functions introduced in Section 5.1. Let $B = \{ x \in R^n \mid \|x\| < r \}$. Suppose that, for $i = 1, 2, \ldots, m$, $V_i(t, x_i)$ satisfies

$$\frac{\partial V_i}{\partial t} + \frac{\partial V_i}{\partial x_i} f_i(t, x_i) \leq -\alpha_i \phi_i^2(x_i) \tag{5.59}$$

$$\left\| \frac{\partial V_i}{\partial x_i} \right\| \leq \beta_i \phi_i(x_i) \tag{5.60}$$

for all $(t, x) \in [0, \infty) \times B$ for some positive constants α_i and β_i, where $\phi_i : R^{n_i} \to R$ are positive definite and continuous. Furthermore, suppose that the interconnection terms $g_i(t, x)$ satisfy the bound

$$\| g_i(t, x) \| \leq \sum_{j=1}^{m} \gamma_{ij} \phi_j(x_j) \tag{5.61}$$

for all $(t, x) \in [0, \infty) \times B$ for some nonnegative constants γ_{ij}. Then, the derivative of $V(t, x) = \sum_{i=1}^{m} d_i V_i(t, x_i)$ along the trajectories of the interconnected system (5.57) satisfies the inequality

$$\dot{V}(t, x) \leq \sum_{i=1}^{m} d_i \left[-\alpha_i \phi_i^2(x_i) + \sum_{j=1}^{m} \beta_i \gamma_{ij} \phi_i(x_i) \phi_j(x_j) \right]$$

The right-hand side is a quadratic form in $\phi_1, \ldots, \phi_m$, which we can rewrite as

$$\dot{V}(t, x) \leq -\tfrac{1}{2} \phi^T (DS + S^T D) \phi$$

where

$$\phi = (\phi_1, \ldots, \phi_m)^T$$
$$D = \mathrm{diag}\,(d_1, \ldots, d_m)$$

and S is an $m \times m$ matrix whose elements are defined by

$$s_{ij} = \begin{cases} \alpha_i - \beta_i \gamma_{ii}, & i = j \\[2mm] -\beta_i \gamma_{ij}, & i \neq j \end{cases} \tag{5.62}$$

If there is a positive diagonal matrix D such that

$$DS + S^T D > 0$$

then $\dot{V}(t, x)$ is negative definite since $\phi(x) = 0$ if and only if $x = 0$; recall that $\phi_i(x_i)$ is a positive definite function of x_i. Thus, a sufficient condition for uniform asymptotic stability of the origin as an equilibrium point of the interconnected system is the existence of a positive diagonal matrix D such that $DS + S^T D$ is positive definite. The matrix S is special in that its off-diagonal elements are nonpositive. The following lemma applies to this class of matrices.

Lemma 5.10 *There exists a positive diagonal matrix D such that $DS + S^T D$ is positive definite if and only if S is an M-matrix; that is, the leading principal minors of S are positive:*

$$
det \begin{bmatrix} s_{11} & s_{12} & \cdots & s_{1k} \\ s_{21} & & & \\ \vdots & & & \\ s_{k1} & \cdots & \cdots & s_{kk} \end{bmatrix} > 0, \quad k = 1, 2, \ldots, m
$$

$\diamond$

Proof: See [48].

The M-matrix condition can be interpreted as a requirement that the diagonal elements of S be "larger as a whole" than the off-diagonal elements. It can be seen (Exercise 5.34) that diagonally dominant matrices with nonpositive off-diagonal elements are M-matrices. The diagonal elements of S are measures of the "degree of stability" for the isolated subsystems in the sense that the constant α_i gives a lower bound on the rate of decrease of the Lyapunov function V_i with respect to $\phi_i^2(x_i)$. The off-diagonal elements of S represent the "strength of the interconnections" in the sense that they give an upper bound on $g_i(t, x)$ with respect to $\phi_j(x_j)$ for $j = 1, \ldots, m$. Thus, the M-matrix condition says that *if the degrees of stability for the isolated subsystems are larger as a whole than the strength of the interconnections, then the interconnected system has a uniformly asymptotically stable equilibrium at the origin*. We summarize our conclusion in the following theorem.

Theorem 5.4 *Consider the system (5.57) and suppose there are positive definite decrescent Lyapunov functions $V_i(t, x_i)$ that satisfy (5.59)–(5.60) and that $g_i(t, x)$ satisfies (5.61) in $[0, \infty) \times B$. Suppose that the matrix S defined by (5.62) is an M-matrix. Then, $x = 0$ is a uniformly asymptotically stable equilibrium point of the system. Moreover, if all the assumptions hold globally and $V_i(t, x_i)$ are radially unbounded, then the origin is globally uniformly asymptotically stable.* $\diamond$

Example 5.10 Consider the fifth-order system

$$
\dot{z}_k = -\rho_k z_k + \sigma, \quad k = 1, 2, 3, 4
$$

$$
\dot{\sigma} = \sum_{k=1}^{4} \eta_k z_k - \zeta \sigma - h(\sigma)
$$

where $\rho_k > 0$, $\zeta > 0$, and η_k are constants and x_k and σ are scalar variables. The function $h : R \to R$ is locally Lipschitz, $h(0) = 0$, and $\sigma h(\sigma) > 0$ for all $\sigma \neq 0$. The system has an equilibrium point at the origin. To investigate stability of this equilibrium point, let us view this system as an interconnection of two isolated subsystems. The first subsystem is

$$\dot{z}_k = -\rho_k z_k, \quad k = 1, 2, 3, 4$$

while the second one is

$$\dot{\sigma} = -\zeta \sigma - h(\sigma)$$

The system can be represented in the form (5.57) with

$$x_1 = (z_1, z_2, z_3, z_4)^T, \quad x_2 = \sigma$$

$$f_1(x_1) = -A_1 x_1, \quad A_1 = \text{diag}(\rho_1, \rho_2, \rho_3, \rho_4)$$

$$f_2(x_2) = -\zeta x_2 - h(x_2)$$

$$g_1(x_2) = \ell^T x_2, \quad \ell^T = [1, 1, 1, 1], \quad g_2(x_1) = \eta^T x_1$$

where $\eta = (\eta_1, \eta_2, \eta_3, \eta_4)^T$. A Lyapunov function for the first isolated subsystem can be taken as $V_1(x_1) = \frac{1}{2} x_1^T x_1$, which satisfies (5.59)–(5.60) with

$$\alpha_1 = \rho_{\min} \stackrel{\text{def}}{=} \min\{\rho_1, \rho_2, \rho_3, \rho_4\}, \quad \beta_1 = 1, \quad \phi_1(x_1) = \|x_1\|_2$$

A Lyapunov function for the second isolated subsystem can be taken as $V_2(x_2) = \frac{1}{2} x_2^2$, which satisfies (5.59)–(5.60) with

$$\alpha_2 = \zeta, \quad \beta_2 = 1, \quad \phi_2(x_2) = |x_2|$$

where we have used the property $\sigma h(\sigma) \geq 0$. The interconnection terms satisfy (5.61) with

$$\gamma_{11} = 0 \qquad \gamma_{12} = \|\ell\|_2 = 2$$
$$\gamma_{21} = \|\eta\|_2 \qquad \gamma_{22} = 0$$

Note that (5.59)–(5.61) are satisfied globally. The matrix S, given by

$$S = \begin{bmatrix} \rho_{\min} & -2 \\ -\|\eta\|_2 & \zeta \end{bmatrix}$$

is an M-matrix if

$$\|\eta\|_2 < \tfrac{1}{2} \zeta \rho_{\min}$$

Thus, we conclude that if the interconnection coefficients η_1 to η_4 are small enough to satisfy the foregoing inequality, then the origin is globally asymptotically stable. Actually, in this example the origin is exponentially stable; see Exercise 5.35.   △

Example 5.11 Consider the second-order system

$$\dot{x}_1 = -x_1 - 1.5x_1^2 x_2^3$$
$$\dot{x}_2 = -x_2^3 + 0.5x_1^2 x_2^2$$

The system can be represented in the form (5.57) with

$$f_1(x_1) = -x_1, \quad g_1(x) = -1.5x_1^2 x_2^3$$

$$f_2(x_2) = -x_2^3, \quad g_2(x) = 0.5x_1^2 x_2^2$$

The first isolated subsystem $\dot{x}_1 = -x_1$ has a Lyapunov function $V_1(x_1) = \frac{1}{2}x_1^2$, which satisfies

$$\frac{\partial V_1}{\partial x_1} f_1(x_1) = -x_1^2 = -\alpha_1 \phi_1^2(x_1)$$

where $\alpha_1 = 1$ and $\phi_1(x_1) = |x_1|$. The second isolated subsystem $\dot{x}_2 = -x_2^3$ has a Lyapunov function $V_2(x_2) = \frac{1}{4}x_2^4$, which satisfies

$$\frac{\partial V_2}{\partial x_2} f_2(x_2) = -x_2^6 = -\alpha_2 \phi_2^2(x_2)$$

where $\alpha_2 = 1$ and $\phi_2(x_2) = |x_2|^3$. The Lyapunov functions satisfy (5.60) with $\beta_1 = \beta_2 = 1$. The interconnection term $g_1(x)$ satisfies the inequality

$$|g_1(x)| = 1.5x_1^2 |x_2|^3 \leq 1.5c_1^2 \phi_2(x_2)$$

for all $|x_1| \leq c_1$. The interconnection term $g_2(x)$ satisfies the inequality

$$|g_2(x)| = 0.5x_1^2 x_2^2 \leq 0.5c_1 c_2^2 \phi_1(x_1)$$

for all $|x_1| \leq c_1$ and $|x_2| \leq c_2$. Thus, if we restrict our attention to a set G defined by

$$G = \{x \in R^2 \mid |x_1| \leq c_1, \ |x_2| \leq c_2\}$$

we can conclude that the interconnection terms satisfy (5.61) with

$$\gamma_{11} = 0 \qquad\qquad \gamma_{12} = 1.5c_1^2$$

$$\gamma_{21} = 0.5c_1 c_2^2 \qquad\qquad \gamma_{22} = 0$$

The matrix

$$S = \begin{bmatrix} 1 & -1.5c_1^2 \\ -0.5c_1 c_2^2 & 1 \end{bmatrix}$$

is an M-matrix if $0.75c_1^3c_2^2 < 1$. This will be the case, for example, when $c_1 = c_2 = 1$. Thus, the origin is asymptotically stable. If we are interested in estimating the region of attraction of the origin, we need to know the composite Lyapunov function $V = d_1V_1 + d_2V_2$; that is, we need to know a positive diagonal matrix D such that

$$DS + S^T D > 0$$

Taking $c_1 = c_2 = 1$, we have

$$DS + S^T D = \begin{bmatrix} 2d_1 & -1.5d_1 - 0.5d_2 \\ -1.5d_1 - 0.5d_2 & 2d_2 \end{bmatrix}$$

which is positive definite for $1 < d_2/d_1 < 9$. Since there is no loss of generality in multiplying a Lyapunov function by a positive constant, we take $d_1 = 1$ and write the composite Lyapunov function as

$$V(x) = \tfrac{1}{2}x_1^2 + \tfrac{1}{4}d_2x_2^4, \quad 1 < d_2 < 9$$

An estimate of the region of attraction is given by

$$\Omega_c = \{x \in R^2 \mid V(x) \le c\}$$

where c is chosen small enough to ensure that Ω_c is inside the rectangle $|x_i| \le 1$. It can be easily seen that the choice of c must satisfy $c \le \min\{1/2, d_2/4\}$. Therefore, to maximize the value of c we want to choose $d_2 \ge 2$. On the other hand, a large value of d_2 will shrink the region of attraction in the direction of the x_2-axis. Therefore, we choose $d_2 = 2$ and $c = \tfrac{1}{2}$ to obtain the following estimate of the region of attraction

$$\Omega_{0.5} = \{x \in R^2 \mid x_1^2 + x_2^4 \le 1\}$$

$$\triangle$$

Example 5.12 The mathematical model of an artificial neural network was presented in Section 1.1.5, and its stability properties were analyzed in Example 3.11 using LaSalle's invariance principle. A key assumption in Example 3.11 is the symmetry requirement $T_{ij} = T_{ji}$, which allows us to represent the right-hand side of the state equation as the gradient of a scalar function. Let us relax this requirement and allow $T_{ij} \ne T_{ji}$. We shall analyze the stability properties of the network by viewing it as an interconnection of subsystems; each subsystem corresponds to one neuron. We find it convenient here to work with the voltages at the amplifier inputs u_i. The equations of motion are

$$C_i \dot{u}_i = \sum_j T_{ij} g_j(u_j) - \frac{1}{R_i} u_i + I_i$$

for $i = 1, 2, \ldots, n$ where $g_i(\cdot)$ are sigmoid functions, I_i are constant current inputs, $R_i > 0$, and $C_i > 0$. We assume that the system has a finite number of isolated equilibrium points. Each equilibrium point u^* satisfies the equation

$$0 = \sum_j T_{ij} g_j(u_j^*) - \frac{1}{R_i} u_i^* + I_i$$

To analyze the stability properties of a given equilibrium point u^*, we shift the equilibrium to the origin. Let $x_i = u_i - u_i^*$.

$$
\begin{aligned}
\dot{x}_i &= \frac{1}{C_i} \dot{u}_i = \frac{1}{C_i} \left[\sum_j T_{ij} g_j(x_j + u_j^*) - \frac{1}{R_i}(x_i + u_i^*) + I_i \right] \\
&= \frac{1}{C_i} \left[\sum_j T_{ij} \left(g_j(x_j + u_j^*) - g_j(u_j^*) \right) - \frac{1}{R_i} x_i \right]
\end{aligned}
$$

Define

$$\eta_i(x_i) = g_i(x_i + u_i^*) - g_i(u_i^*)$$

and rewrite the state equation as

$$\dot{x}_i = \frac{1}{C_i} \left[\sum_j T_{ij} \eta_j(x_j) - \frac{1}{R_i} x_i \right]$$

Assume that $\eta_i(\cdot)$ satisfies the sector condition

$$\sigma^2 k_{i1} \leq \sigma \eta_i(\sigma) \leq \sigma^2 k_{i2}, \quad \text{for } \sigma \in [-r_i, r_i]$$

where k_{i1} and k_{i2} are positive constants. Figure 5.1 shows that such a sector condition is indeed satisfied when

$$g_i(u_i) = \frac{2V_M}{\pi} \tan^{-1} \frac{\lambda \pi u_i}{2V_M}, \quad \lambda > 0$$

We can recast this system in the form (5.57) with

$$
\begin{aligned}
f_i(x_i) &= -\frac{1}{C_i R_i} x_i + \frac{1}{C_i} T_{ii} \eta_i(x_i) \\
g_i(x) &= \frac{1}{C_i} \sum_{j \neq i} T_{ij} \eta_j(x_j)
\end{aligned}
$$

The isolated subsystems are given by

$$\dot{x}_i = f_i(x_i) = -\frac{1}{C_i R_i} x_i + \frac{1}{C_i} T_{ii} \eta_i(x_i)$$

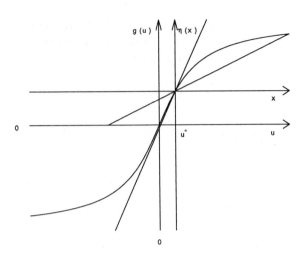

Figure 5.1: The sector nonlinearity $\eta_i(x_i)$ of Example 5.12.

Using

$$V_i(x_i) = \tfrac{1}{2}C_i x_i^2$$

as a Lyapunov function candidate, we obtain

$$\frac{\partial V_i}{\partial x_i} f_i(x_i) = -\frac{1}{R_i}x_i^2 + T_{ii}x_i\eta_i(x_i)$$

If $T_{ii} \leq 0$, then

$$T_{ii}x_i\eta_i(x_i) \leq -|T_{ii}|k_{i1}x_i^2$$

and

$$\frac{\partial V_i}{\partial x_i} f_i(x_i) \leq -\left(\frac{1}{R_i} + |T_{ii}|k_{i1}\right)x_i^2$$

which is negative definite. If $T_{ii} > 0$, then

$$T_{ii}x_i\eta_i(x_i) \leq T_{ii}k_{i2}x_i^2$$

and

$$\frac{\partial V_i}{\partial x_i} f_i(x_i) \leq -\left(\frac{1}{R_i} - T_{ii}k_{i2}\right)x_i^2$$

In this case, we assume that $T_{ii}k_{i2} < 1/R_i$, so that the derivative of V_i is negative definite. To simplify the notation, let

$$\delta_i = \begin{cases} |T_{ii}|k_{i1}, & \text{if } T_{ii} \leq 0 \\[2mm] -T_{ii}k_{i2}, & \text{if } T_{ii} > 0 \end{cases}$$

Then, $V_i(x_i)$ satisfies (5.59)–(5.60) on the interval $[-r_i, r_i]$ with

$$\alpha_i = \left(\frac{1}{R_i} + \delta_i \right), \quad \beta_i = C_i, \quad \phi_i(x_i) = |x_i|$$

where α_i is positive by assumption. The interconnection term $g_i(x)$ satisfies the inequality

$$|g_i(x)| \leq \frac{1}{C_i} \sum_{j \neq i} |T_{ij}| |\eta_j(x_j)| \leq \frac{1}{C_i} \sum_{j \neq i} |T_{ij}| |k_{j2}| x_j|$$

Thus, $g_i(x)$ satisfies (5.61) with $\gamma_{ii} = 0$ and $\gamma_{ij} = k_{j2} |T_{ij}|/C_i$ for $i \neq j$. Now we can form the matrix S as

$$s_{ij} = \begin{cases} \frac{1}{R_i} + \delta_i, & \text{for } i = j \\[2mm] -|T_{ij}| k_{j2} & \text{for } i \neq j \end{cases}$$

The equilibrium point u^* is asymptotically stable if S is an M-matrix. We may estimate the region of attraction by the set

$$\Omega_c = \left\{ x \in R^n \;\middle|\; \sum_{i=1}^{n} d_i V_i(x_i) \leq c \right\}$$

where c is chosen small enough to ensure that Ω_c is inside the set $|x_i| \leq r_i$. This analysis is repeated for each asymptotically stable equilibrium point. The conclusions we could arrive at in this example are more conservative compared to the conclusions we arrived at using LaSalle's invariance principle. First, the magnitude of the interconnection coefficients T_{ij} must be restricted to satisfy the M-matrix condition. Second, we obtain only local estimates of the regions of attractions for the isolated equilibrium points. The union of these estimates does not cover the whole domain of interest. On the other hand, we do not have to assume that $T_{ij} = T_{ji}$. $\triangle$

5.7 Slowly Varying Systems

The system

$$\dot{x} = f(x, u(t)) \tag{5.63}$$

where $x \in R^n$ and $u(t) \in \Gamma \subset R^m$ for all $t \geq 0$ is considered to be slowly varying if $u(t)$ is continuously differentiable and $\|\dot{u}(t)\|$ is "sufficiently" small. The components of $u(t)$ could be input variables or time-varying parameters. In the analysis of (5.63), one usually treats u as a "frozen" parameter and assumes that for each fixed $u = \alpha \in \Gamma$ the frozen system has an isolated equilibrium point defined by

$x = h(\alpha)$. If a property of $x = h(\alpha)$ is uniform in α, then it is reasonable to expect that the slowly varying system (5.63) will possess a similar property. The underlying characteristic of such systems is that the motion caused by changes of initial conditions is much faster than that caused by inputs or time-varying parameters. In this section, we shall see how Lyapunov stability can be used to analyze slowly varying systems.

Suppose $f(x, u)$ is locally Lipschitz on $R^n \times \Gamma$, and for every $u \in \Gamma$ the equation

$$0 = f(x, u)$$

has a continuously differentiable isolated root $x = h(u)$; that is,

$$0 = f(h(u), u)$$

Furthermore, suppose that

$$\left\|\frac{\partial h}{\partial u}\right\| \leq L, \quad \forall\, u \in \Gamma \tag{5.64}$$

To analyze stability properties of the frozen equilibrium point $x = h(\alpha)$, we shift it to the origin via the change of variables $z = x - h(\alpha)$ to obtain the equation

$$\dot{z} = f(z + h(\alpha), \alpha) \stackrel{\text{def}}{=} g(z, \alpha) \tag{5.65}$$

Now we search for a Lyapunov function to show that $z = 0$ is asymptotically stable. Since $g(z, \alpha)$ depends on the parameter α, a Lyapunov function for the system may depend, in general, on α. Suppose we can find a Lyapunov function $V(z, \alpha)$ that satisfies the following conditions:

$$c_1\|z\|^2 \leq V(z, \alpha) \leq c_2\|z\|^2 \tag{5.66}$$

$$\frac{\partial V}{\partial z} g(z, \alpha) \leq -c_3\|z\|^2 \tag{5.67}$$

$$\left\|\frac{\partial V}{\partial z}\right\| \leq c_4\|z\| \tag{5.68}$$

$$\left\|\frac{\partial V}{\partial \alpha}\right\| \leq c_5\|z\|^2 \tag{5.69}$$

for all $z \in D = \{z \in R^n \mid \|z\| < r\}$ and $\alpha \in \Gamma$ where c_i, $i = 1, 2, \ldots, 5$ are positive constants. Inequalities (5.66) and (5.67) state the usual requirements that V be positive definite and decrescent and has a negative definite derivative along the trajectories of the system (5.65). Furthermore, they show that the origin $z = 0$ is exponentially stable. The special requirement here is that these inequalities hold uniformly in α. Inequalities (5.68) and (5.69) are needed to handle the perturbations of (5.65) which will result from the fact that $u(t)$ is not constant, but a time-varying function. The following lemma shows that if the equilibrium point $z = 0$ is exponentially stable uniformly in α, then, under some mild smoothness requirements, there is a Lyapunov function that satisfies (5.66)–(5.69). This is done by deriving a converse Lyapunov function for the system, as in the converse Lyapunov theorems of Section 3.6.

Lemma 5.11 *Consider the system* (5.65) *and suppose* $g(z, \alpha)$ *is continuously differentiable and the Jacobian matrices* $[\partial g/\partial z]$ *and* $[\partial g/\partial \alpha]$ *satisfy*

$$\left\| \frac{\partial g}{\partial z}(z, \alpha) \right\| \leq L_1, \quad \left\| \frac{\partial g}{\partial \alpha}(z, \alpha) \right\| \leq L_2 \|z\|$$

for all $(z, \alpha) \in D \times \Gamma$ *where* $D = \{z \in R^n \mid \|z\| < r\}$. *Let* k, γ, *and* r_0 *be positive constants with* $r_0 < r/k$, *and define* $D_0 = \{z \in R^n \mid \|z\| < r_0\}$. *Assume that the trajectories of the system satisfy*

$$\|z(t)\| \leq k\|z(0)\|e^{-\gamma t}, \quad \forall \ z(0) \in D_0, \ \alpha \in \Gamma, \ t \geq 0$$

Then, there is a function $V : D_0 \times \Gamma \to R$ *that satisfies* (5.66)–(5.69). *Moreover, if all the assumptions hold globally (in* z), *then* $V(z, \alpha)$ *is defined and satisfies* (5.66)–(5.69) *on* $R^n \times \Gamma$. $\diamond$

Proof: Owing to the equivalence of norms, it is sufficient to prove the lemma for the 2-norm. Let $\phi(t, z; \alpha)$ be the solution of (5.65) that starts at $(0, z)$; that is, $\phi(0, z; \alpha) = z$. The notation emphasizes the dependence of the solution on the parameter α. Let

$$V(z, \alpha) = \int_0^T \phi^T(t, z; \alpha)\phi(t, z; \alpha) \, dt$$

where $T = \ln(2k^2)/2\gamma$. Similar to the proof of Theorem 3.12, it can be shown that $V(z, \alpha)$ satisfies (5.66)–(5.68) with $c_1 = [1 - \exp(-2L_1T)]/2L_1$, $c_2 = k^2[1 - \exp(-2\gamma T)]/2\gamma$, $c_3 = \frac{1}{2}$, and $c_4 = 2k\{1 - \exp[-(\gamma - L_1)T]\}/(\gamma - L_1)$. To show that $V(z, \alpha)$ satisfies the last inequality (5.69), note that the sensitivity function $\phi_\alpha(t, z; \alpha)$ satisfies the sensitivity equation

$$\frac{\partial}{\partial t}\phi_\alpha = \frac{\partial g}{\partial z}(\phi(t, z; \alpha), \alpha)\phi_\alpha + \frac{\partial g}{\partial \alpha}(\phi(t, z; \alpha), \alpha), \quad \phi_\alpha(0, z; \alpha) = 0$$

from which we obtain

$$\|\phi_\alpha(t, z; \alpha)\|_2 \ \leq \ \int_0^t L_1\|\phi_\alpha(\tau, z; \alpha)\|_2 \, d\tau + \int_0^t L_2\|\phi(\tau, z; \alpha)\|_2 \, d\tau$$

$$\leq \int_0^t L_1 \|\phi_\alpha(\tau, z; \alpha)\|_2 \, d\tau + \int_0^t L_2 k e^{-\gamma\tau} \, d\tau \|z\|_2$$

$$\leq \int_0^t L_1 \|\phi_\alpha(\tau, z; \alpha)\|_2 \, d\tau + \frac{L_2 k}{\gamma} \|z\|_2$$

Use of the Gronwall-Bellman inequality yields

$$\|\phi_\alpha(t, z; \alpha)\|_2 \leq \frac{L_2 k}{\gamma} \|z\|_2 e^{L_1 t}$$

Hence,

$$\begin{aligned}
\left\| \frac{\partial V}{\partial \alpha} \right\|_2 &= \left\| \int_0^T 2\phi^T(t, z; \alpha)\phi_\alpha(t, z; \alpha) \, dt \right\|_2 \\
&\leq \int_0^T 2ke^{-\gamma t} \|z\|_2 \left(\frac{L_2 k}{\gamma} \right) e^{L_1 t} \|z\|_2 \, dt \\
&\leq \frac{2k^2 L_2}{\gamma(\gamma - L_1)} \left[1 - e^{-(\gamma - L_1)T} \right] \|z\|_2^2 \stackrel{\text{def}}{=} c_5 \|z\|_2^2
\end{aligned}$$

which completes the proof of the lemma.   □

When the frozen system (5.65) is linear, a Lyapunov function satisfying (5.66)–(5.69) can be explicitly determined by solving a parameterized Lyapunov equation. This fact is stated in the following lemma.

Lemma 5.12 *Consider the system $\dot{z} = A(\alpha)z$, where $\alpha \in \Gamma$ and $A(\alpha)$ is continuously differentiable. Suppose the elements of A and their first partial derivatives with respect to α are uniformly bounded; that is,*

$$\|A(\alpha)\|_2 \leq c, \quad \left\| \frac{\partial}{\partial \alpha_i} A(\alpha) \right\|_2 \leq b_i, \quad \forall \, \alpha \in \Gamma, \, \forall \, 1 \leq i \leq m$$

Suppose further that $A(\alpha)$ is Hurwitz uniformly in α; that is,

$$Re[\lambda(A(\alpha))] \leq -\sigma < 0, \quad \forall \, \alpha \in \Gamma$$

Then, the Lyapunov equation

$$PA(\alpha) + A^T(\alpha)P = -I \tag{5.70}$$

has a unique positive definite solution $P(\alpha)$ for every $\alpha \in \Gamma$. Moreover, $P(\alpha)$ is continuously differentiable and satisfies

$$c_1 z^T z \leq z^T P(\alpha)z \leq c_2 z^T z$$

$$\left\| \frac{\partial}{\partial \alpha_i} P(\alpha) \right\|_2 \leq \mu_i, \quad \forall\, 1 \leq i \leq m$$

for all $(z, \alpha) \in R^n \times \Gamma$ where c_1, c_2, and μ_i are positive constants. Consequently, $V(z, \alpha) = z^T P(\alpha) z$ satisfies (5.67)–(5.69) in the 2-norm with $c_3 = 1$, $c_4 = 2c_2$, and $c_5 = \sqrt{\sum_{i=1}^m \mu_i^2}$. $\diamond$

Proof: The uniform Hurwitz property of $A(\alpha)$ implies that the exponential matrix $\exp[tA(\alpha)]$ satisfies

$$\| \exp[tA(\alpha)] \| \leq k(A) e^{-\beta t}, \quad \forall\, t \geq 0, \ \forall\, \alpha \in \Gamma$$

where $\beta > 0$ is independent of α, but $k(A) > 0$ depends on α. For the exponentially decaying bound to hold uniformly in α, we need to use the property that $\|A(\alpha)\|$ is bounded. The set of matrices satisfying $Re[\lambda(A(\alpha))] \leq -\sigma$ and $\|A(\alpha)\| \leq c$ is a compact set, which we denote by S. Let A and B be any two elements of S. Consider [10]

$$\exp[t(A + B)] = \exp[tA] + \int_0^t \exp[(t - \tau)A]B \exp[\tau(A + B)] \, d\tau$$

Using the exponentially decaying bound on $\exp[tA]$, we get

$$\| \exp[t(A + B)] \| \leq k(A) e^{-\beta t} + \int_0^t k(A) e^{-\beta(t - \tau)} \|B\| \, \| \exp[\tau(A + B)] \| \, d\tau$$

Multiply through by $e^{\beta t}$,

$$e^{\beta t} \| \exp[t(A + B)] \| \leq k(A) + k(A)\|B\| \int_0^t e^{\beta \tau} \| \exp[\tau(A + B)] \| \, d\tau$$

Application of the Gronwall-Bellman inequality yields

$$\| \exp[t(A + B)] \| \leq k(A) e^{-(\beta - k(A)\|B\|)t}, \quad \forall\, t \geq 0$$

Hence, there exists a positive constant $\gamma < \beta$ and a neighborhood $\mathcal{N}(A)$ of A such that if $C \in \mathcal{N}(A)$, then

$$\| \exp[tC] \| \leq k(A) e^{-\gamma t}, \quad \forall\, t \geq 0$$

---

[10] This matrix identity follows by writing $\dot{x} = (A + B)x$ as $\dot{x} = Ax + Bx$ and viewing Bx as an input term. Substitution of $x(t) = \exp[t(A + B)]x_0$ in the input term yields

$$\exp[t(A + B)]x_0 = \exp[tA]x_0 + \int_0^t \exp[(t - \tau)A]B \exp[\tau(A + B)]x_0 \, d\tau$$

Since this expression holds for all $x_0 \in R^n$, we arrive at the matrix identity.

Since S is compact, it is covered by a finite number of these neighborhoods. Therefore, we can find a positive constant k independent of α such that

$$\| \exp[tA(\alpha)]\| \leq ke^{-\gamma t}, \quad \forall\, t \geq 0,\; \forall\, \alpha \in \Gamma$$

Consider now the Lyapunov equation (5.70). Existence of a unique positive definite solution for every $\alpha \in \Gamma$ follows from Theorem 3.6. Moreover, the proof of that theorem shows that

$$P(\alpha) = \int_0^\infty \left[e^{tA(\alpha)}\right]^T \left[e^{tA(\alpha)}\right]\, dt$$

Since $A(\alpha)$ is continuously differentiable, so is $P(\alpha)$. We have

$$z^T P(\alpha)z \leq \int_0^\infty k^2 e^{-2\gamma t}\|z\|_2^2\, dt = \frac{k^2}{2\gamma}\|z\|_2^2 \;\Rightarrow\; c_2 = \frac{k^2}{2\gamma}$$

Let $y(t) = e^{tA(\alpha)}z$. Then, $\dot{y} = A(\alpha)y$,

$$-y^T(t)\dot{y}(t) = -y^T(t)A(\alpha)y(t) \leq \|A(\alpha)\|_2 y^T(t)y(t) \leq cy^T(t)y(t)$$

and

$$
\begin{aligned}
z^T P(\alpha)z &= \int_0^\infty y^T(t)y(t)\, dt \;\geq\; \int_0^\infty \frac{-1}{c}y^T(t)\dot{y}(t)\, dt \\
&= \frac{1}{2c}\int_0^\infty \frac{d}{dt}[-y^T(t)y(t)]\, dt \;=\; \frac{1}{2c}\left[-y^T(t)y(t)\right]\big|_0^\infty \\
&= \frac{1}{2c}y^T(0)y(0) \;=\; \frac{1}{2c}z^T z \;\Rightarrow\; c_1 = \frac{1}{2c}
\end{aligned}
$$

Differentiate $P(\alpha)A(\alpha)+A^T(\alpha)P(\alpha) = -I$ partially with respect to any component α_i of α, and denote the derivative of $P(\alpha)$ by $P'(\alpha)$. Then,

$$P'(\alpha)A(\alpha) + A^T(\alpha)P'(\alpha) = -\{P(\alpha)A'(\alpha) + [A'(\alpha)]^T P(\alpha)\}$$

Hence, $P'(\alpha)$ is given by

$$P'(\alpha) = \int_0^\infty \left[e^{tA(\alpha)}\right]^T \{P(\alpha)A'(\alpha) + [A'(\alpha)]^T P(\alpha)\}\left[e^{tA(\alpha)}\right]\, dt$$

It follows that

$$\|P'(\alpha)\|_2 \leq \int_0^\infty k^2 e^{-2\gamma t}2\frac{k^2}{2\gamma}b_i\, dt = \frac{b_i k^4}{2\gamma^2} \;\Rightarrow\; \mu_i = \frac{b_i k^4}{2\gamma^2}$$

which completes the proof of the lemma.   □

It should be noted that the set Γ in Lemma 5.12 is not necessarily compact. When

Γ is compact, the boundedness assumptions follow from the continuous differentiability of $A(\alpha)$.

With $V(z, u)$ as a Lyapunov function candidate, the analysis of (5.63) proceeds as follows. The change of variables $z = x - h(u)$ transforms (5.63) into the form

$$\dot{z} = g(z, u) - \frac{\partial h}{\partial u} \dot{u} \tag{5.71}$$

where the effect of the time variation of u appears as a perturbation of the frozen system (5.65). The derivative of $V(z, u)$ along the trajectories (5.71) is given by

$$
\begin{aligned}
\dot{V} &= \frac{\partial V}{\partial z} \dot{z} + \frac{\partial V}{\partial u} \dot{u}(t) \\
&= \frac{\partial V}{\partial z} g(z, u) + \left[\frac{\partial V}{\partial u} - \frac{\partial V}{\partial z} \frac{\partial h}{\partial u} \right] \dot{u}(t) \\
&\leq -c_3 \|z\|^2 + c_5 \|z\|^2 \|\dot{u}(t)\| + c_4 L \|z\| \, \|\dot{u}(t)\|
\end{aligned}
$$

Setting

$$\gamma(t) = \frac{c_5}{c_4} \|\dot{u}(t)\| \quad \text{and} \quad \delta(t) = L \|\dot{u}(t)\|$$

we can rewrite the last inequality as

$$\dot{V} \leq -c_3 \|z\|^2 + c_4 \gamma(t) \|z\|^2 + c_4 \delta(t) \|z\|$$

which takes the form of inequality (5.41) of Section 5.4. Therefore, by repeating the steps of the comparison lemma application, as conducted in Section 5.4, it can be shown that if $\dot{u}(t)$ satisfies

$$\int_{t_0}^{t} \|\dot{u}(\tau)\| \, d\tau \leq \epsilon_1 (t - t_0) + \eta_1, \quad \text{where } \epsilon_1 < \frac{c_1 c_3}{c_2 c_5} \tag{5.72}$$

and

$$\|z(0)\| < \frac{r}{\rho_1} \sqrt{\frac{c_1}{c_2}}; \quad \sup_{t \geq 0} \|\dot{u}(t)\| \leq \frac{2 c_1 \alpha_1 r}{c_4 \rho_1 L}$$

where α_1 and ρ_1 are defined by

$$\alpha_1 = \frac{1}{2} \left[\frac{c_3}{c_2} - \epsilon_1 \frac{c_5}{c_1} \right] > 0, \quad \rho_1 = \exp \left(\frac{c_5 \eta_1}{2 c_1} \right) \geq 1$$

then $z(t)$ satisfies the inequality

$$\|z(t)\| \leq \sqrt{\frac{c_2}{c_1}} \rho_1 \|z(0)\| e^{-\alpha_1 t} + \frac{c_4 \rho_1 L}{2 c_1} \int_{0}^{t} e^{-\alpha_1 (t - \tau)} \|\dot{u}(\tau)\| \, d\tau \tag{5.73}$$

Depending upon the assumptions for $\|\dot{u}\|$, several conclusions can be drawn from the foregoing inequality. Some of these conclusions are stated in the following theorem.

Theorem 5.5 *Consider the system* (5.71). *Suppose that* $[\partial h/\partial u]$ *satisfies* (5.64), $\|\dot{u}(t)\| \leq \epsilon$ *for all* $t \geq 0$, *and there is a Lyapunov function* $V(z, u)$ *that satisfies* (5.66)–(5.69). *If*

$$\epsilon < \frac{c_1 c_3}{c_2 c_5} \times \frac{r}{r + c_4 L/c_5}$$

then for all $\|z(0)\| < r\sqrt{c_1/c_2}$, *the solutions of* (5.71) *are uniformly bounded for* $t \geq 0$ *and uniformly ultimately bounded with an ultimate bound*

$$b = \frac{c_2 c_4 L \epsilon}{\theta(c_1 c_3 - \epsilon c_2 c_5)}$$

where $\theta \in (0, 1)$ *is an arbitrary constant. If, in addition,* $\dot{u}(t) \to 0$ *as* $t \to \infty$, *then* $z(t) \to 0$ *as* $t \to \infty$. *Finally, if* $h(u) = 0$ *for all* $u \in \Gamma$ *and* $\epsilon < c_3/c_5$, *then* $z = 0$ *is an exponentially stable equilibrium of* (5.71). *Equivalently,* $x = 0$ *is an exponentially stable equilibrium of* (5.63). $\Diamond$

Proof: Since $\|\dot{u}(t)\| \leq \epsilon < c_1 c_3/c_2 c_5$, inequality (5.72) is satisfied with $\epsilon_1 = \epsilon$ and $\eta_1 = 0$. Hence,

$$\alpha_1 = \frac{1}{2}\left[\frac{c_3}{c_2} - \epsilon\frac{c_5}{c_1}\right], \quad \rho_1 = 1$$

Using the given upper bound on ϵ, we have

$$\frac{2c_1\alpha_1 r}{c_4 L} = \frac{c_1 r}{c_4 L}\left[\frac{c_3}{c_2} - \epsilon\frac{c_5}{c_1}\right] > \frac{c_1 r}{c_4 L}\left[\frac{c_3}{c_2} - \frac{c_3}{c_2} \times \frac{r}{r + c_4 L/c_5}\right] = \frac{c_1 c_3}{c_2 c_5} \times \frac{r}{r + c_4 L/c_5} > \epsilon$$

Hence, the inequality $\sup_{t \geq 0}\|\dot{u}(t)\| < 2c_1\alpha_1 r/c_4 L$ is satisfied and from (5.73) we obtain

$$\|z(t)\| \leq \sqrt{\frac{c_2}{c_1}}\|z(0)\|e^{-\alpha_1 t} + \frac{c_4 L}{2c_1}\int_0^t e^{-\alpha_1(t-\tau)}\epsilon \, d\tau$$

$$\leq \sqrt{\frac{c_2}{c_1}}\|z(0)\|e^{-\alpha_1 t} + \frac{c_4 L \epsilon}{2c_1\alpha_1} = \sqrt{\frac{c_2}{c_1}}\|z(0)\|e^{-\alpha_1 t} + b\theta$$

After a finite time, the exponentially decaying term will be less than $(1-\theta)b$, which shows that $z(t)$ will be ultimately bounded by b. If, in addition, $\dot{u}(t) \to 0$ as $t \to \infty$, then it is clear from (5.73) that $z(t) \to 0$ as $t \to \infty$. If $h(u) = 0$ for all $u \in \Gamma$, we can take $L = 0$. Consequently, the upper bound on $\dot{V}$ simplifies to

$$\dot{V} \leq -(c_3 - c_5\epsilon)\|z\|^2$$

which shows that $z = 0$ will be exponentially stable if $\epsilon < c_3/c_5$. $\Box$

Example 5.13 Consider the system

$$\dot{x} = A(\epsilon t)x$$

where $\epsilon > 0$. When ϵ is sufficiently small, we can treat this system as a slowly varying system. It is in the form (5.63) with $u = \epsilon t$ and $\Gamma = [0, \infty)$. For all $u \in \Gamma$, the origin $x = 0$ is an equilibrium point. Hence, this is a special case where $h(u) = 0$. Suppose $Re[\lambda(A(\alpha))] \leq -\sigma < 0$, and $A(\alpha)$ and $A'(\alpha)$ are uniformly bounded for all $\alpha \in \Gamma$. Then, the solution of the Lyapunov equation (5.70) holds the properties stated in Lemma 5.12. Using $V(x, u) = x^T P(u)x$ as a Lyapunov function candidate for $\dot{x} = A(u)x$, we obtain

$$
\begin{aligned}
\dot{V}(t, x) &= x^T [P(u(t))A(u(t)) + A^T(u(t))P(u(t))]x + x^T P'(u(t))\dot{u}(t)x \\
&\leq -x^T x + \epsilon c_5 \|x\|_2^2 = -(1 - \epsilon c_5)\|x\|_2^2
\end{aligned}
$$

where c_5 is an upper bound on $\|P'(\alpha)\|_2$. Therefore, for all $\epsilon < 1/c_5$, the origin $x = 0$ is an exponentially stable equilibrium point of $\dot{x} = A(\epsilon t)x$. $\triangle$

Example 5.14 Consider the pendulum equation

$$
\begin{aligned}
\dot{x}_1 &= x_2 \\
\dot{x}_2 &= -\sin x_1 - x_2 + u(t)
\end{aligned}
$$

where $u(t)$ is a normalized driving torque. Suppose that

$$|u(t)| \leq \frac{1}{\sqrt{2}}, \quad |\dot{u}(t)| \leq \epsilon, \quad \forall \, t \geq 0$$

where ϵ is small. The set Γ is the closed interval $[-1/\sqrt{2}, 1/\sqrt{2}]$. For a constant input torque $u = \alpha$, the equilibrium points are the roots of

$$
\begin{aligned}
0 &= x_2 \\
0 &= -\sin x_1 + \alpha
\end{aligned}
$$

In the set $|x_1| < 3\pi/4$, there is a unique equilibrium point defined by

$$
h(\alpha) = \begin{bmatrix} \sin^{-1}(\alpha) \\ \\ 0 \end{bmatrix}
$$

The change of variables

$$z_1 = x_1 - \sin^{-1}(\alpha), \quad z_2 = x_2$$

transforms the frozen system into

$$\dot{z}_1 = z_2$$
$$\dot{z}_2 = -\eta(z_1, \alpha) - z_2$$

where

$$\eta(z_1, \alpha) = \sin(z_1 + \sin^{-1}(\alpha)) - \alpha = \sin(z_1 + \delta) - \sin \delta$$

and $\delta = \sin^{-1}(\alpha)$. The function $\eta(z_1, \alpha)$ satisfies the inequality

$$z_1 \eta(z_1, \alpha) > 0, \quad \forall \; -\pi - 2\delta < z_1 < \pi - 2\delta$$

Since $|\delta| \leq \pi/4$, the interval $(-\pi - 2\delta, \pi - 2\delta) \supset (-\pi/2, \pi/2)$. Let $\zeta < 1/\sqrt{2}$ be a small positive number and restrict z_1 to a set D_α where

$$z_1 \eta(z_1, \alpha) \geq \zeta z_1^2$$

Since $|\alpha| \leq 1/\sqrt{2}$, the set D_α is nonempty and includes the set $|z_1| \leq \rho$, where ρ can be made arbitrarily close to $\pi/2$ by choosing ζ sufficiently small. This system is a special case of the system studied in Example 3.5. Therefore, a Lyapunov function for the system can be taken as

$$
\begin{aligned}
V(z, \alpha) &= \tfrac{1}{2} z^T P z + \int_0^{z_1} \eta(y, \alpha)\, dy \\
&= \tfrac{1}{2} z^T \begin{bmatrix} \tfrac{1}{2} & \tfrac{1}{2} \\ \tfrac{1}{2} & 1 \end{bmatrix} z + \int_0^{z_1} [\sin(y + \delta) - \sin \delta]\, dy \\
&= \tfrac{1}{4} z_1^2 + \tfrac{1}{2} z_1 z_2 + \tfrac{1}{2} z_2^2 + (1 - \cos z_1) \cos \delta + (\sin z_1 - z_1) \sin \delta
\end{aligned}
$$

Using the facts

$$|y - \sin y| \leq \tfrac{1}{6}|y|^3, \quad |1 - \cos y| \leq \tfrac{1}{2} y^2$$

and $|z_1| \leq 3\pi/2$ for all $z_1 \in D_\alpha$, it can be easily verified that for all $z_1 \in D_\alpha$ the Lyapunov function satisfies (5.66) with

$$c_1 = \tfrac{1}{2}\lambda_{\min}(P), \quad c_2 = \tfrac{1}{2}\lambda_{\max}(P) + \frac{1}{\sqrt{2}}\left(\frac{1}{\sqrt{2}} + \frac{\pi}{4}\right)$$

The derivative of $V(z, \alpha)$ along the trajectories of the frozen system satisfies

$$\frac{\partial V}{\partial z} g(z, \alpha) = -\tfrac{1}{2} z_1 \eta(z_1, \alpha) - \tfrac{1}{2} z_2^2 \leq -\tfrac{1}{2}\zeta z_1^2 - \tfrac{1}{2} z_2^2 \leq -\tfrac{1}{2}\zeta \|z\|_2^2$$

Thus, $V(z, \alpha)$ satisfies (5.67). Finally, inequalities (5.68)–(5.69) follow from

$$\left|\frac{\partial V}{\partial z_1}\right| = \left|\tfrac{1}{2} z_1 + \tfrac{1}{2} z_2 + \eta(z_1, \alpha)\right| \leq \tfrac{3}{2}|z_1| + \tfrac{1}{2}|z_2| \leq \tfrac{1}{2}\sqrt{10}\, \|z\|_2$$

$$\left|\frac{\partial V}{\partial z_2}\right| = \left|\frac{1}{2}z_1 + z_2\right| \leq \frac{1}{2}|z_1| + |z_2| \leq \frac{1}{2}\sqrt{5}\,\|z\|_2$$

$$\left|\frac{\partial V}{\partial \alpha}\right| = \left|\frac{\partial V}{\partial \delta}\frac{\partial \delta}{\partial \alpha}\right| = \left|[(\cos z_1 - 1)\sin \delta + (\sin z_1 - z_1)\cos \delta]\frac{1}{\cos \delta}\right|$$

$$\leq |\cos z_1 - 1| + |\sin z_1 - z_1| \leq \left(\frac{1}{2}z_1^2 + \frac{\pi}{4}z_1^2\right) \leq \left(\frac{1}{2} + \frac{\pi}{4}\right)\|z\|_2^2$$

Thus, all the conditions of Theorem 5.5 are satisfied and we can conclude that, for sufficiently small ϵ, there is a ball around the origin $x = 0$ such that all solutions starting in that ball approach a small ball centered at the slowly moving point $x_1(t) = \sin^{-1}((u(t)))$, $x_2(t) = 0$. The size of this small ball is dependent on ϵ. Furthermore, if $u(t)$ eventually approaches a constant limit u_∞ and $\dot{u}(t) \to 0$, then the solution approaches $x_1 = \sin^{-1}(u_\infty)$, $x_2 = 0$ as $t \to \infty$. $\triangle$

Example 5.15 Consider the system

$$\dot{x} = f(x, u)$$

where $f(x, u)$ is twice continuously differentiable in a domain $D_x \times D_u \subset R^n \times R^m$. Suppose there is a continuously differentiable function $h : D_u \to D_x$ such that

$$0 = f(h(\alpha), \alpha), \quad \forall\ \alpha \in D_u$$

Let

$$A(\alpha) = \left.\frac{\partial f}{\partial x}(x, u)\right|_{x=h(\alpha), u=\alpha}$$

The matrix $A(\alpha)$ is the linearization of the system $\dot{x} = f(x, \alpha)$ about the equilibrium point $x = h(\alpha)$. It is a continuously differentiable function of α. Suppose that $A(\alpha)$ is Hurwitz, uniformly in α, for all $\alpha \in \Gamma$ where $\Gamma \subset D_u$ is a compact set. Then, the matrix $A(\alpha)$ satisfies all the assumptions of Lemma 5.12 for $\alpha \in \Gamma$. Let $P(\alpha)$ be the solution of the Lyapunov equation (5.70). We use $V(z, \alpha) = z^T P(\alpha)z$ as a Lyapunov function candidate for the frozen system

$$\dot{z} = f(z + h(\alpha), \alpha) \overset{\text{def}}{=} g(z, \alpha)$$

Lemma 5.12 shows that $V(z, \alpha)$ satisfies (5.66), (5.68), and (5.69). We only need to verify (5.67). The frozen system can be rewritten as

$$\dot{z} = A(\alpha)z + \Delta g(z, \alpha)$$

where

$$\|\Delta g(z, \alpha)\|_2 = \|g(z, \alpha) - A(\alpha)z\|_2 \leq k_1 \|z\|_2^2$$

in some domain $\{\|z\|_2 < r_1\}$. Thus, the derivative of V along the trajectories of $\dot{z} = g(z, \alpha)$ satisfies

$$\dot{V} \leq -\|z\|_2^2 + 2c_2 k_1 \|z\|_2^3 \leq -\tfrac{1}{2}\|z\|_2^2$$

for $\|z\|_2 < 1/4c_2 k_1$. Hence, there exists $r > 0$ such that $V(z, \alpha)$ satisfies (5.66)–(5.69) for all $(z, \alpha) \in \{\|z\|_2 < r\} \times \Gamma$. Now suppose $u(t)$ is continuously differentiable, $u(t) \in \Gamma$, and $\|\dot{u}(t)\| \leq \epsilon$ for all $t \geq 0$. It follows from Theorem 5.5 that there exist positive constants ρ_1 and ρ_2 such that if $\epsilon < \rho_1$ and $\|x(0) - h(u(0))\|_2 < \rho_2$, then the solution $x(t)$ of $\dot{x} = f(x, u(t))$ will be uniformly bounded for all $t \geq 0$ and $x(t) - h(u(t))$ will be uniformly ultimately bounded by $k\epsilon$ for some $k > 0$. Moreover, if $u(t) \to u_\infty$ and $\dot{u}(t) \to 0$ as $t \to \infty$, then $x(t) \to h(u_\infty)$ as $t \to \infty$. $\triangle$

5.8   Exercises

Exercise 5.1 ([133]) Consider the Lyapunov equation $PA + A^T P = -Q$, where $Q = Q^T > 0$ and A is Hurwitz. Let $\mu(Q) = \lambda_{\min}(Q)/\lambda_{\max}(P)$.

(a) Show that $\mu(kQ) = \mu(Q)$ for any positive constant k.

(b) Let $\hat{Q} = \hat{Q}^T > 0$ have $\lambda_{\min}(\hat{Q}) = 1$. Show that $\mu(I) \geq \mu(\hat{Q})$.

(c) Show that $\mu(I) \geq \mu(Q)$, $\forall\, Q = Q^T > 0$.

Hint: In part (b), let P_1 and P_2 be the solutions of the Lyapunov equation for $Q = I$ and $Q = \hat{Q}$, respectively. Show that

$$P_1 - P_2 = \int_0^\infty \exp(A^T t)(I - \hat{Q})\exp(At)\, dt \leq 0$$

Exercise 5.2 Consider the system $\dot{x} = Ax + Bu$ and let $u = Fx$ be a stabilizing state feedback control; that is, the matrix $(A + BF)$ is Hurwitz. Suppose that, due to physical limitations, we have to use a limiter to limit the value of u_i to $|u_i(t)| \leq L$. The closed-loop system can be represented by $\dot{x} = Ax + BL\,\mathrm{sat}(Fx/L)$, where $\mathrm{sat}(v)$ is a vector whose ith component is the saturation function

$$\mathrm{sat}(v_i) = \begin{cases} -1, & \text{for } v_i < -1 \\ v_i, & \text{for } |v_i| \leq 1 \\ 1, & \text{for } v_i > 1 \end{cases}$$

By adding and subtracting the term BFx, we can rewrite the closed-loop state equation as $\dot{x} = (A + BF)x + Bh(Fx)$ where $h(v) = L\,\mathrm{sat}(v/L) - v$. Thus, the effect of the limiter can be viewed as a perturbation of the nominal system without the limiter.

(a) Show that $|h_i(v)| \leq [\delta/(1+\delta)]|v_i|$, $\forall \, |v_i| \leq L(1+\delta)$.

(b) Let P be the solution of $P(A + BF) + (A + BF)^T P = -I$. Show that the derivative of $V(x) = x^T Px$ along the trajectories of the closed-loop system will be negative definite over the region $|(Fx)_i| \leq L(1+\delta)$, $\forall \, i$, provided $\delta/(1+\delta) < 1/2\|PB\|_2 \, \|F\|_2$.

(c) Show that the origin is asymptotically stable and discuss how you would estimate the region of attraction.

(d) Apply the above result to the case[11]

$$A = \begin{bmatrix} -1 & -1 \\ -1 & 1 \end{bmatrix}, \quad B = \begin{bmatrix} 1 & 0 \\ 0 & 1 \end{bmatrix}, \quad F = \begin{bmatrix} 0 & 1 \\ 1 & -1.9 \end{bmatrix}$$

and $L = 1$. Estimate the region of attraction of the origin.

Exercise 5.3 Consider the system

$$\dot{x} = f(t, x) + Bu, \quad y = Cx, \quad u = -g(t, y)$$

where $f(t, 0) = 0$, $g(t, 0) = 0$, and $\|g(t, y)\| \leq \gamma\|y\|$ for all $t \geq 0$. Suppose that the origin of $\dot{x} = f(t, x)$ is globally exponentially stable and let $V(t, x)$ be a Lyapunov function that satisfies (5.3)–(5.5) globally. Find a bound γ^* on γ such that the origin of the given system is globally exponentially stable for $\gamma < \gamma^*$.

Exercise 5.4 ([47]) Consider the system $\dot{x} = (A + B)x$ and suppose that A is Hurwitz. Let P be the solution of the Lyapunov equation $PA + A^T P = -Q$, $Q = Q^T > 0$. Show that $(A + B)$ is Hurwitz if

$$B^T PQ^{-1}PB < \tfrac{1}{4}Q \quad \text{or} \quad BQ^{-1}B < \tfrac{1}{4}P^{-1}QP^{-1}$$

Hint: In the first case, show that

$$x^T(PB + B^T P)x \leq \tfrac{1}{2}x^T Qx + 2x^T B^T PQ^{-1}PBx, \quad \forall \, x \in R^n$$

starting from

$$x^T \left[\frac{1}{\sqrt{2}}Q^{1/2} - \sqrt{2}Q^{-1/2}PB\right]^T \left[\frac{1}{\sqrt{2}}Q^{1/2} - \sqrt{2}Q^{-1/2}PB\right] x \geq 0$$

A similar idea is used in the second case.

---

[11] This numerical case was studied in [61].

Exercise 5.5 Consider the perturbed system $\dot{x} = Ax + B[u + g(t, x)]$, where $g(t, x)$ is continuously differentiable and satisfies $\|g(t, x)\|_2 \leq k\|x\|_2$, $\forall\, t \geq 0$, $\forall\, x \in B_r$ for some $r > 0$. Let $P = P^T > 0$ be the solution of the Riccati equation

$$PA + A^T P + Q - PBB^T P + 2\alpha P = 0$$

where $Q \geq k^2 I$ and $\alpha > 0$. Show that $u = -B^T Px$ stabilizes the origin of the perturbed system.

Exercise 5.6 ([92]) Consider the perturbed system $\dot{x} = Ax + Bu + Dg(t, y)$, $y = Cx$, where $g(t, y)$ is continuously differentiable and satisfies $\|g(t, y)\|_2 \leq k\|y\|_2$, $\forall\, t \geq 0$, $\forall\, \|y\|_2 \leq r$ for some $r > 0$. Suppose the equation

$$PA + A^T P + \epsilon Q - \frac{1}{\epsilon}PBB^T P + \frac{1}{\gamma}PDD^T P + \frac{1}{\gamma}C^T C = 0$$

where $Q = Q^T > 0$, $\epsilon > 0$, and $0 < \gamma < 1/k$ has a positive definite solution $P = P^T > 0$. Show that $u = -(1/2\epsilon)B^T Px$ stabilizes the origin of the perturbed system.

Exercise 5.7 Consider the system

$$
\begin{aligned}
\dot{x}_1 &= -\alpha x_1 - \omega x_2 + (\beta x_1 - \gamma x_2)(x_1^2 + x_2^2) \\
\dot{x}_2 &= \omega x_1 - \alpha x_2 + (\gamma x_1 + \beta x_2)(x_1^2 + x_2^2)
\end{aligned}
$$

where $\alpha > 0$, β, γ, and $\omega > 0$ are constants.

(a) By viewing this system as a perturbation of the linear system

$$
\begin{aligned}
\dot{x}_1 &= -\alpha x_1 - \omega x_2 \\
\dot{x}_2 &= \omega x_1 - \alpha x_2
\end{aligned}
$$

show that the origin of the perturbed system is exponentially stable with $\{\|x\|_2 \leq r\}$ included in the region of attraction, provided $\beta|$ and $|\gamma|$ are sufficiently small. Find upper bounds on $|\beta|$ and $|\gamma|$ in terms of r.

(b) Using $V(x) = x_1^2 + x_2^2$ as a Lyapunov function candidate for the perturbed system, show that the origin is globally exponentially stable when $\beta < 0$ and exponentially stable with $\{\|x\|_2 < \sqrt{\alpha/\beta}\}$ included in the region of attraction when $\beta \geq 0$.

(c) Compare the results of (a) and (b) and comment on the conservative nature of the result of (a).

Exercise 5.8 Consider the perturbed system $\dot{x} = f(x) + g(x)$. Suppose the origin of the nominal system $\dot{x} = f(x)$ is asymptotically (but not exponentially) stable. Show that, for any $\gamma > 0$, there is a function $g(x)$ satisfying $\|g(x)\| \leq \gamma\|x\|$ in some neighborhood of the origin such that the origin of the perturbed system is unstable.

Exercise 5.9 ([56]) Consider the perturbed system $\dot{x} = f(x) + g(x)$, where $f(x)$ and $g(x)$ are continuously differentiable and $\|g(x)\| \leq \gamma\|x\|$ for all $\|x\| < r$. Suppose

- the origin of the nominal system $\dot{x} = f(x)$ is asymptotically stable and there is a Lyapunov function $V(x)$ that satisfies inequalities (5.20)–(5.22) for all $\|x\| < r$.

- the linearization of $f(x)$ at the origin $(A = [\partial f/\partial x](0))$ is Hurwitz.

(a) Show that there are positive constants ρ and γ_1^* such that, for $\gamma < \gamma_1^*$, the origin of the perturbed system is asymptotically stable and every trajectory starting inside the set $\{\|x\| < \rho\}$ converges to the origin as $t \to \infty$.

(b) Show that there is a positive constant γ^* such that, for $\gamma < \gamma^*$, every trajectory starting inside the set $\{V(x) \leq c\}$ for any $c < \alpha_1(r)$ converges to the origin as $t \to \infty$.

(c) Would (b) hold if the linearization was not Hurwitz? Consider

$$f(x) = \begin{bmatrix} -x_2 - (2x_1 + x_3)^3 \\ x_1 \\ x_2 \end{bmatrix}, \quad g(x) = a\begin{bmatrix} x_1 - x_3 - (2x_1 + x_3)^3 \\ 0 \\ 0 \end{bmatrix}, \quad a \neq 0$$

Hint: For the example of part (c), use $V(x) = x_1^2 + \frac{1}{2}x_2^2 + \frac{1}{2}x_3^2 + x_1x_3$ to show that the origin of $\dot{x} = f(x)$ is asymptotically stable and then apply Theorem 3.14 to obtain a Lyapunov function that satisfies the desired inequalities.

Exercise 5.10 Prove Corollary 5.1.

Exercise 5.11 Prove Corollary 5.2.

Exercise 5.12 Prove Corollary 5.3.

Exercise 5.13 ([62]) Consider the system $\dot{x} = f(t, x)$ and suppose there is a function $V(t, x)$ that satisfies

$$W_1(x) \leq V(t, x) \leq W_2(x), \quad \forall \|x\| \geq r > 0$$

$$\frac{\partial V}{\partial t} + \frac{\partial V}{\partial x}f(t, x) < 0, \quad \forall \|x\| \geq r_1 \geq r$$

where $W_1(x)$ and $W_2(x)$ are continuous, positive definite functions. Show that the solutions of the system are uniformly bounded.
Hint: Notice that $V(t, x)$ is not necessarily positive definite.

Exercise 5.14 Consider the system

$$\dot{x}_1 = -x_1 + \gamma(t)x_2$$
$$\dot{x}_2 = -\gamma(t)x_1 - x_2 + h(t)\cos x_1$$

where $\gamma(t)$ and $h(t)$ are continuous functions, and $|h(t)| \leq H$ for all $t \geq 0$. Show that the solutions of the system are globally uniformly ultimately bounded and estimate the ultimate bound.

Exercise 5.15 Consider the system

$$\dot{x}_1 = -x_1^3 + x_2^5 - \gamma x_2$$
$$\dot{x}_2 = -x_1^3 - x_2^5 + \gamma(x_1 + x_2), \quad 0 \leq \gamma \leq \frac{1}{2}$$

(a) With $\gamma = 0$, show that the origin is globally asymptotically stable. Is it exponentially stable?

(b) With $0 < \gamma \leq \frac{1}{2}$, show that the origin is unstable and the solutions of the system are globally ultimately bounded by an ultimate bound which is a class $\mathcal{K}$ function of γ.

Exercise 5.16 ([14]) Consider the system

$$\dot{x}_1 = x_2$$
$$\dot{x}_2 = -a \sin x_1 - bx_1 - cx_2 - \gamma(cx_1 + 2x_2) + q(t)\cos x_1$$

where a, $b > a$, c, and γ are positive constants, and $q(t)$ is a continuous function of t.

(a) With $q(t) \equiv 0$, use $V(x) = (b + \frac{1}{2}c^2)x_1^2 + cx_1x_2 + x_2^2 + 2a(1 - \cos x_1)$ to show that the origin is globally exponentially stable.

(b) Study stability of the system when $q(t) \neq 0$ and $|q(t)| \leq k$ for all $t \geq 0$.

Exercise 5.17 Consider the system

$$\dot{x}_1 = \left[(\sin x_2)^2 - 1\right] x_1$$
$$\dot{x}_2 = -bx_1 - (1 + b)x_2$$

(a) With $b = 0$, show that the origin is exponentially stable and globally asymptotically stable.

(b) With $b \neq 0$, show that the origin is exponentially stable for sufficiently small $|b|$ but not globally asymptotically stable, no matter how small $|b|$ is.

(c) Discuss the results of parts (a) and (b) in view of the robustness results of Section 5.1, and show that when $b = 0$ the origin is not globally exponentially stable.

Exercise 5.18 ([6]) Consider the system

$$\dot{x}_1 = -x_1 + (x_1 + a)x_2, \quad a \neq 0$$
$$\dot{x}_2 = -x_1(x_1 + a) + bx_2$$

(a) Let $b = 0$. Show that the origin is globally asymptotically stable. Is it exponentially stable?

(b) Let $b > 0$. Show that the origin is exponentially stable for $b < \min\{1, a^2\}$.

(c) Show that the origin is not globally asymptotically stable for any $b > 0$.

(d) Discuss the results of parts (a)–(c) in view of the robustness results of Section 5.1, and show that when $b = 0$ the origin is not globally exponentially stable.

Hint: In part (d), note that the Jacobian matrix of the nominal system is not globally bounded.

Exercise 5.19 Consider the scalar system $\dot{x} = -x/(1 + x^2)$ and $V(x) = x^4$.

(a) Show that inequalities (5.20)–(5.22) are satisfied globally with

$$\alpha_1(r) = \alpha_2(r) = r^4; \quad \alpha_3(r) = \frac{4r^4}{1 + r^2}; \quad \alpha_4(r) = 4r^3$$

(b) Verify that these functions belong to class $\mathcal{K}_\infty$.

(c) Show that the right-hand side of (5.23) approaches zero as $r \to \infty$.

(d) Consider the perturbed system $\dot{x} = -x/(1 + x^2) + \delta$, where δ is a positive constant. Show that whenever $\delta > \frac{1}{2}$, the solution $x(t)$ escapes to ∞ for any initial state $x(0)$.

Exercise 5.20 Consider the scalar system $\dot{x} = -x^3 + e^{-t}$. Show that $x(t) \to 0$ as $t \to \infty$.

Exercise 5.21 For each of the following scalar systems, investigate the input-to-state stability.

$$(1) \quad \dot{x} = -(1 + u)x^3 \qquad\qquad (2) \quad \dot{x} = -(1 + u)x^3 - x^5$$

$$(3) \quad \dot{x} = -x + x^2 u \qquad\qquad (4) \quad \dot{x} = x - x^3 + u$$

Exercise 5.22 For each of the following systems, investigate the input-to-state stability.

(1)
$$\begin{aligned}
\dot{x}_1 &= -x_1 + x_2^2 \\
\dot{x}_2 &= -x_2 + u
\end{aligned}$$

(2)
$$\begin{aligned}
\dot{x}_1 &= (x_1 - x_2 + u)(x_1^2 + x_2^2 - 1) \\
\dot{x}_2 &= (x_1 + x_2 + u)(x_1^2 + x_2^2 - 1)
\end{aligned}$$

(3)
$$\begin{aligned}
\dot{x}_1 &= -x_1 + x_1^2 x_2 \\
\dot{x}_2 &= -x_2 + x_1 + u
\end{aligned}$$

(4)
$$\begin{aligned}
\dot{x}_1 &= -x_1 - x_2 + u_1 \\
\dot{x}_2 &= x_1 - x_2^3 + u_2
\end{aligned}$$

Exercise 5.23 *Consider the system*

$$\begin{aligned}
\dot{\eta} &= f_0(\eta, \xi) \\
\dot{\xi} &= A\xi + Bu
\end{aligned}$$

where $\eta \in R^{n-r}$, $\xi \in R^r$ for some $1 \le r < n$, (A, B) is controllable, and the system $\dot{\eta} = f_0(\eta, \xi)$, with ξ viewed as the input, is locally input-to-state stable. Find a state feedback control $u = \gamma(\eta, \xi)$ that stabilizes the origin of the full system. If $\dot{\eta} = f_0(\eta, \xi)$ is input-to-state stable, design the feedback control such that the origin of the closed-loop system is globally asymptotically stable.

Exercise 5.24 Using Lemma 5.6, show that the origin of the system

$$\begin{aligned}
\dot{x}_1 &= -x_1^3 + x_2 \\
\dot{x}_2 &= -x_2^3
\end{aligned}$$

is asymptotically stable.

Exercise 5.25 Prove another version of Theorem 5.2, where all the assumptions are the same except that inequality (5.29) is replaced by

$$\frac{\partial V}{\partial t} + \frac{\partial V}{\partial x} f(t, x, u) \le -\alpha_3(\|x\|) + \psi(u)$$

where $\alpha_3(\cdot)$ is a class $\mathcal{K}_\infty$ function and $\psi(u)$ is a continuous function of u with $\psi(0) = 0$.

Exercise 5.26 Consider the system

$$\begin{aligned}
\dot{x}_1 &= x_1 \left[\left(\sin \frac{\pi x_2}{2} \right)^2 - 1 \right] \\
\dot{x}_2 &= -x_2 + u
\end{aligned}$$

(a) With $u = 0$, show that the origin is globally asymptotically stable.

(b) Show that for any bounded input $u(t)$, the state $x(t)$ is bounded.

(c) With $u(t) \equiv 1$, $x_1(0) = a$, and $x_2(0) = 1$, show that the solution is $x_1(t) \equiv a$, $x_2(t) \equiv 1$.

(d) Is the system input-to-state stable?

Exercise 5.27 Verify that $D^+W(t)$ satisfies (5.41) when $V(t) = 0$.
Hint: Show that $V(t + h, x(t + h)) \leq \frac{1}{2}c_4h\|g(t, 0)\|^2 + h\, o(h)$, where $o(h)/h \to 0$ as $h \to 0$. Then, use the fact that $\sqrt{c_4/2c_1} \geq 1$.

Exercise 5.28 Consider the linear system of Example 5.9, but change the assumption on $B(t)$ to $\int_0^\infty \|B(t)\|\, dt < \infty$. Show that the origin is exponentially stable.

Exercise 5.29 Consider the linear system of Example 5.9, but change the assumption on $B(t)$ to $\int_0^\infty \|B(t)\|^2\, dt < \infty$. Show that the origin is exponentially stable.
Hint: Use the inequality

$$\int_a^b v(t)\, dt \leq \sqrt{(b - a)\int_a^b v^2(t)\, dt}, \quad \forall\, v(t) \geq 0$$

which follows from the Cauchy-Schwartz inequality.

Exercise 5.30 Consider the system $\dot{x} = A(t)x$ where $A(t)$ is continuous. Suppose the limit $\lim_{t \to \infty} A(t) = \bar{A}$ exists and $\bar{A}$ is Hurwitz. Show that the origin is exponentially stable.

Exercise 5.31 Repeat part(b) of Exercise 5.16 when $q(t)$ is bounded and $q(t) \to 0$ as $t \to \infty$.

Exercise 5.32 Consider the system $\dot{x} = f(t, x)$ where $\|f(t, x) - f(0, x)\|_2 \leq \gamma(t)\|x\|_2$ for all $t \geq 0$, $x \in R^2$, $\gamma(t) \to 0$ as $t \to \infty$,

$$f(0, x) = Ax - (x_1^2 + x_2^2)Bx, \quad A = \begin{bmatrix} -\alpha & -\omega \\ \omega & -\alpha \end{bmatrix}, \quad B = \begin{bmatrix} \beta & \Omega \\ -\Omega & \beta \end{bmatrix}$$

and α, β, ω, Ω are positive constants. Show that the origin is globally exponentially stable.

Exercise 5.33 Consider the perturbed system (5.1). Suppose there is a Lyapunov function $V(t, x)$ that satisfies (5.3)–(5.5), and the perturbation term satisfies $\|g(t, x)\| \leq \delta(t)$, $\forall\, t \geq 0$, $\forall\, x \in D$. Show that for any $\epsilon > 0$ and $\Delta > 0$, there exist $\eta > 0$ and $\rho > 0$ such that whenever $\frac{1}{\Delta}\int_t^{t+\Delta} \delta(\tau)\, d\tau < \eta$, every solution of the perturbed system with $\|x(t_0)\| < \rho$ will satisfy $\|x(t)\| < \epsilon$, $\forall\, t \geq t_0$.
(This result is known as total stability in the presence of perturbation that is bounded in the mean [96].)

Hint: Choose $W = \sqrt{V}$, discretize the time interval with sampling points at $t_0 + i\Delta$ for $i = 0, 1, 2, \ldots$, and show that $W(t_0 + i\Delta)$ satisfies the difference inequality

$$W(t_0 + (i+1)\Delta) \leq e^{-\sigma\Delta} W(t_0 + i\Delta) + k\eta\Delta$$

Exercise 5.34 Let A be an $n \times n$ matrix with $a_{ij} \leq 0$ for all $i \neq j$ and $a_{ii} > \sum_{j \neq i} |a_{ij}|$, $i = 1, 2, \ldots, n$. Show that A is an M-matrix.
Hint: Show that $\sum_{j=1}^{n} a_{ij} > 0$ for $i = 1, \ldots, n$, and use mathematical induction to show that all the leading principal minors are positive.

Exercise 5.35 Suppose the conditions of Theorem 5.4 are satisfied with

$$\phi_i(x_i) = \|x_i\| \quad \text{and} \quad c_{i1}\|x_i\|^2 \leq V_i(t, x_i) \leq c_{i2}\|x_i\|^2$$

Show that the origin is exponentially stable.

Exercise 5.36 ([115]) Study the stability of the origin of the system

$$\begin{aligned}
\dot{x}_1 &= -x_1^3 - 1.5x_1|x_2|^3 \\
\dot{x}_2 &= -x_2^5 + x_1^2 x_2^2
\end{aligned}$$

using composite Lyapunov analysis.

Exercise 5.37 Study the stability of the origin of the system

$$\begin{aligned}
\dot{x}_1 &= x_2 + x_2 x_3^3 \\
\dot{x}_2 &= -x_1 - x_2 + x_1^2 \\
\dot{x}_3 &= x_1 + x_2 - x_3^3
\end{aligned}$$

using composite Lyapunov analysis.

Exercise 5.38 Consider the linear interconnected system

$$\dot{x}_i = A_{ii}x_i + \sum_{j=1; j \neq i}^{m} A_{ij}x_j, \quad i = 1, 2, \ldots, m$$

where x_i are n_i-dimensional vectors and A_{ii} are Hurwitz matrices. Study the stability of the origin using composite Lyapunov analysis.

Exercise 5.39 ([156]) Complex interconnected systems could be subject to structural perturbations which cause groups of subsystems to be connected or disconnected from each other during operation. Such structural perturbations can be represented as

$$\dot{x}_i = f_i(t, x_i) + g_i(t, e_{i1}x_1, \ldots, e_{im}x_m), \quad i = 1, 2, \ldots, m$$

where e_{ij} is a binary variable that takes the value 1 when the jth subsystem acts on the ith subsystem and takes the value 0 otherwise. The origin of the interconnected system is said to be connectively asymptotically stable if it is asymptotically stable for all interconnection patterns, that is, for all possible values of the binary variables e_{ij}. Suppose that all the assumptions of Theorem 5.4 are satisfied, with (5.61) taking the form

$$\|g_i(t, e_{i1}x_1, \ldots, e_{im}x_m)\| \le \sum_{i=1}^{m} e_{ij}\gamma_{ij}\phi_j(x_j)$$

Show that the origin is connectively asymptotically stable.

Exercise 5.40 ([40]) The output $y(t)$ of the linear system

$$\dot{x} = Ax + Bu, \quad y = Cx$$

is required to track a reference input r. Consider the integral controller

$$\dot{z} = r - Cx, \quad u = F_1x + F_2z$$

where we have assumed that the state x can be measured and the matrices F_1 and F_2 can be designed such that the matrix

$$\left[\begin{array}{cc} A + BF_1 & BF_2 \\ -C & 0 \end{array} \right]$$

is Hurwitz.

(a) Show that if $r = \text{constant}$, then $y(t) \to r$ as $t \to \infty$.

(b) Study the tracking properties of the system when $r(t)$ is a slowly varying input.

Exercise 5.41 ([77]) The output $y(t)$ of the nonlinear system

$$\dot{x} = f(x, u), \quad y = h(x)$$

is required to track a reference input r. Consider the integral controller

$$\dot{z} = r - h(x), \quad u = \gamma(x, z, r)$$

where we have assumed that the state x can be measured, the function γ can be designed such that the closed-loop system

$$\begin{aligned} \dot{x} &= f(x, \gamma(x, z, r)) \\ \dot{z} &= r - h(x) \end{aligned}$$

has an exponentially stable equilibrium point $(\bar{x}, \bar{z})$, and the functions f, h, and γ are twice continuously differentiable in their arguments.

(a) Show that if $r = $ constant and the initial state $(x(0), z(0))$ is sufficiently close to $(\bar{x}, \bar{z})$, then $y(t) \to r$ as $t \to \infty$.

(b) Study the tracking properties of the system when $r(t)$ is a slowly varying input.

Exercise 5.42 ([77]) Consider the tracking problem of Exercise 5.41, but assume that we can only measure $y = h(x)$. Consider the observer-based integral controller

$$
\begin{aligned}
\dot{z}_1 &= f(z_1, u) + G(r)[y - h(z_1)] \\
\dot{z}_2 &= r - y \\
u &= \gamma(z_1, z_2, r)
\end{aligned}
$$

Suppose that γ and G can be designed such that the closed-loop system has an exponentially stable equilibrium point $(\bar{x}, \bar{z}_1, \bar{z}_2)$. Study the tracking properties of the system when

(1) $r = $ constant;   **(2)** $r(t)$ is slowly varying.

Exercise 5.43 Consider the linear system $\dot{x} = A(t)x$ where $\|A(t)\| \leq k$ and the eigenvalues of $A(t)$ satisfy $Re[\lambda(t)] \leq -\sigma$ for all $t \geq 0$. Suppose that $\int_0^\infty \|\dot{A}(t)\|^2 \, dt \leq \rho$. Show that the origin of $\dot{x} = A(t)x$ is exponentially stable.

Chapter 6

Input-Output Stability

In most of this book, we use differential equations or state-space models to model nonlinear dynamical systems. A state-space model takes the general form

$$\dot{x}(t) = f(t, x(t), u(t))$$
$$y(t) = h(t, x(t), u(t))$$

where u, y, and x are the input, output, and state variables, respectively. In the state-space approach, we place a lot of emphasis on the behavior of the state variables. This is so true that most of our analysis, so far, has dealt with the state equation, and most of the time without an explicit presence of an external input u. An alternative approach to the mathematical modeling of dynamical systems is the input-output approach.[1] An input-output model relates the output of the system directly to the input, with no knowledge of the internal structure of the system that is represented by the state equation. The system is viewed as a black box that can be accessed only through its input and output terminals. In Section 6.1, we introduce input-output mathematical models and define $\mathcal{L}$ stability, a concept of stability in the input-output sense. In Section 6.2, we study $\mathcal{L}$ stability of nonlinear systems represented by state-space models. In Section 6.3, we introduce the notion of input-to-output stability, which is defined specifically for state-space models. Finally, in Section 6.4 we discuss the calculation of the $\mathcal{L}_2$ gain for a class of autonomous systems.

---

[1] In this chapter, we give just enough of a glimpse of the input-output approach to relate Lyapunov stability to input-output stability and introduce the terminology needed to state a version of the small-gain theorem in Chapter 10. For a comprehensive treatment of the subject, the reader may consult [43], [184], or [146]. The foundation of the input-output approach to nonlinear systems can be found in the work of Sandberg and Zames in the sixties; see, for example, [148], [192], and [193].

6.1 $\mathcal{L}$ Stability

We consider a system whose input-output relation is represented by

$$y = Hu$$

where H is some mapping or operator that specifies y in terms of u. The input u belongs to a space of signals, which is often taken as a normed linear space $\mathcal{L}^m$ of functions $u : [0, \infty) \to R^m$. Examples are the space of piecewise continuous, uniformly bounded functions $(\mathcal{L}^m_\infty)$ with the norm[2]

$$\|u\|_{\mathcal{L}_\infty} = \sup_{t \geq 0} \|u(t)\| < \infty$$

and the space of piecewise continuous, square-integrable functions $(\mathcal{L}^m_2)$ with the norm

$$\|u\|_{\mathcal{L}_2} = \sqrt{\int_0^\infty u^T(t)u(t)\, dt} < \infty$$

More generally, the space $\mathcal{L}^m_p$ for $1 \leq p < \infty$ is defined as the set of all piecewise continuous functions $u : [0, \infty) \to R^m$ such that

$$\|u\|_{\mathcal{L}_p} = \left(\int_0^\infty \|u(t)\|^p\, dt \right)^{1/p} < \infty$$

It can be verified that $\mathcal{L}^m_p$, for each $p \in [1, \infty]$, is a normed linear space.[3]

If we think of $u \in \mathcal{L}^m$ as a "well-behaved" input, the question to ask is whether the output y will be "well behaved" in the sense that $y \in \mathcal{L}^q$ where $\mathcal{L}^q$ is the same normed linear space as $\mathcal{L}^m$, except that the number of output variables q is, in general, different from the number of input variables m. A system that has the property that any "well-behaved" input will generate a "well-behaved" output will be defined as a stable system. However, we cannot define H as a mapping from $\mathcal{L}^m$ to $\mathcal{L}^q$ because we have to deal with systems which are unstable, in that an input $u \in \mathcal{L}^m$ may generate an output y that does not belong to $\mathcal{L}^q$. Therefore, H is usually defined as a mapping from an extended space $\mathcal{L}^m_e$ to an extended space $\mathcal{L}^q_e$, where $\mathcal{L}^m_e$ is defined by

$$\mathcal{L}^m_e = \{u \mid u_\tau \in \mathcal{L}^m, \forall\, \tau \geq 0\}$$

---

[2] To distinguish the norm of u as a vector in the space $\mathcal{L}$ from the norm of $u(t)$ as a vector in R^m, we write the first norm as $\| \cdot \|_{\mathcal{L}}$. Note that the norm $\| \cdot \|$ used in the definition of $\| \cdot \|_{\mathcal{L}_p}$, for any $p \in [1, \infty]$, can be any p-norm in R^m; the number p is not necessarily the same in the two norms. For example, we may define the $\mathcal{L}_\infty$ space with $\|u\|_{\mathcal{L}_\infty} = \sup_{t \geq 0} \|u(t)\|_1$, $\|u\|_{\mathcal{L}_\infty} = \sup_{t \geq 0} \|u(t)\|_2$, or $\|u\|_{\mathcal{L}_\infty} = \sup_{t \geq 0} \|u(t)\|_\infty$. However, it is common to define the $\mathcal{L}_2$ space with the 2-norm in R^m.

[3] See Exercise 2.8 and 2.11.

and u_τ is a truncation of u defined by

$$u_\tau(t) = \left\{ \begin{array}{ll} u(t), & 0 \le t \le \tau \\ 0, & t > \tau \end{array} \right.$$

The extended space $\mathcal{L}_e^m$ is a linear space that contains the unextended space $\mathcal{L}^m$ as a subset. It allows us to deal with unbounded "ever-growing" signals. For example, the signal $u(t) = t$ does not belong to the space $\mathcal{L}_\infty$, but its truncation

$$u_\tau(t) = \left\{ \begin{array}{ll} t, & 0 \le t \le \tau \\ 0, & t > \tau \end{array} \right.$$

belongs to $\mathcal{L}_\infty$ for every finite τ. Hence, $u(t) = t$ belongs to the extended space $\mathcal{L}_{\infty e}$.

A mapping $H : \mathcal{L}_e^m \to \mathcal{L}_e^q$ is said to be causal if the value of the output $(Hu)(t)$ at any time t depends only on the values of the input up to time t. This is equivalent to

$$(Hu)_\tau = (Hu_\tau)_\tau$$

Causality is an intrinsic property of dynamical systems represented by state-space models.

With the space of input and output signals defined, we can now define input-output stability.

Definition 6.1 *A mapping* $H : \mathcal{L}_e^m \to \mathcal{L}_e^q$ *is* $\mathcal{L}$ *stable if there exist a class* $\mathcal{K}$ *function* α, *defined on* $[0, \infty)$, *and a nonnegative constant* β *such that*

$$\|(Hu)_\tau\|_{\mathcal{L}} \le \alpha (\|u_\tau\|_{\mathcal{L}}) + \beta \tag{6.1}$$

for all $u \in \mathcal{L}_e^m$ *and* $\tau \in [0, \infty)$. *It is finite-gain* $\mathcal{L}$ *stable if there exist nonnegative constants* γ *and* β *such that*

$$\|(Hu)_\tau\|_{\mathcal{L}} \le \gamma \|u_\tau\|_{\mathcal{L}} + \beta \tag{6.2}$$

for all $u \in \mathcal{L}_e^m$ *and* $\tau \in [0, \infty)$.

The constant β in (6.1) or (6.2) is called the bias term. It is included in the definition to allow for systems where Hu does not vanish at $u = 0$.[4] When inequality (6.2) is satisfied, we are usually interested in the smallest possible γ for which there is β such that (6.2) is satisfied. When this value of γ is well defined, we shall call it the gain of the system. When inequality (6.2) is satisfied with some $\gamma \ge 0$, we say that the system has an $\mathcal{L}$ gain less than or equal to γ.

---

[4] See Exercise 6.3 for a different role of the bias term.

For causal, $\mathcal{L}$ stable systems, it can be shown by a simple argument that

$$u \in \mathcal{L}^m \Rightarrow Hu \in \mathcal{L}^q$$

and

$$\|Hu\|_{\mathcal{L}} \leq \alpha\left(\|u\|_{\mathcal{L}}\right) + \beta, \quad \forall\, u \in \mathcal{L}^m$$

For causal, finite-gain $\mathcal{L}$ stable systems, the foregoing inequality takes the form

$$\|Hu\|_{\mathcal{L}} \leq \gamma\|u_\tau\|_{\mathcal{L}} + \beta, \quad \forall\, u \in \mathcal{L}^m$$

If $\mathcal{L}$ is the space of uniformly bounded signals ($\mathcal{L}_\infty$), the definition of $\mathcal{L}_\infty$ stability becomes the familiar notion of bounded input-bounded output stability; namely, if the system is $\mathcal{L}_\infty$ stable, then for every bounded input $u(t)$, the output $Hu(t)$ is bounded.

Example 6.1 A memoryless, possibly time-varying, function $h : [0, \infty) \times R \to R$ can be viewed as an operator H that assigns to every input signal $u(t)$ the output signal $y(t) = h(t, u(t))$. We use this simple operator to illustrate the definition of $\mathcal{L}$ stability. Let

$$h(u) = a + b \tanh cu = a + b\, \frac{e^{cu} - e^{-cu}}{e^{cu} + e^{-cu}}$$

for some nonnegative constants a, b, c. Using the fact

$$h'(u) = \frac{4bc}{\left(e^{cu} + e^{-cu}\right)^2} \leq bc, \quad \forall\, u \in R$$

we have

$$|h(u)| \leq a + bc|u|, \quad \forall\, u \in R$$

Hence, H is finite-gain $\mathcal{L}_\infty$ stable with $\gamma = bc$ and $\beta = a$. Furthermore, if $a = 0$, then for each $p \in [1, \infty)$,

$$\int_0^\infty |h(u(t))|^p \, dt \leq (bc)^p \int_0^\infty |u(t)|^p \, dt$$

Hence, for each $p \in [1, \infty]$, the operator H is finite-gain $\mathcal{L}_p$ stable with zero bias and $\gamma = bc$.

Let h be a time-varying function that satisfies

$$|h(t, u)| \leq a|u|, \quad \forall\, t \geq 0, \ \forall\, u \in R$$

for some positive constant a. For each $p \in [1, \infty]$, the operator H is finite-gain $\mathcal{L}_p$ stable with zero bias and $\gamma = a$.

Finally, let

$$h(u) = u^2$$

Since
$$\sup_{t \geq 0} |h(u(t))| \leq \left(\sup_{t \geq 0} |u(t)| \right)^2$$

H is $\mathcal{L}_\infty$ stable with zero bias and $\alpha(r) = r^2$. Notice that H is not finite-gain $\mathcal{L}_\infty$ stable because the function $h(u) = u^2$ cannot be bounded by a straight line of the form $|h(u)| \leq \gamma |u| + \beta$, for all $u \in R$. $\triangle$

Example 6.2 Consider a single-input–single-output system defined by the causal convolution operator
$$y(t) = \int_0^t h(t - \sigma) u(\sigma)\, d\sigma$$

where $h(t) = 0$ for $t < 0$. Suppose $h \in \mathcal{L}_{1e}$; that is, for every $\tau \in [0, \infty)$,
$$\|h_\tau\|_{\mathcal{L}_1} = \int_0^\infty |h_\tau(\sigma)|\, d\sigma = \int_0^\tau |h(\sigma)|\, d\sigma < \infty$$

If $u \in \mathcal{L}_{\infty e}$ and $\tau \geq t$, then
$$
\begin{aligned}
|y(t)| &\leq \int_0^t |h(t - \sigma)|\, |u(\sigma)|\, d\sigma \\
&\leq \int_0^t |h(t - \sigma)|\, d\sigma \sup_{0 \leq \sigma \leq \tau} |u(\sigma)| = \int_0^t |h(s)|\, ds \sup_{0 \leq \sigma \leq \tau} |u(\sigma)|
\end{aligned}
$$

Hence,
$$\|y_\tau\|_{\mathcal{L}_\infty} \leq \|h_\tau\|_{\mathcal{L}_1} \|u_\tau\|_{\mathcal{L}_\infty}, \quad \forall\, \tau \in [0, \infty)$$

This inequality resembles (6.2), but is not the same as (6.2) because the constant γ in (6.2) is required to be independent of τ. While $\|h_\tau\|_{\mathcal{L}_1}$ is finite for every finite τ, it may not be bounded uniformly in τ. For example, $h(t) = e^t$ has $\|h_\tau\|_{\mathcal{L}_1} = (e^\tau - 1)$ which is finite for all $\tau \in [0, \infty)$ but not uniformly bounded in τ. Inequality (6.2) will be satisfied if $h \in \mathcal{L}_1$; that is,
$$\|h\|_{\mathcal{L}_1} = \int_0^\infty |h(\sigma)|\, d\sigma < \infty$$

Then, the inequality
$$\|y_\tau\|_{\mathcal{L}_\infty} \leq \|h\|_{\mathcal{L}_1} \|u_\tau\|_{\mathcal{L}_\infty}, \quad \forall\, \tau \in [0, \infty)$$

shows that the system is finite-gain $\mathcal{L}_\infty$ stable. The condition $\|h\|_{\mathcal{L}_1} < \infty$ actually guarantees finite-gain $\mathcal{L}_p$ stability for each $p \in [1, \infty]$. Consider first the case $p = 1$. For $t \leq \tau < \infty$, we have
$$\int_0^\tau |y(t)|\, dt = \int_0^\tau \left| \int_0^t h(t - \sigma) u(\sigma)\, d\sigma \right|\, dt \leq \int_0^\tau \int_0^t |h(t - \sigma)|\, |u(\sigma)|\, d\sigma\, dt$$

Reversing the order of integration yields

$$\int_0^\tau |y(t)|\, dt \le \int_0^\tau |u(\sigma)| \int_\sigma^\tau |h(t-\sigma)|\, dt\, d\sigma \le \int_0^\tau |u(\sigma)|\, \|h\|_{\mathcal{L}_1}\, d\sigma \le \|h\|_{\mathcal{L}_1} \|u_\tau\|_{\mathcal{L}_1}$$

Thus,

$$\|y_\tau\|_{\mathcal{L}_1} \le \|h\|_{\mathcal{L}_1} \|u_\tau\|_{\mathcal{L}_1}, \quad \forall\, \tau \in [0,\infty)$$

Consider now the case $p \in (1,\infty)$, and let $q \in (1,\infty)$ be defined by $1/p + 1/q = 1$. For $t \le \tau < \infty$, we have

$$
\begin{aligned}
|y(t)| &\le \int_0^t |h(t-\sigma)|\, |u(\sigma)|\, d\sigma \\
&= \int_0^t |h(t-\sigma)|^{1/q} |h(t-\sigma)|^{1/p} |u(\sigma)|\, d\sigma \\
&\le \left(\int_0^t |h(t-\sigma)|\, d\sigma \right)^{1/q} \left(\int_0^t |h(t-\sigma)|\, |u(\sigma)|^p\, d\sigma \right)^{1/p} \\
&\le (\|h_\tau\|_{\mathcal{L}_1})^{1/q} \left(\int_0^t |h(t-\sigma)|\, |u(\sigma)|^p\, d\sigma \right)^{1/p}
\end{aligned}
$$

where the second inequality is obtained by applying Hölder's inequality.[5] Thus,

$$
\begin{aligned}
(\|y_\tau\|_{\mathcal{L}_p})^p &= \int_0^\tau |y(t)|^p\, dt \\
&\le \int_0^\tau (\|h_\tau\|_{\mathcal{L}_1})^{p/q} \left(\int_0^t |h(t-\sigma)|\, |u(\sigma)|^p\, d\sigma \right) dt \\
&= (\|h_\tau\|_{\mathcal{L}_1})^{p/q} \int_0^\tau \int_0^t |h(t-\sigma)|\, |u(\sigma)|^p\, d\sigma\, dt
\end{aligned}
$$

By reversing the order of integration, we obtain

$$
\begin{aligned}
(\|y_\tau\|_{\mathcal{L}_p})^p &\le (\|h_\tau\|_{\mathcal{L}_1})^{p/q} \int_0^\tau |u(\sigma)|^p \int_\sigma^\tau |h(t-\sigma)|\, dt\, d\sigma \\
&\le (\|h_\tau\|_{\mathcal{L}_1})^{p/q} \|h_\tau\|_{\mathcal{L}_1} (\|u_\tau\|_{\mathcal{L}_1})^p = (\|h_\tau\|_{\mathcal{L}_1})^p (\|u_\tau\|_{\mathcal{L}_p})^p
\end{aligned}
$$

Hence,

$$\|y_\tau\|_{\mathcal{L}_p} \le \|h\|_{\mathcal{L}_1} \|u_\tau\|_{\mathcal{L}_p}$$

---

[5]Hölder's inequality states that if $f \in \mathcal{L}_{pe}$ and $g \in \mathcal{L}_{pe}$, where $p \in [0,\infty]$ and $1/p + 1/q = 1$, then

$$\int_0^\tau |f(t)g(t)|\, dt \le \left(\int_0^\tau |f(t)|^p\, dt \right)^{1/p} \left(\int_0^\tau |g(t)|^q\, dt \right)^{1/q}$$

for every $\tau \in [0,\infty)$. See [10].

In summary, if $\|h\|_{\mathcal{L}_1} < \infty$, then for each $p \in [1, \infty]$, the causal convolution operator is finite-gain $\mathcal{L}_p$ stable and (6.2) is satisfied with $\gamma = \|h\|_{\mathcal{L}_1}$ and $\beta = 0$. $\triangle$

Example 6.3 Consider the linear time-invariant system

$$\begin{aligned} \dot{x} &= Ax + Bu, \quad x(0) = x_0 \\ y &= Cx + Du \end{aligned}$$

where A is a Hurwitz matrix. The output $y(t)$ is given by

$$y(t) = Ce^{At}x_0 + \int_0^t Ce^{(t-\tau)A}Bu(\tau)\,d\tau + Du(t)$$

For each fixed $x_0 \in R^n$, the foregoing expression defines an operator H that assigns to each input signal $u(t)$ the corresponding output signal $y(t)$. Due to the Hurwitz property of A, we have

$$\|e^{At}\| \le ke^{-at}, \quad \forall\, t \ge 0$$

for some positive constants k and a.[6] Hence,

$$\|y(t)\| \le k_1 e^{-at} + k_2 \int_0^t e^{-a(t-\tau)}\|u(\tau)\|\,d\tau + k_3\,\|u(t)\| \tag{6.3}$$

where

$$k_1 = k\|C\|\,\|x_0\|, \quad k_2 = k\|B\|\,\|C\|, \quad k_3 = \|D\|$$

Set

$$y_1(t) = k_1 e^{-at}, \quad y_2(t) = k_2 \int_0^t e^{-a(t-\tau)}\|u(\tau)\|\,d\tau, \quad y_3(t) = k_3\,\|u(t)\|$$

Suppose now that $u \in \mathcal{L}_{pe}^m$ for some $p \in [1, \infty]$. Using the results of the previous example, it can be easily verified that

$$\|y_{2\tau}\|_{\mathcal{L}_p} \le \frac{k_2}{a}\|u_\tau\|_{\mathcal{L}_p}$$

It is also straightforward to see that

$$\|y_{3\tau}\|_{\mathcal{L}_p} \le k_3\|u_\tau\|_{\mathcal{L}_p}$$

As for the first term, $y_1(t)$, it can be verified that

$$\|y_{1\tau}\|_{\mathcal{L}_p} \le k_1\rho, \quad \text{where} \quad \rho = \begin{cases} 1, & \text{if } p = \infty \\ \left(\dfrac{1}{ap}\right)^{1/p}, & \text{if } p \in [1, \infty) \end{cases}$$

---

[6] See (3.11).

Thus, by the triangle inequality, (6.2) is satisfied with

$$\gamma = k_3 + \frac{k_2}{a}, \quad \beta = k_1 \rho$$

We conclude that the linear time-invariant system $\{A, B, C, D\}$, with a Hurwitz matrix A, is finite-gain $\mathcal{L}_p$ stable for each $p \in [1, \infty]$. Notice that the bias term β in (6.2) is proportional to $\|x_0\|$. When $x_0 = 0$, (6.2) is satisfied with zero bias. $\triangle$

One drawback of Definition 6.1 is the implicit requirement that inequality (6.1) or (6.2) be satisfied for all signals in the input space $\mathcal{L}^m$. This excludes systems where the input-output relation may be defined only for a subset of the input space. The following example explores this point and motivates the definition of small-signal $\mathcal{L}$ stability that follows the example.

Example 6.4 Consider a single-input–single-output system defined by the nonlinearity

$$y = \tan u$$

The output $y(t)$ is defined only when the input signal is restricted to

$$|u(t)| < \frac{\pi}{2}, \quad \forall \, t \geq 0$$

Thus, the system is not $\mathcal{L}_\infty$ stable in the sense of Definition 6.1. Notice, however, that if we restrict $u(t)$ to the set

$$|u| \leq r < \frac{\pi}{2}$$

then

$$|y| \leq \left(\frac{\tan r}{r} \right) |u|$$

and the system will satisfy the inequality

$$\|y\|_{\mathcal{L}_p} \leq \left(\frac{\tan r}{r} \right) \|u\|_{\mathcal{L}_p}$$

for every $u \in \mathcal{L}_p$ such that $|u(t)| \leq r$ for all $t \geq 0$, where p could be any number in $[1, \infty]$. In the space $\mathcal{L}_\infty$, the requirement $|u(t)| \leq r$ implies that $\|u\|_{\mathcal{L}_\infty} \leq r$, showing that the foregoing inequality holds only for input signals of small norm. Notice, however, that for other $\mathcal{L}_p$ spaces with $p < \infty$, the instantaneous bound on $|u(t)|$ does not necessarily restrict the norm of the input signal. For example, the signal

$$u(t) = re^{-rt/a}, \quad a > 0$$

which belongs to $\mathcal{L}_p$ for each $p \in [1, \infty]$, satisfies the instantaneous bound $|u(t)| \leq r$ while its $\mathcal{L}_p$ norm

$$\|u\|_{\mathcal{L}_p} = r \left(\frac{a}{rp} \right)^{1/p} , \quad 1 \leq p < \infty$$

can be arbitrarily large. $\triangle$

Definition 6.2 *A mapping $H : \mathcal{L}_e^m \to \mathcal{L}_e^q$ is small-signal $\mathcal{L}$ stable (respectively, small-signal finite-gain $\mathcal{L}$ stable) if there is a positive constant r such that inequality (6.1) (respectively, (6.2)) is satisfied for all $u \in \mathcal{L}_e^m$ with $\sup_{0 \leq t \leq \tau} \|u(t)\| \leq r$.*

6.2 $\mathcal{L}$ Stability of State Models

The notion of input-output stability is intuitively appealing. This is probably the reason why most of us were introduced to dynamical system stability in the framework of bounded input-bounded output stability. Since in Lyapunov stability we put a lot of emphasis on studying the stability of equilibrium points and the asymptotic behavior of state variables, one may wonder about what we can see about input-output stability starting from the formalism of Lyapunov stability. In this section, we show how Lyapunov stability tools can be used to establish $\mathcal{L}$ stability of nonlinear systems represented by state-space models.

Consider the system

$$\dot{x} = f(t, x, u), \quad x(0) = x_0 \tag{6.4}$$

$$y = h(t, x, u) \tag{6.5}$$

where $x \in R^n$, $u \in R^m$, $y \in R^q$, $f : [0, \infty) \times D \times D_u \to R^n$ is piecewise continuous in t and locally Lipschitz in (x, u), $h : [0, \infty) \times D \times D_u \to R^q$ is piecewise continuous in t and continuous in (x, u), $D \subset R^n$ is a domain that contains $x = 0$, and $D_u \in R^m$ is a domain that contains $u = 0$. For each fixed $x_0 \in R^n$, the state-space model (6.4)–(6.5) defines an operator H that assigns to each input signal $u(t)$ the corresponding output signal $y(t)$. Suppose $x = 0$ is an equilibrium point of the unforced system

$$\dot{x} = f(t, x, 0) \tag{6.6}$$

The theme of this section is that if the origin of (6.6) is uniformly asymptotically stable (or exponentially stable), then, under some assumptions on f and h, the system (6.4)–(6.5) will be $\mathcal{L}$ stable or small-signal $\mathcal{L}$ stable for a certain class of signal spaces $\mathcal{L}$. We pursue this idea first in the case of exponentially stable origin, and then for the more general case of uniformly asymptotically stable origin.

Theorem 6.1 *Let $D = \{x \in R^n \mid \|x\| < r\}$, $D_u = \{u \in R^m \mid \|u\| < r_u\}$, $f : [0, \infty) \times D \times D_u \to R^n$ be piecewise continuous in t and locally Lipschitz in*

(x, u), and $h : [0, \infty) \times D \times D_u \to R^n$ be piecewise continuous in t and continuous in (x, u). Suppose that

- $x = 0$ is an exponentially stable equilibrium point of (6.6), and there is a Lyapunov function $V(t,x)$ that satisfies

$$c_1\|x\|^2 \leq V(t, x) \leq c_2\|x\|^2 \tag{6.7}$$

$$\frac{\partial V}{\partial t} + \frac{\partial V}{\partial x} f(t, x, 0) \leq -c_3\|x\|^2 \tag{6.8}$$

$$\left\|\frac{\partial V}{\partial x}\right\| \leq c_4\|x\| \tag{6.9}$$

for all $(t, x) \in [0, \infty) \times D$ for some positive constants c_1, c_2, c_3, and c_4.

- f and h satisfy the inequalities

$$\|f(t, x, u) - f(t, x, 0)\| \leq L\|u\| \tag{6.10}$$

$$\|h(t, x, u)\| \leq \eta_1\|x\| + \eta_2\|u\| \tag{6.11}$$

for all $(t, x, u) \in [0, \infty) \times D \times D_u$ for some nonnegative constants L, η_1, and η_2.

Then, for each x_0 with $\|x_0\| < r\sqrt{c_1/c_2}$, the system (6.4)–(6.5) is small-signal finite-gain $\mathcal{L}_p$ stable for each $p \in [1, \infty]$. In particular, for each $u \in \mathcal{L}_{pe}$ with $\sup_{0 \leq t \leq \tau} \|u(t)\| < \min\{r_u, c_1 c_3 r/c_2 c_4 L\}$, the output $y(t)$ satisfies

$$\|y_\tau\|_{\mathcal{L}_p} \leq \gamma\|u_\tau\|_{\mathcal{L}_p} + \beta \tag{6.12}$$

for all $\tau \in [0, \infty)$ with

$$\gamma = \eta_2 + \frac{\eta_1 c_2 c_4 L}{c_1 c_3}, \quad \beta = \eta_1\|x_0\|\sqrt{\frac{c_2}{c_1}}\rho, \quad \text{where } \rho = \begin{cases} 1, & \text{if } p = \infty \\ \left(\frac{2c_2}{c_3 p}\right)^{1/p}, & \text{if } p \in [1, \infty) \end{cases}$$

Furthermore, if the origin is globally exponentially stable and all the assumptions hold globally (with $D = R^n$ and $D_u = R^m$), then the system (6.4)–(6.5) is finite-gain $\mathcal{L}_p$ stable for each $p \in [1, \infty]$ and inequality (6.12) holds for each $x_0 \in R^n$ and $u \in \mathcal{L}_{pe}$. $\diamond$

Proof: The derivative of V along the trajectories of (6.4) satisfies

$$\begin{aligned} \dot{V}(t, x) &= \frac{\partial V}{\partial t} + \frac{\partial V}{\partial x} f(t, x, 0) + \frac{\partial V}{\partial x}[f(t, x, u) - f(t, x, 0)] \\ &\leq -c_3\|x\|^2 + c_4 L\|x\|\,\|u\| \end{aligned}$$

Take $W(t) = \sqrt{V(t, x(t))}$. When $V(t) \neq 0$, use $\dot{W} = \dot{V}/2\sqrt{V}$ and (6.7) to obtain

$$\dot{W} \leq -\frac{1}{2}\left(\frac{c_3}{c_2}\right)W + \frac{c_4 L}{2\sqrt{c_1}}\|u(t)\|$$

When $V(t) = 0$, it can be verified that

$$D^+ W(t) \leq \frac{c_4 L}{2\sqrt{c_1}}\|u(t)\|$$

Hence,

$$D^+ W(t) \leq -\frac{1}{2}\left(\frac{c_3}{c_2}\right)W + \frac{c_4 L}{2\sqrt{c_1}}\|u(t)\|$$

for all values of $V(t)$. By the comparison lemma (Lemma 2.5), $W(t)$ satisfies the inequality

$$W(t) \leq e^{-tc_3/2c_2}W(0) + \frac{c_4 L}{2\sqrt{c_1}}\int_0^t e^{-(t-\tau)c_3/2c_2}\|u(\tau)\| \, d\tau$$

Using (6.7), we obtain

$$\|x(t)\| \leq \sqrt{\frac{c_2}{c_1}}\|x_0\|e^{-tc_3/2c_2} + \frac{c_4 L}{2c_1}\int_0^t e^{-(t-\tau)c_3/2c_2}\|u(\tau)\| \, d\tau$$

It can be easily verified that

$$\|x_0\| < r\sqrt{\frac{c_1}{c_2}} \quad \text{and} \quad \sup_{0 \leq \sigma \leq t}\|u(\sigma)\| < \frac{c_1 c_3 r}{c_2 c_4 L}$$

ensure that $\|x(t)\| < r$; hence, $x(t)$ stays within the domain of validity of the assumptions. Using (6.11), we have

$$\|y(t)\| \leq k_1 e^{-at} + k_2 \int_0^t e^{-a(t-\tau)}\|u(\tau)\| \, d\tau + k_3 \|u(t)\| \qquad (6.13)$$

where

$$k_1 = \sqrt{\frac{c_2}{c_1}}\|x_0\|\eta_1, \quad k_2 = \frac{c_4 L \eta_1}{2c_1}, \quad k_3 = \eta_2, \quad a = \frac{c_3}{2c_2}$$

Inequality (6.13) takes the form of inequality (6.3) of Example 6.3. By repeating the argument used there, we arrive at inequality (6.12). When all the assumptions hold globally, there is no need to restrict $\|x_0\|$ or the instantaneous values of $\|u(t)\|$. Hence, (6.12) is satisfied for each $x_0 \in R^n$ and $u \in \mathcal{L}_{pe}$. □

The use of (the converse Lyapunov) Theorem 3.12 shows the existence of a Lyapunov function satisfying (6.7)–(6.9). Consequently, we have the following corollary.

Corollary 6.1 *Suppose that, in some neighborhood of* $(x = 0, u = 0)$*, the function* $f(t, x, u)$ *is continuously differentiable, the Jacobian matrices* $[\partial f / \partial x]$ *and* $[\partial f / \partial u]$ *are bounded, uniformly in* t*, and* $h(t, x, u)$ *satisfies* (6.11)*. If the origin* $x = 0$ *is an exponentially stable equilibrium point of* (5.25)*, then there is a constant* $r_0 > 0$ *such that for each* x_0 *with* $\|x_0\| < r_0$*, the system* (6.4)–(6.5) *is small-signal finite-gain* $\mathcal{L}_p$ *stable for each* $p \in [1, \infty]$*. Furthermore, if all the assumptions hold globally and the origin* $x = 0$ *is a globally exponentially stable equilibrium point of* (6.11)*, then for each* $x_0 \in R^n$*, the system* (6.4)–(6.5) *is finite-gain* $\mathcal{L}_p$ *stable for each* $p \in [1, \infty]$*.*
$$\diamond$$

Example 6.5 Consider the single-input–single-output first-order system

$$\dot{x} = -x - x^3 + u, \quad x(0) = x_0$$
$$y = \tanh x + u$$

The origin of

$$\dot{x} = -x - x^3$$

is globally exponentially stable, as can be seen by the Lyapunov function $V(x) = \frac{1}{2}x^2$. The function V satisfies (6.7)–(6.9) globally with $c_1 = c_2 = \frac{1}{2}$, $c_3 = c_4 = 1$. The functions f and h satisfy (6.10)–(6.11) globally with $L = \eta_1 = \eta_2 = 1$. Hence, for each $x_0 \in R$ and each $p \in [1, \infty]$, the system is finite-gain $\mathcal{L}_p$ stable. $\triangle$

Example 6.6 Consider the single-input–single-output second-order system

$$\dot{x}_1 = x_2$$
$$\dot{x}_2 = -x_1 - x_2 - a \tanh x_1 + u$$
$$y = x_1$$

where a is a nonnegative constant. The unforced system can be viewed as a perturbation of the linear time-invariant system

$$\dot{x} = Ax, \quad \text{where } A = \begin{bmatrix} 0 & 1 \\ -1 & -1 \end{bmatrix}$$

Since A is Hurwitz, we solve the Lyapunov equation $PA + A^T P = -I$ to obtain

$$P = \begin{bmatrix} 1.5 & 0.5 \\ 0.5 & 1 \end{bmatrix}$$

and use $V(x) = x^T P x$ as a Lyapunov function candidate for the unforced system.

$$\dot{V} = -x_1^2 - x_2^2 - ax_1 \tanh x_1 - 2ax_2 \tanh x_1$$

Using the fact that $x_1 \tanh x_1 \geq 0$ for all $x_1 \in R$, we obtain

$$\dot{V} \leq -\|x\|_2^2 + 2a|x_1|\,|x_2| \leq -(1-a)\|x\|_2^2$$

Thus, for all $a < 1$, V satisfies (6.7)–(6.9) globally with $c_1 = \lambda_{\min}(P)$, $c_2 = \lambda_{\max}(P)$, $c_3 = 1 - a$, $c_4 = 2\|P\|_2 = 2\lambda_{\max}(P)$. The functions f and h satisfy (6.10)–(6.11) globally with $L = \eta_1 = 1$, $\eta_2 = 0$. Hence, for each $x_0 \in R$ and each $p \in [1, \infty]$, the system is finite-gain $\mathcal{L}_p$ stable. $\triangle$

We turn now to the more general case when the origin of (6.6) is uniformly asymptotically stable, and restrict our attention to the study of $\mathcal{L}_\infty$ stability. Recalling the notion of input-to-state stability from Section 5.3, we can state the following theorem.

Theorem 6.2 Let $D = \{x \in R^n \mid \|x\| < r\}$, $D_u = \{u \in R^m \mid \|u\| < r_u\}$, $f : [0, \infty) \times D \times D_u \to R^n$ be piecewise continuous in t and locally Lipschitz in (x, u), and $h : [0, \infty) \times D \times D_u \to R^n$ be piecewise continuous in t and continuous in (x, u). Suppose that

- the system (6.4) is locally input-to-state stable.

- h satisfies the inequality [7]

$$\|h(t, x, u)\| \leq \alpha_1(\|x\|) + \alpha_2(\|u\|) + \eta_3 \tag{6.14}$$

for all $(t, x, u) \in [0, \infty) \times D \times D_u$ for some class $\mathcal{K}$ functions α_1, α_2, and a nonnegative constant η_3.

Then, there is a constant $k_1 > 0$ such that for each x_0 with $\|x_0\| < k_1$, the system (6.4)–(6.5) is small-signal $\mathcal{L}_\infty$ stable. Furthermore, if all the assumptions hold globally (with $D = R^n$ and $D_u = R^m$) and (6.4) is input-to-state stable, then the system (6.4)–(6.5) is $\mathcal{L}_\infty$ stable. $\diamond$

Proof: Recall from Definition 5.2 that if (6.4) is locally input-to-state stable, then there exist a class $\mathcal{KL}$ function β, a class $\mathcal{K}$ function γ, and positive constants k_1 and k_2 such that for any initial state x_0 with $\|x_0\| < k_1$ and any input $u(t)$ with $\sup_{0 \leq \sigma \leq t} \|u(\sigma)\| < k_2$, the solution $x(t)$ satisfies

$$\|x(t)\| \leq \beta(\|x_0\|, t) + \gamma\left(\sup_{0 \leq \sigma \leq t} \|u(\sigma)\|\right) \tag{6.15}$$

---

[7] When h is independent of t and D, D_u are bounded domains, inequality (6.14) follows from continuity of h.

for all $t \geq 0$. Using (6.14), we obtain

$$\|y(t)\| \leq \alpha_1 \left(\beta(\|x_0\|, t) + \gamma \left(\sup_{0 \leq \sigma \leq t} \|u(\sigma)\| \right) \right) + \alpha_2(\|u(t)\|) + \eta_3$$

$$\leq \alpha_1 \left(2\beta(\|x_0\|, t) \right) + \alpha_1 \left(2\gamma \left(\sup_{0 \leq \sigma \leq t} \|u(\sigma)\| \right) \right) + \alpha_2(\|u(t)\|) + \eta_3$$

where we have used the general property of class $\mathcal{K}$ functions

$$\alpha_1(a + b) \leq \alpha_1(2a) + \alpha_1(2b)$$

Thus,

$$\|y_\tau\|_{\mathcal{L}_\infty} \leq \gamma_0 \left(\|u_\tau\|_{\mathcal{L}_\infty} \right) + \beta_0 \qquad (6.16)$$

where $\gamma_0 = \alpha_1 \circ 2\gamma + \alpha_2$, and $\beta_0 = \alpha_1(2\beta(\|x_0\|, 0)) + \eta_3$. When all the assumptions hold globally, inequality (6.15) is satisfied for each x_0 and each $u \in \mathcal{L}_{\infty e}$ and so is (6.16). $\square$

Input-to-state stability of (6.4) can be checked using Theorem 5.2. In view of Lemma 5.4, we have the following corollary.

Corollary 6.2 *Suppose that in some neighborhood of $(x = 0, u = 0)$, the function $f(t, x, u)$ is continuously differentiable and the Jacobian matrices $[\partial f/\partial x]$ and $[\partial f/\partial u]$ are bounded, uniformly in t, and $h(t, x, u)$ satisfies (6.14). If the unforced system (6.6) has a uniformly asymptotically stable equilibrium point at the origin $x = 0$, then the system (6.4)–(6.5) is small-signal $\mathcal{L}_\infty$ stable.* $\diamond$

Example 6.7 Consider the single-input–single-output first-order system

$$\dot{x} = -x - 2x^3 + (1 + x^2)u^2$$
$$y = x^2 + u$$

We have seen in Example 5.7 that the state equation is input-to-state stable. The output function h satisfies (6.14) globally with $\alpha_1(r) = r^2$, $\alpha_2(r) = r$, and $\eta_3 = 0$. Thus, the system is $\mathcal{L}_\infty$ stable. $\triangle$

Example 6.8 Consider the single-input–single-output second-order system

$$\dot{x}_1 = -x_1^3 + g(t)x_2$$
$$\dot{x}_2 = -g(t)x_1 - x_2^3 + u$$
$$y = x_1 + x_2$$

where $g(t)$ is continuous for all $t \geq 0$. Take $V = \frac{1}{2}(x_1^2 + x_2^2)$. Then,

$$\dot{V} = -x_1^4 - x_2^4 + x_2 u$$

Using

$$x_1^4 + x_2^4 \geq \tfrac{1}{2}\|x\|_2^4$$

we obtain

$$\dot{V} \leq -\frac{1-\theta}{2}\|x\|_2^4 - \frac{\theta}{2}\|x\|_2^4 + \|x\|_2|u|, \quad 0 < \theta < 1$$

$$\leq -\frac{1-\theta}{2}\|x\|_2^4, \quad \forall \, \|x\|_2 \geq \left(\frac{2|u|}{\theta}\right)^{1/3}$$

Thus, V satisfies inequalities (5.28)–(5.29) of Theorem 5.2 globally, with $\alpha_1(r) = \alpha_2(r) = \frac{1}{2}r^2$, $\alpha_3(r) = \frac{1-\theta}{2}r^4$, and $\rho(r) = \left(\frac{2r}{\theta}\right)^{1/3}$. Hence, the state equation is input-to-state stable. Furthermore, the function $h = x_1 + x_2$ satisfies (6.14) globally with $\alpha_1(r) = \sqrt{2}r$, $\alpha_2 = 0$, and $\eta_3 = 0$. Thus, the system is $\mathcal{L}_\infty$ stable. △

6.3 Input-to-Output Stability

While showing $\mathcal{L}_\infty$ stability in the proof of Theorem 6.2, we have verified that inequality (6.16) is satisfied with a bias term $\beta_0 = \alpha_1(2\beta(\|x_0\|, 0)) + \eta_3$. In the special case when $h(t, 0, 0) = 0$, it is reasonable to assume that $\eta_3 = 0$. Then, the bias term β_0 is only due to the initial state x_0. This bias term does not reveal the fact that the effect of nonzero initial states decays to zero as t approaches infinity. In particular, from (6.15) it is clear that, instead of (6.16), we can write the inequality

$$\|y_\tau\|_{\mathcal{L}_\infty} \leq \gamma_0\left(\|u_\tau\|_{\mathcal{L}_\infty}\right) + \alpha_1(2\beta(\|x_0\|, t))$$

where

$$\alpha_1(2\beta(\|x_0\|, t)) \to 0, \quad \text{as } t \to \infty$$

showing that the output y is ultimately bounded by a class $\mathcal{K}$ function of the $\mathcal{L}_\infty$ norm of the input. This observation motivates the following definition of input-to-output stability.

Definition 6.3 *The system* (6.4)–(6.5) *is said to be locally input-to-output stable if there exist a class $\mathcal{KL}$ function β, a class $\mathcal{K}$ function γ_0, and positive constants k_1 and k_2 such that, for any initial state $x(t_0)$ with $\|x(t_0)\| < k_1$ and any input $u(t)$ with $\sup_{t \geq t_0} \|u(t)\| < k_2$, the solution $x(t)$ exists and the output $y(t)$ satisfies*

$$\|y(t)\| \leq \beta(\|x(t_0)\|, t - t_0) + \gamma_0\left(\sup_{t_0 \leq \tau \leq t} \|u(\tau)\|\right) \tag{6.17}$$

for all $t \geq t_0 \geq 0$. It is said to be input-to-output stable if $D = R^n$, $D_u = R^m$, and inequality (6.17) *is satisfied for any initial state $x(t_0)$ and any bounded input $u(t)$.*

The following theorem is obvious in view of the discussion preceding Definition 6.3.

Theorem 6.3 *Suppose all the assumptions of Theorem 6.2 are satisfied and $\eta_3 = 0$. Then, the system (6.4)–(6.5) is locally input-to-output stable. If all the assumptions hold globally and (6.4) is input-to-state stable, then the system (6.4)–(6.5) is input-to-output stable.* $\diamond$

It is clear that the systems of Examples 6.7 and 6.8 are input-to-output stable.

6.4 $\mathcal{L}_2$ Gain

$\mathcal{L}_2$ stability plays a special role in systems analysis, since it is natural to work with square-integrable signals which can be viewed as finite-energy signals.[8] In many control problems,[9] the system is represented as an input-output map from a disturbance input u to a controlled output y which is required to be small. With $\mathcal{L}_2$ input signals, the goals of the control design are then to ensure that the input-output map is finite-gain $\mathcal{L}_2$ stable and minimize the $\mathcal{L}_2$ gain of the system. In such problems, it is important not only to be able to find out that the system is finite-gain $\mathcal{L}_2$ stable but also to calculate the $\mathcal{L}_2$ gain of the system or an upper bound on it. In this section, we show how to calculate the $\mathcal{L}_2$ gain for a special class of autonomous systems. We start with linear time-invariant systems.

Theorem 6.4 *Consider the linear time-invariant system*

$$\dot{x} = Ax + Bu \qquad (6.18)$$

$$y = Cx + Du \qquad (6.19)$$

where A is Hurwitz. Let $G(s) = C(sI - A)^{-1}B + D$. Then, the $\mathcal{L}_2$ gain of the system is $\sup_{\omega \in R} \|G(j\omega)\|_2$.[10] $\diamond$

Proof: Due to linearity, we set $x(0) = 0$. From Fourier transform theory,[11] we know that for a causal signal $y \in \mathcal{L}_2$, the Fourier transform $Y(j\omega)$ is given by

$$Y(j\omega) = \int_0^\infty y(t)e^{-j\omega t}\, dt$$

---

[8] If you think of $u(t)$ as current or voltage, then $u^T(t)u(t)$ is proportional to the instantaneous power of the signal, and its integral over all time is a measure of the energy content of the signal.

[9] See the literature on H_∞ control; for example, [15], [44], [51], [81], [178], or [194].

[10] This is the induced 2-norm of the complex matrix $G(j\omega)$, which is equal to $\sqrt{\lambda_{\max}[G^T(-j\omega)G(j\omega)]} = \sigma_{\max}[G(j\omega)]$. This quantity is known as the H_∞ norm of $G(j\omega)$, when $G(j\omega)$ is viewed as an element of the Hardy space H_∞; see [51].

[11] See [43].

and

$$Y(j\omega) = G(j\omega)U(j\omega)$$

Using Parseval's theorem,[12] we can write

$$
\begin{aligned}
\|y\|_{\mathcal{L}_2}^2 &= \int_0^\infty y^T(t)y(t)\, dt = \frac{1}{2\pi}\int_{-\infty}^\infty Y^*(j\omega)Y(j\omega)\, d\omega \\
&= \frac{1}{2\pi}\int_{-\infty}^\infty U^*(j\omega)G^T(-j\omega)G(j\omega)U(j\omega)\, d\omega \\
&\leq \left(\sup_{\omega\in R}\|G(j\omega)\|_2\right)^2 \frac{1}{2\pi}\int_{-\infty}^\infty U^*(j\omega)U(j\omega)\, d\omega \\
&= \left(\sup_{\omega\in R}\|G(j\omega)\|_2\right)^2 \|u\|_{\mathcal{L}_2}^2
\end{aligned}
$$

which shows that the $\mathcal{L}_2$ gain is less than or equal to $\sup_{\omega\in R}\|G(j\omega)\|_2$. Showing that the $\mathcal{L}_2$ gain is equal to $\sup_{\omega\in R}\|G(j\omega)\|_2$ is done by a contradiction argument that is given in Appendix A.9. □

The case of linear time-invariant systems is exceptional in that we can actually find the exact $\mathcal{L}_2$ gain. In more general cases, like the case of the following theorem, we can only find an upper bound on the $\mathcal{L}_2$ gain.

Theorem 6.5 *Consider the autonomous nonlinear system*

$$
\begin{aligned}
\dot{x} &= f(x) + G(x)u, \quad x(0) = x_0 &\qquad (6.20) \\
y &= h(x) &\qquad (6.21)
\end{aligned}
$$

where $f(x)$ is locally Lipschitz and $G(x)$, $h(x)$ are continuous over R^n. The matrix G is $n \times m$ and $h : R^n \to R^q$. The functions f and h vanish at the origin; that is, $f(0) = 0$ and $h(0) = 0$. Let γ be a positive number and suppose there is a continuously differentiable, positive semidefinite function $V(x)$ that satisfies the inequality

$$\mathcal{H}(V,f,G,h,\gamma) \stackrel{\text{def}}{=} \frac{\partial V}{\partial x}f(x) + \frac{1}{2\gamma^2}\frac{\partial V}{\partial x}G(x)G^T(x)\left(\frac{\partial V}{\partial x}\right)^T + \frac{1}{2}h^T(x)h(x) \leq 0 \quad (6.22)$$

for all $x \in R^n$. Then, for each $x_0 \in R^n$, the system (6.20)–(6.21) is finite-gain $\mathcal{L}_2$ stable and its $\mathcal{L}_2$ gain is less than or equal to γ. ◇

---

[12]Parseval's theorem [43] states that for a causal signal $y \in \mathcal{L}_2$,

$$\int_0^\infty y^T(t)y(t)\, dt = \frac{1}{2\pi}\int_{-\infty}^\infty Y^*(j\omega)Y(j\omega)\, d\omega$$

Proof: By completing the squares, we have

$$\frac{\partial V}{\partial x}f(x) + \frac{\partial V}{\partial x}G(x)u = -\frac{1}{2}\gamma^2 \left\| u - \frac{1}{\gamma^2}G^T(x)\left(\frac{\partial V}{\partial x}\right)^T \right\|_2^2 + \frac{\partial V}{\partial x}f(x)$$
$$+ \frac{1}{2\gamma^2}\frac{\partial V}{\partial x}G(x)G^T(x)\left(\frac{\partial V}{\partial x}\right)^T + \frac{1}{2}\gamma^2\|u\|_2^2$$

Substitution of (6.22) yields

$$\frac{\partial V}{\partial x}f(x) + \frac{\partial V}{\partial x}G(x)u \le \frac{1}{2}\gamma^2\|u\|_2^2 - \frac{1}{2}\|y\|_2^2 - \frac{1}{2}\gamma^2 \left\| u - \frac{1}{\gamma^2}G^T(x)\left(\frac{\partial V}{\partial x}\right)^T \right\|_2^2$$

Hence,

$$\frac{\partial V}{\partial x}f(x) + \frac{\partial V}{\partial x}G(x)u \le \frac{1}{2}\gamma^2\|u\|_2^2 - \frac{1}{2}\|y\|_2^2 \tag{6.23}$$

Note that the left-hand side of (6.23) is the derivative of V along the trajectories of the system (6.20). Integration of (6.23) yields

$$V(x(\tau)) - V(x_0) \le \frac{1}{2}\gamma^2 \int_0^\tau \|u(t)\|_2^2 \, dt - \frac{1}{2}\int_0^\tau \|y(t)\|_2^2 \, dt$$

where $x(t)$ is the solution of (6.20) for any $u \in \mathcal{L}_{2e}$. Using $V(x) \ge 0$, we obtain

$$\int_0^\tau \|y(t)\|_2^2 \, dt \le \gamma^2 \int_0^\tau \|u(t)\|_2^2 \, dt + 2V(x_0)$$

Taking the square roots and using the inequality $\sqrt{a^2 + b^2} \le a + b$ for nonnegative numbers a and b, we obtain

$$\|y_\tau\|_{\mathcal{L}_2} \le \gamma\|u_\tau\|_{\mathcal{L}_2} + \sqrt{2V(x_0)} \tag{6.24}$$

which completes the proof.   □

Inequality (6.22) is known as the *Hamilton-Jacobi inequality* (or the *Hamilton-Jacobi equation* when $\le$ is replaced by $=$). The search for a function $V(x)$ that satisfies (6.22) requires basically the solution of a partial differential equation, which might be difficult to solve. If we succeed in finding $V(x)$, we obtain a finite-gain $\mathcal{L}_2$ stability result which, unlike Theorem 6.1, does not require the origin of the unforced system to be exponentially stable. This point is illustrated by the following example.

Example 6.9 Consider the single-input–single-output system

$$
\begin{aligned}
\dot{x}_1 &= x_2 \\
\dot{x}_2 &= -ax_1^3 - kx_2 + u \\
y &= x_2
\end{aligned}
$$

where a and k are positive constants. The unforced system is a special case of the class of systems treated in Example 3.9. In that example, we used the energy-like Lyapunov function $V(x) = \frac{1}{4}ax_1^4 + \frac{1}{2}x_2^2$ to show that the origin is globally asymptotically stable. Using $V(x) = \alpha(\frac{1}{4}ax_1^4 + \frac{1}{2}x_2^2)$ with $\alpha > 0$ as a candidate for the solution of the Hamilton-Jacobi inequality (6.22), it can be shown that

$$
\mathcal{H}(V, f, G, h, \gamma) = \left(-\alpha k + \frac{\alpha^2}{2\gamma^2} + \frac{1}{2} \right) x_2^2
$$

To satisfy (6.22), we need to choose $\alpha > 0$ and $\gamma > 0$ such that

$$
-\alpha k + \frac{\alpha^2}{2\gamma^2} + \frac{1}{2} \leq 0 \tag{6.25}
$$

By simple algebraic manipulation, we can rewrite this inequality as

$$
\gamma^2 \geq \frac{\alpha^2}{2\alpha k - 1}
$$

Since we are interested in the smallest possible γ, we choose α to minimize the right-hand side of the preceding inequality. The minimum value $1/k^2$ is achieved at $\alpha = 1/k$. Thus, choosing $\gamma = 1/k$, we conclude that the system is finite-gain $\mathcal{L}_2$ stable and its $\mathcal{L}_2$ gain is less than or equal to $1/k$. We note that the conditions of Theorem 6.1 are not satisfied in this example since the origin of the unforced system is not exponentially stable. Linearization at the origin yields the matrix

$$
\begin{bmatrix} 0 & 1 \\ 0 & -k \end{bmatrix}
$$

which is not Hurwitz. $\triangle$

The idea of the preceding example is generalized in the following one.

Example 6.10 Consider the nonlinear system (6.20)–(6.21), with $m = q$, and suppose there is a continuously differentiable positive semidefinite function $W(x)$ which satisfies [13]

$$
\frac{\partial W}{\partial x} f(x) \leq -kh^T(x)h(x), \quad k > 0 \tag{6.26}
$$

$$
\frac{\partial W}{\partial x} G(x) = h^T(x) \tag{6.27}
$$

---

[13] A system satisfying (6.26)–(6.27) will be defined in Section 10.3 as an output strictly passive system.

for all $x \in R^n$. Using $V(x) = \alpha W(x)$ with $\alpha > 0$ as a candidate for the solution of the Hamilton-Jacobi inequality (6.22), it can be shown that

$$\mathcal{H}(V, f, G, h, \gamma) = \left(-\alpha k + \frac{\alpha^2}{2\gamma^2} + \frac{1}{2}\right) h^T(x)h(x)$$

To satisfy (6.22), we need to choose $\alpha > 0$ and $\gamma > 0$ such that

$$-\alpha k + \frac{\alpha^2}{2\gamma^2} + \frac{1}{2} \leq 0$$

This inequality is the same as inequality (6.25) of Example 6.9. By repeating the argument used there, it can be shown that the system is finite-gain $\mathcal{L}_2$ stable and its $\mathcal{L}_2$ gain is less than or equal to $1/k$. $\triangle$

Example 6.11 Consider the nonlinear system (6.20)–(6.21), with $m = q$, and suppose there is a continuously differentiable positive semidefinite function $W(x)$ which satisfies [14]

$$\frac{\partial W}{\partial x} f(x) \;\; \leq \;\; 0 \tag{6.28}$$

$$\frac{\partial W}{\partial x} G(x) \;\; = \;\; h^T(x) \tag{6.29}$$

for all $x \in R^n$. The output feedback control

$$u = -ky + v, \quad k > 0$$

results in the closed-loop system

$$\dot{x} \;\; = \;\; f(x) - kG(x)G^T(x)\left(\frac{\partial W}{\partial x}\right)^T + G(x)v \;\; \overset{\text{def}}{=} \;\; f_c(x) + G(x)v$$

$$y \;\; = \;\; h(x) \;\; = \;\; G^T(x)\left(\frac{\partial W}{\partial x}\right)^T$$

It can be easily verified that $W(x)$ satisfies (6.26)–(6.27) of the previous example for the closed-loop system. Hence, the input-output map from v to y is finite-gain $\mathcal{L}_2$ stable and its $\mathcal{L}_2$ gain is less than or equal to $1/k$. This shows, in essence, that the $\mathcal{L}_2$ gain can be made arbitrarily small by choosing the feedback gain k sufficiently large. $\triangle$

---

[14] A system satisfying (6.28)–(6.29) will be defined in Section 10.3 as a passive system. We shall come back to this example in Section 10.3 and look at it from the viewpoint of feedback connection of passive systems.

Example 6.12 Consider the linear time-invariant system

$$\begin{aligned}
\dot{x} &= Ax + Bu \\
y &= Cx
\end{aligned}$$

Suppose there is a positive semidefinite solution P of the Riccati equation

$$PA + A^T P + \frac{1}{\gamma^2} PBB^T P + C^T C = 0 \qquad (6.30)$$

for some $\gamma > 0$. Taking $V(x) = \frac{1}{2} x^T P x$ and using the expression $[\partial V/\partial x] = x^T P$, it can be easily seen that $V(x)$ satisfies the Hamilton-Jacobi equation

$$\mathcal{H}(V, Ax, B, Cx) = x^T P Ax + \frac{1}{2\gamma^2} x^T PB^T BPx + \frac{1}{2} x^T C^T Cx = 0$$

Hence, the system is finite-gain $\mathcal{L}_2$ stable and its $\mathcal{L}_2$ gain is less than or equal to γ. This result gives an alternative method for computing an upper bound on the $\mathcal{L}_2$ gain, as opposed to the frequency-domain calculation of Theorem 6.4. It is interesting to note that the existence of a positive semidefinite solution of (6.30) is a necessary and sufficient condition for the $\mathcal{L}_2$ gain to be less than or equal to γ.[15]

$\triangle$

In Theorem 6.5, we assumed that all the assumptions hold globally. It is clear from the proof of the theorem that if the assumptions hold only on a finite domain D, we shall still arrive at inequality (6.24) as long as the solution of (6.20) stays in D.

Corollary 6.3 *Suppose all the assumptions of Theorem 6.5 are satisfied on a domain $D \subset R^n$ which contains the origin. Then, for any $x_0 \in D$ and any $u \in \mathcal{L}_{2e}$ for which the solution of (6.20) satisfies $x(t) \in D$ for all $t \in [0, \tau]$, we have*

$$\|y_\tau\|_{\mathcal{L}_2} \leq \gamma \|u_\tau\|_{\mathcal{L}_2} + \sqrt{2V(x_0)}$$

$\diamond$

Recalling the definition of input-to-state stability from Section 5.3, we can give the following small-signal $\mathcal{L}_2$ stability result.

Lemma 6.1 *Suppose all the assumptions of Theorem 6.5 are satisfied on a domain $D \subset R^n$ which contains the origin, and the state equation (6.20) is locally input-to-state stable. Then, there exists a positive constant r_0 such that for each x_0 with $\|x_0\|_2 < r_0$, the system (6.20)–(6.21) is small-signal finite-gain $\mathcal{L}_2$ stable and its $\mathcal{L}_2$ gain is less than or equal to γ.*

$\diamond$

---

[15] See [44] for the proof of necessity.

Proof: Recall, from Definition 5.2, that if (6.20) is locally input-to-state stable, then there exist a class $\mathcal{KL}$ function β, a class $\mathcal{K}$ function γ_0, and positive constants k_1 and k_2 such that, for any initial state x_0 with $\|x_0\|_2 < k_1$ and any input $u(t)$ with $\sup_{0 \le \sigma \le t} \|u(\sigma)\|_2 < k_2$, the solution $x(t)$ satisfies

$$\|x(t)\|_2 \le \beta(\|x_0\|_2, t) + \gamma_0 \left(\sup_{0 \le \sigma \le t} \|u(\sigma)\|_2 \right)$$

for all $t \ge 0$. Thus by choosing k_1 and k_2 small enough and $r_0 = k_1$, we can be sure that $x(t) \in D$ for all $t \ge 0$. The lemma follows then from Corollary 6.3. $\square$

Input-to-state stability of (6.20) can be checked using Theorem 5.2. Two easily verifiable cases are given in the following lemma.

Lemma 6.2 *Suppose all the assumptions of Theorem 6.5 are satisfied on a domain $D \subset R^n$ which contains the origin, $f(x)$ is continuously differentiable, and one of the following two conditions is satisfied:*

1. *The matrix $A = [\partial f / \partial x](0)$ is Hurwitz.*

2. *No solution of the unforced system $\dot{x} = f(x)$ can stay identically in $S = \{x \in D \mid h(x) = 0\}$, other than the trivial solution $x(t) \equiv 0$.*

Then, the origin of $\dot{x} = f(x)$ is asymptotically stable and there exists a positive constant r_0 such that for each x_0 with $\|x_0\|_2 < r_0$, the system (6.20)–(6.21) is small-signal finite-gain $\mathcal{L}_2$ stable with $\mathcal{L}_2$ gain less than or equal to γ. $\diamond$

Proof: In case 1, asymptotic stability of the origin of $\dot{x} = f(x)$ follows from Theorem 3.7. In case 2, take $u(t) \equiv 0$. By (6.22), we have

$$\dot{V}(x) = \frac{\partial V}{\partial x} f(x) \le - \tfrac{1}{2} h^T(x) h(x), \quad \forall \, x \in D \tag{6.31}$$

Let $B_r = \{x \in R^n \mid \|x\|_2 \le r\}$ be in the interior of D. We shall show that $V(x)$ is positive definite in B_r. Toward that end, let $\phi(t, x)$ be the solution of $\dot{x} = f(x)$ that starts at $\phi(0, x) = x \in B_r$. By existence and uniqueness of solutions (Theorem 2.2) and continuous dependence of the solution on initial states (Theorem 2.5), there exists $\delta > 0$ such that for each $x \in B_r$ the solution $\phi(t, x)$ stays in D for all $t \in [0, \delta]$. Integrating (6.31) over $[0, \tau]$ for $\tau \le \delta$, we obtain

$$V(\phi(\tau, x)) - V(x) \le - \tfrac{1}{2} \int_0^\tau \|h(\phi(t, x))\|_2^2 \, dt$$

Using $V(\phi(\tau, x)) \ge 0$, we obtain

$$V(x) \ge \tfrac{1}{2} \int_0^\tau \|h(\phi(t, x))\|_2^2 \, dt$$

Define

$$V_a(x) = \tfrac{1}{2} \inf_{\tau \in [0,\delta]} \int_0^\tau \|h(\phi(t,x))\|_2^2 \, dt$$

Then,

$$V(x) \geq V_a(x) \geq 0$$

Suppose now that there is $\bar{x} \neq 0$ such that $V(\bar{x}) = 0$. The foregoing inequality implies that $V_a(\bar{x}) = 0$. Hence,

$$\int_0^\tau \|h(\phi(t,\bar{x}))\|_2^2 \, dt = 0, \quad \forall \, \tau \in [0,\delta] \;\Rightarrow\; h(\phi(t,\bar{x})) \equiv 0, \quad \forall \, t \in [0,\delta]$$

Since during this interval the solution stays in S and, by assumption, the only solution that can stay identically in S is the trivial solution, we conclude that $\phi(t,\bar{x}) \equiv 0 \Rightarrow \bar{x} = 0$. Thus, $V(x)$ is positive definite in B_r. Using $V(x)$ as a Lyapunov function candidate for $\dot{x} = f(x)$, we conclude from (6.31) and LaSalle's invariance principle (Corollary 3.1) that the origin of $\dot{x} = f(x)$ is asymptotically stable. The rest of the proof follows from Lemmas 5.4 and 6.1. $\qquad\square$

Example 6.13 As a variation on the theme of Examples 6.9 and 6.10, consider the system

$$\begin{aligned}
\dot{x}_1 &= x_2 \\
\dot{x}_2 &= -a(x_1 - \tfrac{1}{3}x_1^3) - kx_2 + u \\
y &= x_2
\end{aligned}$$

where $a, k > 0$. The function $V(x) = \alpha\left[a\left(\tfrac{1}{2}x_1^2 - \tfrac{1}{12}x_1^4\right) + \tfrac{1}{2}x_2^2\right]$ with $\alpha > 0$ is positive semidefinite in the set $\{|x_1| \leq \sqrt{6}\}$. Using $V(x)$ as a candidate for the solution of the Hamilton-Jacobi inequality (6.22), it can be shown that

$$\mathcal{H}(V, f, G, h, \gamma) = \left(-\alpha k + \frac{\alpha^2}{2\gamma^2} + \frac{1}{2}\right) x_2^2$$

Repeating the argument used in Example 6.9, it can be easily seen that by choosing $\alpha = \gamma = 1/k$, inequality (6.22) is satisfied for all $x \in R^2$. Since the conditions of Theorem 6.5 are not satisfied globally, we shall investigate small-signal finite-gain stability using Lemma 6.2. We need to show that one of the two conditions of the lemma is satisfied. In this example, both conditions are satisfied. Condition 1 is satisfied since linearization of the unforced system at the origin results in the Hurwitz matrix

$$\begin{bmatrix} 0 & 1 \\ -a & -k \end{bmatrix}$$

Condition 2 is satisfied on the domain $D = \{|x_1| < \sqrt{3}\}$ since

$$x_2(t) \equiv 0 \Rightarrow x_1(t)[3 - x_1^2(t)] \equiv 0 \Rightarrow x_1(t) \equiv 0$$

Thus, we conclude that the system is small-signal finite-gain $\mathcal{L}_2$ stable and its $\mathcal{L}_2$ gain is less than or equal to $1/k$.   △

6.5   Exercises

Exercise 6.1 Show that the series connection of two $\mathcal{L}$ stable (respectively, finite-gain $\mathcal{L}$ stable) systems is $\mathcal{L}$ stable (respectively, finite-gain $\mathcal{L}$ stable).

Exercise 6.2 Show that the parallel connection of two $\mathcal{L}$ stable (respectively, finite-gain $\mathcal{L}$ stable) systems is $\mathcal{L}$ stable (respectively, finite-gain $\mathcal{L}$ stable).

Exercise 6.3 Consider a system defined by the memoryless function $y = u^{1/3}$.

(a) Show that the system is $\mathcal{L}_\infty$ stable with zero bias.

(b) For any positive constant a, show that the system is finite-gain $\mathcal{L}_\infty$ stable with $\gamma = a$ and $\beta = (1/a)^{1/3}$

(c) Compare the two statements.

Exercise 6.4 Consider a memoryless system represented by $y = h(u)$ where $h : R^m \to R^q$ is globally Lipschitz. Investigate $\mathcal{L}_p$ stability for each $p \in [0, \infty]$ when

$$(1) \ \ h(0) = 0; \qquad\qquad (2) \ \ h(0) \neq 0$$

Exercise 6.5 Repeat Example 6.3 for linear time-varying systems.

Exercise 6.6 For each of the following scalar systems, investigate $\mathcal{L}_\infty$ stability and finite-gain $\mathcal{L}_\infty$ stability.

$$(1) \ \ \begin{array}{rcl} \dot{x} & = & -(1+u)x^3 \\ y & = & x \end{array} \qquad\qquad (2) \ \ \begin{array}{rcl} \dot{x} & = & -(1+u)x^3 - x^5 \\ y & = & x + u \end{array}$$

$$(3) \ \ \begin{array}{rcl} \dot{x} & = & -x + x^2 u \\ y & = & \sin x \end{array} \qquad\qquad (4) \ \ \begin{array}{rcl} \dot{x} & = & x - x^3 + u \\ y & = & \tan x \end{array}$$

Exercise 6.7 For each of the following systems, investigate $\mathcal{L}_\infty$ stability and finite-gain $\mathcal{L}_\infty$ stability.

$$
(1) \quad
\begin{aligned}
\dot{x}_1 &= -x_1 + x_2^2 \\
\dot{x}_2 &= -x_2 + u \\
y &= 1/x_1
\end{aligned}
\qquad
(2) \quad
\begin{aligned}
\dot{x}_1 &= (x_1 - x_2 + u)(x_1^2 + x_2^2 - 1) \\
\dot{x}_2 &= (x_1 + x_2 + u)(x_1^2 + x_2^2 - 1) \\
y &= x_1^2 + x_2^2
\end{aligned}
$$

$$
(3) \quad
\begin{aligned}
\dot{x}_1 &= -x_1 + x_1^2 x_2 \\
\dot{x}_2 &= -x_2 + x_1 + u \\
y &= -x_2 + x_1 + u
\end{aligned}
\qquad
(4) \quad
\begin{aligned}
\dot{x}_1 &= -x_1 - x_2 + u_1 \\
\dot{x}_2 &= x_1 - x_2^3 + u_2 \\
y &= x_1(x_2 + u)
\end{aligned}
$$

Exercise 6.8 Suppose all the assumptions of Theorem 6.1 are satisfied except (6.11), which is replaced by

$$\|h(t, x, u)\| \le \eta_1 \|x\| + \eta_2 \|u\| + \eta_3, \quad \eta_3 > 0$$

Show that the system is small-signal finite-gain $\mathcal{L}_\infty$ stable (or finite-gain $\mathcal{L}_\infty$ stable, if the assumptions hold globally) and find the constants γ and β in (6.12).

Exercise 6.9 Suppose all the assumptions of Theorem 6.1 are satisfied except (6.11), which is replaced by (6.14). Show that the system is small-signal $\mathcal{L}_\infty$ stable (or $\mathcal{L}_\infty$ stable, if the assumptions hold globally).

Exercise 6.10 ([173]) Consider the autonomous system

$$
\begin{aligned}
\dot{x} &= f(x, u) \\
y &= h(x, u)
\end{aligned}
$$

where f is locally Lipschitz, h is continuous, $f(0, 0) = 0$, and $h(0, 0) = 0$. Suppose there is a continuously differentiable, positive definite, radially unbounded function $V(x)$ such that

$$\frac{\partial V}{\partial x} f(x, u) \le -W(x) + \psi(u), \quad \forall\, (x, u)$$

where $W(x)$ is continuous, positive definite, and radially unbounded, $\psi(u)$ is continuous, and $\psi(0) = 0$.

(a) Show that the system is $\mathcal{L}_\infty$ stable.

(b) Show that the system is input-to-output stable.

Exercise 6.11 Let $H(s)$ be a Hurwitz strictly proper transfer function, and $h(t) = \mathcal{L}^{-1}\{H(s)\}$ be the corresponding impulse response function. Show that $\sup_{\omega \in R} |H(j\omega)| \le \int_0^\infty |h(t)|\, dt$.

Exercise 6.12 For each of the following systems, show that the system is finite-gain (or small-signal finite-gain) $\mathcal{L}_2$ stable and find an upper bound on the $\mathcal{L}_2$ gain.

(1)
$$
\begin{aligned}
\dot{x}_1 &= x_2 \\
\dot{x}_2 &= -a \sin x_1 - k x_2 + u \\
y &= x_2 \\
& a > 0, \; k > 0
\end{aligned}
$$

(2)
$$
\begin{aligned}
\dot{x}_1 &= -x_2 \\
\dot{x}_2 &= x_1 - x_2 \operatorname{sat}(x_2^2 - x_3^2) + x_2 u \\
\dot{x}_3 &= x_3 \operatorname{sat}(x_2^2 - x_3^2) - x_3 u \\
y &= x_2^2 - x_3^2
\end{aligned}
$$

(3)
$$
\begin{aligned}
\dot{x}_1 &= x_2 \\
\dot{x}_2 &= x_1 - \operatorname{sat}(2 x_1 + x_2) + u \\
y &= x_1
\end{aligned}
$$

(4)
$$
\begin{aligned}
\dot{x}_1 &= x_2 \\
\dot{x}_2 &= -(1 + x_1^2) x_2 - x_1^3 + x_1 u \\
y &= x_1 x_2
\end{aligned}
$$

Exercise 6.13 ([67]) Consider the system

$$
\begin{aligned}
\dot{x} &= f(x) + G(x) u \\
y &= h(x) + J(x) u, \quad x \in R^n, \; u \in R^m, \; y \in R^q
\end{aligned}
$$

where f, G, h and J are smooth functions of x. Suppose there is a positive constant γ such that $\gamma^2 I - J^T(x) J(x) > 0$ and

$$
\mathcal{H} = \frac{\partial V}{\partial x} f + \frac{1}{2}\left[h^T J + \frac{\partial V}{\partial x} G \right] \left(\gamma^2 I - J^T J\right)^{-1} \left[h^T J + \frac{\partial V}{\partial x} G \right]^T + \frac{1}{2} h^T h \le 0
$$

$\forall \, x \in R^n$. Show that the system is finite-gain $\mathcal{L}_2$ stable with $\mathcal{L}_2$ gain less than or equal to γ.
Hint: Set

$$
\begin{aligned}
R(x) &= \gamma^2 I - J^T(x) J(x) = W^T(x) W(x) \\
L(x) &= -\left[W^T(x)\right]^{-1} \left[h^T(x) J(x) + \frac{\partial V}{\partial x} G(x) \right]^T
\end{aligned}
$$

and show that the following inequality holds $\forall \, u \in R^m$

$$
\frac{\partial V}{\partial x} f + \frac{\partial V}{\partial x} G u = -\frac{1}{2}[L + W u]^T [L + W u] + \frac{\gamma^2}{2} u^T u - \frac{1}{2} y^T y + \mathcal{H}
$$

Exercise 6.14 ([178]) Consider the system

$$
\begin{aligned}
\dot{x} &= f(x) + G(x) u + K(x) w, \quad x \in R^n, \; u \in R^m, \; w \in R^r \\
y &= h(x), \quad y \in R^q
\end{aligned}
$$

where u is a control input and w is a disturbance input. The functions f, G, K, and h are smooth, and $f(0) = 0$, $h(0) = 0$. Let $\gamma > 0$. Suppose there is a smooth positive semidefinite function $V(x)$ that satisfies

$$
\frac{\partial V}{\partial x} f(x) + \frac{1}{2} \frac{\partial V}{\partial x} \left[\frac{1}{\gamma^2} K(x) K^T(x) - G(x) G^T(x) \right] \left(\frac{\partial V}{\partial x} \right)^T + \frac{1}{2} h^T(x) h(x) \le 0
$$

$\forall x \in R^n$. Show that with the feedback control $u = -G^T(x) \left(\frac{\partial V}{\partial x}\right)^T$ the closed-loop map from d to $\begin{bmatrix} y \\ u \end{bmatrix}$ is finite-gain $\mathcal{L}_2$ stable with $\mathcal{L}_2$ gain less than or equal to γ.

Exercise 6.15 ([81]) Consider the system (6.20)–(6.21) and assume that $\frac{\partial f}{\partial x}(0)$ is Hurwitz. Suppose $u(t)$ is a T_0-periodic function of time, generated by the linear equation $\dot{u} = Su$, where S is a matrix that has simple eigenvalues on the imaginary axis.

(a) Show that the system

$$\begin{aligned} \dot{x} &= f(x, u) \\ \dot{u} &= Su \end{aligned}$$

has a center manifold $\{x = h_c(u)\}$ in D_u, a neighborhood of $u = 0$, where $h_c(\cdot)$ is continuously differentiable.

(b) Let $x_{ss}(t)$ be the solution of (6.20) when $x(0) = h_c(u(0))$ and suppose $u(t)$ is such that $u(t) \in D_u$ for all $t \geq 0$. Show that $x_{ss}(t)$ is T_0-periodic and, for sufficiently small $\|x(0) - h_c(u(0))\|$, the solution $x(t)$ converges to $x_{ss}(t)$ as $t \to \infty$. In view of these properties we call $x_{ss}(t)$ the steady-state response of the nonlinear system.

(c) Let $y_{ss}(t) = h(x_{ss}(t))$. Show that if the conditions of Theorem 6.5 are satisfied, then

$$\int_{t_0}^{t_0+T_0} y_{ss}^T(t) y_{ss}(t) \, dt \leq \gamma^2 \int_{t_0}^{t_0+T_0} u^T(t) u(t) \, dt, \quad \forall \, t_0 \geq 0$$

Use this inequality to assign a frequency-domain meaning to the $\mathcal{L}_2$ gain of a nonlinear system, in view of frequency response analysis of linear systems.

Exercise 6.16 ([180]) The purpose of this exercise is to show that the $\mathcal{L}_2$ gain of a linear time-invariant system of the form (6.18)–(6.19), with a Hurwitz matrix A, is the same whether the space of functions is defined on $R_+ = [0, \infty)$ or on the whole real line $R = (-\infty, \infty)$. Let $\mathcal{L}_2$ be the space of square integrable functions on R_+ with the norm $\|u\|_{\mathcal{L}_2}^2 = \int_0^\infty u^T(t) u(t) \, dt$ and $\mathcal{L}_{2R}$ be the space of square integrable functions on R with the norm $\|u\|_{\mathcal{L}_{2R}}^2 = \int_{-\infty}^\infty u^T(t) u(t) \, dt$. Let γ_2 and γ_{2R} be the $\mathcal{L}_2$ gains on $\mathcal{L}_2$ and $\mathcal{L}_{2R}$, respectively. Since $\mathcal{L}_2$ is a subset of $\mathcal{L}_{2R}$, it is clear that $\gamma_2 \leq \gamma_{2R}$. We will show that $\gamma_2 = \gamma_{2R}$ by showing that, for every $\epsilon > 0$, there is a signal $u \in \mathcal{L}_2$ such that $y \in \mathcal{L}_2$ and $\|y\|_{\mathcal{L}_2} \geq (1 - \epsilon) \gamma_{2R} \|u\|_{\mathcal{L}_2}$.

(a) Given $\epsilon > 0$ show that we can always choose $\delta > 0$ such that
$(1 - \epsilon/2 - \sqrt{\delta})/\sqrt{1 - \delta} \geq 1 - \epsilon$.

(b) Show that we can always select $u \in \mathcal{L}_{2R}$ and time $t_1 < \infty$ such that

$$\|u\|_{\mathcal{L}_{2R}} = 1, \quad \|y\|_{\mathcal{L}_{2R}} \geq \gamma_{2R} \left(1 - \frac{\epsilon}{2}\right), \quad \int_{-\infty}^{t_1} u^T(t) u(t) \, dt = \delta$$

(c) Let $u(t) = u_1(t) + u_2(t)$, where u_1 vanishes for $t < t_1$ and u_2 vanishes for $t > t_1$. Let $y_1(t)$ be the output corresponding to the input $u_1(t)$. Show that

$$\frac{\|y_1\|_{\mathcal{L}_{2R}}}{\|u_1\|_{\mathcal{L}_{2R}}} \geq \frac{1 - \epsilon/2 - \sqrt{\delta}}{\sqrt{1 - \delta}} \gamma_{2R} \geq (1 - \epsilon)\gamma_{2R}$$

(d) For all $t \geq 0$, define $u(t)$ and $y(t)$ by $u(t) = u_1(t - t_1)$ and $y(t) = y_1(t - t_1)$. Show that both u and y belong to $\mathcal{L}_2$, $y(t)$ is the output corresponding to $u(t)$, and $\|y\|_{\mathcal{L}_2} \geq (1 - \epsilon)\gamma_{2R}\|u\|_{\mathcal{L}_2}$.

Chapter 7

Periodic Orbits

After equilibria, the next simplest steady-state solutions a system can have are periodic solutions. There are two important issues to address concerning periodic solutions: the first issue is predicting their existence and the second one is characterizing their stability. Periodic orbits in the plane are special in that they divide the plane into a region inside the orbit and a region outside it. This makes it possible to obtain criteria for detecting the existence of periodic orbits for second-order systems, which have no generalizations to systems of an order higher than two. The most celebrated of these criteria is the Poincaré-Bendixson theorem, which is presented in Section 7.1. The section also includes a brief account of the Bendixson criterion and the index method, which can be used to rule out the existence of periodic orbits in certain cases. For systems of an order higher than two, we present two methods for predicting the existence of periodic orbits: the averaging method and the describing function method. In each case, the method applies to a special class of nonlinear systems. The averaging method, presented in Section 8.3, applies to a nonlinear system dependent on a small parameter in a special way. The describing function method, presented in Section 10.4, applies to a nonlinear system which can be represented as a feedback connection of a linear dynamical system and a nonlinearity, with the linear system having low-pass filtering characteristics. In Section 7.2, we study stability of periodic solutions and introduce the Poincaré map, a classical tool for characterizing the stability of periodic orbits.

7.1 Second-Order Systems

In this section, we consider second-order autonomous systems of the form

$$\dot{x} = f(x) \tag{7.1}$$

where $f : D \rightarrow R^2$ and $D \subset R^2$ is a domain (open and connected set) that contains the origin $x = 0$. We assume that f is locally Lipschitz in D to ensure that (7.1) has a unique solution through every point in D. Let $\phi(t, y)$ denote the solution of (7.1) that starts at $\phi(0, y) = y$ and denote its maximal interval of existence in forward time by $[0, T^+(y))$ and in backward time by $(-T^-(y), 0]$. In the following discussion, we shall assume that the solution through y can be extended indefinitely in either forward time or backward time; that is, either $T^+(y) = \infty$ or $T^-(y) = \infty$. Of course, we might have $T^+(y) = T^-(y) = \infty$, but we do not require that. If $T^+(y) = \infty$, we define the *positive semiorbit* through y as

$$\gamma^+(y) = \{\phi(t, y) \mid 0 \le t < \infty\}$$

and if $T^-(y) = \infty$, we define the *negative semiorbit* through y as

$$\gamma^-(y) = \{\phi(t, y) \mid -\infty < t \le 0\}$$

Suppose now that $T^+(y) = \infty$ and recall from Section 3.2 that p is a positive limit point of the solution $\phi(t, y)$ (for $t \ge 0$) if there is a sequence $\{t_n\}$, with $t_n \rightarrow \infty$ as $n \rightarrow \infty$, such that $\phi(t_n, y) \rightarrow p$ as $n \rightarrow \infty$. The set of all positive limit points of $\phi(t, y)$ is the positive limit set of $\phi(t, y)$. We have seen in Lemma 3.1 that if $\phi(t, y)$ is bounded, then its positive limit set is a nonempty, compact, invariant set and $\phi(t, y)$ approaches its positive limit set as $t \rightarrow \infty$. We shall refer to this set as the positive limit set of $\gamma^+(y)$. Similarly, when $T^-(y) = \infty$ we define the negative limit set of $\gamma^-(y)$ by replacing ∞ with $-\infty$ in the preceding definition. We can state a lemma similar to Lemma 3.1 for bounded solutions in backward time.

With the requisite terminology in hand, we are ready to state the main result of this section: the Poincaré-Bendixson theorem.

Theorem 7.1 (Poincaré-Bendixson) *Let γ^+ be a bounded positive semiorbit of (7.1) and L^+ be its positive limit set. If L^+ contains no equilibrium points, then it is a periodic orbit. The same result is valid for a negative semiorbit.* $\Diamond$

Proof: Appendix A.10.

The proof uses the fact that a closed curve in R^2 which does not intersect itself must separate R^2 into a region inside the curve and a region outside it. This fact, known as the *Jordan curve theorem*,[1] does not hold in R^n for $n > 2$. Therefore, there is no generalization of the Poincaré-Bendixson theorem to higher-dimensional systems. Exercise 7.8 gives a counterexample in R^3. The intuition behind the theorem is that the possible forms of compact limit sets in the plane are equilibrium points, periodic orbits, and the unions of equilibrium points and the trajectories connecting them. By assuming that L^+ contains no equilibrium points, it must be

---

[1] See [7] for the statement and proof of the Jordan curve theorem.

a periodic orbit

The Poincaré-Bendixson theorem suggests a way to determine the existence of a periodic orbit in the plane. Suppose we can find a closed bounded set M in R^2 such that M contains no equilibrium points of (7.1) and is positively invariant; that is, every solution of (7.1) that starts in M remains in M for all $t \geq 0$. In such a case, we are assured that M contains a bounded positive semiorbit γ^+. Moreover, since M is closed, the positive limit set L^+ is in M. Thus, the existence of a periodic orbit in M follows from the Poincaré-Bendixson theorem. We illustrate this idea in the next two examples.

Example 7.1 Consider the harmonic oscillator

$$\dot{x}_1 = x_2$$
$$\dot{x}_2 = -x_1$$

and the function

$$V(x) = x_1^2 + x_2^2$$

The derivative of $V(x)$ along the trajectories of the system is given by

$$\dot{V}(x) = 2x_1\dot{x}_1 + 2x_2\dot{x}_2 = 0$$

Hence, the trajectory of the system cannot cross the level surface $V(x) = c$ for any positive constant c. Define a set M as the annular region

$$M = \{x \in R^2 \mid c_1 \leq V(x) \leq c_2\}$$

where $c_2 > c_1 > 0$. Clearly, M is bounded and positively invariant. It is also free of equilibrium points since the only equilibrium point of the system is at the origin. Thus, we conclude that there is a periodic orbit in M. $\triangle$

This example emphasizes the fact that the Poincaré-Bendixson theorem assures the existence of a periodic orbit but does not assure its uniqueness. From our earlier study of the harmonic oscillator, we know that it has a continuum of periodic orbits in M.

Example 7.2 Consider the system

$$\dot{x}_1 = x_2 + x_1(1 - x_1^2 - x_2^2)$$
$$\dot{x}_2 = -x_1 + x_2(1 - x_1^2 - x_2^2)$$

and the function

$$V(x) = x_1^2 + x_2^2$$

The derivative of $V(x)$ along the trajectories of the system is given by

$$
\begin{aligned}
\dot{V}(x) &= 2x_1\dot{x}_1 + 2x_2\dot{x}_2 \\
&= 2x_1x_2 + 2x_1^2(1 - x_1^2 - x_2^2) - 2x_1x_2 + 2x_2^2(1 - x_1^2 - x_2^2) \\
&= 2V(x)(1 - V(x))
\end{aligned}
$$

The derivative $\dot{V}$ is positive for $V(x) < 1$ and negative for $V(x) > 1$. Hence, on the level surface $V(x) = c_1$ with $0 < c_1 < 1$ all trajectories will be moving outward, while on the level surface $V(x) = c_2$ with $c_2 > 1$ all trajectories will be moving inward. This shows that the annular region

$$
M = \{x \in R^2 \mid c_1 \le V(x) \le c_2\}
$$

is positively invariant. It is also closed, bounded, and free of equilibrium points since the origin $x = 0$ is the unique equilibrium point. Thus, from the Poincaré-Bendixson theorem, we conclude that there is a periodic orbit in M. Since the above argument is valid for any $c_1 < 1$ and any $c_2 > 1$, we can let c_1 and c_2 approach 1 so that the set M shrinks toward the unit circle. This, in effect, shows that the unit circle is a periodic orbit. We can arrive at the same conclusion by representing the system in the polar coordinates

$$
x_1 = r\cos\theta, \quad x_2 = r\sin\theta
$$

which yields

$$
\dot{r} = r(1 - r^2), \quad \dot{\theta} = -1
$$

This shows not only that the unit circle is a periodic orbit, but also that it is the unique periodic orbit and that all trajectories, other than the trivial solution $x(t) = 0$, spiral toward the unit circle from inside or outside; see Figure 7.1. The uniqueness of the periodic orbit could not be shown by mere application of the Poincaré-Bendixson theorem. $\triangle$

In both examples, the basic idea of using the Poincaré-Bendixson theorem is the same. We show that there is a bounded positive semiorbit γ^+ in M and conclude that its positive limit set L^+ is a periodic orbit. Note, however, that in the harmonic oscillator example every positive semiorbit in M is a periodic orbit itself. In other words, $L^+ = \gamma^+$ and both are periodic. This is not the case in Example 7.2. Except for trajectories starting at the origin or on the unit circle, a positive semiorbit is not periodic. Instead, it spirals toward its positive limit set which is periodic. The periodic orbit in Example 7.2 is a *limit cycle* according to the following definition.

Definition 7.1 *A limit cycle is a closed orbit γ such that γ is the positive limit set of a positive semiorbit $\gamma^+(y)$ or the negative limit set of a negative semiorbit $\gamma^-(y)$ for some $y \notin \gamma$.*

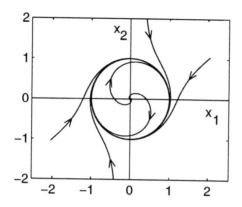

Figure 7.1: Phase portrait of Example 7.2.

The next example is a nontrivial application of the Poincaré-Bendixson theorem to the negative-resistance oscillator of Section 1.1.4.

Example 7.3 The negative-resistance oscillator of Section 1.1.4 is modeled by the second-order differential equation

$$\ddot{v} + \epsilon h'(v)\dot{v} + v = 0$$

where ϵ is a positive constant and $h : R \to R$ satisfies the conditions

$$h(0) = 0, \quad h'(0) < 0$$

$$h(v) \to \infty \quad \text{as} \quad v \to \infty$$

$$h(v) \to -\infty \quad \text{as} \quad v \to -\infty$$

To simplify the analysis, we will impose the additional requirements

$$h(v) = -h(-v)$$

$$h(v) < 0 \text{ for } 0 < v < a, \quad \text{and} \quad h(v) > 0 \text{ for } v > a$$

These additional requirements are satisfied by the typical function of Figure 1.5(b), as well as by the function $h(v) = -v + \frac{1}{3}v^3$ of the Van der Pol oscillator. Choose the state variables as

$$x_1 = v, \quad x_2 = \dot{v} + \epsilon h(v)$$

to obtain the state equation

$$\begin{aligned} \dot{x}_1 &= x_2 - \epsilon h(x_1) \\ \dot{x}_2 &= -x_1 \end{aligned} \tag{7.2}$$

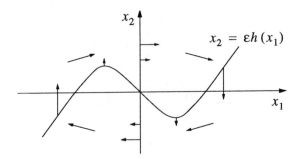

Figure 7.2: Vector field diagram for Example 7.3.

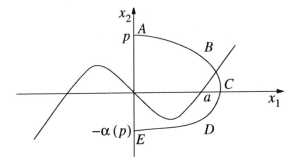

Figure 7.3: The orbit $ABCDE$ of Example 7.3.

which has a unique equilibrium point at the origin. We start our analysis by showing that every nonequilibrium solution rotates around the equilibrium in the clockwise direction. To this end, we divide the state plane into four regions which are determined by the intersection of the two curves

$$x_2 - \epsilon h(x_1) = 0$$
$$-x_1 = 0$$

as shown in Figure 7.2. The figure also shows the general direction of the vector field $f(x)$ of (7.2) in the four regions as well as on the boundaries between them. It is not difficult to see that a solution starting at point $A = (0, p)$ on the upper half of the x_2-axis describes an orbit with an arc of the general nature shown in Figure 7.3. The point E where the arc intersects the lower half of the x_2-axis depends on the starting point A. Let us denote E by $(0, -\alpha(p))$. We will show that if p is chosen large enough, then $\alpha(p) < p$. Consider the function

$$V(x) = \tfrac{1}{2}(x_1^2 + x_2^2)$$

To show that $\alpha(p) < p$, it is enough to show that $V(E) - V(A) < 0$ since

$$V(E) - V(A) = \tfrac{1}{2}[\alpha^2(p) - p^2] \overset{\text{def}}{=} \delta(p)$$

The derivative of $V(x)$ along the trajectories of (7.2) is given by

$$\dot{V}(x) = x_1\dot{x}_1 + x_2\dot{x}_2 = x_1x_2 - \epsilon x_1 h(x_1) - x_1x_2 = -\epsilon x_1 h(x_1)$$

Thus, $\dot{V}(x)$ is positive for $x_1 < a$ and negative for $x_1 > a$. Now,

$$\delta(p) = V(E) - V(A) = \int_{AE} \dot{V}(x(t))\ dt$$

where the right-hand side integral is taken along the arc from A to E. If p is small, the whole arc will lie inside the strip $0 < x_1 < a$. Then, $\delta(p)$ will be positive. As p increases, a piece of the arc will lie outside the strip, that is, the piece BCD in Figure 7.3. In this case, we evaluate the integral in different ways depending on whether the arc is inside or outside the strip $0 < x_1 < a$. We divide the integral into three parts

$$\delta(p) = \delta_1(p) + \delta_2(p) + \delta_3(p)$$

where

$$\delta_1(p) = \int_{AB} \dot{V}(x(t))\ dt$$

$$\delta_2(p) = \int_{BCD} \dot{V}(x(t))\ dt$$

$$\delta_3(p) = \int_{DE} \dot{V}(x(t))\ dt$$

Consider first the term $\delta_1(p)$

$$\delta_1(p) = -\int_{AB} \epsilon x_1 h(x_1)\ dt = -\int_{AB} \epsilon x_1 h(x_1)\ \frac{dt}{dx_1}\ dx_1$$

Substituting for dx_1/dt from (7.2), we obtain

$$\delta_1(p) = -\int_{AB} \epsilon x_1 h(x_1)\frac{1}{x_2 - \epsilon h(x_1)}\ dx_1$$

where, along the arc AB, x_2 is a given function of x_1. Clearly, $\delta_1(p)$ is positive. Note that as p increases, $x_2 - \epsilon h(x_1)$ increases for the arc AB. Hence, $\delta_1(p)$ decreases as $p \to \infty$. Similarly, it can be shown that the third term $\delta_3(p)$ is positive and decreases as $p \to \infty$. Consider now the second term $\delta_2(p)$

$$\delta_2(p) = -\int_{BCD} \epsilon x_1 h(x_1)\ dt = -\int_{BCD} \epsilon x_1 h(x_1)\ \frac{dt}{dx_2}\ dx_2$$

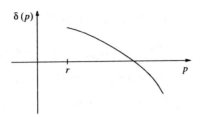

Figure 7.4: A sketch of the function $\delta(p)$ of Example 7.3.

Substituting for dx_2/dt from (7.2), we obtain

$$\delta_2(p) = \int_{BCD} \epsilon h(x_1) \, dx_2$$

where along the arc BCD, x_1 is a given function of x_2. The integral on the right-hand side is negative since $h(x_1) > 0$ and $dx_2 < 0$. As p increases, the arc $ABCDE$ moves to the right and the domain of integration for $\delta_2(p)$ increases. It follows that $\delta_2(p)$ decreases as p increases, and evidently $\lim_{p\to\infty} \delta_2(p) = -\infty$. In summary, we have shown that

- $\delta(p) > 0$, if $p < r$; for some $r > 0$.

- $\delta(p)$ decreases monotonically to $-\infty$ as $p \to \infty$, $p \geq r$.

A sketch of the function $\delta(p)$ is shown in Figure 7.4. It is now clear that by choosing p large enough we can ensure that $\delta(p)$ is negative; hence $\alpha(p) < p$.

Observe that, due to symmetry induced by the fact that $h(\cdot)$ is an odd function, if $(x_1(t), x_2(t))$ is a solution of (7.2) then so is $(-x_1(t), -x_2(t))$. Therefore, if we know that a path $ABCDE$ exists as in Figure 7.3, then the reflection of that path through the origin is another path. Consider $A = (0, p)$ and $E = (0, -\alpha(p))$, where $\alpha(p) < p$. Form a closed curve of the arc $ABCDE$, its reflection through the origin and segments on the x_2-axis connecting these arcs, to form a closed curve; see Figure 7.5. The region enclosed by this closed curve is positively invariant; that is, every trajectory starting inside this region at $t = 0$ will remain in it for all $t \geq 0$. This is a consequence of the directions of the vector fields on the x_2-axis segments and the fact that trajectories do not intersect each other due to uniqueness of solutions. Recall that

$$\dot{V}(x) = -\epsilon x_1 h(x_1) > 0, \quad \text{for } |x_1| < a$$

Let $c > 0$ be small enough that the level surface $V(x) = c$ lies inside the strip $|x_1| < a$. All trajectories on the level surface $V(x) = c$ must move outward. Finally, define a closed set M as the annular region whose outer boundary is the closed curve

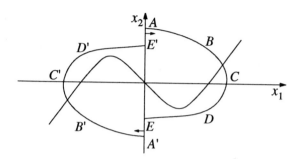

Figure 7.5: The closed curve formed in Example 7.3.

constructed in Figure 7.5 and whose inner boundary is the level surface $V(x) = c$. The set M is closed, bounded, positively invariant, and free of equilibrium points. Thus, from the Poincaré-Bendixson theorem we conclude that there is a closed orbit in M.

Using the same analysis, we can go beyond the Poincaré-Bendixson theorem and show that this closed orbit is unique. Notice that, due to the symmetry property alluded to earlier, the system can have a closed orbit if and only if $\alpha(p) = p$. From Figure 7.4, it is clear that there is only one value of p for which this condition is satisfied. Hence, there is only one closed orbit. Furthermore, we can show that every nonequilibrium solution spirals toward the unique closed orbit. To argue this point, let $p_0 > 0$ be the unique value of p for which $\alpha(p) = p$. Consider a point $(0, p)$ on the x_2-axis with $p > p_0$. As we argued earlier, the trajectory through $(0, p)$ intersects the lower half of the x_2-axis at a point $(0, -\alpha(p))$ where $\alpha(p) < p$. Due to symmetry, the trajectory through $(0, -\alpha(p))$ will meet the upper half of the x_2-axis at a point $(0, \sigma(p))$, where $p_0 \le \sigma(p) < p$. The upper bound follows from the symmetry property, while the lower bound holds since for $\sigma(p)$ to be less than p_0 the trajectory must intersect the closed orbit. The map $p \to \sigma(p)$ is continuous due to the continuous dependence of the solution on initial states. Starting again at the point $(0, \sigma(p))$, the trajectory comes back to the upper half of the x_2-axis at $(0, \sigma^2(p))$ where $p_0 \le \sigma^2(p) < \sigma(p)$. By induction, we generate a sequence $\sigma^n(p)$ which satisfies

$$p_0 \le \sigma^{n+1}(p) < \sigma^n(p), \quad n = 1, 2, \ldots$$

The sequence $\sigma^n(p)$ has a limit $p_1 \ge p_0$. Note that, by continuity of $\sigma(\cdot)$, the limit p_1 satisfies

$$\sigma(p_1) - p_1 = \lim_{n \to \infty} \sigma(\sigma^n(p)) - p_1 = p_1 - p_1 = 0$$

By uniqueness of the closed orbit, it must be that $p_1 = p_0$. This shows that the trajectory of p spirals toward the unique closed orbit as $t \to \infty$. The same thing is true for $p < p_0$. $\triangle$

The next result, known as the *Bendixson criterion*, can be used to rule out the existence of periodic orbits in some cases.

Theorem 7.2 (Bendixson Criterion) *If on a simply connected region[2] $D \subset R^2$ the expression $\partial f_1/\partial x_1 + \partial f_2/\partial x_2$ is not identically zero and does not change sign, then system (7.1) has no periodic orbits lying entirely in D.*   ◇

Proof: On any orbit of (7.1), we have $dx_2/dx_1 = f_2/f_1$. Therefore, on any closed orbit γ we have

$$\int_\gamma f_2(x_1, x_2)\, dx_1 - f_1(x_1, x_2)\, dx_2 = 0$$

This implies, by Green's theorem, that

$$\int\int_S \left(\frac{\partial f_1}{\partial x_1} + \frac{\partial f_2}{\partial x_2} \right) dx_1\, dx_2 = 0$$

where S is the interior of γ. If $\partial f_1/\partial x_1 + \partial f_2/\partial x_2 > 0$ (or < 0) on D, then we cannot find a region $S \subset D$ such that the last equality holds. Hence, there can be no closed orbits entirely in D.   □

Example 7.4 Consider the system

$$\dot{x}_1 = f_1(x_1, x_2) = x_2$$
$$\dot{x}_2 = f_2(x_1, x_2) = ax_1 + bx_2 - x_1^2 x_2 - x_1^3$$

We have

$$\frac{\partial f_1}{\partial x_1} + \frac{\partial f_2}{\partial x_2} = b - x_1^2$$

Hence, there can be no periodic orbits if $b < 0$.   △

We conclude this section with a useful result that relates the existence of periodic orbits and equilibrium points. The result uses the (Poincaré) index of an equilibrium point. Given the second-order system (7.1), let C be a simple closed curve not passing through any equilibrium point of (7.1). Consider the orientation of the vector field $f(x)$ at a point $p \in C$. Letting p traverse C in the counterclockwise direction, the vector $f(x)$ rotates continuously and, upon returning to the original position, must have rotated an angle $2\pi k$ for some integer k, where the angle is measured counterclockwise. The integer k is called the index of the closed curve C. If C is chosen to encircle a single isolated equilibrium point $\bar{x}$, then k is called the index of $\bar{x}$. It is left to the reader (Exercise 7.11) to verify the following lemma by examination of the vector fields.

---

[2] A region $D \subset R^2$ is simply connected if, for every Jordan curve C in D, the inner region of C is a also a subset of D. The interior of any circle is simply connected, but the annular region $0 < c_1 \le x_1^2 + x_2^2 \le c_2$ is not simply connected. Intuitively speaking, simple connectedness is equivalent to the absence of "holes."

Lemma 7.1

(a) *The index of a node, a focus, or a center is +1.*

(b) *The index of a (hyperbolic) saddle is −1.*

(c) *The index of a closed orbit is +1.*

(d) *The index of a closed curve not encircling any equilibrium points is 0.*

(e) *The index of a closed curve is equal to the sum of the indices of the equilibrium points within it.* ◇

As a corollary to this lemma, we have the following.

Corollary 7.1 *Inside any periodic orbit γ, there must be at least one equilibrium point. Suppose the equilibrium points inside γ are hyperbolic, then if N is the number of nodes and foci and S is the number of saddles, it must be that $N - S = 1$.* ◇

Recall that an equilibrium point is hyperbolic if the Jacobian at that point has no eigenvalues on the imaginary axis. If the equilibrium point is not hyperbolic, then its index may differ from ±1; see Exercise 7.12.

The index method is usually useful in ruling out the existence of periodic orbits in certain regions of the plane.

Example 7.5 The system

$$\begin{aligned} \dot{x}_1 &= -x_1 + x_1 x_2 \\ \dot{x}_2 &= x_1 + x_2 - 2x_1 x_2 \end{aligned}$$

has two equilibrium points at $(0,0)$ and $(1,1)$. The Jacobian matrices at these points are

$$\left[\frac{\partial f}{\partial x}\right]_{(0,0)} = \left[\begin{array}{cc} -1 & 0 \\ 1 & 1 \end{array}\right]; \quad \left[\frac{\partial f}{\partial x}\right]_{(1,1)} = \left[\begin{array}{cc} 0 & 1 \\ -1 & -1 \end{array}\right]$$

Hence, $(0,0)$ is a saddle while $(1,1)$ is a focus. The only combination of equilibrium points that can be encircled by a periodic orbit is a single focus. Other possibilities of periodic orbits, like a periodic orbit encircling both equilibria, are ruled out. △

7.2 Stability of Periodic Solutions

In Chapters 3 and 4, we have developed an extensive theory for the behavior of solutions of a dynamical system near an equilibrium point. In this section, we consider the corresponding problem for periodic solutions. If $u(t)$ is a periodic solution of the system

$$\dot{x} = f(t, x) \tag{7.3}$$

what can we say about other solutions which start arbitrarily close to $u(t)$? Will they remain in some neighborhood of $u(t)$ for all t? Will they eventually approach $u(t)$? Such stability properties of the periodic solution $u(t)$ can be characterized and investigated in the sense of Lyapunov. Let

$$y = x - u(t)$$

so that the origin $y = 0$ becomes an equilibrium point for the nonautonomous system

$$\dot{y} = f(t, y + u(t)) - f(t, u(t)) \tag{7.4}$$

The behavior of solutions of (7.3) near $u(t)$ is equivalent to the behavior of solutions of (7.4) near $y = 0$. Therefore, we can characterize stability properties of $u(t)$ from those of the equilibrium $y = 0$. In particular, we say that the periodic solution $u(t)$ is uniformly asymptotically stable if $y = 0$ is a uniformly asymptotically stable equilibrium point for the system (7.4). Similar statements can be made for other stability properties, like uniform stability. Thus, investigating the stability of $u(t)$ has been reduced to studying the stability of an equilibrium point of a nonautonomous system, which we studied in Chapter 3. We shall find this notion of uniform asymptotic stability of periodic solutions in the sense of Lyapunov to be useful when we study nonautonomous systems dependent on small parameters in Chapter 8. The notion, however, is too restrictive when we analyze periodic solutions of autonomous systems. The following example illustrates the restrictive nature of this notion.

Example 7.6 Consider the second-order system

$$
\begin{aligned}
\dot{x}_1 &= x_1\left[\frac{\left(1 - x_1^2 - x_2^2\right)^3}{x_1^2 + x_2^2}\right] - x_2\left[1 + \left(1 - x_1^2 - x_2^2\right)^2\right] \\
\dot{x}_2 &= x_2\left[\frac{\left(1 - x_1^2 - x_2^2\right)^3}{x_1^2 + x_2^2}\right] + x_1\left[1 + \left(1 - x_1^2 - x_2^2\right)^2\right]
\end{aligned}
$$

which is represented in the polar coordinates

$$x_1 = r\cos\theta, \quad x_2 = r\sin\theta$$

by

$$
\begin{aligned}
\dot{r} &= \frac{(1 - r^2)^3}{r} \\
\dot{\theta} &= 1 + (1 - r^2)^2
\end{aligned}
$$

The solution starting at (r_0, θ_0) is given by

$$r(t) = \left[1 - \frac{1 - r_0^2}{\sqrt{1 + 4t\left(1 - r_0^2\right)^2}} \right]^{1/2}$$

$$\theta(t) = \theta_0 + t + \tfrac{1}{4} \ln \left[1 + 4t(1 - r_0^2)^2 \right]$$

From these expressions, we see that the system has a periodic solution

$$\bar{x}_1(t) = \cos t, \quad \bar{x}_2(t) = \sin t$$

The corresponding periodic orbit is the unit circle $r = 1$. All nearby solutions spiral toward this periodic orbit as $t \to \infty$. This is clearly the kind of "asymptotically stable" behavior we expect to see with a periodic orbit. In fact, this periodic orbit has be known classically as a stable limit cycle. However, the periodic solution $\bar{x}(t)$ is not uniformly asymptotically stable in the sense of Lyapunov. Recall that for the solution to be uniformly asymptotically stable, we must have

$$[r(t) \cos \theta(t) - \cos t]^2 + [r(t) \sin \theta(t) - \sin t]^2 \to 0 \quad \text{as} \quad t \to \infty$$

for sufficiently small $[r_0 \cos \theta_0 - 1]^2 + [r_0 \sin \theta_0]^2$. Since $r(t) \to 1$ as $t \to \infty$, we must have

$$|1 - \cos(\theta(t) - t)| \to 0 \quad \text{as} \quad t \to \infty$$

which clearly is not satisfied when $r_0 \neq 1$, since $\theta(t) - t$ is an ever-growing monotonically increasing function of t. $\qquad \triangle$

The point illustrated by this example is true in general. In particular, a nontrivial periodic solution of an autonomous system can never be asymptotically stable in the sense of Lyapunov.[3]

The stability-like properties of the periodic orbit of Example 7.6 can be captured by extending the notion of stability in the sense of Lyapunov from stability of an equilibrium point to stability of an invariant set. Consider the autonomous system

$$\dot{x} = f(x) \tag{7.5}$$

where $f : D \to R^n$ is a continuously differentiable map from a domain $D \subset R^n$ into R^n. Let $M \subset D$ be an invariant set of (7.5). Define an ϵ-neighborhood of M by

$$U_\epsilon = \{x \in R^n \mid \operatorname{dist}(x, M) < \epsilon\}$$

where $\operatorname{dist}(x, M)$ is the minimum distance from x to a point in M; that is,

$$\operatorname{dist}(x, M) = \inf_{y \in M} \|x - y\|$$

---

[3] See [62, Theorem 81.1] for a proof of this statement.

Definition 7.2 *The invariant set M of (7.5) is*

- *stable if, for each $\epsilon > 0$, there is $\delta > 0$ such that*

$$x(0) \in U_\delta \Rightarrow x(t) \in U_\epsilon, \quad \forall\, t \geq 0$$

- *asymptotically stable if it is stable and δ can be chosen such that*

$$x(0) \in U_\delta \Rightarrow \lim_{t \to \infty} \text{dist}(x(t), M) = 0$$

This definition reduces to Definition 3.1 when M is an equilibrium point. Stability and asymptotic stability of invariant sets are interesting concepts in their own sake. We shall apply them here to the specific case when the invariant set M is the closed orbit associated with a periodic solution. Let $u(t)$ be a nontrivial periodic solution of the autonomous system (7.5) with period T, and let γ be the closed orbit defined by

$$\gamma = \{x \in R^n \mid x = u(t),\ 0 \leq t \leq T\}$$

The periodic orbit γ is the image of $u(t)$ in the state space. It is an invariant set whose stability properties are characterized by Definition 7.2. It is common, especially for second-order systems, to refer to asymptotically stable periodic orbits as stable limit cycles

Example 7.7 The harmonic oscillator

$$\dot{x}_1 = x_2$$
$$\dot{x}_2 = -x_1$$

has a continuum of periodic orbits which are concentric circles with a center at the origin. Any one of these periodic orbits is stable. Consider, for example, the periodic orbit γ_c defined by

$$\gamma_c = \{x \in R^2 \mid r = c > 0\}, \quad \text{where } r = \sqrt{x_1^2 + x_2^2}$$

The U_ϵ neighborhood of γ_c is defined by the annular region

$$U_\epsilon = \{x \in R^2 \mid c - \epsilon < r < c + \epsilon\}$$

This annular region itself is an invariant set. Thus, given $\epsilon > 0$, we can take $\delta = \epsilon$ and see that any solution starting in the U_δ neighborhood at $t = 0$ will remain in the U_ϵ neighborhood for all $t \geq 0$. Hence, the periodic orbit γ_c is stable. However, it is not asymptotically stable since a solution starting in a U_δ neighborhood of γ_c does not approach γ_c as $t \to \infty$, no matter how small δ is. $\triangle$

Example 7.8 Consider the system of Example 7.6. It has an isolated periodic orbit

$$\gamma = \{x \in R^2 \mid r = 1\}, \quad \text{where } r = \sqrt{x_1^2 + x_2^2}$$

For $x \notin \gamma$, we have

$$\text{dist}(x, \gamma) = \inf_{y \in \gamma} \|x - y\|_2 = \inf_{y \in \gamma} \sqrt{(x_1 - y_1)^2 + (x_2 - y_2)^2} = |r - 1|$$

Recalling that

$$r(t) = \left[1 - \frac{1 - r_0^2}{\sqrt{1 + 4t(1 - r_0^2)^2}} \right]^{1/2}$$

it can be easily seen that the ϵ-δ requirement for stability is satisfied and

$$\text{dist}(x(t), \gamma) \to 0, \text{ as } t \to \infty$$

Hence, the periodic orbit is asymptotically stable. $\triangle$

Having defined the stability properties of periodic orbits, we can now define the stability properties of periodic solutions.

Definition 7.3 *A nontrivial periodic solution $u(t)$ of (7.5) is*

- *orbitally stable if the closed orbit γ generated by $u(t)$ is stable.*

- *asymptotically orbitally stable if the closed orbit γ generated by $u(t)$ is asymptotically stable.*

Notice that different terminology is used depending on whether we are talking about the periodic solution or about the corresponding periodic orbit. In Example 7.8, we say that the unit circle is an asymptotically stable periodic orbit, but we say that the periodic solution $(\cos t, \sin t)$ is orbitally asymptotically stable.

An important conceptual tool for understanding the stability of periodic orbits is the Poincaré map. It replaces an nth-order continuous-time autonomous system by an $(n-1)$th-order discrete-time system. Let γ be a periodic orbit of the nth-order system (7.5). Let p be a point on γ and H be an $(n-1)$-dimensional hyperplane at p. By a hyperplane, we mean a surface $a^T(x - p) = 0$ for some $a \in R^n$. Suppose that the hyperplane is transversal to γ at p; that is, $a^T f(p) \neq 0$. For $n = 2$, such a hyperplane is any line through p that is not tangent to γ at p. Let $S \subset H$ be a local section such that $p \in S$ and $a^T f(x) \neq 0$ for all $x \in S$. The trajectory starting from p will hit S at p in T seconds, where T is the period of the periodic orbit. Due to the continuity of solutions with respect to initial states, the trajectories starting

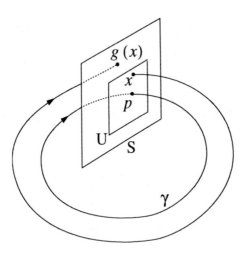

Figure 7.6: Poincaré map.

on S in a sufficiently small neighborhood of p will, in approximately T seconds, intersect S in the vicinity of p. Let $U \subset S$ be a (sufficiently small) neighborhood of p such that γ intersects U only once at p. The Poincaré map $g : U \to S$ is defined for a point $x \in U$ by

$$g(x) = \phi(\tau, x)$$

where $\phi(t, x)$ is the solution of (7.5) that starts at x at time $t = 0$, and $\tau = \tau(x)$ is the time taken for the trajectory starting at x to first return to S; see Figure 7.6. Note that τ generally depends on x and need not be equal to T, the period of γ. The Poincaré map is defined only locally; that is, it need not be defined for all $x \in S$. Suppose that U in the foregoing definition is chosen such that the map is defined for all $x \in U$. Starting with $x^{(0)} \in U$, let $x^{(1)} = g\left(x^{(0)}\right)$. If $x^{(1)} \in U$, the Poincaré map will be defined at $x^{(1)}$. Set $x^{(2)} = g\left(x^{(1)}\right)$. As long as $x^{(k)} \in U$, $x^{(k+1)} = g(x^{(k)})$ will be defined. The sequence $x^{(k)}$ is the solution of the discrete-time system

$$x^{(k+1)} = g\left(x^{(k)}\right) \tag{7.6}$$

It is clear that p is an equilibrium point of (7.6) since $p = g(p)$. Although the vector x is n-dimensional, the solution generated by (7.6) is restricted to the $(n-1)$-dimensional hyperplane H. Hence, it is equivalent to the solution of an $(n - 1)$-dimensional system. This point can be made clearer with a change of variables. Suppose the equation of the hyperplane H is $a^T(x - p) = 0$. Since $a^T f(p) \neq 0$, the vector a must be nonzero. Hence, at least one component of a is nonzero. Since we can rename the components of a, assume without loss of generality that $a_n \neq 0$. Partition a, x, p, and g as

$$
a = \begin{bmatrix} b \\ a_n \end{bmatrix} ; \ x = \begin{bmatrix} y \\ x_n \end{bmatrix} ; p = \begin{bmatrix} q \\ p_n \end{bmatrix} ; \ g = \begin{bmatrix} \eta \\ g_n \end{bmatrix}
$$

Define a nonsingular matrix M (and, subsequently, its inverse) by

$$
M = \begin{bmatrix} I & 0 \\ b^T & a_n \end{bmatrix} ; \ M^{-1} = \begin{bmatrix} I & 0 \\ -\frac{1}{a_n} b^T & \frac{1}{a_n} \end{bmatrix}
$$

where I is the $(n-1) \times (n-1)$ identity matrix. Apply the change of variables

$$
z = M x = \begin{bmatrix} y \\ a^T x \end{bmatrix}
$$

For all $x \in H$, $a^T x = a^T p$. Hence, when expressed in the z-coordinates, the last component of the solution of (7.6) is always constant. The first $(n-1)$ components evolve as the solution of the $(n-1)$-dimensional system

$$
y^{(k+1)} = h\left(y^{(k)} \right) \tag{7.7}
$$

where

$$
h(y) = \eta \left(\begin{bmatrix} y \\ p_n - \frac{1}{a_n} b^T (y - q) \end{bmatrix} \right)
$$

The system (7.7) has an equilibrium point at $y = q$ since

$$
h(q) = \eta \left(\begin{bmatrix} q \\ p_n \end{bmatrix} \right) = \eta(p) = q
$$

There is an intimate relationship between stability properties of the periodic orbit γ and stability properties of q as an equilibrium point for the discrete-time system (7.7). Although we have not formally studied stability of equilibrium points for discrete-time systems, the concepts are completely analogous to the continuous-time case. In particular, Definition 3.1 can be read as a definition of stability and asymptotic stability of the origin of a discrete-time system; the only necessary change is to keep in mind that the time variable t should be taken as an integer. In a few exercises at the end of Section 3.7, we have summarized some facts about Lyapunov stability for discrete-time systems.

Theorem 7.3 *Let γ be a periodic orbit of (7.5). Define the Poincaré map and the discrete-time system (7.7) as explained earlier. If q is an asymptotically stable equilibrium point of (7.7), then γ is asymptotically stable.* $\diamond$

Proof: We need to show that given any U_ϵ neighborhood of γ, there is a U_δ neighborhood such that for all $x \in U_\delta$

$$\phi(t, x) \in U_\epsilon, \quad \forall\, t \geq 0$$

and

$$\lim_{t \to \infty} \text{dist}(\phi(t, x), \gamma) = 0$$

The Poincaré map was defined with respect to a section S at a given point $p \in \gamma$. An initial state near γ need not be in the neighborhood of p. Let

$$x \in U_\delta = \{x \in R^n \mid \text{dist}(x, \gamma) < \delta\}$$

There is a point $p_1 \in \gamma$ such that $\|x - p_1\| < \delta$. By continuity of the solution with respect to initial states, in particular by Theorem 2.5, we have

$$\|\phi(t, x) - \phi(t, p_1)\| \leq \delta \exp(Lt)$$

where L is a Lipschitz constant for f. The solution $\phi(t, p_1)$ is periodic with period T. Therefore, there is a time $t_1 \leq T$ such that $\phi(t_1, p_1) = p$. Consequently,

$$\|\phi(t_1, x) - p\| \leq \delta \exp(LT)$$

Thus, by choosing δ small enough, we can ensure that the trajectory of x will pass within any specified neighborhood of p. Now we want to show that the trajectory will hit the section S of the hyperplane H. Suppose the equation of H is $a^T(x-p) = 0$ and define

$$G(t, x) = a^T[\phi(t, x) - p]$$

Since $\phi(T, p) = p$, we have $G(T, p) = 0$. Moreover,

$$\left.\frac{\partial G}{\partial t}(t, x)\right|_{t=T; x=p} = a^T f(\phi(t, x))\big|_{t=T; x=p} = a^T f(p) \neq 0$$

By the implicit function theorem, there is a unique continuously differentiable map $x \to \tau(x) \in R$ on a neighborhood V of p in R^n such that $\tau(p) = T$ and $G(\tau(x), x) = 0$. This means that any solution starting within the V neighborhood will hit the hyperplane H at time $\tau(x)$. Using continuity of solutions with respect to initial states, it is clear that by choosing the neighborhood V small enough, we can ensure that the solution will hit H within $U \subset S$. By continuity of $\tau(x)$, given any $\Delta \in (0, T)$ we can choose the neighborhood V so small that

$$T - \Delta \leq \tau(x) \leq T + \Delta, \quad \forall\, x \in V \tag{7.8}$$

This argument shows that by starting sufficiently near γ, the trajectory will have a point in $U \subset S$; hence, the Poincaré map will be defined for this point. Now we can consider the discrete-time system (7.7), which has an equilibrium point at q. If q is stable, then given any $\epsilon_1 > 0$ there is $\delta_1 > 0$ such that

$$\|y^{(0)} - q\| < \delta_1 \Rightarrow \|y^{(k)} - q\| < \epsilon_1, \ \forall \ k \geq 0$$

Since

$$x - p = \begin{bmatrix} y - q \\ -\frac{1}{a_n} b^T (y - q) \end{bmatrix}$$

we have

$$c_1 \|y - q\| \leq \|x - p\| \leq c_2 \|y - q\|$$

for some positive constants c_1 and c_2. Therefore, given any $\epsilon_2 > 0$ there is $\delta_2 > 0$ such that

$$\|x^{(0)} - p\| < \delta_2 \Rightarrow \|x^{(k)} - p\| < \epsilon_2, \ \forall \ k \geq 0$$

Choosing ϵ_2 small enough, we can ensure that the sequence $x^{(k)}$ will always be within the neighborhood V. The points of intersection of $\phi(t, x)$ with the section S define a sequence t_k such that

$$T - \Delta \leq t_{k+1} - t_k \leq T + \Delta$$

$$t_k \to \infty \text{ as } k \to \infty$$

and

$$\|\phi(t_k, x) - p\| = \|x^{(k)} - p\| < \epsilon_2, \ \forall \ k \geq 0$$

The bounds on $t_{k+1} - t_k$ follow from (7.8). The fact that $t_k \to \infty$ follows from the lower bound on $t_{k+1} - t_k$. Between two consecutive time points t_k and t_{k+1}, we have

$$\|\phi(t - t_k, x^{(k)}) - \phi(t - t_k, p)\| \leq \|x^{(k)} - p\| \exp[L(T + \Delta)] \qquad (7.9)$$

which follows again from Theorem 2.5. Thus, we can conclude that

$$\text{dist}(\phi(t, x), \gamma) \leq k_1 \epsilon_2, \ \forall \ t \geq 0, \forall \ x \in V$$

for some positive constant k_1, provided the neighborhood V is small enough. By choosing ϵ_2 small enough, we can put $\phi(t, x)$ within any U_ϵ neighborhood of γ. Suppose now that q is an asymptotically stable equilibrium point of (7.7). Then,

$$\|y^{(k)} - q\| \to 0 \text{ as } k \to \infty$$

which implies that

$$\|x^{(k)} - p\| \to 0 \text{ as } k \to \infty$$

For any $\epsilon > 0$, there is $N > 0$ such that

$$\|x^{(k)} - p\| < \epsilon, \quad \forall\, k \geq N$$

It follows from (7.9) that

$$\text{dist}(\phi(t, x), \gamma) < k_1\epsilon, \quad \forall\, t \geq t_N$$

Thus,

$$\text{dist}(\phi(t, x), \gamma) \to 0 \text{ as } t \to \infty$$

□

and the proof is complete.

Example 7.9 Consider the second-order system

$$\dot{x}_1 = x_1 - x_2 - x_1 \left(x_1^2 + x_2^2\right)$$
$$\dot{x}_2 = x_1 + x_2 - x_2 \left(x_1^2 + x_2^2\right)$$

which is represented in the polar coordinates

$$x_1 = r \cos\theta, \quad x_2 = r \sin\theta$$

by

$$\dot{r} = r \left(1 - r^2\right)$$
$$\dot{\theta} = 1$$

The solution starting at (r_0, θ_0) is given by

$$r(t) = \left[1 + \left(\frac{1}{r_0^2} - 1\right) \exp(-2t)\right]^{-1/2}$$
$$\theta(t) = \theta_0 + t$$

From these expressions, we see that the unit circle $r = 1$ is a periodic orbit. To construct a Poincaré map, take $p = (1, 0)$ and the section S as the positive half of the x_1-axis. Starting from a point $(r, 0) \in S$, the first return of the trajectory to S will be in $\tau = 2\pi$. Thus, the Poincaré map (expressed in polar coordinates) is given by

$$g(r) = \left[1 + \left(\frac{1}{r^2} - 1\right) \exp(-4\pi)\right]^{-1/2}$$

which is already a one-dimensional map, so we do not need to go through the reduction procedure explained earlier. The discrete-time system corresponding to this map is given by

$$r_{k+1} = \left[1 + \left(\frac{1}{r_k^2} - 1\right) \exp(-4\pi)\right]^{-1/2}$$

The system has an equilibrium point at $r = 1$, which will be asymptotically stable if the eigenvalues (only one in this case) of the Jacobian at the equilibrium point have magnitudes less than one.

$$
\left. \frac{dg}{dr} \right|_{r=1} = \left. -\frac{1}{2} \left[1 + \left(\frac{1}{r^2} - 1 \right) \exp(-4\pi) \right]^{-3/2} \left(-\frac{2e^{-4\pi}}{r^3} \right) \right|_{r=1}
$$

$$
= e^{-4\pi} < 1
$$

Thus, $r = 1$ is an asymptotically stable equilibrium point of the discrete-time system, and the unit circle is an asymptotically stable periodic orbit. △

The construction of the Poincaré map relies on knowledge of the solution of the differential equation. Therefore, except for trivial examples where the solution is available in a closed form, we cannot construct the Poincaré map analytically. In practice, it has to be generated via computer simulations.[4]

7.3 Exercises

Exercise 7.1 For each of the following systems, use Poincaré-Bendixson's theorem to show that the system has a periodic orbit.

(1)
$$
\dot{x}_1 = x_2
$$
$$
\dot{x}_2 = -x_1 + x_2(1 - 3x_1^2 - 2x_2^2)
$$

(2)
$$
\dot{x}_1 = x_2
$$
$$
\dot{x}_2 = -x_1 + x_2 - 2(x_1 + 2x_2)x_2^2
$$

(3)
$$
\dot{x}_1 = x_1 + x_2 - x_1(|x_1| + |x_2|)
$$
$$
\dot{x}_2 = -2x_1 + x_2 - x_2(|x_1| + |x_2|)
$$

(4)
$$
\ddot{y} + y = \epsilon \dot{y}(1 - y^2 - \dot{y}^2), \quad \epsilon > 0
$$

Exercise 7.2 (Conservative Systems) Consider the second-order system

$$
\dot{x}_1 = x_2
$$
$$
\dot{x}_2 = -g(x_1)
$$

---

[4]Numerical algorithms for generating the Poincaré map can be found, for example, in [132].

where $g(\cdot)$ is continuously differentiable and $zg(z) > 0$ for $z \in (-a, a)$. Consider the energy function

$$V(x) = \tfrac{1}{2}x_2^2 + \int_0^{x_1} g(z)\, dz$$

(a) Show that $V(x)$ remains constant along the trajectories of the system.

(b) Show that, for sufficiently small $\|x(0)\|$, every solution is periodic.

(c) Suppose that $zg(z) > 0$ for $z \in (-\infty, \infty)$ and

$$\int_0^y g(z)\, dz \rightarrow \infty \text{ as } |y| \rightarrow \infty$$

Show that every solution is a periodic solution.

(d) Suppose $g(z) = -g(-z)$ and let $G(y) = \int_0^y g(z)\, dz$. Show that the trajectory through $(A, 0)$ is given by

$$x_2 = \pm\sqrt{2[G(A) - G(x_1)]}$$

(e) Using part (d), show that the period of oscillation of a closed trajectory through $(A, 0)$ is

$$T(A) = 2\sqrt{2} \int_0^A \frac{dy}{[G(A) - G(y)]^{1/2}}$$

(f) Discuss how the trajectory equation in part (d) can be used to construct the phase portrait of the system.

Exercise 7.3 Use the previous exercise to construct the phase portrait and study periodic solutions for each of the following systems. In each case, give the period of oscillation of the periodic orbit through the point $(1, 0)$.

(1) $g(x_1) = \sin x_1$.

(2) $g(x_1) = x_1 + x_1^3$.

(3) $g(x_1) = x_1^3$.

Exercise 7.4 For each of the following systems, show that the system has no limit cycles.

$$\text{(1)} \qquad \dot{x}_1 = x_2$$
$$\dot{x}_2 = g(x_1) + ax_2, \quad a \neq 0$$

(2)
$$\dot{x}_1 = -x_1 + x_1^3 + x_1 x_2^2$$
$$\dot{x}_2 = -x_2 + x_2^3 + x_1^2 x_2$$

(3)
$$\dot{x}_1 = 1 - x_1 x_2$$
$$\dot{x}_2 = x_1$$

(4)
$$\dot{x}_1 = x_1 x_2$$
$$\dot{x}_2 = -x_2$$

(5)
$$\dot{x}_1 = x_2 \cos(x_1)$$
$$\dot{x}_2 = \sin x_1$$

Exercise 7.5 Consider the system

$$\dot{x}_1 = -x_1 + x_2(x_1 + a) - b$$
$$\dot{x}_2 = -cx_1(x_1 + a)$$

where a, b, and c are positive constants with $b > a$. Let

$$D = \left\{ x \in R^2 \mid x_1 < -a \text{ and } x_2 < \frac{x_1 + b}{x_1 + a} \right\}$$

(a) Show that the set D is positively invariant.

(b) Show that there can be no periodic orbits through any point $x \in D$.

Exercise 7.6 Consider the system

$$\dot{x}_1 = ax_1 - x_1 x_2$$
$$\dot{x}_2 = bx_1^2 - cx_2$$

where a, b, and c are positive constants with $c > a$. Let

$$D = \{ x \in R^2 \mid x_2 \geq 0 \}$$

(a) Show that the set D is positively invariant.

(b) Show that there can be no periodic orbits through any point $x \in D$.

Exercise 7.7 ([76]) Consider the system

$$\dot{x}_1 = x_2$$
$$\dot{x}_2 = -[2b - g(x_1)]ax_2 - a^2 x_1$$

where a and b are positive constants, and

$$g(x_1) = \begin{cases} 0, & |x_1| > 1 \\ k, & |x_1| \le 1 \end{cases}$$

(a) Show, using Bendixson's criterion, that there are no periodic orbits if $k < 2b$.

(b) Show, using the Poincaré-Bendixson theorem, that there is a periodic orbit if $k > 2b$.

Exercise 7.8 ([118]) Consider the third-order system, in spherical coordinates,

$$\dot{\theta} = 1, \quad \dot{\phi} = \pi, \quad \dot{\rho} = \rho(\sin\theta + \pi\sin\phi)$$

and let $\theta(0) = \phi(0) = 0$ and $\rho(0) = 1$. Show that the solution is bounded in R^3, but it is not periodic nor does it tend to a periodic solution.

Exercise 7.9 Consider a second-order system and let M be a positively invariant compact set. Show that M either contains a periodic orbit or an equilibrium point.

Exercise 7.10 Consider a second-order system and let

$$M = \{x \in R^2 \mid x_1^2 + x_2^2 \le a^2\}$$

be a positively invariant set. Show that M contains an equilibrium point.

Exercise 7.11 Verify Lemma 7.1 by examination of the vector fields.

Exercise 7.12 ([59]) For each of the following systems, show that the origin is not hyperbolic, find the index of the origin, and verify that it is different from ± 1.

$$(1) \qquad \begin{aligned} \dot{x}_1 &= x_1^2 \\ \dot{x}_2 &= -x_2 \end{aligned}$$

$$(2) \qquad \begin{aligned} \dot{x}_1 &= x_1^2 - x_2^2 \\ \dot{x}_2 &= 2x_1x_2 \end{aligned}$$

Chapter 8

Perturbation Theory and Averaging

Exact closed-form analytic solutions of nonlinear differential equations are possible only for a limited number of special classes of differential equations. In general, we have to resort to approximate methods. There are two distinct categories of approximation methods which engineers and scientists should have at their disposal as they analyze nonlinear systems: (1) numerical solution methods and (2) asymptotic methods. In this and the next chapter, we introduce the reader to some asymptotic methods for the analysis of nonlinear differential equations.[1]

Suppose we are given the state equation

$$\dot{x} = f(t, x, \epsilon)$$

where ϵ is a "small" scalar parameter and, under certain conditions, the equation has an exact solution $x(t, \epsilon)$. Equations of this type are encountered in many applications. The goal of an asymptotic method is to obtain an approximate solution $\tilde{x}(t, \epsilon)$ such that the approximation error $x(t, \epsilon) - \tilde{x}(t, \epsilon)$ is small, in some norm, for small $|\epsilon|$ and the approximate solution $\tilde{x}(t, \epsilon)$ is expressed in terms of equations simpler than the original equation. The practical significance of asymptotic methods is in revealing underlying structural properties possessed by the original state equation for small $|\epsilon|$. We shall see, in Section 8.1, examples where asymptotic methods reveal a weak coupling structure among isolated subsystems or the structure of a weakly nonlinear system. More important, asymptotic methods reveal multiple-time-scale structures inherent in many practical problems. Quite often, the solution of the state equation exhibits the phenomenon that some variables move in time faster than other variables, leading to the classification of variables as "slow" and

---

[1] Numerical solution methods are not studied in this textbook on the premise that most students are introduced to these methods in elementary differential equation courses, and they get their in-depth study of the subject in numerical analysis courses or courses on the numerical solution of ordinary differential equations.

"fast." Both the averaging method of this chapter and the singular perturbation method of the next chapter deal with the interaction of slow and fast variables.

Section 8.1 presents the classical perturbation method of seeking an approximate solution as a finite Taylor expansion of the exact solution. The asymptotic validity of the approximation is established in Section 8.1 on finite time intervals and in Section 8.2 on the infinite-time interval. In Section 8.3, we introduce the averaging method in its simplest form, which is sometimes called "periodic averaging" since the right-hand side function is periodic in time. Section 8.4 gives an application of the averaging method to the study of periodic solutions of weakly nonlinear second-order systems. Finally, we present a more general form of the averaging method in Section 8.5.

8.1   The Perturbation Method

Consider the system

$$\dot{x} = f(t, x, \epsilon) \tag{8.1}$$

where $f : [t_0, t_1] \times D \times [-\epsilon_0, \epsilon_0] \to R^n$ is "sufficiently smooth" in its arguments over a domain $D \subset R^n$. The required smoothness conditions will be spelled out as we proceed. Suppose we want to solve the state equation (8.1) for a given initial state

$$x(t_0) = \eta(\epsilon) \tag{8.2}$$

where, for more generality, we allow the initial state to depend "smoothly" on ϵ. The solution of (8.1)–(8.2) will depend on the parameter ϵ, a point that we shall emphasize by writing the solution as $x(t, \epsilon)$. The goal of the perturbation method is to exploit the "smallness" of the perturbation parameter ϵ to construct approximate solutions that will be valid for sufficiently small $|\epsilon|$. The simplest approximation results by setting $\epsilon = 0$ in (8.1)–(8.2) to obtain the nominal or unperturbed problem

$$\dot{x} = f(t, x, 0), \quad x(t_0) = \eta(0) \tag{8.3}$$

Suppose this problem has a unique solution $x_0(t)$ defined on $[t_0, t_1]$ and $x_0(t) \in D$ for all $t \in [t_0, t_1]$. Suppose further that f and η are continuously differentiable in their arguments for (t, x, ϵ) in $[t_0, t_1] \times D \times [-\epsilon_0, \epsilon_0]$. From continuity of solutions with respect to initial states and parameters (Theorem 2.6) we know that, for sufficiently small $|\epsilon|$, the problem (8.1)-(8.2) has a unique solution $x(t, \epsilon)$ defined on $[t_0, t_1]$. Approximating $x(t, \epsilon)$ by $x_0(t)$ can be justified by using Theorem 2.5 to show that[2]

$$\|x(t, \epsilon) - x_0(t)\| \le k|\epsilon|, \quad \forall |\epsilon| < \epsilon_1, \ \forall t \in [t_0, t_1] \tag{8.4}$$

---

[2] The details of deriving the bound (8.4) are left as an exercise; see Exercise 8.1.

for some $k > 0$ and $\epsilon_1 \leq \epsilon_0$. If we succeed in showing this bound on the approximation error, we then say that the error is of order $O(\epsilon)$ and write

$$x(t, \epsilon) - x_0(t) = O(\epsilon)$$

This order of magnitude notation will be used frequently in this chapter and the next one. It is defined as follows.

Definition 8.1 $\delta_1(\epsilon) = O(\delta_2(\epsilon))$ *if there exist positive constants k and c such that*

$$|\delta_1(\epsilon)| \leq k|\delta_2(\epsilon)|, \quad \forall \; |\epsilon| < c$$

Example 8.1

- $\epsilon^n = O(\epsilon^m)$ for all $n \geq m$, since

$$|\epsilon|^n = |\epsilon|^m |\epsilon|^{n-m} < |\epsilon|^m, \quad \forall \; |\epsilon| < 1$$

- $\epsilon^2/(0.5 + \epsilon) = O(\epsilon^2)$, since

$$\left| \frac{\epsilon^2}{0.5 + \epsilon} \right| < \frac{1}{0.5 - a} |\epsilon|^2, \quad \forall \; |\epsilon| < a < 0.5$$

- $1 + 2\epsilon = O(1)$, since

$$|1 + 2\epsilon| < 1 + 2a, \quad \forall \; |\epsilon| < a$$

- $\exp(-a/\epsilon)$ with positive a and ϵ is $O(\epsilon^n)$ for any positive integer n, since

$$\frac{e^{-a/\epsilon}}{\epsilon^n} \leq \left(\frac{n}{a} \right)^n e^{-n}, \quad \forall \; \epsilon > 0$$

$\triangle$

Suppose for now that we have shown that the approximation error $(x(t, \epsilon) - x_0(t))$ is $O(\epsilon)$, what can we say about the numerical value of this error for a given numerical value of ϵ? Unfortunately, we cannot translate the $O(\epsilon)$ order of magnitude statement into a numerical bound on the error. Knowing that the error is $O(\epsilon)$ means that we know that the norm of the error is less than $k|\epsilon|$ for some positive constant k that is independent of ϵ. However, we do not know the value of k, which might be 1, 10, or any positive number.[3] The fact that k is independent of ϵ guarantees that the bound $k|\epsilon|$ decreases monotonically as $|\epsilon|$ decreases. Therefore, for

---

[3] It should be noted, however, that in a well-formulated perturbation problem where variables are normalized to have dimensionless state variables, time, and perturbation parameter, one should expect the numerical value of k not to be much larger than one. See Example 8.4 for further discussion of normalization, or consult [124] and [89] for more examples.

sufficiently small $|\epsilon|$, the error will be small. More precisely, given any tolerance δ, we know that the norm of the error will be less than δ for all $|\epsilon| < \delta/k$. If this range is too small to cover the numerical values of interest for ϵ, we then need to extend the range of validity by obtaining higher-order approximations. An $O(\epsilon^2)$ approximation will meet the same δ tolerance for all $|\epsilon| < \sqrt{\delta/k_2}$, an $O(\epsilon^3)$ approximation will do it for all $|\epsilon| < (\delta/k_3)^{1/3}$, and so on. Although the constants $k, k_2, k_3, \ldots$ are not necessarily equal, these intervals are increasing in length since the tolerance δ is typically much smaller than one. Another way to look at higher-order approximations is to see that, for a given "sufficiently small" value of ϵ, an $O(\epsilon^n)$ error will be smaller than an $O(\epsilon^m)$ error for $n > m$ since

$$\frac{k_1|\epsilon|^n}{k_2|\epsilon|^m} < 1, \quad \forall\, |\epsilon| < \left(\frac{k_2}{k_1}\right)^{1/(n-m)}$$

Higher-order approximations for solutions of (8.1)–(8.2) can be obtained in a straightforward manner, provided the functions f and η are sufficiently smooth.

Suppose f and η have continuous partial derivatives up to order $N+1$ and N, respectively, with respect to (x, ϵ) for $(t, x, \epsilon) \in [t_0, t_1] \times D \times [-\epsilon_0, \epsilon_0]$. To obtain a higher-order approximation of $x(t, \epsilon)$, we construct a finite Taylor series

$$x(t, \epsilon) = x_0(t) + \sum_{k=1}^{N-1} x_k(t)\epsilon^k + \epsilon^N x_R(t, \epsilon) \tag{8.5}$$

Two things need to be done here. First, we need to know how to calculate the terms x_0, x_1, ..., x_{N-1}; in the process of doing that, it will be shown that these terms are well defined. Second, we need to show that the remainder term x_R is well defined and bounded on $[t_0, t_1]$, which will establish that $\sum_{k=0}^{N-1} x_k(t)\epsilon^k$ is an $O(\epsilon^N)$ (Nth-order) approximation of $x(t, \epsilon)$. Note that, by Taylor's theorem,[4] the smoothness requirement on the initial state $\eta(\epsilon)$ guarantees the existence of a finite Taylor series for $\eta(\epsilon)$; that is,

$$\eta(\epsilon) = \eta(0) + \sum_{k=1}^{N-1} \eta_k \epsilon^k + \epsilon^N \eta_R(\epsilon)$$

Therefore,

$$x_0(t_0) = \eta(0); \quad x_k(t_0) = \eta_k, \quad k = 1, 2, \ldots, N-1$$

Substitution of (8.5) in (8.1) yields

$$\sum_{k=0}^{N-1} \dot{x}_k(t)\epsilon^k + \epsilon^N \dot{x}_R(t, \epsilon) = f(t, x(t, \epsilon), \epsilon) \stackrel{\text{def}}{=} h(t, \epsilon)$$

---

[4] [7, Theorem 5-14].

$$= \sum_{k=0}^{N-1} h_k(t)\epsilon^k + \epsilon^N h_R(t, \epsilon) \qquad (8.6)$$

where the coefficients of the Taylor series of $h(t, \epsilon)$ are functions of the coefficients of the Taylor series of $x(t, \epsilon)$. Since the equation holds for all sufficiently small ϵ, it must hold as an identity in ϵ. Hence, coefficients of like powers of ϵ must be equal. Matching those coefficients, we can derive the equations that must be satisfied by x_0, x_1, and so on. Before we do that, we have to generate the coefficients of the Taylor series of $h(t, \epsilon)$. The zeroth-order term $h_0(t)$ is given by

$$h_0(t) = f(t, x_0(t), 0)$$

Hence, matching coefficients of ϵ^0 in (8.6), we determine that $x_0(t)$ satisfies

$$\dot{x}_0 = f(t, x_0, 0), \quad x_0(t_0) = \eta(0)$$

which, not surprisingly, is the unperturbed problem (8.3). The first-order term $h_1(t)$ is given by

$$
\begin{aligned}
h_1(t) &= \left. \frac{\partial}{\partial \epsilon} f(t, x(t, \epsilon), \epsilon) \right|_{\epsilon=0} \\
&= \left. \left\{ \frac{\partial f}{\partial x}(t, x(t, \epsilon), \epsilon) \frac{\partial x}{\partial \epsilon}(t, \epsilon) + \frac{\partial f}{\partial \epsilon}(t, x(t, \epsilon), \epsilon) \right\} \right|_{\epsilon=0} \\
&= \frac{\partial f}{\partial x}(t, x_0(t), 0) \, x_1(t) + \frac{\partial f}{\partial \epsilon}(t, x_0(t), 0)
\end{aligned}
$$

Matching coefficients of ϵ in (8.6), we find that $x_1(t)$ satisfies

$$\dot{x}_1 = \frac{\partial f}{\partial x}(t, x_0(t), 0) \, x_1 + \frac{\partial f}{\partial \epsilon}(t, x_0(t), 0), \quad x_1(t_0) = \eta_1$$

Define

$$A(t) = \frac{\partial f}{\partial x}(t, x_0(t), 0), \qquad g_1(t, x_0(t)) = \frac{\partial f}{\partial \epsilon}(t, x_0(t), 0)$$

and rewrite the equation for x_1 as

$$\dot{x}_1 = A(t)x_1 + g_1(t, x_0(t)), \quad x_1(t_0) = \eta_1$$

This linear equation has a unique solution defined on $[t_0, t_1]$.

This process can be continued to derive the equations satisfied by x_2, x_3, and so on. This, however, will involve higher-order differentials of f with respect to x, which makes the notation cumbersome. There is no point in writing the equations in a general form. Once the idea is clear, we can generate the equations for the specific problem of interest. Nevertheless, to set the pattern that these equations take, we shall, at the risk of boring some readers, derive the equation for x_2. The second-order coefficient in the Taylor series of $h(t, \epsilon)$ is given by

$$h_2(t) = \frac{1}{2} \frac{\partial^2}{\partial \epsilon^2} h(t, \epsilon) \Big|_{\epsilon=0}$$

Now

$$
\begin{aligned}
\frac{\partial}{\partial \epsilon} h(t, \epsilon) &= \frac{\partial f}{\partial x}(t, x, \epsilon) \frac{\partial x}{\partial \epsilon}(t, \epsilon) + \frac{\partial f}{\partial \epsilon}(t, x, \epsilon) \\
&= \frac{\partial f}{\partial x}(t, x, \epsilon)[x_1(t) + 2\epsilon x_2(t) + \cdots] + \frac{\partial f}{\partial \epsilon}(t, x, \epsilon)
\end{aligned}
$$

To simplify the notation, let

$$\psi(t, x, \epsilon) = \frac{\partial f}{\partial x}(t, x, \epsilon) \, x_1(t)$$

and continue to calculate the second derivative of h with respect to ϵ.

$$
\begin{aligned}
\frac{\partial^2}{\partial \epsilon^2} h(t, \epsilon) &= \frac{\partial \psi}{\partial x}(t, x, \epsilon) \frac{\partial x}{\partial \epsilon}(t, \epsilon) + \frac{\partial}{\partial \epsilon} \frac{\partial f}{\partial x}(t, x, \epsilon) \, x_1(t) \\
&\quad + 2 \frac{\partial f}{\partial x}(t, x, \epsilon) \, x_2(t) + \frac{\partial}{\partial x} \frac{\partial f}{\partial \epsilon}(t, x, \epsilon) \frac{\partial x}{\partial \epsilon}(t, \epsilon) \\
&\quad + \frac{\partial^2 f}{\partial \epsilon^2}(t, x, \epsilon) + \epsilon[\, \cdot \,]
\end{aligned}
$$

Thus,

$$h_2(t) = A(t)x_2(t) + g_2(t, x_0(t), x_1(t))$$

where

$$
\begin{aligned}
g_2(t, x_0(t), x_1(t)) &= \frac{1}{2} \frac{\partial \psi}{\partial x}(t, x_0(t), 0) \, x_1(t) + \frac{\partial}{\partial \epsilon} \frac{\partial f}{\partial x}(t, x_0(t), 0) \, x_1(t) \\
&\quad + \frac{1}{2} \frac{\partial^2 f}{\partial \epsilon^2}(t, x_0(t), 0)
\end{aligned}
$$

Matching coefficients of ϵ^2 in (8.6) yields

$$\dot{x}_2 = A(t)x_2 + g_2(t, x_0(t), x_1(t)), \quad x_2(t_0) = \eta_2$$

In summary, the Taylor series coefficients x_0, x_1, $\ldots$, x_{N-1} are obtained by solving the equations

$$\dot{x}_0 = f(t, x_0, 0), \quad x_0(t_0) = \eta(0) \tag{8.7}$$

$$\dot{x}_k = A(t)x_k + g_k(t, x_0(t), \ldots, x_{k-1}(t)), \quad x_k(t_0) = \eta_k \tag{8.8}$$

for $k = 1, 2, \ldots, N-1$, where $A(t)$ is the Jacobian $[\partial f/\partial x]$ evaluated at $x = x_0(t)$ and $\epsilon = 0$, and the term $g_k(t, x_0(t), x_1(t), \ldots, x_{k-1}(t))$ is a polynomial in $x_1, \ldots, x_{k-1}$ with coefficients depending continuously on t and $x_0(t)$. The assumption that $x_0(t)$ is defined on $[t_0, t_1]$ implies that $A(t)$ is defined on the same interval; hence, the linear equations (8.8) have unique solutions defined on $[t_0, t_1]$. Let us now illustrate the calculation of the Taylor series coefficients by a second-order example.

Example 8.2 Consider the Van der Pol state equation

$$\dot{x}_1 = x_2, \qquad\qquad x_1(0) = \eta_1(\epsilon)$$

$$\dot{x}_2 = -x_1 + \epsilon(1 - x_1^2)x_2, \quad x_2(0) = \eta_2(\epsilon)$$

Suppose we want to construct a finite Taylor series with $N = 3$. Let

$$x_i = x_{i0} + \epsilon x_{i1} + \epsilon^2 x_{i2} + \epsilon^3 x_{iR}, \quad i = 1, 2$$

and

$$\eta_i = \eta_i(0) + \epsilon \eta_{i1} + \epsilon^2 \eta_{i2} + \epsilon^3 \eta_{iR}, \quad i = 1, 2$$

Substitution of the expansions for x_1 and x_2 in the state equation results in

$$\dot{x}_{10} + \epsilon \dot{x}_{11} + \epsilon^2 \dot{x}_{12} + \epsilon^3 \dot{x}_{1R} = x_{20} + \epsilon x_{21} + \epsilon^2 x_{22} + \epsilon^3 x_{2R}$$
$$\dot{x}_{20} + \epsilon \dot{x}_{21} + \epsilon^2 \dot{x}_{22} + \epsilon^3 \dot{x}_{2R} = -x_{10} - \epsilon x_{11} - \epsilon^2 x_{12} - \epsilon^3 x_{1R}$$
$$+ \epsilon \left[1 - (x_{10} + \epsilon x_{11} + \epsilon^2 x_{12} + \epsilon^3 x_{1R})^2 \right]$$
$$\times (x_{20} + \epsilon x_{21} + \epsilon^2 x_{22} + \epsilon^3 x_{2R})$$

Matching coefficients of ϵ^0, we obtain

$$\dot{x}_{10} = x_{20}, \quad x_{10}(0) = \eta_1(0)$$

$$\dot{x}_{20} = -x_{10}, \quad x_{20}(0) = \eta_2(0)$$

which is the unperturbed problem at $\epsilon = 0$. Matching coefficients of ϵ, we obtain

$$\dot{x}_{11} = x_{21}, \qquad\qquad x_{11}(0) = \eta_{11}$$

$$\dot{x}_{21} = -x_{11} + (1 - x_{10}^2)x_{20}, \quad x_{21}(0) = \eta_{21}$$

while matching coefficients of ϵ^2 results in

$$\dot{x}_{12} = x_{22}, \qquad\qquad x_{12}(0) = \eta_{12}$$

$$\dot{x}_{22} = -x_{12} + (1 - x_{10}^2)x_{21} - 2x_{10}x_{11}x_{20}, \quad x_{22}(0) = \eta_{22}$$

The latter two sets of equations are in the form of (8.8) for $k = 1, 2$. △

Having calculated the terms x_0, x_1, ..., x_{N-1}, our task now is to show that $\sum_{k=0}^{N-1} x_k(t)\epsilon^k$ is indeed an $O(\epsilon^N)$ approximation of $x(t,\epsilon)$. Consider the approximation error

$$e = x - \sum_{k=0}^{N-1} x_k(t)\epsilon^k \tag{8.9}$$

Differentiating both sides of (8.9) with respect to t and substituting for the derivatives of x and x_k from (8.1), (8.7), and (8.8), it can be shown that e satisfies the equation

$$\dot{e} = A(t)e + \rho_1(t,e,\epsilon) + \rho_2(t,\epsilon), \quad e(t_0) = \epsilon^N \eta_R(\epsilon) \tag{8.10}$$

where

$$\rho_1(t,e,\epsilon) = f\left(t, e + \sum_{k=0}^{N-1} x_k(t)\epsilon^k, \epsilon\right) - f\left(t, \sum_{k=0}^{N-1} x_k(t)\epsilon^k, \epsilon\right) - A(t)e$$

$$\rho_2(t,\epsilon) = f\left(t, \sum_{k=0}^{N-1} x_k(t)\epsilon^k, \epsilon\right) - f(t, x_0(t), 0)$$
$$- \sum_{k=1}^{N-1} [A(t)x_k(t) + g_k(\cdot)]\epsilon^k$$

By assumption, $x_0(t)$ is bounded and belongs to D for all $t \in [t_0, t_1]$. Hence, there exist $\lambda > 0$ and $\epsilon_1 > 0$ such that for all $\|e\| \leq \lambda$ and $|\epsilon| \leq \epsilon_1$, the functions $x_0(t)$, $\sum_{k=0}^{N-1} x_k(t)\epsilon^k$, and $e + \sum_{k=0}^{N-1} x_k(t)\epsilon^k$ belong to a compact subset of D. It can be easily verified that

$$\rho_1(t, 0, \epsilon) = 0 \tag{8.11}$$

$$\left\| \frac{\partial \rho_1}{\partial e}(t, e, \epsilon) \right\| \leq k_1(\|e\| + |\epsilon|) \tag{8.12}$$

and

$$\|\rho_2(t,\epsilon)\| \leq |\epsilon|^N k_2 \tag{8.13}$$

for all $(t, e, \epsilon) \in [t_0, t_1] \times B_\lambda \times [-\epsilon_1, \epsilon_1]$, for some positive constants k_1 and k_2. Equation (8.10) can be viewed as a perturbation of

$$\dot{e}_0 = A(t)e_0 + \rho_1(t, e_0, \epsilon), \quad e_0(t_0) = 0 \tag{8.14}$$

which has the unique solution $e_0(t, \epsilon) \equiv 0$ for $t \in [t_0, t_1]$. Application of Theorem 2.6 shows that (8.10) has a unique solution defined on $[t_0, t_1]$ for sufficiently small $|\epsilon|$. Application of Theorem 2.5 shows that

$$\|e(t,\epsilon)\| = \|e(t,\epsilon) - e_0(t,\epsilon)\| = O(\epsilon^N)$$

We summarize this conclusion in the following theorem.

Theorem 8.1 *Suppose that*

- *f is continuous and has continuous partial derivatives up to order $N + 1$, with respect to (x, ϵ) for $(t, x, \epsilon) \in [t_0, t_1] \times D \times [-\epsilon_0, \epsilon_0]$.*

- *η and its derivatives up to order N are continuous for $\epsilon \in [-\epsilon_0, \epsilon_0]$.*

- *the nominal problem (8.3) has a unique solution $x_0(t)$ defined on $[t_0, t_1]$ and $x_0(t) \in D$ for all $t \in [t_0, t_1]$.*

Then, there exists $\epsilon^* > 0$ such that $\forall \, |\epsilon| < \epsilon^*$, the problem (8.1)–(8.2) has a unique solution $x(t, \epsilon)$, defined on $[t_0, t_1]$, which satisfies

$$x(t, \epsilon) - \sum_{k=0}^{N-1} x_k(t)\epsilon^k = O(\epsilon^N)$$

$\diamond$

Theorem 8.1, with $N = 1$, is similar to Theorems 2.5 and 2.6 in that it shows continuity of solutions with respect to initial states and parameters. It is a slightly sharper result, however, because it estimates the order of magnitude of the error to be $O(\epsilon)$. Note that when we approximate $x(t, \epsilon)$ by $x_0(t)$ we need not know the value of the parameter ϵ, which could be an unknown parameter that represents deviations of the system's parameters from their nominal values. When we use a higher-order approximation $\sum_{k=0}^{N-1} x_k(t)\epsilon^k$ for $N > 1$, we need to know the value of ϵ to construct the finite Taylor series, even though we do not need it to calculate the terms x_1, x_2, and so on. If we have to know ϵ to construct the Taylor series approximation, we must then compare the computational effort that is needed to approximate the solution via a Taylor series with the effort needed to calculate the exact solution. The exact solution $x(t, \epsilon)$ can be obtained by solving the nonlinear state equation (8.1), while the approximate solution is obtained by solving the nonlinear state equation (8.7) and a number of linear state equations (8.8), depending on the order of the approximation. Since in both cases we have to solve a nonlinear state equation of order n, we must ask ourselves what we will gain by solving (8.7) instead of (8.1). One situation where the Taylor series approximation will be clearly preferable is the case when the solution is sought for several values of ϵ. In the Taylor series approximation, equations (8.7) and (8.8) will be solved only once; then, different Taylor expansions will be constructed for different values of ϵ. Aside from this special (repeated values of ϵ) case, we find the Taylor series approximation to be effective when

- the unperturbed state equation (8.7) is considerably simpler than the ϵ-dependent state equation (8.1),

- ϵ is reasonably small that an "acceptable" approximation can be achieved with a few terms in the series.

In most engineering applications of the perturbation method, adequate approximations are achieved with $N = 2$ or 3, and the process of setting $\epsilon = 0$ simplifies the state equation considerably. In the following two examples, we look at two typical cases where setting $\epsilon = 0$ reduces the complexity of the state equation. In the first example, we consider again the Van der Pol equation of Example 8.2, which represents a wide class of "weakly nonlinear systems" which become linear at $\epsilon = 0$. To construct a Taylor series approximation in this case, we only solve linear equations. In the second example, we look at a system which is an interconnection of subsystems with "weak" or ϵ-coupling. At $\epsilon = 0$, the system decomposes into lower- order decoupled subsystems. To construct a Taylor series approximation in this case, we always solve lower-order decoupled equations as opposed to solving the original higher-order equation (8.1).

Example 8.3 Suppose we want to solve the Van der Pol equation

$$\dot{x}_1 = x_2, \qquad x_1(0) = 1$$

$$\dot{x}_2 = -x_1 + \epsilon(1 - x_1^2)x_2, \qquad x_2(0) = 0$$

over the time interval $[0, \pi]$. We start by setting $\epsilon = 0$ to obtain the linear unperturbed equation

$$\dot{x}_{10} = x_{20}, \qquad x_{10}(0) = 1$$

$$\dot{x}_{20} = -x_{10}, \qquad x_{20}(0) = 0$$

whose solution is given by

$$x_{10}(t) = \cos t, \qquad x_{20}(t) = -\sin t$$

Clearly, all the assumptions of Theorem 8.1 are satisfied and we conclude that the approximation error $x(t, \epsilon) - x_0(t)$ is $O(\epsilon)$. Calculating the exact solution $x(t, \epsilon)$ numerically at three different values of ϵ and using

$$E_0 = \max_{0 \le t \le \pi} \|x(t, \epsilon) - x_0(t)\|_2$$

as a measure of the approximation error, we find that $E_0 = 0.0112$, 0.0589, and 0.1192 for $\epsilon = 0.01$, 0.05, and 0.1, respectively. These numbers show clearly that, for small ϵ, the error is bounded by $k\epsilon$ for some positive constant k. With these numbers, it appears that $k = 1.2$ would work for $\epsilon \le 0.1$. Figure 8.1 shows the exact and approximate trajectories of the first component of the state vector when $\epsilon = 0.1$. Suppose now we want to improve the approximation at $\epsilon = 0.1$. We need

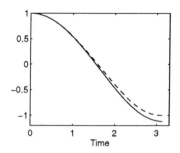

Figure 8.1: $x_1(t, \epsilon)$ (solid) and $x_{10}(t)$ (dashed) of Example 8.3 at $\epsilon = 0.1$.

to calculate x_{11} and x_{21}. We have seen in Example 8.2 that x_{11} and x_{21} satisfy the equation

$$\dot{x}_{11} = x_{21}, \qquad\qquad x_{11}(0) = 0$$

$$\dot{x}_{21} = -x_{11} + [1 - x_{10}^2(t)]x_{20}(t), \quad x_{21}(0) = 0$$

where, in the current problem,

$$[1 - x_{10}^2(t)]x_{20}(t) = -(1 - \cos^2 t)\sin t$$

The solution is

$$x_{11}(t) = -\tfrac{9}{32}\sin t - \tfrac{1}{32}\sin 3t + \tfrac{3}{8}t\cos t$$
$$x_{21}(t) = \tfrac{3}{32}\cos t - \tfrac{3}{32}\cos 3t - \tfrac{3}{8}t\sin t$$

By Theorem 8.1, the second-order approximation

$$\left[\begin{array}{c} x_{10}(t) + \epsilon x_{11}(t) \\ x_{20}(t) + \epsilon x_{21}(t) \end{array} \right]$$

will be $O(\epsilon^2)$ close to the exact solution for sufficiently small ϵ. To compare the approximate solution to the exact one at $\epsilon = 0.1$, we calculate

$$E_1 = \max_{0 \le t \le \pi} \|x(t, 0.1) - x_0(t) - 0.1x_1(t)\|_2 = 0.0057$$

which shows a reduction in the approximation error by almost an order of magnitude. Figure 8.2 shows the approximation errors in the first component of the state vector for the first-order approximation x_0 and the second-order approximation $x_0 + \epsilon x_1$ at $\epsilon = 0.1$. $\triangle$

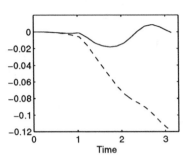

Figure 8.2: Approximation errors $x_1(t, \epsilon) - x_{10}(t)$ (dashed) and $x_1(t, \epsilon) - x_{10}(t) - \epsilon x_{11}(t)$ (solid) of Example 8.3 at $\epsilon = 0.1$.

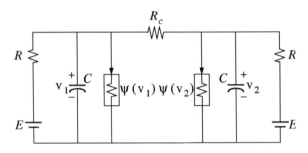

Figure 8.3: Electric circuit of Example 8.4.

Example 8.4 The circuit shown in Figure 8.3 contains nonlinear resistors whose $I - V$ characteristics are given by $i = \psi(v)$. The differential equations for the voltages across the capacitors are

$$C\frac{dv_1}{dt} = \frac{1}{R}(E - v_1) - \psi(v_1) - \frac{1}{R_c}(v_1 - v_2)$$

$$C\frac{dv_2}{dt} = \frac{1}{R}(E - v_2) - \psi(v_2) - \frac{1}{R_c}(v_2 - v_1)$$

The structure of this circuit is such that it has two similar sections, each consisting of an RC section with a nonlinear resistor. The two sections are connected through the resistor R_c. When R_c is "relatively large," the connection between the two sections becomes "weak." In particular, when $R_c = \infty$, the connection is open circuit and the two sections are decoupled from each other. This circuit lends itself to an ϵ-coupling representation where the coupling between the two sections may be parameterized by a small parameter ϵ. At first glance, it appears that a reasonable choice of ϵ is $\epsilon = 1/R_c$. Indeed, with this choice, the coupling terms in the foregoing equations will be multiplied by ϵ. However, such a choice makes ϵ dependent on the absolute value of a physical parameter whose value, no matter how small or large, has no significance by itself without considering the values of other physical parameters in the system. In a well-formulated perturbation problem, the parameter ϵ would be chosen as a ratio between physical parameters that reflects the "true smallness" of ϵ in a relative sense. To choose ϵ this way, we usually start by choosing the state variables and/or the time variable as dimensionless quantities. In our circuit, the clear choice of state variables is v_1 and v_2. Instead of working with v_1 and v_2, we scale them in such a way that the typical extreme values of the scaled variables would be close to ± 1. Due to the weak coupling between the two identical sections, it is reasonable to use the same scaling factor α for both state variables. Define the state variables as $x_1 = v_1/\alpha$ and $x_2 = v_2/\alpha$. Taking a dimensionless time $\tau = t/RC$ and writing $dx/d\tau = \dot{x}$, we obtain the state equation

$$\dot{x}_1 = \frac{E}{\alpha} - x_1 - \frac{R}{\alpha}\psi(\alpha x_1) - \frac{R}{R_c}(x_1 - x_2)$$

$$\dot{x}_2 = \frac{E}{\alpha} - x_2 - \frac{R}{\alpha}\psi(\alpha x_2) - \frac{R}{R_c}(x_2 - x_1)$$

It appears now that a reasonable choice of ϵ is $\epsilon = R/R_c$. Suppose that $R = 1.5 \times 10^3$ Ω, $E = 1.2\ V$, and the nonlinear resistors are tunnel diodes with

$$\psi(v) = 10^{-3} \times \left(17.76v - 103.79v^2 + 229.62v^3 - 226.31v^4 + 83.72v^5\right)$$

With these numbers, we take $\alpha = 1$ and rewrite the state equation as

$$\dot{x}_1 = 1.2 - x_1 - h(x_1) - \epsilon(x_1 - x_2)$$

$$\dot{x}_2 = 1.2 - x_2 - h(x_2) - \epsilon(x_2 - x_1)$$

where $h(v) = 1.5 \times 10^3 \times \psi(v)$. Suppose we want to solve this equation for the initial state

$$x_1(0) = 0.15; \quad x_2(0) = 0.6$$

Setting $\epsilon = 0$, we obtain the decoupled equations

$$\dot{x}_1 = 1.2 - x_1 - h(x_1), \quad x_1(0) = 0.15$$

$$\dot{x}_2 = 1.2 - x_2 - h(x_2), \quad x_2(0) = 0.6$$

which are solved independently of each other. Let $x_{10}(t)$ and $x_{20}(t)$ be the solutions. According to Theorem 8.1, they provide an $O(\epsilon)$ approximation of the exact solution for sufficiently small ϵ. To obtain an $O(\epsilon^2)$ approximation, we set up the equations for x_{11} and x_{21} as

$$\dot{x}_{11} = -[1 + h'(x_{10}(t))]x_{11} - [x_{10}(t) - x_{20}(t)], \quad x_{11}(0) = 0$$

$$\dot{x}_{21} = -[1 + h'(x_{20}(t))]x_{21} - [x_{20}(t) - x_{10}(t)], \quad x_{21}(0) = 0$$

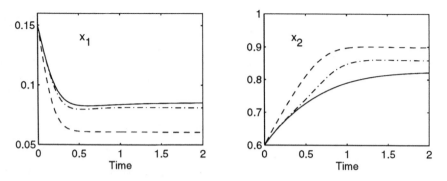

Figure 8.4: Exact solution (solid), first-order approximation (dashed), and second-order approximation (dash/dot) for Example 8.4 at $\epsilon = 0.3$.

where $h'(\cdot)$ is the derivative of $h(\cdot)$. Figure 8.4 shows the exact solution as well as the first-order and second-order approximations for $\epsilon = 0.3$. $\triangle$

A serious limitation of Theorem 8.1 is that the $O(\epsilon^N)$ error bound is valid only on finite (order $O(1)$) time intervals $[t_0, t_1]$. It does not hold on intervals like $[t_0, T/\epsilon]$ nor on the infinite-time interval $[t_0, \infty)$. The reason is that the constant k in the bound $k|\epsilon|^N$ depends on t_1 in such a way that it grows unbounded as t_1 increases. In particular, since the constant k results from application of Theorem 2.5, it has a component of the form $\exp(Lt_1)$. In the next section, we shall see how to employ stability conditions to extend Theorem 8.1 to the infinite interval. With the lack of such stability conditions, the approximation may not be valid for large t even though it is valid on $O(1)$ time intervals. Figure 8.5 shows the exact and approximate solutions for the Van der Pol equation of Example 8.3, at $\epsilon = 0.1$, over a large time interval. For large t, the error $x_1(t, \epsilon) - x_{10}(t)$ is no longer $O(\epsilon)$. More seriously, the error $x_1(t, \epsilon) - x_{10}(t) - \epsilon x_{11}(t)$ grows unbounded, which is a consequence of the term $t \cos t$ in $x_{11}(t)$.

8.2 Perturbation on the Infinite Interval

The perturbation result of Theorem 8.1 can be extended to the infinite time interval $[t_0, \infty)$ if some additional conditions are added to ensure exponential stability of the solution of the nominal system (8.7). A solution $\bar{x}_0(t)$ of (8.7) is exponential stability if there is a constant $c > 0$ such that

$$\|x_0(t_0) - \bar{x}_0(t_0)\| < c \Rightarrow \|x_0(t) - \bar{x}_0(t)\| \leq k e^{-\gamma(t-t_0)}\|x_0(t_0) - \bar{x}_0(t_0)\|, \ \forall \ t \geq t_0$$

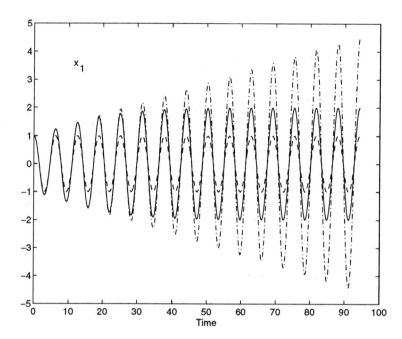

Figure 8.5: Exact solution (solid), first-order approximation (dashed), and second-order approximation (dash/dot) for the Van der Pol equation over a large time interval.

for some positive constants k and γ where $x_0(t)$ is the solution of (8.7) that starts at $x_0(t_0)$. The main condition we add here is a requirement that (8.7) has an exponentially stable equilibrium point $x = p^*$. This ensures that every solution of (8.7) that starts in a (sufficiently small) neighborhood of p^* will be exponentially stable.

Theorem 8.2 *Suppose that*

- *f is continuous and bounded, and has continuous, bounded partial derivatives up to order $N + 1$ with respect to (x, ϵ) for $(t, x, \epsilon) \in [0, \infty) \times D \times [-\epsilon_0, \epsilon_0]$.*

- *η and its derivatives up to order N are continuous for $\epsilon \in [-\epsilon_0, \epsilon_0]$.*

- *$x = p^* \in D$ is an exponentially stable equilibrium point for the nominal system (8.7).*

Then, there exist positive constants ϵ^* and ρ such that for all $|\epsilon| < \epsilon^*$ and all $\|\eta(\epsilon) - p^*\| < \rho$, equations (8.1)–(8.2) have a unique solution $x(t, \epsilon)$, uniformly *bounded on* $[0, \infty)$, *and*

$$x(t, \epsilon) - \sum_{k=0}^{N-1} x_k(t)\epsilon^k = O(\epsilon^N)$$

where $O(\epsilon^N)$ *holds uniformly in t for all* $t \geq t_0 \geq 0$. $\diamond$

Proof: Appendix A.11.

Example 8.5 The electric circuit of Example 8.4 is represented by

$$\dot{x}_1 = 1.2 - x_1 - h(x_1) - \epsilon(x_1 - x_2)$$
$$\dot{x}_2 = 1.2 - x_2 - h(x_2) - \epsilon(x_2 - x_1)$$

where

$$h(v) = 1.5 \left(17.76v - 103.79v^2 + 229.62v^3 - 226.31v^4 + 83.72v^5 \right)$$

At $\epsilon = 0$, the unperturbed system comprises two isolated first-order systems:

$$\dot{x}_1 = 1.2 - x_1 - h(x_1)$$
$$\dot{x}_2 = 1.2 - x_2 - h(x_2)$$

It can be verified that each of the two systems has three equilibrium points at 0.063, 0.285, and 0.884. The Jacobian $-1 + h'(x_i)$ is negative at $x_i = 0.063$ and $x_i = 0.884$ and positive at $x_i = 0.285$. Hence, the equilibria at 0.063 and 0.884 are exponentially stable while the equilibrium at 0.285 is unstable. When the two first-order systems are put together, the composite second-order system will have nine equilibrium points; only four of them will be exponentially stable. These are the equilibrium points $(0.063, 0.063)$, $(0.063, 0.884)$, $(0.884, 0.063)$, and $(0.884, 0.884)$. Theorem 8.2 says that in the neighborhood of each of these equilibria, the approximations calculated in Example 8.4 will be valid for all $t \geq 0$. The simulations shown in Figure 8.4 were taken over a time interval long enough for the solution to reach steady state. In this particular case, the initial state $(0.15, 0.6)$ lies in the domain of attraction of the equilibrium point $(0.085, 0.822)$, which, clearly, is the perturbation of the equilibrium point $(0.063, 0.884)$. Note that, for sufficiently small ϵ, the perturbed system will have nine isolated equilibrium points, each neighboring one of the equilibrium points of the unperturbed system. $\triangle$

The $O(\epsilon^N)$ estimate of Theorem 8.2 is valid only when the equilibrium point $x = p^*$ is exponentially stable. It will not necessarily hold if the equilibrium point is asymptotically stable but not exponentially stable, as illustrated by the following example.

Example 8.6 Consider the first-order system

$$\dot{x} = -x^3 + \epsilon x$$

and suppose $\epsilon > 0$. The origin of the unperturbed system

$$\dot{x} = -x^3$$

is globally asymptotically stable, but not exponentially stable (see Example 3.23). The perturbed system has three equilibrium points at $x = 0$ and $x = \pm\sqrt{\epsilon}$. The equilibrium $x = 0$ is unstable, while the other two equilibria are asymptotically stable. Solving both systems with the same positive initial condition $x(0) = a$, it can be easily seen that

$$x(t, \epsilon) \to \sqrt{\epsilon} \text{ and } x_0(t) \to 0 \text{ as } t \to \infty$$

Since $\sqrt{\epsilon}$ is not $O(\epsilon)$, it is clear that the approximation error $x(t, \epsilon) - x_0(t)$ is not $O(\epsilon)$ for all $t \geq 0$. Nevertheless, since the origin is asymptotically stable, we should be able to make a statement about the asymptotic behavior of the approximation as $t \to \infty$. Indeed we can make a statement, although it will be weaker than the statement of Theorem 8.2. Since the origin of the unperturbed system is asymptotically stable, the solution $x_0(t)$ tends to zero as $t \to \infty$; equivalently, given any $\delta > 0$ there is $T_1 > 0$ such that

$$\|x_0(t)\| < \delta/2, \quad \forall t \geq T_1$$

The solutions of the perturbed system are ultimately bounded by a bound that shrinks with ϵ. Therefore, given any $\delta > 0$ there is $T_2 > 0$ and $\epsilon^* > 0$ such that

$$\|x(t, \epsilon)\| < \delta/2, \quad \forall t \geq T_2, \forall \epsilon < \epsilon^*$$

Combining these two estimates, we can say that for any $\delta > 0$, the approximation error satisfies

$$\|x(t, \epsilon) - x_0(t)\| < \delta, \quad \forall t \geq T, \forall \epsilon < \epsilon^*$$

where $T = \max\{T_1, T_2\}$. On the order $O(1)$ time interval $[0, T]$, we know from the finite time result of Theorem 8.1 that the approximation error is $O(\epsilon)$. Therefore, we can say that for any $\delta > 0$ there is $\epsilon^{**} > 0$ such that

$$\|x(t, \epsilon) - x_0(t)\| < \delta, \quad \forall t \in [0, \infty), \forall \epsilon < \epsilon^{**}$$

The last inequality is equivalent to saying that the approximation error tends to zero as $\epsilon \to 0$, uniformly in t for all $t \geq 0$. This is the best we can show, in general, in the lack of exponential stability. Of course, in this particular example we can obtain both $x_0(t)$ and $x(t, \epsilon)$ in closed form, and we can actually show that the approximation error is $O(\sqrt{\epsilon})$. $\triangle$

8.3   Averaging

The averaging method applies to a system of the form

$$\dot{x} = \epsilon f(t, x, \epsilon)$$

where ϵ is a small positive parameter and $f(t, x, \epsilon)$ is T-periodic in t; that is,

$$f(t + T, x, \epsilon) = f(t, x, \epsilon), \quad \forall \, (t, x, \epsilon) \in [0, \infty) \times D \times [0, \epsilon_0]$$

for some domain $D \subset R^n$. The method approximates the solution of this system by the solution of an "averaged system," obtained by averaging $f(t, x, \epsilon)$ at $\epsilon = 0$. To motivate the averaging method, let us start by examining a scalar example.

Example 8.7 Consider the first-order linear system

$$\dot{x} = \epsilon a(t, \epsilon)x, \quad x(0) = \eta \tag{8.15}$$

where ϵ is a positive parameter and $a(t + T, \epsilon) = a(t, \epsilon)$ for all $t \geq 0$. Assume that a is sufficiently smooth in its arguments. To obtain an approximate solution that is valid for small ϵ, we may apply the perturbation method of Section 8.1. Setting $\epsilon = 0$ results in the unperturbed system

$$\dot{x} = 0, \quad x(0) = \eta$$

which has a constant solution $x_0(t) = \eta$. According to Theorem 8.1, the error of this approximation will be $O(\epsilon)$ on $O(1)$ time intervals. The unperturbed system does not satisfy the conditions of Theorem 8.2. Therefore, it is not clear whether this approximation is valid on time intervals larger than $O(1)$. Since in this example we can write down a closed-form expression for the exact solution, we shall examine the approximation error by direct calculations. The solution of (8.15) is given by

$$x(t, \epsilon) = \exp\left[\epsilon \int_0^t a(\tau, \epsilon) \, d\tau\right] \eta$$

Hence, the approximation error is

$$x(t, \epsilon) - x_0(t) = \left\{\exp\left[\epsilon \int_0^t a(\tau, \epsilon) \, d\tau\right] - 1\right\} \eta$$

To see how the approximation error behaves as t increases, we need to evaluate the integral term in the above expression. The function $a(t, \epsilon)$ is periodic in t. Let its mean be

$$\bar{a}(\epsilon) = \frac{1}{T} \int_0^T a(\tau, \epsilon) \, d\tau$$

We can write $a(t, \epsilon)$ as

$$a(t, \epsilon) = \bar{a}(\epsilon) + [a(t, \epsilon) - \bar{a}(\epsilon)]$$

The term inside the bracket is a T-periodic function of t with zero mean. Therefore, the integral

$$\int_0^t [a(\tau, \epsilon) - \bar{a}(\epsilon)] \; d\tau \stackrel{\text{def}}{=} \Delta(t, \epsilon)$$

is T-periodic and, hence, bounded for all $t \geq 0$. On the other hand, the integration of the term $\bar{a}(\epsilon)$ on $[0, t]$ results in $t\bar{a}(\epsilon)$. Thus,

$$x(t, \epsilon) - x_0(t) = \{\exp\left[\epsilon\bar{a}(\epsilon)t\right]\exp\left[\epsilon\Delta(t, \epsilon)\right] - 1\}\,\eta$$

Except for the case $\bar{a}(\epsilon) = 0$, the approximation error will be $O(\epsilon)$ only on $O(1)$ time intervals. A careful examination of the approximation error suggests that a better approximation of $x(t, \epsilon)$ is $\exp[\epsilon\,\bar{a}(\epsilon)t]\eta$ or even $\exp[\epsilon\,\bar{a}(0)t]\eta$, since $\bar{a}(\epsilon) - \bar{a}(0) = O(\epsilon)$. Let us try $\bar{x}(t, \epsilon) = \exp[\epsilon\bar{a}(0)t]\eta$ as an alternative approximation. The approximation error is given by

$$
\begin{aligned}
x(t, \epsilon) - \bar{x}(t, \epsilon) &= \{\exp\left[\epsilon\bar{a}(\epsilon)t\right]\exp\left[\epsilon\Delta(t, \epsilon)\right] - \exp[\epsilon\bar{a}(0)t]\}\,\eta \\
&= \exp[\epsilon\bar{a}(0)t]\{\exp\left[\epsilon(\bar{a}(\epsilon) - \bar{a}(0))t\right]\exp\left[\epsilon\Delta(t, \epsilon)\right] - 1\}\,\eta
\end{aligned}
$$

Noting that

$$
\begin{aligned}
\exp\left[\epsilon\Delta(t, \epsilon)\right] &= 1 + O(\epsilon), \quad \forall\, t \geq 0 \\
\exp\left[\epsilon(\bar{a}(\epsilon) - \bar{a}(0))t\right] &= \exp[tO(\epsilon^2)] = 1 + O(\epsilon), \quad \forall\, t \in [0, b/\epsilon] \\
\exp[\epsilon\bar{a}(0)t] &= O(1), \quad \forall\, t \in [0, b/\epsilon]
\end{aligned}
$$

for any finite $b > 0$, we conclude that $x(t, \epsilon) - \bar{x}(t, \epsilon) = O(\epsilon)$ on time intervals of order $O(1/\epsilon)$, which confirms the conjecture that the approximation $\bar{x}(t, \epsilon) = \exp[\epsilon\bar{a}(0)t]\eta$ is better than the approximation $x_0(t) = \eta$. Note that $\bar{x}(t, \epsilon)$ is the solution of

$$\dot{x} = \epsilon\bar{a}(0)x, \quad x(0) = \eta \tag{8.16}$$

which is an averaged system whose right-hand side is the average of the right-hand side of (8.15) at $\epsilon = 0$. $\triangle$

In this example, we have arrived at the averaged system (8.16) through our knowledge of the closed-form expression of the exact solution of (8.15). Such closed-form expressions are available only in very special cases. However, the plausibility of averaging is not dependent on the special features of this example. Let us reason the idea of averaging in a different way. The right-hand side of (8.15) is multiplied by a positive constant ϵ. When ϵ is small, the solution x will vary "slowly" with

t relative to the periodic fluctuation of $a(t, \epsilon)$. It is intuitively clear that if the response of a system is much slower than the excitation, then such response will be determined predominantly by the average of the excitation. This intuition has its roots in linear system theory, where we know that if the bandwidth of the system is much smaller than the bandwidth of the input, then the system will act as a low-pass filter that rejects the high-frequency component of the input. If the solution of (8.15) is determined predominantly by the average of the fluctuation of $a(t, \epsilon)$, then it is reasonable, in order to get a first-order approximation with an order $O(\epsilon)$, that the function $a(t, \epsilon)$ be replaced by its average. This two-time-scale interpretation of averaging is not dependent on the special features of Example 8.7, nor is it dependent on the linearity of the system. It is a plausible idea that works in a more general setup, as we shall see in the rest of this chapter.

Consider the system

$$\dot{x} = \epsilon f(t, x, \epsilon) \tag{8.17}$$

where $f : [0, \infty) \times D \times [0, \epsilon_0] \to R^n$ is continuous and bounded and has continuous, bounded derivatives up to the second order with respect to its arguments for $(t, x, \epsilon) \in [0, \infty) \times D \times [0, \epsilon_0]$. Moreover, $f(t, x, \epsilon)$ is T-periodic in t for some $T > 0$. The parameter ϵ is positive and $D \subset R^n$ is a domain. Let $\Gamma = [0, \infty) \times D$. We associate with (8.17) an autonomous averaged system

$$\dot{x} = \epsilon f_{\text{av}}(x) \tag{8.18}$$

where

$$f_{\text{av}}(x) = \frac{1}{T} \int_0^T f(\tau, x, 0) \, d\tau \tag{8.19}$$

The basic problem in the averaging method is to determine in what sense the behavior of the autonomous system (8.18) approximates the behavior of the more complicated nonautonomous system (8.17). We shall address this problem by showing, via a change of variables, that the nonautonomous system (8.17) can be represented as a perturbation of the autonomous system (8.18). Define

$$u(t, x) = \int_0^t h(\tau, x) \, d\tau \tag{8.20}$$

where

$$h(t, x) = f(t, x, 0) - f_{\text{av}}(x) \tag{8.21}$$

Since $h(t, x)$ is T-periodic in t and has zero mean, the function $u(t, x)$ is T-periodic in t. Hence, $u(t, x)$ is bounded for all $(t, x) \in \Gamma$. Moreover, $\partial u / \partial t$ and $\partial u / \partial x$, given by

$$\frac{\partial u}{\partial t} = h(t, x)$$

$$\frac{\partial u}{\partial x} = \int_0^t \frac{\partial h}{\partial x}(\tau, x) \, d\tau$$

are T-periodic in t and bounded on Γ. Here, we have used the fact that $\partial h/\partial x$ is T-periodic in t and has zero mean. Consider the change of variables

$$x = y + \epsilon u(t, y) \tag{8.22}$$

Differentiating both sides with respect to t, we obtain

$$\dot{x} = \dot{y} + \epsilon \frac{\partial u}{\partial t}(t, y) + \epsilon \frac{\partial u}{\partial y}(t, y) \dot{y}$$

Substituting for $\dot{x}$ from (8.17), we find that the new state variable y satisfies the equation

$$
\begin{aligned}
\left[I + \epsilon \frac{\partial u}{\partial y} \right] \dot{y} &= \epsilon f(t, y + \epsilon u, \epsilon) - \epsilon \frac{\partial u}{\partial t} \\
&= \epsilon f(t, y + \epsilon u, \epsilon) - \epsilon f(t, y, 0) + \epsilon f_{\mathrm{av}}(y) \\
&\stackrel{\text{def}}{=} \epsilon f_{\mathrm{av}}(y) + \epsilon p(t, y, \epsilon)
\end{aligned}
$$

where

$$p(t, y, \epsilon) = [f(t, y, \epsilon) - f(t, y, 0)] + [f(t, y + \epsilon u, \epsilon) - f(t, y, \epsilon)]$$

The following Lipschitz inequalities hold on Γ uniformly in t

$$\| f(t, y, \epsilon) - f(t, y, 0) \| \le L_1 \epsilon$$

and

$$\| f(t, y + \epsilon u, \epsilon) - f(t, y, \epsilon) \| \le L_2 \epsilon \| u \|$$

Thus, $p(t, y, \epsilon)$ is $O(\epsilon)$ on Γ. Note also that $p(t, y, \epsilon)$ is T-periodic in t. Since $\partial u/\partial y$ is bounded on Γ, the matrix $I + \epsilon \partial u/\partial y$ is nonsingular for sufficiently small ϵ, and

$$\left[I + \epsilon \frac{\partial u}{\partial y} \right]^{-1} = I + O(\epsilon), \quad \text{on } \Gamma$$

Therefore, the state equation for y is given by

$$\dot{y} = \epsilon f_{\mathrm{av}}(y) + \epsilon^2 q(t, y, \epsilon) \tag{8.23}$$

where $q(t, y, \epsilon)$ is T-periodic in t and bounded on Γ for sufficiently small ϵ. This equation is a perturbation of the averaged system (8.18). By extending the arguments used in the previous two sections, we arrive at the following theorem.

Theorem 8.3 *Let $f(t, x, \epsilon)$ be continuous and bounded, and have continuous and bounded partial derivatives up to the second order with respect to (x, ϵ) for $(t, x, \epsilon) \in [0, \infty) \times D \times [0, \epsilon_0]$. Suppose f is T-periodic in t for some $T > 0$ and ϵ is a positive parameter. Let $x(t, \epsilon)$ and $x_{\mathrm{av}}(t, \epsilon)$ be solutions of (8.17) and (8.18), respectively.*

- If $x_{\mathrm{av}}(t, \epsilon) \in D \; \forall \; t \in [0, b/\epsilon]$ and $\|x(0, \epsilon) - x_{\mathrm{av}}(0, \epsilon)\| = O(\epsilon)$, then

$$\|x(t, \epsilon) - x_{\mathrm{av}}(t, \epsilon)\| = O(\epsilon) \quad \text{on } [0, b/\epsilon]$$

- If $p^* \in D$ is an exponentially stable equilibrium point for the averaged system (8.18), then there exists a positive constant ρ such that if

$$\|x_{\mathrm{av}}(0, \epsilon) - p^*\| < \rho \text{ and } \|x(0, \epsilon) - x_{\mathrm{av}}(0, \epsilon)\| = O(\epsilon)$$

then

$$\|x(t, \epsilon) - x_{\mathrm{av}}(t, \epsilon)\| = O(\epsilon) \quad \text{for all } t \in [0, \infty)$$

- If $p^* \in D$ is an exponentially stable equilibrium point for the averaged system (8.18), then there is a positive constant ϵ^* such that for all $0 < \epsilon < \epsilon^*$, (8.17) has a unique exponentially stable periodic solution of period T in an $O(\epsilon)$ neighborhood of p^*. $\diamond$

Proof: Appendix A.12.

Note that the T-periodic solution $\bar{x}(t, \epsilon)$ may be a trivial periodic solution $\bar{x}(t, \epsilon) = p_\epsilon$ where p_ϵ is an equilibrium point of (8.21). In this case, the theorem ensures that the equilibrium point p_ϵ is exponentially stable. In the special case when $f(t, 0, \epsilon) = 0$ for all $(t, \epsilon) \in [0, \infty) \times [0, \epsilon_0]$, both the original system (8.21) and the averaged system (8.22) have equilibria at $x = 0$. By uniqueness of the T-periodic solution $\bar{x}(t, \epsilon)$, it follows that $\bar{x}(t, \epsilon)$ is the trivial solution $x = 0$.

Example 8.8 Consider the linear system

$$\dot{x} = \epsilon A(t)x$$

where $A(t + T) = A(t)$ and $\epsilon > 0$. Let

$$\bar{A} = \frac{1}{T} \int_0^T A(\tau) \, d\tau$$

The averaged system is given by

$$\dot{x} = \epsilon \bar{A} x$$

It has an equilibrium point at $x = 0$. Suppose the matrix $\bar{A}$ is Hurwitz. Then, it follows from Theorem 8.3 that for sufficiently small ϵ, $\dot{x} = \epsilon A(t)x$ has a unique T-periodic solution in an $O(\epsilon)$ neighborhood of the origin $x = 0$. However, $x = 0$ is an equilibrium point for the system. Hence, the periodic solution is the trivial solution $x(t) = 0$. Consequently, we conclude that for sufficiently small ϵ, $x = 0$ is an exponentially stable equilibrium point for the nonautonomous system $\dot{x} = \epsilon A(t)x$. $\triangle$

Example 8.9 Consider the scalar system

$$\dot{x} = \epsilon(x \sin^2 t - 0.5x^2) = \epsilon f(t, x)$$

The function $f(t, x)$ is π-periodic in t. The averaged function $f_{av}(x)$ is given by

$$f_{av}(x) = \frac{1}{\pi} \int_0^\pi (x \sin^2 t - 0.5x^2)\, dt = 0.5(x - x^2)$$

The averaged system

$$\dot{x} = 0.5\epsilon(x - x^2)$$

has two equilibrium points at $x = 0$ and $x = 1$. The Jacobian df_{av}/dx evaluated at these equilibria is given by

$$\left.\frac{df_{av}}{dx}\right|_{x=0} = (0.5 - x)|_{x=0} = 0.5$$

$$\left.\frac{df_{av}}{dx}\right|_{x=1} = (0.5 - x)|_{x=1} = -0.5$$

Thus, for sufficiently small ϵ, the system has an exponentially stable π-periodic solution in an $O(\epsilon)$ neighborhood of $x = 1$. Moreover, for initial states sufficiently near $x = 1$, solving the averaged system with the same initial state as the original system yields the approximation

$$x(t, \epsilon) - x_{av}(t, \epsilon) = O(\epsilon), \quad \forall\, t \geq 0$$

Suppose we want to calculate a second-order approximation. We need to use the change of variables (8.22) to represent the problem as a standard perturbation problem and then proceed to approximate the solution, as we have done in Section 8.1. Using (8.20), the function $u(t, x)$ is given by

$$u(t, x) = \int_0^t (x \sin^2 \tau - 0.5x^2 - 0.5x + 0.5x^2)\, d\tau$$

$$= x \int_0^t (\sin^2 \tau - 0.5)\, d\tau = -\tfrac{1}{4}x \sin 2t$$

The change of variables (8.22) takes the form

$$x = y - \frac{\epsilon}{4}y \sin 2t = \left(1 - \frac{\epsilon}{4}\sin 2t\right) y$$

Differentiating both sides with respect to t, we obtain

$$\dot{x} = \left(1 - \frac{\epsilon}{4}\sin 2t\right)\dot{y} - \frac{\epsilon}{2}y \cos 2t$$

Hence,

$$\dot{y} = \frac{\epsilon}{1 - (\epsilon/4)\sin 2t}\left(x\sin^2 t - \tfrac{1}{2}x^2 + \tfrac{1}{2}y\cos 2t\right)$$

Substituting x in term of y, and expanding the term $1/[1 - (\epsilon/4)\sin 2t]$ in the power series

$$\frac{1}{1 - (\epsilon/4)\sin 2t} = 1 + \frac{\epsilon}{4}\sin 2t + O(\epsilon^2)$$

we arrive at the equation

$$\dot{y} = \frac{\epsilon}{2}(y - y^2) + \frac{\epsilon^2}{16}(y\sin 4t + 2y^2\sin 2t) + O(\epsilon^3)$$

where the system appears as a perturbation of the averaged system. In order to find a second-order approximation, we need to calculate y_0 and y_1 in the finite Taylor series

$$y = y_0 + \epsilon y_1 + \epsilon^2 y_R$$

We know that $y_0 = x_{\text{av}}$, the solution of the averaged system. The equation for y_1 is given by

$$\dot{y}_1 = \epsilon\left[\left(\tfrac{1}{2} - y_0(t)\right)y_1 + \tfrac{1}{16}y_0(t)\sin 4t + \tfrac{1}{8}y_0^2(t)\sin 2t\right], \quad y_1(0) = 0$$

where we have assumed that the initial state $x(0)$ is independent of ϵ. Using (8.22), we obtain a second-order approximation of x as

$$x = \left(1 - \frac{\epsilon}{4}\sin 2t\right)x_{\text{av}}(t,\epsilon) + \epsilon y_1(t,\epsilon) + O(\epsilon^2)$$

Figure 8.6 shows the solution of the exact system, the solution of the averaged system, and the second-order approximation for the initial state $x(0) = 0.7$ and $\epsilon = 0.3$. The figure illustrates clearly how the solution of the averaged system averages the exact solution. The second-order approximation is almost indistinguishable from the exact solution, but we can see the difference as the solution reaches steady state. $\triangle$

Example 8.10 Consider the suspended pendulum of Section 1.1.1 and assume that the suspension point is subjected to vertical vibrations of small amplitude and high frequency. Suppose the motion of the suspension point is described by $a\sin\omega t$, where a is the amplitude and ω is the frequency. Writing Newton's law in the tangential direction (perpendicular to the rod), the equation of motion is[5]

$$m(l\ddot{\theta} - a\omega^2\sin\omega t\sin\theta) = -mg\sin\theta - k(l\dot{\theta} + a\omega\cos\omega t\sin\theta)$$

---

[5]To derive this equation, write expressions, for the x- and y-coordinates of the bob as $x = l\sin\theta$ and $y = l\cos\theta - a\sin\omega t$. Then, show that the velocity and acceleration of the bob in the tangential direction are $(l\dot{\theta} + a\omega\cos\omega t\sin\theta)$ and $(l\ddot{\theta} - a\omega^2\sin\omega t\sin\theta)$, respectively. The friction force is assumed to be viscous friction proportional to the velocity of the bob with a friction coefficient k.

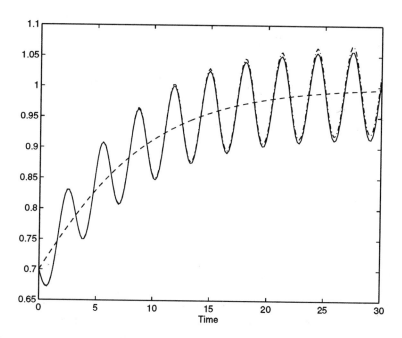

Figure 8.6: The exact (solid), averaged (dashed), and second-order (dash/dot) solutions of Example 8.9 with $\epsilon = 0.3$.

Assume that $a/l \ll 1$ and $\omega_0/\omega \ll 1$, where $\omega_0 = \sqrt{g/l}$ is the frequency of free oscillations of the pendulum in the vicinity of the lower equilibrium position $\theta = 0$. Let $\epsilon = a/l$ and write $\omega_0/\omega = \alpha\epsilon$, where $\alpha = \omega_0 l/\omega a$. Let $\beta = k/m\omega_0$ and change the time scale from t to $\tau = \omega t$. In the new time scale, the equation of motion is

$$\frac{d^2\theta}{d\tau^2} + \alpha\beta\epsilon\frac{d\theta}{d\tau} + (\alpha^2\epsilon^2 - \epsilon\sin\tau)\sin\theta + \alpha\beta\epsilon^2\cos\tau\sin\theta = 0$$

With

$$\begin{aligned} x_1 &= \theta \\ x_2 &= \frac{1}{\epsilon}\frac{d\theta}{d\tau} + \cos\tau\sin\theta \end{aligned}$$

as state variables, the state equation is given by

$$\frac{dx}{d\tau} = \epsilon f(\tau, x) \tag{8.24}$$

where

$$f_1(\tau, x) = x_2 - \cos \tau \sin x_1$$
$$f_2(\tau, x) = -\alpha\beta x_2 - \alpha^2 \sin x_1 + \cos \tau x_2 \cos x_1 - \cos^2 \tau \sin x_1 \cos x_1$$

The function $f(\tau, x)$ is 2π-periodic in τ. We shall study this system by the averaging method. The averaged system is given by

$$\frac{dx}{d\tau} = \epsilon f_{av}(x) \qquad (8.25)$$

where

$$f_{av1}(x) = \frac{1}{2\pi} \int_0^{2\pi} f_1(\tau, x) \, d\tau = x_2$$
$$f_{av2}(x) = \frac{1}{2\pi} \int_0^{2\pi} f_2(\tau, x) \, d\tau = -\alpha\beta x_2 - \alpha^2 \sin x_1 - \tfrac{1}{4} \sin 2x_1$$

In arriving at these expressions, we have used the fact that the average of $\cos \tau$ is zero while the average of $\cos^2 \tau$ is $\frac{1}{2}$. Both the original system (8.24) and the averaged system (8.25) have equilibrium points at $(x_1 = 0, \ x_2 = 0)$ and $(x_1 = \pi, \ x_2 = 0)$, which correspond to the equilibrium positions $\theta = 0$ and $\theta = \pi$. With a fixed suspension point, the equilibrium position $\theta = 0$ is exponentially stable while the equilibrium position $\theta = \pi$ is unstable. Let us see what a vibrating suspension point will do to the system. To apply Theorem 8.3, we analyze the stability properties of the equilibrium points of the averaged system (8.25) via linearization. The Jacobian of $f_{av}(x)$ is given by

$$\frac{\partial f_{av}}{\partial x} = \begin{bmatrix} 0 & 1 \\ -\alpha^2 \cos x_1 - 0.5 \cos 2x_1 & -\alpha\beta \end{bmatrix}$$

At the equilibrium point $(x_1 = 0, x_2 = 0)$, the Jacobian

$$\begin{bmatrix} 0 & 1 \\ -\alpha^2 - 0.5 & -\alpha\beta \end{bmatrix}$$

is Hurwitz for all positive values of α and β. Therefore, Theorem 8.3 says that for sufficiently small ϵ, the original system (8.24) has a unique exponentially stable 2π-periodic solution in an $O(\epsilon)$ neighborhood of the origin. The origin is an equilibrium point for the original system; hence, the periodic solution is the trivial solution $x = 0$. In this case, Theorem 8.3 confirms that for sufficiently small ϵ the origin is an exponentially stable equilibrium point for the original system (8.24). In other words, exponential stability of the lower equilibrium position of the pendulum is preserved under (small amplitude, high frequency) vibration of the suspension point. At the equilibrium point $(x_1 = \pi, x_2 = 0)$, the Jacobian

$$
\begin{bmatrix}
0 & 1 \\
\alpha^2 - 0.5 & -\alpha\beta
\end{bmatrix}
$$

is Hurwitz for $0 < \alpha < 1/\sqrt{2}$ and $\beta > 0$. Noting, again, that $(x_1 = \pi, \ x_2 = 0)$ is an equilibrium point for the original system, application of Theorem 8.3 leads us to the conclusion that if $\alpha < 1/\sqrt{2}$ then the upper equilibrium position $\theta = \pi$ is an exponentially stable equilibrium point for the original system (8.24) for sufficiently small ϵ. This is an intriguing finding because it shows that the unstable upper equilibrium position of the pendulum can be stabilized by vibrating the suspension point vertically with a small amplitude and high frequency.[6] $\triangle$

8.4   Weakly Nonlinear Second-Order Oscillators

Consider the second-order system

$$
\ddot{y} + \omega^2 y = \epsilon g(y, \dot{y}) \tag{8.26}
$$

where $g(\cdot, \cdot)$ is sufficiently smooth and $|g|$ is bounded by $k|y|$ or $k|\dot{y}|$ on compact sets of $(y, \dot{y})$; k is a positive constant. Choosing $x_1 = y$ and $x_2 = \dot{y}/\omega$ as state variables, we obtain the state equation

$$
\begin{aligned}
\dot{x}_1 &= \omega x_2 \\
\dot{x}_2 &= -\omega x_1 + \frac{\epsilon}{\omega} g(x_1, \omega x_2)
\end{aligned}
$$

Representing the system in the polar coordinates

$$
x_1 = r \sin \phi, \quad x_2 = r \cos \phi
$$

we have

$$
\begin{aligned}
\dot{r} &= \frac{1}{r}(x_1 \dot{x}_1 + x_2 \dot{x}_2) \\
&= \frac{1}{r}\left(\omega x_1 x_2 - \omega x_1 x_2 + \frac{\epsilon}{\omega} x_2 g(x_1, \omega x_2) \right) \\
&= \frac{\epsilon}{\omega} g(r \sin \phi, \omega r \cos \phi) \cos \phi \tag{8.27}
\end{aligned}
$$

---

[6]The idea of introducing high-frequency, zero-mean vibrations in the parameters of a dynamic system in order to modify the properties of the system in a desired manner has been generalized into a *principle of vibrational control*; see [110] and [17].

and

$$
\begin{aligned}
\dot{\phi} &= \frac{d}{dt}\tan^{-1}\left(\frac{x_1}{x_2}\right) = \frac{1}{r^2}(x_2\dot{x}_1 - x_1\dot{x}_2) \\
&= \frac{1}{r^2}\left(\omega x_1^2 + \omega x_2^2 - \frac{\epsilon}{\omega}x_1 g(x_1,\omega x_2)\right) \\
&= \omega - \frac{\epsilon}{\omega r}g(r\sin\phi,\omega r\cos\phi)\sin\phi
\end{aligned}
\tag{8.28}
$$

The second term on the right-hand side of (8.28) is $O(\epsilon)$ on bounded sets of r, as a consequence of the assumption that $|g|$ is bounded by $k|y|$ or $k|\dot{y}|$. Hence, the right-hand side of (8.28) is bounded from below for sufficiently small ϵ. Divide (8.27) by (8.28) to obtain

$$
\frac{dr}{d\phi} = \frac{\epsilon g(r\sin\phi,\omega r\cos\phi)\cos\phi}{\omega^2 - (\epsilon/r)g(r\sin\phi,\omega r\cos\phi)\sin\phi}
$$

We rewrite this equation as

$$
\frac{dr}{d\phi} = \epsilon f(\phi,r,\epsilon) \tag{8.29}
$$

where

$$
f(\phi,r,\epsilon) = \frac{g(r\sin\phi,\omega r\cos\phi)\cos\phi}{\omega^2 - (\epsilon/r)g(r\sin\phi,\omega r\cos\phi)\sin\phi}
$$

If we view ϕ as an independent variable, then (8.29) takes the form of (8.17), where $f(\phi,r,\epsilon)$ is 2π-periodic in ϕ. The function $f_{av}(r)$ is given by

$$
f_{av}(r) = \frac{1}{2\pi}\int_0^{2\pi} f(\phi,r,0)\,d\phi = \frac{1}{2\pi\omega^2}\int_0^{2\pi} g(r\sin\phi,\omega r\cos\phi)\cos\phi\,d\phi
$$

Suppose the averaged system

$$
\frac{dr}{d\phi} = \epsilon f_{av}(r) \tag{8.30}
$$

has an equilibrium point r^* where $[\partial f_{av}/\partial r](r^*) < 0$; then, there is $\epsilon^* > 0$ such that $\forall\, 0 < \epsilon < \epsilon^*$ (8.29) has a unique exponentially stable 2π-periodic solution $r = R(\phi,\epsilon)$ in an $O(\epsilon)$ neighborhood of r^*. This, by itself, does not say that (8.26) has a periodic solution with respect to t. More work is needed to reach that conclusion. Substitution of $r = R(\phi,\epsilon)$ in (8.28) yields

$$
\dot{\phi} = \omega - \frac{\epsilon}{\omega R(\phi,\epsilon)}g(R(\phi,\epsilon)\sin\phi,\omega R(\phi,\epsilon)\cos\phi)\sin\phi
$$

Let $\phi^*(t,\epsilon)$ be a solution of this equation. To show that (8.26) has a periodic solution, we need to show that there exists $T = T(\epsilon) > 0$, generally dependent on ϵ, such that

$$
\phi^*(t+T,\epsilon) = 2\pi + \phi^*(t,\epsilon), \quad \forall\, t \geq 0 \tag{8.31}
$$

For then

$$R(\phi^*(t+T,\epsilon),\epsilon) = R(2\pi + \phi^*(t,\epsilon),\epsilon) = R(\phi^*(t,\epsilon),\epsilon)$$

which implies that $R(\phi^*(t,\epsilon),\epsilon)$ is T-periodic in t. The image of the solution $r = R(\phi^*(t,\epsilon),\epsilon)$ in the state plane x_1–x_2 would be a closed orbit in the neighborhood of the circle $r = r^*$. Since the periodic solution $r = R(\phi,\epsilon)$ is exponentially stable, the closed orbit would attract all solutions in its neighborhood; that is, the closed orbit would be a stable limit cycle. Let us turn now to the question of existence of T that satisfies (8.31). Note that, over order $O(1)$ time intervals, the solution of (8.28) can be approximated by ωt with an error of order $O(\epsilon)$; that is, $\phi(t,\epsilon) = t\omega + O(\epsilon)$. At $\epsilon = 0$, (8.31) reduces to

$$(t+T)\omega = 2\pi + t\omega$$

which has the unique solution $T_0 = 2\pi/\omega$. Moreover, the partial derivative of $\phi^*(t+T,\epsilon)$ with respect to T at $\epsilon = 0$ equals $\omega \neq 0$. By the implicit function theorem, we conclude that for sufficiently small ϵ, equation (8.31) has a unique solution $T(\epsilon)$. We can actually show that $T(\epsilon) = 2\pi/\omega + O(\epsilon)$.

Example 8.11 Consider the Van der Pol oscillator

$$\ddot{y} + y = \epsilon\dot{y}(1 - y^2)$$

which is a special case of (8.26) with $\omega = 1$ and $g(y,\dot{y}) = \dot{y}(1 - y^2)$. The function $f_{av}(r)$ is given by

$$
\begin{aligned}
f_{av}(r) &= \frac{1}{2\pi}\int_0^{2\pi} g(r\sin\phi, \omega r\cos\phi)\cos\phi\, d\phi \\
&= \frac{1}{2\pi}\int_0^{2\pi} \left(1 - r^2\sin^2\phi\right) r\cos^2\phi\, d\phi \\
&= \frac{1}{2\pi}\int_0^{2\pi} r\cos^2\phi\, d\phi - \frac{1}{2\pi}\int_0^{2\pi} r^3\sin^2\phi\cos^2\phi\, d\phi \\
&= \frac{1}{2}r - \frac{1}{8}r^3
\end{aligned}
$$

The averaged system

$$\frac{dr}{d\phi} = \epsilon\left(\tfrac{1}{2}r - \tfrac{1}{8}r^3\right)$$

has three equilibrium points at $r = 0$, $r = 2$, and $r = -2$. Since by definition $r \geq 0$, the negative root is rejected. We check stability of the equilibria via linearization. The Jacobian matrix is given by

$$\frac{df_{av}}{dr} = \tfrac{1}{2} - \tfrac{3}{8}r^2$$

and

$$\frac{df_{\mathrm{av}}}{d\,r}\bigg|_{r=0} = \tfrac{1}{2} > 0; \quad \frac{df_{\mathrm{av}}}{d\,r}\bigg|_{r=2} = -1 < 0$$

Thus the equilibrium point $r = 2$ is exponentially stable. Therefore, for sufficiently small ϵ, the Van der Pol oscillator has a stable limit cycle in an $O(\epsilon)$ neighborhood of $r = 2$. The period of the steady-state oscillation is $O(\epsilon)$ close to 2π. This stable limit cycle was observed in Example 1.7 via simulation. $\triangle$

Let us conclude by noting that if the averaged system (8.30) has an equilibrium point r^* where $[\partial f_{\mathrm{av}}/\partial r](r^*) > 0$, then there is $\epsilon^* > 0$ such that $\forall\, 0 < \epsilon < \epsilon^*$ (8.26) has an unstable limit cycle in the neighborhood of the circle $r = r^*$. This can be seen by reversing time in (8.26), that is, replacing t by $\tau = -t$.

8.5   General Averaging

Consider the system

$$\dot{x} = \epsilon f(t, x, \epsilon) \tag{8.32}$$

where $f : [0,\infty) \times D \times [0,\epsilon_0] \to R^n$ is continuous and bounded, and has continuous, bounded partial derivatives up to the second order with respect to (x,ϵ) for $(t,x,\epsilon) \in [0,\infty) \times D \times [0,\epsilon_0]$. The parameter ϵ is positive and $D \subset R^n$ is a domain. Let $\Gamma = [0,\infty) \times D$. The averaging method applies to the system (8.32) in cases more general than the case of $f(t,x,\epsilon)$ being periodic in t. In particular, it applies when the function $f(t,x,0)$ has an average value $f_{\mathrm{av}}(x)$, where the average is defined in a general sense.

Definition 8.2 *A continuous, bounded function $g : \Gamma \to R^n$ is said to have an average $g_{\mathrm{av}}(x)$ if the limit*

$$g_{\mathrm{av}}(x) = \lim_{T \to \infty} \frac{1}{T} \int_t^{t+T} g(\tau, x)\, d\tau$$

exists and

$$\left\| \frac{1}{T} \int_t^{t+T} g(\tau, x)\, d\tau - g_{\mathrm{av}}(x) \right\| \le k\sigma(T), \quad \forall\, (t,x) \in \Gamma$$

where k is a positive constant and $\sigma : [0,\infty) \to [0,\infty)$ is a strictly decreasing, continuous, bounded function such that $\sigma(T) \to 0$ as $T \to \infty$. The function σ is called the convergence function.

Example 8.12

- Let $g(t, x) = \sum_{k=1}^{N} g_k(t, x)$ where $g_k(t, x)$ is periodic in t of period T_k, with $T_i \neq T_j$ when $i \neq j$. The function g is not periodic[7] in t, but it has the average

$$g_{\mathrm{av}}(x) = \sum_{k=1}^{N} g_{k\mathrm{av}}(x)$$

where $g_{k\mathrm{av}}$ is the average of the periodic function $g_k(t, x)$, as defined in Section 8.3. The convergence function σ is of order $O(1/T)$ as $T \to \infty$. We can take it as $\sigma(T) = 1/(T+1)$.

- The average of

$$g(t, x) = \frac{1}{1+t} h(x)$$

is zero, and the convergence function σ can be taken as $\sigma(T) = (1/T) \ln(1+T)$.
△

Suppose now that $f(t, x, 0)$ has the average function $f_{\mathrm{av}}(x)$ on Γ with convergence function $\sigma(\cdot)$. Let

$$h(t, x) = f(t, x, 0) - f_{\mathrm{av}}(x) \tag{8.33}$$

The function $h(t, x)$ has zero average with $\sigma(\cdot)$ as its convergence function. Suppose that the Jacobian matrix $[\partial h/\partial x]$ has zero average on Γ with the same convergence function $\sigma(\cdot)$. Define

$$w(t, x, \eta) = \int_0^t h(\tau, x) \exp[-\eta(t - \tau)] \, d\tau \tag{8.34}$$

for some positive constant η. At $\eta = 0$, the function $w(t, x, 0)$ satisfies

$$\|w(t + \delta, x, 0) - w(t, x, 0)\| = \left\| \int_0^{t+\delta} h(\tau, x) \, d\tau - \int_0^t h(\tau, x) \, d\tau \right\|$$

$$= \left\| \int_t^{t+\delta} h(\tau, x) \, d\tau \right\| \leq k\delta\sigma(\delta) \tag{8.35}$$

This implies, in particular, that

$$\|w(t, x, 0)\| \leq kt\sigma(t), \quad \forall \, (t, x) \in \Gamma$$

---

[7] This function is called almost periodic. An introduction to the theory of almost periodic functions can be found in [65] or [50].

since $w(0, x, 0) = 0$. Integrating the right-hand side of (8.34) by parts, we obtain

$$
\begin{aligned}
w(t, x, \eta) &= w(t, x, 0) - \eta \int_0^t \exp[-\eta(t - \tau)] w(\tau, x, 0) \, d\tau \\
&= \exp(-\eta t) w(t, x, 0) - \eta \int_0^t \exp[-\eta(t - \tau)] \, [w(\tau, x, 0) - w(t, x, 0)] \, d\tau
\end{aligned}
$$

where the second equality is obtained by adding and subtracting

$$
\eta \int_0^t \exp[-\eta(t - \tau)] \, d\tau \; w(t, x, 0)
$$

to the right-hand side. Using (8.35), we obtain

$$
\|w(t, x, \eta)\| \leq kt \exp(-\eta t) \sigma(t) + k\eta \int_0^t \exp[-\eta(t - \tau)](t - \tau)\sigma(t - \tau) \, d\tau \quad (8.36)
$$

This inequality can be used to show that $\eta \|w(t, x, \eta)\|$ is uniformly bounded by a class $\mathcal{K}$ function of η. For example, if $\sigma(t) = 1/(t + 1)$, then

$$
\eta \|w(t, x, \eta)\| \leq k\eta \exp(-\eta t) + k\eta^2 \int_0^t \exp[-\eta(t - \tau)] \, d\tau = k\eta
$$

Defining $\alpha(\eta) = k\eta$, we have $\eta \|w(t, x, \eta)\| \leq \alpha(\eta)$. If $\sigma(t) = 1/(t^r + 1)$ with $0 < r < 1$, then

$$
\begin{aligned}
\eta \|w(t, x, \eta)\| &\leq k\eta t^{(1-r)} e^{-\eta t} + k\eta^2 \int_0^t e^{-\eta(t-\tau)} (t - \tau)^{(1-r)} \, d\tau \\
&\leq k\eta \left(\frac{1-r}{\eta} \right)^{1-r} e^{-(1-r)} + k\eta^2 \int_0^\infty e^{-\eta s} s^{(1-r)} \, ds \\
&\leq k\eta \left(\frac{1-r}{\eta} \right)^{1-r} e^{-(1-r)} + k\eta^2 \frac{\Gamma(2 - r)}{\eta^{(2-r)}} \leq k_1 \eta^r
\end{aligned}
$$

where $\Gamma(\cdot)$ denotes the standard gamma function. Defining $\alpha(\eta) = k_1 \eta^r$, we have $\eta \|w(t, x, \eta)\| \leq \alpha(\eta)$. In general, it can be shown (Exercise 8.18) that there is a class $\mathcal{K}$ function α such that

$$
\eta \|w(t, x, \eta)\| \leq \alpha(\eta), \quad \forall \, (t, x) \in \Gamma \tag{8.37}
$$

Without loss of generality, we can choose $\alpha(\eta)$ such that $\alpha(\eta) \geq c\eta$ for $\eta \in [0, 1]$, where c is a positive constant. The partial derivatives $[\partial w / \partial t]$ and $[\partial w / \partial x]$ are given by

$$
\begin{aligned}
\frac{\partial w}{\partial t} &= h(t, x) - \eta w(t, x, \eta) \\
\frac{\partial w}{\partial x} &= \int_0^t \frac{\partial h}{\partial x}(\tau, x) \exp[-\eta(t - \tau)] \, d\tau
\end{aligned}
$$

Since $[\partial h/\partial x]$ possesses the same properties of h that have been used to arrive at (8.37), it is clear that we can repeat the previous derivations to show that

$$\eta \left\| \frac{\partial w}{\partial x} \right\| \leq \alpha(\eta), \quad \forall \ (t, x) \in \Gamma \tag{8.38}$$

There is no loss of generality in using the same class $\mathcal{K}$ function in both (8.37) and (8.38) since the calculated estimates will differ only in the positive constant that multiplies the η dependent term, so we can define α using the larger of the two constants.

The function $w(t, x, \eta)$ that we have just defined possesses all the key properties of the function $u(t, x)$ of Section 8.3. The only difference is that the function w is parameterized in a parameter η in such a way that the bounds on w and $[\partial w/\partial x]$ are of the form $\alpha(\eta)/\eta$ for some class $\mathcal{K}$ function α. We did not need to parameterize u in terms of any parameter. In fact, $u(t, x)$ is nothing more than the function $w(t, x, \eta)$ evaluated at $\eta = 0$. This should come as no surprise since in the periodic case the convergence function $\sigma(t) = 1/(t + 1)$; hence, $\alpha(\eta) = k\eta$ and the bound $\alpha(\eta)/\eta = k$ will be independent of η.

From this point on, the analysis will be very similar to that of Section 8.3. We define the change of variables

$$x = y + \epsilon w(t, y, \epsilon) \tag{8.39}$$

The term $\epsilon w(t, y, \epsilon)$ is of order $O(\alpha(\epsilon))$; hence, for sufficiently small ϵ, the change of variables (8.39) is well defined since the matrix $[I + \epsilon \partial w/\partial y]$ is nonsingular. In particular,

$$\left[I + \epsilon \frac{\partial w}{\partial y} \right]^{-1} = I + O(\alpha(\epsilon)), \quad \text{on } \Gamma$$

Proceeding as in Section 8.3, we can show that the state equation for y is given by

$$\dot{y} = \epsilon f_{\text{av}}(y) + \epsilon \alpha(\epsilon) q(t, y, \epsilon) \tag{8.40}$$

where $q(t, y, \epsilon)$ is bounded on Γ for sufficiently small ϵ. In arriving at (8.40), we have used the fact that $\alpha(\epsilon) \geq c\epsilon$. Equation (8.40) is a perturbation of the averaged system

$$\dot{x} = \epsilon f(t, x) \tag{8.41}$$

It is similar to (8.23) except for the fact that the coefficient of the q term is $\epsilon \alpha(\epsilon)$ instead of ϵ^2. This observation leads to the following theorem, which is similar to (the first two parts of) Theorem 8.3 except that the estimates $O(\epsilon)$ are replaced by estimates $O(\alpha(\epsilon))$.

Theorem 8.4 Let $f(t, x, \epsilon)$ be continuous and bounded, and have continuous and bounded partial derivatives up to the second order with respect to (x, ϵ) for $(t, x, \epsilon) \in$

$[0, \infty) \times D \times [0, \epsilon_0]$, where $\epsilon > 0$. Suppose that $f(t, x, 0)$ has the average function $f_{av}(x)$ on $[0, \infty) \times D$ and that the Jacobian of $h(t, x) = f(t, x, 0) - f_{av}(x)$ has zero average with the same convergence function as f. Let $x(t, \epsilon)$ and $x_{av}(t, \epsilon)$ be solutions of (8.32) and (8.41), respectively, and $\alpha(\cdot)$ be the class $\mathcal{K}$ function appearing in the estimates (8.37) and (8.38).

- If $x_{av}(t, \epsilon) \in D \ \forall \ t \in [0, b/\epsilon]$ and $\|x(0, \epsilon) - x_{av}(0, \epsilon)\| = O(\alpha(\epsilon))$, then

$$\|x(t, \epsilon) - x_{av}(t, \epsilon)\| = O(\alpha(\epsilon)) \ \text{ on } [0, b/\epsilon]$$

- If $f(t, 0, \epsilon) = 0$ for all $(t, \epsilon) \in [0, \infty) \times [0, \epsilon_0]$ and the origin of the averaged system (8.41) is exponentially stable, then there exist positive constants ϵ^* and ρ such that for all $0 < \epsilon < \epsilon^*$, the origin of the original system (8.32) will be exponentially stable and, for all $\|x_{av}(0, \epsilon)\| < \rho$, the $O(\alpha(\epsilon))$ estimate of the approximation error will be valid for all $t \in [0, \infty)$. $\diamond$

Proof: Appendix A.13.

Example 8.13 Consider the linear system

$$\dot{x} = \epsilon A(t) x$$

where $\epsilon > 0$. Suppose that $A(t)$ and its derivatives up to the second order are continuous and bounded. Moreover, suppose $A(t)$ has an average

$$A_{av} = \lim_{T \to \infty} \frac{1}{T} \int_t^{t+T} A(\tau) \, d\tau$$

in the sense of Definition 8.2. The averaged system is given by

$$\dot{x} = \epsilon A_{av} x$$

Suppose A_{av} is Hurwitz. By Theorem 8.4, we conclude that the origin of the original linear time-varying system is exponentially stable for sufficiently small ϵ. Suppose further that the matrix $A(t) = A_{tr}(t) + A_{ss}(t)$ is the sum of a transient component $A_{tr}(t)$ and a steady-state component $A_{ss}(t)$. The transient component decays to zero exponentially fast; that is,

$$\|A_{tr}(t)\| \leq k_1 \exp(-\alpha t), \quad k_1 > 0, \ \alpha > 0$$

while the elements of the steady-state component are formed of a finite sum of sinusoids with distinct frequencies. The average of the transient component is zero since

$$\frac{1}{T} \int_t^{t+T} \|A_{tr}(\tau)\| \, d\tau \leq \frac{1}{T} \int_t^{t+T} k_1 e^{-\alpha \tau} \, d\tau = \frac{k_1 e^{-\alpha t}}{\alpha T} \left[1 - e^{-\alpha T} \right] \leq \frac{k_2}{T+1}$$

Recalling the first case of Example 8.12, we see that $A(t)$ has an average with convergence function $\sigma(T) = 1/(T+1)$. Hence, the class $\mathcal{K}$ function of Theorem 8.4 is $\alpha(\eta) = k\eta$. Let $x(t, \epsilon)$ and $x_{\text{av}}(t, \epsilon)$ denote solutions of the original and averaged systems which start from the same initial state. By Theorem 8.4,

$$\|x(t, \epsilon) - x_{\text{av}}(t, \epsilon)\| = O(\epsilon), \quad \forall\, t \geq 0$$

$\triangle$

8.6 Exercises

Exercise 8.1 Using Theorem 2.5, verify inequality (8.4).

Exercise 8.2 If $\delta(\epsilon) = O(\epsilon)$, is it $O(\epsilon^{1/2})$? Is it $O(\epsilon^{3/2})$?

Exercise 8.3 If $\delta(\epsilon) = \epsilon^{1/n}$, where $n > 1$ is a positive integer, is there a positive integer N such that $\delta(\epsilon) = O(\epsilon^N)$?

Exercise 8.4 Consider the initial value problem

$$\dot{x}_1 = -(0.2 + \epsilon)x_1 + \frac{\pi}{4} - \tan^{-1} x_1 + \epsilon \tan^{-1} x_2, \quad x_1(0) = \eta_1$$
$$\dot{x}_2 = -(0.2 + \epsilon)x_2 + \frac{\pi}{4} - \tan^{-1} x_2 + \epsilon \tan^{-1} x_1, \quad x_2(0) = \eta_2$$

(a) Find an $O(\epsilon)$ approximation.

(b) Find an $O(\epsilon^2)$ approximation.

(c) Investigate the validity of the approximation on the infinite interval.

(d) Calculate, using a computer program, the exact solution, the $O(\epsilon)$ approximation, and the $O(\epsilon^2)$ approximation for $\epsilon = 0.1$, $\eta_1 = 0.5$, and $\eta_2 = 1.5$ on the time interval $[0, 3]$. Comment on the accuracy of the approximation.

Hint: In parts (a) and (b), it is sufficient to give the equations defining the approximation. You are not required to find an analytic closed-form expression for the approximation.

Exercise 8.5 Repeat Exercise 8.4 for the system

$$\dot{x}_1 = x_2$$
$$\dot{x}_2 = -x_1 - x_2 + \epsilon x_1^3$$

In part (d), let $\epsilon = 0.1$, $\eta_1 = 1.0$, $\eta_2 = 0.0$, and the time interval be $[0, 5]$.

Exercise 8.6 Repeat Exercise 8.4 for the system

$$\dot{x}_1 = -x_1 + x_2$$
$$\dot{x}_2 = \epsilon x_1 - x_2 - \tfrac{1}{3}x_2^3$$

In part (d), let $\epsilon = 0.2$, $\eta_1 = 1.0$, $\eta_2 = 0.0$, and the time interval be $[0, 4]$.

Exercise 8.7 ([150]) Repeat Exercise 8.4 for the system

$$\dot{x}_1 = x_1 - x_1^2 + \epsilon x_1 x_2$$
$$\dot{x}_2 = 2x_2 - x_2^2 - \epsilon x_1 x_2$$

In part (d), let $\epsilon = 0.2$, $\eta_1 = 0.5$, $\eta_2 = 1.0$, and the time interval be $[0, 4]$.

Exercise 8.8 Repeat Exercise 8.4 for the system

$$\dot{x}_1 = -x_1 + x_2(1 + x_1) + \epsilon(1 + x_1)^2$$
$$\dot{x}_2 = -x_1(x_1 + 1)$$

In part (d), let $\epsilon = -0.1$, $\eta_1 = -1$, and $\eta_2 = 2$. Repeat the calculation for $\epsilon = -0.05$ and $\epsilon = -0.2$ and comment on the accuracy of the approximation.

Exercise 8.9 Consider the initial value problem

$$\dot{x}_1 = -x_1 + \epsilon x_2, \quad x_1(0) = \eta$$
$$\dot{x}_2 = -x_2 - \epsilon x_1, \quad x_2(0) = \eta$$

Find an $O(\epsilon)$ approximation. Calculate the exact and approximate solutions at $\epsilon = 0.1$ for two different sets of initial conditions: (1) $\eta = 1$, (2) $\eta = 10$. Comment on the approximation accuracy. Explain any discrepancy with Theorem 8.1.

Exercise 8.10 ([59]) Study, using the averaging method, each of the following scalar systems.

(1) $\dot{x} = \epsilon(x - x^2)\sin^2 t$ (2) $\dot{x} = \epsilon(x \cos^2 t - \tfrac{1}{2}x^2)$

(3) $\dot{x} = \epsilon(-x + \cos^2 t)$ (4) $\dot{x} = -\epsilon x \cos t$

Exercise 8.11 For each of the following systems, show that for sufficiently small $\epsilon > 0$, the origin is exponentially stable.

(1) $\dot{x}_1 = \epsilon x_2$
$$\dot{x}_2 = -\epsilon(1 + 2\sin t)x_2 - \epsilon(1 + \cos t)\sin x_1$$

(2) $\dot{x}_1 = \epsilon[(-1 + 1.5\cos^2 t)x_1 + (1 - 1.5\sin t \cos t)x_2]$
$$\dot{x}_2 = \epsilon[(-1 - 1.5\sin t \cos t)x_1 + (-1 + 1.5\sin^2 t)x_2]$$

(3) $\dot{x} = \epsilon\left(-x \sin^2 t + x^2 \sin t + x e^{-t}\right), \quad \epsilon > 0$

Exercise 8.12 Consider the system $\dot{y} = Ay + \epsilon g(t, y, \epsilon)$, $\epsilon > 0$, where the $n \times n$ matrix A has only simple eigenvalues on the imaginary axis.

(a) Show that $\exp(At)$ and $\exp(-At)$ are bounded for all $t \geq 0$.

(b) Show that the change of variables $y = \exp(At)x$ transforms the system into the form $\dot{x} = \epsilon f(t, x, \epsilon)$ and give an expression for f in terms of g and $\exp(At)$.

Exercise 8.13 ([150]) Study Mathieu's equation $\ddot{y} + (1 + 2\epsilon \cos 2t)y = 0$, $\epsilon > 0$, using the averaging method.
Hint: Use Exercise 8.12.

Exercise 8.14 ([150]) Study the equation $\ddot{y} + y = 8\epsilon(\dot{y})^2 \cos t$ using the averaging method.
Hint: Use Exercise 8.12.

Exercise 8.15 Apply the averaging method to study the existence of limit cycles for each of the following second-order systems. If there is a limit cycle, estimate its location in the state plane and the period of oscillation, and determine whether it is stable or unstable.

(1) $\ddot{y} + y = -\epsilon \dot{y}(1 - y^2)$ **(2)** $\ddot{y} + y = \epsilon \dot{y}(1 - y^2) - \epsilon y^3$

(3) $\ddot{y} + y = -\epsilon \left(1 - \frac{3\pi}{4}|y|\right)\dot{y}$ **(4)** $\ddot{y} + y = -\epsilon \left(1 - \frac{3\pi}{4}|\dot{y}|\right)\dot{y}$

(5) $\ddot{y} + y = -\epsilon(\dot{y} - y^3)$ **(6)** $\ddot{y} + y = \epsilon \dot{y}(1 - y^2 - \dot{y}^2)$

Exercise 8.16 Consider Rayleigh's equation

$$m\frac{d^2u}{dt^2} + ku = \lambda \left[1 - \alpha \left(\frac{du}{dt}\right)^2\right]\frac{du}{dt}$$

where m, k, λ, and α are positive constants.

(a) Using the dimensionless variables $y = u/u^*$, $\tau = t/t^*$, and $\epsilon = \lambda/\lambda^*$, where $(u^*)^2\alpha k = m/3$, $t^* = \sqrt{m/k}$, and $\lambda^* = \sqrt{km}$, show that the equation can be normalized to

$$\ddot{y} + y = \epsilon \left(\dot{y} - \tfrac{1}{3}\dot{y}^3\right)$$

where $\dot{y}$ denotes the derivative of y with respect to τ.

(b) Apply the averaging method to show that the normalized Rayleigh equation has a stable limit cycle. Estimate the location of the limit cycle in the plane $(y, \dot{y})$.

(c) Using a numerical algorithm, obtain the phase portrait of the normalized Rayleigh equation in the plane $(y, \dot{y})$ for

$$\text{(i) } \epsilon = 1, \quad \text{(ii) } \epsilon = 0.1, \quad \text{and} \quad \text{(iii) } \epsilon = 0.01,$$

Compare with the results of part (b).

Exercise 8.17 Consider Duffing's equation

$$m\ddot{y} + c\dot{y} + ky + ka^2 y^3 = A \cos \omega t$$

where A, a, c, k, m and ω are positive constants.

(a) Taking $x_1 = y$, $x_2 = \dot{y}$, $\tau = \omega t$, and $\epsilon = 1/\omega$, show that the equation can be represented as $\frac{dx}{d\tau} = \epsilon f(\tau, x, \epsilon)$.

(b) Show that the system has an exponentially stable periodic solution for sufficiently large ω. Estimate the frequency of oscillation and the location of the periodic orbit in the phase plane.

Exercise 8.18 Verify (8.37).
Hint: Start from (8.36). In majorizing $\sigma(t)$, use the fact that $\sigma(t)$ is bounded for $t \leq 1/\sqrt{\eta}$, while for $t \geq 1/\sqrt{\eta}$ use the inequality $\sigma(t) \leq \sigma(1/\sqrt{\eta})$.

Exercise 8.19 Study, using general averaging, the scalar system

$$\dot{x} = \epsilon \left(\sin^2 t + \sin 1.5t + e^{-t} \right) x$$

Exercise 8.20 ([151]) The output of an nth-order linear time-invariant single-input–single-output system can be represented by $y(t) = \theta^T w(t)$, where θ is a $(2n + 1)$-dimensional vector of constant parameters and $w(t)$ is an auxiliary signal which can be synthesized from the system's input and output, without knowing θ. Suppose that the vector θ is unknown and denote its value by θ^*. In identification experiments, the parameter $\theta(t)$ is updated using an adaptation law of the form $\dot{\theta} = -\epsilon e(t) w(t)$, where $e(t) = [\theta(t) - \theta^*]^T w(t)$ is the error between the actual system's output and the estimated output using $\theta(t)$. Let $\phi(t) = \theta(t) - \theta^*$ denote the parameter error.

(a) Show that $\dot{\phi} = \epsilon A(t)\phi$, where $A(t) = -w(t)w^T(t)$.

(b) Using (general) averaging, derive a condition on $w(t)$ which ensures that, for sufficiently small ϵ, $\theta(t) \to \theta^*$ as $t \to \infty$.

Chapter 9

Singular Perturbations

While the perturbation method of Section 8.1 applies to state equations which depend smoothly on a small parameter ϵ, in this chapter we face a more difficult perturbation problem characterized by discontinuous dependence of system properties on the perturbation parameter ϵ. We shall study the "so-called" standard singular perturbation model

$$\begin{aligned}
\dot{x} &= f(t, x, z, \epsilon) \\
\epsilon \dot{z} &= g(t, x, z, \epsilon)
\end{aligned}$$

where setting $\epsilon = 0$ causes a fundamental and abrupt change in the dynamic properties of the system, as the differential equation $\epsilon \dot{z} = g$ degenerates into the algebraic or transcendental equation

$$0 = g(t, x, z, 0)$$

The essence of the theory developed in this chapter is that the discontinuity of solutions caused by singular perturbations can be avoided if analyzed in separate time scales. This multitime-scale approach is a fundamental characteristic of singular perturbation methods.

In Section 9.1, we define the standard singular perturbation model and illustrate, via examples, some of its physical sources. In Section 9.2, we study the two-time-scale properties of the standard model and give a trajectory approximation result based on the decomposition of the model into reduced (slow) and boundary-layer (fast) models. The intuition behind the time-scale decomposition becomes more transparent with a geometric viewpoint, which we present in Section 9.3. The time-scale decomposition of Section 9.2 is used in Section 9.4 to analyze the stability of equilibria via Lyapunov's method. Section 9.5 extends the trajectory approximation result of Section 9.2 to the infinite-time interval. Finally, we illustrate, via examples, how the results of Sections 9.4 and 9.5 can be used to analyze the robustness of stability to singular perturbations.

9.1   The Standard Singular Perturbation Model

The singular perturbation model of a dynamical system is a state-space model in which the derivatives of some of the states are multiplied by a small positive parameter ϵ; that is,

$$\dot{x} = f(t, x, z, \epsilon), \quad x \in R^n \tag{9.1}$$

$$\epsilon\dot{z} = g(t, x, z, \epsilon), \quad z \in R^m \tag{9.2}$$

We assume that the functions f and g are continuously differentiable in their arguments for $(t, x, z, \epsilon) \in [0, t_1] \times D_1 \times D_2 \times [0, \epsilon_0]$, where $D_1 \subset R^n$ and $D_2 \subset R^m$ are open connected sets. When we set $\epsilon = 0$ in (9.1)–(9.2), the dimension of the state equation reduces from $n+m$ to n because the differential equation (9.2) degenerates into the equation

$$0 = g(t, x, z, 0) \tag{9.3}$$

We shall say that the model (9.1)–(9.2) is in *standard form* if and only if (9.3) has $k \geq 1$ isolated real roots

$$z = h_i(t, x), \quad i = 1, 2, \ldots, k \tag{9.4}$$

for each $(t, x) \in [0, t_1] \times D_1$. This assumption ensures that a well-defined n-dimensional reduced model will correspond to each root of (9.3). To obtain the ith reduced model, we substitute (9.4) into (9.1), at $\epsilon = 0$, to obtain

$$\dot{x} = f(t, x, h(t, x), 0) \tag{9.5}$$

where we have dropped the subscript i from h. It will be clear from the context which root of (9.4) we are using. This model is sometimes called a *quasi-steady-state model* because z, whose velocity $\dot{z} = g/\epsilon$ can be large when ϵ is small and $g \neq 0$, may rapidly converge to a root of (9.3) which is the equilibrium of (9.2). We shall discuss this two-time-scale property of (9.1)–(9.2) in the next section. The model (9.5) is also known as the *slow model*.

Modeling a physical system in the singularly perturbed form may not be easy. It is not always clear how to pick the parameters to be considered as small. Fortunately, in many applications our knowledge of physical processes and components of the system sets us on the right track.[1] The following three examples illustrate

---

[1] More about modeling physical systems in the singularly perturbed form can be found in [95, Chapter 1], [94, Chapter 4], and [32].

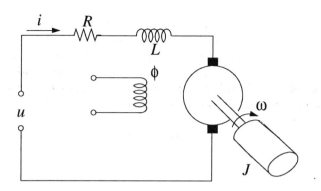

Figure 9.1: Armature-controlled DC motor.

three different "typical" ways of choosing the parameter ϵ. In the first example, ϵ is chosen as a small time constant. This is the most popular source of singularly perturbed models and, historically, the case that motivated interest in singular perturbations. Small time constants, masses, capacitances, and similar "parasitic" parameters which increase the order of a dynamic model are quite common in models of physical systems. In the interest of model simplification, we usually neglect these parasitic parameters to reduce the order of the model. Singular perturbations legitimize this ad hoc model simplification and provide tools for improving oversimplified models. In the second example, the parameter ϵ is the reciprocal of a high-gain parameter in a feedback system. The example represents an important source of singularly perturbed models. The use of high-gain parameters, or more precisely parameters which are driven asymptotically toward infinity, in the design of feedback control systems is quite common. A typical approach to the analysis and design of high-gain feedback systems is to model them in the singularly perturbed form. In the third example, the parameter ϵ is a parasitic resistor in an electric circuit. Although neglecting the parasitic resistor reduces the order of the model, it does it in way that is quite distinct from neglecting a parasitic time constant. Modeling the system in the standard singularly perturbed form involves a careful choice of the state variables.

Example 9.1 An armature-controlled DC motor, shown in Figure 9.1, can be modeled by the second-order state equation

$$
\begin{aligned}
J\frac{d\omega}{dt} &= ki \\
L\frac{di}{dt} &= -k\omega - Ri + u
\end{aligned}
$$

where i, u, R, and L are the armature current, voltage, resistance, and inductance, J

is the moment of inertia, ω is the angular speed, and ki and $k\omega$ are, respectively, the torque and the back e.m.f. (electromotive force) developed with constant excitation flux ϕ. The first state equation is a mechanical torque equation, and the second one is an equation for the electric transient in the armature circuit. Typically, L is "small" and can play the role of our parameter ϵ. This means that, with $\omega = x$ and $i = z$, the motor's model is in the standard form (9.1)–(9.2) whenever $R \neq 0$. Neglecting L, we solve

$$0 = -k\omega - Ri + u$$

to obtain

$$i = \frac{u - k\omega}{R}$$

which is the only root, and substitute it in the torque equation. The resulting model

$$J\dot{\omega} = -\frac{k^2}{R}\omega + \frac{k}{R}u$$

is the commonly used first-order model of the DC motor. As we discussed in Chapter 7, in formulating perturbation models it is preferable to choose the perturbation parameter ϵ as a dimensionless ratio of two physical parameters. To that end, let us define the dimensionless variables

$$\omega_r = \frac{\omega}{\Omega}; \quad i_r = \frac{iR}{k\Omega}; \quad u_r = \frac{u}{k\Omega}$$

and rewrite the state equation as

$$
\begin{aligned}
T_m \frac{d\omega_r}{dt} &= i_r \\
T_e \frac{di_r}{dt} &= -\omega_r - i_r + u_r
\end{aligned}
$$

where $T_m = JR/k^2$ is the mechanical time constant and $T_e = L/R$ is the electrical time constant. Since $T_m \gg T_e$, we let T_m be the time unit; that is, we introduce the dimensionless time variable $t_r = t/T_m$ and rewrite the state equation as

$$
\begin{aligned}
\frac{d\omega_r}{dt_r} &= i_r \\
\frac{T_e}{T_m} \frac{di_r}{dt_r} &= -\omega_r - i_r + u_r
\end{aligned}
$$

This scaling has brought the model into the standard form with a physically meaningful dimensionless parameter

$$\epsilon = \frac{T_e}{T_m} = \frac{Lk^2}{JR^2}$$

$\triangle$

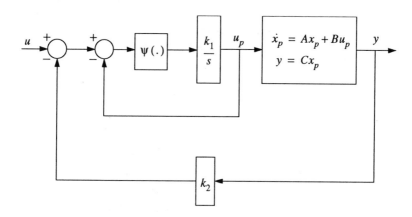

Figure 9.2: Actuator control with high-gain feedback.

Example 9.2 Consider the feedback control system of Figure 9.2; see Figure 3.1. The inner loop represents actuator control with high-gain feedback. The high-gain parameter is the integrator constant k_1. The plant is a single-input–single-output nth-order system represented by the state-space model $\{A, B, C\}$. The nonlinearity $\psi(\cdot)$ is a first quadrant–third quadrant nonlinearity with

$$\psi(0) = 0 \text{ and } y\psi(y) > 0, \ \forall \ y \neq 0$$

The state equation for the closed-loop system is

$$\dot{x}_p = Ax_p + Bu_p$$
$$\frac{1}{k_1}\dot{u}_p = \psi(u - u_p - k_2 C x_p)$$

With $\epsilon = 1/k_1$, $x_p = x$, and $u_p = z$, the model takes the form (9.1)–(9.2). Setting $\epsilon = 0$, or equivalently $k_1 = \infty$, we solve

$$\psi(u - u_p - k_2 C x_p) = 0$$

to obtain

$$u_p = u - k_2 C x_p$$

which is the unique root since $\psi(\cdot)$ vanishes only at its origin. The resulting reduced model is

$$\dot{x}_p = (A - Bk_2 C)x_p + Bu$$

which is the model of the simplified block diagram of Figure 9.3, where the whole inner loop in Figure 9.2 is replaced by a direct connection.   △

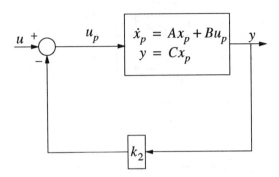

Figure 9.3: Simplified block diagram of Figure 9.2.

Example 9.3 Consider again the electric circuit of Example 8.4, shown in Figure 8.3. The differential equations for the voltages across the capacitors are

$$C\dot{v}_1 = \frac{1}{R}(E - v_1) - \psi(v_1) - \frac{1}{R_c}(v_1 - v_2)$$

$$C\dot{v}_2 = \frac{1}{R}(E - v_2) - \psi(v_2) - \frac{1}{R_c}(v_2 - v_1)$$

In Example 8.4, we analyzed the circuit for a "large" resistor R_c, which was idealized to be open circuit when $1/R_c$ was set to zero. This time, let us study the circuit for a "small" R_c. Setting $R_c = 0$ replaces the resistor with a short-circuit connection that puts the two capacitors in parallel. In a well-defined model for this simplified circuit, the two capacitors in parallel should be replaced by one equivalent capacitor, which means that the model of the simplified circuit will be of order one. To represent this model-order reduction as a singular perturbation, let us start with the seeming choice $\epsilon = R_c$ and rewrite the state equation as

$$\epsilon\dot{v}_1 = \frac{\epsilon}{CR}(E - v_1) - \frac{\epsilon}{C}\psi(v_1) - \frac{1}{C}(v_1 - v_2)$$

$$\epsilon\dot{v}_2 = \frac{\epsilon}{CR}(E - v_2) - \frac{\epsilon}{C}\psi(v_2) - \frac{1}{C}(v_2 - v_1)$$

If this model were in the form (9.1)–(9.2), both v_1 and v_2 would be considered as z variables and (9.3) would be

$$v_1 - v_2 = 0$$

However, the roots of this equation are not isolated, which violates the basic assumption that the roots of (9.3) should be isolated. Therefore, with v_1 and v_2 as z variables, the model is not in the standard form. Let us now try another choice of

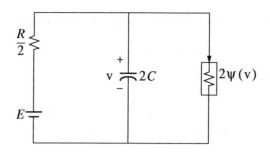

Figure 9.4: Simplified circuit when $R_c = 0$.

the state variables. Take[2]

$$x = \tfrac{1}{2}\left(v_1 + v_2\right); \quad z = \tfrac{1}{2}\left(v_1 - v_2\right)$$

The state equation for the new variables is

$$\dot{x} = \frac{1}{CR}(E - x) - \frac{1}{2C}[\psi(x + z) + \psi(x - z)]$$
$$\epsilon\dot{z} = -\left(\frac{\epsilon}{CR} + \frac{2}{C}\right)z - \frac{\epsilon}{2C}[\psi(x + z) - \psi(x - z)]$$

Now the unique root of (9.3) is $z = 0$, which results in the reduced model

$$\dot{x} = -\frac{1}{CR}(E - x) - \frac{1}{C}\psi(x)$$

This model represents the simplified circuit of Figure 9.4, where each pair of similar parallel branches is replaced by an equivalent single branch. To obtain ϵ as a dimensionless parameter, we normalize x, z, and ψ as

$$x_r = \frac{x}{E}; \quad z_r = \frac{z}{E}; \quad \psi_r(v) = \frac{R}{E}\psi(Ev)$$

and normalize the time variable as $t_r = t/CR$ to obtain the singularly perturbed model

$$\frac{dx_r}{dt_r} = 1 - x_r - \frac{1}{2}[\psi_r(x_r + z_r) + \psi_r(x_r - z_r)]$$
$$\epsilon\frac{dz_r}{dt_r} = -(\epsilon + 2)z_r - \frac{\epsilon}{2}[\psi_r(x_r + z_r) - \psi_r(x_r - z_r)]$$

where $\epsilon = R_c/R$ is dimensionless.   △

---

[2] This choice of state variables follows from a systematic procedure described in [32].

9.2   Time-Scale Properties of the Standard Model

Singular perturbations cause a multitime-scale behavior of dynamic systems characterized by the presence of slow and fast transients in the system's response to external stimuli. Loosely speaking, the slow response is approximated by the reduced model (9.5), while the discrepancy between the response of the reduced model (9.5) and that of the full model (9.1)–(9.2) is the fast transient. To see this point, let us consider the problem of solving the state equation

$$\dot{x} = f(t, x, z, \epsilon), \qquad x(t_0) = \xi(\epsilon) \tag{9.6}$$

$$\epsilon \dot{z} = g(t, x, z, \epsilon), \qquad z(t_0) = \eta(\epsilon) \tag{9.7}$$

where $\xi(\epsilon)$ and $\eta(\epsilon)$ depend smoothly on ϵ and $t_0 \in [0, t_1)$. Let $x(t, \epsilon)$ and $z(t, \epsilon)$ denote the solution of the full problem (9.6)–(9.7). When we define the corresponding problem for the reduced model (9.5), we can only specify n initial conditions since the model is nth order. Naturally, we retain the initial state for x to obtain the reduced problem

$$\dot{x} = f(t, x, h(t, x), 0), \quad x(t_0) = \xi_0 \stackrel{\text{def}}{=} \xi(0) \tag{9.8}$$

Denote the solution of (9.8) by $\bar{x}(t)$. Since the variable z has been excluded from the reduced model and substituted by its "quasi-steady-state" $h(t, x)$, the only information we can obtain about z by solving (9.8) is to compute

$$\bar{z}(t) \stackrel{\text{def}}{=} h(t, \bar{x}(t))$$

which describes the quasi-steady-state behavior of z when $x = \bar{x}$. By contrast to the original variable z starting at t_0 from a prescribed $\eta(\epsilon)$, the quasi-steady-state $\bar{z}$ is not free to start from a prescribed value, and there may be a large discrepancy between its initial value

$$\bar{z}(t_0) = h(t_0, \xi_0)$$

and the prescribed initial state $\eta(\epsilon)$. Thus, $\bar{z}(t)$ cannot be a uniform approximation of $z(t, \epsilon)$. The best we can expect is that the estimate

$$z(t, \epsilon) - \bar{z}(t) = O(\epsilon)$$

will hold on an interval excluding t_0, that is, for $t \in [t_b, t_1]$ where $t_b > t_0$. On the other hand, it is reasonable to expect the estimate

$$x(t, \epsilon) - \bar{x}(t) = O(\epsilon)$$

to hold uniformly for all $t \in [t_0, t_1]$ since

$$x(t_0, \epsilon) - \bar{x}(t_0) = \xi(\epsilon) - \xi(0) = O(\epsilon)$$

If the error $z(t, \epsilon) - \bar{z}(t)$ is indeed $O(\epsilon)$ over $[t_b, t_1]$, then it must be true that during the initial ("boundary-layer") interval $[t_0, t_b]$ the variable z approaches $\bar{z}$. Let us remember that the speed of z can be large since $\dot{z} = g/\epsilon$. In fact, having set $\epsilon = 0$ in (9.2), we have made the transient of z instantaneous whenever $g \neq 0$. From our previous study of the stability of equilibria, it should be clear that we cannot expect z to converge to its quasi-steady-state $\bar{z}$ unless certain stability conditions are satisfied. Such conditions will result from the forthcoming analysis.

It is more convenient in the analysis to perform the change of variables

$$y = z - h(t, x) \tag{9.9}$$

that shifts the quasi-steady-state of z to the origin. In the new variables (x, y), the full problem is

$$\dot{x} = f(t, x, y + h(t, x), \epsilon), \quad x(t_0) = \xi(\epsilon) \tag{9.10}$$

$$\epsilon \dot{y} = g(t, x, y + h(t, x), \epsilon) - \epsilon \frac{\partial h}{\partial t}$$

$$- \epsilon \frac{\partial h}{\partial x} f(t, x, y + h(t, x), \epsilon), \quad y(t_0) = \eta(\epsilon) - h(t_0, \xi(\epsilon)) \tag{9.11}$$

The quasi-steady-state of (9.11) is now $y = 0$, which when substituted in (9.10) results in the reduced model (9.8). To analyze (9.11), let us note that $\epsilon \dot{y}$ may remain finite even when ϵ tends to zero and $\dot{y}$ tends to infinity. We set

$$\epsilon \frac{dy}{dt} = \frac{dy}{d\tau}; \text{ hence, } \frac{d\tau}{dt} = \frac{1}{\epsilon}$$

and use $\tau = 0$ as the initial value at $t = t_0$. The new time variable $\tau = (t - t_0)/\epsilon$ is "stretched"; that is, if ϵ tends to zero, τ tends to infinity even for finite t only slightly larger than t_0 by a fixed (independent of ϵ) difference. In the τ time scale, (9.11) is represented by

$$\frac{dy}{d\tau} = g(t, x, y + h(t, x), \epsilon) - \epsilon \frac{\partial h}{\partial t}$$

$$- \epsilon \frac{\partial h}{\partial x} f(t, x, y + h(t, x), \epsilon), \quad y(0) = \eta(\epsilon) - h(t_0, \xi(\epsilon)) \tag{9.12}$$

The variables t and x in the foregoing equation will be slowly varying since, in the τ time scale, they are given by

$$t = t_0 + \epsilon \tau$$

$$x = x(t_0 + \epsilon \tau, \epsilon)$$

Setting $\epsilon = 0$ freezes these variables at $t = t_0$ and $x = \xi_0$, and reduces (9.12) to the autonomous system

$$\frac{dy}{d\tau} = g(t_0, \xi_0, y + h(t_0, \xi_0), 0), \quad y(0) = \eta(0) - h(t_0, \xi_0) \tag{9.13}$$

which has equilibrium at $y = 0$.[3] The frozen parameters (t_0, ξ_0) in (9.13) depend on the given initial time and initial state for the problem under consideration. In our investigation of the stability of the origin of (9.13), we should allow the frozen parameters to take any values in the region of the slowly-varying parameters (t, x). Assume that the solution of the reduced problem $\bar{x}(t)$ is defined for $t \in [0, t_1]$ and $\|\bar{x}(t)\| \leq r_1$ over $[0, t_1]$. Define the set $B_r = \{x \in R^n \mid \|x\| \leq r\}$, where $r > r_1$. We rewrite (9.13) as

$$\frac{dy}{d\tau} = g(t, x, y + h(t, x), 0) \tag{9.14}$$

where $(t, x) \in [0, t_1] \times B_r$ are treated as fixed parameters. We shall refer to (9.14) as the boundary-layer model or the boundary-layer system. Sometimes, we shall also refer to (9.13) as the boundary-layer model. This should cause no confusion since (9.13) is an evaluation of (9.14) for a given initial time and initial state. The model (9.14) is more suitable when we study stability properties of the boundary-layer system. The crucial stability property we need for the boundary-layer system is exponential stability of its origin, uniformly in the frozen parameters. The following definition states this property precisely.

Definition 9.1 *The equilibrium $y = 0$ of the boundary-layer system (9.14) is exponentially stable, uniformly in $(t, x) \in [0, t_1] \times B_r$, if there exist positive constants k, γ, and ρ_0 such that the solutions of (9.14) satisfy*

$$\|y(\tau)\| \leq k\|y(0)\|\exp(-\gamma\tau), \ \forall \ \|y(0)\| < \rho_0, \ \forall \ (t, x) \in [0, t_1] \times B_r, \ \forall \ \tau \geq 0 \tag{9.15}$$

Inequality (9.15) implies that the solutions of (9.14) satisfy $\|y(\tau)\| < k\rho_0$. Therefore, we shall require that the smoothness properties of the various functions hold for all $y \in B_\rho \overset{\text{def}}{=} \{y \in R^m \mid \|y\| \leq \rho\}$, with $\rho \geq \rho_0 k$. Aside from trivial cases where the solution of the boundary layer model may be known in closed form, verification of exponential stability of the origin will have to be done either by linearization or via search for a Lyapunov function. It can be shown (Exercise 9.5) that if the Jacobian matrix $[\partial g/\partial y]$ satisfies the eigenvalue condition

$$Re\left[\lambda\left\{\frac{\partial g}{\partial y}(t, x, h(t, x), 0)\right\}\right] \leq -c < 0, \ \forall \ (t, x) \in [0, t_1] \times B_r \tag{9.16}$$

then there exist constants k, γ, and ρ_0 for which (9.15) is satisfied. This, of course, is a local result; that is, the constant ρ_0 could be very small. Alternatively, it can be shown (Exercise 9.6) that if there is a Lyapunov function $V(t, x, y)$ which satisfies

$$c_1\|y\|^2 \leq V(t, x, y) \leq c_2\|y\|^2 \tag{9.17}$$

---

[3] Recall from Section 5.7 that if the origin of (9.13) is exponentially stable, uniformly in the frozen parameters (t_0, ξ_0), then it will remain exponentially stable when these parameters are replaced by the slowly varying variables (t, x). This observation motivates the stability analysis of (9.13).

$$\frac{\partial V}{\partial y} g(t, x, y + h(t, x), 0) \leq -c_3 \|y\|^2 \tag{9.18}$$

for $(t, x, y) \in [0, t_1] \times B_r \times B_\rho$, then (9.15) is satisfied with the estimates

$$\rho_0 = \rho\sqrt{c_1/c_2}, \quad k = \sqrt{c_2/c_1}, \quad \gamma = c_3/2c_2 \tag{9.19}$$

Theorem 9.1 *Consider the singular perturbation problem (9.6)–(9.7) and let $z = h(t, x)$ be an isolated root of (9.3). Assume that the following conditions are satisfied for all*

$$[t, x, z - h(t, x), \epsilon] \in [0, t_1] \times B_r \times B_\rho \times [0, \epsilon_0]$$

- *The functions f, g, and their first partial derivatives with respect to (x, z, ϵ) are continuous. The function $h(t, x)$ and the Jacobian $\partial g(t, x, z, 0)/\partial z$ have continuous first partial derivatives with respect to their arguments. The initial data $\xi(\epsilon)$ and $\eta(\epsilon)$ are smooth functions of ϵ.*

- *The reduced problem (9.8) has a unique solution $\bar{x}(t)$, defined on $[t_0, t_1]$, and $\|\bar{x}(t)\| \leq r_1 < r$ for all $t \in [t_0, t_1]$.*

- *The origin of the boundary-layer model (9.14) is exponentially stable, uniformly in (t, x). In particular, the solutions of (9.14) satisfy (9.15) with $\rho_0 \leq \rho/k$.*

Then, there exist positive constants μ and ϵ^* such that for all $\|\eta(0) - h(t_0, \xi(0))\| < \mu$ and $0 < \epsilon < \epsilon^*$, the singular perturbation problem (9.6)–(9.7) has a unique solution $x(t, \epsilon)$, $z(t, \epsilon)$ on $[t_0, t_1]$, and

$$x(t, \epsilon) - \bar{x}(t) = O(\epsilon) \tag{9.20}$$

$$z(t, \epsilon) - h(t, \bar{x}(t)) - \hat{y}(t/\epsilon) = O(\epsilon) \tag{9.21}$$

hold uniformly for $t \in [t_0, t_1]$, where $\hat{y}(\tau)$ is the solution of the boundary-layer model (9.13). Moreover, given any $t_b > t_0$, there is $\epsilon^{**} \leq \epsilon^*$ such that

$$z(t, \epsilon) - h(t, \bar{x}(t)) = O(\epsilon) \tag{9.22}$$

holds uniformly for $t \in [t_b, t_1]$ whenever $\epsilon < \epsilon^{**}$. $\diamond$

Proof: Appendix A.14.

This theorem[4] is known as Tikhonov's theorem. Its proof uses exponential stability of the origin of the boundary-layer model to show that

$$\|y(t, \epsilon)\| \leq k \exp\left[\frac{-\alpha(t - t_0)}{\epsilon}\right] \|y(t_0)\| + \epsilon\delta$$

---

[4] There are other versions of this theorem which use slightly different technical assumptions; see, for example, [95, Chapter 1, Theorem 3.1].

This bound is used in (9.10) to prove (9.20), which is plausible since $\int_0^t \exp(-\alpha s/\epsilon)\, ds$ is $O(\epsilon)$. The proof ends with error analysis of (9.11) in the τ time scale to prove (9.21) and (9.22).

Example 9.4 Consider the singular perturbation problem

$$\dot{x} = z, \qquad\qquad x(0) = \xi_0$$

$$\epsilon \dot{z} = -x - z + u(t), \quad z(0) = \eta_0$$

for the DC motor of Example 9.1. Suppose $u(t) = t$ for $t \geq 0$ and we want to solve the state equation over the interval $[0, 1]$. The unique root of (9.3) is $h(t, x) = -x + t$ and the boundary-layer model (9.14) is

$$\frac{dy}{d\tau} = -y$$

Clearly, the origin of the boundary-layer system is globally exponentially stable. The reduced problem

$$\dot{x} = -x + t, \quad x(0) = \xi_0$$

has the unique solution

$$\bar{x}(t) = t - 1 + (1 + \xi_0)\exp(-t)$$

The boundary-layer problem

$$\frac{dy}{d\tau} = -y, \quad y(0) = \eta_0 + \xi_0$$

has the unique solution

$$\hat{y}(\tau) = (\eta_0 + \xi_0)\exp(-\tau)$$

From Theorem 9.1, we have

$$x - [t - 1 + (1 + \xi_0)\exp(-t)] = O(\epsilon)$$

$$z - \left[(\eta_0 + \xi_0)\exp\left(\frac{-t}{\epsilon}\right) + 1 - (1 + \xi_0)\exp(-t)\right] = O(\epsilon)$$

for all $t \in [0, 1]$. The $O(\epsilon)$ approximation of z clearly exhibits a two-time-scale behavior. It starts with a fast transient $(\eta_0 + \xi_0)\exp(-t/\epsilon)$, which is the so-called "boundary-layer" part of the solution. After the decay of this transient, z remains close to $[1 - (1 + \xi_0)\exp(-t)]$, which is the slow ("quasi-steady-state") part of the solution. The two-time-scale behavior is significant only in z, while x is predominantly slow. In fact, x has a fast (boundary-layer) transient, but it is $O(\epsilon)$. Since this system is linear, we can characterize its two-time-scale behavior via modal analysis. It can be easily seen that the system has one slow eigenvalue λ_1, which is $O(\epsilon)$ close to the eigenvalue of the reduced model, that is, $\lambda_1 = -1 + O(\epsilon)$, and one fast eigenvalue $\lambda_2 = \lambda/\epsilon$, where λ is $O(\epsilon)$ close to the eigenvalue of the boundary-layer model, that is, $\lambda_2 = [-1 + O(\epsilon)]/\epsilon$. The exact solutions of x and z will be linear combinations of the slow mode $\exp(\lambda_1 t)$, the fast mode $\exp(\lambda t/\epsilon)$, and a steady-state component due to the input $u(t) = t$. By actually calculating the modal decomposition, it can be verified that the coefficient of the fast mode in x is $O(\epsilon)$. This can be done for linear systems in general; see Exercise 9.13. $\triangle$

Example 9.5 Consider the singular perturbation problem

$$\dot{x} = Ax + Bz, \qquad x(0) = \xi_0$$

$$\epsilon\dot{z} = \psi(u(t) - z - k_2Cx), \quad z(0) = \eta_0$$

for the high-gain feedback system of Example 9.2. Suppose $u(t) = 1$ for $t \geq 0$ and $\psi(\cdot) = \tan^{-1}(\cdot)$. The unique root of (9.3) is $h(t, x) = 1 - k_2Cx$ and the boundary-layer model (9.14) is

$$\frac{dy}{d\tau} = \tan^{-1}(-y) = -\tan^{-1}(y)$$

The Jacobian

$$\left.\frac{\partial g}{\partial y}\right|_{y=0} = -\left.\frac{1}{1 + y^2}\right|_{y=0} = -1$$

is Hurwitz; hence, the origin of the boundary-layer model is exponentially stable. To estimate the constants k, γ, and ρ_0 in (9.15), we search for a Lyapunov function. Taking $V(y) = \frac{1}{2}y^2$, we have

$$\frac{dV}{dy}g = -y\tan^{-1}y \leq -\frac{\tan^{-1}\rho}{\rho}y^2, \quad \forall\, |y| \leq \rho$$

Hence, (9.15) is satisfied with

$$k = 1; \quad \gamma = \frac{\tan^{-1}\rho}{\rho}; \quad \rho_0 = \rho$$

Since the reduced problem

$$\dot{x} = (A - Bk_2C)x + B, \quad x(0) = \xi_0$$

is linear, it is clear that all the assumptions of Theorem 9.1 are satisfied, and we can proceed to approximate x and z in terms of the solutions of the reduced and boundary-layer problems. $\triangle$

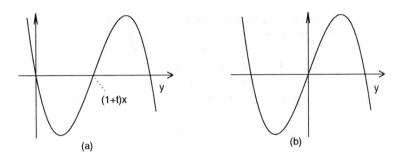

Figure 9.5: RHS of boundary-layer model: (a) $z = -(1+t)x$, (b) $z = 0$.

Example 9.6 Consider the singular perturbation problem

$$\dot{x} = x^2(1+t)/z, \qquad\qquad x(0) = 1$$

$$\epsilon\dot{z} = -[z + (1+t)x]\, z\, [z - (1+t)], \quad z(0) = \eta_0$$

In this case, (9.3) is

$$0 = -[z + (1+t)x]\, z\, [z - (1+t)]$$

and has three isolated roots

$$z = -(1+t)x; \quad z = 0; \quad z = 1+t$$

in the region $\{t \geq 0$ and $x > k > 0\}$. Consider first the root $z = -(1+t)x$. The boundary-layer model (9.14) is

$$\frac{dy}{d\tau} = -y[y - (1+t)x][y - (1+t)x - (1+t)]$$

A sketch of the right-hand side function, Figure 9.5(a), shows that the origin is asymptotically stable with $y < (1+t)x$ as its region of attraction. Taking $V(y) = \frac{1}{2}y^2$, it can be easily verified that V satisfies inequalities (9.17)–(9.18) for $y \leq \rho < (1+t)x$. The reduced problem

$$\dot{x} = -x, \quad x(0) = 1$$

has the unique solution $\bar{x}(t) = \exp(-t)$ for all $t \geq 0$. The boundary-layer problem with $t = 0$ and $x = 1$ is

$$\frac{dy}{d\tau} = -y(y-1)(y-2), \quad y(0) = \eta_0 + 1$$

and has a unique exponentially decaying solution $\hat{y}(\tau)$ for $\eta_0 + 1 \leq 1 - a$, that is, for $\eta_0 \leq -a < 0$ where $a > 0$ can be arbitrarily small. Consider next the root $z = 0$. The boundary-layer model (9.14) is

$$\frac{dy}{d\tau} = -[y + (1+t)x] \, y \, [y - (1+t)]$$

A sketch of the right-hand side function, Figure 9.5(b), shows that the origin is unstable. Hence, Theorem 9.1 does not apply to this case. Finally, the boundary-layer model for the root $z = 1 + t$ is

$$\frac{dy}{d\tau} = -[y + (1+t) + (1+t)x][y + (1+t)]y$$

Similar to the first case, it can be shown that in this case also the origin is exponentially stable uniformly in (t, x). The reduced problem

$$\dot{x} = x^2, \quad x(0) = 1$$

has the unique solution $\bar{x}(t) = 1/(1 - t)$ for all $t \in [0, 1)$. Notice that $\bar{x}(t)$ has a finite escape time at $t = 1$. However, Theorem 9.1 still holds for $t \in [0, t_1]$ with $t_1 < 1$. The boundary-layer problem, with $t = 0$ and $x = 1$,

$$\frac{dy}{d\tau} = -(y + 2)(y + 1)y, \quad y(0) = \eta_0 - 1$$

has a unique exponentially decaying solution $\hat{y}(\tau)$ for $\eta_0 > a > 0$. Among the three roots of (9.3), only two roots, $h = -(1+t)x$ and $h = 1+t$, give rise to valid reduced models. Theorem 9.1 applies to the root $h = -(1 + t)x$ if $\eta_0 < 0$ and to the root $h = 1+t$ if $\eta_0 > 0$. Figures 9.6 and 9.7 show simulation results at $\epsilon = 0.1$. Figure 9.6 shows z for four different values of η_0, two for each reduced model. Figure 9.7 shows the exact and approximate solutions of x and z for $\eta_0 = -0.3$. The trajectories of Figure 9.6 clearly exhibit a two-time-scale behavior. They start with a fast transient of $z(t, \epsilon)$ from η_0 to $\bar{z}(t)$. After the decay of this transient, they remain close to $\bar{z}(t)$. In the case $\eta_0 = -0.3$, the convergence to $\bar{z}(t)$ does not take place within the time interval $[0, 0.2]$. The same case is shown in Figure 9.7 on a longer time interval, where we can see $z(t, \epsilon)$ approaching $\bar{z}(t)$. Figure 9.7 illustrates the $O(\epsilon)$ asymptotic approximation result of Tikhonov's theorem. $\triangle$

9.3 Slow and Fast Manifolds

In this section, we give a geometric view of the two-time-scale behavior of the solutions of (9.1)–(9.2) as trajectories in R^{n+m}. In order to use the concept of invariant

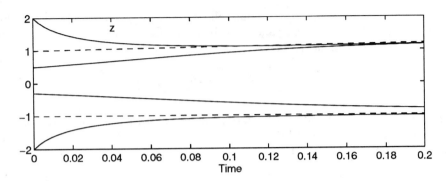

Figure 9.6: Simulation results for z of Example 9.6 at $\epsilon = 0.1$: reduced solution (dashed); exact solution (solid).

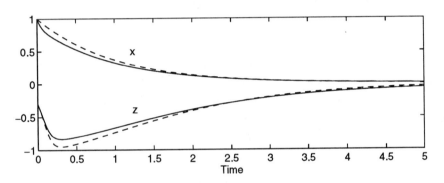

Figure 9.7: Exact (solid) and approximate (dashed) solutions for Example 9.6 at $\epsilon = 0.1$.

manifolds,[5] we restrict our discussion to autonomous systems. Furthermore, to simplify the notation, we take f and g to be independent of ϵ. Thus, we consider the following simpler form of the singularly perturbed system (9.1)–(9.2):

$$\dot{x} = f(x, z) \tag{9.23}$$
$$\epsilon \dot{z} = g(x, z) \tag{9.24}$$

Let $z = h(x)$ be an isolated root of $0 = g(x, z)$ and suppose that the assumptions of Theorem 9.1 are satisfied for this root. The equation $z = h(x)$ describes an n-dimensional manifold in the $(n + m)$-dimensional state space of (x, z). It is an

---

[5] Invariant manifolds have been introduced in Section 4.1.

invariant manifold for the system

$$\dot{x} = f(x, z) \tag{9.25}$$

$$0 = g(x, z) \tag{9.26}$$

for a trajectory of (9.25)–(9.26) that starts in the manifold $z = h(x)$ will remain in the manifold for all future time (for which the solution is defined). The motion in this manifold is described by the reduced model

$$\dot{x} = f(x, h(x))$$

A consequence of Theorem 9.1 is that trajectories of (9.23)–(9.24) which start in an $O(\epsilon)$ neighborhood of $z = h(x)$ will remain within an $O(\epsilon)$ neighborhood of $z = h(x)$. This motivates the question: Is there an analog of the invariant manifold $z = h(x)$ for $\epsilon > 0$? It turns out that, under the assumptions of Theorem 9.1, there is a nearby invariant manifold for (9.23)–(9.24) which lies within an $O(\epsilon)$ neighborhood of $z = h(x)$. We seek the invariant manifold for (9.23)–(9.24) in the form

$$z = H(x, \epsilon) \tag{9.27}$$

where H is a sufficiently smooth (that is, sufficiently many times continuously differentiable) function of x and ϵ. The expression (9.27) defines an n-dimensional manifold, dependent on ϵ, in the $(n + m)$-dimensional state space of (x, z). For $z = H(x, \epsilon)$ to be an invariant manifold of (9.23)–(9.24), it must be true that

$$z(0, \epsilon) - H(x(0, \epsilon), \epsilon) = 0 \Rightarrow z(t, \epsilon) - H(x(t, \epsilon), \epsilon) \equiv 0, \quad \forall\, t \in J \subset [0, \infty)$$

where J is any time interval over which the solution $[x(t, \epsilon), z(t, \epsilon)]$ exists. Differentiating both sides of (9.27) with respect to t, multiplying through by ϵ, and substituting for $\dot{x}$, $\epsilon \dot{z}$, and z from (9.23), (9.24), and (9.27), respectively, we obtain the *manifold condition*

$$0 = g(x, H(x, \epsilon)) - \epsilon \frac{\partial H}{\partial x} f(x, H(x, \epsilon)) \tag{9.28}$$

which $H(x, \epsilon)$ must satisfy for all x in the region of interest and all $\epsilon \in [0, \epsilon_0]$. At $\epsilon = 0$, the partial differential equation (9.28) degenerates into

$$0 = g(x, H(x, 0))$$

which shows that $H(x, 0) = h(x)$. Since $0 = g(x, z)$ may have more than one isolated root $z = h(x)$, we may seek an invariant manifold for (9.23)–(9.24) in the neighborhood of each root. It can be shown[6] that there exist $\epsilon^* > 0$ and a function

---

[6] We will not prove the existence of the invariant manifold here. A proof can be done by a variation of the proof of (the center manifold) Theorem 4.1, which is given in Appendix A.7; see [28, Section 2.7]. A proof under the basic assumptions of Theorem 9.1 can be found in [93].

$H(x, \epsilon)$ satisfying the manifold condition (9.28) for all $\epsilon \in [0, \epsilon^*]$ and

$$H(x, \epsilon) - h(x) = O(\epsilon)$$

for all $\|x\| < r$. The invariant manifold $z = H(x, \epsilon)$ is called a *slow manifold* for (9.23)–(9.24). For each slow manifold, there corresponds a slow model

$$\dot{x} = f(x, H(x, \epsilon)) \tag{9.29}$$

which describes *exactly* the motion on that manifold.

In most cases, we cannot solve the manifold condition (9.28) exactly, but we can approximate $H(x, \epsilon)$ arbitrarily closely as a Taylor series at $\epsilon = 0$. The approximation procedure starts by substituting in (9.28) a Taylor series for $H(x, \epsilon)$,

$$H(x, \epsilon) = H_0(x) + \epsilon H_1(x) + \epsilon^2 H_2(x) + \cdots$$

and calculating $H_0(x)$, $H_1(x)$, and so on, by equating terms of like powers of ϵ. This requires the functions f and g to be continuously differentiable in their arguments a sufficient number of times. It is clear that $H_0(x) = H(x, 0) = h(x)$. The equation for $H_1(x)$ is

$$\frac{\partial g}{\partial z}(x, h(x))H_1(x) = \frac{\partial h}{\partial x} f(x, h(x))$$

and has a unique solution if the Jacobian $[\partial g/\partial z]$ at $z = h(x)$ is nonsingular. The nonsingularity of the Jacobian is implied by the eigenvalue condition (9.16). Similar to H_1, the equations for higher-order terms will be linear and solvable if the Jacobian $[\partial g/\partial z]$ is nonsingular.

To introduce the notion of a fast manifold, we examine (9.23)–(9.24) in the $\tau = t/\epsilon$ time scale. At $\epsilon = 0$, $x(\tau) \equiv x(0)$ while $z(\tau)$ evolves according to

$$\frac{dz}{d\tau} = g(x(0), z)$$

approaching the equilibrium point $z = h(x(0))$. This motion describes trajectories (x, z) in R^{n+m} which, for every given $x(0)$, lie in a fast manifold F_x defined by $x = x(0) = $ constant and rapidly descend to the manifold $z = h(x)$. For ϵ larger than zero but small, the fast manifolds are "foliations" of solutions rapidly approaching the slow manifold. Let us illustrate this picture by two second-order examples.

Example 9.7 Consider the singularly perturbed system

$$\begin{aligned} \dot{x} &= -x + z \\ \epsilon \dot{z} &= \tan^{-1}(1 - z - x) \end{aligned}$$

At $\epsilon = 0$, the slow manifold is $z = h(x) = 1 - x$. The corresponding slow model

$$\dot{x} = -2x + 1$$

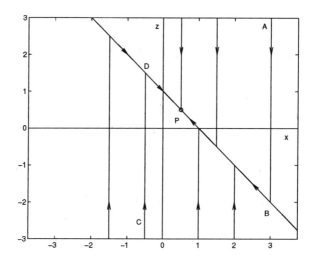

Figure 9.8: Approximate phase portrait of Example 9.7.

has an asymptotically stable equilibrium at $x = \frac{1}{2}$. Therefore trajectories on the manifold $z = 1 - x$ will be heading toward the point $P = (\frac{1}{2}, \frac{1}{2})$, as indicated by the arrow heads in Figure 9.8. Notice that $(\frac{1}{2}, \frac{1}{2})$ is an equilibrium point of the full system. The fast manifolds at $\epsilon = 0$ are parallel to the z-axis, with the trajectories heading toward the slow manifold $z = 1 - x$. With this information, we can construct an approximate phase portrait of the system. For example, a trajectory starting at point A will move down vertically until it hits the manifold $z = 1 - x$ at point B. From B, the trajectory moves along the manifold toward the equilibrium point P. Similarly, a trajectory starting at point C will move up vertically to point D and then along the manifold to the equilibrium point P. For $\epsilon > 0$ but small, the phase portrait of the system will be close to the approximate picture we have drawn at $\epsilon = 0$. Figure 9.9 shows the phase portrait for $\epsilon = 0.1$. The proximity of the two portraits is noticeable.   △

Example 9.8 Consider the Van der Pol equation

$$\frac{d^2v}{ds^2} - \mu(1 - v^2)\frac{dv}{ds} + v = 0$$

and assume that $\mu \gg 1$. With

$$x = -\frac{1}{\mu}\frac{dv}{ds} + v - \frac{1}{3}v^3; \quad z = v$$

as state variables, $t = s/\mu$ as the time variable, and $\epsilon = 1/\mu^2$ the system is repre-

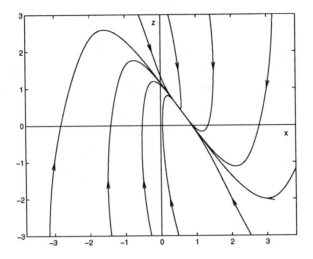

Figure 9.9: Phase portrait of Example 9.7 for $\epsilon = 0.1$.

sented by the standard singularly perturbed model

$$\begin{aligned}\dot{x} &= z \\ \epsilon\dot{z} &= -x + z - \tfrac{1}{3}z^3\end{aligned}$$

We already know by the Poincaré-Bendixson theorem (Example 7.2) that the Van der Pol equation has a stable limit cycle. What we would like to do here is to use singular perturbations to have a better estimate of the location of the limit cycle. At $\epsilon = 0$, we need to solve for the roots $z = h(x)$ of

$$0 = -x + z - \tfrac{1}{3}z^3$$

The curve $-x + z - \tfrac{1}{3}z^3 = 0$, the slow manifold at $\epsilon = 0$, is sketched in Figure 9.10. For $x < -\tfrac{2}{3}$, there is only one root on the branch AB. For $-\tfrac{2}{3} < x < \tfrac{2}{3}$, there are three roots, one on each of the branches AB, BC, and CD. For $x > \tfrac{2}{3}$, there is one root on the branch CD. For roots on the branch AB, the Jacobian

$$\frac{\partial g}{\partial z} = 1 - z^2 < 0, \quad \text{for } z^2 > 1$$

Thus, roots on the branch AB (excluding a neighborhood of point B) are exponentially stable. The same is true for roots on the branch CD (excluding a neighborhood of point C). On the other hand, roots on the branch BC are unstable because they lie in the region $z^2 < 1$. Let us construct an approximate phase portrait using singular perturbations. We divide the state plane into three regions, depending on

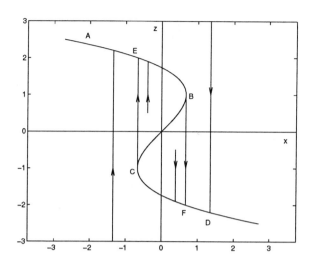

Figure 9.10: Approximate phase portrait of the Van der Pol oscillator.

the value of x. Trajectories starting in the region $x < -\frac{2}{3}$ will move parallel to the z-axis approaching the branch AB of the slow manifold. Trajectories starting in the region $-\frac{2}{3} < x < \frac{2}{3}$ will again be parallel to the z-axis, approaching either the branch AB or the branch CD, depending on the initial value of z. If the initial point is over the branch BC, the trajectory will approach AB; otherwise, it will approach CD. Finally, trajectories starting in the region $x > \frac{2}{3}$ will approach the branch CD. For trajectories on the slow manifold itself, they will move along the manifold. The direction of motion can be determined by inspection of the vector field sign and is indicated in Figure 9.10. In particular, since $\dot{x} = z$, trajectories on the branch AB will be sliding down, while those on the branch CD will be climbing up. There is no point to talk about motion on the branch BC since there are no reduced models corresponding to the unstable roots on that branch. So far, we have formed an approximate phase portrait everywhere except the branch BC and the neighborhood of points B and C. We cannot use singular perturbation theory to predict the phase portrait in these regions. Let us investigate what happens in the neighborhood of B when ϵ is larger than zero but small. Trajectories sliding along the branch AB toward B are actually sliding along the exact slow manifold $z = H(x, \epsilon)$. Since the trajectory is moving toward B, we must have $g < 0$. Consequently, the exact slow manifold must lie above the branch AB. Inspection of the vector field diagram in the neighborhood of B shows that the trajectory crosses the vertical line through B (that is, $x = \frac{2}{3}$) at a point above B. Once the trajectory crosses this line, it belongs to the region of attraction of a stable root on the branch CD; therefore the trajectory moves rapidly in a vertical line toward the branch CD.

By a similar argument, it can be shown that a trajectory moving along the branch CD will cross the vertical line through C at a point below C and then will move vertically toward the branch AB. This completes the picture of the approximate portrait. Trajectories starting at any point are attracted to one of the two branches AB or CD, which they approach vertically. Once on the slow manifold, the trajectory will move toward the closed curve $E - B - F - C - E$, if not already on it, and will cycle through it. The exact limit cycle of the Van der Pol oscillator will lie within an $O(\epsilon)$ neighborhood of this closed curve. The phase portrait for $\epsilon = 0.1$, shown in Figure 9.11, confirms this prediction.

We can also estimate the period of oscillation of the periodic solution. The closed curve $E - B - F - C - E$ has two slow sides and two fast ones. Neglecting the time of the fast transients from B to F and from C to E, we estimate the period of oscillation by $t_{EB} + t_{FC}$. The time t_{EB} can be estimated from the reduced model

$$\dot{x} \;=\; z$$
$$0 \;=\; -x + z - \tfrac{1}{3}z^3$$

Differentiating the second equation with respect to t and equating the expressions for $\dot{x}$ from the two equations, we obtain the equation

$$\dot{z} = \frac{z}{1 - z^2}$$

which, when integrated from E to B, yields $t_{EB} = \tfrac{3}{2} - \ln 2$. The time t_{FC} can be estimated similarly and, due to symmetry, $t_{EB} = t_{FC}$. Thus, the period of oscillation is approximated for small ϵ by $3 - 2\ln 2$. $\triangle$

9.4   Stability Analysis

We consider the autonomous singularly perturbed system

$$\dot{x} \;=\; f(x, z) \qquad\qquad (9.30)$$
$$\epsilon\dot{z} \;=\; g(x, z) \qquad\qquad (9.31)$$

and assume that the origin $(x = 0, z = 0)$ is an isolated equilibrium point and the functions f and g are locally Lipschitz in an open connected set that contains the origin. Consequently,

$$f(0,0) = 0, \quad g(0,0) = 0$$

We want to analyze stability of the origin by examining the reduced and boundary-layer models. Let $z = h(x)$ be an isolated root of

$$0 = g(x, z)$$

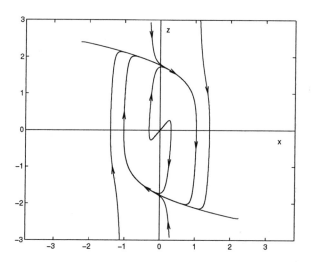

Figure 9.11: Phase portrait of the Van der Pol oscillator for $\epsilon = 0.1$.

defined for all $x \in D_1 \subset R^n$, where D_1 is an open connected set that contains $x = 0$. Suppose $h(0) = 0$. If $z = h(x)$ is the only root of $0 = g$, then it must vanish at the origin since $g(0,0) = 0$. If there are two or more isolated roots, then one of them must vanish at $x = 0$ and that is the one we must work with. It is more convenient to work in the (x,y)-coordinates, where

$$y = z - h(x)$$

because this change of variables shifts the equilibrium of the boundary-layer model to the origin. In the new coordinates, the singularly perturbed system is

$$\dot{x} \;=\; f(x, y + h(x)) \tag{9.32}$$

$$\epsilon \dot{y} \;=\; g(x, y + h(x)) - \epsilon \, \frac{\partial h}{\partial x} f(x, y + h(x)) \tag{9.33}$$

Assuming that $\|h(x)\| \leq \zeta(\|x\|) \ \forall \ x \in D_1$, where ζ is a class $\mathcal{K}$ function, the map $y = z - h(x)$ is stability preserving; that is, the origin of (9.30)–(9.31) is asymptotically stable if and only if the origin of (9.32)–(9.33) is asymptotically stable. The reduced system

$$\dot{x} = f(x, h(x)) \tag{9.34}$$

has equilibrium at $x = 0$, and the boundary-layer system

$$\frac{dy}{d\tau} = g(x, y + h(x)) \tag{9.35}$$

where $\tau = t/\epsilon$ and x is treated as a fixed parameter, has equilibrium at $y = 0$. The main theme of our analysis is to assume that, for each of the two systems, the origin is asymptotically stable and that we have a Lyapunov function that satisfies the conditions of Lyapunov's theorem. In the case of the boundary-layer system, we require asymptotic stability of the origin to hold uniformly in the frozen parameter x. We have already defined what this means in the case of an exponentially stable origin (Definition 9.1). More generally, we say that the origin of (9.35) is asymptotically stable uniformly in x if the solutions of (9.35) satisfy

$$\|y(\tau)\| \le \beta(y(0), \tau), \quad \forall \ \tau \ge 0, \ \forall \ x \in D_1$$

where β is a class $\mathcal{KL}$ function. This conditions will be implied by the conditions we shall impose on the Lyapunov function for (9.35). Viewing the full singularly perturbed system (9.32)–(9.33) as an interconnection of the reduced and boundary-layer systems, we form a composite Lyapunov function candidate for the full system as a linear combination of the Lyapunov functions for the reduced and boundary-layer systems. We then proceed to calculate the derivative of the composite Lyapunov function along the trajectories of the full system and verify, under reasonable growth conditions on f and g, that the composite Lyapunov function will satisfy the conditions of Lyapunov's theorem for sufficiently small ϵ.

Let $V(x)$ be a Lyapunov function for the reduced system (9.34) such that

$$\frac{\partial V}{\partial x} f(x, h(x)) \le -\alpha_1 \psi_1^2(x) \tag{9.36}$$

$\forall \ x \in D_1$, where $\psi_1 : R^n \to R$ is a positive definite function; that is, $\psi_1(0) = 0$ and $\psi_1(x) > 0$ for all $x \in D_1 - \{0\}$. Let $W(x, y)$ be a Lyapunov function for the boundary-layer system (9.35) such that

$$\frac{\partial W}{\partial y} g(x, y + h(x)) \le -\alpha_2 \psi_2^2(y) \tag{9.37}$$

$\forall \ (x, y) \in D_1 \times D_2$, where $D_2 \subset R^m$ is an open connected set that contains $y = 0$, and $\psi_2 : R^m \to R$ is a positive definite function; that is, $\psi_2(0) = 0$ and $\psi_2(y) > 0$ for all $y \in D_2 - \{0\}$. We allow the Lyapunov function W to depend on x since x is a parameter of the system and Lyapunov functions may, in general, depend on the system's parameters. Since x is not a true constant parameter, we have to keep track of the effect of the dependence of W on x. To ensure that the origin of (9.35) is asymptotically stable uniformly in x, we assume that $W(x, y)$ satisfies

$$W_1(y) \le W(x, y) \le W_2(y), \quad \forall \ (x, y) \in D_1 \times D_2 \tag{9.38}$$

for some positive definite continuous functions W_1 and W_2. Now consider the composite Lyapunov function candidate

$$\nu(x, y) = (1 - d)V(x) + dW(x, y), \quad 0 < d < 1 \tag{9.39}$$

where the constant d is to be chosen. Calculating the derivative of ν along the trajectories of the full system (9.32)–(9.33), we obtain

$$
\begin{aligned}
\dot{\nu} &= (1-d)\frac{\partial V}{\partial x}f(x, y+h(x)) + \frac{d}{\epsilon}\frac{\partial W}{\partial y}g(x, y+h(x))\\
&\quad - d\frac{\partial W}{\partial y}\frac{\partial h}{\partial x}f(x, y+h(x)) + d\frac{\partial W}{\partial x}f(x, y+h(x))\\
&= (1-d)\frac{\partial V}{\partial x}f(x, h(x)) + \frac{d}{\epsilon}\frac{\partial W}{\partial y}g(x, y+h(x))\\
&\quad + (1-d)\frac{\partial V}{\partial x}[f(x, y+h(x)) - f(x, h(x))]\\
&\quad + d\left[\frac{\partial W}{\partial x} - \frac{\partial W}{\partial y}\frac{\partial h}{\partial x}\right]f(x, y+h(x))
\end{aligned}
$$

We have represented the derivative $\dot{\nu}$ as the sum of four terms. The first two terms are the derivatives of V and W along the trajectories of the reduced and boundary-layer systems. These two terms are negative definite in x and y, respectively, by inequalities (9.36) and (9.37). The other two terms represent the effect of the interconnection between the slow and fast dynamics, which is neglected at $\epsilon = 0$. These terms are, in general, indefinite. The first of these two terms

$$
\frac{\partial V}{\partial x}[f(x, y+h(x)) - f(x, h(x))]
$$

represents the effect of the deviation of (9.32) from the reduced system (9.34). The other term

$$
\left[\frac{\partial W}{\partial x} - \frac{\partial W}{\partial y}\frac{\partial h}{\partial x}\right]f(x, y+h(x))
$$

represents the deviation of (9.33) from the boundary-layer system (9.34), as well as the effect of freezing x during the boundary-layer analysis. Suppose that these perturbation terms satisfy

$$
\frac{\partial V}{\partial x}[f(x, y+h(x)) - f(x, h(x))] \le \beta_1\psi_1(x)\psi_2(y) \tag{9.40}
$$

and

$$
\left[\frac{\partial W}{\partial x} - \frac{\partial W}{\partial y}\frac{\partial h}{\partial x}\right]f(x, y+h(x)) \le \beta_2\psi_1(x)\psi_2(y) + \gamma\psi_2^2(y) \tag{9.41}
$$

for some nonnegative constants β_1, β_2, and γ. Using inequalities (9.36), (9.37), (9.40), and (9.41), we obtain

$$
\begin{aligned}
\dot{\nu} &\le -(1-d)\alpha_1\psi_1^2(x) - \frac{d}{\epsilon}\alpha_2\psi_2^2(y) + (1-d)\beta_1\psi_1(x)\psi_2(y)\\
&\quad + d\beta_2\psi_1(x)\psi_2(y) + d\gamma\psi_2^2(y)\\
&= -\psi^T(x, y)\Lambda\psi(x, y)
\end{aligned}
$$

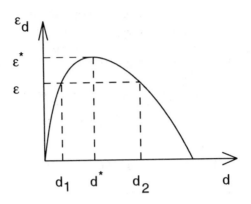

Figure 9.12: Upper bound on ϵ.

where

$$\psi(x,y) = \begin{bmatrix} \psi_1(x) \\ \psi_2(y) \end{bmatrix}$$

and

$$\Lambda = \begin{bmatrix} (1-d)\alpha_1 & -\frac{1}{2}(1-d)\beta_1 - \frac{1}{2}d\beta_2 \\ -\frac{1}{2}(1-d)\beta_1 - \frac{1}{2}d\beta_2 & d\left(\frac{\alpha_2}{\epsilon} - \gamma\right) \end{bmatrix}$$

The right-hand side of the last inequality is a quadratic form in ψ. The quadratic form is negative definite when

$$d(1-d)\alpha_1 \left(\frac{\alpha_2}{\epsilon} - \gamma\right) > \frac{1}{4}[(1-d)\beta_1 + d\beta_2]^2$$

which is equivalent to

$$\epsilon < \frac{\alpha_1\alpha_2}{\alpha_1\gamma + \frac{1}{4d(1-d)}[(1-d)\beta_1 + d\beta_2]^2} \stackrel{\text{def}}{=} \epsilon_d \tag{9.42}$$

The dependence of ϵ_d on d is sketched in Figure 9.12. It can be easily seen that the maximum value of ϵ_d occurs at $d^* = \beta_1/(\beta_1 + \beta_2)$ and is given by

$$\epsilon^* = \frac{\alpha_1\alpha_2}{\alpha_1\gamma + \beta_1\beta_2} \tag{9.43}$$

It follows that the origin of (9.32)–(9.33) is asymptotically stable for all $\epsilon < \epsilon^*$. Theorem 9.2 summarizes our findings.

Theorem 9.2 *Consider the singularly perturbed system* (9.32)–(9.33). *Assume there are Lyapunov functions* $V(x)$ *and* $W(x,y)$ *which satisfy* (9.36)–(9.38) *and* (9.40)–(9.41). *Let* ϵ_d *and* ϵ^* *be defined by* (9.42) *and* (9.43). *Then, the origin of* (9.32)–(9.33) *is asymptotically stable for all* $0 < \epsilon < \epsilon^*$. *Moreover,* $\nu(x,y)$, *defined by* (9.39), *is a Lyapunov function for* $\epsilon \in (0, \epsilon_d)$. $\diamond$

The stability analysis that led to Theorem 9.2 delineates a procedure for constructing Lyapunov functions for the singularly perturbed system (9.32)–(9.33). We start by studying the origins of the reduced and boundary-layer systems, searching for Lyapunov functions $V(x)$ and $W(x,y)$ that satisfy (9.36)–(9.38). Then inequalities (9.40)–(9.41), which we shall refer to as the interconnection conditions, are checked. Several choices of V and W may be tried before one finds the desired Lyapunov functions. As a guideline in that search, notice that the interconnection conditions will be satisfied if

$$\left\| \frac{\partial V}{\partial x} \right\| \leq k_1 \psi_1(x); \quad \| f(x, h(x)) \| \leq k_2 \psi_1(x)$$

$$\| f(x, y + h(x)) - f(x, h(x)) \| \leq k_3 \psi_2(y)$$

$$\left\| \frac{\partial W}{\partial y} \right\| \leq k_4 \psi_2(y); \quad \left\| \frac{\partial W}{\partial x} \right\| \leq k_5 \psi_2(y)$$

A Lyapunov function $V(x)$ that satisfies (9.36) and $\|\partial V/\partial x\| \leq k_1 \psi_1(x)$ is known as a *quadratic-type* Lyapunov function, and ψ_1 is called a *comparison function*. Thus, the search would be successful if we could find quadratic-type Lyapunov functions V and W with comparison functions ψ_1 and ψ_2 such that $f(x, h(x))$ could be majorized by $\psi_1(x)$ and $[f(x, y+h(x)) - f(x, h(x))]$ could be majorized by $\psi_2(y)$. If we succeed in finding V and W, we can conclude that the origin is asymptotically stable for $\epsilon < \epsilon^*$. For a given $\epsilon < \epsilon^*$, there is a range (d_1, d_2), illustrated in Figure 9.12, such that for any $d \in (d_1, d_2)$, the function $\nu(x, y) = (1 - d)V(x) + dW(x, y)$ is a valid Lyapunov function. The freedom in choosing d can be used to achieve other objectives, like improving estimates of the region of attraction.

Example 9.9 The second-order system

$$\dot{x} = f(x, z) = x - x^3 + z$$
$$\epsilon \dot{z} = g(x, z) = -x - z$$

has a unique equilibrium point at the origin. Let $y = z - h(x) = z + x$ and rewrite the system as

$$\dot{x} = -x^3 + y$$
$$\epsilon \dot{y} = -y + \epsilon(-x^3 + y)$$

For the reduced system

$$\dot{x} = -x^3$$

we take $V(x) = \frac{1}{4}x^4$, which satisfies (9.36) with $\psi_1(x) = |x|^3$ and $\alpha_1 = 1$. For the boundary-layer system

$$\frac{dy}{d\tau} = -y$$

we take $W(y) = \frac{1}{2}y^2$, which satisfies (9.38) with $\psi_2(y) = |y|$ and $\alpha_2 = 1$. As for the interconnection conditions (9.40)–(9.41), we have

$$\frac{\partial V}{\partial x}[f(x, y + h(x)) - f(x, h(x))] = x^3 y \leq \psi_1\psi_2$$

and

$$\frac{\partial W}{\partial y} f(x, y + h(x)) = y(-x^3 + y) \leq \psi_1\psi_2 + \psi_2^2$$

Note that $\partial W/\partial x = 0$. Hence, (9.40)–(9.41) are satisfied with $\beta_1 = \beta_2 = \gamma = 1$. Therefore, the origin is asymptotically stable for $\epsilon < \epsilon^* = 0.5$. In fact, since all the conditions are satisfied globally and $\nu(x, y) = (1 - d)V(x) + dW(y)$ is radially unbounded, the origin is globally asymptotically stable for $\epsilon < 0.5$. To see how conservative this bound is, let us note that the characteristic equation of the linearization at the origin is

$$\lambda^2 + \left(\frac{1}{\epsilon} - 1\right)\lambda = 0$$

which shows that the origin is unstable for $\epsilon > 1$. Since our example is a simple second-order system, we may calculate the derivative of the Lyapunov function

$$\nu(x, y) = \frac{1-d}{4}x^4 + \frac{d}{2}y^2$$

along the trajectories of the full singularly perturbed system and see if we can get an upper bound on ϵ less conservative than the one provided by Theorem 9.2.

$$\begin{aligned}
\dot{\nu} &= (1-d)x^3(-x^3 + y) - \frac{d}{\epsilon}y^2 + dy(-x^3 + y) \\
&= -(1-d)x^6 + (1-2d)x^3 y - d\left(\frac{1}{\epsilon} - 1\right)y^2
\end{aligned}$$

It is apparent that the choice $d = \frac{1}{2}$ cancels the cross-product terms and yields

$$\dot{\nu} = -\frac{1}{2}x^6 - \frac{1}{2}\left(\frac{1}{\epsilon} - 1\right)y^2$$

which is negative definite for all $\epsilon < 1$. This estimate is indeed less conservative than that of Theorem 9.2. In fact, it is the actual range of ϵ for which the origin is asymptotically stable.   △

Example 9.10 The system

$$
\begin{aligned}
\dot{x} &= -x + z \\
\epsilon\dot{z} &= \tan^{-1}(1 - x - z)
\end{aligned}
$$

has an equilibrium point at $(\frac{1}{2}, \frac{1}{2})$. The change of variables

$$
\tilde{x} = x - \tfrac{1}{2}; \quad \tilde{z} = z - \tfrac{1}{2}
$$

shifts the equilibrium to the origin. To simplify the notation, let us drop the tilde and write the state equation as

$$
\begin{aligned}
\dot{x} &= -x + z \\
\epsilon\dot{z} &= -\tan^{-1}(x + z)
\end{aligned}
$$

The equation

$$
0 = -\tan^{-1}(x + z)
$$

has a unique root $z = h(x) = -x$. We apply the change of variables $y = z + x$ to obtain

$$
\begin{aligned}
\dot{x} &= -2x + y \\
\epsilon\dot{y} &= -\tan^{-1} y + \epsilon(-2x + y)
\end{aligned}
$$

For the reduced system, we take $V(x) = \frac{1}{2}x^2$, which satisfies (9.36) with $\alpha_1 = 2$ and $\psi_1(x) = |x|$. For the boundary-layer system, we take $W(y) = \frac{1}{2}y^2$ and (9.38) takes the form

$$
\frac{dW}{dy}[-\tan^{-1} y] = -y\tan^{-1} y \le -\frac{\tan^{-1}\rho}{\rho}y^2
$$

for all $y \in D_2 = \{y \mid |y| < \rho\}$. Thus, (9.38) is satisfied with $\alpha_2 = (\tan^{-1}\rho)/\rho$ and $\psi_2(y) = |y|$. The interconnection conditions (9.40)–(9.41) are satisfied globally with $\beta_1 = 1$, $\beta_2 = 2$, and $\gamma = 1$. Hence, the origin is asymptotically stable for all $\epsilon < \epsilon^* = (\tan^{-1}\rho)/2\rho$. In fact, the origin is exponentially stable since ν is quadratic in (x, y) and the negative definite upper bound on $\dot{\nu}$ is quadratic in (x, y). $\triangle$

The Lyapunov analysis we have just presented can be extended to nonautonomous systems. We shall not give the details here;[7] instead, we consider the case of exponential stability and use converse Lyapunov theorems to prove a result of conceptual importance.

---

[7] A detailed treatment of the nonautonomous case can be found in [95, Section 7.5].

Theorem 9.3 *Consider the singularly perturbed system*

$$\dot{x} = f(t, x, z, \epsilon) \tag{9.44}$$

$$\epsilon\dot{z} = g(t, x, z, \epsilon) \tag{9.45}$$

Assume that the following assumptions are satisfied for all

$$(t, x, \epsilon) \in [0, \infty) \times B_r \times [0, \epsilon_0]$$

- $f(t, 0, 0, \epsilon) = 0$ *and* $g(t, 0, 0, \epsilon) = 0$.

- *The equation*

$$0 = g(t, x, z, 0)$$

  has an isolated root $z = h(t, x)$ *such that* $h(t, 0) = 0$.

- *The functions* f, g, *and* h *and their partial derivatives up to order* 2 *are bounded for* $z - h(t, x) \in B_\rho$.

- *The origin of the reduced system*

$$\dot{x} = f(t, x, h(t, x), 0)$$

  is exponentially stable.

- *The origin of the boundary-layer system*

$$\frac{dy}{d\tau} = g(t, x, y + h(t, x), 0)$$

  is exponentially stable, uniformly in (t, x).

Then, there exists $\epsilon^* > 0$ *such that for all* $\epsilon < \epsilon^*$, *the origin of* (9.44)–(9.45) *is exponentially stable.* ◇

Proof: By Theorem 3.12, there is a Lyapunov function $V(t, x)$ for the reduced system which satisfies

$$c_1\|x\|^2 \leq V(t, x) \leq c_2\|x\|^2$$

$$\frac{\partial V}{\partial t} + \frac{\partial V}{\partial x} f(t, x, h(t, x), 0) \leq -c_3\|x\|^2$$

$$\left\|\frac{\partial V}{\partial x}\right\| \leq c_4\|x\|$$

for some positive constants c_i, $i = 1, \ldots, 4$, and for $x \in B_{r_0}$ where $r_0 \leq r$. By Lemma 5.11, there is a Lyapunov function $W(t, x, y)$ for the boundary-layer system which satisfies

$$b_1\|y\|^2 \leq W(t, x, y) \leq b_2\|y\|^2$$

$$\frac{\partial W}{\partial y} g(t, x, y + h(t, x), 0) \leq -b_3 \|y\|^2$$

$$\left\| \frac{\partial W}{\partial y} \right\| \leq b_4 \|y\|$$

$$\left\| \frac{\partial W}{\partial t} \right\| \leq b_5 \|y\|^2; \quad \left\| \frac{\partial W}{\partial x} \right\| \leq b_6 \|y\|^2$$

for some positive constants b_i, $i = 1, \ldots, 6$, and for $y \in B_{\rho_0}$ where $\rho_0 \leq \rho$. Apply the change of variables

$$. \ y = z - h(t, x)$$

to transform (9.44)–(9.45) into

$$\dot{x} = f(t, x, y + h(t, x), \epsilon) \tag{9.46}$$

$$\epsilon \dot{y} = g(t, x, y + h(t, x), \epsilon) - \epsilon \frac{\partial h}{\partial t}$$

$$- \epsilon \frac{\partial h}{\partial x} f(t, x, y + h(t, x), \epsilon) \tag{9.47}$$

We are going to use

$$\nu(t, x, y) = V(t, x) + W(t, x, y)$$

as a Lyapunov function candidate for the system (9.46)–(9.47). In preparation for that, let us note the following estimates in the neighborhood of the origin. Since f and g vanish at the origin for all $\epsilon \in [0, \epsilon_0]$, they are Lipschitz in ϵ linearly in the state (x, y). In particular,

$$\|f(t, x, y + h(t, x), \epsilon) - f(t, x, y + h(t, x), 0)\| \leq \epsilon L_1(\|x\| + \|y\|)$$

$$\|g(t, x, y + h(t, x), \epsilon) - g(t, x, y + h(t, x), 0)\| \leq \epsilon L_2(\|x\| + \|y\|)$$

Also,

$$\|f(t, x, y + h(t, x), 0) - f(t, x, h(t, x), 0)\| \leq L_3 \|y\|$$

$$\|f(t, x, h(t, x), 0)\| \leq L_4 \|x\|$$

$$\left\| \frac{\partial h}{\partial t} \right\| \leq k_1 \|x\|; \quad \left\| \frac{\partial h}{\partial x} \right\| \leq k_2$$

where we have used the fact that $f(t, x, h(t, x), 0)$ and $h(t, x)$ vanish at $x = 0$ for all t. Using these estimates and the properties of the functions V and W, it can be verified that the derivative of ν along the trajectories of (9.46)–(9.47) satisfies inequality of the form

$$\dot{\nu} \leq -a_1 \|x\|^2 + \epsilon a_2 \|x\|^2 - \frac{a_3}{\epsilon} \|y\|^2 + a_4 \|y\|^2$$

$$+ a_5 \|x\| \ \|y\| + a_6 \|x\| \ \|y\|^2 + a_7 \|y\|^3$$

For all $\|y\| \leq \rho_0$, this inequality simplifies to

$$\dot{\nu} \leq -a_1\|x\|^2 + \epsilon a_2\|x\|^2 - \frac{a_3}{\epsilon}\|y\|^2 + a_8\|y\|^2 + 2a_9\|x\|\,\|y\|$$

$$= -\begin{bmatrix} \|x\| \\ \|y\| \end{bmatrix}^T \begin{bmatrix} a_1 - \epsilon a_2 & -a_9 \\ -a_9 & (a_3/\epsilon) - a_8 \end{bmatrix} \begin{bmatrix} \|x\| \\ \|y\| \end{bmatrix}$$

Thus, there exists $\epsilon^* > 0$ such that for all $0 < \epsilon < \epsilon^*$, we have

$$\dot{\nu} \leq -2\gamma\nu$$

for some $\gamma > 0$. It follows that

$$\nu(t, x(t), y(t)) \leq \exp[-2\gamma(t - t_0)]\nu(t_0, x(t_0), y(t_0))$$

and, from the properties of V and W,

$$\left\| \begin{array}{c} x(t) \\ y(t) \end{array} \right\| \leq K_1 \exp[-\gamma(t - t_0)] \left\| \begin{array}{c} x(t_0) \\ y(t_0) \end{array} \right\|$$

Since $y = z - h(t, x)$ and $\|h(t, x)\| \leq k_2\|x\|$, we obtain

$$\left\| \begin{array}{c} x(t) \\ z(t) \end{array} \right\| \leq K_2 \exp[-\gamma(t - t_0)] \left\| \begin{array}{c} x(t_0) \\ z(t_0) \end{array} \right\|$$

which completes the proof of the theorem.   □

Theorem 9.3 is conceptually important because it establishes robustness of exponential stability to unmodeled fast (high frequency) dynamics. Quite often in the analysis of dynamical systems, we use reduced-order models obtained by neglecting small "parasitic" parameters which increase the order of the model. This model order reduction can be represented as a singular perturbation problem where the full singularly perturbed model represents the actual system with the parasitic parameters and the reduced model is the simplified model used in the analysis. It is quite reasonable to assume that the boundary-layer model has an exponentially stable origin. In fact, if the dynamics associated with the parasitic elements were unstable we should not have neglected them in the first place. The technicalities of assuming exponential stability instead of only asymptotic stability, or assuming that exponential stability holds uniformly, are quite reasonable in most applications. It is enough to mention that all these technicalities will automatically hold when the fast dynamics are linear. Suppose that we analyzed stability of the origin of the reduced-order model and concluded that it is exponentially stable. Theorem 9.3 assures us that the origin of the actual system will be exponentially stable, provided the neglected fast dynamics are sufficiently fast. The following example illustrates how this robustness property arises in control design.

Example 9.11 Consider the feedback stabilization of the system

$$
\begin{aligned}
\dot{x} &= f(t, x, v) \\
\epsilon \dot{z} &= Az + Bu \\
v &= Cz
\end{aligned}
$$

where $f(t, 0, 0) = 0$ and A is a Hurwitz matrix. The system has an open-loop equilibrium point at the origin, and the control task is to design a state feedback control law to stabilize the origin. The linear part of this model represents actuator dynamics which are, typically, much faster than the plant dynamics represented by the nonlinear equation $\dot{x} = f$. To simplify the design problem, we may neglect the actuator dynamics by setting $\epsilon = 0$ and substituting $v = -CA^{-1}Bu$ in the plant equation. To simplify the notation, let us assume that $-CA^{-1}B = I$ and write the reduced-order model as

$$
\dot{x} = f(t, x, u)
$$

We use this model to design a state feedback control law $u = \gamma(t, x)$ such that the origin of the closed-loop model

$$
\dot{x} = f(t, x, \gamma(t, x))
$$

is exponentially stable. We shall refer to this model as the nominal closed-loop system. Suppose we have designed such a control law. Will this control stabilize the actual system with the actuator dynamics included? When the control is applied to the actual system, the closed-loop equation is

$$
\begin{aligned}
\dot{x} &= f(t, x, Cz) \\
\epsilon \dot{z} &= Az + B\gamma(t, x)
\end{aligned}
$$

We have a singular perturbation problem, where the full singularly perturbed model is the actual closed-loop system and the reduced model is the nominal closed-loop system. By design, the origin of the reduced model is exponentially stable. The boundary-layer model

$$
\frac{dy}{d\tau} = Ay
$$

is independent of (t, x) and its origin is exponentially stable since A is a Hurwitz matrix. Assuming that f and γ are smooth enough to satisfy the conditions of Theorem 9.3, we conclude that the origin of the actual closed-loop system is exponentially stable for sufficiently small ϵ. This result legitimizes the ad hoc model simplification process of neglecting the actuator dynamics. $\triangle$

9.5   Singular Perturbation on the Infinite Interval

Theorem 9.1 is valid only on order $O(1)$ time intervals. This fact can be easily seen from the proof of the theorem. In particular, it is established in (A.36) that

$$\|x(t, \epsilon) - \bar{x}(t)\| \le \epsilon k_3 [1 + t_1 - t_0] \exp[L_6(t_1 - t_0)]$$

For any finite t_1, the foregoing estimate is $O(\epsilon)$ but is not $O(\epsilon)$ uniformly in t for all $t \ge t_0$. For the latter statement to hold, we need to show that

$$\|x(t, \epsilon) - \bar{x}(t)\| \le \epsilon k, \quad \forall\, t \in [t_0, \infty)$$

This can be done if some additional conditions are added to ensure exponential stability of the solutions of the reduced problem. The following theorem extends Theorem 9.1 to the infinite-time interval.

Theorem 9.4 *Consider the singular perturbation problem (9.6)–(9.7) and let $z = h(t, x)$ be an isolated root of (9.3). Assume that the following conditions are satisfied for all*

$$(t, x, z - h(t, x), \epsilon) \in [0, \infty) \times B_r \times B_\rho \times [0, \epsilon_0]$$

- *The functions f, g, and their first partial derivatives with respect to (x, z, ϵ) are continuous and bounded. The function $h(t, x)$ and the Jacobian $[\partial g(t, x, z, 0)/\partial z]$ have bounded first partial derivatives with respect to their arguments. The Jacobian $[\partial f(t, x, h(t, x), 0)/\partial x]$ has bounded first partial derivatives with respect to x. The initial data $\xi(\epsilon)$ and $\eta(\epsilon)$ are smooth functions of ϵ.*

- *The origin of the reduced model (9.5) is exponentially stable.*

- *The origin of the boundary-layer model (9.14) is exponentially stable, uniformly in (t, x). In particular, the solutions of (9.14) satisfy (9.15) with $\rho_0 \le \rho/k$.*

Then, there exist positive constants μ_1, μ_2, and ϵ^* such that for all

$$\|\xi(0)\| < \mu_1, \quad \|\eta(0) - h(t_0, \xi(0))\| < \mu_2, \quad \text{and} \quad 0 < \epsilon < \epsilon^*$$

the singular perturbation problem (9.6)–(9.7) has a unique solution $x(t, \epsilon)$, $z(t, \epsilon)$ defined for all $t \ge t_0 \ge 0$, and

$$x(t, \epsilon) - \bar{x}(t) = O(\epsilon) \tag{9.48}$$

$$z(t, \epsilon) - h(t, \bar{x}(t)) - \hat{y}(t/\epsilon) = O(\epsilon) \tag{9.49}$$

hold uniformly for $t \in [t_0, \infty)$, *where* $\bar{x}(t)$ *and* $\hat{y}(\tau)$ *are the solutions of the reduced and boundary-layer problems* (9.8) *and* (9.13). *Moreover, given any* $t_b > t_0$, *there is* $\epsilon^{**} \leq \epsilon^*$ *such that*

$$z(t, \epsilon) - h(t, \bar{x}(t)) = O(\epsilon) \tag{9.50}$$

holds uniformly for $t \in [t_b, \infty)$ *whenever* $\epsilon < \epsilon^{**}$.   ◇

Proof: Appendix A.15.

Example 9.12 Consider the singular perturbation problem

$$\dot{x} = 1 - x - \frac{1}{2}[\psi(x+z) + \psi(x-z)], \qquad x(0) = \xi_0$$

$$\epsilon\dot{z} = -(\epsilon+2)z - \frac{\epsilon}{2}[\psi(x+z) - \psi(x-z)], \quad z(0) = \eta_0$$

for the electric circuit of Example 9.3, and assume that

$$\psi(v) = a\left[\exp\left(\frac{v}{b}\right) - 1\right], \quad a > 0, \ b > 0$$

We have dropped the subscript r as we copied these equations from Example 9.3. The differentiability and boundedness assumptions of Theorem 9.4 are satisfied on any compact set of (x, z). The reduced model

$$\dot{x} = 1 - x - a\left[\exp\left(\frac{x}{b}\right) - 1\right]$$

has a unique equilibrium point at $x = p^*$, where p^* is the unique root of

$$0 = 1 - p^* - a\left[\exp\left(\frac{p^*}{b}\right) - 1\right]$$

It can be easily seen that $0 < p^* < 1$. The Jacobian

$$\left.\frac{df}{dx}\right|_{x=p^*} = -1 - \frac{a}{b}\exp\left(\frac{p^*}{b}\right) < -1$$

is negative; hence, the equilibrium point $x = p^*$ is exponentially stable. The change of variables $\tilde{x} = x - p^*$ shifts the equilibrium to the origin. The boundary-layer model

$$\frac{dz}{d\tau} = -2z$$

is independent of x, and its origin is exponentially stable. Thus, all the conditions of Theorem 9.4 are satisfied and the estimates (9.48)–(9.50), with $h = 0$, hold for all $t \geq 0$.   △

Example 9.13 Consider the adaptive control of a plant represented by the second-order transfer function

$$\tilde{P}(s) = \frac{k_p}{(s - a_p)(\epsilon s + 1)}$$

where a_p, $k_p > 0$ and $\epsilon > 0$ are unknown parameters. The parameter ϵ represents a small "parasitic" time constant. Suppose we have neglected ϵ and simplified the transfer function to

$$P(s) = \frac{k_p}{s - a_p}$$

We may now proceed to design the adaptive controller for this first-order transfer function. It is shown in Section 1.1.6 that a model reference adaptive controller is given by

$$
\begin{aligned}
u(t) &= \theta_1(t)r(t) + \theta_2(t)y_p(t) \\
\dot{\theta}_1 &= -\gamma[y_p(t) - y_m(t)]r(t) \\
\dot{\theta}_2 &= -\gamma[y_p(t) - y_m(t)]y_p(t)
\end{aligned}
$$

where y_p, u, r, and y_m are the plant output, the control input, the reference input, and the reference model output, respectively. With (the first-order model of) the plant and the reference model represented by

$$\dot{y}_p = a_p y_p + k_p u$$

and

$$\dot{y}_m = a_m y_m + k_m r, \qquad k_m > 0$$

it is shown in Section 1.1.6 that the closed-loop adaptive control system is represented by the third-order state equation

$$
\begin{aligned}
\dot{e}_o &= a_m e_o + k_p \phi_1 r(t) + k_p \phi_2 (e_o + y_m(t)) \\
\dot{\phi}_1 &= -\gamma e_o r(t) \\
\dot{\phi}_2 &= -\gamma e_o (e_o + y_m(t))
\end{aligned}
$$

where

$$e_o = y_p - y_m, \quad \phi_1 = \theta_1 - \theta_1^*, \quad \phi_2 = \theta_2 - \theta_2^*$$

and $\theta_1^* = k_m/k_p$, $\theta_2^* = (a_m - a_p)/k_p$. Define

$$x = [\ e_o \quad \phi_1 \quad \phi_2\ ]^T$$

as the state vector and rewrite the state equation as

$$\dot{x} = f_0(t, x)$$

where $f_0(t, 0) = 0$. We shall refer to this third-order state equation as the nominal adaptive control system. This is the model we use in the stability analysis. We assume that the origin of this model is exponentially stable.[8] When the adaptive controller is applied to the actual system, the closed-loop system will be different from this nominal model. Let us represent the situation as a singular perturbation problem. The actual second-order model of the plant can be represented by the singularly perturbed model

$$\dot{y}_p = a_p y_p + k_p z$$
$$\epsilon \dot{z} = -z + u$$

By repeating the derivations of Sections 1.1.6, it can be seen that the actual adaptive control system is represented by the singularly perturbed model

$$\dot{x} = f_0(t, x) + K(z - h(t, x))$$
$$\epsilon \dot{z} = -z + h(t, x)$$

where

$$h(t, x) = u = (\theta_1^* + \phi_1)r(t) + (\theta_2^* + \phi_2)(e_o + y_m(t)), \quad K = [k_p, 0, 0]^T$$

The signal $y_m(t)$ is the output of a Hurwitz transfer function driven by $r(t)$. Therefore, it has the same smoothness and boundedness properties of $r(t)$. In particular, if $r(t)$ has continuous and bounded derivatives up to order N, the same will be true for $y_m(t)$. Let us analyze this singularly perturbed system. At $\epsilon = 0$, we have $z = h(t, x)$ and the reduced model is

$$\dot{x} = f_0(t, x)$$

which is the closed-loop model of the nominal adaptive control system. We have assumed that the origin of this model is exponentially stable. The boundary-layer model

$$\frac{dy}{d\tau} = -y$$

is independent of (t, x) and its origin is exponentially stable. If the reference input $r(t)$ is bounded and has a bounded derivative $\dot{r}(t)$, then all the assumptions of

---

[8] It is shown in Section 13.4 that this will be the case under a persistence of excitation condition; see, in particular, Example 13.15 where it is shown that the origin will be exponentially stable if $r(t) = 1 + a \sin \omega t$. A word of caution at this point: note that our analysis in this example assumes that $r(t)$ is fixed and studies the asymptotic behavior of the system for small ϵ. As we fix the value of ϵ at some small numerical value, our underlying assumption puts a constraint on $r(t)$; in particular, on the input frequency ω. If we start to increase the input frequency ω, we may reach a point where the conclusions of this example are no longer valid because a high-frequency input may violate the slowly varying nature of the slow variables x. For example, the signal $\dot{r}(t)$, which is of order $O(\omega)$, may violate our assumption that $\dot{r}$ is of order $O(1)$ with respect to ϵ.

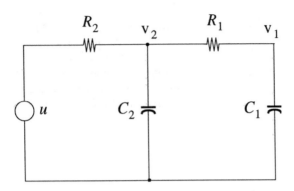

Figure 9.13: Exercises 9.1 and 9.2.

Theorem 9.4 are satisfied on any compact set of (x, z). Let $\bar{x}$ denote the solution of the nominal adaptive control system and $x(t, \epsilon)$ denote the solution of the actual adaptive control system, both starting from the same initial state. By Theorem 9.4, we conclude that there exists $\epsilon^* > 0$ such that for all $0 < \epsilon < \epsilon^*$,

$$x(t, \epsilon) - \bar{x}(t) = O(\epsilon)$$

where $O(\epsilon)$ holds uniformly in t for all $t \geq t_0$. Once again, this result shows robustness to unmodeled fast dynamics. It is different, however, from the robustness result we have seen in Example 9.11. In that example, the origin $(x = 0, z = 0)$ is an equilibrium point of the full singularly perturbed system, as required in Theorem 9.3, and we have shown that it is exponentially stable for sufficiently small ϵ. In the current example, the origin $(x = 0, z = 0)$ is not an equilibrium point of the full singularly perturbed system, although $x = 0$ is an equilibrium point of the reduced system. In this situation, the best we can expect to show for the actual adaptive system is closeness of its response to that of the nominal system and uniform ultimate boundedness of its solutions; we have achieved both. Notice that, since $\bar{x}(t)$ tends to zero as t tends to infinity, the solutions of the actual adaptive system are uniformly ultimately bounded with an $O(\epsilon)$ ultimate bound. $\triangle$

9.6   Exercises

Exercise 9.1 Consider the RC circuit of Figure 9.13 and suppose that the capacitor C_2 is small relative to C_1 while $R_1 = R_2 = R$. Represent the system in the standard singularly perturbed form.

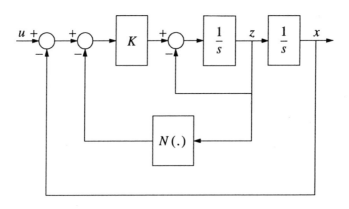

Figure 9.14: Exercise 9.4.

Exercise 9.2 Consider the RC circuit of Figure 9.13 and suppose that the resistor R_1 is small relative to R_2, while $C_1 = C_2 = C$. Represent the system in the standard singularly perturbed form.

Exercise 9.3 Consider the tunnel diode circuit of Section 1.1.2 and suppose that the inductance L is relatively small so that the time constant L/R is much smaller than the time constant CR. Represent the system as a standard singularly perturbed model with $\epsilon = L/CR^2$.

Exercise 9.4 ([95]) The feedback system of Figure 9.14 has a high-gain amplifier K and a nonlinear element $N(\cdot)$. Represent the system as a standard singularly perturbed model with $\epsilon = 1/K$.

Exercise 9.5 Show that if the Jacobian $[\partial g/\partial y]$ satisfies the eigenvalue condition (9.16), then there exist constants k, γ, and ρ_0 for which inequality (9.15) is satisfied.

Exercise 9.6 Show that if there is a Lyapunov function satisfying (9.17)–(9.18), then inequality (9.15) is satisfied with the estimates (9.19).

Exercise 9.7 Consider the singular perturbation problem

$$\begin{aligned}
\dot{x} &= x^2 + z, & x(0) &= \xi \\
\epsilon\dot{z} &= x^2 - z + 1, & z(0) &= \eta
\end{aligned}$$

(a) Find an $O(\epsilon)$ approximation of x and z on the time interval $[0, 1]$.

(b) Let $\xi = \eta = 0$. Simulate x and z for

$$(1)\ \epsilon = 0.1 \quad \text{and} \quad (2)\ \epsilon = 0.05$$

and compare with the approximation derived in part (a). In carrying out the computer simulation note that the system has a finite escape time shortly after $t = 1$.

Exercise 9.8 Consider the singular perturbation problem

$$
\begin{aligned}
\dot{x} &= x + z, \quad x(0) = \xi \\
\epsilon \dot{z} &= -\frac{2}{\pi} \tan^{-1}\left(\frac{\pi}{2}(2x + z)\right), \quad z(0) = \eta
\end{aligned}
$$

(a) Find an $O(\epsilon)$ approximation of x and z on the time interval $[0, 1]$.

(b) Let $\xi = \eta = 1$. Simulate x and z for

$$(1)\ \epsilon = 0.2 \quad \text{and} \quad (2)\ \epsilon = 0.1$$

and compare with the approximation derived in part (a).

Exercise 9.9 Consider the singularly perturbed system

$$
\begin{aligned}
\dot{x} &= z \\
\epsilon \dot{z} &= -x - \epsilon z - \exp(z) + 1 + u(t)
\end{aligned}
$$

Find the reduced and boundary-layer models and analyze the stability properties of the boundary-layer model.

Exercise 9.10 ([95]) Consider the singularly perturbed system

$$
\begin{aligned}
\dot{x} &= \frac{x^2 t}{z} \\
\epsilon \dot{z} &= -(z + xt)(z - 2)(z - 4)
\end{aligned}
$$

(a) How many reduced models can this system have?

(b) Investigate boundary-layer stability for each reduced model.

(c) Let $x(0) = 1$ and $z(0) = a$. Find an $O(\epsilon)$ approximation of x and z on the time interval $[0, 1]$ for all values of a in the interval $[-2, 6]$.

Exercise 9.11 ([95]) Find the exact slow manifold of the singularly perturbed system

$$
\begin{aligned}
\dot{x} &= x z^3 \\
\epsilon \dot{z} &= -z - x^{4/3} + \frac{4}{3}\epsilon x^{16/3}
\end{aligned}
$$

Exercise 9.12 ([95]) How many slow manifolds does the following system have? Which of these manifolds will attract trajectories of the system?

$$\dot{x} = -xz$$
$$\epsilon \dot{z} = -(z - \sin^2 x)(z - e^{ax})(z - 2e^{2ax}), \quad a > 0$$

Exercise 9.13 ([95]) Consider the linear autonomous singularly perturbed system

$$\dot{x} = A_{11}x + A_{12}z$$
$$\epsilon \dot{z} = A_{21}x + A_{22}z$$

where $x \in R^n$, $z \in R^m$, and A_{22} is a Hurwitz matrix.

(a) Show that for sufficiently small ϵ, the system has an exact slow manifold $z = -L(\epsilon)x$, where L satisfies the algebraic equation

$$-\epsilon L(A_{11} - A_{12}L) = A_{21} - A_{22}L$$

(b) Show that the change of variables

$$\eta = z + L(\epsilon)x$$

transforms the system into a block triangular form.

(c) Show that the eigenvalues of the system cluster into a group of n slow eigenvalues of order $O(1)$ and m fast eigenvalues of order $O(1/\epsilon)$.

(d) Let $H(\epsilon)$ be the solution of the linear equation

$$\epsilon(A_{11} - A_{12}L)H - H(A_{22} + \epsilon LA_{12}) + A_{12} = 0$$

Show that the similarity transformation

$$\begin{bmatrix} \xi \\ \eta \end{bmatrix} = \begin{bmatrix} I - \epsilon HL & -\epsilon H \\ L & I \end{bmatrix} \begin{bmatrix} x \\ z \end{bmatrix}$$

transforms the system into the block modal form

$$\dot{\xi} = A_s(\epsilon)\xi; \quad \epsilon\dot{\eta} = A_f(\epsilon)\eta$$

where the eigenvalues of A_s and A_f/ϵ are, respectively, the slow and fast eigenvalues of the full singularly perturbed system.

(e) Show that the component of the fast mode in x is $O(\epsilon)$.

(f) Give an independent proof of Tikhonov's theorem in the current case.

Exercise 9.14 Consider the linear singularly perturbed system

$$\dot{x} = A_{11}x + A_{12}z + B_1u(t), \quad x(0) = \xi$$
$$\epsilon\dot{z} = A_{21}x + A_{22}z + B_2u(t), \quad z(0) = \eta$$

where $x \in R^n$, $z \in R^m$, $u \in R^p$, A_{22} is Hurwitz, and $u(t)$ is uniformly bounded for all $t \geq 0$. Let $\bar{x}(t)$ be the solution of the reduced system

$$\dot{x} = A_0 x + B_0 u(t), \quad x(0) = \xi$$

where $A_0 = A_{11} - A_{12}A_{22}^{-1}A_{21}$, and $B_0 = B_1 - A_{12}A_{22}^{-1}B_2$.

(a) Show that $x(t, \epsilon) - \bar{x}(t) = O(\epsilon)$ on any compact interval $[0, t_1]$.

(b) Show that if A_0 is Hurwitz, then $x(t, \epsilon) - \bar{x}(t) = O(\epsilon)$ for all $t \geq 0$.

Hint: Use the transformation of the previous Exercise.

Exercise 9.15 Consider the singularly perturbed system

$$\dot{x}_1 = x_2$$
$$\dot{x}_2 = -x_2 + z$$
$$\epsilon\dot{z} = \tan^{-1}(1 - x_1 - z)$$

(a) Find the reduced and boundary-layer models.

(b) Analyze the stability properties of the boundary-layer model.

(c) Let $x_1(0) = x_2(0) = z(0) = 0$. Find an $O(\epsilon)$ approximation of the solution. Using a numerical algorithm, calculate the exact and approximate solutions over the time interval $[0, 10]$ for $\epsilon = 0.1$.

(d) Show that the system has a unique equilibrium point and analyze its stability using the singular perturbation approach. Is the equilibrium point asymptotically stable? Is it globally asymptotically stable? Is it exponentially stable? Calculate an upper bound ϵ^* on ϵ for which your stability analysis is valid.

(e) Investigate the validity of the approximation on the infinite time interval.

Exercise 9.16 Repeat Exercise 9.15 for the singularly perturbed system

$$\dot{x} = -2x + x^2 + z$$
$$\epsilon\dot{z} = x - x^2 - z$$

In part (c), let $x(0) = z(0) = 1$ and the time interval be $[0, 5]$.

Exercise 9.17 Repeat Exercise 9.15 for the singularly perturbed system

$$\dot{x} = xz^3$$
$$\epsilon\dot{z} = -2x^{4/3} - 2z$$

In part (c), let $x(0) = z(0) = 1$ and the time interval be $[0, 1]$.

Exercise 9.18 Repeat Exercise 9.15 for the singularly perturbed system

$$\dot{x} = -x^3 + \tan^{-1}(z)$$
$$\epsilon\dot{z} = -x - z$$

In part (c), let $x(0) = -1$, $z(0) = 2$ and the time interval be $[0, 2]$.

Exercise 9.19 Repeat Exercise 9.15 for the singularly perturbed system

$$\dot{x} = -x + z_1 + z_2 + z_1 z_2$$
$$\epsilon\dot{z}_1 = -z_1$$
$$\epsilon\dot{z}_2 = -z_2 - (x + z_1 + xz_1)$$

In part (c), let $x(0) = z_1(0) = z_2(0) = 1$ and the time interval be $[0, 2]$.

Exercise 9.20 Consider the field-controlled DC motor of Exercise 1.12. Let $v_a = V = $ constant, and $v_f = U = $ constant.

(a) Show that the system has a unique equilibrium point at

$$I_f = \frac{U}{R_f}, \quad I_a = \frac{c_3 V_a}{c_3 R_a + c_1 c_2 U^2 / R_f^2}, \quad \Omega = \frac{c_2 V_a U / R_f}{c_3 R_a + c_1 c_2 U^2 / R_f^2}$$

We will use (I_f, I_a, Ω) as a nominal operating point.

(b) It is typical that the armature circuit time constant $T_a = L_a/R_a$ is much smaller than the field circuit time constant $T_f = L_f/R_f$ and the mechanical time constant. Therefore, the system can be modeled as a singularly perturbed system with i_f and Ω as the slow variables and i_a as the fast variable. Taking $x_1 = i_f/I_f$, $x_2 = \omega/\Omega$, $z = i_a/I_a$, $u = v_f/U$, and $\epsilon = T_a/T_f$, and using $t' = t/T_f$ as the time variable, show that the singularly perturbed model is given by

$$\dot{x}_1 = -x_1 + u$$
$$\dot{x}_2 = a(x_1 z - x_2)$$
$$\epsilon\dot{z} = -z - bx_1 x_2 + c$$

where $a = L_f c_3/R_f J$, $b = c_1 c_2 U^2/c_3 R_a R_f^2$, $c = V_a/I_a R_a$, and $(\dot{\cdot})$ denotes the derivative with respect to t'.

(c) Find the reduced and boundary-layer models.

(d) Analyze the stability properties of the boundary-layer model.

(e) Find an $O(\epsilon)$ approximation of x and z.

(f) Investigate the validity of the approximation on the infinite time interval.

(f) Using a numerical algorithm, calculate the exact and approximate solutions for a unit step input at u and zero initial states over the time interval $[0, 10]$ for $\epsilon = 0.2$ and $\epsilon = 0.1$. Use the numerical data $c_1 = c_2 = \sqrt{2} \times 10^{-2}\ N - m/A$, $c_3 = 6 \times 10^{-6}\ N - m - s/\text{rad}$, $J = 10^{-6}\ N - m - s^2/\text{rad}$, $R_a = R_f = 1\Omega$, $L_f = 0.2\ H$, $V_a = 1\ V$, and $V_f = 0.2\ V$.

Exercise 9.21 ([95]) Consider the singularly perturbed system

$$\dot{x} = -\eta(x) + az$$
$$\epsilon\dot{z} = -\frac{x}{a} - z$$

where a is a positive constant and $\eta(\cdot)$ is a smooth nonlinear function which satisfies

$$\eta(0) = 0 \quad \text{and} \quad x\eta(x) > 0, \quad \text{for } x \in (-\infty, b) - \{0\}$$

for some $b > 0$. Investigate the stability of the origin for small ϵ using the singular perturbation approach.

Exercise 9.22 ([95]) The singularly perturbed system

$$\dot{x} = -2x^3 + z^2$$
$$\epsilon\dot{z} = x^3 - \tan z$$

has an isolated equilibrium point at the origin.

(a) Show that asymptotic stability of the origin cannot be established by linearization.

(b) Using the singular perturbation approach, show that the origin is asymptotically stable for $\epsilon \in (0, \epsilon^*)$. Estimate ϵ^* and the region of attraction.

Exercise 9.23 ([95]) Let the assumptions of Theorem 9.2 hold with $\psi_1(x) = \|x\|$ and $\psi_2(y) = \|y\|$ and suppose, in addition, that $V(x)$ and $W(x, y)$ satisfy

$$k_1\|x\|^2 \leq V(x) \leq k_2\|x\|^2$$
$$k_3\|y\|^2 \leq W(x, y) \leq k_4\|y\|^2$$

$\forall\ (x, y) \in D_1 \times D_2$, where k_1 to k_4 are positive constants. Show that the conclusions of Theorem 9.2 hold with exponential stability replacing asymptotic stability.

Exercise 9.24 ([172]) Consider the singularly perturbed system

$$\dot{x} = f(x, y)$$
$$\epsilon\dot{y} = Ay + \epsilon g_1(x, y)$$

where A is Hurwitz and f and g_1 are sufficiently smooth functions that vanish at the origin. Suppose there is a Lyapunov function $V(x)$ such that $\frac{\partial V}{\partial x} f(x, 0) \le -\alpha_1 \phi(x)$ in the domain of interest, where $\alpha_1 > 0$ and $\phi(x)$ is positive definite. Let P be the solution of the Lyapunov equation $PA + A^T P = -I$ and take $W(y) = y^T Py$.

(a) Suppose f and g_1 satisfy the inequalities

$$\|g_1(x, 0)\|_2 \le k_1 \phi^{1/2}(x), \quad k_1 \ge 0$$

$$\frac{\partial V}{\partial x}[f(x, y) - f(x, 0)] \le k_2 \phi^{1/2}(x)\|y\|_2, \quad k_2 \ge 0$$

in the domain of interest. Using the Lyapunov function candidate $\nu(x, y) = (1 - d)V(x) + dW(y)$, $0 < d < 1$ and the analysis preceding Theorem 9.2, show that the origin is asymptotically stable for sufficiently small ϵ.

(b) As an alternative to Theorem 9.2, suppose f and g_1 satisfy the inequalities

$$\|g_1(x, 0)\|_2 \le k_3 \phi^a(x), \quad k_3 \ge 0, \quad 0 < a \le \tfrac{1}{2}$$

$$\frac{\partial V}{\partial x}[f(x, y) - f(x, 0)] \le k_4 \phi^b(x)\|y\|_2^c, \quad k_4 \ge 0, \quad 0 < b < 1, \quad c = \frac{1 - b}{a}$$

in the domain of interest. Using the Lyapunov function candidate $\nu(x, y) = V(x) + (y^T Py)^\gamma$ where $\gamma = 1/2a$, show that the origin is asymptotically stable for sufficiently small ϵ.

Hint: Use Young's inequality

$$uw \le \frac{1}{\mu}|u|^p + \mu^{\frac{1}{p-1}}|w|^{\frac{p}{p-1}}, \forall\, u \ge 0,\ w \ge 0,\ \mu > 0,\ p > 1$$

to show that $\dot{\nu} \le -c_1 \phi - c_2 \|y\|_2^{2\gamma}$. Then show that the coefficients c_1 and c_2 can be made positive for sufficiently small ϵ.

(c) Give an example where the interconnection conditions of part (b) are satisfied but not those of part (a).

Exercise 9.25 ([91]) Consider the multiparameter singularly perturbed system

$$\dot{x} = f(x, z_1, \ldots, z_m)$$
$$\epsilon_i \dot{z}_i = \eta_i(x) + \sum_{j=1}^{m} a_{ij} z_j, \quad i = 1, \ldots, m$$

where x is an n-dimensional vector, z_i's are scalar variables, and ϵ_i's are small positive parameters. Let $\epsilon = \max_i \epsilon_i$. This equation can be rewritten as

$$\dot{x} = f(x, z)$$
$$\epsilon D \dot{z} = \eta(x) + Az$$

where z and η are m-dimensional vectors whose components are z_i and η_i, respectively, A is an $m \times m$ matrix whose elements are a_{ij}, and D is an $m \times m$ diagonal matrix whose ith diagonal element is ϵ_i/ϵ. The diagonal elements of D are positive and bounded by one. Suppose that the origin of the reduced system

$$\dot{x} = f(x, -A^{-1}\eta(x))$$

is asymptotically stable and there is a Lyapunov function $V(x)$ which satisfies the conditions of Theorem 9.2. Suppose further that there is a diagonal matrix P with positive elements such that

$$PA + A^T P = -Q, \quad Q > 0$$

Using

$$\nu(x, z) = (1 - d)V(x) + d(z + A^{-1}\eta(x))^T PD(z + A^{-1}\eta(x)), \quad 0 < d < 1$$

as a Lyapunov function candidate, analyze the stability of the origin. State and prove a theorem similar to Theorem 9.2 for the multiparameter case. Your conclusion should allow the parameters ϵ_i's to be arbitrary, subject only to a requirement that they be sufficiently small.

Exercise 9.26 ([95]) The singularly perturbed system

$$\dot{x}_1 = (a + x_2)x_1 + 2z, \quad a > 0$$
$$\dot{x}_2 = bx_1^2, \quad b > 0$$
$$\epsilon \dot{z} = -x_1 x_2 - z$$

has an equilibrium set $\{x_1 = 0, \ z = 0\}$. Study the asymptotic behavior of the solution, for small ϵ, using LaSalle's invariance principle.
Hint: The asymptotic behavior of the reduced model has been studied in Example 3.10. Use a composite Lyapunov function and proceed as in Section 9.4. Notice, however, that Theorem 9.2 does not apply to the current problem.

Exercise 9.27 ([90]) Consider the linear autonomous singularly perturbed system

$$\dot{x} = A_{11}x + A_{12}z + B_1 u$$
$$\epsilon \dot{z} = A_{21}x + A_{22}z + B_2 u$$
$$y = C_1 x + C_2 z + Du$$

where u and y are the input and output vectors, respectively. Suppose that A_{22} is Hurwitz. The reduced model, obtained by setting $\epsilon = 0$, may be used as a simplified model for the purpose of designing a stabilizing output feedback controller. Let the reduced model be represented by

$$\dot{x} = A_0 x + B_0 u, \qquad y = C_0 x + D_0 u$$

and suppose that

$$\dot{v} = F v + G y, \qquad u = H v + J y$$

is a stabilizing output feedback controller; that is, the closed-loop system composed of the reduced model and the controller has an asymptotically stable origin. Study the robustness of stability to singular perturbations when (1) $J = 0$ and (2) $J \neq 0$. In the second case, consider the following example in your study:

$$\begin{aligned} \dot{x} &= x + z + u \\ \epsilon \dot{z} &= x - z + u \\ y &= x + z \end{aligned}$$

with the controller $u = 2y$.

Exercise 9.28 Consider the singularly perturbed system

$$\begin{aligned} \dot{x} &= f(x, u(t)) + B(u(t))[z - g(x, u(t))] \\ \epsilon \dot{z} &= -z + g(x, u(t)), \qquad x \in R^n, \; z \in R^m, \; u \in R^q \end{aligned}$$

where where $B(u)$ and $G(x, u)$ are continuously differentiable and $f(x, u)$ is twice continuously differentiable in a domain $D_x \times D_u \subset R^n \times R^q$. Suppose there is a continuously differentiable function $h : D_u \to D_x$ such that $0 = f(h(\alpha), \alpha)$, $\forall \, \alpha \in D_u$, and $A(\alpha) = \frac{\partial f}{\partial x}(h(\alpha), \alpha)$ is Hurwitz, uniformly in α, for all $\alpha \in \Gamma$ where $\Gamma \subset D_u$ is a compact set. Furthermore, suppose that $u(t)$ is continuously differentiable, $u(t) \in \Gamma$, and $\|\dot{u}(t)\| \leq \mu$ for all $t \geq 0$.

(a) Verify that the change of variables $w = x - h(u(t))$ and $y = z - g(x, u(t))$ transforms the system into

$$\begin{aligned} \dot{w} &= f(w + h(u), u) + B(u)y - \frac{\partial h}{\partial u}\dot{u} \\ \epsilon \dot{y} &= -y - \epsilon \frac{\partial g}{\partial x}[f(w + h(u), u) + B(u)y] - \epsilon \frac{\partial g}{\partial u}\dot{u} \end{aligned}$$

(b) Let $P(\alpha)$ be the unique positive definite solution of the Lyapunov equation $P(\alpha)A(\alpha) + A^T(\alpha)P(\alpha) = -I$. Verify that $V(w, \alpha) = w^T P(\alpha) w$ satisfies inequalities (5.66)–(5.69) in a neighborhood of $w = 0$.

(c) Let $\nu(w, y, u) = V(w, u) + \frac{1}{2}y^T y$. Verify that there exist $\epsilon^* > 0$ and $\mu^* > 0$ such that for all $0 < \epsilon < \epsilon^*$ and $0 \leq \mu < \mu^*$, the inequality

$$\dot{\nu} \leq -\alpha\nu + \beta\sqrt{\nu}\,\|\dot{u}\|$$

is satisfied in a neighborhood of $(w, y) = (0, 0)$ for some $\alpha > 0$ and $\beta \geq 0$.

(d) Show that there exist positive constants ρ_1, ρ_2, ρ_3, ρ_4 such that if $0 < \epsilon < \rho_1$, $\mu < \rho_2$, $\|x(0) - h(u(0))\| < \rho_3$, and $\|z(0) - g(x(0), u(0))\| < \rho_4$, then $x(t)$ and $z(t)$ are uniformly bounded for all $t \geq 0$, and $x(t) - h(u(t))$ is uniformly ultimately bounded by $k\mu$ for some $k > 0$.

(e) Show that if (in addition to the previous assumptions) $u(t) \to u_\infty$ and $\dot{u}(t) \to 0$ as $t \to \infty$, then $x(t) \to h(u_\infty)$ as $t \to \infty$.

Hint: In part (b), use Lemma 5.12 and in parts (d) and (e), apply the comparison lemma

Exercise 9.29 Apply Theorem 9.4 to study the asymptotic behavior of the system

$$\begin{aligned}
\dot{x} &= -x + z - \sin t \\
\epsilon\dot{z} &= -z + \sin t
\end{aligned}$$

as $t \to \infty$.

Chapter 10

Analysis of Feedback Systems

In this chapter, we present some tools for the analysis of feedback systems, represented by the feedback connection of Figure 10.1. The interest in feedback systems stems from two facts. First, many physical systems can be represented by the feedback connection of Figure 10.1. For control engineers, this should be an obvious statement since feedback remains the basic concept in automatic control systems, whereby certain variables are measured, processed, and fed back to manipulate the control inputs. The interest in feedback systems, however, is not limited to control systems. Many physical systems which are not originally modeled in the feedback form may be represented in this form to take advantage of the special tools available for feedback systems. Second, the feedback structure brings in new challenges and opportunities, which have made the analysis of feedback systems a hot research topic over many years. The challenge is exemplified by the fact that the feedback connection of two stable systems could be unstable. In the face of such challenge comes the opportunity to take advantage of the feedback structure itself to come up with conditions that will guard against instability. Both the small gain approach of Section 10.2 and the passivity approach of Section 10.3 provide such conditions. In the first case, it is done by ensuring that a signal attenuates as it flows through the feedback loop, while in the latter it is done by ensuring passivity of the system. Another opportunity that arises with feedback systems is the possibility of modeling the system in such a way that the component in the forward path is a linear time-invariant system while the component in the feedback path is a memoryless nonlinearity (as shown in Figure 10.2); thus, localizing the effect of nonlinearity to the feedback path and taking advantage of linearity in dealing with the forward path. The absolute stability techniques of Section 10.1 and the describing function method of Section 10.4 exploit this structure and come up with graphical methods that build on the classical linear feedback analysis tools like Nyquist criterion and Nyquist plot.

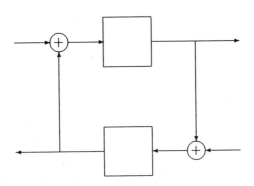

Figure 10.1: Feedback connection.

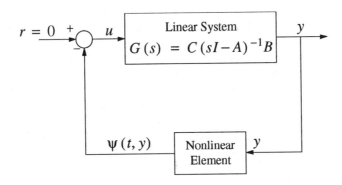

Figure 10.2: Feedback connection of a linear system and a nonlinear element.

10.1 Absolute Stability

Many nonlinear physical systems can be represented as a feedback connection of a linear dynamical system and a nonlinear element, as shown in Figure 10.2. The process of representing a system in this form depends on the particular system involved. For instance, in the case in which a control system's only nonlinearity is in the form of a relay or actuator/sensor nonlinearity, there is no difficulty in representing the system in the feedback form of Figure 10.2. In other cases, the representation may be less obvious. We assume that the external input $r = 0$ and study the behavior of the unforced system, represented by

$$\dot{x} = Ax + Bu \tag{10.1}$$

$$y = Cx \tag{10.2}$$

$$u = -\psi(t, y) \tag{10.3}$$

where $x \in R^n$, $u, y \in R^p$, (A, B) is controllable, (A, C) is observable, and $\psi : [0, \infty) \times R^p \rightarrow R^p$ is a memoryless, possibly time-varying nonlinearity which is piecewise continuous in t and locally Lipschitz in y. The transfer function matrix of the linear system is given by

$$G(s) = C(sI - A)^{-1}B \tag{10.4}$$

which is a square strictly proper transfer function.[1] The controllability-observability assumptions ensure that $\{A, B, C\}$ is a minimal realization of $G(s)$. From linear system theory, we know that for any rational strictly proper $G(s)$, a minimal realization always exists. The nonlinearity $\psi(\cdot, \cdot)$ will be required to satisfy a *sector condition*. To describe the sector condition, let us start with the case when $G(s)$ is a single-input–single-output transfer function; that is, $p = 1$. In this case, $\psi : [0, \infty) \times R \rightarrow R$ satisfies a sector condition (or is a sector nonlinearity) if there are constants α, β, a, and b (with $\beta > \alpha$ and $a < 0 < b$) such that

$$\alpha y^2 \leq y\psi(t, y) \leq \beta y^2, \quad \forall\, t \geq 0, \; \forall\, y \in [a, b] \tag{10.5}$$

If (10.5) holds for all $y \in (-\infty, \infty)$, we say that the sector condition holds globally. Figure 10.3 shows sketches of sector nonlinearities when (10.5) holds globally as well as when it holds only on a finite domain. Sometimes, the $\leq$ sign on one or both sides of (10.5) may be strengthened to a $<$ sign. When (10.5) holds globally, it is customary to say that $\psi(\cdot, \cdot)$ belongs to a sector $[\alpha, \beta]$. If strict inequality is required on either side of (10.5), we say that $\psi(\cdot, \cdot)$ belongs to a sector $(\alpha, \beta]$, $[\alpha, \beta)$, or (α, β), with obvious implications. It is simple to show that (10.5) implies the inequality

$$[\psi(t, y) - \alpha y][\psi(t, y) - \beta y] \leq 0, \quad \forall\, t \geq 0, \; \forall\, y \in [a, b] \tag{10.6}$$

In the multivariable case when u and y are p-dimensional vectors, the form of the sector condition is more involved. Let us consider first the case when the nonlinearity $\psi(t, y)$ is decoupled in the sense that $\psi_i(t, y)$ depends only on y_i; that is,

$$\psi(t, y) = \begin{bmatrix} \psi_1(t, y_1) \\ \psi_2(t, y_2) \\ \vdots \\ \psi_p(t, y_p) \end{bmatrix} \tag{10.7}$$

---

[1] Many of the results of this section can be stated in the more general case when $y = Cx + Du$; that is, when $G(s)$ is proper but not strictly proper. The analysis will be basically the same, except for the need to verify that the feedback connection is well posed. For convenience, we shall deal with the case $D \neq 0$ only in the exercises.

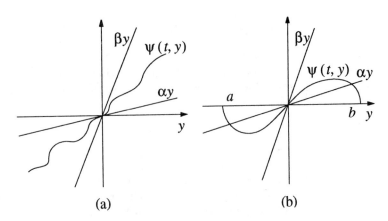

Figure 10.3: (a) Global sector nonlinearities; (b) local sector nonlinearities.

Suppose each component $\psi_i(\cdot, \cdot)$ satisfies the sector condition (10.5) with constants α_i, β_i, a_i, and b_i. Taking

$$K_{\min} = \text{diag}(\alpha_1, \alpha_2, \ldots, \alpha_p), \quad K_{\max} = \text{diag}(\beta_1, \beta_2, \ldots, \beta_p)$$

and

$$\Gamma = \{y \in R^p \mid a_i \leq y_i \leq b_i\}$$

it can be easily seen that

$$[\psi(t, y) - K_{\min}y]^T[\psi(t, y) - K_{\max}y] \leq 0, \quad \forall\, t \geq 0,\ \forall\, y \in \Gamma \qquad (10.8)$$

Note that $K_{\max} - K_{\min}$ is a positive definite symmetric (diagonal) matrix. Inequality (10.8) may hold for more general multivariable nonlinearities, with different matrices $K_{\min}$ and $K_{\max}$. For example, suppose that $\psi(t, y)$ satisfies the inequality

$$\|\psi(t, y) - Ly\|_2 \leq \gamma\|y\|_2, \quad \forall\, t \geq 0,\ \forall\, y \in \Gamma \subset R^p$$

where the interior of Γ is connected and contains the origin. Taking

$$K_{\min} = L - \gamma I; \quad K_{\max} = L + \gamma I$$

we can write

$$[\psi(t, y) - K_{\min}y]^T[\psi(t, y) - K_{\max}y] = \|\psi(t, y) - Ly\|_2^2 - \gamma^2\|y\|_2^2 \leq 0$$

Note, again, that $K_{\max} - K_{\min}$ is a positive definite symmetric (diagonal) matrix. The following definition of a sector nonlinearity covers many nonlinearities of interest, both in the single variable as well as the multivariable cases.

Definition 10.1 *A memoryless nonlinearity* $\psi : [0, \infty) \times R^p \to R^p$ *is said to satisfy a sector condition if*

$$[\psi(t, y) - K_{\min}y]^T[\psi(t, y) - K_{\max}y] \leq 0, \quad \forall \, t \geq 0, \; \forall \, y \in \Gamma \subset R^p \tag{10.9}$$

for some real matrices $K_{\min}$ *and* $K_{\max}$, *where* $K = K_{\max} - K_{\min}$ *is a positive definite symmetric matrix and the interior of* Γ *is connected and contains the origin. If* $\Gamma = R^p$, *then* $\psi(\cdot, \cdot)$ *satisfies the sector condition globally, in which case it is said that* $\psi(\cdot, \cdot)$ *belongs to a sector* $[K_{\min}, K_{\max}]$. *If* (10.9) *holds with strict inequality, then* $\psi(\cdot, \cdot)$ *is said to belong to a sector* $(K_{\min}, K_{\max})$.

For all nonlinearities satisfying the sector condition (10.9), the origin $x = 0$ is an equilibrium point of the system (10.1)–(10.3). The problem of interest here is to study the stability of the origin, not for a given nonlinearity but rather for a class of nonlinearities that satisfy a given sector condition. If we succeed in showing that the origin is asymptotically stable for all nonlinearities in the sector, the system is said to be absolutely stable. The problem was originally formulated by Lure, and is sometimes called *Lure's problem*. Traditionally, absolute stability has been defined for the case when the origin is globally asymptotically stable. To keep up with that tradition, we shall use the phrase "absolute stability" when the sector condition is satisfied globally and the origin is globally asymptotically stable. Otherwise, we shall use the phrase "absolute stability with a finite domain."

Definition 10.2 *Consider the system* (10.1)–(10.3), *where* $\psi(\cdot, \cdot)$ *satisfies a sector condition per Definition 10.1. The system is absolutely stable if the origin is globally uniformly asymptotically stable for any nonlinearity in the given sector. It is absolutely stable with a finite domain if the origin is uniformly asymptotically stable.*

We shall investigate asymptotic stability of the origin using two Lyapunov function candidates. The first one is a simple quadratic function

$$V(x) = x^T P x; \quad P = P^T > 0$$

and the second one is a function of the form

$$V(x) = x^T P x + \eta \int_0^y \psi^T(\sigma) K \, d\sigma; \quad P = P^T > 0, \; \eta \geq 0$$

which is known as a Lure-type Lyapunov function. In the latter case, we assume that the nonlinearity ψ is time invariant and satisfies some conditions to ensure that the integral is well defined and nonnegative. In both cases, we start the derivation with P undetermined. Then, we determine conditions under which there exists P such that the derivative of $V(x)$ along the trajectories of the system is negative definite for all nonlinearities that satisfy a given sector condition. In Sections 10.1.1

and 10.1.2, these conditions take the form of a frequency-domain condition on a certain transfer function, while in Section 10.1.3 the search for P is replaced by an optimization problem whose solution is sought numerically. The frequency-domain conditions of Sections 10.1.1 and 10.1.2 use the concept of a positive real transfer function and a key lemma that relates that concept to the existence of Lyapunov functions.

Definition 10.3 *A $p \times p$ proper rational transfer function matrix $Z(s)$ is called positive real if*

- *all elements of $Z(s)$ are analytic for $Re[s] > 0$,*

- *any pure imaginary pole of any element of $Z(s)$ is a simple pole and the associated residue matrix of $Z(s)$ is positive semidefinite Hermitian, and*

- *for all real ω for which $j\omega$ is not a pole of any element of $Z(s)$, the matrix $Z(j\omega) + Z^T(-j\omega)$ is positive semidefinite.*

The transfer function $Z(s)$ is called strictly positive real [2] *if $Z(s-\epsilon)$ is positive real for some $\epsilon > 0$.*

When $p = 1$, the frequency-domain condition of Definition 10.3 reduces to $ReZ(j\omega) \geq 0$, $\forall\ \omega \in R$, which is equivalent to the condition that the Nyquist plot of of $Z(j\omega)$ lies in the closed right-half complex plane.

The following lemma gives an equivalent characterization of strictly positive real transfer functions.

Lemma 10.1 *Let $Z(s)$ be a $p \times p$ proper rational transfer function matrix, and suppose $\det[Z(s) + Z^T(-s)]$ is not identically zero.* [3] *Then, $Z(s)$ is strictly positive real if and only if*

- *$Z(s)$ is Hurwitz; that is, poles of all elements of $Z(s)$ have negative real parts,*

- *$Z(j\omega) + Z^T(-j\omega) > 0$, $\forall\ \omega \in R$, and*

- *one of the following three conditions is satisfied:*

  1. $Z(\infty) + Z^T(\infty) > 0$;

  2. $Z(\infty) + Z^T(\infty) = 0$ and $\lim_{\omega \to \infty} \omega^2[Z(j\omega) + Z^T(-j\omega)] > 0$;

---

[2] The definition of strictly positive real transfer functions is not uniform in the literature; see [182] for various definitions and the relationship between them.

[3] Equivalently, $Z(s) + Z^T(-s)$ has a normal rank p over the field of rational functions of s.

3. $Z(\infty)+Z^T(\infty) \geq 0$ (*but not zero nor nonsingular*) *and there exist positive constants* σ_0 *and* ω_0 *such that*

$$\omega^2 \sigma_{\min}[Z(j\omega) + Z^T(-j\omega)] \geq \sigma_0, \quad \forall \, |\omega| \geq \omega_0$$

◇

Proof: Appendix A.16.

Example 10.1 The transfer function $Z(s) = 1/s$ is positive real since it has no poles in $Re[s] > 0$, a simple pole at $s = 0$ whose residue is 1, and

$$Re[Z(j\omega)] = Re \left[\frac{1}{j\omega}\right] = 0, \quad \forall \, \omega \neq 0$$

It is not strictly positive real since $1/(s - \epsilon)$ has a pole in $Re[s] > 0$ for any $\epsilon > 0$. The transfer function $Z(s) = 1/(s + a)$ with $a > 0$ is positive real, since it has no poles in $Re[s] \geq 0$ and

$$Re[Z(j\omega)] = \frac{a}{\omega^2 + a^2} > 0, \quad \forall \, \omega \in R$$

Since this is so for every $a > 0$, we see that for any $\epsilon \in (0, a)$ the transfer function $Z(s - \epsilon) = 1/(s + a - \epsilon)$ will be positive real. Hence, $Z(s) = 1/(s + a)$ is strictly positive real. The same conclusion can be drawn from Lemma 10.1 by noting that

$$\lim_{\omega \to \infty} \omega^2 Re[Z(j\omega)] = \lim_{\omega \to \infty} \frac{\omega^2 a}{\omega^2 + a^2} = a > 0$$

The transfer function

$$Z(s) = \frac{1}{s^2 + s + 1}$$

is not positive real since

$$Re[Z(j\omega)] = \frac{1 - \omega^2}{(1 - \omega^2)^2 + \omega^2} < 0, \quad \forall \, \omega > 1$$

In fact, any scalar transfer function of a relative degree[4] higher than one is not positive real because the Nyquist plot of $Z(j\omega)$ will not be confined to the right-half complex plane. Consider the 2×2 transfer function matrix

$$Z(s) = \frac{1}{s + 1} \begin{bmatrix} 1 & 1 \\ 1 & 1 \end{bmatrix}$$

---

[4] The relative degree of a proper rational function $Z(s) = n(s)/d(s)$ is deg d - deg n.

We cannot apply Lemma 10.1 because $\det[Z(s) + Z^T(-s)] \equiv 0 \; \forall \; s$. However, $Z(s)$ is strictly positive real, as can be seen by checking the conditions of Definition 10.3. Note that, for $\epsilon < 1$, the poles of the elements of $Z(s - \epsilon)$ will be in $Re[s] < 0$ and

$$Z(j\omega - \epsilon) + Z^T(-j\omega - \epsilon) = \frac{2(1 - \epsilon)}{\omega^2 + (1 - \epsilon)^2} \begin{bmatrix} 1 & 1 \\ 1 & 1 \end{bmatrix}$$

will be positive semidefinite for all $\omega \in R$. Similarly, it can be seen that the 2×2 transfer function matrix

$$Z(s) = \frac{1}{s + 1} \begin{bmatrix} s & 1 \\ -1 & 2s + 1 \end{bmatrix}$$

is strictly positive real. This time, however, $\det[Z(s) + Z^T(-s)]$ is not identically zero and we can apply Lemma 10.1 to arrive at the same conclusion by noting that $Z(\infty) + Z^T(\infty)$ is nonsingular and

$$Z(j\omega) + Z^T(-j\omega) = \frac{2}{\omega^2 + 1} \begin{bmatrix} \omega^2 & -j\omega \\ j\omega & 2\omega^2 + 1 \end{bmatrix}$$

is positive definite for all $\omega \in R$. Finally, the 2×2 transfer function matrix

$$Z(s) = \begin{bmatrix} \frac{s}{s+1} & \frac{1}{s+2} \\ \frac{-1}{s+2} & \frac{2}{s+1} \end{bmatrix}$$

has

$$Z(\infty) + Z^T(\infty) = \begin{bmatrix} 2 & 0 \\ 0 & 0 \end{bmatrix}$$

It can be verified that

$$Z(j\omega) + Z^T(-j\omega) = \begin{bmatrix} \frac{2\omega^2}{1+\omega^2} & \frac{-2j\omega}{4+\omega^2} \\ \frac{2j\omega}{4+\omega^2} & \frac{4}{1+\omega^2} \end{bmatrix}$$

is positive definite for all $\omega \in R$. Moreover, the minimum eigenvalue of $\omega^2[Z(j\omega) + Z^T(-j\omega)]$ approaches 2 as $\omega \to \infty$. Hence, by Lemma 10.1 we conclude that $Z(s)$ is strictly positive real. $\triangle$

Strictly positive real transfer functions will arise in our analysis as a consequence of the following lemma, known as the *Kalman-Yakubovich-Popov lemma* or the *positive real lemma*.

Lemma 10.2 Let $Z(s) = \mathcal{C}(sI - \mathcal{A})^{-1}\mathcal{B} + \mathcal{D}$ be a $p \times p$ transfer function matrix, where $\mathcal{A}$ is Hurwitz, $(\mathcal{A}, \mathcal{B})$ is controllable, and $(\mathcal{A}, \mathcal{C})$ is observable. Then, $Z(s)$ is strictly positive real if and only if there exist a positive definite symmetric matrix P, matrices W and L, and a positive constant ϵ such that

$$PA + \mathcal{A}^T P = -L^T L - \epsilon P \tag{10.10}$$

$$P\mathcal{B} = \mathcal{C}^T - L^T W \tag{10.11}$$

$$W^T W = \mathcal{D} + \mathcal{D}^T \tag{10.12}$$

$\diamond$

Proof: Appendix A.17.

10.1.1 Circle Criterion

Consider the system (10.1)–(10.3) and suppose that A is Hurwitz and the nonlinearity $\psi(\cdot, \cdot)$ satisfies the sector condition (10.9) with $K_{\min} = 0$; that is,

$$\psi^T(t, y)[\psi(t, y) - Ky] \leq 0, \quad \forall\, t \geq 0, \ \forall\, y \in \Gamma \subset R^p \tag{10.13}$$

where K is a positive definite symmetric matrix. Consider a Lyapunov function candidate

$$V(x) = x^T P x$$

where P is a positive definite symmetric matrix to be chosen. The derivative of $V(x)$ along the trajectories of the system (10.1)–(10.3) is given by

$$\dot{V}(t, x) = x^T(PA + A^T P)x - 2x^T PB\psi(t, y)$$

Since $-2\psi^T(\psi - Ky) \geq 0$, its addition to the right-hand side of the last equality gives an upper bound on $\dot{V}(t, x)$. Therefore,

$$
\begin{aligned}
\dot{V}(t, x) &\leq x^T(PA + A^T P)x - 2x^T PB\psi(t, y) - 2\psi^T(t, y)[\psi(t, y) - Ky] \\
&= x^T(PA + A^T P)x + 2x^T(C^T K - PB)\psi(t, y) - 2\psi^T(t, y)\psi(t, y)
\end{aligned}
$$

We will show that the right-hand side is negative definite by completing a square term. Toward that goal, suppose there are matrices $P = P^T > 0$ and L and a constant $\epsilon > 0$ such that

$$PA + A^T P = -L^T L - \epsilon P \tag{10.14}$$

$$PB = C^T K - \sqrt{2}L^T \tag{10.15}$$

Then,

$$
\begin{aligned}
\dot{V}(t,x) &\leq -\epsilon x^T P x - x^T L^T L x + 2\sqrt{2} x^T L^T \psi(t,y) - 2\psi^T(t,y)\psi(t,y) \\
&= -\epsilon x^T P x - [Lx - \sqrt{2}\psi(t,y)]^T [Lx - \sqrt{2}\psi(t,y)] \\
&\leq -\epsilon x^T P x
\end{aligned}
$$

Thus, we can show that $\dot{V}(t,x)$ is negative definite provided we can find P, L, and ϵ that satisfy (10.14)–(10.15). Using Lemma 10.2, we can see that this is the case if and only if

$$
Z(s) = I + KC(sI - A)^{-1}B
$$

is strictly positive real. Notice that the pair (A, KC) is observable since (A, C) is observable and K is nonsingular. To summarize, we have shown the following lemma.

Lemma 10.3 *Consider the system* (10.1)–(10.3), *where A is Hurwitz, (A,B) is controllable, (A,C) is observable, and $\psi(\cdot,\cdot)$ satisfies the sector condition* (10.13) *globally. Then, the system is absolutely stable if $Z(s) = I + KG(s)$ is strictly positive real. If* (10.13) *is satisfied only on a set $\Gamma \subset R^p$, then the same condition on $Z(s)$ ensures that the system is absolutely stable with a finite domain.* $\diamond$

The restriction on A to be Hurwitz can be removed by an idea known as *loop transformation* or *pole shifting*. Suppose we start with a system where A is not Hurwitz and the nonlinearity $\psi(\cdot,\cdot)$ satisfies the more general sector condition (10.9). Figure 10.4 shows an equivalent representation of the system where a constant-gain negative feedback $K_{min}y$ is applied around the linear component of the system. The effect of this feedback is offset by subtracting $K_{min}y$ from the output of the nonlinearity. This process defines a new linear system and a new nonlinearity. The new linear system is represented by the transfer function

$$
G_T(s) = G(s)\left[I + K_{min}G(s)\right]^{-1}
$$

or equivalently by the state-space model

$$
\begin{aligned}
\dot{x} &= (A - BK_{min}C)x + Bu \\
y &= Cx
\end{aligned}
$$

The new nonlinearity is given by

$$
\psi_T(t,y) = \psi(t,y) - K_{min}y
$$

It can be easily seen that if $\psi(\cdot,\cdot)$ satisfies the sector condition (10.9), then $\psi_T(\cdot,\cdot)$ satisfies the sector condition (10.13) with $K = K_{max} - K_{min}$. Therefore, if $(A - BK_{min}C)$ is a Hurwitz matrix, we can apply Lemma 10.3 to conclude that the

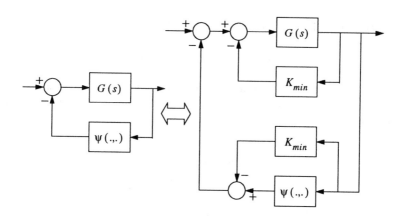

Figure 10.4: Loop transformation.

system will be absolutely stable if $Z_T(s) = I + KG_T(s)$ is strictly positive real. Noting that

$$\begin{aligned} Z_T(s) &= I + (K_{\max} - K_{\min})G(s)[I + K_{\min}G(s)]^{-1} \\ &= [I + K_{\max}G(s)][I + K_{\min}G(s)]^{-1} \end{aligned}$$

we arrive at the following theorem, which includes Lemma 10.3 as a special case.

Theorem 10.1 *Consider the system* (10.1)–(10.3), *where* (A, B) *is controllable,* (A, C) *is observable, and* $\psi(\cdot, \cdot)$ *satisfies the sector condition* (10.9) *globally. Then, the system is absolutely stable if*

$$G_T(s) = G(s)[I + K_{\min}G(s)]^{-1}$$

is Hurwitz and

$$Z_T(s) = [I + K_{\max}G(s)][I + K_{\min}G(s)]^{-1}$$

is strictly positive real. If (10.9) *is satisfied only on a set* $\Gamma \subset R^p$, *then the conditions given on* $G_T(s)$ *and* $Z_T(s)$ *ensure that the system is absolutely stable with a finite domain.* $\diamond$

We shall refer to this theorem as the *multivariable circle criterion*, although the reason for using this name will not be clear until we specialize to the scalar case $p = 1$. Theorem 10.1 reduces to Lemma 10.3 when $K_{\min} = 0$ and $G(s)$ is Hurwitz. Notice that Lemma 10.1 can be used to characterize the strict positive realness of $Z_T(s)$. In fact, since $Z_T(\infty) = I$, the lemma shows that $Z_T(s)$ is strictly positive real if and only if $Z_T(s)$ is Hurwitz and

$$Z_T(j\omega) + Z_T^T(-j\omega) > 0, \quad \forall \, \omega \in R$$

Example 10.2 Consider the system (10.1)–(10.3) and suppose $G(s)$ is Hurwitz. Let

$$\gamma_1 = \sup_{w \in R} \sigma_{\max}[G(j\omega)] = \sup_{w \in R} \|G(j\omega)\|_2$$

where $\sigma_{\max}[\cdot]$ denotes the maximum singular value of a complex matrix. The constant γ_1 is finite since $G(s)$ is Hurwitz. Suppose that the nonlinearity $\psi(\cdot, \cdot)$ satisfies the inequality

$$\|\psi(t, y)\|_2 \leq \gamma_2 \|y\|_2, \quad \forall\, t \geq 0, \, \forall\, y \in R^p \tag{10.16}$$

then it satisfies the sector condition (10.9) with

$$K_{\min} = -\gamma_2 I; \quad K_{\max} = \gamma_2 I$$

To apply Theorem 10.1, we need to show that

$$G_T(s) = G(s)[I - \gamma_2 G(s)]^{-1}$$

is Hurwitz and

$$Z_T(s) = [I + \gamma_2 G(s)][I - \gamma_2 G(s)]^{-1}$$

is strictly positive real. For the first point, note that[5]

$$\sigma_{\min}[I - \gamma_2 G(j\omega)] \geq 1 - \gamma_1 \gamma_2$$

If $\gamma_1 \gamma_2 < 1$, the plot of $\det[I - \gamma_2 G(j\omega)]$ does not go through nor encircle the origin. Hence, by the multivariable Nyquist criterion,[6] $[I - \gamma_2 G(s)]^{-1}$ is Hurwitz; consequently, $G_T(s)$ and $Z_T(s)$ are Hurwitz. Using Lemma 10.1, we see that $Z_T(s)$ is strictly positive real if

$$Z_T(j\omega) + Z_T^T(-j\omega) > 0, \quad \forall\, \omega \in R$$

The left-hand side of this inequality is given by

$$
\begin{aligned}
Z_T(j\omega) + Z_T^T(-j\omega) &= [I + \gamma_2 G(j\omega)][I - \gamma_2 G(j\omega)]^{-1} \\
&\quad + [I - \gamma_2 G^T(-j\omega)]^{-1}[I + \gamma_2 G^T(-j\omega)] \\
&= [I - \gamma_2 G^T(-j\omega)]^{-1} \left[2I - 2\gamma_2^2 G^T(-j\omega)G(j\omega)\right] \\
&\quad \times [I - \gamma_2 G(j\omega)]^{-1}
\end{aligned}
$$

---

[5] The following properties of singular values of a complex matrix are used:

$$\det G \neq 0 \Leftrightarrow \sigma_{\min}[G] > 0$$
$$\sigma_{\max}[G^{-1}] = 1/\sigma_{\min}[G], \text{ if } \sigma_{\min}[G] > 0$$
$$\sigma_{\min}[I + G] \geq 1 - \sigma_{\max}[G]$$
$$\sigma_{\max}[G_1 G_2] \leq \sigma_{\max}[G_1]\sigma_{\max}[G_2]$$

[6] See [27, pp. 160–161] for a statement of the multivariable Nyquist criterion.

Hence, $Z_T(j\omega) + Z_T^T(-j\omega)$ is positive definite for all ω if and only if

$$\sigma_{\min}\left[I - \gamma_2^2 G^T(-j\omega)G(j\omega)\right] > 0, \quad \forall\, \omega \in R$$

Now, for $\gamma_1 \gamma_2 < 1$, we have

$$\begin{aligned}
\sigma_{\min}[I - \gamma_2^2 G^T(-j\omega)G(j\omega)] &\geq 1 - \gamma_2^2 \sigma_{\max}[G^T(-j\omega)]\sigma_{\max}[G(j\omega)] \\
&\geq 1 - \gamma_1^2\gamma_2^2 > 0
\end{aligned}$$

Hence, all the conditions of Theorem 10.1 are satisfied and we can conclude that the system is absolutely stable if $\gamma_1\gamma_2 < 1$. This is a robustness result which shows that closing the loop around a Hurwitz transfer function with a nonlinearity satisfying (10.16), with a sufficiently small γ_2, does not destroy the stability of the system.[7]

$\triangle$

In the scalar case $p = 1$, the conditions of Theorem 10.1 can be verified graphically by examining the Nyquist plot of $G(j\omega)$. The sector condition for a single-input–single-output nonlinearity takes the form (10.5). The conditions of Theorem 10.1 are

1. $G_T(s) = G(s)/[1 + \alpha G(s)]$ is Hurwitz

2. $Z_T(s) = [1 + \beta G(s)]/[1 + \alpha G(s)]$ is strictly positive real

where all transfer functions are scalar. To verify that $Z_T(s)$ is strictly positive real, we can use Lemma 10.1 which states that $Z_T(s)$ is strictly positive real if and only if $Z_T(s)$ is Hurwitz and

$$Re\left[\frac{1 + \beta G(j\omega)}{1 + \alpha G(j\omega)}\right] > 0, \quad \forall\, \omega \in R \qquad (10.17)$$

To relate condition (10.17) to the Nyquist plot of $G(j\omega)$, we have to distinguish between three different cases, depending on the sign of α. Consider first the case when $\beta > \alpha > 0$. In this case, condition (10.17) can be rewritten as

$$Re\left[\frac{\frac{1}{\beta} + G(j\omega)}{\frac{1}{\alpha} + G(j\omega)}\right] > 0, \quad \forall\, \omega \in R \qquad (10.18)$$

For a point q on the Nyquist plot of $G(j\omega)$, the two complex numbers $[1/\beta + G(j\omega)]$ and $[1/\alpha + G(j\omega)]$ can be represented by the lines connecting q to $-1/\beta + j0$ and $-1/\alpha + j0$, respectively, as shown in Figure 10.5. The real part of the ratio of two complex numbers is positive when the angle difference between the two numbers is less than $\pi/2$; that is, the angle $(\theta_1 - \theta_2)$ in Figure 10.5 is less than $\pi/2$. If we

---

[7] The inequality $\gamma_1\gamma_2 < 1$ can be derived also from the small-gain theorem; see Example 10.8.

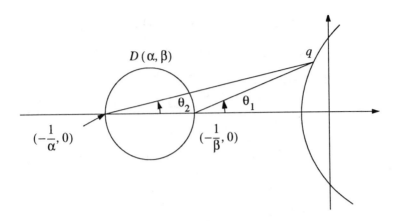

Figure 10.5: Graphical representation of the circle criterion.

define $D(\alpha, \beta)$ to be the closed disk in the complex plane whose diameter is the line segment connecting the points $-1/\alpha + j0$ and $-1/\beta + j0$, then it is simple to see that the angle $(\theta_1 - \theta_2)$ is less than $\pi/2$ when q is outside the disk $D(\alpha, \beta)$. Since (10.18) is required to hold for all ω, all points on the Nyquist plot of $G(j\omega)$ must be strictly outside the disk $D(\alpha, \beta)$. On the other hand, to verify that $G_T(s)$ is Hurwitz, which implies that $Z_T(s)$ is Hurwitz, we can use the Nyquist criterion which states that $G_T(s)$ is Hurwitz if and only if the Nyquist plot of $G(j\omega)$ does not intersect the point $-1/\alpha + j0$ and encircles it exactly m times in the counterclockwise direction, where m is the number of poles of $G(s)$ in the open right-half complex plane.[8] Therefore, the conditions of Theorem 10.1 are satisfied if the Nyquist plot of $G(j\omega)$ does not enter the disk $D(\alpha, \beta)$ and encircles it m times in the counterclockwise direction. Consider, next, the case when $\beta > 0$ and $\alpha = 0$. For this case, the conditions of Theorem 10.1 are: $G(s)$ is Hurwitz and

$$Re[1 + \beta G(j\omega)] > 0, \quad \forall \, \omega \in R$$

The latter condition can be rewritten as

$$Re[G(j\omega)] > -\frac{1}{\beta}, \quad \forall \, \omega \in R$$

which is equivalent to the graphical condition that the Nyquist plot of $G(j\omega)$ lies to the right of the vertical line defined by $Re[s] = -1/\beta$. Finally, consider the case

---

[8] When $G(s)$ has poles on the imaginary axis, the Nyquist path is indented in the right-half plane, as usual.

when $\alpha < 0 < \beta$. In this case, condition (10.17) is equivalent to

$$Re\left[\frac{\frac{1}{\beta} + G(j\omega)}{\frac{1}{\alpha} + G(j\omega)}\right] < 0, \quad \forall\, \omega \in R \qquad (10.19)$$

where the inequality sign is reversed because, as we go from (10.17) to (10.19), we multiply by α/β which is now negative. Repeating previous arguments, it can be easily seen that for (10.19) to hold, the Nyquist plot of $G(j\omega)$ must lie inside the disk $D(\alpha, \beta)$. Consequently, the Nyquist plot cannot encircle the point $-1/\alpha + j0$. Therefore, from the Nyquist criterion we see that $G(s)$ must be Hurwitz for $G_T(s)$ to be so. The stability criteria for the three cases are summarized in the following theorem, which is known as the *circle criterion*.

Theorem 10.2 *Consider a scalar system of the form* (10.1)–(10.3)*, where* $\{A, B, C\}$ *is a minimal realization of* $G(s)$ *and* $\psi(\cdot, \cdot)$ *satisfies the sector condition* (10.5) *globally. Then the system is absolutely stable if one of the following conditions is satisfied, as appropriate:*

1. *If* $0 < \alpha < \beta$*, the Nyquist plot of* $G(j\omega)$ *does not enter the disk* $D(\alpha, \beta)$ *and encircles it m times in the counterclockwise direction, where m is the number of poles of* $G(s)$ *with positive real parts.*

2. *If* $0 = \alpha < \beta$*,* $G(s)$ *is Hurwitz and the Nyquist plot of* $G(j\omega)$ *lies to the right of the vertical line defined by* $Re[s] = -1/\beta$*.*

3. *If* $\alpha < 0 < \beta$*,* $G(s)$ *is Hurwitz and the Nyquist plot of* $G(j\omega)$ *lies in the interior of the disk* $D(\alpha, \beta)$*.*

If the sector condition (10.5) *is satisfied only on an interval* $[a, b]$*, then the foregoing conditions ensure that the system is absolutely stable with a finite domain.* $\diamond$

The circle criterion allows us to investigate absolute stability using only the Nyquist plot of $G(j\omega)$. This is important because the Nyquist plot can be determined directly from experimental data. Given the Nyquist plot of $G(j\omega)$, we can determine permissible sectors for which the system is absolutely stable. The following two examples illustrate the use of the circle criterion.

Example 10.3 Let

$$G(s) = \frac{4}{(s+1)(\frac{1}{2}s+1)(\frac{1}{3}s+1)}$$

The Nyquist plot of $G(j\omega)$ is shown in Figure 10.6. Since $G(s)$ is Hurwitz, we can allow α to be negative and apply the third case of the circle criterion. So, we need to determine a disk $D(\alpha, \beta)$ that encloses the Nyquist plot. Clearly, the choice of the disk is not unique. Suppose we decide to locate the center of the

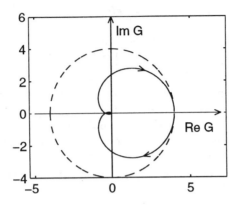

Figure 10.6: Nyquist plot for Example 10.3.

disk at the origin of the complex plane. This means that we shall work with a disk $D(-\gamma_2, \gamma_2)$, where the radius $\gamma_2 > 0$ is to be chosen. The Nyquist plot will be inside this disk if $|G(j\omega)| < \gamma_2$. In particular, if we set $\gamma_1 = \sup_{\omega \in R} |G(j\omega)|$, then γ_2 must be chosen to satisfy $\gamma_1 \gamma_2 < 1$. This is the same condition we found in Example 10.2. It is not hard to see that $|G(j\omega)|$ is maximum at $\omega = 0$ and $\gamma_1 = 4$. Thus, γ_2 must be less than 0.25. Hence, we can conclude that the system is absolutely stable for all nonlinearities in the sector $[-0.25 + \epsilon, 0.25 - \epsilon]$, where $\epsilon > 0$ can be arbitrarily small. Inspection of the Nyquist plot and the disk $D(-0.25, 0.25)$ in Figure 10.6 suggests that the choice to locate the center at the origin may not be the best choice. By locating the center at another point, we might be able to obtain a disk that encloses the Nyquist plot more tightly. For example, let us locate the center at the point $1.5 + j0$. The maximum distance from this point to the Nyquist plot is 2.834. Hence, choosing the radius of the disk to be 2.9 ensures that the Nyquist plot is inside the disk $D(-1/4.4, 1/1.4)$, and we can conclude that the system is absolutely stable for all nonlinearities in the sector $[-0.227, 0.714]$. Comparison of this sector with the previous one (see Figure 10.7) shows that by giving in a little bit on the lower bound of the sector, we achieve a significant improvement in the upper bound. Clearly, there is still room for optimizing the choice of the center of the disk but we shall not pursue it. The point we wanted to show is that the graphical representation used in the circle criterion gives us a closer look at the problem, compared with the use of norm inequalities as in Example 10.2, which allows us to obtain less conservative estimates of the sector. Another direction we can pursue in applying the circle criterion is to restrict α to zero and apply the second case of the circle criterion. The Nyquist plot lies to the right of the vertical line $Re[s] = -0.857$. Hence, we can conclude that the system is absolutely stable for all nonlinearities in the sector $[0, 1.166]$. This sector

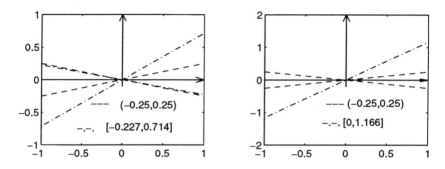

Figure 10.7: Sectors for Example 10.3.

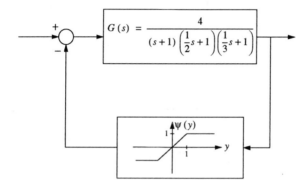

Figure 10.8: Feedback connection with saturation nonlinearity.

is sketched in Figure 10.7, together with the previous two sectors. It gives the best estimate of β, which is achieved at the expense of limiting the nonlinearity to be a first quadrant–third quadrant nonlinearity. To appreciate how this flexibility in using the circle criterion could be useful in applications, let us suppose that we are interested in studying the stability of the system of Figure 10.8, which includes a limiter or saturation nonlinearity (a typical nonlinearity in feedback control systems due to constraints on physical variables). The saturation nonlinearity belongs to a sector $[0, 1]$. Therefore, it is included in the sector $[0, 1.166]$, but not in the sector $(-0.25, 0.25)$ or $[-0.227, 0.714]$. Thus, based on the application of the second case of the circle criterion, we can conclude that the feedback system of Figure 10.8 has a globally asymptotically stable equilibrium at the origin. $\triangle$

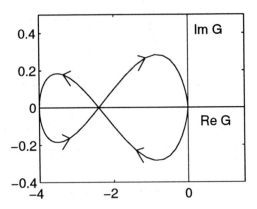

Figure 10.9: Nyquist plot for Example 10.4.

Example 10.4 Let

$$G(s) = \frac{4}{(s - 1)(\frac{1}{2}s + 1)(\frac{1}{3}s + 1)}$$

This transfer function is not Hurwitz, since it has a pole in the open right-half plane. So, we must restrict α to be positive and apply the first case of the circle criterion. The Nyquist plot of $G(j\omega)$ is shown in Figure 10.9. From the circle criterion, we know that the Nyquist plot must encircle the disk $D(\alpha, \beta)$ once in the counterclockwise direction. Inspection of the Nyquist plot shows that a disk can be encircled by the Nyquist plot only if it is totally inside one of the two lobes formed by the Nyquist plot in the left-half plane. A disk inside the right lobe is encircled once in the clockwise direction. Hence, it does not satisfy the circle criterion. A disk inside the left lobe is encircled once in the counterclockwise direction. Thus, we need to choose α and β to locate the disk $D(\alpha, \beta)$ inside the left lobe. Let us locate the center of the disk at the point $-3.2 + j0$, about halfway between the two ends of the lobe on the real axis. The minimum distance from this center to the Nyquist plot is 0.1688. Hence, choosing the radius to be 0.168, we conclude that the system is absolutely stable for all nonlinearities in the sector $[0.2969, 0.3298]$.

$\triangle$

In Examples 10.2–10.4, we have considered cases where the sector condition is satisfied globally. In the next example, the sector condition is satisfied only on a finite interval.

Example 10.5 Consider the feedback connection of Figure 10.2, where the linear system is represented by the transfer function

$$G(s) = \frac{s + 2}{(s + 1)(s - 1)}$$

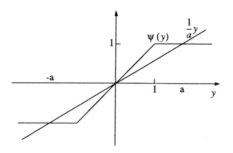

Figure 10.10: Sector for Example 10.5.

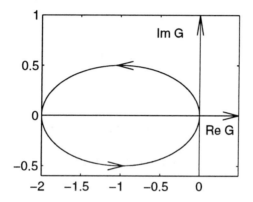

Figure 10.11: Nyquist plot for Example 10.5.

and the nonlinear element is $\psi(y) = \text{sat}(y)$. The nonlinearity belongs globally to the sector $[0, 1]$. However, since $G(s)$ is not Hurwitz, we must apply the first case of the circle criterion, which requires the sector condition (10.5) to hold with a positive α. Thus, we cannot conclude absolute stability using the circle criterion.[9] The best we can hope for is to show absolute stability with a finite domain. Figure 10.10 shows that on the interval $[-a, a]$, the nonlinearity $\psi(\cdot)$ satisfies the sector condition (10.5) with $\alpha = 1/a$ and $\beta = 1$. Since $G(s)$ has a pole with positive real part, the Nyquist plot of $G(j\omega)$, shown in Figure 10.11, must encircle the disk $D(\alpha, 1)$ once in the counterclockwise direction. It can be verified, analytically, that condition (10.17) is satisfied for $\alpha > 0.5359$. Thus, choosing $\alpha = 0.55$, the sector condition (10.5) is satisfied on the interval $[-1.818, 1.818]$ and the disk $D(0.55, 1)$ is encircled once by the Nyquist plot in the counterclockwise direction. Thus, from the first case of the

---

[9] In fact, the origin is not globally asymptotically stable; see Exercise 3.8.

circle criterion, we conclude that the system is absolutely stable with a finite domain. We can also use the quadratic Lyapunov function $V(x) = x^T P x$ to estimate the region of attraction. Consider the state-space model

$$\begin{aligned}
\dot{x}_1 &= x_2 \\
\dot{x}_2 &= x_1 + u \\
y &= 2x_1 + x_2 \\
u &= -\psi(y)
\end{aligned}$$

The loop transformation

$$u = -\alpha y - \psi(y) + \alpha y = -\alpha y - \psi_T(y)$$

brings the system into the form

$$\begin{aligned}
\dot{x} &= A_T x - B\psi_T(y) \\
y &= Cx
\end{aligned}$$

where

$$A_T = \begin{bmatrix} 0 & 1 \\ -0.1 & -0.55 \end{bmatrix}, \quad B = \begin{bmatrix} 0 \\ 1 \end{bmatrix}, \quad C = \begin{bmatrix} 2 & 1 \end{bmatrix}$$

and $\psi_T(\cdot)$ satisfies the sector condition (10.13) with

$$K = 1 - \alpha = 0.45$$

To find the Lyapunov function, we need to find P, L, and ϵ that satisfy (10.14)–(10.15), with A replaced by A_T. The existence of a solution is guaranteed by Lemma 10.2. It can be verified that[10]

$$\epsilon = 0.02, \quad P = \begin{bmatrix} 0.4946 & 0.4834 \\ 0.4834 & 1.0774 \end{bmatrix}, \quad L = \begin{bmatrix} 0.2946 & -0.4436 \end{bmatrix}$$

satisfy (10.14)–(10.15). Thus, $V(x) = x^T P x$ is a Lyapunov function for the system. We estimate the region of attraction by

$$\Omega_c = \{x \in R^2 \mid V(x) \le c\}$$

where $c \le \min_{\{|y|=1.818\}} V(x) = 0.3445$ to ensure that Ω_c is contained in the set $\{|y| \le 1.818\}$. Taking $c = 0.34$ gives the estimate shown in Figure 10.12. $\triangle$

---

[10]The value of ϵ is chosen such that $Z_T(s - 0.5\epsilon)$ is positive real and $[(\epsilon/2)I + A]$ is Hurwitz. Equation (10.15) is used to substitute for L in (10.14), which is then solved as a quadratic equation in P; see Exercise 10.11.

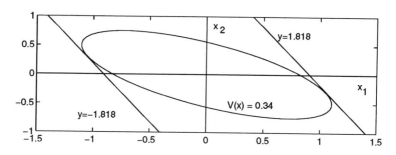

Figure 10.12: Region of attraction for Example 10.5.

10.1.2   Popov Criterion

Consider again the system (10.1)–(10.3), and suppose that A is Hurwitz and $\psi(\cdot)$ is a time-invariant nonlinearity that satisfies the sector condition (10.9) with $K_{\min} = 0$; that is,

$$\psi^T(y)[\psi(y) - Ky] \leq 0, \quad \forall\, y \in \Gamma \subset R^p \tag{10.20}$$

where K is a positive definite symmetric matrix. Suppose further that $K\psi(y)$ is the gradient of a scalar function[11] and

$$\int_0^y \psi^T(\sigma)K \, d\sigma \geq 0, \quad \forall\, y \in \Gamma \subset R^p \tag{10.21}$$

This will be the case, for example, when $\psi(y)$ is a decoupled nonlinearity of the form (10.7) and each component $\psi_i(\cdot)$ satisfies the sector condition (10.5) with $\alpha = 0$; that is, $\psi(\cdot)$ satisfies the sector condition (10.20) with

$$K = \text{diag}(\beta_1, \beta_2, \ldots, \beta_p) \quad \text{and} \quad \Gamma = \{y \in R^p \mid a_i \leq y_i \leq b_i\}$$

Another example will be any nonlinearity that satisfies (10.20) with $K = \beta I$, $\beta > 0$, and whose Jacobian matrix $[\partial\psi/\partial y]$ is symmetric and integral $\int_0^y \psi^T(\sigma) \, d\sigma$ is nonnegative. Consider a Lyapunov function candidate of the Lure-type

$$V(x) = x^T P x + 2\eta \int_0^y \psi^T(\sigma)K \, d\sigma$$

where $\eta \geq 0$ is to be chosen. The derivative of $V(x)$ along the trajectories of the system (10.1)–(10.3) is given by

$$\dot{V}(x) = x^T(PA + A^T P)x - 2x^T PB\psi(y) + 2\eta\psi^T(y)KC[Ax - B\psi(y)]$$

[11] Recall from our discussion of the variable gradient method in Section 3.1 that a vector $g(y)$ is the gradient of a scalar function if and only if the Jacobian matrix $[\partial g/\partial y]$ is symmetric, and in this case the integral $\int_0^y g^T(x) \, dx$ exists and is independent of the path joining the origin to y.

Since $-2\psi^T(\psi - Ky) \geq 0$, its addition to the right-hand side of the last equality gives an upper bound on $V(x)$. Therefore,

$$
\begin{aligned}
\dot{V}(x) \;\leq\;& x^T(PA + A^T P)x - 2x^T PB\psi(y) \\
& +2\eta\psi^T(y)KC[Ax - B\psi(y)] - 2\psi^T(y)[\psi(y) - Ky] \\
=\;& x^T(PA + A^T P)x - 2x^T(PB - \eta A^T C^T K - C^T K)\psi(y) \\
& -\psi^T(y)(2I + \eta KCB + \eta B^T C^T K)\psi(y)
\end{aligned}
$$

where we have used the fact that $\psi^T M x = x^T M^T \psi$ for any $p \times p$ matrix M. Choose η such that

$$
2I + \eta KCB + \eta B^T C^T K \geq 0
$$

which can be always done.[12] Let

$$
2I + \eta KCB + \eta B^T C^T K = W^T W
$$

Now, suppose there are matrices $P = P^T > 0$ and L and a constant $\epsilon > 0$ such that

$$
\begin{aligned}
PA + A^T P &= -L^T L - \epsilon P & (10.22) \\
PB &= C^T K + \eta A^T C^T K - L^T W & (10.23)
\end{aligned}
$$

Then,

$$
\dot{V}(x) \leq -\epsilon x^T Px - [Lx - W\psi(y)]^T[Lx - W\psi(y)] \leq -\epsilon x^T Px
$$

which is negative definite. The question of existence of P, L, and ϵ satisfying (10.22)–(10.23) can be studied using Lemma 10.2. Consider the transfer function

$$
\begin{aligned}
Z(s) &= I + \eta KCB + (KC + \eta KCA)(sI - A)^{-1}B \\
&= \mathcal{D} + \mathcal{C}(sI - \mathcal{A})^{-1}\mathcal{B}
\end{aligned}
$$

where

$$
\mathcal{A} = A, \ \mathcal{B} = B, \ \mathcal{C} = KC + \eta KCA, \ \mathcal{D} = I + \eta KCB
$$

Note that

$$
Z(\infty) + Z^T(\infty) = W^T W
$$

Suppose that η is chosen such that $(1 + \eta\lambda_i) \neq 0$ for all eigenvalues of A; that is, $-1/\eta$ is not an eigenvalue of A. Then, the pair $(A, KC + \eta KCA)$ is observable since (A, C) is observable.[13] Thus, all the conditions of Lemma 10.2 are satisfied and we

---

[12]By choosing η small enough, for example.

[13]Note that if $(A, KC + \eta KCA)$ is not observable, then there is an eigenvalue λ_i and associated eigenvector v_i of A for which

$$
(1 + \eta\lambda_i)Cv_i = 0 \ \Rightarrow \ Cv_i = 0
$$

which implies that (A, C) is not observable.

conclude that there are P, L, and ϵ satisfying (10.22)–(10.23) if and only if $Z(s)$ is strictly positive real. The transfer function $Z(s)$ can be expressed as

$$
\begin{aligned}
Z(s) &= I + \eta KC(sI - A + A)(sI - A)^{-1}B + KC(sI - A)^{-1}B \\
&= I + (1 + \eta s)KG(s)
\end{aligned}
$$

Our conclusion is summarized in the following theorem, which is known as the multivariable Popov criterion.

Theorem 10.3 *Consider the system* (10.1)–(10.3) *where A is Hurwitz, (A, B) is controllable, (A, C) is observable, and $\psi(\cdot)$ is a time-invariant nonlinearity that satisfies the sector condition* (10.20) *globally with a positive definite symmetric K. Suppose that $K\psi(y)$ is the gradient of a scalar function and* (10.21) *is satisfied globally. Then, the system is absolutely stable if there is $\eta \geq 0$, with $-1/\eta$ not an eigenvalue of A such that*

$$
Z(s) = I + (1 + \eta s)KG(s)
$$

is strictly positive real. If (10.13) *and* (10.21) *are satisfied only on a set $\Gamma \subset R^p$, then the same condition on $Z(s)$ ensures that the system is absolutely stable with a finite domain.* ◇

As we have done in the circle criterion, the restriction on A to be Hurwitz can be removed by performing a loop transformation. We will not repeat this idea in general, but it will be illustrated by examples and exercises. Note that with $\eta = 0$, Theorem 10.3 reduces to Lemma 10.3. This shows that the conditions of Theorem 10.3 are weaker than those of Lemma 10.3. With $\eta > 0$, absolute stability can be established under less stringent conditions. In the scalar case $p = 1$, we can test the strict positive realness of $Z(s)$ graphically. Choose η such that $Z(\infty) > 0$. Then, by Lemma 10.1, $Z(s)$ is strictly positive real if and only if

$$
Re[1 + (1 + j\eta\omega)kG(j\omega)] > 0, \quad \forall \, \omega \in R
$$

which is equivalent to

$$
\frac{1}{k} + Re[G(j\omega)] - \eta\omega Im[G(j\omega)] > 0, \quad \forall \, \omega \in R \tag{10.24}
$$

where $G(j\omega) = Re[G(j\omega)] + jIm[G(j\omega)]$. If we plot $Re[G(j\omega)]$ versus $\omega Im[G(j\omega)]$ with ω as a parameter, then condition (10.24) is satisfied if the plot lies to the right of the line that intercepts the point $-1/k + j0$ with a slope $1/\eta$; see Figure 10.13. Such a plot is known as a Popov plot, in contrast to a Nyquist plot which is a plot of $Re[G(j\omega)]$ versus $Im[G(j\omega)]$.

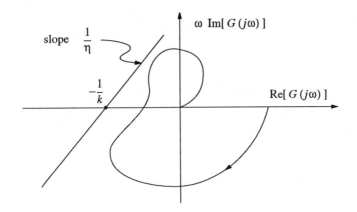

Figure 10.13: Popov plot.

Example 10.6 Consider the second-order system

$$\dot{x}_1 = x_2$$
$$\dot{x}_2 = -x_2 - h(y)$$
$$y = x_1$$

This system would fit the form (10.1)–(10.3) if we took $\psi(\cdot) = h(\cdot)$, but the matrix A would not be Hurwitz. Adding and subtracting the term αy to the right-hand side of the second state equation, where $\alpha > 0$, and defining $\psi(y) = h(y) - \alpha y$, the system takes the form (10.1)–(10.3), with

$$A = \begin{bmatrix} 0 & 1 \\ -\alpha & -1 \end{bmatrix}, \ B = \begin{bmatrix} 0 \\ 1 \end{bmatrix}, \ C = \begin{bmatrix} 1 & 0 \end{bmatrix}$$

Assume that $h(\cdot)$ belongs to a sector $[\alpha, \beta]$, where $\beta > \alpha$. Then, $\psi(\cdot)$ belongs to the sector $[0, k]$, where $k = \beta - \alpha$. We have $Z(\infty) = 1$, independent of η. Condition (10.24) takes the form

$$\frac{1}{k} + \frac{\alpha - \omega^2 + \eta\omega^2}{(\alpha - \omega^2)^2 + \omega^2} > 0, \ \ \forall \, \omega \in R$$

For all finite positive values of α and k, this inequality is satisfied by choosing $\eta \geq 1$. Hence, the system is absolutely stable for all nonlinearities $h(\cdot)$ in the sector $[\alpha, \beta]$, where α can be arbitrarily small and β can be arbitrarily large. Figure 10.14 shows the Popov plot of $G(j\omega)$ for $\alpha = 1$. The plot is drawn only for $\omega \geq 0$ since $Re[G(j\omega)]$ and $\omega Im[G(j\omega)]$ are even functions of ω. The Popov plot asymptotically approaches the line through the origin of slope unity from the right side. Therefore,

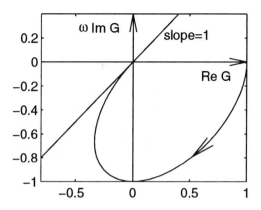

Figure 10.14: Popov plot for Example 10.6.

it lies to the right of any line of slope ≤ 1 that intersects the real axis to the left of the origin. To see the advantage of having $\eta > 0$, let us take $\eta = 0$ and apply the circle criterion. From the second case of Theorem 10.2, the system is absolutely stable if the Nyquist plot of $G(j\omega)$ lies to the right of the vertical line defined by $Re[s] = -1/k$. Since a portion of the Nyquist plot lies in the left-half plane, k cannot be arbitrarily large. The maximum permissible value of k can be determined analytically from the condition

$$\frac{1}{k} + \frac{\alpha - \omega^2}{(\alpha - \omega^2)^2 + \omega^2} > 0, \quad \forall \, \omega \in R$$

which yields $k < 1 + 2\sqrt{\alpha}$. Thus, using the circle criterion we can only conclude that the system is absolutely stable for all nonlinearities $h(\cdot)$ in the sector $[\alpha, 1 + \alpha + 2\sqrt{\alpha} - \epsilon]$, where $\alpha > 0$ and $\epsilon > 0$ can be arbitrarily small. $\triangle$

10.1.3 Simultaneous Lyapunov Functions

Consider the system (10.1)–(10.3) with a decoupled nonlinearity $\psi(\cdot, \cdot)$. Suppose that each component of $\psi(\cdot, \cdot)$ satisfies the sector condition (10.5) with constants α_i, β_i, a_i, and b_i. As we saw before, the nonlinearity $\psi(\cdot, \cdot)$ satisfies the sector condition (10.8) with positive diagonal matrices $K_{\min}$ and $K_{\max}$. We want to study asymptotic stability of the origin $x = 0$ for all nonlinearities satisfying (10.8). Consider a Lyapunov function candidate

$$V(x) = x^T P x$$

where $P = P^T > 0$ is to be chosen. The derivative of $V(x)$ along the trajectories of the system (10.1)–(10.3) is given by

$$\dot{V}(t,x) = x^T(PA + A^TP)x - 2x^TPB\psi(t,y) \tag{10.25}$$

Our goal is to find, if possible, a matrix P such that the right-hand side of (10.25) is negative definite for all nonlinearities $\psi(\cdot,\cdot)$ which satisfy the sector condition (10.8). In Section 10.1.1, we found a sufficient condition for the existence of such a matrix P in terms of the strict positive realness of a certain transfer function. This time, we take a different approach to the problem where we search directly for P. The key property we have at hand is the fact that each component of $\psi(\cdot,\cdot)$ satisfies the inequality

$$\alpha_i y_i^2 \le y_i \psi_i(t,y_i) \le \beta_i y_i^2 \tag{10.26}$$

For a given trajectory $x(t)$ of (10.1)–(10.3), let us define the time-varying gains $k_i(t)$ by[14]

$$\psi_i(t,y_i(t)) = k_i(t)y_i(t) \tag{10.27}$$

For different trajectories of the system, the time-varying gains defined in (10.27) are different. Irrespective of the particular trajectory traced, it is always true that

$$\alpha_i \le k_i(t) \le \beta_i \tag{10.28}$$

because $\psi_i(\cdot,\cdot)$ satisfies (10.26). The derivative $\dot{V}(t,x)$ can be rewritten as

$$\begin{aligned}
\dot{V}(t,x) &= x^T(PA + A^TP)x - 2x^TPBK(t)y \\
&= x^T[P(A - BK(t)C) + (A - BK(t)C)^TP]x \tag{10.29}
\end{aligned}$$

where

$$K(t) = \text{diag}[k_1(t), k_2(t), \ldots, k_p(t)]$$

It is clear that if the right-hand side of (10.29) is negative definite for all time-varying gains satisfying (10.28), then the right-hand side of (10.25) is negative definite for all nonlinearities satisfying (10.26). More specifically, if there is a positive constant ϵ such that

$$x^T[P(A - BK(t)C) + (A - BK(t)C)^TP]x < -\epsilon x^T x \tag{10.30}$$

for all time-varying gains satisfying (10.28), then

$$x^T(PA + A^TP)x - 2x^TPB\psi(t,y) < -\epsilon x^T x$$

---

[14]Equation (10.27) defines $k_i(t)$ uniquely whenever $y_i(t) \neq 0$. At $y_i(t) = 0$, we choose $k_i(t)$ arbitrarily such that (10.28) is satisfied.

for all nonlinearities satisfying (10.26). With this observation, our problem reduces to studying (10.30) for all time-varying gains satisfying (10.28). The gains $k_1(\cdot)$ to $k_p(\cdot)$ take values in the convex hypercube

$$H = \{k \in R^p \mid \alpha_i \le k_i \le \beta_i\}$$

We denote by $k^{(1)}$, $k^{(2)}$, ... the 2^p vertices of H. Corresponding to each vertex, let

$$A^{(i)} = A - BK^{(i)}C$$

where the diagonal elements of $K^{(i)}$ are taken at the vertex $k^{(i)}$. Let

$$w(x, k) = x^T[P(A - BKC) + (A - BKC)^T P]x \tag{10.31}$$

where $K = \text{diag}(k_1, k_2, \ldots, k_p)$. It is obvious that there exist $P = P^T > 0$ and $\epsilon > 0$ such that

$$w(x, k(t)) < -\epsilon x^T x$$

for all time-varying gains satisfying (10.28) if and only if

$$w(x, k) < -\epsilon x^T x, \quad \forall\, k \in H$$

Since $w(x, k)$ is linear in k, it attains its maximum value over H at one of the vertices of H. Hence, the foregoing condition is equivalent to

$$w(x, k^{(i)}) < -\epsilon x^T x, \quad \forall\, i = 1, 2, \ldots, 2^p$$

Therefore, we need to conduct our search for P only at the vertices of H. We summarize this finding in the following theorem.

Theorem 10.4 *Consider the system* (10.1)–(10.3), *where* $\psi(\cdot, \cdot)$ *is a decoupled non-linearity which satisfies the sector condition* (10.8) *globally. Then, there exist a positive definite symmetric matrix P and a positive constant ϵ such that*

$$x^T(PA + A^T P)x - 2x^T PB\psi(t, y) < -\epsilon x^T x, \quad \forall\, x \in R^n$$

for all $\psi(\cdot, \cdot)$ *if and only if*

$$PA^{(i)} + \left(A^{(i)}\right)^T P < -\epsilon I, \quad \forall\, i = 1, 2, \ldots, 2^p \tag{10.32}$$

Thus, condition (10.32) *ensures absolute stability of the system. Moreover, if the sector condition* (10.8) *is satisfied only on a set $\Gamma \subset R^p$, then condition* (10.32) *ensures absolute stability with a finite domain.* ◇

Proof: We have already shown the sufficiency of (10.32). Necessity follows from the fact that within the class of nonlinearities satisfying the sector condition (10.8), we can choose $\psi(t, y) = K^{(i)}y$, where $K^{(i)}$ is a diagonal matrix whose elements are taken at the ith vertex. If the sector condition is satisfied on a set Γ, the whole argument will still be valid, except that x will be limited to a domain about the origin defined by Γ.   □

While condition (10.32) involves only the vertices of H, it does not merely require that $A^{(i)}$ be Hurwitz for all i; the same P must work for all i. In other words, we must have a simultaneous quadratic Lyapunov function for all vertices.[15] The existence of P and ϵ satisfying (10.32) can be established by solving a special type of minimax problem. Let

$$\mathcal{A} = \text{block diag}[A^{(1)}, A^{(2)}, \ldots, A^{(2^p)}]$$

and

$$\mathcal{P} = \text{block diag}[P, P, \ldots, P]$$

Then, (10.32) can be rewritten as[16]

$$\mathcal{P}\mathcal{A} + \mathcal{A}^T\mathcal{P} < -\epsilon I$$

Since P is an $n \times n$ symmetric matrix ($p_{ij} = p_{ji}$), it has only $n(n+1)/2$ undetermined parameters; let $r = n(n + 1)/2$. Denote these r parameters by $z_1, z_2, \ldots, z_r$ and stack them in a vector z. Define the scalar function

$$\eta(z, v) = v^T(\mathcal{P}\mathcal{A} + \mathcal{A}^T\mathcal{P})v$$

Since this function is linear in the elements of $\mathcal{P}$, it can be written as

$$\eta(z, v) = v^T \sum_{i=1}^{r} z_i L_i v$$

Let

$$\phi(z) = \max_{\|v\|_2=1} \eta(z, v) = \lambda_{\max}(\mathcal{P}\mathcal{A} + \mathcal{A}^T\mathcal{P}) \tag{10.33}$$

---

[15] The concept of simultaneous Lyapunov functions appears in a number of problems, some of which will be covered in the exercises. A good reference on the subject is [22].

[16] By rewriting the equation in this form, the problem of simultaneous Lyapunov functions has been recast as a problem of structured Lyapunov functions, where in our case $\mathcal{P}$ is restricted to be block diagonal with equal diagonal blocks. The need to work with structured Lyapunov functions appears in a number of stability problems; see [22] for further discussion of structured Lyapunov functions. We encounter structured Lyapunov functions in the stability analysis of interconnected systems (Section 5.6) and multiparameter singular perturbations (Exercise 9.25).

where $\lambda_{\max}(\cdot)$ denotes the maximum eigenvalue of the respective matrix. There is no loss of generality in scaling P such that $|p_{ij}| \leq 1$. Consequently,

$$z \in \mathcal{Z} = \{z \in R^r \mid |z_i| \leq 1\}$$

The connection between condition (10.32) and the function $\phi(z)$ is established in the following lemma.

Lemma 10.4 *Let $\phi(z)$ be defined by (10.33) and assume that at least one of the $A^{(i)}$ matrices is known to be Hurwitz. Then, there are $P = P^T > 0$ and $\epsilon > 0$ such that*

$$PA^{(i)} + \left[A^{(i)}\right]^T P < -\epsilon I, \quad \forall \; i = 1, 2, \ldots, 2^p$$

if and only if

$$\phi^* = \min_{z \in \mathcal{Z}} \phi(z) < -\epsilon$$

$\diamond$

Proof: By continuous dependence of the eigenvalues of a matrix on its parameters,[17] the function $\phi(z)$ is continuous. Hence, it attains a minimum over the compact set $\mathcal{Z}$. The necessity is obvious. To prove sufficiency, let z^* be a point at which $\phi(z^*) < -\epsilon$, and let P^* and $\mathcal{P}^*$ be the corresponding matrices. Then,

$$-\epsilon > \phi(z^*) = \lambda_{\max}(\mathcal{P}^* \mathcal{A} + \mathcal{A}^T \mathcal{P}^*)$$

Hence,

$$P^* A^{(i)} + \left[A^{(i)}\right]^T P^* < -\epsilon I, \quad \forall \; i = 1, 2, \ldots, 2^p$$

Since one of the matrices $A^{(i)}$ is known to be Hurwitz, it follows from Theorem 3.6 that P^* is positive definite. $\qquad\square$

Testing condition (10.32) has thus been reduced to the minimization of $\phi(z)$ over $\mathcal{Z}$, which will be formulated next as a minimax problem. Let z and w be any two points in $\mathcal{Z}$. Let $v(w)$ be an eigenvector of $\sum_{i=1}^r w_i L_i$ corresponding to the maximum eigenvalue, with $\|v(w)\|_2 = 1$. Let $g(w)$ be an r-dimensional vector, whose ith component is given by $g_i(w) = v^T(w) L_i v(w)$. Then, it can be easily seen that $\phi(w) = w^T g(w)$ and

$$
\begin{aligned}
\phi(z) \;&=\; \max_{\|v\|_2 = 1} \eta(z, v) \\[2mm]
&\geq\; \eta(z, v(w)) \;=\; v^T(w) \sum_{i=1}^r z_i L_i v(w) \\[2mm]
&=\; \sum_{i=1}^r z_i v^T(w) L_i v(w) \;=\; \sum_{i=1}^r z_i g_i(w)
\end{aligned}
$$

---

[17] See [57, Section 7.2].

Therefore,

$$\phi(z) \geq z^T g(w), \quad \forall \; z, w \in \mathcal{Z} \tag{10.34}$$

This inequality enables us to search for the minimizing policy using a cutting-plane algorithm.[18] We start by choosing ϵ as a small positive number[19] and search for a minimizing policy. Note that our interest is not in finding the minimum policy itself; rather, it is enough to find any point z for which $\phi(z) < -\epsilon$. Therefore, we may stop the search once we hit such a point. Suppose we start the search at a point $z^{(1)}$. If $\phi(z^{(1)}) < -\epsilon$, we stop. If $\phi(z^{(1)}) \geq -\epsilon$, we compute $g^{(1)} = g(z^{(1)})$. According to (10.34), all points in the half-space $z^T g^{(1)} \geq 0$ will have $\phi(z) \geq 0$. Therefore, we eliminate this half-space and continue searching in the other half $z^T g^{(1)} < 0$. We choose a new point $z^{(2)}$ which minimizes $z^T g^{(1)}$ over $\mathcal{Z}$. After k iterations, suppose that $z^{(1)}, z^{(2)}, \ldots, z^{(k)}$ are the first k iterates and set $g^{(i)} = g(z^{(i)})$. If $\phi(z^{(k)}) < -\epsilon$, we stop. Otherwise, we choose a new point $z^{(k+1)}$ as the minimizing policy of the minimax problem

$$s_k = \min_{z \in \mathcal{Z}} \max_{1 \leq i \leq k} \left\{ z^T g^{(i)} \right\} \tag{10.35}$$

The following lemma summarizes some observations on this minimax problem.

Lemma 10.5

- $-\sigma \leq s_k \leq 0$, where $\sigma = \left(r \sum_{i=1}^{r} \| L_i \|_2^2 \right)^{1/2}$

- s_k is monotonically nondecreasing; that is, $s_{k+1} \geq s_k$.

- $\phi^* \geq s_k, \; \forall \; k$.

- If $\phi(z^{(k)}) \geq -\epsilon, \; \forall \; k$, then

$$s_k \to s_\infty \geq -\epsilon, \; as \; k \to \infty$$

- If $(\hat{x}^{(k)}, \hat{y}^{(k)})$ is a solution of the linear programming problem

$$\left. \begin{array}{ll} \text{minimize} & -y \\ \text{subject to} & x^T g^{(i)} + y \leq l^T g^{(i)}, \quad i = 1, 2, \ldots, k \\ & 0 \leq x_i \leq 2, \qquad\qquad i = 1, 2, \ldots, r \\ & 0 \leq y \leq \sigma \end{array} \right\} \tag{10.36}$$

where l is an r-dimensional vector of ones, then $\hat{x}^{(k)} - l$ is a solution of the minimax problem (10.35) and $s_k = -\hat{y}^{(k)}$. $\diamond$

---

[18] See [107].

[19] The value of ϵ is a measure of the positive definiteness of the matrix $-(\mathcal{P}\mathcal{A} + \mathcal{A}^T\mathcal{P})$. For example, if we choose $\epsilon = 10^{-4}$, then this matrix will be positive definite and its minimum eigenvalue will be greater than 10^{-4}. Moreover, since the elements of $\mathcal{P}$ have magnitudes less than or equal to one, the number $1/\epsilon$ is a measure of the condition number of the matrix $-(\mathcal{P}\mathcal{A} + \mathcal{A}^T\mathcal{P})$.

Proof: The first two items are obvious. The third item follows immediately from (10.34) and (10.35). Suppose that $\phi(z^{(k)}) \geq -\epsilon$ for all k. The sequence s_k is monotonically nondecreasing and bounded from above by zero. Hence, $s_k \to s_\infty \leq 0$ as $k \to \infty$. We want to show that $s_\infty \geq -\epsilon$. Suppose $s_\infty = -\epsilon - \alpha$ for some $\alpha > 0$. By definition,

$$s_k = \max_{1 \leq i \leq k} \left\{ \left[z^{(k+1)} \right]^T g^{(i)} \right\} \geq \max_{1 \leq i \leq j} \left\{ \left[z^{(k+1)} \right]^T g^{(i)} \right\}, \text{ for } j \leq k$$

Therefore,

$$\left[z^{(k+1)} \right]^T g^{(j)} \leq s_k \leq -\epsilon - \alpha, \quad \forall \, k \geq j$$

but

$$\left[z^{(j)} \right]^T g^{(j)} = \phi(z^{(j)}) \geq -\epsilon, \quad \forall \, j$$

Hence,

$$\left[z^{(k+1)} - z^{(j)} \right]^T g^{(j)} \leq -\alpha, \quad \forall \, k \geq j$$

Consequently,

$$\left\| z^{(k+1)} - z^{(j)} \right\|_2 \geq \frac{\alpha}{\beta}, \quad \forall \, k \geq j$$

where $\beta = \left(\sum_{i=1}^r \|L_i\|_2^2 \right)^{1/2}$ is an upper bound on $\|g(z)\|_2$ for all $z \in \mathcal{Z}$. The last inequality is impossible since $z^{(k)}$ is a bounded sequence, so it must have at least one accumulation point. Thus, $s_\infty \geq -\epsilon$. Consider now the linear programming problem (10.36). If $(\hat{x}, \hat{y})$ is a solution of (10.36), then

$$-\hat{y} \geq \max_{1 \leq i \leq k} (\hat{x} - l)^T g^{(i)} \geq s_k$$

since $\hat{x} - l \in \mathcal{Z}$. Alternatively, if $z^{(k+1)}$ is a solution to the minimax problem (10.35), then $x = z^{(k+1)} + l$ and $y = -s_k$ is a feasible solution of (10.36); this implies that $-\hat{y} \leq s_k$. Thus, $-\hat{y} = s_k$ and $\hat{x} - l$ is a solution of (10.35).   □

In view of this lemma, if the condition $\phi(z^{(k)}) < -\epsilon$ is not reached in a finite number of steps, then $s_k \to s_\infty \geq -\epsilon$, indicating that there is no P that satisfies (10.32) for the given ϵ. The fact that s_k is monotonically nondecreasing can be used to devise a stopping criterion. For example, we may stop when s_k exceeds some threshold $-(1 + \delta)\epsilon$ where δ is a small positive number. For given ϵ and δ, the search algorithm for finding P can now be summarized as follows.

Step 0: Take $z^{(1)} \in \mathcal{Z}$; compute $\phi(z^{(1)})$; if $\phi(z^{(1)}) < -\epsilon$, stop; otherwise, continue; compute $g^{(1)}$; set $k = 1$.

Step 1: Solve the linear programming problem (10.36); if $s_k = -\hat{y}^{(k)} \geq -(1+\delta)\epsilon$, stop; otherwise, continue.

Step 2: Take $z^{(k+1)} = \hat{x}^{(k)} - l$; if $\phi(z^{(k+1)}) < -\epsilon$, stop; otherwise, compute $g^{(k+1)}$, replace k by $k+1$, and go to step 1.

The algorithm terminates in a finite number of steps. The termination point is either:

- $\phi(z^{(k)}) < -\epsilon$; there is P satisfying (10.32), or

- $s_k \geq -(1+\delta)\epsilon$; there is no P that satisfies

$$PA^{(i)} + \left[A^{(i)}\right]^T P < -(1+\delta)\epsilon I, \quad \forall \, i = 1, 2, \ldots, 2^p$$

Example 10.7 Consider again the system of Example 10.5 with the sector $[0.55, 1]$. There are two vertices with corresponding matrices

$$A^{(1)} = A - 0.55BC = \begin{bmatrix} 0 & 1 \\ -0.1 & -0.55 \end{bmatrix}, \quad A^{(2)} = A - BC = \begin{bmatrix} 0 & 1 \\ -1 & -1 \end{bmatrix}$$

The system is absolutely stable if we can find $P = P^T > 0$ that satisfies (10.32) for some $\epsilon > 0$. Taking $\epsilon = \delta = 10^{-6}$, we start the iteration at $z^{(1)} = [\,1 \quad 0 \quad 1\,]^T$. The algorithm stops in five steps at the negative value $\phi(z^{(6)}) = -0.0199$, corresponding to $z^{(6)} = [\,0.4245 \quad 0.3688 \quad 1.0\,]^T$. Thus, the desired matrix P is

$$P = \begin{bmatrix} 0.4245 & 0.3688 \\ 0.3688 & 1.0 \end{bmatrix}$$

and the system is absolutely stable. The monotonically increasing variable s_k takes the values

$$-1.3477 \quad -1.031 \quad -0.1948 \quad -0.0463 \quad -0.039$$

Let us try now the sector $[0, 1]$, for which the system is not absolutely stable. The matrix $A^{(2)}$ is the same as in the previous case, while $A^{(1)} = A$. We apply the algorithm with the same values of ϵ, δ, and $z^{(1)}$. The algorithm stops in four steps at $s_4 = -2.96 \times 10^{-22}$, showing that there is no P for the given values of ϵ and δ. $\triangle$

10.2   Small-Gain Theorem

The formalism of input-output stability is particularly useful in studying the stability of interconnections of dynamical systems, since the notion of the gain of a

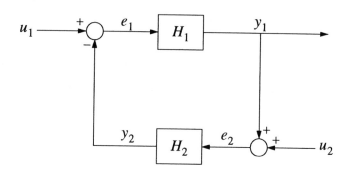

Figure 10.15: Feedback connection.

system allows us to track how the norm of a signal increases or decreases as it passes through a system. This is particularly so for the feedback connection of Figure 10.15. Here we have two systems: $H_1 : \mathcal{L}_e^m \to \mathcal{L}_e^q$ and $H_2 : \mathcal{L}_e^q \to \mathcal{L}_e^m$. Suppose both systems are finite-gain $\mathcal{L}$ stable;[20] that is,

$$\|y_{1\tau}\|_{\mathcal{L}} \leq \gamma_1 \|e_{1\tau}\|_{\mathcal{L}} + \beta_1, \quad \forall \, e_1 \in \mathcal{L}_e^m, \; \forall \, \tau \in [0, \infty) \tag{10.37}$$

$$\|y_{2\tau}\|_{\mathcal{L}} \leq \gamma_2 \|e_{2\tau}\|_{\mathcal{L}} + \beta_2, \quad \forall \, e_2 \in \mathcal{L}_e^q, \; \forall \, \tau \in [0, \infty) \tag{10.38}$$

Suppose further that the feedback system is *well defined* in the sense that for every pair of inputs $u_1 \in \mathcal{L}_e^m$ and $u_2 \in \mathcal{L}_e^q$, there exist unique outputs $e_1, y_2 \in \mathcal{L}_e^m$ and $e_2, y_1 \in \mathcal{L}_e^q$.[21] The question of interest is whether the feedback connection, when viewed as a mapping from the input pair (u_1, u_2) to the output pair (e_1, e_2), is $\mathcal{L}$ stable.[22] The following theorem, known as *the small-gain theorem*, gives a sufficient condition for finite-gain $\mathcal{L}$ stability of the feedback connection.

Theorem 10.5 *Under the preceding assumptions, if*

$$\gamma_1 \gamma_2 < 1$$

then for all $u_1 \in \mathcal{L}_e^m$ and $u_2 \in \mathcal{L}_e^q$,

$$\|e_{1\tau}\|_{\mathcal{L}} \leq \frac{1}{1 - \gamma_1 \gamma_2} (\|u_{1\tau}\|_{\mathcal{L}} + \gamma_2 \|u_{2\tau}\|_{\mathcal{L}} + \beta_2 + \gamma_2 \beta_1) \tag{10.39}$$

---

[20] In this section, we present a version of the classical small-gain theorem which applies to finite-gain $\mathcal{L}$ stability. For more general versions which apply to $\mathcal{L}$ stability, see[108, 85].

[21] Sufficient conditions for existence and uniqueness of solutions are available in the literature. The most common approach uses the contraction mapping principle; see, for example, [43, Theorem III.3.1]. A more recent approach that makes use of existence and uniqueness of the solution of state equations can be found in [85].

[22] See Exercise 10.17 for an explanation of why we have to consider both inputs and outputs in studying stability of the feedback connection.

$$\|e_{2\tau}\|_{\mathcal{L}} \leq \frac{1}{1-\gamma_1\gamma_2}(\|u_{2\tau}\|_{\mathcal{L}} + \gamma_1\|u_{1\tau}\|_{\mathcal{L}} + \beta_1 + \gamma_1\beta_2) \qquad (10.40)$$

for all $\tau \in [0,\infty)$. If $u_1 \in \mathcal{L}^m$ and $u_2 \in \mathcal{L}^q$, then $e_1, y_2 \in \mathcal{L}^m$, and $e_2, y_1 \in \mathcal{L}^q$ and the norms of e_1 and e_2 are bounded by the right-hand side above with nontruncated functions. $\diamond$

Proof: Assuming existence of the solution, we can write

$$e_{1\tau} = u_{1\tau} - (H_2 e_2)_{\tau}, \qquad e_{2\tau} = u_{2\tau} + (H_1 e_1)_{\tau}$$

Then,

$$
\begin{aligned}
\|e_{1\tau}\|_{\mathcal{L}} &\leq \|u_{1\tau}\|_{\mathcal{L}} + \|(H_2 e_2)_{\tau}\|_{\mathcal{L}} \leq \|u_{1\tau}\|_{\mathcal{L}} + \gamma_2\|e_{2\tau}\|_{\mathcal{L}} + \beta_2 \\
&\leq \|u_{1\tau}\|_{\mathcal{L}} + \gamma_2\left(\|u_{2\tau}\|_{\mathcal{L}} + \gamma_1\|e_{1\tau}\|_{\mathcal{L}} + \beta_1\right) + \beta_2 \\
&= \gamma_1\gamma_2\|e_{1\tau}\|_{\mathcal{L}} + \left(\|u_{1\tau}\|_{\mathcal{L}} + \gamma_2\|u_{2\tau}\|_{\mathcal{L}} + \beta_2 + \gamma_2\beta_1\right)
\end{aligned}
$$

Since $\gamma_1\gamma_2 < 1$,

$$\|e_{1\tau}\|_{\mathcal{L}} \leq \frac{1}{1-\gamma_1\gamma_2}(\|u_{1\tau}\|_{\mathcal{L}} + \gamma_2\|u_{2\tau}\|_{\mathcal{L}} + \beta_2 + \gamma_2\beta_1)$$

The bound on $\|e_{2\tau}\|_{\mathcal{L}}$ is shown similarly. If $u_1 \in \mathcal{L}^m$ and $u_2 \in \mathcal{L}^q$, the norms $\|u_1\|_{\mathcal{L}}$ and $\|u_2\|_{\mathcal{L}}$ are finite. Then, $\|e_{1\tau}\|_{\mathcal{L}}$ is bounded for all τ, uniformly in τ, since $\|u_{i\tau}\|_{\mathcal{L}} \leq \|u_i\|_{\mathcal{L}}$ for $i = 1, 2$. Thus, $\|e_1\|_{\mathcal{L}}$ is finite. A similar argument applies to $\|e_2\|_{\mathcal{L}}$, $\|y_1\|_{\mathcal{L}}$, and $\|y_2\|_{\mathcal{L}}$. $\square$

The small-gain theorem simply says that the feedback connection of two input-output stable systems, as in Figure 10.15, will be input-output stable provided the product of the system gains is less than one. The feedback connection of Figure 10.15 provides a convenient setup for studying robustness issues in dynamical systems. Quite often, dynamical systems subject to model uncertainties can be represented in the form of a feedback connection with H_1, say, as a stable nominal system and H_2 as a stable perturbation. Then, the requirement $\gamma_1\gamma_2 < 1$ is satisfied whenever γ_2 is small enough. Therefore, the small-gain theorem provides a conceptual framework for understanding many of the robustness results that arise in the study of dynamical systems, especially when feedback is used. Many of the robustness results that we can derive using Lyapunov stability techniques can be interpreted as special cases of the small-gain theorem.

Example 10.8 Consider the feedback connection of Figure 10.15. Let H_1 be a linear time-invariant system with a Hurwitz square transfer function matrix $G(s) = C(sI - A)^{-1}B$. Let H_2 be a memoryless function $e_2 = \psi(t, y_2)$ which satisfies

$$\|\psi(t, y)\|_2 \leq \gamma_2\|y\|_2, \quad \forall\, t \geq 0, \,\forall\, y \in R^m$$

This system was treated in Example 10.2 using the multivariable circle criterion. From Theorem 6.4, we know that H_1 is finite-gain $\mathcal{L}_2$ stable and its $\mathcal{L}_2$ gain is given by

$$\gamma_1 = \sup_{w \in R} \|G(j\omega)\|_2$$

It can be easily seen that H_2 is finite-gain $\mathcal{L}_2$ stable and its $\mathcal{L}_2$ gain is less than or equal to γ_2. Assuming the feedback system is well defined, we conclude by the small-gain theorem that it will be finite-gain $\mathcal{L}_2$ stable if $\gamma_1 \gamma_2 < 1$.   △

Example 10.9 Consider the system

$$
\begin{aligned}
\dot{x} &= f(t, x, v + d_1(t)) \\
\epsilon \dot{z} &= Az + B[u + d_2(t)] \\
v &= Cz
\end{aligned}
$$

where f is a smooth function of its arguments, A is a Hurwitz matrix, $-CA^{-1}B = I$, ϵ is a small positive parameter, and d_1, d_2 are disturbance signals. The linear part of this model represents actuator dynamics which are, typically, much faster than the plant dynamics represented here by the nonlinear equation $\dot{x} = f$. The disturbance signals d_1 and d_2 enter the system at the input of the plant and the input of the actuator, respectively. Suppose the disturbances d_1 and d_2 belong to a signal space $\mathcal{L}$, where $\mathcal{L}$ could be any $\mathcal{L}_p$ space, and the control goal is to attenuate the effect of this disturbance on the state x. This goal can be met if feedback control can be designed so that the closed-loop input-output map from (d_1, d_2) to x is finite-gain $\mathcal{L}$ stable and the $\mathcal{L}$ gain is less than some given tolerance $\delta > 0$. To simplify the design problem, it is common to neglect the actuator dynamics by setting $\epsilon = 0$ and substituting $v = -CA^{-1}B(u + d_2) = u + d_2$ in the plant equation to obtain the reduced-order model

$$\dot{x} = f(t, x, u + d)$$

where $d = d_1 + d_2$. Assuming the state variables are available for measurement, we use this model to design a state feedback control law $u = \gamma(t, x)$ to meet the design objective. Suppose we have succeeded in designing a smooth state feedback control $u = \gamma(t, x)$ such that

$$\|x\|_{\mathcal{L}} \leq \gamma \|d\|_{\mathcal{L}} + \beta \tag{10.41}$$

for some $\gamma < \delta$. Will this control meet the design objective when applied to the actual system with the actuator dynamics included? This is a question of robustness of the controller with respect to the unmodeled actuator dynamics.[23] When the

---

[23] In Example 9.11 we investigated a similar robustness problem, in the context of stabilization, using singular perturbation theory.

control is applied to the actual system, the closed-loop equation is given by

$$\dot{x} = f(t, x, Cz + d_1(t))$$
$$\epsilon\dot{z} = Az + B[\gamma(t, x) + d_2(t)]$$

Let us assume that $d_2(t)$ is differentiable and $\dot{d}_2 \in \mathcal{L}$. The change of variables

$$\eta = z + A^{-1}B[\gamma(t, x) + d_2(t)]$$

brings the closed-loop system into the form

$$\dot{x} = f(t, x, \gamma(t, x) + d(t) + C\eta)$$
$$\epsilon\dot{\eta} = A\eta + \epsilon A^{-1}B[\dot{\gamma} + \dot{d}_2(t)]$$

where

$$\dot{\gamma} = \frac{\partial\gamma}{\partial t}(t, x) + \frac{\partial\gamma}{\partial x}(t, x)f(t, x, \gamma(t, x) + d + C\eta)$$

It is not difficult to see that the closed-loop system can be represented in the form of Figure 10.15 with H_1 defined by

$$\dot{x} = f(t, x, \gamma(t, x) + e_1)$$
$$y_1 = \dot{\gamma} = \frac{\partial\gamma}{\partial t}(t, x) + \frac{\partial\gamma}{\partial x}(t, x)f(t, x, \gamma(t, x) + e_1)$$

H_2 defined by

$$\dot{\eta} = \frac{1}{\epsilon}A\eta + A^{-1}Be_2$$
$$y_2 = -C\eta$$

and

$$u_1 = d_1 + d_2 = d, \quad u_2 = \dot{d}_2$$

Notice that, in this representation, the system H_1 is the nominal reduced-order closed-loop system while H_2 represents the effect of the unmodeled dynamics. Setting $\epsilon = 0$ opens the loop and the overall closed-loop system reduces to the nominal one. Let us assume that the feedback function $\gamma(t, x)$ satisfies the inequality

$$\left\| \frac{\partial\gamma}{\partial t}(t, x) + \frac{\partial\gamma}{\partial x}(t, x)f(t, x, \gamma(t, x) + e_1) \right\| \leq c_1\|x\| + c_2\|e_1\| \qquad (10.42)$$

for all (t, x, e_1), where c_1 and c_2 are nonnegative constants. Using (10.41) and (10.42), it can be shown that

$$\|y_1\|_{\mathcal{L}} \leq \gamma_1\|e_1\|_{\mathcal{L}} + \beta_1$$

where
$$\gamma_1 = c_1\gamma + c_2, \quad \beta_1 = c_1\beta$$

System H_2 is a linear time-invariant system with a Hurwitz matrix A. Therefore,

$$\left\|e^{At}\right\| \leq ke^{-at} \Rightarrow \left\|e^{At/\epsilon}\right\| \leq ke^{-at/\epsilon}$$

for some positive constants k and a. From Example 6.3, we see that the system is finite-gain $\mathcal{L}_p$ stable for any $p \in [1, \infty]$ and

$$\|y_2\|_{\mathcal{L}} \leq \gamma_2\|e_2\|_{\mathcal{L}} + \beta_2 \stackrel{\text{def}}{=} \epsilon\gamma_f\|e_2\|_{\mathcal{L}} + \beta_2$$

where

$$\gamma_f = \frac{k\|A^{-1}B\|\ \|C\|}{a}, \quad \beta_2 = \rho k\|C\|\ \|\eta_0\|$$

and ρ, which depends on p of the space $\mathcal{L}_p$, is defined in Example 6.3. Now,

$$\gamma_1\gamma_2 = \epsilon\gamma_1\gamma_f < 1, \quad \forall\, \epsilon < \frac{1}{\gamma_1\gamma_f}$$

Thus, assuming the feedback connection is well defined, we conclude from the small-gain theorem that the input-output map from (u_1, u_2) to (e_1, e_2) is $\mathcal{L}$ stable. From (10.39), we have

$$\|e_1\|_{\mathcal{L}} \leq \frac{1}{1 - \epsilon\gamma_1\gamma_f}[\|u_1\|_{\mathcal{L}} + \epsilon\gamma_f\|u_2\|_{\mathcal{L}} + \epsilon\gamma_f\beta_1 + \beta_2]$$

Using

$$\|x\|_{\mathcal{L}} \leq \gamma\|e_1\|_{\mathcal{L}} + \beta$$

which follows from (10.41), and the definition of u_1 and u_2, we obtain

$$\|x\|_{\mathcal{L}} \leq \frac{\gamma}{1 - \epsilon\gamma_1\gamma_f}[\|d\|_{\mathcal{L}} + \epsilon\gamma_f\|\dot{d}_2\|_{\mathcal{L}} + \epsilon\gamma_f\beta_1 + \beta_2] + \beta \qquad (10.43)$$

It is interesting to note that the right-hand side of (10.43) approaches

$$\gamma\|d\|_{\mathcal{L}} + \beta + \gamma\beta_2$$

as $\epsilon \to 0$, which shows that for sufficiently small ϵ the upper bound on the $\mathcal{L}$ gain of the map from d to x, under the actual closed-loop system, will be close to the corresponding quantity under the nominal closed-loop system. $\triangle$

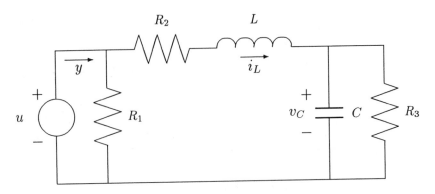

Figure 10.16: RLC circuit of Example 10.10.

10.3   Passivity Approach

An alternative approach to the stability analysis of feedback systems is the passivity approach. If, in the feedback connection of Figure 10.15, both components H_1 and H_2 are passive in the sense that they do not generate energy of their own, then it is intuitively clear that the feedback system will be passive. If one of the two feedback components dissipates energy, then the feedback system will dissipate energy. To carry this intuition into precise statements, we need a precise definition of what we mean by a passive system. We shall define passivity for two types of systems: dynamical systems represented by state-space models, and memoryless functions. We start with dynamical systems. The following RLC circuit example motivates the definition, which follows the example.

Example 10.10 Consider the RLC circuit shown in Figure 10.16. The circuit features a voltage source connected to a passive RLC network. We will take the voltage u as the input to the system and the current y as its output. The product uy is the power flow into the passive network. Taking the current through the inductor, x_1, and the voltage across the capacitor, x_2, as the state variables, we can write the state model

$$
\begin{aligned}
L\dot{x}_1 &= u - R_2 x_1 - x_2 \\
C\dot{x}_2 &= x_1 - \frac{1}{R_3} x_2 \\
y &= x_1 + \frac{1}{R_1} u
\end{aligned}
$$

The energy stored in the system is given by $V(x) = \frac{1}{2}Lx_1^2 + \frac{1}{2}Cx_2^2$. Since the network is passive, the energy absorbed by the network over any period of time $[0, t]$ must be greater than or equal to the increase in the energy stored in the network over the same period; that is,

$$\int_0^t u(s)y(s) \ ds \geq V(x(t)) - V(x(0)) \qquad (10.44)$$

If (10.44) holds with strict inequality, then the difference between the absorbed energy and the increase in the stored energy must be the energy dissipated in the resistive components of the network. Since (10.44) must hold for every $t \geq 0$, it must be true that the instantaneous power inequality

$$u(t)y(t) \geq \dot{V}(x(t)) \qquad (10.45)$$

holds for all t; that is, the power flow into the network must be greater than or equal to the rate of change of the energy stored in the network. We can arrive at inequality (10.45) by calculating the derivative of V along the trajectories of the system. We have

$$
\begin{aligned}
\dot{V} &= Lx_1\dot{x}_1 + Cx_2\dot{x}_2 = x_1(u - R_2x_1 - x_2) + x_2\left(x_1 - \frac{1}{R_3}x_2\right) \\
&= x_1(u - R_2x_1) - \frac{1}{R_3}x_2^2 = \left(x_1 + \frac{1}{R_1}u\right)u - \frac{1}{R_1}u^2 - R_2x_1^2 - \frac{1}{R_3}x_2^2 \\
&= uy - \frac{1}{R_1}u^2 - R_2x_1^2 - \frac{1}{R_3}x_2^2
\end{aligned}
$$

Thus,

$$uy = \dot{V} + \frac{1}{R_1}u^2 + R_2x_1^2 + \frac{1}{R_3}x_2^2 \geq \dot{V}$$

The quadratic nonnegative terms in the foregoing equation represent the dissipation rate. The dissipation rate takes different forms, which we illustrate by various special cases of the network.

Case 1: Take $R_1 = R_3 = \infty$ and $R_2 = 0$. Then,

$$uy = \dot{V}$$

In this case, there is no energy dissipation in the network; that is, the system is lossless.

Case 2: Take $R_2 = 0$ and $R_3 = \infty$. Then,

$$uy = \dot{V} + \frac{1}{R_1}u^2$$

The dissipation rate is proportional to u^2. There is no energy dissipation if and only if $u(t)$ is identically zero.

Case 3: Take $R_1 = R_3 = \infty$. Then,

$$uy = \dot{V} + R_2 y^2$$

where we have used the fact that in this case $y = x_1$. The dissipation rate is proportional to y^2. There is no energy dissipation if and only if $y(t)$ is identically zero.

Case 4: Take $R_1 = \infty$. Then,

$$uy = \dot{V} + R_2 x_1^2 + \frac{1}{R_3} x_2^2$$

The dissipation rate is a positive definite function of the state x. There is no energy dissipation if and only if $x(t)$ is identically zero.

Case 5: Take $R_1 = \infty$, $R_2 = 0$. Then,

$$uy = \dot{V} + \frac{1}{R_3} x_2^2$$

The dissipation rate is a positive semidefinite function of the state. Notice, however, that from the second state equation, we have

$$x_2(t) \equiv 0 \Rightarrow x_1(t) \equiv 0$$

irrespective of the input u. Therefore, like the previous case, there is no energy dissipation if and only if $x(t)$ is identically zero.

These five cases illustrate four basic forms of the dissipation rate: no dissipation, strict dissipation when the input is not identically zero, strict dissipation when the output is not identically zero, and strict dissipation when the state is not identically zero. These four basic forms will be captured in Definition 10.4. It is clear that combinations of these forms are also possible. For example, for the complete circuit when all resistors are present, we have

$$uy = \dot{V} + \frac{1}{R_1} u^2 + R_2 x_1^2 + \frac{1}{R_3} x_2^2$$

whose dissipation rate is the sum of a quadratic term in the input and a positive definite function of the state.   △

Consider a dynamical system represented by

$$\dot{x} = f(x, u) \tag{10.46}$$
$$y = h(x, u) \tag{10.47}$$

where $f : R^n \times R^m \to R^n$ is locally Lipschitz, $h : R^n \times R^m \to R^m$ is continuous, $f(0,0) = 0$, and $h(0,0) = 0$. The system has the same number of inputs and outputs.

Definition 10.4 *The system* (10.46)–(10.47) *is said to be passive if there exists a continuously differentiable positive semidefinite function* $V(x)$ *(called the storage function) such that*

$$u^T y \geq \frac{\partial V}{\partial x} f(x, u) + \epsilon u^T u + \delta y^T y + \rho \psi(x), \quad \forall \, (x, u) \in R^n \times R^m \qquad (10.48)$$

where ϵ, δ, *and* ρ *are nonnegative constants, and* $\psi(x)$ *is a positive semidefinite function of* x *such that*

$$\psi(x(t)) \equiv 0 \Rightarrow x(t) \equiv 0 \qquad (10.49)$$

for all solutions of (10.46) *and any* $u(t)$ *for which the solution exists. The term* $\rho\psi(x)$ *is called the state dissipation rate. Furthermore, the system is said to be*

- *lossless if* (10.48) *is satisfied with equality and* $\epsilon = \delta = \rho = 0$; *that is,*

$$u^T y = \frac{\partial V}{\partial x} f(x, u)$$

- *input strictly passive if* $\epsilon > 0$;

- *output strictly passive if* $\delta > 0$;

- *state strictly passive if* $\rho > 0$;

If more than one of the constants ϵ, δ, ρ *are positive, we combine names. For example, if both* ϵ *and* δ *are positive, we say the system is input and output strictly passive.*

In the case of output strictly passive systems, we will be interested in the additional property that

$$y(t) \equiv 0 \Rightarrow x(t) \equiv 0 \qquad (10.50)$$

for all solutions of (10.46) when $u = 0$. Equivalently, no solutions of $\dot{x} = f(x, 0)$ can stay identically in $S = \{x \in R^n \mid h(x, 0) = 0\}$, other than the trivial solution $x(t) \equiv 0$. It is important to see the difference between (10.50) and the condition (10.49) imposed on the state dissipation rate. Condition (10.49) is required to hold for any u, while (10.50) is required to hold only at $u = 0$. Recall Case 3 of Example 10.10, where the system

$$\begin{aligned} L\dot{x}_1 &= u - R_2 x_1 - x_2 \\ C\dot{x}_2 &= x_1 \\ y &= x_1 \end{aligned}$$

is output strictly passive. It can be easily seen that

$$y(t) \equiv 0 \Rightarrow x_1(t) \equiv 0 \Rightarrow x_2 \equiv u(t) \equiv \text{a constant}$$

Thus, when $u = 0$, condition (10.50) is satisfied. However, if we allow $u \neq 0$, then the condition $y(t) \equiv 0$ can be maintained by any constant input $u \neq 0$ which results in a constant $x_2 \neq 0$. This situation corresponds to DC inputs where no DC current can flow in the circuit due to the series inductance, while energy could be stored on the capacitor. The condition (10.50) can be interpreted as an observability condition. Recall that for the linear system

$$\begin{aligned} \dot{x} &= Ax \\ y &= Cx \end{aligned}$$

observability is equivalent to

$$y(t) = Ce^{At}x(0) \equiv 0 \;\Leftrightarrow\; x(0) = 0 \;\Leftrightarrow\; x(t) \equiv 0$$

For easy reference, we define (10.50) as an observability property of the system.

Definition 10.5 *The system* (10.46)–(10.47) *is said to be zero-state observable if no solution of* $\dot{x} = f(x, 0)$ *can stay identically in* $S = \{x \in R^n \mid h(x, 0) = 0\}$, *other than the trivial solution* $x(t) \equiv 0$.

The following lemma relates passivity to some of the stability notions we are familiar with.

Lemma 10.6 *Consider the system* (10.46)–(10.47).

- *If the system is passive with a positive definite storage function* $V(x)$, *then the origin of* $\dot{x} = f(x, 0)$ *is stable.*

- *If the system is output strictly passive, then it is finite-gain* $\mathcal{L}_2$ *stable.*

- *If the system is output strictly passive with a positive definite storage function* $V(x)$, *and zero-state observable, then the origin of* $\dot{x} = f(x, 0)$ *is asymptotically stable.*

- *If the system is state strictly passive with a positive definite storage function* $V(x)$, *then the origin of* $\dot{x} = f(x, 0)$ *is asymptotically stable.*

Furthermore, if $V(x)$ *is radially unbounded in any one of the last two cases, the origin will be globally asymptotically stable.*   ◇

Proof: The first item is obvious from $\dot{V} \leq u^T y$, which reduces to $\dot{V}(x) \leq 0$ when $u = 0$. The requirement that $V(x)$ be positive definite qualifies it to be a Lyapunov function. To prove the second item, start from the inequality

$$\begin{aligned} \dot{V} &\leq u^T y - \delta y^T y = -\frac{1}{2\delta}(u - \delta y)^T(u - \delta y) + \frac{1}{2\delta}u^T u - \frac{\delta}{2}y^T y \\ &\leq \frac{1}{2\delta}u^T u - \frac{\delta}{2}y^T y \end{aligned}$$

integrating both sides over $[0, \tau]$ yields

$$\int_0^\tau y^T(t)y(t)\ dt \le \frac{1}{\delta^2}\int_0^\tau u^T(t)u(t)\ dt - \frac{2}{\delta}[V(x(\tau)) - V(x(0))]$$

Hence,

$$\|y_\tau\|_{\mathcal{L}_2} \le \frac{1}{\delta}\|u_\tau\|_{\mathcal{L}_2} + \sqrt{\frac{2}{\delta}V(x(0))}$$

where we have used $V(x) \ge 0$ and $\sqrt{a^2 + b^2} \le a + b$ for nonnegative numbers a and b. To prove the third item, use $V(x)$ as a Lyapunov function candidate. We have

$$\dot{V} \le -\delta h^T(x, 0)h(x, 0) \le 0$$

By zero-state observability, the only solution of $\dot{x} = f(x, 0)$ that can stay identically in $S = \{x \in R^n \mid h(x, 0) = 0\}$ is the trivial solution. Hence, by LaSalle's invariance principle (Corollary 3.1), the origin is asymptotically stable. If $V(x)$ is radially unbounded, we conclude by Corollary 3.2 that the origin is globally asymptotically stable. The proof of the fourth item is exactly the same as that of the third item. $\square$

Requiring the storage function to be positive definite is not a severe restriction. In fact, in the last two items of the lemma, it can be shown that any storage function that will satisfy the definition of passivity must be positive definite in a neighborhood of the origin.[24]

The following two examples show passive systems that we have encountered earlier in the book.

Example 10.11 A linear time-invariant system with a minimal realization and a strictly positive real transfer function $G(s) = C(sI - A)^{-1}B + D$ is state strictly passive.[25] This fact can be shown using the Kalman-Yakubovich-Popov lemma (Lemma 10.2). From the lemma, there are a positive definite symmetric matrix P, matrices W and L, and a positive constant ϵ such that

$$\begin{aligned}
PA + A^T P &= -L^T L - \epsilon P \\
PB &= C^T - L^T W \\
W^T W &= D + D^T
\end{aligned}$$

Taking $V(x) = \frac{1}{2}x^T P x$ as a candidate for a storage function, we have

$$u^T y - \frac{\partial V}{\partial x}(Ax + Bu) = u^T(Cx + Du) - x^T P(Ax + Bu)$$

---

[24] See Exercise 10.20.

[25] It can be also shown that if $G(s)$ is positive real, then the system is passive; see Exercise 10.21.

$$
\begin{aligned}
&= \; u^T C x + \tfrac{1}{2} u^T (D + D^T) u - \tfrac{1}{2} x^T (PA + A^T P) x - x^T P B u \\
&= \; u^T (B^T P + W^T L) x + \tfrac{1}{2} u^T W^T W u \\
&\quad + \tfrac{1}{2} x^T L^T L x + \tfrac{1}{2} \epsilon x^T P x - x^T P B u \\
&= \; \tfrac{1}{2} (Lx + Wu)^T (Lx + Wu) + \tfrac{1}{2} \epsilon x^T P x \; \geq \; \tfrac{1}{2} \epsilon x^T P x
\end{aligned}
$$

Hence, inequality (10.48) is satisfied with $\rho \psi(x) = \tfrac{1}{2} \epsilon x^T P x$. Since P is positive definite, we conclude that the system is state strictly passive. $\triangle$

Example 10.12 Consider the m-input–m-output system [26]

$$
\begin{aligned}
\dot{x} &= f(x) + G(x) u \\
y &= h(x)
\end{aligned}
$$

where f is locally Lipschitz, G and h are continuous, $f(0) = 0$, and $h(0) = 0$. Suppose there is a continuously differentiable positive semidefinite function $V(x)$ such that

$$
\begin{aligned}
\frac{\partial V}{\partial x} f(x) &\leq 0 \\
\frac{\partial V}{\partial x} G(x) &= h^T(x)
\end{aligned}
$$

Then,

$$
u^T y - \frac{\partial V}{\partial x} [f(x) + G(x) u] = u^T h(x) - \frac{\partial V}{\partial x} f(x) - h^T(x) u = - \frac{\partial V}{\partial x} f(x) \geq 0
$$

which shows that the system is passive. If we have the stronger condition

$$
\begin{aligned}
\frac{\partial V}{\partial x} f(x) &\leq -k h^T(x) h(x), \quad k > 0 \\
\frac{\partial V}{\partial x} G(x) &= h^T(x)
\end{aligned}
$$

then

$$
u^T y - \frac{\partial V}{\partial x} [f(x) + G(x) u] \geq -k y^T y
$$

and the system is output strictly passive. If, in addition, $V(x)$ is positive definite and the system is zero-state observable, then the origin of $\dot{x} = f(x)$ will be asymptotically stable. $\triangle$

We turn now to the definition of passivity of a memoryless function. We start with an example to motivate the definition.

---

[26] $\mathcal{L}_2$ stability of this system was studied in Examples 6.10 and 6.11.

Example 10.13 Consider a voltage source connected to a linear resistor with positive resistance R. From Ohm's law, we have

$$y = \frac{1}{R}u$$

We can view this memoryless function as a system with the voltage u as input and the current y as output. The product uy is the power flow into the passive resistance. Unlike the electric network of Example 10.11, the current network has no energy storing elements. Therefore, the power flow uy must be always nonnegative. In fact,

$$uy = \frac{1}{R}u^2 = Ry^2 = \frac{1}{2R}u^2 + \frac{R}{2}y^2$$

which shows that the system is strictly passive with a dissipation rate that can be expressed as a quadratic function of the input, a quadratic function of the output, or the sum of two such functions. Thus, the system can be defined to be "input strictly passive," "output strictly passive," or "input and output strictly passive" as in the case of dynamical systems. Of course, there is no counterpart for "state strictly passive" since the system is memoryless. $\triangle$

Consider the nonlinear time-varying function

$$y = h(t, u) \tag{10.51}$$

where $h : R \times R^m \to R^m$ is piecewise continuous in t and locally Lipschitz in u.

Definition 10.6 *The system* (10.51) *is said to be passive if*

$$u^T y \geq \epsilon u^T u + \delta y^T y \quad \forall\ u \in R^m \tag{10.52}$$

where ϵ and δ are nonnegative constants. Furthermore, the system is said to be

- *input strictly passive if $\epsilon > 0$;*

- *output strictly passive if $\delta > 0$;*

We are now ready to analyze stability of the feedback system of Figure 10.15 when both components H_1 and H_2 are passive. We shall give three theorems. The first theorem deals with finite-gain $\mathcal{L}_2$ stability. The second theorem deals with asymptotic stability of the origin of the unforced system when both H_1 and H_2 are dynamical systems. The third theorem also deals with asymptotic stability, but with a memoryless system H_2.

Theorem 10.6 *Suppose the feedback system of Figure 10.15 is well defined in the sense that for every pair of inputs u_1, $u_2 \in \mathcal{L}_e^m$ there exist unique outputs e_1, e_2, y_1, $y_2 \in \mathcal{L}_e^m$. Let H_1 and H_2 be passive systems; that is, they satisfy Definition 10.4 or Definition 10.5 with*

$$e_i^T y_i \geq \frac{\partial V_i}{\partial x_i} f_i(x_i, e_i) + \epsilon_i e_i^T e_i + \delta_i y_i^T y_i, \quad \text{for } i = 1, 2 \qquad (10.53)$$

We write (10.53) for the case of a dynamical system, with the understanding that if H_i is a memoryless function we will take $V_i = 0$. Then, the closed-loop input-output map from (u_1, u_2) to (y_1, y_2) is finite gain $\mathcal{L}_2$ stable if

$$\epsilon_1 + \delta_2 > 0 \quad \text{and} \quad \epsilon_2 + \delta_1 > 0 \qquad (10.54)$$

$\diamond$

Proof: From $e_1 = u_1 - y_2$ and $e_2 = u_2 + y_1$, we have

$$\begin{aligned}
e_1^T y_1 + e_2^T y_2 &= (u_1 - y_2)^T y_1 + (u_2 + y_1)^T y_2 = u_1^T y_1 + u_2^T y_2 \quad (10.55) \\
e_1^T e_1 &= u_1^T u_1 - 2u_1^T y_2 + y_2^T y_2 \qquad (10.56) \\
e_2^T e_2 &= u_2^T u_2 + 2u_2^T y_1 + y_1^T y_1 \qquad (10.57)
\end{aligned}$$

Adding the inequalities (10.53) for $i = 1, 2$ and using (10.55)–(10.57), we obtain

$$\frac{\partial V}{\partial x} f \leq -y^T L y - u^T M u + u^T N y$$

where

$$\frac{\partial V}{\partial x} = \left[\frac{\partial V_1}{\partial x_1} \ \frac{\partial V_2}{\partial x_2} \right], \ f = \left[\begin{array}{c} f_1 \\ f_2 \end{array} \right], \ u = \left[\begin{array}{c} u_1 \\ u_2 \end{array} \right], \ y = \left[\begin{array}{c} y_1 \\ y_2 \end{array} \right]$$

$$L = \left[\begin{array}{cc} (\epsilon_2 + \delta_1)I & 0 \\ 0 & (\epsilon_1 + \delta_2)I \end{array} \right], \ M = \left[\begin{array}{cc} \epsilon_1 I & 0 \\ 0 & \epsilon_2 I \end{array} \right], \ N = \left[\begin{array}{cc} I & 2\epsilon_1 I \\ -2\epsilon_2 I & I \end{array} \right]$$

By assumption (10.54), the matrix L is positive definite. Using this fact, together with $u^T M u \geq 0$, we obtain

$$\begin{aligned}
\frac{\partial V}{\partial x} &\leq -a\|y\|_2^2 + b\|u\|_2\|y\|_2 \\
&= -\frac{1}{2a}(b\|u\|_2 - a\|y\|_2)^2 + \frac{b^2}{2a}\|u\|_2^2 - \frac{a}{2}\|y\|_2^2 \\
&\leq \frac{b^2}{2a}\|u\|_2^2 - \frac{a}{2}\|y\|_2^2
\end{aligned}$$

where $a = \lambda_{\min}(L) > 0$ and $b = \|N\|_2$. Integrating over $[0, \tau]$, using $V(x) \geq 0$, and taking the square roots, we arrive at

$$\|y_\tau\|_{\mathcal{L}_2} \leq \frac{b}{a}\|u_\tau\|_{\mathcal{L}_2} + \sqrt{2V(x(0))/a}$$

which completes the proof of the theorem. □

Condition (10.54) is satisfied if

- both H_1 and H_2 are input strictly passive,

- both H_1 and H_2 are output strictly passive, or

- one component is passive while the other one is input and output strictly passive.

Theorem 10.7 *Consider the feedback system of Figure 10.15 where H_1 and H_2 are dynamical systems of the form (10.46)–(10.47); that is,*

$$\dot{x}_i = f_i(x_i, e_i)$$
$$y_i = h_i(x_i, e_i)$$

for $i = 1, 2$. Suppose the feedback system has a well-defined state-space model

$$\dot{x} = f(x, u)$$
$$y = h(x, u)$$

where

$$x = \begin{bmatrix} x_1 \\ x_2 \end{bmatrix}, \quad u = \begin{bmatrix} u_1 \\ u_2 \end{bmatrix}, \quad y = \begin{bmatrix} y_1 \\ y_2 \end{bmatrix}$$

f is locally Lipschitz, h is continuous, $f(0,0) = 0$, and $h(0,0) = 0$. Let H_1 and H_2 be passive systems with positive definite storage functions $V_1(x_1)$ and $V_2(x_2)$; that is,

$$e_i^T y_i \geq \frac{\partial V_i}{\partial x_i} f_i(x_i, e_i) + \epsilon_i e_i^T e_i + \delta_i y_i^T y_i + \rho_i \psi_i(x_i), \quad \text{for } i = 1, 2$$

Then, the origin of

$$\dot{x} = f(x, 0) \tag{10.58}$$

is stable and all solutions starting sufficiently near the origin are bounded for all $t \geq 0$. If $V_1(x_1)$ and $V_2(x_2)$ are radially unbounded, then all solutions of (10.58) are bounded. The origin of (10.58) is asymptotically stable in any one of the following cases:

Case 1: $\rho_1 > 0$ and $\rho_2 > 0$;

Case 2: $\rho_1 > 0$, $\epsilon_1 + \delta_2 > 0$, and H_2 is zero-state observable.

Case 3: $\rho_2 > 0$, $\epsilon_2 + \delta_1 > 0$, and H_1 is zero-state observable;

Case 4: $\epsilon_1 + \delta_2 > 0$, $\epsilon_2 + \delta_1 > 0$, and both H_1 and H_2 are zero-state observable.

Furthermore, if, in any one of the four cases, $V_1(x_1)$ and $V_2(x_2)$ are radially unbounded, the origin will be globally asymptotically stable.   ◇

Proof: Take $u_1 = u_2 = 0$. In this case, $e_1 = -y_2$ and $e_2 = y_1$. Using $V(x) = V_1(x_1) + V_2(x_2)$ as a Lyapunov function candidate for the feedback system, we have

$$
\begin{aligned}
\dot{V} &= \frac{\partial V_1}{\partial x_1} f_1(x_1, e_1) + \frac{\partial V_2}{\partial x_2} f_2(x_2, e_2) \\
&\leq \sum_{i=1}^{2} [e_i^T y_i - \epsilon_i e_i^T e_i - \delta_i y_i^T y_i - \rho_i \psi_i(x_i)] \\
&= -y_2^T y_1 - \epsilon_1 y_2^T y_2 - \delta_1 y_1^T y_1 - \rho_1 \psi_1(x_1) \\
&\quad + y_1^T y_2 - \epsilon_2 y_1^T y_1 - \delta_2 y_2^T y_2 - \rho_2 \psi_2(x_2) \\
&= -\rho_1 \psi_1(x_1) - \rho_2 \psi_2(x_2) - (\epsilon_2 + \delta_1) y_1^T y_1 - (\epsilon_1 + \delta_2) y_2^T y_2
\end{aligned}
$$

If all we know is that the two systems are passive, that is, all ϵ_i, δ_i, ρ_i could be zero, we have $\dot{V} \leq 0$, which shows that the origin is stable and there is $c > 0$ such that all solutions starting in $\{V(x) \leq c\}$ remain bounded for all $t \geq 0$. If $V(x)$ is radially unbounded, c can be arbitrarily large. To show asymptotic stability using LaSalle's invariance principle (Corollary 3.1), we need to show that $\dot{V}(x(t)) \equiv 0$ implies $x(t) \equiv 0$. In Case 1, this follows from property (10.49) of $\psi_i(x_i)$. In Case 2,

$$
\rho_1 > 0 \;\Rightarrow\; \psi_1(x_1(t)) \equiv 0 \;\Rightarrow\; x_1(t) \equiv 0
$$

and

$$
\epsilon_1 + \delta_2 > 0 \;\Rightarrow\; y_2(t) \equiv 0 \;\Rightarrow\; e_1(t) \equiv 0
$$

Now,

$$
x_1(t) \equiv 0 \text{ and } e_1(t) \equiv 0 \;\Rightarrow\; y_1(t) \equiv 0 \;\Rightarrow\; e_2(t) \equiv 0
$$

By zero-state observability of H_2, we conclude that $x_2(t) \equiv 0$. Case 3 is analogous to Case 2. Finally, in Case 4,

$$
\epsilon_1 + \delta_2 > 0 \;\Rightarrow\; y_2(t) \equiv 0 \text{ and } \epsilon_2 + \delta_1 > 0 \;\Rightarrow\; y_1(t) \equiv 0
$$

By zero-state observability of H_1 and H_2, we conclude that $x(t) \equiv 0$. If $V(x)$ is radially unbounded, global asymptotic stability follows from Corollary 3.2.   □

It is straightforward to see that the asymptotic stability condition of Theorem 10.7 is satisfied if [27]

- both H_1 and H_2 are state strictly passive,

---

[27] There are additional cases which do not follow directly from the statement of Theorem 10.7, but can be deduced from its proof; see Exercises 10.25 and 10.26.

- one component is state and input strictly passive and the other one is zero-state observable.

- one component is state strictly passive and the other one is output strictly passive and zero-state observable, or

- any one of the cases listed after Theorem 10.6 holds and both H_1 and H_2 are zero-state observable.

Theorem 10.8 *Consider the feedback system of Figure 10.15 where H_1 is a dynamical system of the form*

$$\begin{aligned} \dot{x}_1 &= f_1(x_1, e_1) \\ y_1 &= h_1(x_1, e_1) \end{aligned}$$

and H_2 is a memoryless function of the form

$$y_2 = h_2(t, e_2)$$

Suppose the feedback system has a well-defined state-space model

$$\begin{aligned} \dot{x} &= f(t, x, u) \\ y &= h(t, x, u) \end{aligned}$$

where

$$x = x_1, \ u = \left[\begin{array}{c} u_1 \\ u_2 \end{array} \right], \ y = \left[\begin{array}{c} y_1 \\ y_2 \end{array} \right]$$

f is piecewise continuous in t and locally Lipschitz in (x, u), h is piecewise continuous in t and continuous in (x, u), $f(t, 0, 0) = 0$, and $h(t, 0, 0) = 0$. Let H_1 be a passive system with a positive definite storage function $V_1(x_1)$ and H_2 be a passive system; that is,

$$e_1^T y_1 \geq \frac{\partial V_1}{\partial x_1} f_1(x_1, e_1) + \epsilon_1 e_1^T e_1 + \delta_1 y_1^T y_1 + \rho_1 \psi_1(x_1)$$

$$e_2^T y_2 \geq \epsilon_2 e_2^T e_2 + \delta_2 y_2^T y_2$$

Then, the origin of

$$\dot{x} = f(t, x, 0) \tag{10.59}$$

is uniformly stable and all solutions starting sufficiently near the origin are bounded for all $t \geq 0$. If $V_1(x_1)$ is radially unbounded, then all solutions of (10.59) are bounded. The origin of (10.59) is uniformly asymptotically stable if H_1 is state strictly passive with a positive definite $\psi_1(x_1)$. If H_2 is time invariant, then the origin of (10.59) is asymptotically stable in either of the following cases:

Case 1: $\rho_1 > 0$;

Case 2: $\epsilon_2 + \delta_1 > 0$ and H_1 is zero-state observable.

Furthermore, if $V_1(x_1)$ is radially unbounded, the origin will be globally asymptotically stable. $\diamond$

Proof: Take $u_1 = u_2 = 0$ and proceed as in the proof of Theorem 10.7, using $V(x) = V_1(x_1)$, to obtain

$$\dot{V} \le -\rho_1 \psi_1(x_1) - (\epsilon_2 + \delta_1) y_1^T y_1 - (\epsilon_1 + \delta_2) y_2^T y_2$$

The uniform stability and boundedness conclusions follow from $\dot{V} \le 0$. If $\rho_1 > 0$ and $\psi_1(x)$ is a positive definite function, we apply Theorem 3.8 to conclude uniform asymptotic stability. In the time-invariant case, the closed-loop system is autonomous and we can apply LaSalle's invariance principle, as in the proof of Theorem 10.7, to complete the proof. $\square$

Both Theorems 10.7 and 10.8 require the feedback system to have a well-defined state-space model. This condition can be easily checked from the models of H_1 and H_2. In particular, in the case of Theorem 10.7 it can be easily verified that a state-space model for the feedback system will exist if the equations

$$e_1 = u_1 - h_2(x_2, e_2) \tag{10.60}$$

$$e_2 = u_2 + h_1(x_1, e_1) \tag{10.61}$$

have a unique solution (e_1, e_2) for every (x_1, x_2, u_1, u_2). The functions f and h of the closed-loop state model will satisfy the required smoothness conditions if the functions f_i and h_i of the feedback components H_1 and H_2 are sufficiently smooth. In the case of Theorem 10.8, we have a similar situation if the equations

$$e_1 = u_1 - h_2(t, e_2) \tag{10.62}$$

$$e_2 = u_2 + h_1(x_1, e_1) \tag{10.63}$$

have a unique solution (e_1, e_2) for every (x_1, t, u_1, u_2). The existence of solutions for (10.60)–(10.61) and (10.62)–(10.63) is pursued further in Exercises 10.28 and 10.29.

Example 10.14 Consider the feedback system of Figure 10.15. Let H_1 be a linear, time-invariant, m-input–m-output system represented by the minimal realization

$$\dot{x} = Ax + Be_1$$

$$y_1 = Cx$$

Let H_2 be the memoryless function $y_2 = h(t, e_2)$, where h is piecewise continuous in t and locally Lipschitz in e_2. Suppose $G(s) = C(sI - A)^{-1}B$ is strictly positive real and

$$u^T h(t, u) \geq 0, \quad \forall\, u \in R^m, \ \forall\, t \geq 0$$

The preceding inequality shows that H_2 is passive. We have seen in Example 10.11 that H_1 is state strictly passive with storage function $V(x) = \frac{1}{2}x^T P x$ and state dissipation rate $\rho\psi(x) = \frac{1}{2}\epsilon x^T P x$, where $P > 0$ and $\epsilon > 0$. It can be easily seen that the feedback system has a well-defined state-space model. Thus, from Theorem 10.8 we conclude that the origin of the closed-loop system is globally uniformly asymptotically stable. This is an absolute stability result similar to the ones we have seen in Section 10.1. In fact, this result can be derived as a limiting case of the multivariable circle criterion by considering the sector $[0, kI]$ with $k > 0$ and then taking the limit $k \to \infty$. It can be also derived directly by repeating the proof of Lemma 10.3; see Exercise 10.4. $\triangle$

Example 10.15 Consider the feedback system of Figure 10.15. Let H_1 be an m-input–m-output dynamical system given by

$$\dot{x} = f(x) + G(x)e_1$$
$$y_1 = h(x)$$

where f is locally Lipschitz, G and h are continuous, $f(0) = 0$, and $h(0) = 0$. Let H_2 be the linear time-invariant function $y_2 = ke_2$, $k > 0$. Suppose there is a continuously differentiable positive semidefinite function $V(x)$ such that

$$\frac{\partial V}{\partial x} f(x) \leq 0$$
$$\frac{\partial V}{\partial x} G(x) = h^T(x)$$

We have seen in Example 10.12 that H_1 is passive. On the other hand, it is clear that H_2 is input and output strictly passive. Thus, the conditions of Theorem 10.6 are satisfied with $\epsilon_2 > 0$ and $\delta_2 > 0$, and we conclude that the closed-loop input-output map from (u_1, u_2) to (y_1, y_2) is finite-gain $\mathcal{L}_2$ stable. This same conclusion was obtained in Example 6.11 using Theorem 6.5. Suppose further that $V(x)$ is positive definite and H_1 is zero-state-observable. Then, the conditions of Case 2 of Theorem 10.8 are satisfied and we conclude that the origin of

$$\dot{x} = f(x) - kG(x)h(x)$$

is asymptotically stable. It will be globally asymptotically stable if $V(x)$ is radially unbounded. $\triangle$

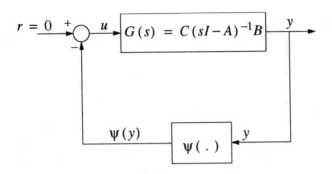

Figure 10.17: The feedback connection of Section 10.4.

10.4   The Describing Function Method

Consider a nonlinear system represented by a feedback connection of a linear time-invariant dynamical system and a nonlinear element, as shown in Figure 10.17. We assume the external input $r = 0$ and study the behavior of the autonomous system

$$
\begin{aligned}
\dot{x} &= Ax + Bu \\
y &= Cx \\
u &= -\psi(y)
\end{aligned}
$$

where $x \in R^n$, $u, y \in R$, (A, B) is controllable, (A, C) is observable, and $\psi : R \to R$ is a memoryless time-invariant nonlinearity. The scalar transfer function of the linear system is given by

$$G(s) = C(sI - A)^{-1}B$$

We are interested in investigating the existence of periodic solutions. A periodic solution satisfies $y(t + 2\pi/\omega) = y(t)$ for all t, where ω is the frequency of oscillation. We shall use a general method for finding periodic solutions, known as *the method of harmonic balance*. The idea of the method is to represent a periodic solution by a Fourier series and seek a frequency ω and a set of Fourier coefficients which satisfy the system's equation. Suppose that $y(t)$ is periodic and let

$$y(t) = \sum_{k=-\infty}^{\infty} a_k \exp(jk\omega t)$$

be its Fourier series, where a_k are complex coefficients,[28] $a_k = \bar{a}_{-k}$ and $j = \sqrt{-1}$. Since $\psi(\cdot)$ is a time-invariant nonlinearity, $\psi(y(t))$ is periodic with the same fre-

---

[28] A bar over a complex variable denotes its complex conjugate.

quency ω and can be written as

$$\psi(y(t)) = \sum_{k=-\infty}^{\infty} c_k \exp(jk\omega t)$$

where each complex coefficient c_k is a function of all a_i's. For $y(t)$ to be a solution of the feedback system, it must satisfy the differential equation

$$d(p)y(t) + n(p)\psi(y(t)) = 0$$

where p is the differential operator $p(\cdot) = d(\cdot)/dt$ and $n(s)$ and $d(s)$ are the numerator and denominator polynomials of $G(s)$. Since

$$p \exp(jk\omega t) = \frac{d}{dt} \exp(jk\omega t) = jk\omega \exp(jk\omega t)$$

we have

$$d(p) \sum_{k=-\infty}^{\infty} a_k \exp(jk\omega t) = \sum_{k=-\infty}^{\infty} d(jk\omega)a_k \exp(jk\omega t)$$

and

$$n(p) \sum_{k=-\infty}^{\infty} c_k \exp(jk\omega t) = \sum_{k=-\infty}^{\infty} n(jk\omega)c_k \exp(jk\omega t)$$

Substitution of these expressions back into the differential equation yields

$$\sum_{k=-\infty}^{\infty} [d(jk\omega)a_k + n(jk\omega)c_k] \exp(jk\omega t) = 0$$

Using the orthogonality of the functions $\exp(jk\omega t)$ for different values of k, we find that the Fourier coefficients must satisfy

$$G(jk\omega)c_k + a_k = 0 \qquad (10.64)$$

for all integers k. Because $G(jk\omega) = \bar{G}(-jk\omega)$, $a_k = \bar{a}_{-k}$, and $c_k = \bar{c}_{-k}$, we need only look at (10.64) for $k \geq 0$. Equation (10.64) is an infinite-dimensional equation, which we can hardly solve. We need to find a finite-dimensional approximation of (10.64). Noting that the transfer function $G(s)$ is strictly proper, that is, $G(j\omega) \to 0$ as $\omega \to \infty$, it is reasonable to assume that there is an integer $q > 0$ such that for all $k > q$, $|G(jk\omega)|$ is small enough to replace $G(jk\omega)$ (and consequently a_k) by 0. This approximation reduces (10.64) to a finite-dimensional problem

$$G(jk\omega)\hat{c}_k + \hat{a}_k = 0, \quad k = 0, 1, 2, \ldots, q \qquad (10.65)$$

where the Fourier coefficients are written with a hat accent to emphasize that a solution of (10.65) is only an approximation to the solution of (10.64). In essence, we can proceed to solve (10.65). However, the complexity of the problem will grow with q and, for a large q, the finite-dimensional problem (10.65) might still be difficult to solve. The simplest problem results if we can choose $q = 1$. This, of course, requires the transfer function $G(s)$ to have sharp "low-pass filtering" characteristics to allow us to approximate $G(jk\omega)$ by 0 for all $k > 1$. Even though we know $G(s)$, we cannot judge whether this is a good approximation since we do not know the frequency of oscillation ω. Nevertheless, the classical describing function method makes this approximation and sets $\hat{a}_k = 0$ for $k > 1$ to reduce the problem to one of solving the two equations

$$G(0)\hat{c}_0(\hat{a}_0, \hat{a}_1) + \hat{a}_0 = 0 \qquad (10.66)$$

$$G(j\omega)\hat{c}_1(\hat{a}_0, \hat{a}_1) + \hat{a}_1 = 0 \qquad (10.67)$$

Notice that (10.66)–(10.67) define one real equation (10.66) and one complex equation (10.67) in two real unknowns, ω and $\hat{a}_0$, and a complex unknown $\hat{a}_1$. When expressed as real quantities, they define three equations in four unknowns. This is expected since the time origin is arbitrary for an autonomous system, so if $(\hat{a}_0, \hat{a}_1)$ satisfies the equation then $(\hat{a}_0, \hat{a}_1 e^{j\theta})$ will give another solution for arbitrary real θ. To take care of this nonuniqueness, we take the first harmonic of $y(t)$ to be $a \sin \omega t$, with $a \geq 0$; that is, we choose the time origin such that the phase of the first harmonic is zero. Using

$$a \sin \omega t = \frac{a}{2j}[\exp(j\omega t) - \exp(-j\omega t)] \Rightarrow \hat{a}_1 = \frac{a}{2j}$$

we rewrite (10.66)–(10.67) as

$$G(0)\hat{c}_0\left(\hat{a}_0, \frac{a}{2j}\right) + \hat{a}_0 = 0 \qquad (10.68)$$

$$G(j\omega)\hat{c}_1\left(\hat{a}_0, \frac{a}{2j}\right) + \frac{a}{2j} = 0 \qquad (10.69)$$

Since (10.68) does not depend on ω, it may be solved for $\hat{a}_0$ as a function of a. Note that if $\psi(\cdot)$ is an odd function, that is,

$$\psi(-y) = -\psi(y)$$

then $\hat{a}_0 = \hat{c}_0 = 0$ is a solution of (10.68) because

$$\hat{c}_0 = \frac{\omega}{2\pi} \int_0^{2\pi/\omega} \psi(\hat{a}_0 + a \sin \omega t) \, dt$$

For convenience, let us restrict our attention to nonlinearities with odd symmetry and take $\hat{a}_0 = \hat{c}_0 = 0$. Then, we can rewrite (10.69) as

$$G(j\omega)\hat{c}_1\left(0, \frac{a}{2j}\right) + \frac{a}{2j} = 0 \qquad (10.70)$$

The coefficient $\hat{c}_1(0, a/2j)$ is the complex Fourier coefficient of the first harmonic at the output of the nonlinearity when its input is the sinusoidal signal $a \sin \omega t$. It is given by

$$\hat{c}_1(0, a/2j) = \frac{\omega}{2\pi} \int_0^{2\pi/\omega} \psi(a \sin \omega t) \exp(-j\omega t)\ dt$$

$$= \frac{\omega}{2\pi} \int_0^{2\pi/\omega} [\psi(a \sin \omega t) \cos \omega t - j\psi(a \sin \omega t) \sin \omega t]\ dt$$

The first term under the integral sign is an odd function while the second term is an even function. Therefore, the integration of the first term over one complete cycle is zero, and the integral simplifies to

$$\hat{c}_1(0, a/2j) = -j\frac{\omega}{\pi} \int_0^{\pi/\omega} \psi(a \sin \omega t) \sin \omega t\ dt$$

Define a function $\Psi(a)$ by

$$\Psi(a) = \frac{\hat{c}_1(0, a/2j)}{a/2j} = \frac{2\omega}{\pi a} \int_0^{\pi/\omega} \psi(a \sin \omega t) \sin \omega t\ dt \qquad (10.71)$$

so that (10.70) can be rewritten as

$$[G(j\omega)\Psi(a) + 1]a = 0 \qquad (10.72)$$

Since we are not interested in a solution with $a = 0$, we can solve (10.72) completely by finding all solutions of

$$G(j\omega)\Psi(a) + 1 = 0 \qquad (10.73)$$

Equation (10.73) is known as the *first-order harmonic balance equation*, or simply the *harmonic balance equation*. The function $\Psi(a)$ defined by (10.71) is called *the describing function* of the nonlinearity ψ. It is obtained by applying a sinusoidal signal $a \sin \omega t$ at the input of the nonlinearity and calculating the ratio of the Fourier coefficient of the first harmonic at the output to a. It can be thought of as an "equivalent gain" of a linear time-invariant element whose response to $a \sin \omega t$ is $\Psi(a)a \sin \omega t$. This equivalent gain concept (sometimes called *equivalent linearization*) can be applied to more general time-varying nonlinearities or nonlinearities with memory, like hysteresis and backlash.[29] In that general context, the describing function might be complex and dependent on both a and ω. We shall only deal with describing functions of odd, time-invariant, memoryless nonlinearities for which $\Psi(a)$ is real, dependent only on a, and given by the expression

$$\Psi(a) = \frac{2}{\pi a} \int_0^\pi \psi(a \sin \theta) \sin \theta \, d\theta \qquad (10.74)$$

which is obtained from (10.71) by changing the integration variable from t to $\theta = \omega t$.

The describing function method states that if (10.73) has a solution (a_s, ω_s), then there is "probably" a periodic solution of the system with frequency and amplitude (at the input of the nonlinearity) close to ω_s and a_s. Conversely, if (10.73) has no solutions, then the system "probably" does not have a periodic solution. More analysis is needed to replace the word "probably" with "certainly," and to quantify the phrase "close to ω_s and a_s," when there is a periodic solution. We shall postpone these investigations until a later point in the section. For now, we would like to look more closely at the calculation of the describing function and the question of solving the harmonic balance equation (10.73). The following three examples illustrate the calculation of describing functions for odd nonlinearities.

Example 10.16 Consider the signum nonlinearity

$$\psi(y) = \text{sgn}(y) = \begin{cases} 1, & \text{if } y > 0 \\ 0, & \text{if } y = 0 \\ -1, & \text{if } y < 0 \end{cases}$$

The describing function is given by

$$\Psi(a) = \frac{2}{\pi a} \int_0^\pi \psi(a \sin \theta) \sin \theta \, d\theta = \frac{2}{\pi a} \int_0^\pi \sin \theta \, d\theta = \frac{4}{\pi a}$$

$\triangle$

Example 10.17 Consider the *piecewise-linear* nonlinearity of Figure 10.18. If a sinusoidal input to this nonlinearity has amplitude $a \le \delta$, the nonlinearity will act as a linear gain. The output will be a sinusoid with amplitude $s_1 a$. Hence, the describing function is $\Psi(a) = s_1$, independent of a. When $a > \delta$, we divide the integral on the right-hand side of (10.74) into pieces, with each piece corresponding to a linear portion of $\psi(\cdot)$. Furthermore, using the odd symmetry of the output waveform, we simplify the integration to

$$\Psi(a) = \frac{2}{\pi a} \int_0^\pi \psi(a \sin \theta) \sin \theta \, d\theta = \frac{4}{\pi a} \int_0^{\pi/2} \psi(a \sin \theta) \sin \theta \, d\theta$$

---

[29] See [76] or [13].

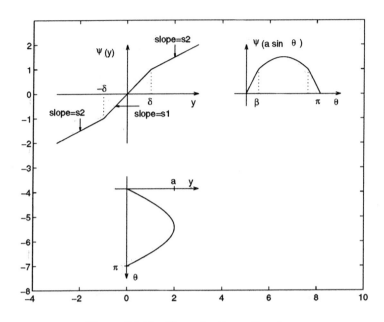

Figure 10.18: Piecewise-linear function.

$$
\begin{aligned}
&= \frac{4}{\pi a} \int_0^\beta a s_1 \sin^2 \theta \, d\theta \\
&\quad + \frac{4}{\pi a} \int_\beta^{\pi/2} [\delta s_1 + s_2(a \sin \theta - \delta)] \sin \theta \, d\theta, \quad \beta = \sin^{-1}\left(\frac{\delta}{a}\right) \\
&= \frac{2s_1}{\pi}\left(\beta - \frac{1}{2}\sin 2\beta\right) + \frac{4\delta(s_1 - s_2)}{\pi a}\left(\cos\beta - \cos\frac{\pi}{2}\right) \\
&\quad + \frac{2s_2}{\pi}\left(\frac{\pi}{2} - \frac{1}{2}\sin\pi - \beta + \frac{1}{2}\sin 2\beta\right) \\
&= \frac{2(s_1 - s_2)}{\pi}\left(\beta + \frac{\delta}{a}\cos\beta\right) + s_2
\end{aligned}
$$

Thus,

$$
\Psi(a) = \frac{2(s_1 - s_2)}{\pi}\left[\sin^{-1}\left(\frac{\delta}{a}\right) + \frac{\delta}{a}\sqrt{1 - \left(\frac{\delta}{a}\right)^2}\right] + s_2
$$

A sketch of the describing function is shown in Figure 10.19. By selecting specific values for δ and the slopes s_1 and s_2, we can obtain the describing function of several common nonlinearities. For example, the saturation nonlinearity is a special case

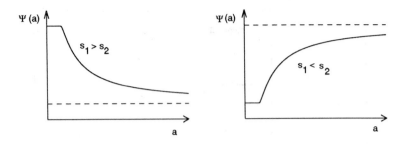

Figure 10.19: Describing function for the piecewise nonlinearity of Figure 10.18.

of the piecewise-linear nonlinearity of Figure 10.18 with $\delta = 1$, $s_1 = 1$, and $s_2 = 0$. Hence, its describing function is given by

$$
\Psi(a) = \begin{cases} 1, & \text{if } 0 \leq a \leq 1 \\[2mm] \frac{2}{\pi}\left[\sin^{-1}\left(\frac{1}{a}\right) + \frac{1}{a}\sqrt{1 - \left(\frac{1}{a}\right)^2}\right], & \text{if } a > 1 \end{cases}
$$

$\triangle$

Example 10.18 Consider an odd nonlinearity that satisfies the sector condition

$$
\alpha y^2 \leq y\psi(y) \leq \beta y^2
$$

for all $y \in R$. The describing function $\Psi(a)$ satisfies the lower bound

$$
\Psi(a) = \frac{2}{\pi a}\int_0^\pi \psi(a\sin\theta)\sin\theta \; d\theta \geq \frac{2\alpha}{\pi}\int_0^\pi \sin^2\theta \; d\theta = \alpha
$$

and the upper bound

$$
\Psi(a) = \frac{2}{\pi a}\int_0^\pi \psi(a\sin\theta)\sin\theta \; d\theta \leq \frac{2\beta}{\pi}\int_0^\pi \sin^2\theta \; d\theta = \beta
$$

Therefore,

$$
\alpha \leq \Psi(a) \leq \beta, \quad \forall \, a \geq 0
$$

$\triangle$

Since the describing function $\Psi(a)$ is real, (10.73) can be rewritten as

$$
\{Re[G(j\omega)] + jIm[G(j\omega)]\}\,\Psi(a) + 1 = 0
$$

where $Re[\cdot]$ and $Im[\cdot]$ denote the real and imaginary parts of a complex function. This equation is equivalent to the two real equations

$$1 + \Psi(a)Re[G(j\omega)] = 0 \tag{10.75}$$
$$Im[G(j\omega)] = 0 \tag{10.76}$$

Since (10.76) is independent of a, we can solve it first for ω to determine the possible frequencies of oscillation. For each solution ω, we solve (10.75) for a. Note that the possible frequencies of oscillation are determined solely by the transfer function $G(s)$; they are independent of the nonlinearity $\psi(\cdot)$. The nonlinearity determines the corresponding value of a, that is, the possible amplitude of oscillation. This procedure can be carried out analytically for low-order transfer functions, as illustrated by the following examples.

Example 10.19 Let

$$G(s) = \frac{1}{s(s+1)(s+2)}$$

and consider two nonlinearities: the signum nonlinearity and the saturation nonlinearity. By simple manipulation, we can write $G(j\omega)$ as

$$G(j\omega) = \frac{-3\omega - j\left(2 - \omega^2\right)}{9\omega^3 + \omega\left(2 - \omega^2\right)^2}$$

Equation (10.76) takes the form

$$\frac{\left(2 - \omega^2\right)}{9\omega^3 + \omega\left(2 - \omega^2\right)^2} = 0$$

which has one positive root $\omega = \sqrt{2}$. Note that for each positive root of (10.76) there is a negative root of equal magnitude. We only consider the positive roots. Note also that a root at $\omega = 0$ would be of no interest because it would not give rise to a nontrivial periodic solution. Evaluating $Re[G(j\omega)]$ at $\omega = \sqrt{2}$ and substituting it in (10.75), we obtain $\Psi(a) = 6$. All this information has been gathered without specifying the nonlinearity $\psi(\cdot)$. Consider now the signum nonlinearity. We found in Example 10.16 that $\Psi(a) = 4/\pi a$. Therefore, $\Psi(a) = 6$ has a unique solution $a = 2/3\pi$. Now we can say that the nonlinear system formed of $G(s)$ and the signum nonlinearity will "probably" oscillate with a frequency close to $\sqrt{2}$ and an amplitude (at the input of the nonlinearity) close to $2/3\pi$. Consider next the saturation nonlinearity. We found in Example 10.17 that $\Psi(a) \leq 1$ for all a. Therefore, $\Psi(a) = 6$ has no solutions. Therefore, we expect that the nonlinear system formed of $G(s)$ and the saturation nonlinearity will not have sustained oscillations. $\triangle$

Example 10.20 Let
$$G(s) = \frac{-s}{s^2 + 0.8s + 8}$$
and consider two nonlinearities: the saturation nonlinearity and a dead-zone non-linearity that is a special case of the piecewise-linear nonlinearity of Example 10.17 with $s_1 = 0$, $s_2 = 0.5$, and $\delta = 1$. We can write $G(j\omega)$ as

$$G(j\omega) = \frac{-0.8\omega^2 - j\omega\left(8 - \omega^2\right)}{0.64\omega^2 + \left(8 - \omega^2\right)^2}$$

Equation (10.76) has a unique positive root $\omega = 2\sqrt{2}$. Evaluating $Re[G(j\omega)]$ at $\omega = 2\sqrt{2}$ and substituting it in (10.75), we obtain $\Psi(a) = 0.8$. For the saturation nonlinearity, the describing function is

$$\Psi(a) = \begin{cases} 1, & \text{if } 0 \le a \le 1 \\ \frac{2}{\pi}\left[\sin^{-1}\left(\frac{1}{a}\right) + \frac{1}{a}\sqrt{1 - \left(\frac{1}{a}\right)^2}\right], & \text{if } a > 1 \end{cases}$$

Hence, $\Psi(a) = 0.8$ has a unique solution $a = 1.455$. Therefore, we expect that the nonlinear system formed of $G(s)$ and the saturation nonlinearity will oscillate with frequency close to $2\sqrt{2}$ and amplitude (at the input of the nonlinearity) close to 1.455. For the dead-zone nonlinearity, the describing function $\Psi(a)$ is less than 0.8 for all a. Hence, $\Psi(a) = 0.8$ has no solutions, and we expect that the nonlinear system formed of $G(s)$ and the dead-zone nonlinearity will not have sustained os-cillations. In this particular example, we can confirm this no oscillation conjecture by showing that the system is absolutely stable for a class of sector nonlinearities which includes the given dead-zone nonlinearity. It can be easily checked that

$$Re[G(j\omega)] \ge -1.25, \quad \forall\, \omega \in R$$

Therefore, from the circle criterion (Theorem 10.2) we know that the system is absolutely stable for a sector $[0, \beta]$ with $\beta < 0.8$. The given dead-zone nonlinearity belongs to this sector. Hence, the origin of the state space is globally asymptotically stable and the system cannot have sustained oscillations. $\triangle$

Example 10.21 Consider Raleigh's equation
$$\ddot{z} + z = \epsilon\left(\dot{z} - \tfrac{1}{3}\dot{z}^3\right)$$

where ϵ is a positive parameter. This equation can be represented in the feedback form of Figure 10.2. Let $u = -\tfrac{1}{3}\dot{z}^3$ and rewrite the system's equation as

$$\begin{aligned} \ddot{z} - \epsilon\dot{z} + z &= \epsilon u \\ u &= -\tfrac{1}{3}\dot{z}^3 \end{aligned}$$

The first equation defines a linear system. Taking $y = \dot{z}$ to be the output of the system, its transfer function is

$$G(s) = \frac{\epsilon s}{s^2 - \epsilon s + 1}$$

The second equation defines a nonlinearity $\psi(y) = \frac{1}{3}y^3$. The two equations together represent the system in the feedback form of Figure 10.2. The describing function of $\psi(y) = \frac{1}{3}y^3$ is given by

$$\Psi(a) = \frac{2}{3\pi a} \int_0^\pi (a \sin \theta)^3 \sin \theta \, d\theta = \frac{1}{4}a^2$$

The function $G(j\omega)$ can be written as

$$G(j\omega) = \frac{j\epsilon\omega[(1 - \omega^2) + j\epsilon\omega]}{(1 - \omega^2)^2 + \epsilon^2\omega^2}$$

The equation $Im[G(j\omega)] = 0$ yields $\omega(1 - \omega^2) = 0$; hence, there is a unique positive solution $\omega = 1$. Then,

$$1 + \Psi(a)Re[G(j\omega)] = 0 \implies a = 2$$

Therefore, we expect that Raleigh's equation has a periodic solution of frequency near 1 rad/sec and that the amplitude of oscillation in $\dot{z}$ is near 2. $\triangle$

For higher-order transfer functions, solving the harmonic balance equation (10.73) analytically might be very complicated. Of course, we can always resort to numerical methods for solving (10.73). However, the power of the describing function method is not in solving (10.73) analytically or numerically; rather, it is the graphical solution of (10.73) that made the method popular. Equation (10.73) can be rewritten as

$$G(j\omega) = - \frac{1}{\Psi(a)} \tag{10.77}$$

or

$$\frac{1}{G(j\omega)} = -\Psi(a) \tag{10.78}$$

Equation (10.77) suggests that we can solve (10.73) by plotting the Nyquist plot of $G(j\omega)$ for $\omega > 0$ and the locus of $-1/\Psi(a)$ for $a \geq 0$. Intersections of these loci give the solutions of (10.73). Since $\Psi(a)$ is real for odd nonlinearities, the locus of $-1/\Psi(a)$ in the complex plane will be confined to the real axis. Equation (10.78) suggests a similar procedure by plotting the inverse Nyquist plot of $G(j\omega)$ (that is, the locus in the complex plane of $1/G(j\omega)$ as ω varies) and the locus of $-\Psi(a)$. The important role of Nyquist plots in classical control theory made this graphical implementation of the describing function method a popular tool with control engineers as they faced nonlinearities in control systems.

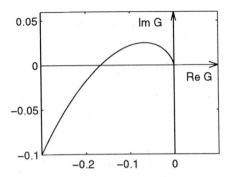

Figure 10.20: Nyquist plot for Example 10.22.

Example 10.22 Consider again the transfer function $G(s)$ of Example 10.19. The Nyquist plot of $G(j\omega)$ is shown in Figure 10.20. It intersects the real axis at $(-1/6, 0)$. For odd nonlinearities, (10.73) will have a solution if the locus of $-1/\Psi(a)$ on the real axis includes this point of intersection. $\triangle$

Let us turn now to the question of justifying the describing function method. Being an approximate method for solving the infinite-dimensional equation (10.64), the describing function method can be justified by providing estimates of the error caused by the approximation. In the interest of simplicity, we shall pursue this analysis only for nonlinearities with the following two features:[30]

- Odd nonlinearity, that is, $\psi(y) = -\psi(-y)$, $\forall\, y \neq 0$.

- Single-valued nonlinearity with a slope between α and β; that is,

$$\alpha(y_2 - y_1) \leq [\psi(y_2) - \psi(y_1)] \leq \beta(y_2 - y_1)$$

  for all real numbers y_1 and $y_2 > y_1$.

A nonlinearity $\psi(\cdot)$ with these features belongs to a sector $[\alpha, \beta]$. Hence, from Example 10.18, its describing function satisfies $\alpha \leq \Psi(a) \leq \beta$ for all $a \geq 0$. It should be noted, however, that the slope restriction is not the same as the sector condition. A nonlinearity may satisfy the foregoing slope restriction with bounds α and β, and could belong to a sector $[\bar{\alpha}, \bar{\beta}]$ with different bounds $\bar{\alpha}$ and $\bar{\beta}$.[31] We emphasize that in the forthcoming analysis, we should use the slope bounds α and β, not the sector boundaries $\bar{\alpha}$ and $\bar{\beta}$.

---

[30] See [18], [112], and [170] for describing function theory for more general nonlinearities.
[31] Verify that $[\bar{\alpha}, \bar{\beta}] \subset [\alpha, \beta]$.

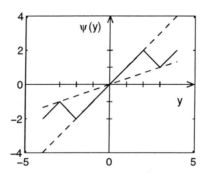

Figure 10.21: Nonlinearity of Example 10.23.

Example 10.23 Consider the piecewise odd nonlinearity

$$\psi(y) = \begin{cases} y, & \text{for } 0 \leq y \leq 2 \\ 4 - y, & \text{for } 2 \leq y \leq 3 \\ y - 2, & \text{for } y \geq 3 \end{cases}$$

shown in Figure 10.21. The nonlinearity satisfies the slope restriction

$$-1 \leq \frac{\psi(y_2) - \psi(y_1)}{y_2 - y_1} \leq 1$$

as well as the sector condition

$$\frac{1}{3} \leq \frac{\psi(y)}{y} \leq 1$$

In the forthcoming analysis, we should take $\alpha = -1$ and $\beta = 1$.   △

We shall restrict our attention to the question of existence of half-wave symmetric periodic solutions;[32] that is, periodic solutions that only have odd harmonics. This is a reasonable restriction in view of the odd symmetry of ψ. The Fourier coefficients of the odd harmonics of a periodic solution $y(t)$ satisfy (10.64) for $k = 1, 3, 5, \ldots$. The basic idea of the error analysis is to split the periodic solution $y(t)$ into a first harmonic $y_1(t)$ and higher harmonics $y_h(t)$. We choose the time origin such that the phase of the first harmonic is zero; that is, $y_1(t) = a \sin \omega t$. Thus,

$$y(t) = a \sin \omega t + y_h(t)$$

---

[32] This restriction is made only for convenience. See [111] for a more general analysis that does not make this assumption.

Using this representation, the Fourier coefficients of the first harmonic of $y(t)$ and $\psi(y(t))$ are

$$a_1 = \frac{a}{2j}$$

$$c_1 = \frac{\omega}{\pi} \int_0^{\pi/\omega} \psi(a \sin \omega t + y_h(t)) \exp(-j\omega t) \, dt$$

From (10.64), with $k = 1$ we have

$$G(j\omega)c_1 + a_1 = 0$$

Introducing the function

$$\Psi^*(a, y_h) = \frac{c_1}{a_1} = j \frac{2\omega}{\pi a} \int_0^{\pi/\omega} \psi(a \sin \omega t + y_h(t)) \exp(-j\omega t) \, dt$$

we can rewrite the equation as

$$\frac{1}{G(j\omega)} + \Psi^*(a, y_h) = 0 \qquad\qquad (10.79)$$

Adding $\Psi(a)$ to both sides of (10.79), we can rewrite it as

$$\frac{1}{G(j\omega)} + \Psi(a) = \delta\Psi \qquad\qquad (10.80)$$

where

$$\delta\Psi = \Psi(a) - \Psi^*(a, y_h)$$

When $y_h = 0$, $\Psi^*(a, 0) = \Psi(a)$; hence, $\delta\Psi = 0$ and (10.80) reduces to the harmonic balance equation

$$\frac{1}{G(j\omega)} + \Psi(a) = 0 \qquad\qquad (10.81)$$

Therefore, the harmonic balance equation (10.81) is an approximate version of the exact equation (10.80). The error term $\delta\Psi$ cannot be found exactly, but its size can often be estimated. Our next step is to find an upper bound on $\delta\Psi$. To that end, let us define two functions $\rho(\omega)$ and $\sigma(\omega)$. Start by drawing the locus of $1/G(j\omega)$ in the complex plane. On the same graph paper, draw a (critical) circle with the interval $[-\beta, -\alpha]$ on the real axis as a diameter. Notice that the locus of $-\Psi(a)$ lies inside this circle on the real axis since $\alpha \leq \Psi(a) \leq \beta$. Now consider an ω such that the points on the locus $1/G$ corresponding to $k\omega$ ($k > 1$ and odd) lie outside the critical circle, as shown in Figure 10.22. The distance from any one of these points

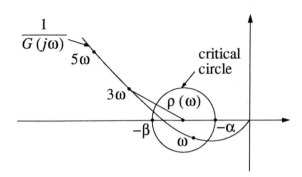

Figure 10.22: Finding $\rho(\omega)$.

to the center of the critical circle is

$$\left| \frac{\alpha + \beta}{2} + \frac{1}{G(jk\omega)} \right|$$

Define

$$\rho(\omega) = \inf_{k>1;k \text{ odd}} \left| \frac{\alpha + \beta}{2} + \frac{1}{G(jk\omega)} \right| \tag{10.82}$$

Note that we have defined $\rho(\omega)$ only for ω at which $1/G(jk\omega)$ lies outside the critical circle for all $k = 3, 5, \ldots$; that is, for ω in the set

$$\Omega = \{\omega \mid \rho(\omega) > \tfrac{1}{2}(\beta - \alpha)\}$$

On any connected subset Ω' of Ω, define

$$\sigma(\omega) = \frac{\left(\frac{\beta - \alpha}{2} \right)^2}{\rho(\omega) - \frac{\beta - \alpha}{2}} \tag{10.83}$$

The positive quantity $\sigma(\omega)$ is an upper bound on the error term $\delta\Psi$, as stated in the following lemma.

Lemma 10.7 *Under the stated assumptions,*

$$\frac{\omega}{\pi} \int_0^{2\pi/\omega} y_h^2(t) \, dt \leq \left[\frac{2\sigma(\omega)a}{\beta - \alpha} \right]^2, \quad \forall \, \omega \in \Omega' \tag{10.84}$$

$$|\delta\Psi| \leq \sigma(\omega), \quad \forall \, \omega \in \Omega' \tag{10.85}$$

Proof: Appendix A.18.

The proof of this lemma is based on writing an equation for $y_h(t)$ in the form $y_h = T(y_h)$ and showing that $T(\cdot)$ is a contraction mapping. This allows us to calculate the upper bound (10.84), which is then used to calculate the upper bound on the error term (10.85). The slope restrictions on the nonlinearity ψ are used in showing that $T(\cdot)$ is a contraction mapping.

Using the bound (10.85) in (10.80), we see that a necessary condition for the existence of a half-wave symmetric periodic solution with $\omega \in \Omega'$ is

$$\left| \frac{1}{G(j\omega)} + \Psi(a) \right| \leq \sigma(\omega)$$

Geometrically, this condition states that the point $-\Psi(a)$ must be contained in a circle with a center at $1/G(j\omega)$ and radius $\sigma(\omega)$. For each $\omega \in \Omega' \subset \Omega$, we can draw such an error circle. The envelope of all error circles over the connected set Ω' forms an uncertainty band. The reason for choosing a subset of Ω is that, as ω approaches the boundary of Ω, the error circles become arbitrarily large and cease to give any useful information. The subset Ω' should be chosen with the objective of drawing a narrow band. If $G(j\omega)$ has sharp low-pass filtering characteristics, the uncertainty band can be quite narrow over Ω'. Note that $\rho(\omega)$ is a measure of the low-pass filtering characteristics of $G(j\omega)$; for the smaller $|G(jk\omega)|$ for $k > 1$, the larger $\rho(\omega)$, as seen from (10.82). A large $\rho(\omega)$ results in a small radius $\sigma(\omega)$ for the error circle, as seen from (10.83).

We are going to look at intersections of the uncertainty band with the locus of $-\Psi(a)$. If no part of the band intersects the $-\Psi(a)$ locus, then clearly (10.80) has no solution with $\omega \in \Omega'$. If the band intersects the locus completely, as in Figure 10.23, then we expect that there is a solution. This is indeed true provided we exclude some degenerate cases. Actually, we can find error bounds by examining the intersection. Let a_1 and a_2 be the amplitudes corresponding to the intersections of the boundary of the uncertainty band with the $-\Psi(a)$ locus. Let ω_1 and ω_2 be the frequencies corresponding to the error circles of radii $\sigma(\omega_1)$ and $\sigma(\omega_2)$, which are tangent to the $-\Psi(a)$ locus on either side of it. Define a rectangle Γ in the (ω, a) plane by

$$\Gamma = \{(\omega, a) \mid \omega_1 < \omega < \omega_2,\ a_1 < a < a_2\}$$

The rectangle Γ contains the point (ω_s, a_s) for which the loci of $1/G$ and $-\Psi$ intersect, that is, the solution of the harmonic balance equation (10.81). It turns out that if certain regularity conditions hold, then it is possible to show that (10.80) has a solution in the closure of Γ. These regularity conditions are

$$\left. \frac{d}{da} \Psi(a) \right|_{a=a_s} \neq 0; \quad \left. \frac{d}{d\omega} Im[G(j\omega)] \right|_{\omega=\omega_s} \neq 0$$

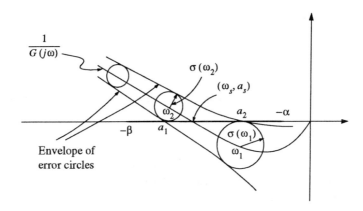

Figure 10.23: A complete intersection.

A complete intersection between the uncertainty band and the $-\Psi(a)$ locus can now be precisely defined as taking place when the $1/G(j\omega)$ locus intersects the $-\Psi(a)$ locus and a finite set Γ can be defined, as shown, such that (w_s, a_s) is the unique intersection point in Γ and the regularity conditions hold.

Finally, notice that at high frequencies for which all harmonics (including the first) have the corresponding $1/G(j\omega)$ points outside the critical circle, we do not need to draw the uncertainty band. Therefore, we define a set

$$\tilde{\Omega} = \left\{ \omega \;\middle|\; \left| \frac{\alpha + \beta}{2} + \frac{1}{G(jk\omega)} \right| > \frac{\beta - \alpha}{2}, \; k = 1, 3, 5, \ldots \right\}$$

and take the smallest frequency in $\tilde{\Omega}$ as the largest frequency in Ω', then decrease ω until the error circles become uncomfortably large.

The main result of this section is the following theorem on the justification of the describing function method.

Theorem 10.9 *Consider the feedback connection of Figure 10.2, where the nonlinearity $\psi(\cdot)$ is memoryless, time invariant, odd, and single valued with slopes between α and β. Draw the loci of $1/G(j\omega)$ and $-\Psi(a)$ in the complex plane and construct the critical circle and the band of uncertainty as described earlier. Then,*

- *the system has no half-wave symmetric periodic solutions with fundamental frequency $\omega \in \tilde{\Omega}$.*

- *the system has no half-wave symmetric periodic solutions with fundamental frequency $\omega \in \Omega'$ if the corresponding error circle does not intersect the $-\Psi(a)$ locus.*

- *for each complete intersection defining a set Γ in the (ω, a) plane, there is at least one half-wave symmetric periodic solution*

$$y(t) = a \sin \omega t + y_h(t)$$

with (ω, a) in $\bar{\Gamma}$ and $y_h(t)$ satisfies the bound (10.84). $\diamond$

Proof: Appendix A.19.

Note that the theorem gives a sufficient condition for oscillation and a sufficient condition for nonoscillation. Between the two conditions, there is an area of ambiguity where we cannot reach conclusions of oscillation or nonoscillation.

Example 10.24 Consider again

$$G(s) = \frac{-s}{s^2 + 0.8s + 8}$$

together with the saturation nonlinearity. We have seen in Example 10.20 that the harmonic balance equation has a unique solution $w_s = 2\sqrt{2} \approx 2.83$ and $a_s = 1.455$. The saturation nonlinearity satisfies the slope restrictions with $\alpha = 0$ and $\beta = 1$. Therefore, the critical circle is centered at -0.5 and its radius is 0.5. The function $1/G(j\omega)$ is given by

$$\frac{1}{G(j\omega)} = -0.8 + j\frac{8 - \omega^2}{\omega}$$

Hence, the locus of $1/G(j\omega)$ lies on the line $Re[s] = -0.8$, as shown in Figure 10.24. The radius of the error circle $\sigma(\omega)$ has been calculated for eight frequencies starting with $\omega = 2.65$ and ending with $\omega = 3.0$, with uniform increments of 0.05. The centers of the error circles are spread on the line $Re[s] = -0.8$ inside the critical circle. The value of $\sigma(\omega)$ at $\omega = 2.65$ is 0.0388 and at $\omega = 3.0$ is 0.0321, with monotonic change between the two extremes. In all cases, the closest harmonic to the critical circle is the third harmonic, so that the infimum in (10.82) is achieved at $k = 3$. The boundaries of the uncertainty band are almost vertical. The intersection of the uncertainty band with $-\Psi(a)$ locus correspond to the points $a_1 = 1.377$ and $a_2 = 1.539$. The error circle corresponding to $\omega = 2.85$ is almost tangent to the real axis from the lower side, so we take $\omega_2 = 2.85$. The error circle corresponding to $\omega = 2.8$ is the closest circle to be tangent to the real axis from the upper side. This means that $\omega_1 > 2.8$. Trying $\omega = 2.81$, we have obtained a circle that is almost tangent to the real axis. Therefore, we define the set Γ as

$$\Gamma = \{(\omega, a) \mid 2.81 < \omega < 2.85, \ 1.377 < a < 1.539\}$$

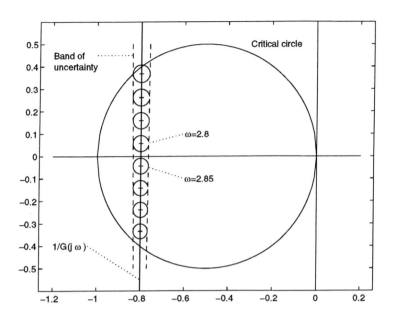

Figure 10.24: Uncertainty band for Example 10.24.

There is only one intersection point in Γ. We need to check the regularity conditions. The derivative

$$\frac{d}{da}\Psi(a) = \frac{2}{\pi}\frac{d}{da}\left[\sin^{-1}\left(\frac{1}{a}\right) + \frac{1}{a}\sqrt{1 - \left(\frac{1}{a}\right)^2}\right] = -\frac{4}{\pi a^3}\sqrt{a^2 - 1}$$

is different from zero at $a = 1.455$, and

$$\frac{d}{d\omega}Im[G(j\omega)]\bigg|_{\omega=\sqrt{8}} = \frac{2}{(0.8)^2} \neq 0$$

Thus, by Theorem 10.9, we conclude that the system indeed has a periodic solution. Moreover, we conclude that the frequency of oscillation ω belongs to the interval $[2.81, 2.85]$, and the amplitude of the first harmonic of the oscillation at the input of the nonlinearity belongs to the interval $[1.377, 1.539]$. From the bound (10.84) of Lemma 10.7, we know also that the higher-harmonic component $y_h(t)$ satisfies

$$\frac{\omega}{\pi}\int_0^{2\pi/\omega} y_h^2(t)\, dt \leq 0.0123, \quad \forall\, (\omega, a) \in \Gamma$$

which shows that the waveform of the oscillating signal at the nonlinearity input is fairly close to its first harmonic $a\sin\omega t$. Δ

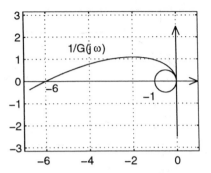

Figure 10.25: Inverse Nyquist plot and critical circle for Example 10.25.

Example 10.25 Reconsider Example 10.19 with

$$G(s) = \frac{1}{s(s+1)(s+2)}$$

and the saturation nonlinearity. The nonlinearity satisfies the slope restriction with $\alpha = 0$ and $\beta = 1$. The inverse Nyquist plot of $G(j\omega)$, shown in Figure 10.25, lies outside the critical circle for all $\omega > 0$. Hence, $\tilde{\Omega} = (0, \infty)$ and we conclude that there is no oscillation. $\triangle$

10.5 Exercises

Exercise 10.1 Consider the transfer function

$$Z(s) = \frac{\omega_n^2(s+\alpha)}{s^2 + 2\zeta\omega_n s + \omega_n^2}$$

where $\omega_n > 0$ and $0 < \zeta < 1$. Show that $Z(s)$ is strictly positive real if and only if $0 < \alpha < 2\zeta\omega_n$.

Exercise 10.2 ([2]) Consider the square transfer function matrix $G(s) = \mathcal{C}(sI - \mathcal{A})^{-1}\mathcal{B} + \mathcal{D}$ where $(\mathcal{A}, \mathcal{B}, \mathcal{C})$ is a minimal realization.

(a) Suppose $\mathcal{A}$ has no eigenvalues on the imaginary axis. Show that $G(s)$ is positive real if and only if there exist matrices P, L, and W, with P symmetric positive definite, such that (10.10)–(10.12) are satisfied with $\epsilon = 0$.
    Hint: Read the proof of Lemma 10.2.

(b) Suppose $\mathcal{A}$ has only pure imaginary eigenvalues. Show that $G(s)$ is positive real if and only if there exists a symmetric positive definite matrix P such that (10.10)–(10.12) are satisfied with $\epsilon = 0$, $L = 0$, and $W = 0$.
Hint: In this case, $Z(s)$ takes the form $Z(s) = \sum_i \frac{1}{s^2 + \omega_i^2}(F_i s + G_i)$ where $\omega_i \neq \omega_k$ for $i \neq k$.

(c) Using the results of (a) and (b), show that, for any matrix $\mathcal{A}$, $G(s)$ is positive real if and only is there exist matrices P, L, and W, with P symmetric positive definite, such that (10.10)–(10.12) are satisfied with $\epsilon = 0$.

Exercise 10.3 Prove Lemma 10.3 in the case when $G(s) = C(sI - A)^{-1}B + D$ with $D \neq 0$.

Exercise 10.4 Consider the feedback connection of Figure 10.2. Suppose $G(s)$ is strictly positive real and $\psi(t, y)$ satisfies

$$\psi(t, 0) = 0, \quad y^T \psi(t, y) \geq 0, \quad \forall\, y \in R, \; \forall\, t \geq 0$$

Show that the origin is globally exponentially stable.
Hint: Repeat the proof of Lemma 10.3.

Exercise 10.5 Consider the absolute stability problem for a scalar transfer function $G(s)$, which is proper but not strictly proper. Let

$$\dot{x} = Ax + Bu, \quad y = Cx + du, \quad d \neq 0$$

be a minimal state-space realization of $G(s)$. Suppose that A is Hurwitz and the time-varying nonlinearity $\psi(\cdot, \cdot)$ belongs to a sector $[0, \beta]$.

(a) Show that $1 + \beta d > 0$ is a necessary condition for absolute stability.

(b) Suppose $1 + \beta d > 0$. With $V(x) = x^T P x$ as a Lyapunov function candidate, show that the system is absolutely stable if $Re[G(j\omega)] > -1/\beta$, $\forall\, \omega \in R$.

Exercise 10.6 Study absolute stability, using the circle criterion, for each of the scalar transfer functions given below. In each case, find a sector $[\alpha, \beta]$ for which the system is absolutely stable.

(1) $G(s) = \frac{s}{s^2 - s + 1}$ 　　　　(2) $G(s) = \frac{1}{(s+1)(s+2)}$

(3) $G(s) = \frac{1}{(s+1)^2}$ 　　　　(4) $G(s) = \frac{s^2 - 0.5}{(s+1)(s^2+1)}$

(5) $G(s) = \frac{1-s}{(s+1)^2}$ 　　　　(6) $G(s) = \frac{s+2}{(s+1)(s-1)}$

(7) $G(s) = \frac{1}{(s+1)^3}$ 　　　　(8) $G(s) = \frac{s+2}{(s+1)^3}$

Exercise 10.7 Consider the feedback connection of Figure 10.2 with

$$G(s) = \frac{2s}{(s^2 + s + 1)}$$

(a) Show that the system is absolutely stable for nonlinearities in the sector $[0, 1]$.

(b) Show that the system has no limit cycles when $\psi(y) = \text{sat}(y)$.

Exercise 10.8 Consider the system

$$\begin{aligned}
\dot{x}_1 &= -x_1 - h(x_1 + x_2) \\
\dot{x}_2 &= x_1 - x_2 - 2h(x_1 + x_2)
\end{aligned}$$

where $h(\cdot)$ is a smooth function satisfying

$$yh(y) \geq 0, \ \forall \ y \in R, \quad h(y) = \begin{cases} c, & y \geq a_2 \\ 0, & |y| \leq a_1 \\ -c & y \leq -a_2 \end{cases}$$

$$|h(y)| \leq c, \quad \text{for } a_1 < y < a_2 \text{ and } -a_2 < y < -a_1$$

(a) Show that the origin is the unique equilibrium point.

(b) Show, using the circle criterion, that the origin is globally asymptotically stable.

Exercise 10.9 ([181]) Consider the system

$$\begin{aligned}
\dot{x}_1 &= x_2 \\
\dot{x}_2 &= -(\mu^2 + a^2 - q\cos\omega t)x_1 - 2\mu x_2
\end{aligned}$$

where μ, a, q, and ω are positive constants. Represent the system in the form of Figure 10.2 with $\psi(t, y) = qy\cos\omega t$ and use the circle criterion to derive conditions on the parameters μ, a, q, and ω which will ensure that the origin is exponentially stable.

Exercise 10.10 Consider the linear time-varying system $\dot{x} = [A + BE(t)C]x$ where A is Hurwitz, $\|E(t)\|_2 \leq 1$, $\forall \ t \geq 0$, and $\sup_{\omega \in R} \sigma_{max}[C(j\omega I - A)^{-1}B] < 1$. Show that the origin is uniformly asymptotically stable.

Exercise 10.11 Consider (10.10)–(10.12) and suppose that $\mathcal{D}$ is nonsingular. Show that P satisfies the Riccati equation

$$P[(\epsilon/2)I + \mathcal{A}] + [(\epsilon/2)I + \mathcal{A}^T]P + (\mathcal{C}^T - P\mathcal{B})(\mathcal{D} + \mathcal{D}^T)^{-1}(\mathcal{C} - \mathcal{B}^T P) = 0$$

Exercise 10.12 Consider the closed-loop system of Exercise 5.2, with the limiter included.

(a) Show that the system can be represented in the form of Figure 10.2 with $G(s) = -F(sI - A - BF)^{-1}B$ and $\psi(y) = L \, \text{sat}(y/L) - y$.

(b) Derive a condition for asymptotic stability of the origin using the multivariable circle criterion.

(c) Apply the result to the numerical case of Exercise 5.2, part (d), and estimate the region of attraction.

Exercise 10.13 In this exercise, we derive a version of the Popov criterion for a scalar transfer function $G(s)$ with all poles in the open left-half plane, except for a simple pole on the imaginary axis having a positive residue. The system is represented by

$$
\begin{aligned}
\dot{z} &= Az - B\psi(y) \\
\dot{v} &= -\psi(y) \\
y &= Cz + dv, \quad d > 0
\end{aligned}
$$

where A is Hurwitz, (A, B) is controllable, (A, C) is observable, and $\psi(\cdot)$ belongs to a sector $(0, k]$. Let $V(z, v) = z^T P z + a(y - Cz)^2 + b \int_0^y \psi(\sigma) \, d\sigma$, where $P = P^T > 0$, $a > 0$, and $b \geq 0$.

(a) Show that V is positive definite and radially unbounded.

(b) Show that $\dot{V}$ satisfies the inequality

$$
\dot{V} \leq z^T (PA + A^T P)z - 2z^T(PB - w)\psi(y) - \gamma\psi^2(y)
$$

where

$$
w = adC^T + \frac{1}{2}bA^T C^T; \quad \gamma = \frac{2ad}{k} + b(d + CB)
$$

Assume that b is chosen to ensure that $\gamma \geq 0$.

(c) Show that the system is absolutely stable if

$$
\frac{1}{k} + R_e[(1 + j\omega\eta)G(j\omega)] > 0, \quad \forall \, \omega \in R, \text{ where } \eta = \frac{b}{2ad}
$$

Exercise 10.14 Repeat Exercise 10.6 using the Popov criterion.

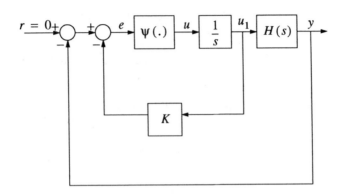

Figure 10.26: Exercise 10.15.

Exercise 10.15 ([76]) The feedback control system shown in Figure 10.26 represents a control system where $H(s)$ is the (scalar) transfer function of the plant and the inner loop models the actuator. Let

$$H(s) = \frac{s+6}{(s+2)(s+3)}$$

and suppose that $K \geq 0$ and $\psi(\cdot)$ belongs to a sector $(0, \beta]$ where β could be arbitrarily large but finite.

(a) Show that the system can be represented as the feedback connection of Figure 10.2 with $G(s) = [H(s) + K]/s$.

(b) Using the version of the Popov criterion in Exercise 10.13, find a lower bound K_c such that the system is absolutely stable for all $K > K_c$.

Exercise 10.16 ([74]) Let $\dot{x} = A(\nu)x$ define a family of linear systems where the elements of $A(\nu)$ are multilinear functions of the p-dimensional parameter vector ν; that is, $a_{ij}(\nu)$ is linear in ν_k with all other components of ν held fixed. The set of allowable ν is a convex hypercube $V = \{\nu \mid \alpha_i \leq \nu_i \leq \beta_i\}$. Denote the 2^p vertices of V by $\nu^{(1)}, \nu^{(2)}, \ldots,$ and the corresponding state equation matrices by $A^{(i)} = A(\nu^{(i)})$. Show that $A(\nu)$ is Hurwitz for all $\nu \in V$ if there exists a real symmetric matrix $P > 0$ such that

$$PA^{(i)} + \left(A^{(i)}\right)^T P < -\epsilon I, \quad \forall\, i = 1, 2, \ldots, 2^p$$

Exercise 10.17 Consider the feedback connection of Figure 10.15, where H_1 and H_2 are linear time-invariant systems represented by the transfer functions

$$H_1(s) = \frac{s-1}{s+1}, \quad H_2(s) = \frac{1}{s-1}$$

Find the closed-loop transfer functions from $\begin{bmatrix} u_1 \\ u_2 \end{bmatrix}$ to $\begin{bmatrix} y_1 \\ y_2 \end{bmatrix}$ and from $\begin{bmatrix} u_1 \\ u_2 \end{bmatrix}$ to $\begin{bmatrix} e_1 \\ e_2 \end{bmatrix}$. Use these transfer functions to discuss why we have to consider both inputs (u_1, u_2) and both outputs (e_1, e_2) (or (y_1, y_2)) in studying stability of the feedback connection.

Exercise 10.18 Under the conditions of the small gain theorem, show that the input-output map $(u_1, u_2) \to (y_1, y_2)$ is $\mathcal{L}$-stable and derive inequalities similar to (10.39)–(10.40) for $y_{1\tau}$ and $y_{2\tau}$.

Exercise 10.19 Let $d_2(t) = a \sin \omega t$ in Example 10.9, where a and ω are positive constants.

(a) Show that, for sufficiently small ϵ, the state of the closed-loop system is uniformly bounded.

(b) Investigate the effect of increasing ω.

Exercise 10.20 Show that any storage function that satisfies the definition of passivity for a system that is output strictly passive and zero-state observable, or a system that is state strictly passive, must be positive definite in some neighborhood of the origin.
Hint: Read the proof of Lemma 6.2.

Exercise 10.21 Consider the linear time-invariant system

$$\dot{x} = \mathcal{A}x + \mathcal{B}u, \quad y = \mathcal{C}x + \mathcal{D}u$$

where $(\mathcal{A}, \mathcal{B}, \mathcal{C})$ is a minimal realization and $Z(s) = \mathcal{C}(sI - \mathcal{A})^{-1}\mathcal{B} + \mathcal{D}$ is positive real. Show that the system is passive.
Hint: Use Exercise 10.2.

Exercise 10.22 Consider the system

$$\begin{aligned} \dot{x}_1 &= x_2 \\ \dot{x}_2 &= -h(x_1) - ax_2 + u \\ y &= \alpha x_1 + x_2 \end{aligned}$$

where $0 < \alpha < a$ and $yh(y) > 0$, $\forall\, y \neq 0$. Show that the system is state strictly passive.
Hint: Use $V(x)$ of Example 3.5 as a storage function.

Exercise 10.23 Show that if a system is input strictly passive and finite-gain $\mathcal{L}_2$-stable, then it is input and output strictly passive.

Exercise 10.24 Show that the parallel connection of two passive (respectively, input strictly passive, output strictly passive, state strictly passive) dynamical systems is passive (respectively, input strictly passive, output strictly passive, state strictly passive).

Exercise 10.25 ([68]) Consider the feedback system of Figure 10.15 where H_1 and H_2 are passive dynamical systems of the form (10.46)–(10.47). Suppose the feedback connection has a well-defined state-space model, and the series connection $H_1(-H_2)$, with input e_2 and output y_1, is zero-state observable. Show that the origin is asymptotically stable if H_2 is input strictly passive or H_1 is output strictly passive.

Exercise 10.26 ([68]) Consider the feedback system of Figure 10.15 where H_1 and H_2 are passive dynamical systems of the form (10.46)–(10.47). Suppose the feedback connection has a well-defined state-space model, and the series connection $H_2 H_1$, with input e_1 and output y_2, is zero-state observable. Show that the origin is asymptotically stable if H_1 is input strictly passive or H_2 is output strictly passive.

Exercise 10.27 ([68]) As a generalization of the concept of passivity, a dynamical system of the form (10.46)–(10.47) is said to be dissipative with respect to a supply rate $w(u, y)$ if there is a positive definite storage function $V(x)$ such that $\dot{V} \leq w$. Consider the feedback system of Figure 10.15 where H_1 and H_2 are zero-state observable, dynamical systems of the form (10.46)–(10.47). Suppose each of H_1 and H_2 is dissipative with storage function $V_i(x_i)$ and supply rate $w_i(u_i, y_i) = y_i^T Q_i y_i + 2 y_i^T S_i u_i + u_i^T R_i u_i$, where Q_i, S_i, R_i are real matrices with Q_i and R_i symmetric. Show that the origin is stable (respectively, asymptotically stable) if the matrix

$$\hat{Q} = \begin{bmatrix} Q_1 + \alpha R_2 & -S_1 + \alpha S_2^T \\ -S_1^T + \alpha S_2 & R_1 + \alpha Q_2 \end{bmatrix}$$

is negative semidefinite (respectively, negative definite) for some $\alpha > 0$.

Exercise 10.28 Consider the feedback system of Figure 10.15 where H_1 and H_2 have state-space models

$$\dot{x}_i = f_i(x_i) + G_i(x_i)e_i, \qquad y_i = h_i(x_i) + J_i(x_i)e_i$$

for $i = 1, 2$. Show that the feedback system has a well-defined state-space model if the matrix $I + J_2(x_2)J_1(x_1)$ is nonsingular for all x_1 and x_2.

Exercise 10.29 Consider equations (10.60)–(10.61) and (10.62)–(10.63), and suppose $h_1 = h_1(x_1)$, independent of e_1. Show, in each case, that the equations have a unique solution (e_1, e_2).

Exercise 10.30 ([129]) Consider the equations of motion of an m-link robot, described in Exercise 1.3.

(a) Using the total energy $V = \frac{1}{2}\dot{q}^T M(q)\dot{q} + P(q)$ as a storage function, show that the map from u to $\dot{q}$ is passive.

(b) With $u = -K_d\dot{q} + v$ where K_d is a positive diagonal constant matrix, show that the map from v to $\dot{q}$ is output strictly passive.

(c) Show that $u = -K_d\dot{q}$, where K_d is a positive diagonal constant matrix, makes the origin $(q = 0, \dot{q} = 0)$ asymptotically stable. Under what additional conditions will it be globally asymptotically stable?

Exercise 10.31 ([161]) Consider the equations of motion of an m-link robot, described in Exercise 1.3. Let $u = M(q)\ddot{q}_r + C(q,\dot{q})\dot{q}_r + g(q) - K_p e + v$, where $\dot{q}_r = \dot{q}_d - \Lambda e$, $\sigma = \dot{q} - \dot{q}_r = \dot{e} + \Lambda e$, $e = q - q_d$, $q_d(t)$ is a smooth desired trajectory in R^m that is bounded and has bounded derivatives up to the second order, and K_p and Λ are positive diagonal constant matrices.

(a) Using $V = \frac{1}{2}\sigma^T M(q)\sigma + \frac{1}{2}e^T K_p e$ as a storage function, show that the map from v to σ is passive.

(b) With $v = -K_d\sigma + w$, where K_d is a positive diagonal constant matrix, show that the map from w to σ is output strictly passive.

(c) Show that $v = -K_d\sigma$, where K_d is a positive diagonal constant matrix, ensures that q and $\dot{q}$ are bounded, and $e(t)$ converges to zero as t tends to infinity.
Hint: Use Lemma 4.2.

Exercise 10.32 ([134]) Euler equations for a rotating rigid spacecraft are given by

$$\begin{aligned}
J_1\dot{\omega}_1 &= (J_2 - J_3)\omega_2\omega_3 + u_1 \\
J_2\dot{\omega}_2 &= (J_3 - J_1)\omega_3\omega_1 + u_2 \\
J_3\dot{\omega}_3 &= (J_1 - J_2)\omega_1\omega_2 + u_3
\end{aligned}$$

where, ω_1 to ω_3 are the components of the angular velocity vector along the principal axes, u_1 to u_3 are the torque inputs applied about the principal axes, and J_1 to J_3 are the principal moments of inertia.

(a) Show that the map from $u = [u_1, u_2, u_3]^T$ to $\omega = [\omega_1, \omega_2, \dot{\omega}_3]^T$ is lossless.

(b) Show that $u = -K\omega$, where K is a positive definite constant matrix, makes the origin $\omega = 0$ globally asymptotically stable.

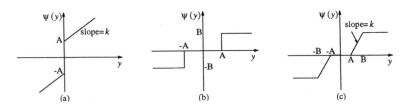

Figure 10.27: Exercise 10.33.

Exercise 10.33 For each odd nonlinearity $\psi(y)$ on the following list, verify the given expression of the describing function $\Psi(a)$.

(**1**) $\psi(y) \;=\; y^5; \quad \Psi(a) = 5a^4/8$

(**2**) $\psi(y) \;=\; y^3|y|; \quad \Psi(a) = 32a^3/15\pi$

(**3**) $\psi(y) \quad : \quad$ Figure 10.27(a); $\Psi(a) = K + \dfrac{4A}{\pi a}$

(**4**) $\psi(y) \quad : \quad$ Figure 10.27(b)

$$\Psi(a) \;=\; \begin{cases} 0, & \text{for } a \le A \\ (4B/\pi a)[1 - (A/a)^2]^{1/2}, & \text{for } a \ge A \end{cases}$$

(**5**) $\psi(y) \quad : \quad$ Figure 10.27(c)

$$\Psi(a) \;=\; \begin{cases} 0, & \text{for } a \le A \\ K[1 - N(a/A)], & \text{for } A \le a \le B \\ K[N(a/B) - N(a/A)], & \text{for } a \ge B \end{cases}$$

where

$$N(x) = \frac{2}{\pi}\left[\sin^{-1}\left(\frac{1}{x}\right) + \frac{1}{x}\sqrt{1 - \left(\frac{1}{x}\right)^2}\; \right]$$

Exercise 10.34 Consider the feedback connection of Figure 10.17 with

$$G(s) = \frac{1 - s}{s(s + 1)}$$

Using the describing function method, investigate the possibility of existence of periodic solutions and the possible frequency and amplitude of oscillation for each of the following nonlinearities.

(1) $\psi(y) = y^5$.

(2) The nonlinearity of Exercise 10.33, part (5), with $A = 1$, $B = \frac{3}{2}$, and $K = 2$.

Exercise 10.35 Consider the feedback connection of Figure 10.17 with

$$G(s) = \frac{1}{(s+1)^6} \quad \text{and} \quad \psi(y) = \text{sgn}(y)$$

Using the describing function method, investigate the possibility of existence of periodic solutions and the possible frequency and amplitude of oscillation.

Exercise 10.36 Repeat Exercise 10.35 with

$$G(s) = \frac{s+6}{s(s+2)(s+3)} \quad \text{and} \quad \psi(y) = \text{sgn}(y)$$

Exercise 10.37 Repeat Exercise 10.35 with

$$G(s) = \frac{s}{s^2 - s + 1} \quad \text{and} \quad \psi(y) = y^5$$

Exercise 10.38 Consider the feedback connection of Figure 10.17 with

$$G(s) = \frac{5(s+0.25)}{s^2(s+2)^2}$$

Using the describing function method, investigate the possibility of existence of periodic solutions and the possible frequency and amplitude of oscillation for each of the following nonlinearities.

(1) The nonlinearity of Exercise 10.33, part (3), with $A = 1$ and $K = 2$.

(2) The nonlinearity of Exercise 10.33, part (4), with $A = 1$ and $B = 1$.

(3) The nonlinearity of Exercise 10.33, part (5), with $A = 1$, $B = \frac{3}{2}$, and $K = 2$.

Exercise 10.39 Consider the feedback connection of Figure 10.17 with

$$G(s) = \frac{2bs}{s^2 - bs + 1}, \quad \psi(y) = \text{sat}(y)$$

Using the describing function method, show that for sufficiently small $b > 0$ the system has a periodic solution. Confirm your conclusion by applying Theorem 10.9 and estimate the frequency and amplitude of oscillation.

Exercise 10.40 ([113]) Consider the feedback connection of Figure 10.17 with

$$G(s) = \frac{2(s-1)}{s^3(s+1)} \quad \text{and} \quad \psi(y) = \text{sat}(y)$$

Study the existence of periodic solutions using Theorem 10.9. For each oscillation, if any, find a frequency interval $[\omega_1, \omega_2]$ and an amplitude interval $[a_1, a_2]$ such that the frequency of oscillation is in $[\omega_1, \omega_2]$ and the amplitude of the first harmonic is in $[a_1, a_2]$.

Exercise 10.41 Repeat Exercise 10.40 with

$$G(s) = \frac{-s}{s^2 + 0.8s + 8} \quad \text{and} \quad \psi(y) = \tfrac{1}{2}\sin y$$

Exercise 10.42 Repeat Exercise 10.40 with

$$G(s) = \frac{-s}{s^2 + 0.8s + 8}$$

and $\psi(y)$ the nonlinearity of Example 10.23.

Exercise 10.43 Repeat Exercise 10.40 with

$$G(s) = \frac{-24}{s^2(s+1)^3} \quad \text{and} \quad \psi(y) = \begin{cases} y^3 + \frac{y}{2}, & \text{for } 0 \le y \le 1 \\ 2y - \frac{1}{2}, & \text{for } y \ge 1 \end{cases}$$

where $\psi(y)$ is an odd nonlinearity.

Chapter 11

Feedback Control

The last three chapters of the book deal with the design of feedback control. Various tools of nonlinear control design are introduced, including linearization, gain scheduling, exact feedback linearization, integral control, Lyapunov redesign, backstepping, sliding mode control, and adaptive control. Most of the nonlinear analysis tools we have learned so far come into play in these three chapters, solidifying our understanding of these tools. This chapter starts with a section on control problems which serves as an introduction to all three chapters. This is followed by two sections on design via linearization and gain scheduling. Exact feedback linearization is presented in Chapter 12, and Lyapunov-based design tools are presented in Chapter 13.

11.1 Control Problems

There are many control tasks which require the use of feedback control. Depending on the goal of control design, there are several formulations of the control problem. The tasks of stabilization, tracking, and disturbance rejection/attenuation (and various combinations of them) lead to a number of control problems. In each problem, we may have a state feedback version where all state variables can be measured, or an output feedback version where only an output vector whose dimension is typically less than the dimension of the state can be measured. In a typical control problem, there are additional goals for the design, like meeting certain requirements on the transient response or certain constraints on the control input. These requirements could be conflicting and the designer has to trade off various conflicting requirements. The desire to optimize this design tradeoff leads to various optimal control formulations. When model uncertainty is taken into consideration, issues of sensitivity and robustness come into play. The attempt to design feedback control to cope with a wide range of model uncertainty leads to either robust or adaptive control problem formulations. In robust control, the model uncertainty is characterized as perturbations of the nominal model. You may think of the nominal model as a point in space and the perturbed models as points in a ball that contains the nominal model. A robust control design tries to meet the control objective for any model in the "ball of uncertainty." Adaptive control, on the other hand, parameterizes the uncertainty in terms of certain unknown parameters and tries to use feedback to learn these parameters on line, that is, during the operation of the system. In a more elaborate adaptive scheme, the controller might be learning certain unknown nonlinear functions rather than just learning some unknown parameters. There are also problem formulations that mix robust and adaptive control. In this section, we describe the control problems we shall encounter in this chapter and the next two chapters. We will limit our discussions to the basic tasks of stabilization, tracking, and disturbance rejection. We start with the stabilization problem, both state feedback and output feedback versions. Then, we describe tracking and disturbance rejection problems. Some robust and adaptive control problems will be described in Chapter 13, as needed.

The state feedback stabilization problem for the system

$$\dot{x} = f(t, x, u)$$

is the problem of designing a feedback control law

$$u = \gamma(t, x)$$

such that the origin $x = 0$ is a uniformly asymptotically stable equilibrium point of the closed-loop system

$$\dot{x} = f(t, x, \gamma(t, x))$$

Once we know how to solve this problem, we can stabilize the system with respect to an arbitrary point p by translating p to the origin via the change of variables $y = x - p$. Furthermore, p does not have to be an equilibrium point of the open-loop system; see Exercise 11.1. The feedback control law $u = \gamma(t, x)$ is usually called "static feedback" because it is a memoryless function of x. Sometimes, we use a dynamic state feedback control

$$u = \gamma(t, x, z)$$

where z is the solution of a dynamical system driven by x; that is,

$$\dot{z} = g(t, x, z)$$

Common examples of dynamic state feedback control arise when we use integral control (see Sections 11.2.2 and 12.3.3) or adaptive control (see Section 1.1.6).

The output feedback stabilization problem for the system

$$\dot{x} = f(t, x, u)$$
$$y = h(t, x, u)$$

is the problem of designing a static output feedback control law

$$u = \gamma(t, y)$$

or a dynamic output feedback control law

$$u = \gamma(t, y, z)$$
$$\dot{z} = g(t, y, z)$$

such that the origin is a uniformly asymptotically stable equilibrium point of the closed-loop system. In the case of dynamic feedback control, the origin to be stabilized is $(x = 0, z = 0)$. Dynamic feedback control is more common in output feedback schemes since the lack of measurement of some of the state variables is usually compensated for by including "observers" or "observer-like" components in the feedback controller.

Naturally, the feedback stabilization problem is much simpler when the system is linear and time invariant:

$$\dot{x} = Ax + Bu$$
$$y = Cx + Du$$

In this case, the state feedback control $u = Kx$ preserves linearity of the open-loop system, and the origin of the closed-loop system

$$\dot{x} = (A + BK)x$$

is asymptotically stable if and only if the matrix $A + BK$ is Hurwitz. Thus, the state feedback stabilization problem reduces to a problem of designing a matrix K to assign the eigenvalues of $A + BK$ in the open left-half complex plane. Linear control theory[1] confirms that the eigenvalues of $A + BK$ can be arbitrarily assigned (subject only to the constraint that complex eigenvalues are in conjugate pairs) provided the pair (A, B) is controllable. Even if some eigenvalues of A are not controllable, stabilization is still possible provided the uncontrollable eigenvalues have negative real parts. In this case, the pair (A, B) is called stabilizable, and the uncontrollable (open-loop) eigenvalues of A will be (closed-loop) eigenvalues of

---

[1] See, for example, [98], [29], [41], or [142].

$A + BK$. If we can only measure the output y, we can use dynamic compensation, like the observer-based controller

$$u = K\hat{x}$$
$$\dot{\hat{x}} = A\hat{x} + Bu + H(C\hat{x} + Du - y)$$

to stabilize the system. Here, the feedback gain K is designed as in state feedback, such that $A + BK$ is Hurwitz, while the observer gain H is designed such that $A + HC$ is Hurwitz. The closed-loop eigenvalues will consist of the eigenvalues of $A + BK$ and the eigenvalues of $A + HC$.[2] The stabilization of $A + HC$ is dual to the stabilization of $A + BK$ and requires observability (or at least detectability) of the pair (A, C).

For a general nonlinear system, the problem is more difficult and less understood. The most practical way to approach the stabilization problem for nonlinear systems is to appeal to the neat results available in the linear case, that is, via linearization. In Section 11.2, a feedback control law is designed by linearizing the system about the origin and designing a stabilizing linear feedback control for the linearization. The validity of this idea comes from Lyapunov's indirect method stated in Theorem 3.7. Clearly, this approach is local; that is, it can only guarantee asymptotic stability of the origin and cannot, in general, prescribe a region of attraction nor achieve global asymptotic stability. In Section 11.3, we describe *gain scheduling*, a technique that aims at extending the region of validity of linearization by solving the stabilization problem at different operating points and allowing the controller to move from one design to another in a smooth or abrupt way. In Chapter 12, another linearization idea is presented. There, we deal with a special class of nonlinear systems which can be transformed into linear systems via feedback and (possibly) a change of variables. After this transformation, a stabilizing linear state feedback control is designed for the linear system. This linearization approach is different from the first one in that no approximation is used; it is exact. This exactness, however, assumes perfect knowledge of the state equation and uses that knowledge to cancel the nonlinearities of the system. Since perfect knowledge of the state equation and exact mathematical cancelation of terms are almost impossible, the implementation of this approach will almost always result in a closed-loop system, which is a perturbation of a nominal system whose origin is exponentially stable. The validity of the method draws upon Lyapunov theory for perturbed systems, which was discussed in Chapter 5 specifically regarding the exponential stability of the origin being a robust property of the system.

When a linear system is stabilized by feedback, the origin of the closed-loop system is globally asymptotically stable. This is not the case for nonlinear systems,

---

[2] This fact is usually referred to as the "separation principle," since the assignment of the closed-loop eigenvalues can be carried out in separate tasks for the state feedback and observer problems.

where different stabilization notions can be introduced. If the nonlinear system is stabilized via linearization, then the origin of the closed-loop system will be asymptotically stable. Without further analysis of the system, the region of attraction of the origin will be unknown. In this case, we say the feedback control achieves *local stabilization*. If the feedback control guarantees that a certain set is included in the region of attraction or if an estimate of the region of attraction is given, we say that the feedback control achieves *regional stabilization*. If the origin of the closed-loop system is globally asymptotically stable, we say that the control achieves *global stabilization*. If feedback control does not achieve global stabilization but can be designed such that any given compact set (no matter how large) can be included in the region of attraction, we say that the feedback control achieves *semiglobal stabilization*. These four stabilization notions are illustrated by the following example.

Example 11.1 Suppose we want to stabilize the scalar system

$$\dot{x} = x^2 + u$$

using state feedback. Linearization at the origin results in the linear system $\dot{x} = u$, which can be stabilized by $u = -kx$ with $k > 0$. When this control is applied to the nonlinear system, it results in

$$\dot{x} = -kx + x^2$$

whose linearization at the origin is $\dot{x} = -kx$. Thus, by Theorem 3.7, the origin is asymptotically stable and we say that $u = -kx$ achieves local stabilization. In this simple example, it is not hard to see that the region of attraction is the set $\{x < k\}$. With this estimate, we say that $u = -kx$ achieves regional stabilization. By increasing k, we can expand the region of attraction. In fact, given any compact set $B_r = \{|x| \leq r\}$, we can include it in the region of attraction by choosing $k > r$. Hence, $u = -kx$ achieves semiglobal stabilization. It is important to notice that $u = -kx$ does not achieve global stabilization. In fact, for any finite k, there is a part of the state space (that is, $x > k$) which is not in the region of attraction. While semiglobal stabilization can include any compact set in the region of attraction, the control law is dependent on the given set and will not necessarily work with a bigger set. For a given r, we can choose $k > r$. Once k is fixed and the controller is implemented, if the initial state happens to be in the region $\{x > k\}$, the solution $x(t)$ will diverge to infinity. Global stabilization can be achieved by the nonlinear control law

$$u = -x^2 - kx$$

which cancels the open-loop nonlinearity and yields the linear closed-loop system $\dot{x} = -kx$.   △

We turn now to the description of a more general control problem; namely, the tracking problem in the presence of disturbance. Here, we have a system modeled by

$$\dot{x} = f(t, x, u, w)$$
$$y = h(t, x, u, w)$$
$$y_m = h_m(t, x, u, w)$$

where x is the state, u is the control input, w is a disturbance input, y is the controlled output, and y_m is the measured output. The basic goal of the control problem is to design the control input so that the controlled output y tracks a reference signal y_R; that is,

$$e(t) = y(t) - y_R(t) \approx 0, \quad \forall\, t \geq t_0$$

where t_0 is the time at which control starts. Since the initial value of y depends on the initial state $x(t_0)$, meeting this requirement for all $t \geq t_0$ would clearly require presetting the initial state at time t_0, which is quite unusual in practice. Therefore, we usually seek an asymptotic output tracking goal where the tracking error e approaches zero as t tends to infinity; that is,

$$e(t) \to 0 \quad \text{as} \quad t \to \infty$$

This asymptotic output tracking is feasible in the absence of input disturbance w or for certain types of disturbance inputs, in which case we say that we have achieved asymptotic disturbance rejection. For a general time-varying disturbance input $w(t)$, it might not be feasible to achieve asymptotic disturbance rejection. In such cases, we may attempt to achieve disturbance attenuation, which can take the form of a requirement to achieve ultimate boundedness of the tracking error with prescribed tolerance; that is,

$$\|e(t)\| \leq \epsilon, \quad \forall\, t \geq T$$

where ϵ is a prespecified (small) positive number. Alternatively, we may consider attenuating the closed-loop input-output map from the disturbance input w to the tracking error e. For example, if we consider w as an $\mathcal{L}_2$ signal, then we may state our goal as to minimize the $\mathcal{L}_2$ gain of the closed-loop input-output map from w to e, or at least make this gain less than a prescribed tolerance.[3]

An important class of tracking problems are those problems where the reference signal y_R is constant. In such cases, it is common to refer to y_R as the set point and to the problem as a regulation problem. Here, the main control problem is usually to stabilize the closed-loop system at an equilibrium point where $y = y_R$.

---

[3] This is the formulation of the H_∞ control problem. See, for example, [194, 51, 44, 15, 81, 179].

Feedback control laws for the tracking problem are classified in the same way we have seen in the stabilization problem. We speak of state feedback if x can be measured; that is, $y_m = x$; otherwise we speak of output feedback. Also, the feedback control law can be static or dynamic. The control law may achieve local, regional, semiglobal, or global tracking. The new element here is that these phrases refer not only to the size of the initial state, but to the size of the exogenous signals y_R and w as well. For example, in a typical problem, local tracking means tracking is achieved for sufficiently small initial states and sufficiently small exogenous signals, while global tracking means tracking is achieved for any initial state and any (y, w) in a prescribed class of exogenous signals.

11.2 Design via Linearization

We illustrate the design via linearization approach by considering the stabilization and regulation problems. In each case, we start with state feedback control and then present output feedback control.

11.2.1 Stabilization

For state feedback stabilization, consider the system

$$\dot{x} = f(x, u) \tag{11.1}$$

where $f(0,0) = 0$ and $f(x, u)$ is continuously differentiable in a domain $D_x \times D_u \subset R^n \times R^p$ that contains the origin $(x = 0, \ u = 0)$. We want to design a state feedback control $u = \gamma(x)$ to stabilize the system. Linearization of (11.1) about $(x = 0, \ u = 0)$ results in the linear system

$$\dot{x} = Ax + Bu \tag{11.2}$$

where

$$A = \left. \frac{\partial f}{\partial x}(x, u) \right|_{x=0, u=0} \quad ; \quad B = \left. \frac{\partial f}{\partial u}(x, u) \right|_{x=0, u=0}$$

Assume the pair (A, B) is controllable, or at least stabilizable. Design a matrix K to assign the eigenvalues of $A + BK$ to desired locations in the open left-half complex plane. Now apply the linear state feedback control $u = Kx$ to the nonlinear system (11.1). The closed-loop system is

$$\dot{x} = f(x, Kx) \tag{11.3}$$

Clearly, the origin is an equilibrium point of the closed-loop system. The linearization of (11.3) about the origin $x = 0$ is given by

$$\dot{x} = \left[\frac{\partial f}{\partial x}(x, Kx) + \frac{\partial f}{\partial u}(x, Kx) \ K \right]_{x=0} x = (A + BK)x$$

Since $A+BK$ is Hurwitz, it follows from Theorem 3.7 that the origin is an asymptotically stable equilibrium point of the closed-loop system (11.3). Actually, according to Theorem 3.11, the origin is exponentially stable. As a byproduct of the linearization approach, we can always find a Lyapunov function for the closed-loop system. Let Q be any positive-definite symmetric matrix and solve the Lyapunov equation

$$P(A + BK) + (A + BK)^T P = -Q$$

for P. Since $(A + BK)$ is Hurwitz, the Lyapunov equation has a unique positive definite solution (Theorem 3.6). The quadratic function $V(x) = x^T P x$ is a Lyapunov function for the closed-loop system in the neighborhood of the origin. We can use $V(x)$ to estimate the region of attraction of the origin.

Example 11.2 Consider the pendulum equation

$$\ddot{\theta} = -a \sin \theta - b\dot{\theta} + cT$$

where $a = g/l > 0$, $b = k/m \geq 0$, $c = 1/ml^2 > 0$, θ is the angle subtended by the rod and the vertical axis, and T is a torque applied to the pendulum. View the torque as the control input and suppose we want to stabilize the pendulum at an angle $\theta = \delta$. Choose the state variables as

$$x_1 = \theta - \delta; \quad x_2 = \dot{\theta}$$

so that the desired equilibrium point is located at the origin of the state space. The state equation is

$$\begin{aligned} \dot{x}_1 &= x_2 \\ \dot{x}_2 &= -a\sin(x_1 + \delta) - bx_2 + cT \end{aligned}$$

For the system to maintain equilibrium at the origin, the torque must have a steady-state component T_{ss} that satisfies

$$0 = -a \sin \delta + cT_{ss}$$

Define a control variable u as $u = T - T_{ss}$ and rewrite the state equation as

$$\begin{aligned} \dot{x}_1 &= x_2 \\ \dot{x}_2 &= -a[\sin(x_1 + \delta) - \sin \delta] - bx_2 + cu \end{aligned}$$

This equation is in the standard form (11.1) where $f(0,0) = 0$. Linearization of the system at the origin results in

$$A = \begin{bmatrix} 0 & 1 \\ -a\cos(x_1 + \delta) & -b \end{bmatrix}_{x_1=0} = \begin{bmatrix} 0 & 1 \\ -a\cos\delta & -b \end{bmatrix}; \quad B = \begin{bmatrix} 0 \\ c \end{bmatrix}$$

The pair (A, B) is controllable. Taking $K = [k_1 \ k_2]$, it can be easily verified that $A + BK$ is Hurwitz for

$$k_1 < \frac{a \cos \delta}{c}, \quad k_2 < \frac{b}{c}$$

The torque control law is given by

$$T = \frac{a \sin \delta}{c} + Kx = \frac{a \sin \delta}{c} + k_1(\theta - \delta) + k_2 \dot{\theta}$$

We leave it to the reader (Exercise 11.2) to continue with the Lyapunov analysis of the closed-loop system.   △

For output feedback stabilization, consider the system

$$\dot{x} = f(x, u) \tag{11.4}$$
$$y = h(x) \tag{11.5}$$

where $f(0, 0) = 0$, $h(0) = 0$, and $f(x, u)$, $h(x)$ are continuously differentiable in a domain $D_x \times D_u \subset R^n \times R^p$ that contains the origin ($x = 0$, $u = 0$). We want to design an output feedback control law (using only measurements of y) to stabilize the system. Linearization of (11.4)–(11.5) about ($x = 0$, $u = 0$) results in the linear system

$$\dot{x} = Ax + Bu \tag{11.6}$$
$$y = Cx \tag{11.7}$$

where A and B are defined after (11.2) and

$$C = \frac{\partial h}{\partial x}(x)\bigg|_{x=0}$$

Assume (A, B) is stabilizable and (A, C) is detectable, and design a linear dynamic output feedback controller

$$\dot{z} = Fz + Gy \tag{11.8}$$
$$u = Lz + My \tag{11.9}$$

such that the closed-loop matrix

$$\begin{bmatrix} A + BMC & BL \\ GC & F \end{bmatrix} \tag{11.10}$$

is Hurwitz. An example of such design is the observer-based controller, where

$$F = A + BK + HC, \quad G = -H, \quad L = K, \quad M = 0$$

and K and H are designed such that $A + BK$ and $A + HC$ are Hurwitz. When the controller (11.8)–(11.9) is applied to the nonlinear system (11.4)–(11.5), it results in the closed-loop system

$$\dot{x} = f(x, Lz + Mh(x)) \tag{11.11}$$
$$\dot{z} = Fz + Gh(x) \tag{11.12}$$

It can be verified that the origin ($x = 0$, $z = 0$) is an equilibrium point of the closed-loop system (11.11)–(11.12) and linearization about the origin results in the Hurwitz matrix (11.10). Thus, once again, we conclude that the origin is an exponentially stable equilibrium point of the closed-loop system (11.11)–(11.12). A Lyapunov function for the closed-loop system can be obtained by solving a Lyapunov equation for the Hurwitz matrix (11.10).

Example 11.3 Reconsider the pendulum equation of Example 11.2, and suppose we measure the angle θ but do not measure the angular velocity $\dot{\theta}$. An output variable y can be taken as $y = x_1 = \theta - \delta$, and the state feedback controller of Example 11.2 can be implemented using the observer

$$\dot{\hat{x}} = A\hat{x} + Bu + H(\hat{x}_1 - y)$$

Taking $H = [h_1 \; h_2]^T$, it can be verified that $A + HC$ will be Hurwitz if

$$-h_1 + b > 0, \quad -h_1 b - h_2 + a \cos \delta > 0$$

The torque control law is given by

$$T = \frac{a \sin \delta}{c} + K\hat{x}$$

$\triangle$

11.2.2   Regulation via Integral Control

In Example 11.2, we considered the problem of regulating the angle θ to a constant value δ. We reduced the problem to a stabilization problem by shifting the desired equilibrium point to the origin. While this approach is sound when the parameters of the system are known precisely, it could be unacceptable under parameter perturbations. In this section, we present an integral control approach which ensures robust regulation under all parameter perturbations which do not destroy stability of the closed-loop system. Before we present this approach, let us examine Example 11.2 to see the effect of parameter perturbations. The control law we arrived at in Example 11.2 is given by

$$T = \frac{a \sin \delta}{c} + Kx$$

It comprises the steady-state component $T_{ss} = (a/c)\sin\delta$, which assigns the equilibrium value of θ, say θ_{ss}, at the desired angle δ and the feedback component Kx which makes $A + BK$ Hurwitz. While the calculation of both components depends on the parameters of the system, the feedback part can be designed to be robust to a wide range of parameter perturbations. In particular, if all we know about the parameters a, b, c is an upper bound on the ratio a/c, that is, $a/c \leq \rho$, together with the fact that a and c are positive and b is nonnegative, we can ensure that $A + BK$ will be Hurwitz by choosing k_1 and k_2 to satisfy

$$k_1 < -\rho, \quad k_2 < 0$$

The calculation of T_{ss}, on the other hand, could be sensitive to parameter perturbations. Suppose T_{ss} is calculated using nominal values a_0 and c_0 of a and c, respectively. The equilibrium point of the closed-loop system is given by

$$\sin\theta_{ss} = \frac{cT_{ss}}{a} = \frac{a_0 c}{a c_0}\sin\delta$$

If $\delta = 0$ or $\delta = \pi$ (that is, the pendulum is stabilized at one of the open-loop equilibrium points), $T_{ss} = 0$ and the approach used in Example 11.2 will be robust to parameter perturbations. For other values of δ, the error in the steady-state angle could be of the order of the error in the parameters. For example, if $\delta = 45°$, $c \approx c_0$, and $a \approx 1.1a_0$, we have $\theta_{ss} \approx 40°$, an error of about 10%.

Consider the system

$$\dot{x} = f(x, u) \tag{11.13}$$

$$y = h(x) \tag{11.14}$$

where $f(x, u)$ and $h(x)$ are continuously differentiable in a domain $D_x \times D_u \subset R^n \times R^p$. Here x is the state, u is the control input, and $y \in R^p$ is the controlled output. Let $y_R \in R^p$ be a constant reference. We want to design state feedback control such that

$$y(t) \to y_R \quad \text{as} \quad t \to \infty$$

In addition to the measurement of x, we assume that we can physically measure the controlled output y. Quite often, y is included in x, in which case this additional assumption is redundant. However, when this is not the case, we cannot use the equation $y = h(x)$ to compute y from the measured x for such computation will be subject to errors due to model uncertainty. In particular, suppose the nominal model of $h(x)$ is $h_0(x)$. If an integral controller is built on a computed output $h_0(x)$, it will regulate $h_0(x)$ to y_R rather than regulating y to y_R, which will defeat the purpose of using integral control. The regulation task will be achieved by stabilizing the system at an equilibrium point where $y = y_R$. For the system to maintain such an equilibrium condition, it must be true that there exists a pair $(x_{ss}, u_{ss}) \in D_x \times D_u$ such that

$$0 \;=\; f(x_{ss}, u_{ss}) \tag{11.15}$$

$$0 \;=\; h(x_{ss}) - y_R \tag{11.16}$$

We assume that equations (11.15)–(11.16) have a unique solution $(x_{ss}, u_{ss}) \in D_x \times D_u$. To proceed with the design of the controller, we integrate the tracking error $e = y - y_R$:

$$\dot{\sigma} = e$$

and augment the integrator with the system (11.13) to obtain

$$\dot{x} \;=\; f(x, u) \tag{11.17}$$

$$\dot{\sigma} \;=\; h(x) - y_R \tag{11.18}$$

For multi-output systems $(p > 1)$, the integrator equation $\dot{\sigma} = e$ represents a stack of p integrators where each component of the tracking error vector is integrated. The control u will be designed as a feedback function of (x, σ) such that the closed-loop system has an equilibrium point $(\bar{x}, \bar{\sigma})$ with $\bar{x} = x_{ss}$. Assuming for the moment that this task is indeed possible, we linearize (11.17)–(11.18) about $(x = x_{ss},\ \sigma = \bar{\sigma},\ u = u_{ss})$ to obtain

$$\dot{\xi} = \begin{bmatrix} A & 0 \\ C & 0 \end{bmatrix} \xi + \begin{bmatrix} B \\ 0 \end{bmatrix} v \overset{\text{def}}{=} \mathcal{A}\xi + \mathcal{B}v, \quad \text{where } \xi = \begin{bmatrix} x - x_{ss} \\ \sigma - \bar{\sigma} \end{bmatrix}, \quad v = u - u_{ss}$$

where

$$A = \left. \frac{\partial f}{\partial x}(x, u) \right|_{x = x_{ss}, u = u_{ss}}, \quad B = \left. \frac{\partial f}{\partial u}(x, u) \right|_{x = x_{ss}, u = u_{ss}}, \quad C = \left. \frac{\partial h}{\partial x}(x) \right|_{x = x_{ss}}$$

The matrices A, B, C are, in general, dependent on y_R. Suppose now that (A, B) is controllable (respectively, stabilizable) and [4]

$$\text{rank} \begin{bmatrix} A & B \\ C & 0 \end{bmatrix} = n + p \tag{11.19}$$

Then, $(\mathcal{A}, \mathcal{B})$ is controllable (respectively, stabilizable).[5] Design a matrix K such that $\mathcal{A} + \mathcal{B}K$ is Hurwitz. The matrix K may depend on y_R. Partition K as

---

[4] The rank condition (11.19) implies that the linear state model (A, B, C) has no transmission zeros at the origin.

[5] See Exercise 11.12.

$K = [K_1 \ K_2]$, where K_1 is $p \times n$; thus, K_2 must be nonsingular.[6] The state feedback control should be taken as

$$u = K_1(x - x_{ss}) + K_2(\sigma - \bar{\sigma}) + u_{ss}$$

One degree of freedom, introduced by the integral control, is the fact that $\bar{\sigma}$ can be assigned by feedback. Choosing $\bar{\sigma}$ as

$$\bar{\sigma} = K_2^{-1}(u_{ss} - K_1 x_{ss}) \stackrel{\text{def}}{=} \sigma_{ss}$$

the state feedback control simplifies to

$$u = K_1 x + K_2 \sigma \qquad (11.20)$$

When the control (11.20) is applied to (11.17)–(11.18), it results in the closed-loop system

$$\dot{x} = f(x, K_1 x + K_2 \sigma) \qquad (11.21)$$
$$\dot{\sigma} = h(x) - y_R \qquad (11.22)$$

It can be easily verified that (x_{ss}, σ_{ss}) is indeed an equilibrium point of the closed-loop system. In fact, it is the unique equilibrium point in the domain of interest since there is a unique pair $(x_{ss}, u_{ss}) \in D_x \times D_u$ that satisfies the equilibrium equations (11.15)–(11.16). Linearization of (11.21)–(11.22) about this point yields

$$\dot{\xi} = \mathcal{A}_c \xi$$

where

$$\mathcal{A}_c = \begin{bmatrix} \frac{\partial f}{\partial x} + \frac{\partial f}{\partial u} K_1 & \frac{\partial f}{\partial u} K_2 \\ \frac{\partial h}{\partial x} & 0 \end{bmatrix}_{x = x_{ss}, \sigma = \sigma_{ss}} = \mathcal{A} + \mathcal{B}K$$

Thus, (x_{ss}, σ_{ss}) is exponentially stable and all solutions starting sufficiently close to it approach it as t tends to infinity. Consequently, $y(t) - y_R \to 0$ as $t \to \infty$.

In summary, assuming $(\mathcal{A}, \mathcal{B})$ is stabilizable and the rank condition (11.19) is satisfied, the state feedback control is given by

$$u = K_1 x + K_2 \sigma$$
$$\dot{\sigma} = e = y - y_R$$

where $K = [K_1 \ K_2]$ is designed such that $\mathcal{A} + \mathcal{B}K$ is Hurwitz.

---

[6] If K_2 is singular, it can be shown that $\mathcal{A} + \mathcal{B}K$ will be singular, which contradicts the fact that $\mathcal{A} + \mathcal{B}K$ is Hurwitz.

Example 11.4 Reconsider the pendulum equation of Exercise 11.2, where the task is to stabilize the pendulum at $\theta = \delta$. Taking

$$x_1 = \theta - \delta, \quad x_2 = \dot{\theta}, \quad y = \theta - \delta, \quad u = T$$

we obtain the state model

$$\begin{aligned} \dot{x}_1 &= x_2 \\ \dot{x}_2 &= -a \sin(x_1 + \delta) - bx_2 + cu \\ y &= x_1 = e \end{aligned}$$

Taking $y_R = 0$, it can be easily seen that the desired equilibrium is

$$x_{ss} = \begin{bmatrix} 0 \\ 0 \end{bmatrix}, \quad u_{ss} = \frac{a}{c} \sin \delta$$

The matrices A, B, C are given by

$$A = \begin{bmatrix} 0 & 1 \\ -a\cos\delta & -b \end{bmatrix}; \quad B = \begin{bmatrix} 0 \\ c \end{bmatrix}; \quad C = \begin{bmatrix} 1 & 0 \end{bmatrix}$$

Noting that $c > 0$, it can be easily verified that (A, B) is controllable and the rank condition (11.19) is satisfied. Taking $K = [k_1 \ k_2 \ k_3]$, it can be verified, using the Routh-Hurwitz criterion, that $A + BK$ will be Hurwitz if

$$b - k_2 c > 0, \quad (b - k_2 c)(a \cos \delta - k_1 c) + k_3 c > 0, \quad -k_3 c > 0$$

Suppose we do not know the exact values of the parameters $a > 0$, $b \geq 0$, $c > 0$, but we know an upper bound ρ_1 on a and a lower bound ρ_2 on c. Then, the choice

$$k_2 < 0, \quad k_3 < 0, \quad k_1 < -\frac{\rho_1}{\rho_2}\left(1 + \frac{k_3}{k_2 \rho_1}\right) \qquad (11.23)$$

ensures that $A + BK$ will be Hurwitz. The torque feedback law is given by

$$T = k_1(\theta - \delta) + k_2\dot{\theta} + k_3\sigma$$

where

$$\dot{\sigma} = \theta - \delta$$

which is the classical PID (proportional-integral-derivative) controller. Comparison of this feedback law with the one derived in Example 11.2 shows that we no longer calculate the steady-state torque needed to maintain the equilibrium position. Regulation will be achieved for all parameter perturbations which satisfy the bounds $a \leq \rho_1$ and $c \geq \rho_2$. $\qquad \triangle$

The robustness of regulation under integral control can be intuitively explained as follows. The feedback controller creates an asymptotically stable equilibrium point. At this point, all signals must be constant. For the integrator $\dot{\sigma} = e$ to have a constant output σ, its input e must be zero. Thus, the inclusion of the integrator forces the tracking error to be zero at equilibrium. Parameter perturbations will change the equilibrium point, but the condition $e = 0$ at equilibrium will be maintained. Thus, as long as the perturbed equilibrium point remains asymptotically stable, regulation will be achieved. However, since our stabilization result is local, we have to require the initial conditions $(x(0), \sigma(0))$ to be sufficiently close to the equilibrium point (x_{ss}, σ_{ss}). This requirement will, in general, restrict the range of parameter perturbations. Later on, we shall see nonlocal integral control designs which remove this restriction.[7]

The foregoing state feedback controller can be extended to output feedback via the use of observers. Suppose we cannot measure x, but we can measure the controlled output y. Suppose further that (A, C) is detectable and design the observer gain H, possibly dependent on y_R, such that $A + HC$ is Hurwitz. Then, the observer-based controller is given by

$$u = K_1 \hat{x} + K_2 \sigma \tag{11.24}$$

$$\dot{\sigma} = e = y - y_R \tag{11.25}$$

$$\dot{\hat{x}} = A\hat{x} + Bu + H(C\hat{x} - y) \tag{11.26}$$

We leave it as an exercise for the reader (Exercise 11.14) to verify the asymptotic regulation properties of this controller. Notice that we use an observer to estimate x; as for σ, it is already available for feedback since it is obtained by integrating the regulation error $e = y - y_R$.

11.3 Gain Scheduling

The basic limitation of the design via linearization approach is the fact that the controller is guaranteed to work only in the neighborhood of a single operating (equilibrium) point. In this section, we introduce gain scheduling, a technique that can extend the validity of the linearization approach to a range of operating points. In many situations, it is known how the dynamics of a system change with its operating points. It might even be possible to model the system in such a way that the operating points are parameterized by one or more variables which we call *scheduling variables*. In such situations, we may linearize the system at several equilibrium points, design a linear feedback controller at each point, and implement the resulting family of linear controllers as a single controller whose parameters are

---

[7]See Section 12.3.3.

changed by monitoring the scheduling variables. Such a controller is called a *gain scheduled controller.*

The concept of gain scheduling originated in connection with flight control systems.[8] In this application, the nonlinear equations of motion of an airplane or a missile are linearized about selected operating points which capture the key modes of operation throughout the flight envelope. Linear controllers are designed to achieve desired stability and performance requirements for the linearizations of the system about the selected operating points. The parameters of the controllers are then interpolated as functions of gain scheduling variables; typical variables are dynamic pressure, Mach number, altitude, and angle of attack. Finally, the gain scheduled controller is implemented on the nonlinear system. We start with a simple example to illustrate the idea of gain scheduling.

Example 11.5 Consider a tank system where the cross-sectional area β varies with height. The system can be modeled by the equation

$$\frac{d}{dt}\left(\int_0^h \beta(y)\ dy\right) = q - c\sqrt{2h}$$

where h is the liquid height in the tank, q is the input flow and c is a positive constant. Taking $x = h$ as the state variable and $u = q$ as the control input, the state model is given by

$$\dot{x} = \frac{1}{\beta(x)}\left(u - c\sqrt{2x}\right) \stackrel{\text{def}}{=} f(x, u)$$

Suppose we want to design a controller such that x tracks a reference signal y_R. We define $y = x$ as the controlled output and use y_R as the scheduling variable. When $y_R = \alpha = $ positive constant, the output y should be maintained at the value α. This requires the input u to be maintained at $\bar{u}(\alpha) = c\sqrt{2\alpha}$. Thus, for every value of α in the operating range, the desired operating point is defined by $x = y = \alpha$ and $u = \bar{u}(\alpha)$. Linearization of the nonlinear state equation about $x = \alpha$ and $u = \bar{u}(\alpha)$ results in the family of parameterized linear models

$$\dot{x}_\delta = a(\alpha)x_\delta + b(\alpha)u_\delta$$

where

$$
\begin{aligned}
a(\alpha) &= \left.\frac{\partial f}{\partial x}\right|_{x=\alpha, u=c\sqrt{2\alpha}} = \left[\frac{1}{\beta(x)}\left(\frac{-c}{\sqrt{2x}}\right) - \frac{\beta'(x)}{\beta^2(x)}\left(u - c\sqrt{2x}\right)\right]_{x=\alpha, u=c\sqrt{2\alpha}} \\
&= -\frac{c\sqrt{2\alpha}}{2\alpha\beta(\alpha)}
\end{aligned}
$$

---

[8] Examples of gain scheduling application to flight control problems can be found in [167, 168, 78, 125]. Examples in automotive and process control can be found in [11, 75, 84].

$$b(\alpha) = \left.\frac{\partial f}{\partial u}\right|_{x=\alpha, u=c\sqrt{2x}} = \frac{1}{\beta(\alpha)}, \quad x_\delta = x - \alpha, \quad u_\delta = u - c\sqrt{2\alpha}$$

Consider the PI (proportional-integral) controller

$$u_\delta = k_1 e + k_2 \sigma, \quad \dot{\sigma} = e = x - y_R = x_\delta - r_\delta$$

where $r_\delta = y_R - \alpha$. The closed-loop linear system has the characteristic equation

$$s^2 - (a + bk_1)s - bk_2 = 0$$

For a desired characteristic equation of the form

$$s^2 + 2\zeta\omega_n s + \omega_n^2, \quad \omega_n > 0, \ 0 < \zeta < 1$$

the gains k_1 and k_2 are taken as

$$k_1(\alpha) = -\frac{2\zeta\omega_n + a(\alpha)}{b(\alpha)}, \quad k_2(\alpha) = -\frac{\omega_n^2}{b(\alpha)}$$

A gain scheduled controller can be obtained by scheduling k_1 and k_2 as functions of y_R; that is, α is replaced by y_R so that the gains vary directly with the desired height. Sometimes it may be possible to simplify the scheduling task. For example, suppose the numerical values are such that $|a(\alpha)| << 2\zeta\omega_n$ over the operating range. If we reparameterize the PI controller as

$$u_\delta = k\left(e + \frac{1}{T}\sigma\right), \quad \text{where } k = k_1, \ T = \frac{k_1}{k_2}$$

we obtain

$$k(\alpha) = -\frac{2\zeta\omega_n + a(\alpha)}{b(\alpha)} \approx -\frac{2\zeta\omega_n}{b(\alpha)} = -2\zeta\omega_n\beta(\alpha)$$

$$T(\alpha) = \frac{2\zeta\omega_n + a(\alpha)}{\omega_n^2} \approx \frac{2\zeta}{\omega_n}$$

We take $k(\alpha) = -2\zeta\omega_n\beta(\alpha)$ and $T = 2\zeta/\omega_n$, so that we only need to schedule the gain k. The closed-loop linear system is given by

$$\dot{\xi} = A_\ell(\alpha)\xi + B_\ell r_\delta, \quad y_\delta = C_\ell\xi$$

where $\xi = [x_\delta \ \sigma]^T$, $y_\delta = x_\delta$, and

$$A_\ell(\alpha) = \begin{bmatrix} a(\alpha) - 2\zeta\omega_n & -\omega_n^2 \\ 1 & 0 \end{bmatrix}, \quad B_\ell = \begin{bmatrix} 2\zeta\omega_n \\ -1 \end{bmatrix}, \quad C_\ell = \begin{bmatrix} 1 & 0 \end{bmatrix}$$

The closed-loop transfer function from the command input r_δ to the output y_δ is given by

$$\frac{2\zeta\omega_n s + \omega_n^2}{s^2 + [2\zeta\omega_n - a(\alpha)]s + \omega_n^2}$$

Now, leaving aside this linear analysis, let us consider the gain scheduled PI controller

$$u = k(y_R)\left(e + \frac{1}{T}\sigma\right), \quad \dot{\sigma} = e = x - y_R$$

where the gain k is scheduled as a function of the reference signal y_R. When this control is applied to the nonlinear state equation, it results in the closed-loop system

$$\begin{aligned}
\dot{x} &= \frac{1}{\beta(x)}\left[k(y_R)\left(x - y_R + \frac{1}{T}\sigma\right) - c\sqrt{2x}\right] \\
\dot{\sigma} &= x - y_R \\
y &= x
\end{aligned}$$

When $y_R = \alpha = $ positive constant, the system has an equilibrium point at $x = \alpha$, $\sigma = \bar{\sigma}(\alpha) \stackrel{\text{def}}{=} -c\sqrt{2\alpha}/\omega_n^2\beta(\alpha)$. The corresponding equilibrium value of u is $c\sqrt{2\alpha}$.[9] This shows that the closed-loop nonlinear system achieves the desired operating point for every α in the operating range. Linearization of the closed-loop system about $(x, \sigma) = (\alpha, \bar{\sigma}(\alpha))$ and $y_R = \alpha$ results in the linearized closed-loop equation

$$\dot{\rho}_\delta = A_n(\alpha)\rho_\delta + B_n(\alpha)r_\delta, \quad y_\delta = C_n\rho_\delta$$

where

$$A_n(\alpha) = \begin{bmatrix} a(\alpha) - 2\zeta\omega_n & -\omega_n^2 \\ 1 & 0 \end{bmatrix}, \quad B_n(\alpha) = \begin{bmatrix} 2\zeta\omega_n + \gamma(\alpha) \\ -1 \end{bmatrix}, \quad C_n = \begin{bmatrix} 1 & 0 \end{bmatrix}$$

$\gamma(\alpha) = k'(\alpha)\bar{\sigma}(\alpha)/T\beta(\alpha)$, and $\rho_\delta = [(x - \alpha), (\sigma - \bar{\sigma}(\alpha))]^T$. The closed-loop transfer function from the command input r_δ to the output y_δ is given by

$$\frac{[2\zeta\omega_n + \gamma(\alpha)]s + \omega_n^2}{s^2 + [2\zeta\omega_n - a(\alpha)]s + \omega_n^2}$$

Let us note the difference between the two linear state models represented by (A_ℓ, B_ℓ, C_ℓ) and (A_n, B_n, C_n). The first model is the closed-loop model of the family of parameterized linear models used in the design of the controller, while the second model is the linearization of the closed-loop nonlinear system (under the gain scheduled controller) about the desired operating point. Ideally, we would like

---

[9] Notice the role played by integral control in alleviating the need to use the equilibrium value of u in the control. This is the same observation we used in the previous section.

these two models to be equivalent, for then we would know that the local behavior of the closed-loop system near the desired operating point matches the behavior predicted by the design model. Comparison of the two models shows that $A_n = A_\ell$ and $C_n = C_\ell$ but $B_n \neq B_l$, resulting in a different zero location in the closed-loop transfer function. Despite this difference, the two transfer functions share the same poles and property of zero steady-state tracking error to step inputs. If the only objective of control design is to assign the closed-loop poles and achieve zero steady-state tracking error to step inputs, we can say that the gain scheduled controller meets this objective. On the other hand, if other performance issues are of concern, like the transient part of the step response which is affected by the zero location, then we have to study the effect of the zero shift by linear analysis and/or simulation of the model (A_n, B_n, C_n). Alternatively, we may modify the gain scheduled controller with the objective of arriving at a linear model (A_n, B_n, C_n) that is equivalent to the linear model (A_ℓ, B_ℓ, C_ℓ) for every α in the operating range. Looking back at the expression for B_n, it is not hard to see that the term $\gamma(\alpha)$, which is not present in B_ℓ, is due to differentiation of the scheduled gain $k(y_R)$ with respect to y_R. This term has no counterpart in the model (A_ℓ, B_ℓ, C_ℓ) because in that model the gain was fixed at $k(\alpha)$. We will present two ideas to modify the gain scheduled controller in such a way that the effect of partial differentiation with respect to the scheduling variable cancels out at the operating point.[10] The first idea exploits the freedom that exists in choosing the steady-state value of σ. Consider the gain scheduled controller

$$u = k(y_R)\left[e + \frac{1}{T}(\sigma - \bar{\sigma}(y_R))\right] + \bar{u}(y_R), \quad \dot{\sigma} = e = x - y_R$$

where $\bar{u}(y_R) = c\sqrt{2y_R}$ and $\bar{\sigma}(y_R)$ is arbitrary. The controller we started with is a special case of this when $\bar{\sigma}(y_R)$ is chosen such that

$$\frac{k(y_R)}{T}\bar{\sigma}(y_R) = \bar{u}(y_R)$$

Instead of making this choice, let us keep $\bar{\sigma}(y_R)$ undetermined and proceed to calculate the linear model (A_n, B_n, C_n), which is now given by

$$A_n(\alpha) = \left[\begin{array}{cc} a(\alpha) - 2\zeta\omega_n & -\omega_n^2 \\ 1 & 0 \end{array}\right], \quad B_n(\alpha) = \left[\begin{array}{c} 2\zeta\omega_n + \phi(\alpha) \\ -1 \end{array}\right], \quad C_n = \left[\begin{array}{cc} 1 & 0 \end{array}\right]$$

where

$$\phi(\alpha) = \frac{1}{\beta(\alpha)}\left[-\frac{k(\alpha)}{T}\frac{\partial\bar{\sigma}}{\partial\alpha}(\alpha) + \frac{\partial\bar{u}}{\partial\alpha}(\alpha)\right]$$

---

[10] These two ideas are applications of general algorithms presented, respectively, in [101] and [88].

Choosing $\bar{\sigma}$ to satisfy the equation

$$\frac{k(\alpha)}{T}\frac{\partial\bar{\sigma}}{\partial\alpha}(\alpha) = \frac{\partial\bar{u}}{\partial\alpha}(\alpha)$$

ensures that (A_n, B_n, C_n) coincides with (A_ℓ, B_ℓ, C_ℓ). The second idea replaces the controller

$$u = k(y_R)\left(e + \frac{1}{T}\sigma\right), \quad \dot{\sigma} = e$$

by the controller

$$u = k(y_R)e + \frac{1}{T}z, \quad \dot{z} = k(y_R)e \qquad (11.27)$$

which, for a constant k, can be interpreted as commuting the gain k with the integrator. We leave it as an exercise for the reader (Exercise 11.21) to verify that this gain scheduled controller results in a triple (A_n, B_n, C_n) which is equivalent to the triple (A_ℓ, B_ℓ, C_ℓ). $\triangle$

In view of this example, we can describe the development of a gain scheduled controller for tracking control of a nonlinear system by the following steps.

1. Linearize the nonlinear model about the family of operating (equilibrium) points, parameterized by the scheduling variables.

2. Design a parameterized family of linear controllers to achieve the specified performance for the parameterized family of linear systems at each operating point.

3. Construct a gain scheduled controller such that, at each constant operating point,

   - the controller provides a constant control value yielding zero error,
   - the linearization of the closed-loop nonlinear system at each operating point is the same as the feedback connection of the parameterized linear system and the corresponding linear controller.

4. Check nonlocal performance of the gain scheduled controller for the nonlinear model by simulation.

The second step can be achieved by solving the design problem for the family of linear models parameterized by the scheduling variables, as we have done in the foregoing example, or by solving the problem only at a finite number of operating points using the same controller structure for all of them but allowing the controller parameters to change from one operating point to the other; then, the controller parameters are interpolated at intermediate operating points to produce the parameterized family of linear controllers. This interpolation process is usually ad hoc in nature and relies on physical insight.[11] As we have seen in the foregoing example, the choice of a gain scheduled controller in the third step is not unique. Meeting the linearization property stated in the third step is dependent on the controller structure. Also, deciding to what extent we have to insist on complete agreement between the linearization of the closed-loop nonlinear system and the parameterized closed-loop linear system depends on the control objective and specifications. We will not pursue these issues in a general setup.[12] Instead, we will show how to develop a gain scheduled controller for the observer-based integral controller of the previous section.

Consider the system

$$\dot{x} = f(x, u, w) \tag{11.28}$$

$$y = h(x) \tag{11.29}$$

where $f(x, u, w)$ and $h(x)$ are twice continuously differentiable functions in a domain $D_x \times D_u \times D_w \subset R^n \times R^p \times R^q$. Here, x is the state, u is the control input, w is a measured exogenous signal used for scheduling, and $y \in R^p$ is a measured output. Let $y_R \in D_r \subset R^p$ be a reference signal. We want to design an output feedback controller that achieves small tracking error $e = y - y_R$ in response to the exogenous input

$$v = \begin{bmatrix} y_R \\ w \end{bmatrix} \in D_v \overset{\text{def}}{=} D_r \times D_w$$

Toward that end, we require the closed-loop system to have an asymptotically stable equilibrium point, with zero steady-state tracking error, when $v = \alpha = $ constant. Let us partition α as $\alpha = [\alpha_r^T, \; \alpha_w^T]^T$. We assume that there exists a unique pair of continuously differentiable functions $\mathcal{X} : D_v \to D_x$ and $\mathcal{U} : D_v \to D_u$ such that

$$0 = f(\mathcal{X}(\alpha), \mathcal{U}(\alpha), \alpha_w) \tag{11.30}$$

$$0 = h(\mathcal{X}(\alpha)) - \alpha_r \tag{11.31}$$

for all $\alpha \in D_v$. This assumption guarantees that, for each $\alpha \in D_v$, there is an equilibrium point at which zero steady-state tracking error can be achieved. Linearization of (11.28)–(11.29) about $x = \mathcal{X}(\alpha)$, $u = \mathcal{U}(\alpha)$, and $w = \alpha_w$ results in the parameterized family of linear systems

$$\dot{x}_\delta = A(\alpha)x_\delta + B(\alpha)u_\delta + E(\alpha)w_\delta \tag{11.32}$$

$$y_\delta = C(\alpha)x_\delta \tag{11.33}$$

where

$$A(\alpha) = \left. \frac{\partial f}{\partial x}(x, u, w) \right|_{\mathcal{X}(\alpha), \mathcal{U}(\alpha), \alpha_w} , \quad B(\alpha) = \left. \frac{\partial f}{\partial u}(x, u, w) \right|_{\mathcal{X}(\alpha), \mathcal{U}(\alpha), \alpha_w}$$

---

[11] For examples of this interpolation process, see [78, 125].

[12] Consult [101, 88] for further elaboration on these issues.

$$C(\alpha) = \left.\frac{\partial h}{\partial x}(x)\right|_{\mathcal{X}(\alpha)}, \quad E(\alpha) = \left.\frac{\partial f}{\partial w}(x, u, w)\right|_{\mathcal{X}(\alpha), \mathcal{U}(\alpha), \alpha_w}$$

$$x_\delta = x - \mathcal{X}(\alpha), \quad u_\delta = u - \mathcal{U}(\alpha), \quad w_\delta = w - \alpha_w$$

Our smoothness assumptions on f, h, $\mathcal{X}$, and $\mathcal{U}$ ensure that the matrices A, B, C, and E are continuously differentiable functions of α. In anticipation of the use of integral control, define the matrices

$$\mathcal{A}(\alpha) = \left[\begin{array}{cc} A(\alpha) & 0 \\ C(\alpha) & 0 \end{array}\right], \quad \mathcal{B}(\alpha) = \left[\begin{array}{c} B(\alpha) \\ 0 \end{array}\right]$$

We assume that the pairs $(\mathcal{A}(\alpha), \mathcal{B}(\alpha))$ and $(A(\alpha), C(\alpha))$ are, respectively, controllable and observable for every $\alpha \in D_v$.[13] For any compact subset $S \subset D_v$, we can design matrices $K(\alpha)$ and $H(\alpha)$, continuously differentiable in α, such that $\mathcal{A}(\alpha) + \mathcal{B}(\alpha)K(\alpha)$ and $A(\alpha) + H(\alpha)C(\alpha)$ are Hurwitz uniformly in α; that is, the real parts of all eigenvalues are less than a negative number, independent of α, for all $\alpha \in S$.[14] It follows that the matrix

$$\mathcal{A}_c(\alpha) = \left[\begin{array}{ccc} A(\alpha) & B(\alpha)K_2(\alpha) & B(\alpha)K_1(\alpha) \\ C(\alpha) & 0 & 0 \\ -H(\alpha)C(\alpha) & B(\alpha)K_2(\alpha) & A(\alpha) + B(\alpha)K_1(\alpha) + H(\alpha)C(\alpha) \end{array}\right] \tag{11.34}$$

where K is partitioned as $K = [K_1 \ K_2]$, is Hurwitz uniformly in α for all $\alpha \in S$. Consider the observer-based integral controller derived in the previous section:

$$u_\delta = K_1(\alpha)\hat{x}_\delta + K_2(\alpha)\sigma \tag{11.35}$$

$$\dot{\sigma} = e = y - y_R \tag{11.36}$$

$$\dot{\hat{x}}_\delta = A(\alpha)\hat{x}_\delta + B(\alpha)u_\delta + H(\alpha)[C(\alpha)\hat{x}_\delta - y_\delta] \tag{11.37}$$

When this controller is applied to (11.32)–(11.33), it results in the parameterized family of linear closed-loop systems

$$\dot{\xi} = A_\ell(\alpha)\xi + B_\ell(\alpha)r_\delta + E_\ell(\alpha)w_\delta \tag{11.38}$$

$$y_\delta = C_\ell(\alpha)\xi \tag{11.39}$$

where

$$A_\ell(\alpha) = \mathcal{A}_c(\alpha), \quad B_\ell(\alpha) = \left[\begin{array}{c} 0 \\ -I \\ 0 \end{array}\right], \quad E_\ell(\alpha) = \left[\begin{array}{c} E(\alpha) \\ 0 \\ 0 \end{array}\right], \quad \xi = \left[\begin{array}{c} x_\delta \\ \sigma \\ \hat{x}_\delta \end{array}\right]$$

---

[13] Recall that controllability of $(\mathcal{A}, \mathcal{B})$ is implied by controllability of (A, B) and the rank condition (11.19).

[14] One method to compute $K(\alpha)$ and $H(\alpha)$ is outlined in Exercise 11.22.

$$C_\ell(\alpha) = \begin{bmatrix} C(\alpha) & 0 & 0 \end{bmatrix}, \quad r_\delta = y_R - \alpha_r$$

It is clear that when $r_\delta = 0$ and $w_\delta = 0$, this system has an asymptotically stable equilibrium point at the origin. Moreover, the steady-state tracking error is zero, as it can be seen from the integrator equation (11.36).

Now, let us consider the gain scheduled controller

$$
\begin{align}
u &= K_1(v)\hat{x} + K_2(v)\sigma \tag{11.40} \\
\dot{\sigma} &= e \tag{11.41} \\
\dot{\hat{x}} &= A(v)\hat{x} + B(v)u + H(v)[C(v)\hat{x} - v] \tag{11.42}
\end{align}
$$

where the matrices A, B, C, K_1, K_2, and H are scheduled as functions of the scheduling variable v. When this controller is applied to the nonlinear system (11.28)–(11.29), it results in the closed-loop system

$$\dot{\rho} = g_o(\rho, v) \tag{11.43}$$

where

$$\rho = \begin{bmatrix} x \\ \sigma \\ \hat{x} \end{bmatrix}$$

and

$$g_o(\rho, v) = \begin{bmatrix} f(x, K_1(v)\hat{x} + K_2(v)\sigma, w) \\ h(x) - y_R \\ \{A(v) + B(v)K_1(v) + H(v)C(v)\}\hat{x} + B(v)K_2(v)\sigma - H(v)h(x) \end{bmatrix}$$

When $v = \alpha = \text{constant}$, the system (11.43) has an equilibrium point $\bar{\rho}(\alpha)$ defined by [15]

$$\bar{\rho}(\alpha) = \begin{bmatrix} \mathcal{X}(\alpha) \\ \bar{\sigma}(\alpha) \\ \bar{x}(\alpha) \end{bmatrix} = \begin{bmatrix} \mathcal{X}(\alpha) \\ K_2^{-1}(\alpha)\{\mathcal{U}(\alpha) - K_1(\alpha)\bar{x}(\alpha)\} \\ -\{A(\alpha) + H(\alpha)C(\alpha)\}^{-1}\{B(\alpha)\mathcal{U}(\alpha) - H(\alpha)\alpha_r\} \end{bmatrix}$$

Moreover, due to uniqueness of the pair of functions $(\mathcal{X}(\alpha), \mathcal{U}(\alpha))$ which satisfy the equilibrium equations (11.30)–(11.31), the equilibrium point $\bar{\rho}(\alpha)$ is unique. The corresponding equilibrium values of u and y are $u = \mathcal{U}(\alpha)$ and $y = h(\mathcal{X}(\alpha)) = y_R$, respectively. Thus, the closed-loop system (11.43) achieves the desired operating point for every $\alpha \in S$. Linearization of (11.43) about $\rho = \bar{\rho}(\alpha)$, $y_R = \alpha_r$, and $w = \alpha_w$ results in the linearized closed-loop equation

$$
\begin{align}
\dot{\rho}_\delta &= A_n(\alpha)\rho_\delta + B_n(\alpha)r_\delta + E_n(\alpha)w_\delta \tag{11.44} \\
y_\delta &= C_n(\alpha)\rho_\delta \tag{11.45}
\end{align}
$$

---

[15]$(A+HC)$ is nonsingular since it is Hurwitz, and K_2 is nonsingular since $(A+BK)$ is Hurwitz; see Section 11.2.2 for further explanation of the latter point.

where $\rho_\delta = \rho - \bar{\rho}(\alpha)$. It can be verified that

$$A_n(\alpha) = A_\ell(\alpha) = \mathcal{A}_c(\alpha), \quad C_n(\alpha) = C_\ell(\alpha)$$

but, in general, $B_n(\alpha) \neq B_\ell(\alpha)$ and $E_n(\alpha) \neq E_\ell(\alpha)$ due to partial differentiation of the scheduled variables with respect to v. The fact that $\mathcal{A}_c(\alpha)$ is Hurwitz, uniformly in α for all $\alpha \in S$, shows that the gain scheduled controller (11.40)–(11.42) achieves the objective of producing an asymptotically stable equilibrium point, with zero steady-state tracking error, when $v = \alpha = $ constant. However, the closed-loop transfer functions from r_δ and w_δ to the output y_δ could be dramatically different from the corresponding transfer functions of the design model (11.38)–(11.39). Therefore, we must check the performance of the model (11.44)–(11.45) via analysis and/or simulation.

If the performance of the model (11.44)–(11.45) is acceptable, we may go ahead and implement the gain scheduled controller (11.40)–(11.42) despite the difference between the model (11.44)–(11.45) and the design model (11.38)–(11.39). It turns out, however, that we can do better than that. As in Example 11.5, we can modify the gain scheduled controller to achieve equivalence between the linearized model of the closed-loop system and the linear design model. We will present a modification that uses the second of the two ideas presented in Example 11.5. Recall that the idea was to commute the scheduled variable and the integrator; that is, to move the integrator from the input side of the controller to its output side. This is basically what we would like to do for the controller (11.40)–(11.42). However, the current situation is complicated by the presence of the observer equation, which can be written as

$$\dot{\hat{x}} = [A(v) + B(v)K_1(v) + H(v)C(v)]\hat{x} + B(v)K_2(v)\sigma - H(v)y$$

after substitution of u using (11.40). This observer equation has two driving inputs: σ and y. While σ is the output of an integrator and it makes sense to talk about moving that integrator to the output side of the observer, y is not the output of an integrator. This difficulty can be overcome if we can measure $\dot{y}$, the derivative of y. For, then, we can represent the observer equation as

$$\dot{\hat{x}} = F(v)\hat{x} + G(v)\lambda, \quad \dot{\lambda} = \psi$$

where

$$F = A + BK_1 + HC, \quad G = \begin{bmatrix} BK_2 & -H \end{bmatrix}, \quad \psi = \begin{bmatrix} e \\ \dot{y} \end{bmatrix}$$

For a fixed-gain controller, that is, with $v = \alpha$, the controller transfer function from ψ to u is given by

$$\{K_1(\alpha)[sI - F(\alpha)]^{-1}G(\alpha) + K_2(\alpha)[I \quad 0]\}\frac{1}{s}$$

which is equivalent to

$$\frac{1}{s}\{K_1(\alpha)[sI - F(\alpha)]^{-1}G(\alpha) + K_2(\alpha)[I \quad 0]\}$$

The latter transfer function can be realized as

$$\dot{\varphi} = F(\alpha)\varphi + G(\alpha)\psi$$
$$\dot{\eta} = K_1(\alpha)\varphi + K_2(\alpha)e$$
$$u = \eta$$

Scheduling the matrices A, B, C, K_1, K_2, and H in this realization as functions of the scheduling variable v, we obtain the gain scheduled controller

$$\dot{\varphi} = [A(v) + B(v)K_1(v) + H(v)C(v)]\varphi + B(v)K_2(v)e - H(v)\theta \quad (11.46)$$
$$\dot{\eta} = K_1(v)\varphi + K_2(v)e \quad (11.47)$$
$$u = \eta \quad (11.48)$$
$$\theta = \dot{y} \quad (11.49)$$

When this controller is applied to the nonlinear system (11.28)–(11.29), it results in the closed-loop system

$$\dot{z} = g(z, v) \quad (11.50)$$

where

$$z = \begin{bmatrix} x \\ \eta \\ \varphi \end{bmatrix}, \quad g(z, v) = \begin{bmatrix} f(x, \eta, w) \\ K_1(v)\varphi + K_2(v)\{h(x) - y_R\} \\ g_3(z, v) \end{bmatrix}$$

and

$$g_3(z, v) = [A(v) + B(v)K_1(v) + H(v)C(v)]\varphi + B(v)K_2(v)[h(x) - y_R]$$
$$- H(v)\frac{\partial h}{\partial x}(x)f(x, \eta, w)$$

When $v = \alpha = $ constant, the system (11.50) has an equilibrium point $\bar{z}(\alpha)$ defined by

$$\bar{z}(\alpha) = \begin{bmatrix} \mathcal{X}(\alpha) \\ \mathcal{U}(\alpha) \\ 0 \end{bmatrix} \quad (11.51)$$

Once again, it can be shown (Exercise 11.23) that this equilibrium point is unique. The corresponding value of y is $y = h(\mathcal{X}(\alpha)) = y_R$. Linearization of (11.50) about $z = \psi_m$, $y_R = \alpha_r$, and $w = \alpha_w$ results in the linearized closed-loop equation [16]

$$\dot{z}_\delta = A_m(\alpha)z_\delta + B_m(\alpha)r_\delta + E_m(\alpha)w_\delta \quad (11.52)$$
$$y_\delta = C_m(\alpha)z_\delta \quad (11.53)$$

---

[16] While calculating the matrices A_m, B_m, and E_m, note that partial derivatives which appear as coefficients of φ, $[h - y_R]$, or f vanish at the operating point.

where

$$
A_m = \begin{bmatrix} A & B & 0 \\ K_2 C & 0 & K_1 \\ -HCA + BK_2 C & -HCB & A + BK_1 + HC \end{bmatrix}, \quad B_m = \begin{bmatrix} 0 \\ -K_2 \\ -BK_2 \end{bmatrix}
$$

$$
C_m = \begin{bmatrix} C & 0 & 0 \end{bmatrix}, \quad E_m = \begin{bmatrix} E \\ 0 \\ -HCE \end{bmatrix}, \quad z_\delta = z - \bar{z}(\alpha)
$$

We leave it as an exercise for the reader (Exercise 11.24) to verify that the matrix $P(\alpha)$, defined by

$$
P = \begin{bmatrix} I & 0 & 0 \\ 0 & K_2 & K_1 \\ -HC & BK_2 & A + BK_1 + HC \end{bmatrix} \tag{11.54}
$$

is nonsingular and

$$
P^{-1} A_m P = A_\ell, \quad P^{-1} B_m = B_\ell, \quad P^{-1} E_m = E_\ell, \quad C_m P = C_\ell \tag{11.55}
$$

Hence, the linear model (11.52)–(11.53) is equivalent to the linear model (11.38)–(11.39).

If measurement of $\dot{y}$ is not available, we may still use the gain scheduled controller (11.46)–(11.48) with θ taken as the estimate of $\dot{y}$, given by

$$
\epsilon \dot{\zeta} = -\zeta + y \tag{11.56}
$$

$$
\theta = \frac{1}{\epsilon}(-\zeta + y) \tag{11.57}
$$

where ϵ is a "sufficiently small" positive constant and the filter (11.56) is always initialized such that

$$
\zeta(0) - y(0) = O(\epsilon) \tag{11.58}
$$

Since y is measured, we can always meet this initialization condition. Furthermore, whenever the system is initialized from an equilibrium point, the condition (11.58) is automatically satisfied since, at equilibrium, $y = \zeta$. The linear system (11.56)–(11.57) acts as a derivative approximator when ϵ is sufficiently small, as it can be seen from its transfer function

$$
\frac{s}{\epsilon s + 1} I
$$

which approximates the differentiator transfer function sI for frequencies much smaller than $1/\epsilon$. The use of the derivative approximator (11.56)–(11.57) can be justified by application of singular perturbation results from Chapter 9. To derive a singularly perturbed model for the closed-loop system, we start by using θ as the state of the additional filter instead of ζ; that is, we perform a change of variables according to (11.57). The state equation for θ is

$$\epsilon\dot{\theta} = -\theta + \dot{y} \tag{11.59}$$

Note that (11.59) cannot be implemented to compute θ since it is driven by $\dot{y}$, which is not available. For implementation, we must use the filter (11.56)–(11.57) but, for analysis, it is more convenient to work with (11.59). The closed-loop system under the gain scheduled controller (11.46)–(11.48) and (11.56)–(11.58) is described by the singularly perturbed model

$$\dot{z} = g(z, v) - NH(v)\left[\theta - \frac{\partial h}{\partial x}(x)f(x, \eta, w)\right] \tag{11.60}$$

$$\epsilon\dot{\theta} = -\theta + \frac{\partial h}{\partial x}(x)f(x, \eta, w) \tag{11.61}$$

where $N = [0\ 0\ I]^T$. When $v = \alpha = $ constant, the system (11.60)–(11.61) has an equilibrium point at $z = \bar{z}(\alpha)$ and $\theta = 0$. This shows that the additional filter (11.56)–(11.57) does not change the equilibrium condition. Linearization of (11.60)–(11.61) about $z = \bar{z}(\alpha), \theta = 0, y_R = \alpha_r$, and $w = \alpha_w$ results in the linear singularly perturbed system

$$\dot{z}_\delta = [A_m(\alpha) + NH(\alpha)C_m(\alpha)A_m(\alpha)]z_\delta - NH(\alpha)\theta$$
$$+ B_m(\alpha)r_\delta + [E_m(\alpha) + NH(\alpha)C_m(\alpha)E_m(\alpha)]w_\delta \tag{11.62}$$

$$\epsilon\dot{\theta} = -\theta + C_m(\alpha)A_m(\alpha)z_\delta + C_m(\alpha)E_m(\alpha)w_\delta \tag{11.63}$$

$$y_\delta = C_m(\alpha)z_\delta \tag{11.64}$$

It is easy to see that the reduced model, obtained by setting $\epsilon = 0$ and eliminating θ, is exactly (11.52)–(11.53). Noting that $A_m(\alpha)$ is Hurwitz, we can use singular perturbation theory to show that for any bounded inputs r_δ and w_δ and any bounded initial conditions $z_\delta(0)$ and $\theta(0)$, we have

$$\|z_\delta(t, \epsilon) - z_{\delta s}(t)\| \leq k\epsilon, \quad \forall\, t \geq 0 \tag{11.65}$$

where $z_\delta(t, \epsilon)$ and $z_{\delta s}(t)$ are, respectively, solutions of (11.62) and (11.52) which start from the same initial conditions. The details of the singular perturbation argument are outlined in Exercise 9.14. The approximation result (11.65) shows that, for sufficiently small ϵ, the performance of the linearized closed-loop system when $\dot{y}$ is approximated by filter (11.56)–(11.57) recovers the performance of the linearized closed-loop system when $\dot{y}$ is used. Note that for (11.65) to hold, the initial conditions $z_\delta(0)$ and $\theta(0)$ must be bounded uniformly in ϵ. Noting the scaling factor $1/\epsilon$ in (11.57) shows that the initialization condition (11.58) is needed to ensure uniform boundedness of $\theta(0)$.

In summary, the design of a gain scheduled, observer-based, integral controller for tracking control can be carried out as follows:

1. Solve the equilibrium equations (11.30)–(11.31) to compute $\mathcal{X}(\alpha)$ and $\mathcal{U}(\alpha)$.

2. Compute the matrices $A(\alpha)$, $B(\alpha)$, $C(\alpha)$, $E(\alpha)$, $\mathcal{A}(\alpha)$, and $\mathcal{B}(\alpha)$, all of which are defined after (11.32)–(11.33).

3. Design continuously differentiable matrices $K(\alpha)$ and $H(\alpha)$ such that the matrices $\mathcal{A}(\alpha) + \mathcal{B}(\alpha)K(\alpha)$ and $A(\alpha) + H(\alpha)C(\alpha)$ are Hurwitz, uniformly in α, for all α in a compact set S.

4. If $\dot{y}$ is available for feedback, implement the gain scheduled controller (11.46)–(11.49); otherwise, implement the gain scheduled controller (11.46)–(11.48) and (11.56)–(11.58).

So far, our analysis of the closed-loop system under the gain scheduled controller has focused on the local behavior in the neighborhood of a constant operating point. Can we say more about the behavior of the nonlinear system? What if the scheduling variable is not constant? In applications of gain scheduling, the practice has been that you can schedule on time-varying variables as long as they are slow enough relative to the dynamics of the system. This practice can be justified by analyzing the closed-loop system (11.50) as a slowly varying system. Stability analysis of slowly varying systems is presented in Section 5.7. Example 5.15 of that section deals specifically with a class of systems that includes the closed-loop system (11.50) as a special case. Using the results of Example 5.15, we can arrive at the following conclusions. Suppose the scheduling variable $v(t)$ is continuously differentiable and $\|\dot{v}(t)\| \leq \mu$ for all $t \geq 0$. Then, there exist positive constants ρ_1 and ρ_2 such that if $\mu < \rho_1$ and $\|z(0) - \bar{z}(v(0))\| < \rho_2$, the solution $z(t)$ of $\dot{z} = g(z, v(t))$ will be uniformly bounded for all $t \geq 0$. Moreover, after some finite time, $z(t) - \bar{z}(v(t))$ will be of order $O(\mu)$. This implies that $x(t) - \mathcal{X}(v(t))$ will be $O(\mu)$. Hence,

$$e(t) = h(x(t), v(t)) = h(\mathcal{X}(v(t)) + O(\mu), v(t)) = O(\mu)$$

since $h(\mathcal{X}(v), v) = 0$. In other words, the tracking error will, eventually, be of the order of the derivative of the scheduling variable. Furthermore, if $v(t) \to v_\infty$ and $\dot{v}(t) \to 0$ as $t \to \infty$, then $z(t) \to \bar{z}(v_\infty)$ as $t \to \infty$. Consequently, $e(t) \to 0$ as $t \to \infty$. Similar conclusions can be stated for the closed-loop system (11.60)–(11.61).[17]

Example 11.6 Consider the second-order system

$$\dot{x}_1 = \tan x_1 + x_2$$
$$\dot{x}_2 = x_1 + u$$
$$y = x_2$$

---

[17]See Exercise 9.28.

where y is the only measured signal. We want y to track a reference signal y_R. We use y_R as the scheduling variable. When $y_R = \alpha = \text{constant}$, the equilibrium equations (11.30)–(11.31) have the unique solution

$$X(\alpha) = \begin{bmatrix} -\tan^{-1}\alpha \\ \alpha \end{bmatrix}, \quad U(\alpha) = \tan^{-1}\alpha$$

The family of parameterized linear systems (11.32)–(11.33) is defined by

$$A(\alpha) = \begin{bmatrix} 1+\alpha^2 & 1 \\ 1 & 0 \end{bmatrix}, \quad B = \begin{bmatrix} 0 \\ 1 \end{bmatrix}, \quad C = \begin{bmatrix} 0 & 1 \end{bmatrix}, \quad E = 0$$

It can be easily checked that $(A(\alpha), B)$ is controllable and $(A(\alpha), C)$ is observable for all α. The matrix $K(\alpha)$ is designed as

$$K(\alpha) = \begin{bmatrix} -(1+\alpha^2)(3+\alpha^2) - 3 - \frac{1}{1+\alpha^2} & -3-\alpha^2 & \frac{1}{1+\alpha^2} \end{bmatrix}$$

to assign the eigenvalues of $A(\alpha) + BK(\alpha)$ at $-1, -\frac{1}{2} \pm j\frac{\sqrt{3}}{2}$. The matrix $H(\alpha)$ is designed as

$$H(\alpha) = \begin{bmatrix} -10 - (4+\alpha^2)(1+\alpha^2) \\ -(4+\alpha^2) \end{bmatrix}$$

to assign the eigenvalues of $A(\alpha) + H(\alpha)C$ at $-\frac{3}{2} \pm j\frac{3\sqrt{3}}{2}$. We have chosen the eigenvalues independent of α for convenience, but that is not necessary. We could have allowed the eigenvalues to depend on α as long as their real parts are less than a negative number independent of α. Since $\dot{y}$ is not available, we implement the gain scheduled controller (11.46)–(11.48) and (11.56)–(11.58) with $\epsilon = 0.01$. Figure 11.1 shows the response of the closed-loop system to a sequence of step changes in the reference signal. A step change in the reference signal resets the equilibrium point of the system, and the initial state of the system at time 0_+ is the equilibrium state at time 0_-. If the initial state is within the region of attraction of the new equilibrium point, the system reaches steady state at that point. Since our controller is based on linearization, it guarantees only local stabilization. Therefore, in general, step changes in the reference signal will have to be limited. Reaching a large value of the reference signal can be a done by a sequence of step changes, as in the figure, allowing enough time for the system to settle down after each step change. Another method to change the reference set point is to move slowly from one set point to another. Figure 11.2 shows the response of the closed-loop system to a slow ramp that takes the set point from zero to one over a period of 100 seconds. This response is consistent with our conclusions about the behavior of gain scheduled controllers under slowly varying scheduling variables. The same figure shows the response to a faster ramp signal. As the slope of the ramp increases, tracking performance deteriorates. If we keep increasing the slope of the ramp, the

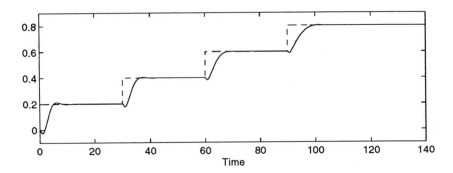

Figure 11.1: The reference (dashed) and output (solid) signals of the gain scheduled controller of Example 11.6.

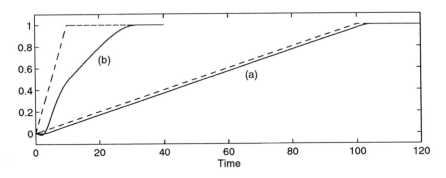

Figure 11.2: The reference (dashed) and output (solid) signals of the gain scheduled controller of Example 11.6 with ramp reference: (a) ramp slope $= 0.01$; (b) ramp slope $= 0.1$.

system will eventually go unstable. To appreciate what we gain by gain scheduling, Figure 11.3 shows the response of the closed-loop system to the same sequence of step changes of Figure 11.1 when a fixed-gain controller evaluated at $\alpha = 0$ is used. For small reference inputs, the response is as good as the one with the gain scheduled controller but, as the reference signal increases, the performance deteriorates and the system goes unstable. Finally, to see why we may have to modify the gain scheduled controller (11.40)–(11.42) that we started with, Figure 11.4 shows the response of the closed-loop system under this controller to the same sequence of step changes of Figure 11.1. While stability and zero steady-state tracking error are achieved, as predicted by our analysis, the transient response deteriorates rapidly as the reference signal increases. This is due to additional zeros in the closed-loop transfer function. Such bad transient behavior could lead to instability as it could

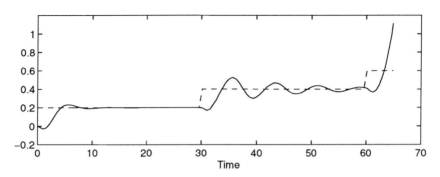

Figure 11.3: The reference (dashed) and output (solid) signals of the fixed gain controller of Example 11.6.

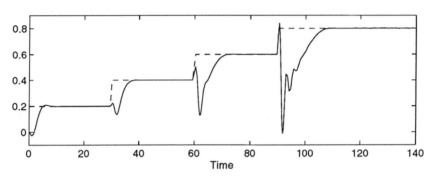

Figure 11.4: The reference (dashed) and output (solid) signals of the unmodified gain scheduled controller of Example 11.6.

take the state of the system out of the finite region of attraction, although such instability was not observed in this example. △

11.4 Exercises

Exercise 11.1 Suppose we want to stabilize the system $\dot{x} = f(x, u)$, $x \in R^n$, $u \in R^p$, with respect to an arbitrary point $p \in R^n$.

(a) Show that this is possible only if there is a vector $q \in R^p$ such that $0 = f(p, q)$.

(b) Show that the change of variables $u = q + v$; $x = p + y$ reduces the problem to stabilization with respect to the origin.

Exercise 11.2 Consider the closed-loop system of Example 11.2. For convenience, assume $a = c = 1$, $\delta = \pi/4$, $b = 0$, and $k_1 = k_2 = -1$. Find a Lyapunov function for the system, and use it to estimate the region of attraction of the origin.

Exercise 11.3 Consider the system

$$
\begin{aligned}
\dot{x}_1 &= a(x_2 - x_1) \\
\dot{x}_2 &= bx_1 - x_2 - x_1 x_3 + u \\
\dot{x}_3 &= x_1 + x_1 x_2 - 2ax_3
\end{aligned}
$$

where a and b are positive constants. Using linearization, design a state feedback control law to stabilize the origin.

Exercise 11.4 Using linearization, design a state feedback control law to stabilize the origin of the system

$$
\begin{aligned}
\dot{x}_1 &= \exp(x_2)u \\
\dot{x}_2 &= x_1 + x_2^2 + \exp(x_2)u \\
\dot{x}_3 &= x_1 - x_2
\end{aligned}
$$

Estimate the region of attraction of the closed-loop system.

Exercise 11.5 Figure 11.5 shows a schematic diagram of a magnetic suspension system, where a ball of magnetic material is suspended by means of an electromagnet whose current is controlled by feedback from the, optically measured, ball position [187, pp. 192–200]. This system has the basic ingredients of systems constructed to levitate mass, used in gyroscopes, accelerometers, and fast trains. The equation of motion of the ball is

$$
m\ddot{y} = -k\dot{y} + mg + F(y, i)
$$

where m is the mass of the ball, $y \geq 0$ is the vertical (downward) position of the ball measured from a reference point ($y = 0$ when the ball is next to the coil), k is a viscous friction coefficient, g is the acceleration due to gravity, $F(y, i)$ is the force generated by the electromagnet, and i is its electric current. The inductance of the electromagnet depends on the position of the ball, and can be modeled as

$$
L(y) = L_1 + \frac{L_0}{1 + y/a}
$$

where L_1, L_0, and a are positive constants. This model represents the case that the inductance has its highest value when the ball is next to the coil, and decreases to a constant value as the ball is removed to $y = \infty$. With $E(y, i) = \frac{1}{2}L(y)i^2$ as the energy stored in the electromagnet, the force $F(y, i)$ is given by

$$
F(y, i) = \frac{\partial E}{\partial y} = -\frac{L_0 i^2}{2a(1 + y/a)^2}
$$

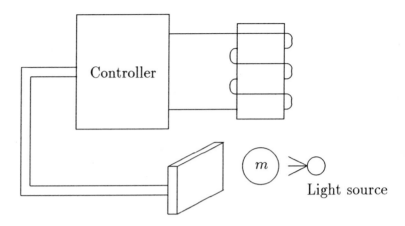

Figure 11.5: Magnetic suspension system of Exercise 11.5.

When the electric circuit of the coil is driven by a voltage source with voltage v, Kirchhoff's voltage law gives the relationship $v = \dot{\phi} + Ri$, where R is the series resistance of the circuit and $\phi = L(y)i$ is the magnetic flux linkage.

(a) Using $x_1 = y$, $x_2 = \dot{y}$, and $x_3 = i$ as state variables, and $u = v$ as control input, show that the state equation is given by

$$
\begin{aligned}
\dot{x}_1 &= x_2 \\
\dot{x}_2 &= g - \frac{k}{m}x_2 - \frac{L_0 a x_3^2}{2m(a + x_1)^2} \\
\dot{x}_3 &= \frac{1}{L(x_1)}\left[-Rx_3 + \frac{L_0 a x_2 x_3}{(a + x_1)^2} + u\right]
\end{aligned}
$$

(b) Suppose it is desired to balance the ball at a desired position $y_R > 0$. Find the steady-state values I_{ss} and V_{ss} of i and v, respectively, which are necessary to maintain such balance.

(c) Show that the equilibrium point obtained by taking $u = V_{ss}$ is unstable.

(d) Using linearization, design a state feedback control law to stabilize the ball at the desired position, that is, to make the equilibrium point asymptotically stable.

(e) Study the performance of the closed-loop system using computer simulation. In particular, study the transient behavior and the effect of $\pm 20\%$ parameter perturbations from nominal values for all parameters of the system.

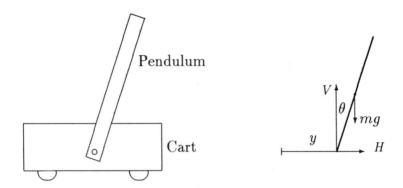

Figure 11.6: Inverted pendulum of Exercise 11.6.

(f) Starting with the ball at equilibrium, move it a small distance up (and then down) and let it go. Repeat this experiment, gradually increasing the amount of initial disturbance. Using simulation, determine the largest range of initial disturbance for which the ball will return to equilibrium.

In parts (e) and (f), use the numerical data: $m = 0.01$ kg, $k = 0.001$ N/m/sec, $g = 9.81$ m/sec^2, $a = 0.05$ m, $L_0 = 0.01$ H, $L_1 = 0.02$ H, $R = 10\,\Omega$, and $y_R = 0.05$ m. Unless otherwise specified, use the same numerical data in all forthcoming exercises that apply to the magnetic suspension system.

Exercise 11.6 Consider the inverted pendulum of Figure 11.6 [98]. The pivot of the pendulum is mounted on a cart which can move in a horizontal direction. The cart is driven by a motor that exerts a horizontal force F on the cart. This force is the only control input to the system. By manipulating F, we would like to control the position of the pendulum. From the physical description of the system, it is clear that balancing the pendulum in a vertical position is open-loop unstable. By feedback we can stabilize the pendulum in a vertical position, or even command it to do some desired maneuvers. This will be the task of this exercise and a few others. The figure shows also the forces acting on the pendulum, which are the force mg at the center of gravity, a horizontal reaction force H, and a vertical reaction force V at the pivot. Writing horizontal and vertical Newton's laws at the center of gravity of the pendulum yields

$$m \frac{d^2}{dt^2}(y + L\sin\theta) = H, \qquad m \frac{d^2}{dt^2}(L\cos\theta) = V - mg$$

Taking moments about the center of gravity yields the torque equation

$$I\ddot{\theta} = VL\sin\theta - HL\cos\theta$$

while a horizontal Newton's law for the cart yields

$$M\ddot{y} = F - H - k\dot{y}$$

Here m is the mass of the pendulum, M is the mass of the cart, L is the distance from the center of gravity to the pivot, I is the moment of inertia of the pendulum with respect to the center of gravity, k is a friction coefficient, y is the displacement of the pivot, θ is the angular rotation of the pendulum (measured clockwise), and g is the acceleration due to gravity. Carrying out the indicated differentiation and eliminating V and H result in the equations of motion

$$I\ddot{\theta} = mgL\sin\theta - mL^2\ddot{\theta} - mL\ddot{y}\cos\theta$$

$$M\ddot{y} = F - m\left(\ddot{y} + L\ddot{\theta}\cos\theta - L\dot{\theta}^2\sin\theta\right) - k\dot{y}$$

Solving these equations for $\ddot{\theta}$ and $\ddot{y}$, we obtain

$$\begin{bmatrix} \ddot{\theta} \\ \ddot{y} \end{bmatrix} = \frac{1}{\Delta(\theta)} \begin{bmatrix} m+M & -mL\cos\theta \\ -mL\cos\theta & I+mL^2 \end{bmatrix} \begin{bmatrix} mgL\sin\theta \\ F + mL\dot{\theta}^2\sin\theta - k\dot{y} \end{bmatrix}$$

where

$$\Delta(\theta) = (I+mL^2)(m+M) - m^2L^2\cos^2\theta \geq (I+mL^2)M + m > 0$$

(a) Using $x_1 = \theta$, $x_2 = \dot{\theta}$, $x_3 = y$, and $x_4 = \dot{y}$ as state variables and $u = F$ as control input, write down the state equation.

(b) Show that the open-loop system has an equilibrium set.

(c) Suppose we want to stabilize the pendulum at the vertical position ($\theta = 0$). Find an open-loop equilibrium point at which $\theta = 0$, and show that it is unstable.

(d) Linearize the nonlinear state equation at the desired equilibrium point, and verify that the linearized state equation is controllable.

(e) Using linearization, design a state feedback control law to stabilize the system at the desired equilibrium point.

(f) Study the performance of the closed-loop system using computer simulation. In particular, study the transient behavior and the effect of $\pm 10\%$ parameter perturbations from nominal values for all parameters of the system.

(g) Starting with the pendulum at equilibrium, move it a small angle to the right (and then to the left) and let it go. Repeat this experiment, gradually increasing the amount of initial disturbance. Using simulation, determine the largest range of initial disturbance for which the pendulum will return to equilibrium.

In parts (e), (f), and (g), use the numerical data: $m = 0.1$ kg, $M = 1$ kg, $k = 0.1$ N/m/sec, $I = 0.025/3$ kg m^2, $g = 9.81$ m/sec^2, and $L = 0.5$ m. Unless otherwise specified, use the same numerical data in all forthcoming exercises that apply to the inverted pendulum.

Exercise 11.7 Reconsider the system of Exercise 11.3, but assume you can only measure x_1 and x_3. Using linearization, design an output feedback control law to stabilize the origin.

Exercise 11.8 Reconsider the system of Exercise 11.4, but assume you can only measure x_3. Using linearization, design an output feedback control law to stabilize the origin.

Exercise 11.9 Reconsider the magnetic suspension system of Exercise 11.5, but assume you can only measure the ball position y and the current i.

(a) Using linearization, design an output feedback control law to stabilize the ball at a desired position $y_R > 0$.

(b) Repeat parts (e) and (f) of Exercise 11.5.

(c) Compare the performance of this controller with the one designed in Exercise 11.5.

Exercise 11.10 Reconsider the magnetic suspension system of Exercise 11.5, but assume you can only measure the ball position y.

(a) Using linearization, design an output feedback control law to stabilize the ball at a desired position $y_R > 0$.

(b) Repeat parts (e) and (f) of Exercise 11.5.

(c) Compare the performance of this controller with the one designed in Exercise 11.5.

Exercise 11.11 Reconsider the inverted pendulum of Exercise 11.6, but assume you can only measure the angle θ and the cart position y.

(a) Using linearization, design an output feedback control law to stabilize the pendulum at $\theta = 0$.

(b) Repeat parts (f) and (g) of Exercise 11.6.

(c) Compare the performance of this controller with the one designed in Exercise 11.6.

Exercise 11.12 Let

$$\mathcal{A} = \begin{bmatrix} A & 0 \\ C & 0 \end{bmatrix}, \quad \mathcal{B} = \begin{bmatrix} B \\ 0 \end{bmatrix}$$

where A, B, C satisfy the rank condition (11.19). Show that $(\mathcal{A}, \mathcal{B})$ is controllable (respectively, stabilizable) if and only if (A, B) is controllable (respectively, stabilizable)

Exercise 11.13 Consider the magnetic suspension system of Exercise 11.5.

(a) Redesign the linear state feedback control of Exercise 11.5 including integral control.

(b) Repeat parts (f) and (e) of Exercise 11.5.

(c) Compare the performance of this controller with the one designed in Exercise 11.5.

Exercise 11.14 Apply the observer-based controller (11.24)–(11.26) to the nonlinear system (11.13)–(11.14). Verify that the closed-loop system has an equilibrium point at which the regulation error is zero, and show that this equilibrium point is exponentially stable.

Exercise 11.15 Consider the pendulum of Example 11.2. Assume you measure the angle θ, but you do not measure the angular velocity $\dot{\theta}$. Using linearization, design an output feedback integral controller to stabilize the pendulum at an angle $\theta = \delta$.

Exercise 11.16 Consider the magnetic suspension system of Exercises 11.5 and 11.9.

(a) Redesign the linear output feedback control of Exercise 11.9 including integral control.

(b) Repeat parts (e) and (f) of Exercise 11.5.

(c) Compare the performance of this controller with the one designed in Exercise 11.9.

Exercise 11.17 Consider the magnetic suspension system of Exercises 11.5 and 11.10.

(a) Redesign the linear output feedback control of Exercise 11.10 including integral control.

(b) Repeat parts (e) and (f) of Exercise 11.5.

(c) Compare the performance of this controller with the one designed in Exercise 11.10.

Exercise 11.18 Reconsider the magnetic suspension system of Exercise 11.5, but assume you can only measure the current i.

(a) Can you design a linear output feedback control law to stabilize the ball at the desired position?

(b) Can you design a linear output feedback integral controller to stabilize the ball at the desired position?

Exercise 11.19 Consider the inverted pendulum of Exercise 11.6. Suppose it is desired to stabilize the pendulum at angle $\theta = \theta_r$, where $-\pi/2 < \theta_r < \pi/2$, and assume that you can only measure the angle θ and the cart position y.

(a) Using linearization, design an output feedback integral controller to stabilize the pendulum at the desired position.

(b) Repeat part (f) of Exercise 11.6.

Exercise 11.20 A field-controlled DC motor is described in Exercise 1.12. When the field circuit is driven by a current source, we can view the field current as the control input and model the system by the second-order state model

$$
\begin{aligned}
\dot{x}_1 &= -\theta_1 x_1 - \theta_2 x_2 u + \theta_3 \\
\dot{x}_2 &= -\theta_4 x_2 + \theta_5 x_1 u \\
y &= x_2
\end{aligned}
$$

where x_1 is the armature current, x_2 is the speed, u is the field current, and θ_1 to θ_5 are positive constants. It is required to design a speed control system so that y asymptotically tracks a constant speed reference y_R. It is assumed that $y_R^2 < \theta_3^2 \theta_5/4\theta_1\theta_2\theta_4$ and the domain of operation is restricted to $x_1 > \theta_3/2\theta_1$.

(a) Find the steady-state input u_{ss} needed to maintain the output at the constant speed y_R. Verify that the control $u = u_{ss}$ results in an exponentially stable equilibrium point.

(b) Study the performance of the system using computer simulation. In particular, study the transient behavior and the effect of $\pm 20\%$ parameter perturbations from nominal values for all parameters of the system.

(c) Starting with the motor at rest $(y = 0)$ apply a small step change in the reference signal and watch the motor tracking the reference speed. Repeat this experiment, gradually increasing the amount of step change. Using simulation, determine the largest range of initial step for which the motor will reach steady-state at the desired speed.

(d) Using linearization, design a state feedback integral controller to achieve the desired speed regulation. Repeat parts (b) and (c), and compare the performance of this feedback control to the open-loop control of part (a).

(e) Suppose you measure the speed x_2 but you do not measure the armature current x_1. Repeat part (d) using an observer to estimate the armature current. Repeat parts (b) and (c), and compare the performance of this controller with the one designed in part (d).

In parts (b) through (e), use the numerical data: $\theta_1 = 60$, $\theta_2 = 0.5$, $\theta_3 = 40$, $\theta_4 = 6$, and $\theta_5 = 4 \times 10^4$. Unless otherwise specified, use the same numerical data in all forthcoming exercises that apply to this system.

Exercise 11.21 With reference to Example 11.5, show that the controller (11.27) results in a triple (A_n, B_n, C_n) which is equivalent to the triple (A_ℓ, B_ℓ, C_ℓ).

Exercise 11.22 Consider the linear system $\dot{x} = A(\alpha)x + B(\alpha)u$, where $A(\alpha)$ and $B(\alpha)$ are continuously differentiable functions of the constant vector α, and $\alpha \in \Gamma$, a compact subset of R^m. Let $W(\alpha)$ be the controllability Gramian, defined by

$$W(\alpha) = \int_0^T \exp[-A(\alpha)\sigma]B(\alpha)B^T(\alpha)\exp[-A^T(\alpha)\sigma] \, d\sigma$$

for some $T > 0$, independent of α. Suppose (A, B) is controllable, uniformly in α, in the sense that there are positive constants c_1 and c_2, independent of α, such that

$$c_1 I \le W(\alpha) \le c_2 I, \quad \forall \, \alpha \in \Gamma$$

Let

$$Q(\alpha) = \int_0^T e^{-2c\sigma} \exp[-A(\alpha)\sigma]B(\alpha)B^T(\alpha)\exp[-A^T(\alpha)\sigma] \, d\sigma, \quad c > 0$$

(a) Show that $c_1 e^{-2cT} I \le Q(\alpha) \le c_2 I$, for all $\alpha \in \Gamma$.

(b) Let $u = K(\alpha)x \stackrel{\text{def}}{=} -\frac{1}{2}B^T(\alpha)P(\alpha)x$, where $P(\alpha) = Q^{-1}(\alpha)$. Using $V = x^T P(\alpha)x$ as a Lyapunov function candidate for $\dot{x} = [A(\alpha) + B(\alpha)K(\alpha)]x$, show that $\dot{V} \le -2cV$.

(c) Show that $[A(\alpha) + B(\alpha)K(\alpha)]$ is Hurwitz uniformly in α for all $\alpha \in \Gamma$.

Exercise 11.23 Show that $\bar{z}(\alpha)$, defined by (11.51), is the unique equilibrium point of (11.50) for all $\alpha \in D_v$.

Exercise 11.24 Show that $P(\alpha)$, defined by (11.54), is nonsingular and satisfies (11.55).

CHAPTER 11. DESIGN VIA LINEARIZATION

Exercise 11.25 Consider the pendulum of Example 11.2. Assume you can measure both the angle θ and the angular velocity $\dot{\theta}$.

(a) Design a gain scheduled, state feedback, integral controller so that the angle θ tracks a reference angle θ_r.

(b) Study the performance of the gain scheduled controller by computer simulation. Use the numerical data: a $= 10$, $b = 0.1$, and $c = 10$.

Exercise 11.26 Consider the pendulum of Example 11.2. Assume you measure the angle θ, but you do not measure the angular velocity $\dot{\theta}$.

(a) Design a gain scheduled, observer-based, integral controller so that the angle θ tracks a reference angle θ_r.

(b) Study the performance of the gain scheduled controller by computer simulation. Use the numerical data: a $= 10$, $b = 0.1$, and $c = 10$.

Exercise 11.27 Consider the magnetic suspension system of Exercise 11.5. Assume you measure x_1 and x_3, but you do not measure x_2.

(a) Design a gain scheduled, observer-based, integral controller so that the ball position x_1 tracks a reference position y_R.

(b) Study the performance of the gain scheduled controller by computer simulation.

Exercise 11.28 Consider the field controlled DC motor of Exercise 11.20. Assume you measure the speed x_2, but you do not measure the armature current x_1.

(a) Design a gain scheduled, observer-based, integral controller so that the speed x_2 tracks a reference speed y_R.

(b) Study the performance of the gain scheduled controller by computer simulation.

Chapter 12

Exact Feedback Linearization

In this chapter, we consider a class of nonlinear systems of the form

$$\dot{x} = f(x) + G(x)u$$
$$y = h(x)$$

and pose the question of whether there exist a state feedback control

$$u = \alpha(x) + \beta(x)v$$

and a change of variables

$$z = T(x)$$

that transform the nonlinear system into an equivalent linear system. We start in Section 12.1 with *input-state linearization*, where the full state equation is linearized. In Section 12.2, we introduce the notion of *input-output linearization*, where the emphasis is on linearizing the input-output map from u to y even if the state equation is only partially linearized. State feedback control of feedback linearizable systems is discussed in Section 12.3, where we deal with stabilization, tracking, and regulation via integral control problems. Finally, in Section 12.4 we present the differential geometric approach to feedback linearization which allows us to characterize the class of feedback linearizable systems by geometric conditions.

12.1 Input-State Linearization

To introduce the idea of exact feedback linearization, let us start with the pendulum stabilization problem of Example 11.2. Inspection of the state equation

$$\dot{x}_1 = x_2$$
$$\dot{x}_2 = -a[\sin(x_1 + \delta) - \sin \delta] - bx_2 + cu$$

519

shows that we can choose u as

$$u = \frac{a}{c}[\sin(x_1 + \delta) - \sin \delta] + \frac{v}{c}$$

to cancel the nonlinear term $a[\sin(x_1 + \delta) - \sin \delta]$. This cancellation results in the linear system

$$\begin{aligned}
\dot{x}_1 &= x_2 \\
\dot{x}_2 &= -bx_2 + v
\end{aligned}$$

Thus, the stabilization problem for the nonlinear system has been reduced to a stabilization problem for a controllable linear system. We can proceed to design a stabilizing linear state feedback control

$$v = k_1 x_1 + k_2 x_2$$

to locate the eigenvalues of the closed-loop system

$$\begin{aligned}
\dot{x}_1 &= x_2 \\
\dot{x}_2 &= k_1 x_1 + (k_2 - b)x_2
\end{aligned}$$

in the open left-half plane. The overall state feedback control law is given by

$$u = \left(\frac{a}{c}\right)[\sin(x_1 + \delta) - \sin \delta] + \frac{1}{c}(k_1 x_1 + k_2 x_2)$$

How general is this idea of nonlinearity cancellation? Clearly, we should not expect to be able to cancel nonlinearities in every nonlinear system. There must be a certain structural property of the system that allows us to perform such cancellation. It is not hard to see that to cancel a nonlinear term $\alpha(x)$ by subtraction, the control u and the nonlinearity $\alpha(x)$ must always appear together as a sum $u + \alpha(x)$. To cancel a nonlinear term $\gamma(x)$ by division, the control u and the nonlinearity $\gamma(x)$ must always appear as a product $\gamma(x)u$. If the matrix $\gamma(x)$ is nonsingular in the domain of interest, then it can be cancelled by $u = \beta(x)v$, where $\beta(x) = \gamma^{-1}(x)$ is the inverse of the matrix $\gamma(x)$. Therefore, the ability to use feedback to convert a nonlinear state equation into a controllable linear state equation by cancelling nonlinearities requires the nonlinear state equation to have the structure

$$\dot{x} = Ax + B\beta^{-1}(x)[u - \alpha(x)] \tag{12.1}$$

where A is $n \times n$, B is $n \times p$, the pair (A, B) is controllable, and the functions $\alpha : R^n \to R^p$ and $\beta : R^n \to R^{p \times p}$ are defined in a domain $D_x \subset R^n$ that contains the origin. The matrix $\beta(x)$ is assumed to be nonsingular for every $x \in D_x$. Notice that β^{-1} here denotes the inverse of the matrix $\beta(x)$ for every x, and not the inverse map of the function $\beta(x)$. If the state equation takes the form (12.1), then we can linearize it via the state feedback

$$u = \alpha(x) + \beta(x)v \tag{12.2}$$

to obtain the linear state equation

$$\dot{x} = Ax + Bv \tag{12.3}$$

For stabilization, we design $v = Kx$ such that $A + BK$ is Hurwitz. The overall nonlinear stabilizing state feedback control is

$$u = \alpha(x) + \beta(x)Kx \tag{12.4}$$

Suppose the nonlinear state equation does not have the structure of (12.1); does this mean we cannot linearize the system via feedback? The answer is no. Recall that the state model of a system is not unique. It depends on the choice of the state variables. Even if the state equation does not have the structure (12.1) for one choice of state variables, it might do so for another choice. Consider, for example, the system

$$\begin{aligned} \dot{x}_1 &= a \sin x_2 \\ \dot{x}_2 &= -x_1^2 + u \end{aligned}$$

We cannot simply choose u to cancel the nonlinear term $a \sin x_2$. However, if we first change the variables by the transformation

$$\begin{aligned} z_1 &= x_1 \\ z_2 &= a \sin x_2 = \dot{x}_1 \end{aligned}$$

then z_1 and z_2 satisfy

$$\begin{aligned} \dot{z}_1 &= z_2 \\ \dot{z}_2 &= a \cos x_2 \left(-x_1^2 + u \right) \end{aligned}$$

and the nonlinearities can be cancelled by the control

$$u = x_1^2 + \frac{1}{a \cos x_2} v$$

which is well defined for $-\pi/2 < x_2 < \pi/2$. The state equation in the new coordinates (z_1, z_2) can be found by inverting the transformation to express (x_1, x_2) in terms of (z_1, z_2); that is,

$$\begin{aligned} x_1 &= z_1 \\ x_2 &= \sin^{-1}\left(\frac{z_2}{a} \right) \end{aligned}$$

which is well defined for $-a < z_2 < a$. The transformed state equation is given by

$$\dot{z}_1 = z_2$$
$$\dot{z}_2 = a\cos\left(\sin^{-1}\left(\frac{z_2}{a}\right)\right)\left(-z_1^2 + u\right)$$

When a change of variables $z = T(x)$ is used to transform the state equation from the x-coordinates to the z-coordinates, the map T must be invertible; that is, it must have an inverse map $T^{-1}(\cdot)$ such that $x = T^{-1}(z)$ for all $z \in T(D_x)$, where D_x is the domain of T. Moreover, since the derivatives of z and x should be continuous, we require that both $T(\cdot)$ and $T^{-1}(\cdot)$ be continuously differentiable. A continuously differentiable map with a continuously differentiable inverse is known as a *diffeomorphism*. Now we have all the elements we need to define input-state linearizable systems.

Definition 12.1 *A nonlinear system*

$$\dot{x} = f(x) + G(x)u \tag{12.5}$$

where $f : D_x \to R^n$ and $G : D_x \to R^{n \times p}$ are sufficiently smooth[1] on a domain $D_x \subset R^n$, is said to be input-state linearizable if there exists a diffeomorphism $T : D_x \to R^n$ such that $D_z = T(D_x)$ contains the origin and the change of variables $z = T(x)$ transforms the system (12.5) into the form

$$\dot{z} = Az + B\beta^{-1}(x)[u - \alpha(x)] \tag{12.6}$$

with (A, B) controllable and $\beta(x)$ nonsingular for all $x \in D_x$.

Setting
$$\alpha_0(z) = \alpha\left(T^{-1}(z)\right) \quad \text{and} \quad \beta_0(z) = \beta\left(T^{-1}(z)\right)$$
we can write equation (12.6) as

$$\dot{z} = Az + B\beta_0^{-1}(z)[u - \alpha_0(z)] \tag{12.7}$$

which takes the form (12.1). It is more convenient, however, to express α and β in the x-coordinates since the state feedback control is implemented in these coordinates.

Suppose we are given an input-state linearizable system (12.5). Let $z = T(x)$ be a change of variables that brings the system into the form (12.6). We have

$$\dot{z} = \frac{\partial T}{\partial x}\dot{x} = \frac{\partial T}{\partial x}[f(x) + G(x)u] \tag{12.8}$$

---

[1] By "sufficiently smooth" we mean that all the partial derivatives, that will appear later on, are defined and continuous.

On the other hand, from (12.6),

$$\dot{z} = Az + B\beta^{-1}(x)[u - \alpha(x)] = AT(x) + B\beta^{-1}(x)[u - \alpha(x)] \qquad (12.9)$$

From (12.8) and (12.9), we see that the equality

$$\frac{\partial T}{\partial x}[f(x) + G(x)u] = AT(x) + B\beta^{-1}(x)[u - \alpha(x)]$$

must hold for all x and u in the domain of interest. By first taking $u = 0$, we split
the equation into two:

$$\frac{\partial T}{\partial x}f(x) = AT(x) - B\beta^{-1}(x)\alpha(x) \qquad (12.10)$$

$$\frac{\partial T}{\partial x}G(x) = B\beta^{-1}(x) \qquad (12.11)$$

Therefore, we conclude that any function $T(\cdot)$ that transforms (12.5) into the form
(12.6) must satisfy the partial differential equations (12.10)–(12.11). Alternatively,
if there is a map $T(\cdot)$ that satisfies (12.10)–(12.11) for some α, β, A, and B with the
desired properties, then it can be easily seen that the change of variable $z = T(x)$
transforms (12.5) into (12.6). Hence, the existence of T, α, β, A, and B that
satisfy the partial differential equations (12.10)–(12.11) is a necessary and sufficient
condition for the system (12.5) to be input-state linearizable.

When a nonlinear system is input-state linearizable, the map $z = T(x)$ that
transforms the system into the form (12.6) is not unique. Probably the easiest way
to see this point is to notice that if we apply the linear state transformation $\zeta = Mz$,
with a nonsingular M, to (12.6) then the state equation in the ζ-coordinates will
be

$$\dot{\zeta} = MAM^{-1}\zeta + MB\beta^{-1}(x)[u - \alpha(x)]$$

which is still of the form (12.6), but with different A and B matrices. Therefore,
the composition of the transformations $z = T(x)$ and $\zeta = Mz$ gives a new transformation that transforms the system into the special structure of (12.6). The
nonuniqueness of T can be exploited to simplify the partial differential equations
(12.10)–(12.11). To illustrate the idea without complications, we will restrict our
discussions to single-input systems ($p = 1$) and write the single-column input matrix G as g . In this case, for any controllable pair (A, B) we can find a nonsingular matrix M that transforms (A, B) into a controllable canonical form;[2] that is,

---

[2] See, for example, [98], [29], [41], or [142].

$MAM^{-1} = A_c + B_c\lambda^T$ and $MB = B_c$, where

$$A_c = \begin{bmatrix} 0 & 1 & 0 & \cdots & 0 \\ 0 & 0 & 1 & \cdots & 0 \\ \vdots & & \ddots & & \vdots \\ \vdots & & & 0 & 1 \\ 0 & \cdots & \cdots & 0 & 0 \end{bmatrix}_{n \times n} \quad \text{and} \quad B_c = \begin{bmatrix} 0 \\ 0 \\ \vdots \\ 0 \\ 1 \end{bmatrix}_{n \times 1} \tag{12.12}$$

The term $B_c\lambda^T\zeta = B_c\lambda^T MT(x)$ can be included in the term $B_c\beta^{-1}(x)\alpha(x)$. There-fore, without loss of generality, we can assume that the matrices A and B in (12.10)–(12.11) are the canonical form matrices A_c and B_c. Let

$$T(x) = \begin{bmatrix} T_1(x) \\ T_2(x) \\ \vdots \\ T_{n-1}(x) \\ T_n(x) \end{bmatrix}$$

It can be easily verified that

$$A_c T(x) - B_c\beta^{-1}(x)\alpha(x) = \begin{bmatrix} T_2(x) \\ T_3(x) \\ \vdots \\ T_n(x) \\ -\alpha(x)/\beta(x) \end{bmatrix} \quad \text{and} \quad B_c\beta^{-1}(x) = \begin{bmatrix} 0 \\ 0 \\ \vdots \\ 0 \\ 1/\beta(x) \end{bmatrix}$$

where α and β are scalar functions. Using these expressions in (12.10)–(12.11) simplifies the partial differential equations. Equation (12.10) simplifies to

$$\frac{\partial T_1}{\partial x} f(x) = T_2(x)$$

$$\frac{\partial T_2}{\partial x} f(x) = T_3(x)$$

$$\vdots$$

$$\frac{\partial T_{n-1}}{\partial x} f(x) = T_n(x)$$

$$\frac{\partial T_n}{\partial x} f(x) = -\alpha(x)/\beta(x)$$

The first $n - 1$ equations show that the components T_2 to T_n of T are determined functions of the first component T_1. The last equation defines α/β in terms of T_1.

Equation (12.11) simplifies to

$$\frac{\partial T_1}{\partial x} g(x) = 0$$

$$\frac{\partial T_2}{\partial x} g(x) = 0$$

$$\vdots$$

$$\frac{\partial T_{n-1}}{\partial x} g(x) = 0$$

$$\frac{\partial T_n}{\partial x} g(x) = 1/\beta(x) \neq 0$$

We need to search for a function $T_1(x)$ that satisfies

$$\frac{\partial T_i}{\partial x} g(x) = 0, \quad i = 1, 2, \ldots, n-1; \quad \frac{\partial T_n}{\partial x} g(x) \neq 0 \tag{12.13}$$

where

$$T_{i+1}(x) = \frac{\partial T_i}{\partial x} f(x), \quad i = 1, 2, \ldots, n-1$$

If there is a function $T_1(x)$ that satisfies (12.13), then β and α are given by

$$\beta(x) = \frac{1}{(\partial T_n/\partial x) g(x)}; \quad \alpha(x) = -\frac{(\partial T_n/\partial x) f(x)}{(\partial T_n/\partial x) g(x)} \tag{12.14}$$

Thus, we have reduced the problem to solving (12.13) for T_1. We shall see in Section 12.4 that the existence of a function T_1 that satisfies (12.13) can be characterized by a necessary and sufficient condition in terms of the functions f, g, and their partial derivatives. Furthermore, if $T_1(x)$ satisfies (12.13) in a domain D, then for each $x_0 \in D$ there is a neighborhood N of x_0 such that the restriction of $T(x)$ to N is a diffeomorphism. For now, let us illustrate, via examples, how to solve for T_1. Before we work out the examples, we would like to explain an additional condition that we will impose on T_1. In all examples, we assume that our goal is to stabilize the system with respect to an open-loop equilibrium point x^*; that is, $f(x^*) = 0$. We would like to choose the map T such that the point x^* is mapped into the origin; that is, $T(x^*) = 0$. This can be achieved by solving (12.13) for T_1 subject to the condition $T_1(x^*) = 0$. Notice that the other components of T vanish at x^* since $f(x^*) = 0$.

Example 12.1 As a first example, we reconsider the system

$$\dot{x} = \begin{bmatrix} a \sin x_2 \\ -x_1^2 \end{bmatrix} + \begin{bmatrix} 0 \\ 1 \end{bmatrix} u \overset{\text{def}}{=} f(x) + gu$$

and see how we can arrive at the transformation we used earlier by solving (12.13). The open-loop system has an equilibrium point at $x = 0$. We want to find $T_1(x)$ that satisfies the conditions

$$\frac{\partial T_1}{\partial x}g = 0; \quad \frac{\partial T_2}{\partial x}g \neq 0$$

with $T_1(0) = 0$, where

$$T_2(x) = \frac{\partial T_1}{\partial x}f(x)$$

From the condition $[\partial T_1/\partial x]g = 0$, we have

$$\frac{\partial T_1}{\partial x}g = \frac{\partial T_1}{\partial x_2} = 0$$

So, T_1 must be independent of x_2. Therefore,

$$T_2(x) = \frac{\partial T_1}{\partial x_1}a \sin x_2$$

The condition

$$\frac{\partial T_2}{\partial x}g = \frac{\partial T_2}{\partial x_2} = \frac{\partial T_1}{\partial x_1}a \cos x_2 \neq 0$$

is satisfied in the domain where $\cos x_2 \neq 0$ by any choice of $T_1 = T_1(x_1)$ such that $(\partial T_1/\partial x_1) \neq 0$. Taking $T_1(x) = x_1$ results in the transformation we used earlier. Other choices of T_1 could have been made. For example, $T_1(x_1) = x_1 + x_1^3$ would give another change of variables that transforms the system into the form (12.6).
$$\triangle$$

Example 12.2 A synchronous generator connected to an infinite bus may be represented by a third-order model of the form (Exercise 1.7)

$$\dot{x} = f(x) + gu$$

with

$$f(x) = \begin{bmatrix} x_2 \\ -a[(1 + x_3)\sin(x_1 + \delta) - \sin \delta] - bx_2 \\ -cx_3 + d[\cos(x_1 + \delta) - \cos \delta] \end{bmatrix} ; \quad g = \begin{bmatrix} 0 \\ 0 \\ 1 \end{bmatrix}$$

where a, b, c, d, and δ are positive constants. The open-loop system has equilibrium at $x = 0$. We want to find $T_1 = T_1(x)$ that satisfies the conditions

$$\frac{\partial T_1}{\partial x}g = 0; \quad \frac{\partial T_2}{\partial x}g = 0; \quad \frac{\partial T_3}{\partial x}g \neq 0$$

with $T_1(0) = 0$, where

$$T_2(x) = \frac{\partial T_1}{\partial x} f(x); \quad T_3(x) = \frac{\partial T_2}{\partial x} f(x)$$

From the condition $[\partial T_1/\partial x]g = 0$, we have

$$\frac{\partial T_1}{\partial x} g = \frac{\partial T_1}{\partial x_3} = 0$$

We choose T_1 independent of x_3. Therefore,

$$T_2(x) = \frac{\partial T_1}{\partial x_1} x_2 - \frac{\partial T_1}{\partial x_2}\{a[(1 + x_3)\sin(x_1 + \delta) - \sin \delta] + bx_2\}$$

From the condition $[\partial T_2/\partial x]g = 0$ we have

$$\frac{\partial T_2}{\partial x} g = \frac{\partial T_2}{\partial x_3} = -a \sin(x_1 + \delta)\frac{\partial T_1}{\partial x_2} = 0$$

We choose T_1 independent of x_2. Therefore, T_2 simplifies to

$$T_2(x) = \frac{\partial T_1}{\partial x_1} x_2$$

and

$$T_3(x) = \frac{\partial T_2}{\partial x_1} x_2 - \frac{\partial T_1}{\partial x_1}\{a[(1 + x_3)\sin(x_1 + \delta) - \sin \delta] + bx_2\}$$

Hence,

$$\frac{\partial T_3}{\partial x} g = \frac{\partial T_3}{\partial x_3} = -a \sin(x_1 + \delta)\frac{\partial T_1}{\partial x_1}$$

and the condition $[\partial T_3/\partial x]g \neq 0$ is satisfied in the domain $0 < x_1 + \delta < \pi$ with any choice of $T_1 = T_1(x_1)$ such that $(\partial T_1/\partial x_1) \neq 0$ on this domain. The simple choice $T_1 = x_1$ satisfies this requirement as well as the condition $T_1(0) = 0$. Thus, we choose $T_1 = x_1$ and find from the previous expressions that the change of variables $z = T(x)$ is given by

$$
\begin{aligned}
z_1 &= T_1(x) &&= x_1 \\
z_2 &= T_2(x) &&= x_2 \\
z_3 &= T_3(x) &&= -a[(1 + x_3)\sin(x_1 + \delta) - \sin \delta] - bx_2
\end{aligned}
$$

The inverse of this transformation $x = T^{-1}(z)$ is defined for all $0 < z_1 + \delta < \pi$ and given by

$$
\begin{aligned}
x_1 &= z_1 \\
x_2 &= z_2 \\
x_3 &= -1 - \frac{z_3 + bz_2 - a\sin\delta}{a\sin(z_1 + \delta)}
\end{aligned}
$$

The functions β and α are given by

$$\beta(x) = \frac{1}{(\partial T_3/\partial x)\, g} = \frac{1}{(\partial T_3/\partial x_3)} = \frac{-1}{a \sin(x_1 + \delta)}$$

$$
\begin{aligned}
\alpha(x) &= -\frac{(\partial T_3/\partial x)\, f(x)}{(\partial T_3/\partial x)\, g} \\
&= \frac{-a(1 + x_3)\cos(x_1 + \delta) f_1(x) - b f_2(x) - a \sin(x_1 + \delta) f_3(x)}{a \sin(x_1 + \delta)}
\end{aligned}
$$

The state equation in the z-coordinates is

$$
\begin{aligned}
\dot{z}_1 &= z_2 \\
\dot{z}_2 &= z_3 \\
\dot{z}_3 &= -a \sin(x_1 + \delta)[u - \alpha(x)]
\end{aligned}
$$

which can be linearized by the linearizing feedback control

$$u = \alpha(x) - \frac{1}{a \sin(x_1 + \delta)} v$$

Notice that the state equation in the z-coordinates is valid only in the domain $0 < z_1 + \delta < \pi$, which is the domain over which the change of variables $z = T(x)$ is a well-defined diffeomorphism. $\triangle$

Example 12.3 A single link manipulator with flexible joints and negligible damping may be represented by a fourth-order model of the form (Exercise 1.4)

$$\dot{x} = f(x) + g u$$

with

$$f(x) = \begin{bmatrix} x_2 \\ -a \sin x_1 - b(x_1 - x_3) \\ x_4 \\ c(x_1 - x_3) \end{bmatrix}; \quad g = \begin{bmatrix} 0 \\ 0 \\ 0 \\ d \end{bmatrix}$$

where a, b, c, and d are positive constants. The open-loop system has equilibrium at $x = 0$. We want to find $T_1 = T_1(x)$ that satisfies the conditions

$$\frac{\partial T_i}{\partial x} g = 0, \quad i = 1, 2, 3; \quad \frac{\partial T_4}{\partial x} g \neq 0$$

with $T_1(0) = 0$, where

$$T_{i+1}(x) = \frac{\partial T_i}{\partial x} f(x), \quad i = 1, 2, 3$$

From the condition $[\partial T_1/\partial x]g = 0$, we have $(\partial T_1/\partial x_4) = 0$, so we must choose T_1 independent of x_4. Therefore,

$$T_2(x) = \frac{\partial T_1}{\partial x_1}x_2 + \frac{\partial T_1}{\partial x_2}[-a\sin x_1 - b(x_1 - x_3)] + \frac{\partial T_1}{\partial x_3}x_4$$

From the condition $[\partial T_2/\partial x]g = 0$, we have

$$\frac{\partial T_2}{\partial x_4} = 0 \Rightarrow \frac{\partial T_1}{\partial x_3} = 0$$

So, we choose T_1 independent of x_3. Therefore, T_2 simplifies to

$$T_2(x) = \frac{\partial T_1}{\partial x_1}x_2 + \frac{\partial T_1}{\partial x_2}[-a\sin x_1 - b(x_1 - x_3)]$$

and

$$T_3(x) = \frac{\partial T_2}{\partial x_1}x_2 + \frac{\partial T_2}{\partial x_2}[-a\sin x_1 - b(x_1 - x_3)] + \frac{\partial T_2}{\partial x_3}x_4$$

Finally,

$$\frac{\partial T_3}{\partial x_4} = 0 \Rightarrow \frac{\partial T_2}{\partial x_3} = 0 \Rightarrow \frac{\partial T_1}{\partial x_2} = 0$$

and we choose T_1 independent of x_2. Hence,

$$T_4(x) = \frac{\partial T_3}{\partial x_1}x_2 + \frac{\partial T_3}{\partial x_2}[-a\sin x_1 - b(x_1 - x_3)] + \frac{\partial T_3}{\partial x_3}x_4$$

and the condition $[\partial T_4/\partial x]g \neq 0$ is satisfied whenever $(\partial T_1/\partial x_1) \neq 0$. Therefore, we take $T_1(x) = x_1$. The change of variables

$$
\begin{aligned}
z_1 &= T_1(x) &&= x_1 \\
z_2 &= T_2(x) &&= x_2 \\
z_3 &= T_3(x) &&= -a\sin x_1 - b(x_1 - x_3) \\
z_4 &= T_4(x) &&= -ax_2\cos x_1 - b(x_2 - x_4)
\end{aligned}
$$

transforms the state equation into

$$
\begin{aligned}
\dot{z}_1 &= z_2 \\
\dot{z}_2 &= z_3 \\
\dot{z}_3 &= z_4 \\
\dot{z}_4 &= -(a\cos z_1 + b + c)z_3 + a(z_2^2 - c)\sin z_1 + bdu
\end{aligned}
$$

which is of the form (12.7). Unlike the previous example, in the current one the state equation in the z-coordinates is valid globally because $z = T(x)$ is a global diffeomorphism; that is, $T(\cdot)$ and $T^{-1}(\cdot)$ are defined globally and $T(R^n) = R^n$. $\triangle$

Example 12.4 A field-controlled DC motor with negligible shaft damping may be represented by a third-order model of the form (Exercise 1.12)

$$\dot{x} = f(x) + gu$$

with

$$f(x) = \begin{bmatrix} -ax_1 \\ -bx_2 + \rho - cx_1x_3 \\ \theta x_1 x_2 \end{bmatrix}; \quad g = \begin{bmatrix} 1 \\ 0 \\ 0 \end{bmatrix}$$

where a, b, c, θ, and ρ are positive constants. The open-loop system has an equilibrium set at $x_1 = 0$ and $x_2 = \rho/b$. We take the desired operating point as $x^* = (0, \rho/b, \omega_0)^T$, where ω_0 is a desired set point for the angular velocity x_3. We want to find $T_1 = T_1(x)$ that satisfies the conditions

$$\frac{\partial T_1}{\partial x}g = 0; \quad \frac{\partial T_2}{\partial x}g = 0; \quad \frac{\partial T_3}{\partial x}g \neq 0$$

with $T_1(x^*) = 0$, where

$$T_2(x) = \frac{\partial T_1}{\partial x}f(x); \quad T_3(x) = \frac{\partial T_2}{\partial x}f(x)$$

From the condition

$$\frac{\partial T_1}{\partial x}g = \frac{\partial T_1}{\partial x_1} = 0$$

we see that T_1 must be independent of x_1. Therefore,

$$T_2(x) = \frac{\partial T_1}{\partial x_2}[-bx_2 + \rho - cx_1x_3] + \frac{\partial T_1}{\partial x_3}\theta x_1 x_2$$

From the condition $[\partial T_2/\partial x]g = 0$, we have

$$cx_3\frac{\partial T_1}{\partial x_2} = \theta x_2 \frac{\partial T_1}{\partial x_3}$$

which is satisfied if T_1 takes the form

$$T_1 = c_1[\theta x_2^2 + cx_3^2] + c_2$$

for some constants c_1 and c_2. We choose $c_1 = 1$ and, to satisfy the condition $T_1(x^*) = 0$, we take

$$c_2 = -\theta\left(x_2^*\right)^2 - c\left(x_3^*\right)^2 = -\theta(\rho/b)^2 - c\omega_0^2$$

With this choice of T_1, T_2 simplifies to

$$T_2(x) = 2\theta x_2(\rho - bx_2)$$

Consequently, T_3 is given by

$$T_3(x) = 2\theta(\rho - 2bx_2)(-bx_2 + \rho - cx_1x_3)$$

Hence,

$$\frac{\partial T_3}{\partial x}g = \frac{\partial T_3}{\partial x_1} = -2c\theta(\rho - 2bx_2)x_3$$

and the condition $[\partial T_3/\partial x]g \neq 0$ is satisfied whenever $x_2 \neq \rho/2b$ and $x_3 \neq 0$. Assuming $x_3^* > 0$, we define a domain

$$D_x = \{x \in R^3 \mid x_2 > \frac{\rho}{2b} \text{ and } x_3 > 0\}$$

which contains the point x^*. It can be easily verified (Exercise 12.6) that the map $z = T(x)$ is a diffeomorphism on D_x and the state equation in the z-coordinates is well defined in the domain

$$D_z = T(D_x) = \left\{z \in R^3 \mid z_1 > \theta\phi^2(z_2) - \theta(\rho/b)^2 - c\omega_0^2 \text{ and } z_2 < \frac{\theta\rho^2}{2b}\right\}$$

where $\phi(\cdot)$ is the inverse of the map $2\theta x_2(\rho - bx_2)$, which is well defined for $x_2 > \rho/2b$. The domain D_z contains the origin $z = 0$. $\triangle$

12.2 Input-Output Linearization

When certain output variables are of interest, as in tracking control problems, the state model is described by state and output equations. Linearizing the state equation, as we have done in the previous section, does not necessarily linearize the output equation. For example, if the system

$$\dot{x}_1 = a \sin x_2$$
$$\dot{x}_2 = -x_1^2 + u$$

has an output $y = x_2$, then the change of variables and state feedback control

$$z_1 = x_1, \quad z_2 = a \sin x_2, \quad \text{and} \quad u = x_1^2 + \frac{1}{a \cos x_2}v$$

yield

$$\dot{z}_1 = z_2$$
$$\dot{z}_2 = v$$
$$y = \sin^{-1}\left(\frac{z_2}{a}\right)$$

While the state equation is linear, solving a tracking control problem for y is still complicated by the nonlinearity of the output equation. Inspection of both the state and output equations in the x-coordinates shows that, if we use the state feedback control $u = x_1^2 + v$, we can linearize the input-output map from u to y, which will be described by the linear model

$$\begin{aligned}
\dot{x}_2 &= v \\
y &= x_2
\end{aligned}$$

We can now proceed to solve the tracking control problem using linear control theory. This discussion shows that sometimes it is more beneficial to linearize the input-output map even at the expense of leaving part of the state equation nonlinear. This is the input-output linearization problem which we will address in this section. One catch about input-output linearization is that the linearized input-output map may not account for all the dynamics of the system. In the foregoing example, the full system is described by

$$\begin{aligned}
\dot{x}_1 &= a \sin x_2 \\
\dot{x}_2 &= v \\
y &= x_2
\end{aligned}$$

Note that the state variable x_1 is not connected to the output y. In other words, the linearizing feedback control has made x_1 unobservable from y. When we design tracking control, we should make sure that the variable x_1 is well behaved; that is, stable or bounded in some sense. A naive control design that uses only the linear input-output map may result in an ever-growing signal $x_1(t)$. For example, suppose we design a linear control to stabilize the output y at a constant value y_R. The corresponding solution of the x_1 equation is $x_1(t) = x_1(0) + t\,a \sin y_R$. So, for any $y_R \neq 0$, the variable $x_1(t)$ will grow unbounded. This internal stability issue will be addressed in the context of input-output linearization.

Consider the single-input–single-output system

$$\begin{aligned}
\dot{x} &= f(x) + g(x)u \\
y &= h(x)
\end{aligned}$$

where f, g, and h are sufficiently smooth in a domain $D \subset R^n$. The simplest case of input-output linearization arises when the system is both input-state and input-output linearizable. Starting with an input-state linearizable system, let $T_1(x)$ be a solution of (12.13). Suppose the output function $h(x)$ happens to be equal to $T_1(x)$. For example, in the robotic manipulator example 12.3, we chose $T_1(x) = x_1$; this could indeed be the output of interest, since in this problem we are usually interested in controlling the angle x_1. If $h(x) = T_1(x)$, then the change of variables

$z = T(x)$ and state feedback control $u = \alpha(x) + \beta(x)v$ yield the system

$$\dot{z} = A_c z + B_c v$$
$$y = C_c z$$

where (A_c, B_c, C_c) is a canonical form representation of a chain of n integrators; that is, A_c and B_c take the form (12.12) and

$$C_c = \begin{bmatrix} 1 & 0 & \dots & 0 & 0 \end{bmatrix}_{1 \times n} \tag{12.15}$$

In this system, both the state and output equations are linear. For a given output function $h(x)$, we can find out whether or not $h(x)$ satisfies the condition (12.13) by direct verification; we do not need to solve partial differential equations. The condition (12.13) can be interpreted as a restriction on the way the derivatives of y depend on u. To see this point, set $\psi_1(x) = h(x)$. The derivative $\dot{y}$ is given by

$$\dot{y} = \frac{\partial \psi_1}{\partial x}[f(x) + g(x)u]$$

If $[\partial \psi_1 / \partial x]g(x) = 0$, then

$$\dot{y} = \frac{\partial \psi_1}{\partial x}f(x) \stackrel{\text{def}}{=} \psi_2(x)$$

If we continue to calculate the second derivative of y, denoted by $y^{(2)}$, we obtain

$$y^{(2)} = \frac{\partial \psi_2}{\partial x}[f(x) + g(x)u]$$

Once again, if $[\partial \psi_2 / \partial x]g(x) = 0$, then

$$y^{(2)} = \frac{\partial \psi_2}{\partial x}f(x) \stackrel{\text{def}}{=} \psi_3(x)$$

Repeating this process, we see that if $h(x) = \psi_1(x)$ satisfies (12.13); that is,

$$\frac{\partial \psi_i}{\partial x}g(x) = 0, \quad i = 1, 2, \dots, n-1; \quad \frac{\partial \psi_n}{\partial x}g(x) \neq 0$$

where

$$\psi_{i+1}(x) = \frac{\partial \psi_i}{\partial x}f(x), \quad i = 1, 2, \dots, n-1$$

then u does not appear in the equations of y, $\dot{y}$, ..., $y^{(n-1)}$ and appears in the equation of $y^{(n)}$ with a nonzero coefficient

$$y^{(n)} = \frac{\partial \psi_n}{\partial x}f(x) + \frac{\partial \psi_n}{\partial x}g(x)u$$

This equation shows clearly that the system is input-output linearizable since the state feedback control

$$u = \frac{1}{\frac{\partial \psi_n}{\partial x} g(x)} \left[- \frac{\partial \psi_n}{\partial x} f(x) + v \right]$$

reduces the input-output map to

$$y^{(n)} = v$$

which is a chain of n integrators. What if u appears in the equation of one of the derivatives $\dot{y}, \ldots, y^{(n-1)}$? If the coefficient of u (when it appears) is nonzero, then we can again linearize the input-output map. In particular, if $h = \psi_1(x)$ satisfies

$$\frac{\partial \psi_i}{\partial x} g(x) = 0, \quad i = 1, 2, \ldots, r - 1; \quad \frac{\partial \psi_r}{\partial x} g(x) \neq 0$$

for some $1 \leq r < n$, then the equation of $y^{(r)}$ is given by

$$y^{(r)} = \frac{\partial \psi_r}{\partial x} f(x) + \frac{\partial \psi_r}{\partial x} g(x) u$$

and the state feedback control

$$u = \frac{1}{\frac{\partial \psi_r}{\partial x} g(x)} \left[- \frac{\partial \psi_r}{\partial x} f(x) + v \right]$$

linearizes the input-output map to the chain of r integrators

$$y^{(r)} = v$$

In this case, the integer r is called the relative degree of the system, according to the following definition.

Definition 12.2 *A nonlinear system*

$$\dot{x} = f(x) + g(x)u \tag{12.16}$$
$$y = h(x) \tag{12.17}$$

where $f : D \to R^n$, $g : D \to R^n$, and $h : D \to R$ are sufficiently smooth on a domain $D \subset R^n$, is said to have relative degree r, $1 \leq r \leq n$, in a region $D_0 \subset D$ if

$$\frac{\partial \psi_i}{\partial x} g(x) = 0, \quad i = 1, 2, \ldots, r - 1; \quad \frac{\partial \psi_r}{\partial x} g(x) \neq 0 \tag{12.18}$$

for all $x \in D_0$ where

$$\psi_1(x) = h(x) \quad \text{and} \quad \psi_{i+1}(x) = \frac{\partial \psi_i}{\partial x} f(x), \quad i = 1, 2, \ldots, r - 1 \tag{12.19}$$

If the system (12.16)–(12.17) has relative degree r, then it is input-output linearizable. If it has relative degree n, then it is both input-state and input-output linearizable.

Example 12.5 Consider the controlled van der Pol equation

$$\dot{x}_1 = x_2$$
$$\dot{x}_2 = -x_1 + \epsilon(1 - x_1^2)x_2 + u, \quad \epsilon > 0$$

with output $y = x_1$. Calculating the derivatives of the output, we obtain

$$\dot{y} = \dot{x}_1 = x_2$$
$$\ddot{y} = \dot{x}_2 = -x_1 + \epsilon(1 - x_1^2)x_2 + u$$

Hence, the system has relative degree 2 in R^2. Therefore, it is both input-state and input-output linearizable. If the output is $y = x_2$, then

$$\dot{y} = -x_1 + \epsilon(1 - x_1^2)x_2 + u$$

and the system has relative degree 1 in R^2. If the output is $y = x_1 + x_2^2$, then

$$\dot{y} = x_2 + x_2[-x_1 + \epsilon(1 - x_1^2)x_2 + u]$$

and the system has relative degree 1 in $D_0 = \{x \in R^2 \mid x_2 \neq 0\}$. $\triangle$

Example 12.6 Consider the system

$$\dot{x}_1 = x_1$$
$$\dot{x}_2 = x_2 + u$$
$$y = x_1$$

Calculating the derivatives of y, we obtain

$$\dot{y} = \dot{x}_1 = x_1 = y$$

Consequently, $y^{(n)} = y = x_1$ for all $n \geq 1$. In this case, the system does not have a well-defined relative degree. In this simple example, it is not difficult to see why this is so; the output $y(t) = x_1(t) = e^t x_1(0)$ is independent of the input u. $\triangle$

Example 12.7 Consider the field-controlled DC motor of Exercise 1.12 and Example 12.4. The state equation is given by

$$\dot{x}_1 = -ax_1 + u$$
$$\dot{x}_2 = -bx_2 + \rho - cx_1x_3$$
$$\dot{x}_3 = \theta x_1 x_2$$

where x_1, x_2, and x_3 are the field current, armature current, and angular velocity, respectively. For speed control problems, we choose the output as $y = x_3$. The derivatives of the output are given by

$$\dot{y} = \dot{x}_3 = \theta x_1 x_2$$
$$\ddot{y} = \theta x_1 \dot{x}_2 + \theta \dot{x}_1 x_2 = (\cdot) + \theta x_2 u$$

where $(\cdot)$ contains terms which are functions of x. The system has relative degree 2 in the region $D_0 = \{x \in R^3 \mid x_2 \neq 0\}$. $\triangle$

Example 12.8 Consider a linear system represented by the transfer function

$$H(s) = \frac{b_m s^m + b_{m-1} s^{m-1} + \cdots + b_0}{s^n + a_{n-1} s^{n-1} + \cdots + a_0}$$

where $m < n$ and $b_m \neq 0$. A state model for the system can be taken as

$$\dot{x} = Ax + Bu$$
$$y = Cx$$

where

$$A = \begin{bmatrix} 0 & 1 & 0 & \cdots & & & \cdots & 0 \\ 0 & 0 & 1 & \cdots & & & \cdots & 0 \\ \vdots & & \ddots & & & & & \vdots \\ & & & \ddots & & & & \\ & & & & \ddots & & & \vdots \\ \vdots & & & & \ddots & & & 0 \\ 0 & & & & & 0 & 1 \\ -a_0 & -a_1 & \cdots & \cdots & -a_m & \cdots & \cdots & -a_{n-1} \end{bmatrix}_{n \times n}, \quad B = \begin{bmatrix} 0 \\ 0 \\ \vdots \\ \vdots \\ 0 \\ 1 \end{bmatrix}_{n \times 1}$$

$$C = \begin{bmatrix} b_0 & b_1 & \cdots & \cdots & b_m & 0 & \cdots & 0 \end{bmatrix}_{1 \times n}$$

This linear state model is a special case of (12.16)–(12.17) where $f(x) = Ax$, $g = B$, and $h(x) = Cx$. To check the relative degree of the system, we calculate the derivatives of the output. The first derivative is given by

$$\dot{y} = CAx + CBu$$

If $m = n-1$, then $CB = b_{n-1} \neq 0$ and the system has relative degree 1. Otherwise, $CB = 0$ and we continue to calculate the second derivative of y. Noting that CA is a row vector obtained by shifting the elements of C one position to the right while

CA^2 is obtained by shifting the elements of C two positions to the right, and so on, we see that

$$CA^{i-1}B = 0, \quad \text{for } i = 1, 2, \ldots, n - m - 1, \quad \text{and} \quad CA^{n-m-1}B = b_m \neq 0$$

Thus, u appears first in the equation of $y^{(n-m)}$, which is given by

$$y^{(n-m)} = CA^{n-m}x + CA^{n-m-1}Bu$$

and the relative degree of the system is $n - m$ (the difference between the degrees of the denominator and numerator polynomials of $H(s)$).[3] $\triangle$

To probe further into the control of input-output linearizable systems and issues of internal stability, let us start with the linear system of the foregoing example. The transfer function $H(s)$ can be written as

$$H(s) = \frac{N(s)}{D(s)}$$

where $\deg D = n$ and $\deg N = m < n$. The relative degree $r = n - m$. By Euclidean division, we can write $D(s)$ as

$$D(s) = Q(s)N(s) + R(s)$$

where $Q(s)$ and $R(s)$ are the quotient and remainder polynomials, respectively. From Euclidean division rules, we know that

$$\deg Q = n - m = r, \quad \deg R < m$$

and the leading coefficient of $Q(s)$ is $1/b_m$. With this representation of $D(s)$, we can rewrite $H(s)$ as

$$H(s) = \frac{N(s)}{Q(s)N(s) + R(s)} = \frac{\frac{1}{Q(s)}}{1 + \frac{1}{Q(s)}\frac{R(s)}{N(s)}}$$

Thus, $H(s)$ can be represented as a negative feedback connection with $1/Q(s)$ in the forward path and $R(s)/N(s)$ in the feedback path; see Figure 12.1. The rth-order transfer function $1/Q(s)$ has no zeros and can be realized by the rth-order state vector

$$\xi = \begin{bmatrix} y, & \dot{y}, & \ldots, & y^{(r-1)} \end{bmatrix}^T$$

---

[3] The terminology "relative degree" of a nonlinear system is consistent with the use of the term relative degree in linear control theory, which is defined as $n - m$.

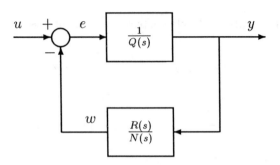

Figure 12.1: Feedback representation of $H(s)$.

to obtain the state model

$$\dot\xi = (A_c + B_c\lambda^T)\xi + B_c b_m e$$
$$y = C_c\xi$$

where (A_c, B_c, C_c) is a canonical form representation of a chain of r integrators as in (12.12) and (12.15), $\lambda \in R^r$, and e is the signal at the output of the feedback summation node. Let (A_0, B_0, C_0) be a minimal realization of the transfer function $R(s)/N(s)$; that is,

$$\dot\eta = A_0\eta + B_0 y$$
$$w = C_0\eta$$

where w is the feedback signal; that is, $e = u - w$. Notice that the eigenvalues of A_0 are the zeros of the polynomial $N(s)$, that is, the zeros of the transfer function $H(s)$. From the feedback connection, we see that $H(s)$ can be realized by the state model

$$\dot\eta = A_0\eta + B_0 C_c\xi \qquad\qquad (12.20)$$
$$\dot\xi = A_c\xi + B_c(\lambda^T\xi - b_m C_0\eta + b_m u) \qquad\qquad (12.21)$$
$$y = C_c\xi \qquad\qquad (12.22)$$

Using the special structure of (A_c, B_c, C_c), it is straightforward to verify that

$$y^{(r)} = \lambda^T\xi - b_m C_0\eta + b_m u$$

The (input-output linearizing) state feedback control

$$u = \frac{1}{b_m}[-\lambda^T\xi + b_m C_0\eta + v]$$

results in the system

$$
\begin{aligned}
\dot{\eta} &= A_0\eta + B_0 C_c\xi \\
\dot{\xi} &= A_c\xi + B_c v \\
y &= C_c\xi
\end{aligned}
$$

whose input-output map is a chain of r integrators, and whose state subvector η is unobservable from the output y. Suppose we want to stabilize the output at a constant reference y_R. This requires stabilizing ξ at $\xi^* = (y_R, 0, \ldots, 0)^T$. Shifting the equilibrium point to the origin by the change of variables $\zeta = \xi - \xi^*$ reduces the problem to a stabilization problem for $\dot{\zeta} = A_c\zeta + B_c v$. Taking $v = K\zeta = K(\xi - \xi^*)$, where K is chosen such that $A_c + B_c K$ is Hurwitz, completes the design of the control law as

$$
u = \frac{1}{b_m}[-\lambda^T\xi + b_m C_0\eta + K(\xi - \xi^*)]
$$

The corresponding closed-loop system is given by

$$
\begin{aligned}
\dot{\eta} &= A_0\eta + B_0 C_c(\xi^* + \zeta) \\
\dot{\zeta} &= (A_c + B_c K)\zeta
\end{aligned}
$$

Since $(A_c + B_c K)$ is Hurwitz, for any initial state $\zeta(0)$ we have $\zeta(t) \to 0$ as $t \to \infty$. Consequently, $y(t) \to y_R$ as $t \to \infty$. What about η? Equation (12.20) is driven by $y = C_c\xi$ as input. To ensure that $\eta(t)$ will be bounded for all possible waveforms of $y(t)$ and all possible initial states $\eta(0)$, we must require A_0 to be Hurwitz. Equivalently, the zeros of $H(s)$ must lie in the open left-half plane. A transfer function having all zeros in the open-left half plane is called *minimum phase*. From a pole placement viewpoint, the state feedback control we have just designed via the input-output linearization procedure assigns the closed-loop eigenvalues into two groups: r eigenvalues are assigned in the open-left half plane as the eigenvalues of $A_c + B_c K$, and $n - r$ eigenvalues are assigned at the open-loop zeros.[4]

Our analysis of the linear system of Example 12.8 sheds some light on the meaning of the state feedback control that reduces the input-output map to a chain of integrators, and on how to characterize internal stability. The key tool that allowed us to develop this understanding is the state model (12.20)–(12.22). Our next task is to develop a nonlinear version of (12.20)–(12.22) for the nonlinear system (12.16)–(12.17), which has relative degree r. The ξ variables are taken the same as in the linear case, since the linear input-output map will still be a chain of r integrators. We would like to choose the η variables to produce a nonlinear version of (12.20). The key feature of (12.20) is the absence of the control input u.

---

[4] It should be noted that stabilizing the output at a constant reference does not require the system to be minimum phase. This requirement is a consequence of our choice to assign some of the closed-loop eigenvalues at the open-loop zeros.

A change of variables that would transform (12.16)–(12.17) into a nonlinear version
of (12.20)–(12.22) can be taken as

$$
z = T(x) = \begin{bmatrix} \phi_1(x) \\ \vdots \\ \phi_{n-r}(x) \\ --- \\ \psi_1(x) \\ \vdots \\ \psi_r(x) \end{bmatrix} \overset{\text{def}}{=} \begin{bmatrix} \phi(x) \\ --- \\ \psi(x) \end{bmatrix} \overset{\text{def}}{=} \begin{bmatrix} \eta \\ --- \\ \xi \end{bmatrix} \tag{12.23}
$$

where ψ_1 to ψ_r are given by (12.19), and ϕ_1 to ϕ_{n-r} are chosen such that $T(x)$ is
a diffeomorphism on a domain $D_x \subset D_0$ and

$$
\frac{\partial \phi_i}{\partial x} g(x) = 0, \quad \text{for } 1 \le i \le n - r, \ \forall \ x \in D_x \tag{12.24}
$$

It will be shown in Section 12.4 that if (12.16)–(12.17) has relative degree $r < n$
in a domain D_0, then for every $x_0 \in D_0$, there is a neighborhood N of x_0 and
functions ϕ_1 to ϕ_{n-r} that satisfy (12.24) and make $T(x)$ a diffeomorphism on N.
The condition (12.24) ensures that when we calculate

$$
\dot{\eta} = \frac{\partial \phi}{\partial x}[f(x) + g(x)u]
$$

the u term cancels out. It is now easy to verify that the change of variables (12.23)
transforms (12.16)–(12.17) into

$$
\dot{\eta} = f_0(\eta, \xi) \tag{12.25}
$$
$$
\dot{\xi} = A_c \xi + B_c \beta^{-1}(x)[u - \alpha(x)] \tag{12.26}
$$
$$
y = C_c \xi \tag{12.27}
$$

where $\xi \in R^r$, $\eta \in R^{n-r}$, (A_c, B_c, C_c) is a canonical form representation of a chain
of r integrators,

$$
f_0(\eta, \xi) = \frac{\partial \phi}{\partial x} f(x) \bigg|_{x = T^{-1}(z)} \tag{12.28}
$$

$$
\beta(x) = \frac{1}{(\partial \psi_r / \partial x) g(x)} \quad \text{and} \quad \alpha(x) = -\frac{(\partial \psi_r / \partial x) f(x)}{(\partial \psi_r / \partial x) g(x)} \tag{12.29}
$$

In (12.26), we have kept α and β expressed in the original coordinates. These
expressions are uniquely determined by (12.29) in terms of f, g, and h. They are
independent of the choice of $\phi(x)$. They can be expressed in the new coordinates
by setting

$$
\alpha_0(\eta, \xi) = \alpha\left(T^{-1}(z)\right) \quad \text{and} \quad \beta_0(\eta, \xi) = \beta\left(T^{-1}(z)\right)
$$

which, of course, will depend on the choice of $\phi(x)$. In this case, (12.6) can be rewritten as

$$\dot{\xi} = A_c \xi + B_c \beta_0^{-1}(\eta, \xi)[u - \alpha_0(\eta, \xi)]^T$$

If x^* is an open-loop equilibrium point of (12.16), then (η^*, ξ^*), defined by

$$\eta^* = \phi(x^*), \quad \xi^* = \left[\begin{array}{cccc} h(x^*) & 0 & \ldots & 0 \end{array} \right]$$

is an equilibrium point of (12.25)–(12.26). If y vanishes at $x = x^*$, that is, $h(x^*) = 0$, we can transform x^* into the origin point ($\eta = 0$, $\xi = 0$) by choosing $\phi(x)$ such that $\phi(x^*) = 0$.

Equations (12.25)–(12.27) are said to be in the *normal form*. This form decomposes the system into an external part ξ and an internal part η. The external part is linearized by the state feedback control

$$u = \alpha(x) + \beta(x)v$$

while the internal part is made unobservable by the same control. The internal dynamics are described by (12.25). Setting $\xi = 0$ in (12.25) results in

$$\dot{\eta} = f_0(\eta, 0) \tag{12.30}$$

which is called the *zero dynamics*, a name that matches nicely with the fact that for linear systems (12.30) is given by $\dot{\eta} = A_0 \eta$, where the eigenvalues of A_0 are the zeros of the transfer function $H(s)$. The system is said to be *minimum phase* if (12.30) has an asymptotically stable equilibrium point in the domain of interest. In particular, if $T(x)$ is chosen such that the origin ($\eta = 0$, $\xi = 0$) is an equilibrium point of (12.25)–(12.27), then the system is said to be minimum phase if the origin of the zero dynamics (12.30) is asymptotically stable. It is useful to know that the zero dynamics can be characterized in the original x-coordinates. Notice that

$$y(t) \equiv 0 \Rightarrow \xi(t) \equiv 0 \Rightarrow u(t) \equiv \alpha(x(t))$$

Thus, keeping the output identically zero implies that the solution of the state equation must be confined to the set

$$Z^* = \{x \in D_0 \mid \psi_1(x) = \psi_2(x) = \ldots = \psi_r(x) = 0\}$$

and the input must be

$$u = u^*(x) \stackrel{\text{def}}{=} \alpha(x)|_{x \in Z^*}$$

The restricted motion of the system is described by

$$\dot{x} = f^*(x) \stackrel{\text{def}}{=} [f(x) + g(x)\alpha(x)]_{x \in Z^*}$$

Example 12.9 Consider the controlled van der Pol equation

$$\dot{x}_1 = x_2$$
$$\dot{x}_2 = -x_1 + \epsilon(1 - x_1^2)x_2 + u$$
$$y = x_2$$

We have seen in Example 12.5 that the system has relative degree 1 in R^2. Taking $\xi = y$ and $\eta = x_1$, we see that the system is already in the normal form. The zero dynamics are given by $\dot{x}_1 = 0$, which does not have an asymptotically stable equilibrium point. Hence, the system is not minimum phase. $\triangle$

Example 12.10 Consider the system

$$\dot{x}_1 = -x_1 + \frac{2 + x_3^2}{1 + x_3^2} u$$
$$\dot{x}_2 = x_3$$
$$\dot{x}_3 = x_1 x_3 + u$$
$$y = x_2$$

which has an open-loop equilibrium point at the origin. The derivatives of the output are given by

$$\dot{y} = \dot{x}_2 = x_3$$
$$\ddot{y} = \dot{x}_3 = x_1 x_3 + u$$

Hence, the system has relative degree 2 in R^3. We have $\psi_1(x) = x_2$ and $\psi_2(x) = x_3$. Using (12.29), we obtain

$$\beta = 1 \quad \text{and} \quad \alpha(x) = -x_1 x_3$$

To characterize the zero dynamics, restrict x to

$$Z^* = \{x \in R^3 \mid x_2 = x_3 = 0\}$$

and take $u = u^*(x) = 0$. This process yields

$$\dot{x}_1 = -x_1$$

which shows that the system is minimum phase. To transform the system into the normal form, we want to choose a function $\phi(x)$ such that

$$\phi(0) = 0, \quad \frac{\partial \phi}{\partial x} g(x) = 0$$

and

$$T(x) = \begin{bmatrix} \phi(x) & x_2 & x_3 \end{bmatrix}^T$$

is a diffeomorphism on some domain containing the origin. The partial differential equation

$$\frac{\partial \phi}{\partial x_1} \cdot \frac{2 + x_3^2}{1 + x_3^2} + \frac{\partial \phi}{\partial x_3} = 0$$

can be solved by separation of variables to obtain

$$\phi(x) = -x_1 + x_3 + \tan^{-1} x_3$$

which satisfies the condition $\phi(0) = 0$. The mapping $T(x)$ is a global diffeomorphism, as can be seen by the fact that for any $z \in R^3$, the equation $T(x) = z$ has a unique solution. Thus, the normal form

$$\dot{\eta} = \left(-\eta + \xi_2 + \tan^{-1} \xi_2\right) \left(1 + \frac{2 + \xi_2^2}{1 + \xi_2^2} \xi_2\right)$$

$$\dot{\xi}_1 = \xi_2$$

$$\dot{\xi}_2 = \left(-\eta + \xi_2 + \tan^{-1} \xi_2\right) \xi_2 + u$$

$$y = \xi_1$$

is defined globally. $\triangle$

Example 12.11 The field-controlled DC motor of Example 12.7 has relative degree 2 in $D_0 = \{x \in R^3 \mid x_2 \neq 0\}$. The functions ψ_1 and ψ_2 are given by $\psi_1(x) = x_3$ and $\psi_2(x) = \theta x_1 x_2$. Using (12.29), we obtain

$$\beta = \frac{1}{\theta x_2} \quad \text{and} \quad \alpha(x) = -\frac{\theta x_2(-ax_1) + \theta x_1(-bx_2 + \rho - cx_1 x_3)}{\theta x_2}$$

To characterize the zero dynamics, restrict x to

$$Z^* = \{x \in D_0 \mid x_3 = 0 \text{ and } x_1 x_2 = 0\} = \{x \in D_0 \mid x_3 = 0 \text{ and } x_1 = 0\}$$

and take $u = u^*(x) = 0$ to obtain

$$\dot{x}_2 = -bx_2 + \rho$$

The zero dynamics have an asymptotically stable equilibrium point at $x_2 = \rho/b$. Hence, the system is minimum phase. To transform the system into the normal form, we want to find a function $\phi(x)$ such that $[\partial \phi / \partial x]g = \partial \phi / \partial x_1 = 0$ and $T = [\phi(x), x_3, \theta x_1 x_2]^T$ is a diffeomorphism on some domain $D_x \subset D_0$. The choice $\phi(x) = x_2 - \rho/b$ satisfies $\partial \phi / \partial x_1 = 0$, makes $T(x)$ a diffeomorphism on $D_x = \{x \in R^3 \mid x_2 > 0\}$, and transforms the equilibrium point of the zero dynamics to the origin. $\triangle$

Example 12.12 Consider the single-input–single-output nonlinear system represented by the nth-order differential equation

$$y^{(n)} = p\left(z, z^{(1)}, \ldots, z^{(m-1)}, y, y^{(1)}, \ldots, y^{(n-1)}\right)$$
$$+ q\left(z, z^{(1)}, \ldots, z^{(m-1)}, y, y^{(1)}, \ldots, y^{(n-1)}\right) z^{(m)}, \quad m < n \quad (12.31)$$

where z is the input, y is the output, $p(\cdot)$ and $q(\cdot)$ are sufficiently smooth functions in a domain of interest, and $q(\cdot) \neq 0$. This nonlinear input-output model reduces to the transfer function model of Example 12.8 for linear systems. We will extend the dynamics of this system by adding a series of m integrators at the input side and define $u = z^{(m)}$ as the control input of the extended system.[5] The extended system is of order $(n + m)$. A state model of the extended system can be obtained by taking the state variables as

$$\zeta = \begin{bmatrix} z \\ z^{(1)} \\ \vdots \\ z^{(m-1)} \end{bmatrix}, \quad \xi = \begin{bmatrix} y \\ y^{(1)} \\ \vdots \\ \vdots \\ y^{(n-1)} \end{bmatrix}, \quad \text{and } x = \begin{bmatrix} \zeta \\ \xi \end{bmatrix}$$

The state model is given by

$$\dot{\zeta} = A_u \zeta + B_u u$$
$$\dot{\xi} = A_c \xi + B_c [p(x) + q(x)u]$$
$$y = C_c \xi$$

where (A_c, B_c, C_c) is a canonical form representation of a chain of n integrators as in (12.12) and (12.15), and (A_u, B_u) is a controllable canonical pair that represents a chain of m integrators as in (12.12). Let $D \in R^{n+m}$ be a domain over which p and q are sufficiently smooth and $q \neq 0$. Using the special structure of (A_c, B_c, C_c), it can be easily seen that

$$y^{(i)} = C_c A_c^i \xi, \quad \text{for } 1 \le i \le n - 1, \quad \text{and } y^{(n)} = p(x) + q(x)u$$

Hence, the system has relative degree n. To find the zero dynamics, notice that $\psi_i(x) = \xi_i$. Hence, $Z^* = \{x \in R^{n+m} \mid \xi = 0\}$ and $u^*(x) = -p(x)/q(x)$ evaluated at $\xi = 0$. Thus, the zero dynamics are given by

$$\dot{\zeta} = A_u \zeta + B_u u^*(x)$$

---

[5] In this example we show that the extended system is input-output linearizable, which allows us to design feedback control using input-output linearization techniques, like the tracking controller of the next section. When such control is applied to the original system, the m integrators become part of the dynamics of the controller.

Recalling the definition of ζ, it can be easily seen that $\zeta_1 = z$ satisfies the mth-order differential equation

$$0 = p\left(z, z^{(1)}, \ldots, z^{(m-1)}, 0, 0, \ldots, 0\right) + q\left(z, z^{(1)}, \ldots, z^{(m-1)}, 0, 0, \ldots, 0\right) z^{(m)}$$
(12.32)

which is the same equation obtained from (12.31) upon setting $y(t) \equiv 0$. For linear systems, (12.32) reduces to a linear differential equation that corresponds to the numerator polynomial of the transfer function. The minimum phase property of the system can be determined by studying (12.32). To transform the system into the normal form, we note that ξ is already a vector of y and its derivatives up to $y^{(n-1)}$. So, we only need to find a function $\phi = \phi(\zeta, \xi) : R^{n+m} \to R^m$ such that

$$\frac{\partial \phi}{\partial \zeta} B_u + \frac{\partial \phi}{\partial \xi} B_c q(x) = 0$$

which is equivalent to

$$\frac{\partial \phi_i}{\partial \zeta_m} + \frac{\partial \phi_i}{\partial \xi_n} q(x) = 0, \quad \text{for } 1 \le i \le m$$
(12.33)

In some special cases, there are obvious solutions for these partial differential equations. For example, if q is constant, ϕ can be taken as

$$\phi_i = \zeta_i - \frac{1}{q}\xi_{n-m+i}, \quad \text{for } 1 \le i \le m$$

Another case is pursued in Exercise 12.11. $\triangle$

12.3 State Feedback Control

12.3.1 Stabilization

Consider an input-state linearizable system

$$\dot{z} = Az + B\beta^{-1}(x)[u - \alpha(x)], \quad z = T(x)$$

where $T(x)$ is a diffeomorphism on a domain $D_x \subset R^n$, $D_z = T(D_x)$ contains the origin, (A, B) is controllable, $\beta(x)$ is nonsingular for all $x \in D_x$, and $\alpha(x)$ and $\beta(x)$ are continuously differentiable. Design K such that $A + BK$ is Hurwitz. The feedback control

$$u = \alpha(x) + \beta(x)KT(x)$$

results in the linear closed-loop system

$$\dot{z} = (A + BK)z$$

This beautiful result, however, is based on exact mathematical cancellation of the nonlinear terms α and β. Exact cancellation is almost impossible for several practical reasons such as model simplification, parameter uncertainty, and computational errors. Most likely, the controller will be implementing functions $\hat{\alpha}$, $\hat{\beta}$, and $\hat{T}$ which are approximations of α, β, and T; that is to say, the actual controller will be implementing the feedback control law

$$u = \hat{\alpha}(x) + \hat{\beta}(x)K\hat{T}(x)$$

The closed-loop system under this feedback is

$$\dot{z} = Az + B\beta^{-1}(x)[\hat{\alpha}(x) + \hat{\beta}(x)K\hat{T}(x) - \alpha(x)]$$

By adding and subtracting the term BKz, we can rewrite this equation as

$$\dot{z} = (A + BK)z + B\delta(z)$$

where

$$\delta(z) = \beta^{-1}(x)\{\hat{\alpha}(x) - \alpha(x) + [\hat{\beta}(x) - \beta(x)]KT(x) + \hat{\beta}(x)K[\hat{T}(x) - T(x)]\}\Big|_{x=T^{-1}(z)}$$

Thus, the closed-loop system appears as a perturbation of the nominal system

$$\dot{z} = (A + BK)z$$

Since the matrix $A + BK$ is Hurwitz, we expect from the robustness results of Chapter 5 that no serious problem will result from a small error term $\delta(z)$. Let us analyze the stability of the closed-loop system. Let $P = P^T > 0$ be the solution of the Lyapunov equation

$$P(A + BK) + (A + BK)^T P = -Q$$

where $Q = Q^T > 0$. Suppose that, in a neighborhood of the origin, the error term $\delta(z)$ satisfies $\|\delta(z)\|_2 \leq \gamma_1\|z\|_2 + \gamma_2$ for some nonnegative constants γ_1 and γ_2. This is a reasonable requirement that follows from continuous differentiability of the nonlinear functions. Using $V(z) = z^T P z$ as a Lyapunov function candidate for the closed-loop system, we have

$$\begin{aligned}
\dot{V}(z) &= -z^T Q z + 2z^T P B\delta(z) \\
&\leq -\lambda_{\min}(Q)\|z\|_2^2 + 2\|PB\|_2\gamma_1\|z\|_2^2 + 2\|PB\|_2\gamma_2\|z\|_2
\end{aligned}$$

If $\gamma_1 < \lambda_{\min}(Q)/4\|PB\|_2$, we have

$$\dot{V}(z) \leq -\tfrac{1}{2}\lambda_{\min}(Q)\|z\|_2^2 + 2\|PB\|_2\gamma_2\|z\|_2 \leq -\tfrac{1}{4}\lambda_{\min}(Q)\|z\|_2^2, \ \ \forall \ \|z\|_2 \geq \frac{8\|PB\|_2\gamma_2}{\lambda_{\min}(Q)}$$

which shows that the solutions of the system will be ultimately bounded with an ultimate bound proportional to γ_2. Moreover, if $\delta(0) = 0$, we can take $\gamma_2 = 0$. In such a case, the origin will be exponentially stable if $\gamma_1 < \lambda_{\min}(Q)/4\|PB\|_2$.

Example 12.13 Consider the pendulum equation

$$\dot{x}_1 = x_2$$
$$\dot{x}_2 = -a\sin(x_1 + \delta_1) - bx_2 + cu$$

where $x_1 = \theta - \delta_1$, $x_2 = \dot{\theta}$, and $u = T$ is a torque input. The goal is to stabilize the pendulum at the angle $\theta = \delta_1$. A linearizing-stabilizing feedback control is given by

$$u = \left(\frac{a}{c}\right)\sin(x_1 + \delta_1) + \left(\frac{1}{c}\right)(k_1 x_1 + k_2 x_2)$$

where k_1 and k_2 are chosen such that

$$A + BK = \left[\begin{array}{cc} 0 & 1 \\ k_1 & k_2 - b \end{array}\right]$$

is Hurwitz. Suppose that, due to uncertainties in the parameters a and c, the actual control is

$$u = \left(\frac{\hat{a}}{\hat{c}}\right)\sin(x_1 + \delta_1) + \left(\frac{1}{\hat{c}}\right)(k_1 x_1 + k_2 x_2)$$

where $\hat{a}$ and $\hat{c}$ are estimates of a and c. The closed-loop system is given by

$$\dot{x}_1 = x_2$$
$$\dot{x}_2 = k_1 x_1 + (k_2 - b)x_2 + \delta(x)$$

where

$$\delta(x) = \left(\frac{\hat{a}c - a\hat{c}}{\hat{c}}\right)\sin(x_1 + \delta_1) + \left(\frac{c - \hat{c}}{\hat{c}}\right)(k_1 x_1 + k_2 x_2)$$

The error term $\delta(x)$ satisfies the bound $|\delta(x)| \leq \gamma_1 \|x\|_2 + \gamma_2$ globally, where

$$\gamma_1 = \left|\frac{\hat{a}c - a\hat{c}}{\hat{c}}\right| + \left|\frac{c - \hat{c}}{\hat{c}}\right|\sqrt{k_1^2 + k_2^2}, \quad \gamma_2 = \left|\frac{\hat{a}c - a\hat{c}}{\hat{c}}\right||\sin\delta_1|$$

The constants γ_1 and γ_2 are measures of the size of the error in estimating the parameters a and c. Let

$$P = \left[\begin{array}{cc} p_{11} & p_{12} \\ p_{12} & p_{22} \end{array}\right]$$

be the solution of the Lyapunov equation $P(A + BK) + (A + BK)^T P = -I$. If

$$\gamma_1 < \frac{1}{4\sqrt{p_{12}^2 + p_{22}^2}}$$

then the solutions of the system are globally ultimately bounded by a bound that is proportional to γ_2. If $\sin\delta_1 = 0$, the foregoing bound on γ_1 ensures global exponential stability of the origin. $\triangle$

The Lyapunov analysis we have just carried out to study the effect of the perturbation term $B\delta(z)$ is quite general. It actually applies to any additive perturbation term $\Delta(z)$ on the right-hand side of the closed-loop equation; that is, we can perform the same analysis when the closed-loop system is given by

$$\dot{z} = (A + BK)z + \Delta(z)$$

The error term $B\delta(z)$, however, has a special structure; it is in the range space of the input matrix B. Such an error is said to satisfy the *matching condition*. There is more we can do about the robustness of the system when the error has this special structure, as we shall see in Sections 13.1 and 13.3 of the next chapter.

Consider now a partially feedback linearizable system of the form

$$\begin{align}
\dot{\eta} &= f_0(\eta, \xi) \tag{12.34} \\
\dot{\xi} &= A\xi + B\beta^{-1}(x)[u - \alpha(x)] \tag{12.35}
\end{align}$$

where

$$z = \begin{bmatrix} \eta \\ \xi \end{bmatrix} = T(x) \overset{\text{def}}{=} \begin{bmatrix} T_1(x) \\ T_2(x) \end{bmatrix}$$

$T(x)$ is a diffeomorphism on a domain $D_x \subset R^n$, $D_z = T(D_x)$ contains the origin, (A, B) is controllable, $\beta(x)$ is nonsingular for all $x \in D_x$, $f_0(0, 0) = 0$, and $f_0(\eta, \xi)$, $\alpha(x)$, and $\beta(x)$ are continuously differentiable. The form (12.34)–(12.35) is clearly motivated by the normal form (12.25)–(12.27) of input-output linearizable systems. However, (12.27) is dropped since the output y plays no role in the state feedback stabilization problem. Moreover, we do not have to restrict our discussions to single-input systems or to a pair (A, B) in the controllable canonical form. So, we proceed to discuss the more general system (12.34)–(12.35) and our conclusions will apply to the normal form (12.25)–(12.27). The state feedback control

$$u = \alpha(x) + \beta(x)v$$

reduces (12.34)–(12.35) to the "triangular" system

$$\begin{align}
\dot{\eta} &= f_0(\eta, \xi) \tag{12.36} \\
\dot{\xi} &= A\xi + Bv \tag{12.37}
\end{align}$$

Equation (12.37) can be easily stabilized by $v = K\xi$, where K is designed such that $A + BK$ is Hurwitz. Asymptotic stability of the origin of the full closed-loop system

$$\begin{align}
\dot{\eta} &= f_0(\eta, \xi) \tag{12.38} \\
\dot{\xi} &= (A + BK)\xi \tag{12.39}
\end{align}$$

will now follow from the concept of input-to-state stability, introduced in Section 5.3. Lemma 5.6 confirms that the origin of (12.38)–(12.39) will be asymptotically stable if the system (12.38), with ξ as input, is locally input-to-state stable. Furthermore, Lemma 5.4 shows that (12.38) will be locally input-to-state stable if the origin of the unforced system

$$\dot{\eta} = f_0(\eta, 0) \tag{12.40}$$

is asymptotically stable. Thus, a minimum phase input-output linearizable system can be stabilized by the state feedback control

$$u = \alpha(x) + \beta(x) K T_2(x) \tag{12.41}$$

Note that this control law depends on $\alpha(x)$, $\beta(x)$, and $T_2(x)$, but not on $T_1(x)$. This is significant for input-output linearizable systems because α, β, and T_2 are uniquely determined by the functions f, g, and h, while T_1 depends on the choice of $\phi(x)$ that satisfies the partial differential equation (12.24). Robustness to model uncertainties can be handled as in our earlier treatment of fully linearizable systems; we leave it to Exercise 12.19.

What about global stabilization? From Lemma 5.6, we know that the origin of (12.38)–(12.39) will be globally asymptotically stable if (12.38) is input-to-state stable. However, input-to-state stability does not follow from global asymptotic, or even exponential, stability of the origin of (12.40), as we saw in Example 5.8. Consequently, knowing that an input-output linearizable system is "globally" minimum phase does not automatically guarantee that the control (12.41) will globally stabilize the system. It will be globally stabilizing if the origin of (12.40) is globally exponentially stable and $f_0(\eta, \xi)$ is globally Lipschitz in (η, ξ), since in this case Lemma 5.5 confirms that the system (12.38) will be input-to-state stable. Otherwise, we have to establish the input-to-state stability of (12.38) by further analysis. Global Lipschitz conditions are sometimes referred to as *linear growth conditions*. The following two examples illustrate some of the difficulties that may arise in the absence of linear growth conditions.

Example 12.14 Consider the second-order system

$$
\begin{aligned}
\dot{\eta} &= -\eta + \eta^2 \xi \\
\dot{\xi} &= v
\end{aligned}
$$

While the origin of $\dot{\eta} = -\eta$ is globally exponentially stable, the system $\dot{\eta} = -\eta + \eta^2 \xi$ is not input-to-state stable. This fact can be seen by repeating the argument used in Example 5.8. It is, of course, locally input-to-state stable. Thus, the linear control $v = -\gamma\xi$, with $\gamma > 0$, will stabilize the origin of the full system. In fact, the origin will be exponentially stable. Taking $\nu = \eta\xi$ and noting that

$$\dot{\nu} = \eta\dot{\xi} + \dot{\eta}\xi = -\gamma\eta\xi - \eta\xi + \eta^2\xi^2 = -(1+\gamma)\nu + \nu^2$$

we see that the hyperbola $\eta\xi = 1+\gamma$ consists of two trajectories. By explicit solution of the closed-loop state equation, it can be shown that the region of attraction is exactly the set $\{\eta\xi < 1+\gamma\}$. Hence, the origin is not globally asymptotically stable. Notice, however, that the region of attraction expands as γ increases. In fact by choosing γ large enough, we can include any compact set in the region of attraction. Thus, the linear feedback control $v = -\gamma\xi$ can achieve semiglobal stabilization. $\triangle$

If the origin of $\dot{\eta} = f_0(\eta, 0)$ is globally asymptotically stable, one might think that the triangular system (12.36)–(12.37) can be globally stabilized, or at least semiglobally stabilized, by designing the linear feedback control $v = K\xi$ to assign the eigenvalues of $(A + BK)$ far to the left in the complex plane so that the solution of $\dot{\xi} = (A + BK)\xi$ decays to zero arbitrarily fast. Then, the solution of $\dot{\xi} = f_0(\eta, \xi)$ will quickly approach the solution of $\dot{\eta} = f_0(\eta, 0)$, which is well behaved since its origin is globally asymptotically stable. It may even appear that this strategy is the one used to achieve semiglobal stabilization in the preceding example. The following example shows why such strategy may fail.[6]

Example 12.15 Consider the third-order system

$$\dot{\eta} = -\tfrac{1}{2}(1 + \xi_2)\eta^3$$
$$\dot{\xi}_1 = \xi_2$$
$$\dot{\xi}_2 = v$$

The linear feedback control

$$v = -\gamma^2\xi_1 - 2\gamma\xi_2 \stackrel{\text{def}}{=} K\xi$$

assigns the eigenvalues of

$$A + BK = \left[\begin{array}{cc} 0 & 1 \\ -\gamma^2 & -2\gamma \end{array}\right]$$

at $-\gamma$ and $-\gamma$. The exponential matrix

$$e^{(A+BK)t} = \left[\begin{array}{cc} (1 + \gamma t)e^{-\gamma t} & te^{-\gamma t} \\ -\gamma^2 te^{-\gamma t} & (1 - \gamma t)e^{-\gamma t} \end{array}\right]$$

shows that as $\gamma \to \infty$, the solution $\xi(t)$ will decay to zero arbitrarily fast. Notice, however, that the coefficient of the (2,1) element of the exponential matrix is a quadratic function of γ. It can be shown that the absolute value of this element reaches a maximum value γ/e at $t = 1/\gamma$. While this term can be made to decay to zero arbitrarily fast by choosing γ large, its transient behavior exhibits a peak of the order of γ. This phenomenon is known as *the peaking phenomenon*.[7] The

---

[6] See, however, Exercise 12.17 for a special case where this strategy will work.

[7] To read more about the peaking phenomenon, see [169]. For an illustration of the peaking phenomenon in high-gain observers, see [46].

interaction of this peaking phenomenon with nonlinear growth in $\dot{\eta} = -\frac{1}{2}(1+\xi_2)\eta^3$ could destabilize the system. In particular, for the initial states $\eta(0) = \eta_0$, $\xi_1(0) = 1$, and $\xi_2(0) = 0$, we have $\xi_2(t) = -\gamma^2 t e^{-\gamma t}$ and

$$\dot{\eta} = -\frac{1}{2}\left(1 - \gamma^2 t e^{-\gamma t}\right)\eta^3$$

During the peaking period, the coefficient of η^3 is positive, causing $|\eta(t)|$ to grow. Eventually, the coefficient of η^3 will become negative, but that might not happen soon enough since the system might have a finite escape time. Indeed, the solution

$$\eta^2(t) = \frac{\eta_0^2}{1 + \eta_0^2[t + (1+\gamma t)e^{-\gamma t} - 1]}$$

shows that if $\eta_0^2 > 1$, the system will have a finite escape time if γ is chosen large enough. $\triangle$

We will come back to the triangular system (12.36)–(12.37) in Section 13.2 and show how to design v as a nonlinear function of ξ and η to achieve global stabilization. This will be done using the backstepping design procedure. We will even deal with cases where (12.36) is not input-to-state stable.

We conclude this section by discussing a pitfall of feedback linearization.[8] The basic philosophy of feedback linearization is to cancel the nonlinear terms of the system. Aside from the issues of whether or not we can cancel the nonlinear terms, effect of uncertainties, implementation factors, and so on, we should examine the philosophy itself: Is it a good idea to cancel nonlinear terms? Our motivation to do so has been mathematically driven. We wanted to linearize the system to make it more tractable and to use the relatively well-developed linear control theory. From a performance viewpoint, however, a nonlinear term could be "good" or "bad" and the decision whether we should use feedback to cancel a nonlinear term is, in reality, problem dependent. Let us use an example to illustrate this point.

Example 12.16 Consider the scalar system

$$\dot{x} = ax - bx^3 + u$$

where a and b are positive constants. A linearizing-stabilizing feedback control can be taken as

$$u = -(\gamma + a)x + bx^3, \quad \gamma > 0$$

which results in the closed-loop system $\dot{x} = -\gamma x$. This feedback control cancels the nonlinear term $-bx^3$, but this term provides "nonlinear damping." In fact, without

---

[8]Another pitfall lies in the sensitivity of the relative degree and the minimum phase property to parameter perturbations. To read more about this issue, see [152] and [83].

any feedback control, this nonlinear damping would guarantee boundedness of the solutions despite the fact that the origin is unstable. So, why should we cancel it? If we simply use the linear control

$$u = -(\gamma + a)x, \quad \gamma > 0$$

we will obtain the closed-loop system

$$\dot{x} = -\gamma x - bx^3$$

whose origin is globally asymptotically stable and its trajectories approach the origin faster than the trajectories of $\dot{x} = -\gamma x$. Moreover, linear control is simpler to implement than the linearizing control and uses less control effort. $\triangle$

The point we want to convey is that feedback linearization theory provides us with a very valuable tool to represent the system in a form where nonlinear terms enter the state equation at the same point as the control input. The next chapter describes robust control techniques which make use of this special structure. The same structure allows us to use feedback to cancel the nonlinearity. While this could be the way to go in many cases, it should not be an automatic choice. We should try our best to understand the effect of the nonlinear terms and decide whether or not cancellation is appropriate. Admittedly, this is not an easy task and the preceding example might be deceiving in that regard.

12.3.2   Tracking

Consider a single-input–single-output, input-output linearizable system represented in the normal form (12.25)–(12.27):

$$\dot{\eta} = f_0(\eta, \xi)$$
$$\dot{\xi} = A_c\xi + B_c \frac{1}{\beta(x)}[u - \alpha(x)]$$
$$y = C_c\xi$$

Without loss of generality, we assume that $f_0(0,0) = 0$. We want to design a tracking control such that the output y asymptotically tracks a reference signal $y_R(t)$. We assume that

- $y_R(t)$ and its derivatives up to $y_R^{(r)}(t)$ are bounded for all $t \geq 0$, and the rth derivative $y_R^{(r)}(t)$ is a piecewise continuous function of t;

- the signals $y_R, \ldots, y_R^{(r)}$ are available on line.

When the system has relative degree $r = n$, it has no nontrivial zero dynamics. In this case, the η variable and its equation are dropped but the rest of the development remains the same. Let

$$
\mathcal{Y}_R = \begin{bmatrix} y_R \\ \vdots \\ y_R^{(r-1)} \end{bmatrix}, \quad e = \begin{bmatrix} \xi_1 - y_R \\ \vdots \\ \xi_r - y_R^{(r-1)} \end{bmatrix} = \xi - \mathcal{Y}_R
$$

The change of variables $e = \xi - \mathcal{Y}_R$ yields

$$
\begin{aligned}
\dot{\eta} &= f_0(\eta, e + \mathcal{Y}_R) \\
\dot{e} &= A_c e + B_c \left\{ \frac{1}{\beta(x)} [u - \alpha(x)] - y_R^{(r)} \right\}
\end{aligned}
$$

The state feedback control

$$
u = \alpha(x) + \beta(x) \left[v + y_R^{(r)} \right]
$$

reduces the normal form to the cascade system

$$
\begin{aligned}
\dot{\eta} &= f_0(\eta, e + \mathcal{Y}_R) \\
\dot{e} &= A_c e + B_c v
\end{aligned}
$$

Our control objective can be met by any design of v that stabilizes the $\dot{e}$-equation while maintaining η bounded for all $t \geq 0$. With the linear control $v = Ke$, where K is designed such that $A_c + B_c K$ is Hurwitz, the complete state feedback control is given by [9]

$$
u = \alpha(x) + \beta(x) \left\{ K[T_2(x) - \mathcal{Y}_R] + y_R^{(r)} \right\} \tag{12.42}
$$

and the closed-loop system is given by

$$
\begin{aligned}
\dot{\eta} &= f_0(\eta, e + \mathcal{Y}_R) \tag{12.43} \\
\dot{e} &= (A_c + B_c K)e \tag{12.44}
\end{aligned}
$$

For minimum phase systems, the equation $\dot{\eta} = f_0(\eta, \xi)$ is locally input-to-state stable. Hence, for sufficiently small initial states $e(0)$ and $\eta(0)$ and sufficiently small $\mathcal{Y}_R(t)$, $\eta(t)$ will be bounded for all $t \geq 0$. Thus, the state feedback control (12.42) solves the local tracking problem. To extend the validity of the control to global tracking problems, where $\mathcal{Y}_R(t)$ can be any bounded function of t, we face the same issues we encountered in global stabilization. A sufficient condition to ensure global tracking is to require the system $\dot{\eta} = f_0(\eta, \xi)$ to be input-to-state stable.

---

[9] As in the previous section, T_2 comprises the last r components of the diffeomorphism $T(x)$ that transforms the system into the normal form.

Example 12.17 Consider the pendulum equation

$$\dot{x}_1 = x_2$$
$$\dot{x}_2 = -a \sin x_1 - bx_2 + cu$$
$$y = x_1$$

The system has relative degree 2 in R^2 and is already represented in the normal form. It has no nontrivial zero dynamics, so it is minimum phase by default. We want the output y to track a reference signal $y_R(t)$, with bounded derivatives $\dot{y}_R$ and $\ddot{y}_R$. Taking

$$e_1 = x_1 - y_R, \quad e_2 = x_2 - \dot{y}_R$$

we obtain

$$\dot{e}_1 = e_2$$
$$\dot{e}_2 = -a \sin x_1 - bx_2 + cu - \ddot{y}_R$$

The state feedback control (12.42) is given by

$$u = \frac{1}{c}[a \sin x_1 + bx_2 + \ddot{y}_R + k_1 e_1 + k_2 e_2]$$

where $K = [k_1, k_2]$ is designed to assign the eigenvalues of $A + BK$ at desired locations in the open left-half plane. Since all the assumptions hold globally, this control achieves global tracking. Figure 12.2 shows the response of the system when $a = b = c = 1$, $k_1 = 1600$, and $k_2 = 40$ to some reference signal. The solid curve is both the reference signal and output signal in the nominal case; they are identical. Here, tracking is achieved for all t and not just asymptotically because $x(0) = y_R(0)$. If $x(0) \neq y_R(0)$, tracking will be achieved asymptotically; this is shown by the dashed curve. Finally, the dotted curve shows the response of the system under parameter perturbations: a, b, and c are perturbed to $a = 1/1.05$; $b = 1/4$, and $c = 1/2.1$, which correspond to a 5%, −50%, and 100% change in ℓ, k, and m, respectively. $\triangle$

12.3.3 Regulation via Integral Control

Consider a single-input–single-output, input-output linearizable system represented in the normal form (12.25)–(12.27)

$$\dot{\eta} = f_0(\eta, \xi)$$
$$\dot{\xi} = A_c \xi + B_c \frac{1}{\beta(x)}[u - \alpha(x)]$$
$$y = C_c \xi$$

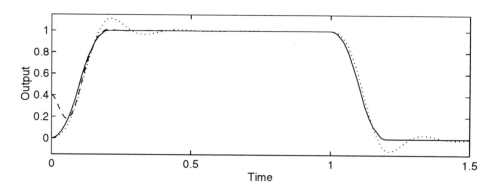

Figure 12.2: Simulation of the tracking control of Example 12.17.

Without loss of generality, we assume that $f_0(0,0) = 0$. We want to design a tracking control such that the output y asymptotically tracks a constant reference y_R. To ensure zero steady-state tracking error in the presence of uncertainties, we want to use integral control. We assume that y_R is available on line and y is physically measured. This is a new assumption that we did not need in the tracking control of the previous section. All our controllers in this chapter are designed under the assumption that the state x is physically measured. The extra assumption here is that y should also be physically measured; it is not good enough to calculate y using $y = h(x)$. The need to measure y in integral control has been discussed in Section 11.2. We integrate the tracking error $e = y - y_R$ using the integrator $\dot\sigma = e$, and augment the integrator with the system to obtain

$$\dot\eta = f_0(\eta, e + y_R)$$
$$\dot\xi_a = \mathcal{A}\xi_a + \mathcal{B}\frac{1}{\beta(x)}[u - \alpha(x)]$$

where

$$\mathcal{A} = \begin{bmatrix} A_c & 0 \\ C_c & 0 \end{bmatrix}, \mathcal{B} = \begin{bmatrix} B_c \\ 0 \end{bmatrix}, \mathcal{Y}_R = \begin{bmatrix} y_R \\ 0 \\ \vdots \\ 0 \end{bmatrix}, e = \xi - y_R, \text{ and } \xi_a = \begin{bmatrix} e \\ \sigma \end{bmatrix}$$

Notice that augmenting the integrator with the system does not change the structure of the normal form. The new pair of matrices $(\mathcal{A}, \mathcal{B})$ is controllable, as it can be easily verified that the rank condition (11.19) is satisfied. From this point on, the derivation of the control proceeds exactly as in the previous section. Design a matrix K such that $\mathcal{A} + \mathcal{B}K$ is Hurwitz. Partition K as $K = [K_1 \ K_2]$, where K_1 is $1 \times r$

and K_2 is 1×1. The state feedback control

$$u = \alpha(x) + \beta(x)\{K_1[T_2(x) - \mathcal{Y}_R] + K_2\sigma\} \qquad (12.45)$$

results in the closed-loop system

$$\begin{aligned} \dot{\eta} &= f_0(\eta, e + \mathcal{Y}_R) \\ \dot{\xi}_a &= (\mathcal{A} + \mathcal{B}K)\xi_a \end{aligned}$$

which takes the same form as the closed-loop system (12.43)–(12.44). Therefore, we refer the reader to the paragraph following (12.43)–(12.44) for the completion of the discussion.

The robustness of the integral controller is due to the fact that whenever the system is at equilibrium, the input to the integrator must be zero. This point was explained in Section 11.2, and will be emphasized again in the next example.

Example 12.18 The tank system of Example 11.5 is represented by the first-order state model

$$\dot{x} = \frac{1}{\beta(x)}\left(u - c\sqrt{2x}\right), \quad y = x$$

The system has relative degree $r = n = 1$. It is already in the normal form and has no nontrivial zero dynamics. Augmenting

$$\dot{\sigma} = e = y - y_R$$

with the state equation, we obtain the second-order augmented system

$$\dot{\xi}_a = \begin{bmatrix} 0 & 0 \\ 1 & 0 \end{bmatrix}\xi_a + \begin{bmatrix} 1 \\ 0 \end{bmatrix}\frac{1}{\beta(y)}\left(u - c\sqrt{2y}\right)$$

The matrix K is designed as

$$K = \begin{bmatrix} -2\zeta\omega_n & -\omega_n^2 \end{bmatrix}, \quad 0 < \zeta < 1, \quad \omega_n > 0$$

to assign the characteristic polynomial of $\mathcal{A} + \mathcal{B}K$ as $s^2 + 2\zeta\omega_n s + \omega_n^2$. The state feedback control (12.45) is given by [10]

$$u = c\sqrt{2y} - \beta(y)(2\zeta\omega_n e + \omega_n^2\sigma)$$

---

[10] It is interesting to compare this controller with one of the gain scheduled controllers derived in Example 11.5, which takes the form

$$u = c\sqrt{2y_R} - \beta(y_R)(2\zeta\omega_n e + \omega_n^2\sigma)$$

upon choosing $\bar{\sigma} = 0$.

and results in the linear closed-loop system

$$\dot{\xi}_a = (A + BK)\xi_a = \begin{bmatrix} -2\zeta\omega_n & -\omega_n^2 \\ 1 & 0 \end{bmatrix} \xi_a$$

The controller achieves regulation for all values of $y_R > 0$ and all initial states $x(0)$ in the region of interest, restricted only by the physical constraint $x \geq 0$. To investigate the robustness of the zero steady-state regulation error, let us suppose that, due to parameter uncertainty, the implemented controller is

$$u = \hat{c}\sqrt{2y} - \hat{\beta}(y)(2\zeta\omega_n e + \omega_n^2\sigma)$$

where $\hat{c}$ and $\hat{\beta}$ are estimates of c and β. It is assumed that $\hat{c}$ and $\hat{\beta}(y)$ are always positive. The closed-loop system under this controller is given by

$$
\begin{aligned}
\dot{e} &= \frac{1}{\beta(y)}\left[\hat{c}\sqrt{2y} - \hat{\beta}(y)(2\zeta\omega_n e + \omega_n^2\sigma) - c\sqrt{2y}\right] \\
\dot{\sigma} &= e
\end{aligned}
$$

The system has a unique equilibrium point at

$$e = 0, \quad \bar{\sigma} = \frac{(\hat{c} - c)\sqrt{2y_R}}{\omega_n^2\hat{\beta}(y_R)}$$

While this equilibrium point is different from the nominal equilibrium point at the origin, the value of e at equilibrium is still zero. To analyze the stability of the perturbed equilibrium point, we shift it to the origin via the change of variables $\sigma_\delta = \sigma - \bar{\sigma}$. Taking $\xi_\delta = [e, \sigma_\delta]^T$, it can be verified that ξ_δ satisfies the state equation

$$\dot{\xi}_\delta = (A + BK)\xi_\delta + B\delta(\xi_\delta)$$

where the perturbation term δ is given by

$$
\begin{aligned}
\delta(\xi_\delta) &= \frac{1}{\beta(y)}\left\{(\hat{c} - c)\left(\sqrt{2y} - \sqrt{2y_R}\right) + [\beta(y) - \hat{\beta}(y)](2\zeta\omega_n e + \omega_n^2\sigma_\delta)\right. \\
&\quad \left. + \omega_n^2[\hat{\beta}(y_R) - \hat{\beta}(y)]\bar{\sigma}\right\}
\end{aligned}
$$

Assuming that

$$\left|\frac{\beta(y) - \hat{\beta}(y)}{\beta(y)}\right| \leq \rho$$

and y, y_R, and $\beta(y)$ are bounded from below by positive constants, it can be shown that

$$|\delta(\xi_\delta)| \leq \gamma\|\xi_\delta\|_2, \quad \text{where } \gamma = L_1|\hat{c} - c| + L_2\rho$$

for some nonnegative constants L_1 and L_2. Standard Lyapunov analysis can be used to show that if γ is sufficiently small, the perturbed equilibrium point will be exponentially stable. To appreciate what we gain by integral control, let us consider another tracking control design that does not use integral control. The tracking controller of the previous section takes the form

$$u = c\sqrt{2y} - \beta(y)ke, \quad k > 0$$

and yields the nominal closed-loop system $\dot{e} = -ke$. The perturbed control

$$u = \hat{c}\sqrt{2y} - \hat{\beta}(y)ke$$

results in the closed-loop system

$$\dot{e} = \frac{1}{\beta(y)}\left[\hat{c}\sqrt{2y} - \hat{\beta}(y)ke - c\sqrt{2y}\right]$$

It can be easily seen that whenever $(\hat{c} - c)\sqrt{2y_R} \neq 0$, the right-hand side does not vanish at $e = 0$. Hence, the perturbed equilibrium point will not yield zero steady-state tracking error. $\triangle$

12.4   Differential Geometric Approach

We start by recalling some differential geometric tools;[11] then we apply these tools to input-output linearization in Section 12.4.2 and to input-state linearization in Section 12.4.3. For clarity of exposition, and convenience, we restrict our presentation to single-input–single-output systems.[12]

12.4.1   Differential Geometric Tools

All functions we shall encounter in the forthcoming definitions are assumed to be smooth in their arguments; that is, they have continuous partial derivatives to any order.

Vector Field: A mapping $f : D \to R^n$, where $D \subset R^n$ is a domain, is said to be a vector field on D. A vector field is an n-dimensional column.

Covector Field: The transpose of a vector field is said to be a covector field. A covector field is an n-dimensional row.

---

[11] The notation and some of the examples used to illustrate it are taken from [80].

[12] Interested readers will find a comprehensive treatment of the differential geometric approach to feedback linearization in [80] and [126], where multivariable systems are treated.

Inner Product: If f and w are, respectively, a vector field and a covector field on D, then the inner product $< w, f >$ is defined as

$$< w, f >= w(x)f(x) = \sum_{i=1}^{n} w_i(x)f_i(x)$$

Differential: Let $h : D \to R$. The differential of h is a covector field, defined by

$$dh = \frac{\partial h}{\partial x} = \left[\frac{\partial h}{\partial x_1}, \ldots, \frac{\partial h}{\partial x_n} \right]$$

Lie Derivative: Let $h : D \to R$ and $f : D \to R^n$. The Lie derivative of h with respect to f or along f, written as $L_f h$, is defined by

$$L_f h(x) = \frac{\partial h}{\partial x} f(x)$$

This is the familiar notion of the derivative of h along the trajectories of the system $\dot{x} = f(x)$. The new notation is convenient when we repeat the calculation of the derivative with respect to the same vector field or a new one. For example, the following notation is used:

$$L_g L_f h(x) = \frac{\partial(L_f h)}{\partial x} g(x)$$

$$L_f^2 h(x) = L_f L_f h(x) = \frac{\partial(L_f h)}{\partial x} f(x)$$

$$L_f^k h(x) = L_f L_f^{k-1} h(x) = \frac{\partial(L_f^{k-1} h)}{\partial x} f(x)$$

$$L_f^0 h(x) = h(x)$$

Example 12.19 Let

$$h(x) = x_1^2 + x_2, \quad f(x) = \begin{bmatrix} x_2 \\ -\sin x_1 - x_2 \end{bmatrix}, \quad g = \begin{bmatrix} 0 \\ 1 \end{bmatrix}$$

Then,

$$L_f h = \frac{\partial h}{\partial x} f(x) = \begin{bmatrix} 2x_1 & 1 \end{bmatrix} \begin{bmatrix} x_2 \\ -\sin x_1 - x_2 \end{bmatrix} = 2x_1 x_2 - \sin x_1 - x_2$$

$$\begin{aligned} L_f^2 h = \frac{\partial(L_f h)}{\partial x} f(x) &= \begin{bmatrix} 2x_2 - \cos x_1 & 2x_1 - 1 \end{bmatrix} \begin{bmatrix} x_2 \\ -\sin x_1 - x_2 \end{bmatrix} \\ &= (2x_2 - \cos x_1)x_2 + (2x_1 - 1)(-\sin x_1 - x_2) \end{aligned}$$

$$L_g L_f h(x) = \frac{\partial (L_f h)}{\partial x} g(x) = \begin{bmatrix} 2x_2 - \cos x_1 & 2x_1 - 1 \end{bmatrix} \begin{bmatrix} 0 \\ 1 \end{bmatrix} = 2x_1 - 1$$

$$L_g h = \frac{\partial h}{\partial x} g(x) = \begin{bmatrix} 2x_1 & 1 \end{bmatrix} \begin{bmatrix} 0 \\ 1 \end{bmatrix} = 1$$

Finally, $L_f L_g h = 0$. $\qquad\qquad\qquad\qquad\qquad\qquad\qquad\qquad\qquad \triangle$

Using the Lie derivative notation, we can restate the definition of relative degree in a simpler form. The reader can verify that the single-input–single-output system

$$\begin{aligned} \dot{x} &= f(x) + g(x)u \\ y &= h(x) \end{aligned}$$

has relative degree r in a domain D_0 if

$$L_g L_f^i h(x) = 0, \text{ for } 0 \le i \le r - 2 \text{ and } L_g L_f^{r-1} h(x) \ne 0, \quad \forall\, x \in D_0$$

Lie Bracket: Let f and g be two vector fields on $D \subset R^n$. The Lie bracket of f and g, written as $[f, g]$, is a third vector field defined by

$$[f, g](x) = \frac{\partial g}{\partial x} f(x) - \frac{\partial f}{\partial x} g(x)$$

where $[\partial g / \partial x]$ and $[\partial f / \partial x]$ are Jacobian matrices. We may repeat bracketing of g with f. The following notation is used to simplify this process:

$$\begin{aligned} ad_f^0 g(x) &= g(x) \\ ad_f g(x) &= [f, g](x) \\ ad_f^k g(x) &= [f, ad_f^{k-1} g](x) \end{aligned}$$

Example 12.20 Let

$$f(x) = \begin{bmatrix} x_2 \\ -\sin x_1 - x_2 \end{bmatrix}, \quad g = \begin{bmatrix} 0 \\ x_1 \end{bmatrix}$$

Then,

$$\begin{aligned} [f, g](x) &= \begin{bmatrix} 0 & 0 \\ 1 & 0 \end{bmatrix} \begin{bmatrix} x_2 \\ -\sin x_1 - x_2 \end{bmatrix} - \begin{bmatrix} 0 & 1 \\ -\cos x_1 & -1 \end{bmatrix} \begin{bmatrix} 0 \\ x_1 \end{bmatrix} \\ &= \begin{bmatrix} -x_1 \\ x_1 + x_2 \end{bmatrix} \stackrel{\text{def}}{=} ad_f g \end{aligned}$$

$$ad_f^2 g = [f, ad_f g]$$

$$= \begin{bmatrix} -1 & 0 \\ 1 & 1 \end{bmatrix} \begin{bmatrix} x_2 \\ -\sin x_1 - x_2 \end{bmatrix} - \begin{bmatrix} 0 & 1 \\ -\cos x_1 & -1 \end{bmatrix} \begin{bmatrix} -x_1 \\ x_1 + x_2 \end{bmatrix}$$

$$= \begin{bmatrix} -x_1 - 2x_2 \\ x_1 + x_2 - \sin x_1 - x_1 \cos x_1 \end{bmatrix}$$

$\triangle$

Example 12.21 If f and g are constant vector fields, then $[f, g] = 0$. $\triangle$

Example 12.22 If $f(x) = Ax$ and g is a constant vector field, then

$$ad_f g(x) = [f, g](x) = -Ag \quad \text{and} \quad ad_f^2 g = [f, ad_f g] = -A(-Ag) = A^2 g$$

$\triangle$

We leave it to the reader (Exercise 12.25) to verify the following properties of Lie brackets:

- **Bilinear:** Let f_1, f_2, g_1, and g_2 be vector fields and r_1 and r_2 be real numbers. Then,

$$[r_1 f_1 + r_2 f_2, g_1] = r_1[f_1, g_1] + r_2[f_2, g_1]$$
$$[f_1, r_1 g_1 + r_2 g_2] = r_1[f_1, g_1] + r_2[f_1, g_2]$$

- **Skew commutative:** $[f, g] = -[g, f]$

- **Jacobi identity:** If f and g are vector fields and h is a real-valued function, then

$$L_{[f,g]} h(x) = L_f L_g h(x) - L_g L_f h(x)$$

Diffeomorphism: A mapping $T : D \rightarrow R^n$ is a diffeomorphism on D if it is invertible on D; that is, there exists a function $T^{-1}(x)$ such that $T^{-1}(T(x)) = x$ for all $x \in D$, and both $T(x)$ and $T^{-1}(x)$ are continuously differentiable. If the Jacobian matrix $[\partial T/\partial x]$ is nonsingular at a point $x_0 \in D$, then it follows from the inverse function theorem [13] that there is a neighborhood N of x_0 such that T restricted to N is a diffeomorphism on N. T is said to be a global diffeomorphism if it is a diffeomorphism on R^n and $T(R^n) = R^n$. T is a global diffeomorphism if and only if [14]

1. $[\partial T/\partial x]$ is nonsingular for all $x \in R^n$;

2. T is proper;[15] that is, $\lim_{\|x\| \rightarrow \infty} \|T(x)\| = \infty$

---

[13] See [7, Theorem 7-5].

[14] See [188] or [149] for a proof of this statement.

[15] Note that a radially unbounded Lyapunov function is a proper map.

Exercise 12.26 gives a counter example to show that nonsingularity of the Jacobian $[\partial T / \partial x]$ for all $x \in R^n$ is not sufficient to make T a global diffeomorphism.

Distribution: Let $f_1, f_2, \ldots, f_k$ be vector fields on $D \subset R^n$. At any fixed point $x \in D$, $f_1(x), f_2(x), \ldots, f_k(x)$ are vectors in R^n and

$$\Delta(x) = \text{span}\{f_1(x), f_2(x), \ldots, f_k(x)\}$$

is a subspace of R^n. To each point $x \in R^n$, we assign a subspace $\Delta(x)$. We will refer to this assignment by

$$\Delta = \text{span}\{f_1, f_2, \ldots, f_k\}$$

which we call a distribution. In other words, Δ is the collection of all vector spaces $\Delta(x)$ for $x \in D$. Note that the dimension of $\Delta(x)$, that is,

$$\dim(\Delta(x)) = \text{rank } [f_1(x), f_2(x), \ldots, f_k(x)]$$

may vary with x, but if $\Delta = \text{span}\{f_1, \ldots, f_r\}$, where $\{f_1(x), \ldots, f_r(x)\}$ are linearly independent for all $x \in D$, then $\dim(\Delta(x)) = r$ for all $x \in D$. In this case, we say that Δ is a nonsingular distribution on D, generated by $f_1, \ldots, f_r$. It follows from the smoothness of the vector fields that every $g \in \Delta$ can be expressed as

$$g(x) = \sum_{i=1}^{r} c_i(x) f_i(x)$$

where $c_i(x)$ are smooth functions defined on D.

Involutive Distribution: A distribution Δ is involutive if

$$g_1 \in \Delta \quad \text{and} \quad g_2 \in \Delta \quad \Rightarrow \quad [g_1, g_2] \in \Delta$$

If Δ is a nonsingular distribution on D, generated by $f_1, \ldots, f_r$, then it can be verified (Exercise 12.27) that Δ is involutive if and only if

$$[f_i, f_j] \in \Delta, \quad \forall \ 1 \leq i, j \leq r$$

Example 12.23 Let $D = R^3$ and $\Delta = \text{span}\{f_1, f_2\}$, where

$$f_1 = \begin{bmatrix} 2x_2 \\ 1 \\ 0 \end{bmatrix}, \quad f_2 = \begin{bmatrix} 1 \\ 0 \\ x_2 \end{bmatrix}$$

It can be verified that $\dim(\Delta(x)) = 2$ for all $x \in D$. We have

$$[f_1, f_2] = \frac{\partial f_2}{\partial x} f_1 - \frac{\partial f_1}{\partial x} f_2 = \begin{bmatrix} 0 \\ 0 \\ 1 \end{bmatrix}$$

$[f_1, f_2] \in \Delta$ if and only if rank $[f_1(x), f_2(x), [f_1, f_2](x)] = 2$, for all $x \in D$. However,

$$\text{rank } [f_1(x), f_2(x), [f_1, f_2](x)] = \text{rank} \begin{bmatrix} 2x_2 & 1 & 0 \\ 1 & 0 & 0 \\ 0 & x_2 & 1 \end{bmatrix} = 3, \ \forall \ x \in D$$

Hence, Δ is not involutive. $\triangle$

Example 12.24 Let $D = \{x \in R^3 \mid x_1^2 + x_3^2 \neq 0\}$ and $\Delta = \text{span}\{f_1, f_2\}$, where

$$f_1 = \begin{bmatrix} 2x_3 \\ -1 \\ 0 \end{bmatrix}, \ f_2 = \begin{bmatrix} -x_1 \\ -2x_2 \\ x_3 \end{bmatrix}$$

It can be verified that $\dim(\Delta(x)) = 2$ for all $x \in D$. We have

$$[f_1, f_2] = \frac{\partial f_2}{\partial x} f_1 - \frac{\partial f_1}{\partial x} f_2 = \begin{bmatrix} -4x_3 \\ 2 \\ 0 \end{bmatrix}$$

and

$$\text{rank } [f_1(x), f_2(x), [f_1, f_2](x)] = \text{rank} \begin{bmatrix} 2x_3 & -x_1 & -4x_3 \\ -1 & -2x_2 & 2 \\ 0 & x_3 & 0 \end{bmatrix} = 2, \ \forall \ x \in D$$

Hence, $[f_1, f_2] \in \Delta$. Since $[f_2, f_1] = -[f_1, f_2]$, we conclude that Δ is involutive. $\triangle$

Codistribution: Codistributions are dual objects of distributions. They are defined using covector fields and possess properties analogous to those of distributions. One particular codistribution of interest is the annihilator of a distribution Δ, written as $\Delta^{\perp}$ and defined by

$$\Delta^{\perp}(x) = \left\{ w \in (R^n)^* \mid < w, v >= 0, \ \forall \ v \in \Delta(x) \right\}$$

where $(R^n)^*$ is the n-dimensional space of row vectors.

Complete Integrability: Let Δ be a nonsingular distribution on D, generated by $f_1, \ldots, f_r$. Then, Δ is said to be completely integrable if for each $x_0 \in D$, there exists a neighborhood N of x_0 and $n - r$ real-valued smooth functions $h_1(x), \ldots, h_{n-r}(x)$ such that $h_1(x), \ldots, h_{n-r}(x)$ satisfy the partial differential equations

$$\frac{\partial h_j}{\partial x} f_i(x) = 0, \quad \forall \ 1 \leq i \leq r \text{ and } 1 \leq j \leq n - r$$

and the covector fields $dh_j(x)$ are linearly independent for all $x \in D$. Equivalently,

$$\Delta^{\perp} = \text{span}\{dh_1, \ldots, dh_{n-r}\}$$

A key result from differential geometry (for which we have introduced all this terminology) is **Frobenius' theorem**,[16] which states that *a nonsingular distribution is completely integrable if and only if it is involutive.*

Example 12.25 Consider the set of partial differential equations

$$2x_3\frac{\partial\phi}{\partial x_1} - \frac{\partial\phi}{\partial x_2} = 0$$

$$-x_1\frac{\partial\phi}{\partial x_1} - 2x_2\frac{\partial\phi}{\partial x_2} + x_3\frac{\partial\phi}{\partial x_3} = 0$$

To investigate the existence of a smooth function $\phi(x)$ that satisfies these equations, we define the vector fields

$$f_1 = \begin{bmatrix} 2x_3 \\ -1 \\ 0 \end{bmatrix}, \quad f_2 = \begin{bmatrix} -x_1 \\ -2x_2 \\ x_3 \end{bmatrix}$$

so that the foregoing set of partial differential equations can be written as

$$\frac{\partial\phi}{\partial x}f_i(x) = 0, \quad \text{for } i = 1, 2$$

We have seen in Example 12.24 that $\Delta = \text{span}\{f_1, f_2\}$ is nonsingular with $r = 2$, and involutive on $D = \{x \in R^3 \mid x_1^2 + x_3^2 \neq 0\}$. Thus, by Frobenius' theorem, Δ is completely integrable on D. Consequently, for each $x_0 \in D$, there exist a neighborhood N of x_0 and a real-valued smooth function $\phi(x)$ with $d\phi(x) \neq 0$ that satisfies the given set of partial differential equations. Notice that there is only one function ϕ since $n - r = 3 - 2 = 1$. $\triangle$

12.4.2   Input-Output Linearization

Consider the single-input–single-output system

$$\dot{x} = f(x) + g(x)u \qquad (12.46)$$

$$y = h(x) \qquad (12.47)$$

where $f : D \to R^n$, $g : D \to R^n$, and $h : D \to R$ are smooth, $D \subset R^n$ is a domain, and the system has relative degree r on D; that is,

$$L_g L_f^i h(x) = 0, \text{ for } 0 \leq i \leq r - 2 \text{ and } L_g L_f^{r-1}h(x) \neq 0, \ \forall \, x \in D \qquad (12.48)$$

The following two lemmas pertain to the system (12.46)–(12.47).

---

[16]For the proof of Frobenius' theorem, see [80].

Lemma 12.1 *For all $x \in D$ and all integers k and j such that $k \geq 0$ and $0 \leq j \leq r - 1$, we have*

$$L_{ad_f^j g} L_f^k h(x) = \begin{cases} 0, & 0 \leq j + k < r - 1 \\ (-1)^j L_g L_f^{r-1} h(x) \neq 0, & j + k = r - 1 \end{cases} \tag{12.49}$$

$\diamond$

Proof: Prove it by induction of j. For $j = 0$, (12.49) holds by the definition of relative degree. Assume now that (12.49) holds for some j and prove it for $j + 1$. Recall from the Jacobi identity that

$$L_{[f,\beta]} \lambda(x) = L_f L_\beta \lambda(x) - L_\beta L_f \lambda(x)$$

for any real-valued function λ and any vector fields f and β. Taking $\lambda = L_f^k h$ and $\beta = ad_f^j g$, we obtain

$$L_{ad_f^{j+1} g} L_f^k h(x) = L_{[f, ad_f^j g]} L_f^k h(x) = L_f L_{ad_f^j g} L_f^k h(x) - L_{ad_f^j g} L_f^{k+1} h(x)$$

We note that the first term on the right-hand side vanishes since

$$j + k + 1 \leq r - 1 \Rightarrow j + k < r - 1 \Rightarrow L_f L_{ad_f^j g} L_f^k h(x) = 0$$

Moreover,

$$L_{ad_f^j g} L_f^{k+1} h(x) = \begin{cases} 0, & 0 \leq j + k + 1 < r - 1 \\ (-1)^j L_g L_f^{r-1} h(x) \neq 0, & j + k + 1 = r - 1 \end{cases}$$

by the assumption that (12.49) holds for j. Thus

$$L_{ad_f^{j+1} g} L_f^k h(x) = \begin{cases} 0, & 0 \leq j + k + 1 < r - 1 \\ (-1)^{j+1} L_g L_f^{r-1} h(x) \neq 0, & j + k + 1 = r - 1 \end{cases}$$

which completes the proof of the lemma. $\square$

Lemma 12.2 *For all $x \in D$,*

- *the row vectors $dh(x), dL_f h(x), \ldots, dL_f^{r-1} h(x)$ are linearly independent;*
- *the column vectors $g(x), ad_f g(x), \ldots, ad_f^{r-1} g(x)$ are linearly independent.*

◇

Proof: We have

$$
\begin{bmatrix} dh(x) \\ \vdots \\ dL_f^{r-1}h(x) \end{bmatrix}
\begin{bmatrix} g(x) & \cdots & ad_f^{r-1}g(x) \end{bmatrix} =
$$

$$
\begin{bmatrix}
L_g h(x) & L_{ad_f g} h(x) & \cdots & \cdots & L_{ad_f^{r-1} g} h(x) \\
L_g L_f h(x) & & & L_{ad_f^{r-2} g} L_f h(x) & * \\
\vdots & & & & \vdots \\
L_g L_f^{r-1} h(x) & * & \cdots & & *
\end{bmatrix}
$$

From the previous lemma, the right-hand side matrix takes the form

$$
\begin{bmatrix}
0 & \cdots & \cdots & 0 & \diamond \\
0 & & & \diamond & * \\
\vdots & & & & \vdots \\
0 & \diamond & & & * \\
\diamond & * & \cdots & & *
\end{bmatrix}
$$

where ◇ denotes a nonzero element. Thus, the matrix is nonsingular, which proves
the lemma, for if any of the two matrices on the left-hand side has rank less than
r, their product must be singular. □

This lemma shows that $r \leq n$. We are now ready to prove the main result on
input-output linearization; namely, a single-input-single-output system of relative
degree $r < n$ can be (locally) transformed into the normal form (12.25)–(12.27).
The result is stated in the following theorem.

Theorem 12.1 *Consider the system (12.46)–(12.47) and suppose it has relative
degree $r < n$ in D. Then, for every $x_0 \in D$, there exist a neighborhood N of x_0 and
smooth functions $\phi_1(x), \ldots, \phi_{n-r}(x)$ such that*

$$L_g \phi_i(x) = 0, \quad \text{for } 1 \leq i \leq n-r, \ \forall \ x \in N$$

and the mapping

$$
T(x) =
\begin{bmatrix}
\phi_1(x) \\
\vdots \\
\phi_{n-r}(x) \\
--- \\
h(x) \\
\vdots \\
L_f^{r-1}h(x)
\end{bmatrix}
$$

restricted to N, is a diffeomorphism on N. ◇

Proof: The distribution $\Delta = \text{span}\{g\}$ is nonsingular, involutive, and has dimension 1.[17] By Frobenius' theorem, Δ is completely integrable. Hence, for every $x_0 \in D$, there exist a neighborhood N_1 of x_0 and $n-1$ smooth functions $\phi_1(x), \ldots, \phi_{n-1}(x)$, with linearly independent differentials such that

$$L_g \phi_i(x) = 0, \quad \text{for } 1 \leq i \leq n-1, \; \forall \; x \in N_1$$

Equivalently,

$$\Delta^\perp = \text{span}\{d\phi_1, \ldots, d\phi_{n-1}\}$$

on N_1. Since

$$L_g L_f^i h(x) = 0, \quad \text{for } 0 \leq i \leq r-2$$

and $dh(x), \ldots, dL_f^{r-2}h(x)$ are linearly independent, we can use $h, \ldots, L_f^{r-2}h$ as part of these $n-1$ functions. In particular, we take them as $\phi_{n-r+1}, \ldots, \phi_{n-1}$. Since $L_g L_f^{r-1}h(x) \neq 0$, we have

$$dL_f^{r-1}h(x_0) \notin \Delta^\perp(x_0) \; \Rightarrow \; \text{rank}\left[\frac{\partial T}{\partial x}(x_0)\right] = n \; \Rightarrow \; \frac{\partial T}{\partial x}(x_0) \text{ is nonsingular}$$

which shows that there is a neighborhood N_2 of x_0 such that $T(x)$, restricted to N_2, is a diffeomorphism on N_2. Taking $N = N_1 \cap N_2$ completes the proof of the theorem. □

12.4.3 Input-State Linearization

Consider the single-input system

$$\dot{x} = f(x) + g(x)u \tag{12.50}$$

where $f : D \to R^n$ and $g : D \to R^n$ are smooth functions on a domain $D \subset R^n$. The following theorem gives necessary and sufficient conditions for (12.50) to be input-state linearizable near a point $x_0 \in D$.

Theorem 12.2 *The system (12.50) is input-state linearizable if and only if there is a domain $D_x \subset D$ such that*

1. *the matrix $\mathcal{G}(x) = [g(x), ad_f g(x), \ldots, ad_f^{n-1}g(x)]$ has rank n for all $x \in D_x$;*

2. *the distribution $\mathcal{D} = \text{span}\{g, ad_f g, \ldots, ad_f^{n-2}g\}$ is involutive in D_x.* ◇

---

[17]Note that any nonsingular distribution of dimension 1 is automatically involutive.

Proof: The system (12.50) is input-state linearizable if and only if there is a real-valued function $h(x)$ such that the system

$$\dot{x} = f(x) + g(x)u$$
$$y = h(x)$$

has relative degree n in D_x; that is, $h(x)$ satisfies

$$L_g L_f^i h(x) = 0, \text{ for } 0 \le i \le n-2 \text{ and } L_g L_f^{n-1} h(x) \ne 0, \ \forall x \in D_x \qquad (12.51)$$

The *only if* part follows from our discussion in Section 12.1, while the *if* part follows from applying the change of variables $z = T(x) = [h(x), L_f h(x), \ldots, L_f^{n-1} h(x)]^T$. Lemma 12.2 confirms that the Jacobian matrix $[\partial T/\partial x](x)$ is nonsingular for all $x \in D_x$. Hence, for each $x_0 \in D_x$, there is a neighborhood N of x_0 such that $T(x)$, restricted to N, is a diffeomorphism. Thus, to prove the theorem we need to show that the existence of $h(x)$ satisfying (12.51) is equivalent to conditions 1 and 2.
Necessity: Suppose there is $h(x)$ satisfying (12.51). Lemma 12.2 shows that rank $\mathcal{G} = n$. Then, $\mathcal{D}$ is nonsingular and has dimension $n-1$. From, (12.49), with $k = 0$ and $r = n$, we have

$$L_g h(x) = L_{ad_f g} h(x) = \cdots = L_{ad_f^{n-2} g} h(x) = 0$$

which can be written as

$$dh(x)[g(x), ad_f g(x), \ldots, ad_f^{n-2} g(x)] = 0$$

This equation implies that $\mathcal{D}^\perp = \text{span}\{dh\}$. Hence, $\mathcal{D}$ is completely integrable and it follows from Frobenius' theorem that $\mathcal{D}$ is involutive.
Sufficiency: Suppose conditions 1 and 2 are satisfied. Then, $\mathcal{D}$ is nonsingular and has dimension $n-1$. By Frobenius' theorem, there exists $h(x)$ satisfying

$$L_g h(x) = L_{ad_f g} h(x) = \cdots = L_{ad_f^{n-2} g} h(x) = 0$$

Using the Jacobi identity, it can be verified that

$$L_g h(x) = L_g L_f h(x) = \cdots = L_g L_f^{n-2} h(x) = 0$$

Furthermore,

$$dh(x)\mathcal{G}(x) = dh(x)[g(x), ad_f g(x), \ldots, ad_f^{n-1} g(x)] = [0, \ldots, 0, L_{ad_f^{n-1} g} h(x)]$$

Since rank $\mathcal{G} = n$ and $dh(x) \ne 0$, it must be true that $L_{ad_f^{n-1} g} h(x) \ne 0$. Using the Jacobi identity, it can be verified that $L_g L_f^{n-1} h(x) \ne 0$, which completes the proof of the theorem. $\square$

Example 12.26 Reconsider the system of Example 12.1:

$$f(x) = \begin{bmatrix} a\sin x_2 \\ -x_1^2 \end{bmatrix}, \quad g = \begin{bmatrix} 0 \\ 1 \end{bmatrix}$$

$$ad_f g = [f, g] = -\frac{\partial f}{\partial x} g = \begin{bmatrix} -a\cos x_2 \\ 0 \end{bmatrix}$$

The matrix

$$\mathcal{G} = [g, ad_f g] = \begin{bmatrix} 0 & -a\cos x_2 \\ 1 & 0 \end{bmatrix}$$

has rank 2 for all x such that $\cos x_2 \neq 0$. The distribution $\mathcal{D} = \text{span}\{g\}$ is involutive. Hence, the conditions of Theorem 12.2 are satisfied in the domain $D_x = \{x \in R^2 \mid \cos x_2 \neq 0\}$. $\triangle$

Example 12.27 Reconsider the system of Example 12.4:

$$f(x) = \begin{bmatrix} -ax_1 \\ -bx_2 + \rho - cx_1 x_3 \\ \theta x_1 x_2 \end{bmatrix}, \quad g = \begin{bmatrix} 1 \\ 0 \\ 0 \end{bmatrix}$$

$$ad_f g = [f, g] = \begin{bmatrix} a \\ cx_3 \\ -\theta x_2 \end{bmatrix}; \quad ad_f^2 g = [f, ad_f g] = \begin{bmatrix} a^2 \\ (a+b)cx_3 \\ (b-a)\theta x_2 - \theta \rho \end{bmatrix}$$

Consider

$$\mathcal{G} = [g, ad_f g, ad_f^2 g] = \begin{bmatrix} 1 & a & a^2 \\ 0 & cx_3 & (a+b)cx_3 \\ 0 & -\theta x_2 & (b-a)\theta x_2 - \theta\rho \end{bmatrix}$$

The determinant of $\mathcal{G}$ is given by

$$\det \mathcal{G} = c\theta(-\rho + 2bx_2)x_3$$

Hence, $\mathcal{G}$ has rank 3 for $x_2 \neq \rho/2b$ and $x_3 \neq 0$. The distribution $\mathcal{D} = \text{span}\{g, ad_f g\}$ is involutive if $[g, ad_f g] \in \mathcal{D}$. We have

$$[g, ad_f g] = \frac{\partial(ad_f g)}{\partial x} g = \begin{bmatrix} 0 & 0 & 0 \\ 0 & 0 & c \\ 0 & -\theta & 0 \end{bmatrix} \begin{bmatrix} 1 \\ 0 \\ 0 \end{bmatrix} = \begin{bmatrix} 0 \\ 0 \\ 0 \end{bmatrix}$$

Hence, $\mathcal{D}$ is involutive and the conditions of Theorem 12.2 are satisfied in the domain

$$D_x = \{x \in R^3 \mid x_2 > \frac{\rho}{2b} \text{ and } x_3 > 0\}$$

which was determined in Example 12.4. $\triangle$

12.5   Exercises

Exercise 12.1 Show that the state equation of the m-link robot of Exercise 1.3 is input-state linearizable.

Exercise 12.2 Show that the state equation of Exercise 10.32 is input-state linearizable.

Exercise 12.3 ([80]) For each of the following systems, show that the system is input-state linearizable. Give the transformation that transforms the system into the form (12.6) and specify the domain over which it is valid.

(1)
$$\begin{aligned}
\dot{x}_1 &= \exp(x_2)u \\
\dot{x}_2 &= x_1 + x_2^2 + \exp(x_2)u \\
\dot{x}_3 &= x_1 - x_2
\end{aligned}$$

(2)
$$\begin{aligned}
\dot{x}_1 &= x_3(1 + x_2) \\
\dot{x}_2 &= x_1 + (1 + x_2)u \\
\dot{x}_3 &= x_2(1 + x_1) - x_3 u
\end{aligned}$$

Hint: In part (1) try $T_1 = T_1(x_3)$ and in part (2) try $T_1 = T_1(x_1)$.

Exercise 12.4 An articulated vehicle (a semitrailer-like vehicle) can be modeled by the state equation

$$\begin{aligned}
\dot{x}_1 &= \tan(x_3) \\
\dot{x}_2 &= -\frac{\tan(x_2)}{a\cos(x_3)} + \frac{1}{b\cos(x_2)\cos(x_3)}\tan(u) \\
\dot{x}_3 &= \frac{\tan(x_2)}{a\cos(x_3)}
\end{aligned}$$

where a and b are positive constants. Show that system is input-state linearizable. Find the domain of validity of the exact linear model.

Exercise 12.5 Consider the system

$$\begin{aligned}
\dot{x}_1 &= -x_1 + x_2 - x_3 \\
\dot{x}_2 &= -x_1 x_3 - x_2 + u \\
\dot{x}_3 &= -x_1 + u
\end{aligned}$$

(a) Is the system input-state linearizable?

(b) If yes, find a feedback control and a change of variables that linearize the state equation.

Exercise 12.6 Verify that the map $z = T(x)$ in Example 12.4 is a diffeomorphism on D_x, and the state equation in the z coordinates is well defined on $D_z = T(D_x)$.

Exercise 12.7 Consider the third-order model of a synchronous generator connected to an infinite bus from Example 12.2. Consider two possible choices of the output:

$$(1) \ \ y = h(x) = x_1; \quad (2) \ \ y = h(x) = x_1 + \gamma x_2, \ \gamma \neq 0$$

In each case, study the relative degree of the system and transform it into the normal form. Specify the region over which the transformation is valid. If there are non-trivial zero dynamics, find whether or not the system is minimum phase.

Exercise 12.8 Consider the system

$$
\begin{aligned}
\dot{x}_1 &= -x_1 + x_2 - x_3 \\
\dot{x}_2 &= -x_1 x_3 - x_2 + u \\
\dot{x}_3 &= -x_1 + u \\
y &= x_3
\end{aligned}
$$

(a) Is the system input-output linearizable?

(b) If yes, transform it into the normal form and specify the region over which the transformation is valid.

(c) Is the system minimum phase?

Exercise 12.9 Consider the inverted pendulum of Exercise 11.6 and let θ be the output. Is the system input-output linearizable? Is it minimum phase?

Exercise 12.10 Consider the system of Example 11.6. Is the system input-output linearizable? Is it minimum phase?

Exercise 12.11 With reference to Example 12.12, consider the partial differential equations (12.33). Suppose $q(x)$ is independent of ζ_m and ξ_n. Show that $\phi_i = \zeta_i$ for $1 \leq i \leq m-1$, and $\phi_m = \zeta_m - \xi_n/q(x)$ satisfy the partial differential equations.

Exercise 12.12 Consider the pendulum of Example 11.2 with the numerical data of Exercise 11.2. Design a stabilizing state feedback control via exact feedback linearization, locating the eigenvalues of the closed-loop system at $-1/2 \pm j\sqrt{3}/2$. Compare the performance of the closed-loop system with that of Exercise 11.2.

Exercise 12.13 Show that the system

$$
\begin{aligned}
\dot{x}_1 &= a(x_2 - x_1), \quad a > 0 \\
\dot{x}_2 &= bx_1 - x_2 - x_1 x_3 + u, \quad b > 0 \\
\dot{x}_3 &= x_1 + x_1 x_2 - 2ax_3
\end{aligned}
$$

is input-state linearizable and design a state feedback control to globally stabilize the origin.

Exercise 12.14 Consider the system

$$\dot{x}_1 = x_2$$
$$\dot{x}_2 = a \sin x_1 - bu \cos x_1$$

where a and b are positive constants.

(a) Show that the system is input-state linearizable.

(b) Using feedback linearization, design a state feedback controller to stabilize the system at $x_1 = \theta$, where $0 \leq \theta < \pi/2$. Can you make this equilibrium point globally asymptotically stable?

Exercise 12.15 Consider the link manipulator of Example 12.3. Suppose the parameters a, b, c, and d are not known exactly, but we know estimates $\hat{a}$, $\hat{b}$, $\hat{c}$, and $\hat{d}$ of them. Design a linearizing state feedback control in terms of $\hat{a}$, $\hat{b}$, $\hat{c}$, and $\hat{d}$ and represent the closed-loop system as a perturbation of a nominal linear system.

Exercise 12.16 Consider the magnetic suspension system of Exercise 11.5.

(a) Show that the system is input-state linearizable.
    Hint: Try $T_1 = T_1(x_1)$.

(b) Using feedback linearization, design a state feedback control law to stabilize the ball at a desired position $y_R > 0$.

(c) Repeat parts (f) and (e) of Exercise 11.5.

(d) Compare the performance of this controller with the one designed in Exercise 11.5.

Exercise 12.17 Consider a special case of the system (12.34)–(12.35) where $f_0(\eta, \xi)$ depends only on ξ_1, and $(A, B) = (A_c, B_c)$ is a controllable canonical form that represents a chain of r integrators. Such system is said to be in a special normal form. Assume that the origin of $\dot{\eta} = f_0(\eta, 0)$ is globally asymptotically stable and there is a radially unbounded Lyapunov function $V_0(\eta)$ such that

$$\frac{\partial V_0}{\partial \eta} f_0(\eta, 0) \leq -W(\eta), \quad \frac{\partial V_0}{\partial \eta} [f_0(\eta, \xi_1) - f_0(\eta, 0)] \leq k|\xi_1|$$

for all (η, ξ_1), where $W(\eta)$ is a positive definite function of η and k is a nonnegative constant.

(a) Show that the control $u = \alpha(x) + \beta(x)v$ and the change of variables

$$z_1 = \xi_1, \ z_2 = \epsilon \xi_2, \ldots, z_r = \epsilon^{r-1}\xi_r, \quad w = \epsilon^r v$$

bring the system into the form

$$\dot{\eta} = f_0(\eta, z_1)$$
$$\epsilon\dot{z} = A_c z + B_c w$$

(b) Let K be chosen such that $A_c + B_c K$ is Hurwitz, and P be the positive definite solution of the Lyapunov equation $P(A_c + B_c K) + (A_c + B_c K)^T P = -I$. Taking $w = Kz$, and using $V(\eta, z) = V_0(\eta) + \sqrt{z^T P z}$ as a Lyapunov function candidate for the closed-loop system, show that, for sufficiently small ϵ, the origin ($\eta = 0$, $z = 0$) is asymptotically stable and the set $\{V(\eta, z) \leq c\}$, with an arbitrary $c > 0$, is included in the region of attraction.

(c) Show that the feedback control achieves semiglobal stabilization; that is, initial states (η_0, ξ_0) in any compact subset of R^n can be included in the region of attraction.

(d) In view of Example 12.15, investigate whether the current controller exhibits a peaking phenomenon, and if so, explain why is it possible to achieve semiglobal stabilization despite the presence of peaking.

Exercise 12.18 Consider the system

$$\dot{x}_1 = x_2 + x_1 x_2 - x_2^2 + u$$
$$\dot{x}_2 = x_1 x_2 - x_2^2 + u$$
$$\dot{x}_3 = x_1 + x_1 x_2 - x_2^2 - (x_3 - x_1)^3 + u$$
$$y = x_1 - x_2$$

(a) Show that the system has a globally defined special normal form.

(b) Show that the origin of the zero dynamics is globally asymptotically stable.

(c) Design a state feedback control law to achieve semiglobal stabilization of the origin.

Hint: See Exercise 12.17.

Exercise 12.19 Consider the system (12.34)–(12.35), where (12.34) is input-to-state stable, together with the feedback control law (12.41). Suppose that, due to model uncertainty, the actual controller will be implementing the feedback control law $u = \hat{\alpha}(x) + \hat{\beta}(x)K\hat{T}_2(x)$.

(a) Show that the closed-loop system can be represented as

$$\dot{\eta} = f_0(\eta, \xi)$$
$$\dot{\xi} = (A + BK)\xi + \delta(\eta, \xi)$$

and give an expression for $\delta(\eta, \xi)$.

(b) Show that if $\|\delta(\eta, \xi)\| \leq \gamma_1 \|\xi\|$ with sufficiently small γ_1, then the origin ($\eta = 0$, $\xi = 0$) is asymptotically stable.

(c) Show that if $\|\delta(\eta, \xi)\| \leq \gamma_1 \|\xi\| + \gamma_2$ with sufficiently small γ_1, then the state (η, ξ) is ultimately bounded by an ultimate bound that is a class $\mathcal{K}$ function of γ_2.

(d) Suppose now that $\|\delta(\eta, \xi)\| \leq \gamma_{11} \|\eta\| + \gamma_{12} \|\xi\|$. Under what additional conditions can you conclude that the origin ($\eta = 0$, $\xi = 0$) will be asymptotically stable for sufficiently small γ_{11} and γ_{12}?

Exercise 12.20 Consider the magnetic suspension system of Exercise 11.5.

(a) Show that, with the ball position y as the output, the system is input-output linearizable.

(b) Using feedback linearization, design a state feedback control law so that the output y asymptotically tracks $y_R(t) = 0.05 + 0.01 \sin t$.

(c) Simulate the closed-loop system.

Exercise 12.21 Consider the field controlled DC motor of Exercise 11.20. Suppose it is desired that the output y asymptotically tracks a time-varying reference signal $y_R(t)$, where both $y_R(t)$ and $\dot{y}_R(t)$ are continuous and bounded for all $t \geq 0$. As in Exercise 11.20, assume that the domain of operation is restricted to $x_1 > \theta_3/2\theta_1$.

(a) Show that the system is input-output linearizable and has relative degree one.

(b) Show that it is minimum phase.

(c) Using feedback linearization, design a state feedback control to achieve the desired tracking.

(d) Study the performance of the system using computer simulation. Take $y_R(t) = 100 + 50 \sin t$. Study the effect of $\pm 20\%$ parameter perturbations from nominal values for all parameters of the system.

Exercise 12.22 Consider the system

$$\begin{aligned} \dot{x}_1 &= x_2 + 2x_1^2 \\ \dot{x}_2 &= x_3 + u \\ \dot{x}_3 &= x_1 - x_3 \\ y &= x_1 \end{aligned}$$

Design a state feedback control law such that the output y asymptotically tracks the reference signal $y_R(t) = \sin t$.

Exercise 12.23 Consider the magnetic suspension system of Exercises 11.5 and 12.16.

(a) Using feedback linearization, design an integral state feedback control law to stabilize the ball at a desired position $y_R > 0$.

(b) Repeat parts (f) and (e) of Exercise 11.5.

(c) Compare the performance of this controller with the one designed in Exercise 12.16.

Exercise 12.24 Consider the field controlled DC motor of Exercise 11.20.

(a) Using feedback linearization and integral control, design a state feedback control law to regulate the speed to a constant reference y_R.

(b) Repeat parts (b) and (c) of Exercise 11.20.

(c) Compare the performance of this controller to the one designed in part (d) of that exercise.

Hint: See parts (a) and (b) of Exercise 12.21.

Exercise 12.25 Verify the bilinear, skew commutative, and Jacobi identity properties of Lie brackets.

Exercise 12.26 ([188]) Let $f(x) = [e^{x_1}, x_2 e^{-x_1}]^T$. Show that $[\partial f/\partial x]$ is nonsingular for all $x \in R^2$, but f is not a global diffeomorphism.

Exercise 12.27 Let Δ be a nonsingular distribution on D, generated by $f_1, \ldots, f_r$. Show that Δ is involutive if and only if $[f_i, f_j] \in \Delta, \ \forall \ 1 \leq i, j \leq r$.

Exercise 12.28 Let

$$f_1(x) = \begin{bmatrix} x_1 \\ 1 \\ 0 \\ x_3 \end{bmatrix}, \quad f_2(x) = \begin{bmatrix} -e^{x_2} \\ 0 \\ 0 \\ 0 \end{bmatrix}$$

$D = R^4$ and $\Delta = \text{span}\{f_1, f_2\}$. Show that Δ is involutive.

Exercise 12.29 Consider the set of partial differential equations

$$\frac{\partial \lambda}{\partial x_1} - \frac{\partial \lambda}{\partial x_3} = 0$$

$$(x_1 + x_2)\frac{\partial \lambda}{\partial x_1} + (2x_1 + x_3)\frac{\partial \lambda}{\partial x_2} - (x_1 + x_2)\frac{\partial \lambda}{\partial x_3} = 0$$

Study the existence of $\lambda(x)$ that satisfies the equations.

Exercise 12.30 Repeat Exercise 12.29 for

$$\frac{\partial \lambda}{\partial x_1} - \frac{\partial \lambda}{\partial x_3} = 0$$

$$(x_1 + x_2 + x_3^2)\frac{\partial \lambda}{\partial x_1} + (2x_1 + x_3)\frac{\partial \lambda}{\partial x_2} - (x_1 + x_2)\frac{\partial \lambda}{\partial x_3} = 0$$

Exercise 12.31 For each of the following systems, investigate input-state linearizability using Theorem 12.2.

(1) The system of Example 12.2;

(2) The system of Example 12.3;

(3) The system of Exercise 12.4;

(4) The system of Exercise 12.5;

(5) The system of Exercise 12.13;

(6) The system of Exercise 12.16;

(7) The system of Exercise 12.18;

(8) The system of Exercise 11.6.

Chapter 13

Lyapunov-Based Design

Lyapunov's method, introduced originally as an analysis tool, turns out to be a useful tool in feedback design. Many feedback control techniques are based on the idea of designing the feedback control in such a way that a Lyapunov function, or more specifically the derivative of a Lyapunov function, has certain properties that guarantee boundededness of trajectories and convergence to an equilibrium point or an equilibrium set. In this chapter, we present four such methods. We start in Section 13.1 with Lyapunov redesign, a technique that uses a Lyapunov function of a nominal system to design an additional control component to robustify the design to a class of large uncertainties that satisfy the matching condition; that is, the uncertain terms enter the state equation at the same point as the control input. We show how Lyapunov redesign can be used to achieve robust stabilization, and to introduce nonlinear damping that guarantees boundedness of trajectories even when no upper bound on the uncertainty is known. The matching condition assumption can be relaxed via the backstepping technique, introduced in Section 13.2. Backstepping is a recursive procedure that interlaces the choice of a Lyapunov function with the design of feedback control. It breaks a design problem for the full system into a sequence of design problems for lower-order (even scalar) systems. By exploiting the extra flexibility that exists with lower order and scalar systems, backstepping can often solve stabilization, tracking, and robust control problems under conditions less restrictive than those encountered in other methods. In Section 13.3, we use backstepping and Lyapunov redesign to introduce sliding mode control, a popular technique for robust control under the matching condition. In sliding mode control, trajectories are forced to reach a sliding manifold in finite time, and stay on the manifold for all future time. Using a lower-order model, the sliding manifold is designed to achieve the desired control objective. Finally, in Section 13.4, we show the use of Lyapunov's method in the design of adaptive control by presenting a direct model reference adaptive controller. The analysis

577

of the adaptive control system makes an interesting application of the invariance theorems of Section 4.3.

13.1   Lyapunov Redesign

13.1.1   Robust Stabilization

Consider the system

$$\dot{x} = f(t, x) + G(t, x)[u + \delta(t, x, u)] \tag{13.1}$$

where $x \in R^n$ is the state and $u \in R^p$ is the control input. The functions f, G, and δ are defined for $(t, x, u) \in [0, \infty) \times D \times R^p$, where $D \subset R^n$ is a domain that contains the origin. We assume that f, G, and δ are piecewise continuous in t and locally Lipschitz in x and u so that with any feedback control $u = \psi(t, x)$, that is piecewise continuous in t and locally Lipschitz in x, the closed-loop system will have a unique solution through every point $(t_0, x_0) \in [0, \infty) \times D$. The functions f and G are known precisely, while the function δ is an unknown function which lumps together various uncertain terms due to model simplification, parameter uncertainty, and so on. The uncertain term δ satisfies an important structural property; namely, it enters the state equation exactly at the point where the control variable enters. We shall refer to this structural property by saying that the uncertain term satisfies the *matching condition*. A nominal model of the system can be taken as

$$\dot{x} = f(t, x) + G(t, x)u \tag{13.2}$$

We proceed to design a stabilizing state feedback controller using this nominal model. Suppose we have succeeded to design a feedback control law $u = \psi(t, x)$ such that the origin of the nominal closed-loop system

$$\dot{x} = f(t, x) + G(t, x)\psi(t, x) \tag{13.3}$$

is uniformly asymptotically stable. Suppose further that we know a Lyapunov function for (13.3); that is, we have a continuously differentiable function $V(t, x)$ that satisfies the inequalities

$$\alpha_1(\|x\|) \leq V(t, x) \leq \alpha_2(\|x\|) \tag{13.4}$$

$$\frac{\partial V}{\partial t} + \frac{\partial V}{\partial x}[f(t, x) + G(t, x)\psi(t, x)] \leq -\alpha_3(\|x\|) \tag{13.5}$$

for all $(t, x) \in [0, \infty) \times D$, where α_1, α_2, and α_3 are class $\mathcal{K}$ functions. Assume that with $u = \psi(t, x) + v$, the uncertain term δ satisfies the inequality

$$\|\delta(t, x, \psi(t, x) + v)\| \leq \rho(t, x) + k\|v\|, \quad 0 \leq k < 1 \tag{13.6}$$

where $\rho : [0, \infty) \times D \to R$ is a nonnegative continuous function. The estimate (13.6) is the only information we need to know about the uncertain term δ. The function ρ is a measure of the size of the uncertainty. It is important to emphasize that we shall not require ρ to be small; we shall only require it to be known. Our goal in this section is to show that with the knowledge of the Lyapunov function V, the function ρ, and the constant k in (13.6), we can design an additional feedback control $v = \gamma(t, x)$ such that the overall control $u = \psi(t, x) + \gamma(t, x)$ stabilizes the actual system (13.1) in the presence of the uncertainty. (Actually, due to a practical reason that will be described later, this control law will stabilize the system with respect to a small neighborhood of the origin.) The design of $\gamma(t, x)$ is called *Lyapunov redesign*.

Before we carry on with the Lyapunov redesign technique, let us illustrate that the feedback linearization problem of the previous chapter fits into the framework of the current problem.

Example 13.1 Consider the input-state linearizable system

$$\dot{x} = f(x) + G(x)u$$

where $f : D_x \to R^n$ and $G : D_x \to R^{n \times p}$ are smooth functions on a domain $D_x \subset R^n$ and there is a diffeomorphism $T : D_x \to R^n$ such that $D_z = T(D_x)$ contains the origin and $T(x)$ satisfies the partial differential equations (12.10)–(12.11)

$$\frac{\partial T}{\partial x} f(x) = AT(x) - B\beta^{-1}(x)\alpha(x)$$
$$\frac{\partial T}{\partial x} G(x) = B\beta^{-1}(x)$$

Consider also the perturbed system

$$\dot{x} = f(x) + \Delta_f(x) + [G(x) + \Delta_G(x)]u$$

with smooth perturbations that satisfy the matching conditions

$$\Delta_f(x) = G(x)\Delta_1(x), \quad \Delta_G(x) = G(x)\Delta_2(x)$$

on D_x.[1] The perturbed system can be represented in the form (13.1); that is,

$$\dot{z} = f_1(z) + G_1(z)[u + \delta(z, u)]$$

---

[1] It can be easily seen that the perturbed system is still input-state linearizable with the same diffeomorphism $T(x)$, provided $I + \Delta_2$ is nonsingular. Condition (13.7) implies that $I + \Delta_2$ is nonsingular.

where $z = T(x)$, $f_1(z) = Az - B\beta_0^{-1}(z)\alpha_0(z)$, $G_1(z) = B\beta_0^{-1}(z)$, $\delta(z, u) = \delta_1(z) + \delta_2(z)u$, and $\alpha_0(z)$, $\beta_0(z)$, $\delta_1(z)$, and $\delta_2(z)$ are, respectively, $\alpha(x)$, $\beta(x)$, $\Delta_1(x)$, and $\Delta_2(x)$, evaluated at $x = T^{-1}(z)$. Since the nominal system is input-state linearizable, we can take the nominal stabilizing feedback control as

$$\psi(z) = \alpha_0(z) + \beta_0(z)Kz$$

where K is chosen such that $(A + BK)$ is Hurwitz. A Lyapunov function for the nominal closed-loop system

$$\dot{z} = (A + BK)z$$

can be taken as $V(z) = z^T P z$, where P is the solution of the Lyapunov equation

$$P(A + BK) + (A + BK)^T P = -I$$

With $u = \psi(z) + v$, the uncertain term $\delta(z, u)$ satisfies the inequality

$$\|\delta(z, \psi(z) + v)\| \le \|\delta_1(z) + \delta_2(z)\alpha_0(z) + \delta_2(z)\beta_0(z)Kz\| + \|\delta_2(z)\| \, \|v\|$$

Thus, to satisfy (13.6) we need the inequalities

$$\|\delta_2(z)\| \le k < 1 \qquad (13.7)$$

and

$$\|\delta_1(z) + \delta_2(z)\alpha_0(z) + \delta_2(z)\beta_0(z)Kz\| \le \rho(z) \qquad (13.8)$$

to hold over a domain that contains the origin for some continuous function $\rho(z)$. Inequality (13.7) is restrictive because it puts a definite limit on the perturbation δ_2. Inequality (13.8), on the other hand, is not restrictive because we do not require ρ to be small. It is basically a choice of a function ρ to estimate the growth of the left-hand side of (13.8). $\triangle$

Consider now the system (13.1) and apply the control $u = \psi(t, x) + v$. The closed-loop system

$$\dot{x} = f(t, x) + G(t, x)\psi(t, x) + G(t, x)[v + \delta(t, x, \psi(t, x) + v)] \qquad (13.9)$$

is a perturbation of the nominal closed-loop system (13.3). Let us calculate the derivative of $V(t, x)$ along the trajectories of (13.9). For convenience, we will not write the argument of the various functions.

$$\dot{V} = \frac{\partial V}{\partial t} + \frac{\partial V}{\partial x}(f + G\psi) + \frac{\partial V}{\partial x}G(v + \delta) \le -\alpha_3(\|x\|) + \frac{\partial V}{\partial x}G(v + \delta)$$

Set $w^T = [\partial V/\partial x]G$ and rewrite the last inequality as

$$\dot{V} \le -\alpha_3(\|x\|) + w^T v + w^T \delta$$

The first term on the right-hand side is due to the nominal closed-loop system. The second and third terms represent, respectively, the effect of the control v and the uncertain term δ on $\dot{V}$. Due to the matching condition, the uncertain term δ appears on the right-hand side exactly at the same point where v appears. Consequently, it is possible to choose v to cancel the (destabilizing) effect of δ on $\dot{V}$. We shall now explore two different methods for choosing v so that $w^T v + w^T \delta \leq 0$. Suppose that inequality (13.6) is satisfied with $\| \cdot \|_2$; that is,

$$\|\delta(t, x, \psi(t, x) + v)\|_2 \leq \rho(t, x) + k\|v\|_2, \quad 0 \leq k < 1$$

We have

$$w^T v + w^T \delta \leq w^T v + \|w\|_2 \, \|\delta\|_2 \leq w^T v + \|w\|_2[\rho(t, x) + k\|v\|_2]$$

Choose

$$v = -\frac{\eta(t, x)}{1 - k} \cdot \frac{w}{\|w\|_2} \tag{13.10}$$

where $\eta(t, x) \geq \rho(t, x)$ for all $(t, x) \in [0, \infty) \times D$. Then,

$$
\begin{aligned}
w^T v + w^T \delta &\leq -\frac{\eta}{1-k}\|w\|_2 + \rho\|w\|_2 + \frac{k\eta}{1-k}\|w\|_2 \\
&= -\eta\left(\frac{1}{1-k} - \frac{k}{1-k}\right)\|w\|_2 + \rho\|w\|_2 \\
&\leq -\rho\|w\|_2 + \rho\|w\|_2 = 0
\end{aligned}
$$

Hence, with the control (13.10), the derivative of $V(t, x)$ along the trajectories of the closed-loop system (13.9) is negative definite.

As an alternative idea, suppose that inequality (13.6) is satisfied with $\| \cdot \|_\infty$; that is,

$$\|\delta(t, x, \psi(t, x) + v)\|_\infty \leq \rho(t, x) + k\|v\|_\infty, \quad 0 \leq k < 1$$

We have

$$w^T v + w^T \delta \leq w^T v + \|w\|_1 \, \|\delta\|_\infty \leq w^T v + \|w\|_1[\rho(t, x) + k\|v\|_\infty]$$

Choose

$$v = -\frac{\eta(t, x)}{1 - k} \, \text{sgn}(w) \tag{13.11}$$

where $\eta(t, x) \geq \rho(t, x)$ for all $(t, x) \in [0, \infty) \times D$ and $\text{sgn}(w)$ is a p-dimensional vector whose ith component is $\text{sgn}(w_i)$. Then,

$$
\begin{aligned}
w^T v + w^T \delta &\leq -\frac{\eta}{1-k}\|w\|_1 + \rho\|w\|_1 + \frac{k\eta}{1-k}\|w\|_1 \\
&= -\eta\left(\frac{1}{1-k} - \frac{k}{1-k}\right)\|w\|_1 + \rho\|w\|_1 \\
&\leq -\rho\|w\|_1 + \rho\|w\|_1 = 0
\end{aligned}
$$

Hence, with the control (13.11), the derivative of $V(t, x)$ along the trajectories of the closed-loop system (13.9) is negative definite. Notice that the control laws (13.10) and (13.11) coincide for single-input systems ($p = 1$).

The control laws (13.10) and (13.11) are discontinuous functions of the state x. This discontinuity causes some theoretical as well as practical problems. Theoretically, we have to change the definition of the control law to avoid division by zero. We also have to examine the question of existence and uniqueness of solutions more carefully since the feedback functions are not locally Lipschitz in x. Practically, implementation of such discontinuous controllers is characterized by the phenomenon of *chattering* where, due to imperfections in switching devices or computational delays, the control has fast switching fluctuations across the switching surface.[2] Instead of trying to work out all these problems, we shall choose the easy and more practical route of approximating the discontinuous control law by a continuous one. The development of such approximation is similar for both control laws. Therefore, we continue with the control (13.10) and leave the parallel development of a continuous approximation of (13.11) to Exercises 13.2 and 13.3.

Consider the feedback control law

$$
v = \begin{cases}
- \frac{\eta(t,x)}{1-k} \cdot \frac{w}{\|w\|_2}, & \text{if } \eta(t, x)\|w\|_2 \geq \epsilon \\[3mm]
- \frac{\eta^2(t,x)}{1-k} \cdot \frac{w}{\epsilon}, & \text{if } \eta(t, x)\|w\|_2 < \epsilon
\end{cases}
\tag{13.12}
$$

With (13.12), the derivative of V along the trajectories of the closed-loop system (13.9) will be negative definite whenever $\eta(t, x)\|w\|_2 \geq \epsilon$. We only need to check $\dot{V}$ when $\eta(t, x)\|w\|_2 < \epsilon$. In this case,

$$
\begin{aligned}
\dot{V} &\leq -\alpha_3(\|x\|_2) + w^T \left[-\frac{\eta^2}{1-k} \cdot \frac{w}{\epsilon} + \delta \right] \\[2mm]
&\leq -\alpha_3(\|x\|_2) - \frac{\eta^2}{(1-k)\epsilon}\|w\|_2^2 + \rho\|w\|_2 + k\|w\|_2\|v\|_2 \\[2mm]
&= -\alpha_3(\|x\|_2) - \frac{\eta^2}{(1-k)\epsilon}\|w\|_2^2 + \rho\|w\|_2 + \frac{k\eta^2}{(1-k)\epsilon}\|w\|_2^2 \\[2mm]
&= -\alpha_3(\|x\|_2) - \frac{\eta^2}{\epsilon}\|w\|_2^2 + \rho\|w\|_2 \leq -\alpha_3(\|x\|_2) - \frac{\eta^2}{\epsilon}\|w\|_2^2 + \eta\|w\|_2
\end{aligned}
$$

The term

$$
- \frac{\eta^2}{\epsilon}\|w\|_2^2 + \eta\|w\|_2
$$

attains a maximum value $\epsilon/4$ at $\eta\|w\|_2 = \epsilon/2$. Therefore,

$$
\dot{V} \leq -\alpha_3(\|x\|_2) + \frac{\epsilon}{4}
$$

---

[2]See Section 13.3 for further discussion of chattering.

whenever $\eta(t,x)\|w\|_2 < \epsilon$. On the other hand, when $\eta(t,x)\|w\|_2 \geq \epsilon$, $\dot{V}$ satisfies

$$\dot{V} \leq -\alpha_3(\|x\|_2) \leq -\alpha_3(\|x\|_2) + \frac{\epsilon}{4}$$

Thus, the inequality

$$\dot{V} \leq -\alpha_3(\|x\|_2) + \frac{\epsilon}{4}$$

is satisfied irrespective of the value of $\eta(t,x)\|w\|_2$. Choose $\epsilon < 2\alpha_3(\alpha_2^{-1}(\alpha_1(r)))$ and set $\mu(\epsilon) = \alpha_3^{-1}(\epsilon/2) < \alpha_2^{-1}(\alpha_1(r))$. Then,

$$\dot{V} \leq -\tfrac{1}{2}\alpha_3(\|x\|_2), \quad \forall\ \mu(\epsilon) \leq \|x\|_2 < r$$

This inequality shows that $\dot{V}$ is negative outside a ball of radius $\mu(\epsilon)$. By Theorem 5.1 and its corollaries, we arrive at the following theorem, which shows that the solutions of the closed-loop system are uniformly ultimately bounded with an ultimate bound $b(\epsilon)$ which is a class $\mathcal{K}$ function of ϵ.

Theorem 13.1 *Consider the system (13.1) and let $D = \{x \in R^n \mid \|x\|_2 < r\}$. Let $\psi(t,x)$ be a stabilizing feedback control for the nominal system (13.1) with a Lyapunov function $V(t,x)$ that satisfies (13.4)–(13.5) in 2-norm for all $t \geq 0$ and all $x \in D$, with some class $\mathcal{K}$ functions $\alpha_1(\cdot)$, $\alpha_2(\cdot)$, and $\alpha_3(\cdot)$ defined on $[0,r)$. Suppose that the uncertain term δ satisfies (13.6) in 2-norm for all $t \geq 0$ and all $x \in D$. Let v be given by (13.12) and choose $\epsilon < 2\alpha_3(\alpha_2^{-1}(\alpha_1(r)))$. Then, for any $\|x(t_0)\|_2 < \alpha_2^{-1}(\alpha_1(r))$, there exists a finite time t_1 such that the solution of the closed-loop system (13.9) satisfies*

$$\|x(t)\|_2 \leq \beta(\|x(t_0)\|_2, t - t_0), \quad \forall\ t_0 \leq t < t_1 \tag{13.13}$$

$$\|x(t)\|_2 \leq b(\epsilon), \quad \forall\ t \geq t_1 \tag{13.14}$$

where $\beta(\cdot,\cdot)$ is a class $\mathcal{KL}$ function and $b(\epsilon)$ is a class $\mathcal{K}$ function defined by

$$b(\epsilon) = \alpha_1^{-1}(\alpha_2(\mu(\epsilon))) = \alpha_1^{-1}(\alpha_2(\alpha_3^{-1}(\epsilon/2)))$$

If all the assumptions hold globally and $\alpha_1(\cdot)$ belongs to class $\mathcal{K}_\infty$, then (13.13)–(13.14) hold for any initial state $x(t_0)$. Moreover, if $\alpha_i(r) = k_i r^c$ for some positive constants k_i and c, then $\beta(r,s) = kr\exp(-\gamma s)$ with positive constants k and γ. $\diamond$

Thus, in general, the continuous Lyapunov redesign (13.12) does not stabilize the origin as its discontinuous counterpart (13.10) does. Nevertheless, it guarantees uniform ultimate boundedness of the solutions. The ultimate bound $b(\epsilon)$ is a class $\mathcal{K}$ function of ϵ. Therefore, it can be made arbitrarily small by choosing ϵ small enough. In the limit, as $\epsilon \to 0$, we recover the performance of the discontinuous controller. Notice that there is no analytical reason to require ϵ to be very small. The only analytical restriction on ϵ is the requirement $\epsilon < 2\alpha_3(\alpha_2^{-1}(\alpha_1(r)))$. This requirement is satisfied for any ϵ when the assumptions hold globally and α_i ($i = 1, 2, 3$) are class $\mathcal{K}_\infty$ functions. Of course, from a practical viewpoint, we would like to make ϵ as small as feasible because we would like to drive the state of the system to a neighborhood of the origin which is as small as it could be. Exploiting the smallness of ϵ in the analysis, we can arrive at a sharper result when the uncertainty δ vanishes at the origin. Suppose there is a ball $D_0 = \{x \in R^n \mid \|x\|_2 < r_0\} \subset D$ such that the following inequalities are satisfied for all $x \in D_0$.

$$\alpha_3(\|x\|_2) \geq \phi^2(x) \tag{13.15}$$

$$\eta(t, x) \geq \eta_0 > 0 \tag{13.16}$$

$$\rho(t, x) \leq \rho_1 \phi(x) \tag{13.17}$$

where $\phi : R^n \to R$ is a positive definite function of x. Choosing $\epsilon < b^{-1}(r_0)$ ensures that the trajectories of the closed-loop systems will be confined to D_0 after a finite time. When $\eta(t, x)\|w\|_2 < \epsilon$, the derivative $\dot{V}$ satisfies

$$
\begin{aligned}
\dot{V} &\leq -\alpha_3(\|x\|_2) - \frac{\eta^2}{\epsilon}\|w\|_2^2 + \rho\|w\|_2 \\
&\leq -\phi^2(x) - \frac{\eta_0^2}{\epsilon}\|w\|_2^2 + \rho_1\phi(x)\|w\|_2 \\
&\leq -\frac{1}{2}\phi^2(x) - \frac{1}{2}\begin{bmatrix} \phi(x) \\ \|w\|_2 \end{bmatrix}^T \begin{bmatrix} 1 & -\rho_1 \\ -\rho_1 & 2\eta_0^2/\epsilon \end{bmatrix} \begin{bmatrix} \phi(x) \\ \|w\|_2 \end{bmatrix}
\end{aligned}
$$

The matrix of the quadratic form will be positive definite if $\epsilon < 2\eta_0^2/\rho_1^2$. Thus, choosing $\epsilon < 2\eta_0^2/\rho_1^2$ we have $\dot{V} \leq -\frac{1}{2}\phi^2(x)$. Since $\dot{V} \leq -\phi^2(x) \leq -\frac{1}{2}\phi^2(x)$ when $\eta(t, x)\|w\|_2 \geq \epsilon$, we conclude that

$$\dot{V} \leq -\frac{1}{2}\phi^2(x)$$

which shows that the origin is uniformly asymptotically stable.

Corollary 13.1 *Assume inequalities (13.15)–(13.17) are satisfied, in addition to all the assumptions of Theorem 13.1. Then, for all $\epsilon < \min\{2\eta_0^2/\rho_1^2, \ b^{-1}(r_0)\}$, the origin of the closed-loop system (13.9) is uniformly asymptotically stable. If $\alpha_i(r) = k_i r^c$, then the origin is exponentially stable.* $\diamond$

Corollary 13.1 is particularly useful when the origin of the nominal closed-loop system (13.2) is exponentially stable and the perturbation $\delta(t, x, u)$ is Lipschitz in x and u and vanishes at ($x = 0$, $u = 0$). In this case, $\phi(x)$ is proportional to $\|x\|_2$ and the uncertain term satisfies (13.6) with $\rho(x)$ that satisfies (13.17). In general, the condition (13.17) may require more than just a vanishing perturbation at the origin. For example if, in a scalar case, $\phi(x) = |x|^3$, then a perturbation term x cannot be bounded by $\rho_1 \phi(x)$.

The stabilization result of Corollary 13.1 is dependent on the choice of η to satisfy (13.16). It can be shown (Exercise 13.5) that if η does not satisfy (13.16), the feedback control might not be stabilizing. When η satisfies (13.16), the feedback control law (13.12) acts in the region $\eta \|w\|_2 < \epsilon$ as a high-gain feedback control $v = -\gamma w$ with $\gamma \geq \eta_0^2/\epsilon(1 - k)$. Such high-gain feedback control law can stabilize the origin when (13.15)–(13.17) are satisfied (Exercise 13.6).

Example 13.2 Let us continue Example 13.1 on feedback linearizable systems. Suppose inequality (13.7) is satisfied in $\| \cdot \|_2$. Suppose further that

$$\|\delta_1(z) + \delta_2(z)\alpha_0(z) + \delta_2(z)\beta_0(z)Kz\|_2 \leq \rho_1 \|z\|_2$$

for all $z \in B_r \subset D_z$. Then, (13.8) is satisfied with $\rho(z) = \rho_1 \|z\|_2$. We take the control v as in (13.12) with $\eta(z) = 1 + \rho_1 \|z\|_2$, $w^T = 2z^T PB$, and $\epsilon < \min\{2/\rho_1^2,\ 2r^2 \lambda_{\min}(P)/\lambda_{\max}(P)\}$. It can be verified that all the assumptions of Theorem 13.1 and Corollary 13.1 are satisfied with $\alpha_1(r) = \lambda_{\min}(P)r^2$, $\alpha_2(r) = \lambda_{\max}(P)r^2$, $\alpha_3(r) = r^2$, $\phi(z) = \|z\|_2$, and $r_0 = r$. Thus, with the overall feedback control $u = \psi(z) + v$, the origin of the perturbed closed-loop system is exponentially stable. If all the assumptions hold globally, the origin will be globally exponentially stable. $\triangle$

Example 13.3 Reconsider the pendulum equation of Example 12.13, with $\delta_1 = \pi$,

$$\dot{x}_1 = x_2$$
$$\dot{x}_2 = a \sin x_1 - bx_2 + cu$$

We want to stabilize the pendulum at the open-loop equilibrium point $x = 0$. This system is input-state linearizable with $T(x) = x$. A nominal stabilizing feedback control can be taken as

$$\psi(x) = -\left(\frac{\hat{a}}{\hat{c}}\right)\sin x_1 + \left(\frac{1}{\hat{c}}\right)(k_1 x_1 + k_2 x_2)$$

where $\hat{a}$ and $\hat{c}$ are the nominal values of a and c, respectively, and k_1 and k_2 are chosen such that

$$A + BK = \begin{bmatrix} 0 & 1 \\ k_1 & k_2 - b \end{bmatrix}$$

is Hurwitz. With $u = \psi(x) + v$, the uncertainty term δ is given by

$$\delta = \frac{1}{\hat{c}}\left[\left(\frac{a\hat{c} - \hat{a}c}{\hat{c}}\right)\sin x_1 + \left(\frac{c - \hat{c}}{\hat{c}}\right)(k_1 x_1 + k_2 x_2)\right] + \left(\frac{c - \hat{c}}{\hat{c}}\right)v$$

Hence,

$$|\delta| \le \rho_1 \|x\|_2 + k|v|, \quad \forall\, x \in R^2 \;\; \forall\, v \in R$$

where

$$k = \left|\frac{c - \hat{c}}{\hat{c}}\right|, \quad \rho_1 = \frac{\gamma_1}{\hat{c}}, \quad \gamma_1 = \left|\frac{\hat{a}c - a\hat{c}}{\hat{c}}\right| + \left|\frac{c - \hat{c}}{\hat{c}}\right|\sqrt{k_1^2 + k_2^2}$$

Assuming $k < 1$ and taking v as in the previous example, the control law $u = \psi(x) + v$ makes the origin globally exponentially stable. In Example 12.13, we analyzed the same system under the control $u = \psi(x)$. Comparison of the results shows exactly the contribution of the additional control component v. In Example 12.13, we were able to show that the control $u = \psi(x)$ stabilizes the system when γ_1 is restricted to satisfy

$$\gamma_1 < \frac{1}{4\sqrt{p_{12}^2 + p_{22}^2}}$$

This restriction has now been completely removed, provided we know γ_1. $\triangle$

Example 13.4 Once again, consider the pendulum equation of the previous example. This time, suppose that the suspension point of the pendulum is subjected to a time-varying, bounded, horizontal acceleration. For simplicity, neglect friction ($b = 0$). The state equation is given by

$$\begin{aligned}
\dot{x}_1 &= x_2 \\
\dot{x}_2 &= a\sin x_1 + cu + h(t)\cos x_1
\end{aligned}$$

where $h(t)$ is the (normalized) horizontal acceleration of the suspension point. We have $|h(t)| \le H$ for all $t \ge 0$. The nominal model and the nominal stabilizing control can be taken as in the previous example (with $b = 0$). The uncertain term δ satisfies

$$|\delta| \le \rho_1 \,\|x\|_2 + k\,|v| + H$$

where ρ_1 and k are the same as in the previous example. This time, we have $\rho(x) = \rho_1\|x\|_2 + H$ which does not vanish at $x = 0$. The choice of $\eta(x)$ in the control law (13.12) must satisfy $\eta(x) \ge \rho_1\|x\|_2 + H$. We take $\eta(x) = \rho_1\|x\|_2 + \eta_0$ with $\eta_0 \ge H$. In the previous example, we arbitrarily set $\eta(0) = 1$. This is the only modification we have to make to accommodate the nonvanishing disturbance term $h(t)\cos x_1$. Since $\rho(0) \ne 0$, Corollary 13.1 does not apply. We can only conclude, by Theorem 13.1, that the solutions of the closed-loop system are uniformly ultimately bounded with an ultimate bound that is proportional to $\sqrt{\epsilon}$. $\triangle$

13.1.2   Nonlinear Damping

Reconsider the system (13.1) with $\delta(t, x, u) = \Gamma(t, x)\delta_0(t, x, u)$; that is,

$$\dot{x} = f(t, x) + G(t, x)[u + \Gamma(t, x)\delta_0(t, x, u)] \tag{13.18}$$

As before, we assume that f and G are known while $\delta_0(t, x, u)$ is an uncertain term. The function $\Gamma(t, x)$ is known. We assume that f, G, Γ, and δ_0 are piecewise continuous in t and locally Lipschitz in x and u for all $(t, x, u) \in [0, \infty) \times R^n \times R^p$. We assume also that δ_0 is uniformly bounded for all (t, x, u). Let $\psi(t, x)$ be a nominal stabilizing feedback control law such that the origin of the nominal closed-loop system (13.3) is globally uniformly asymptotically stable and there is a known Lyapunov function $V(t, x)$ that satisfies (13.4)–(13.5) for all $(t, x) \in [0, \infty) \times R^n$ with class $\mathcal{K}_\infty$ functions α_1, α_2, and α_3. If an upper bound on $\|\delta_0(t, x, u)\|$ is known, we can design the control component v, as before, to ensure robust global stabilization. In this section, we show that even when no upper bound on δ_0 is known, we can design the control component v to ensure boundedness of the trajectories of the closed-loop system. Toward that end, let $u = \psi(t, x) + v$ and recall that the derivative of V along the trajectories of the closed-loop system satisfies

$$\dot{V} = \frac{\partial V}{\partial t} + \frac{\partial V}{\partial x}(f + G\psi) + \frac{\partial V}{\partial x}G(v + \Gamma\delta_0) \leq -\alpha_3(\|x\|) + w^T(v + \Gamma\delta_0)$$

where $w^T = [\partial V/\partial x]G$. Taking

$$v = -\gamma w \|\Gamma(t, x)\|_2^2, \quad \gamma > 0 \qquad (13.19)$$

we obtain

$$\dot{V} \leq -\alpha_3(\|x\|) - \gamma\|w\|_2^2\|\Gamma\|_2^2 + \|w\|_2\|\Gamma\|_2 k_0$$

where k_0 is an (unknown) upper bound on $\|\delta_0\|$. The term

$$-\gamma\|w\|_2^2\|\Gamma\|_2^2 + \|w\|_2\|\Gamma\|_2 k_0$$

attains a maximum value $k_0^2/4\gamma$ at $\|w\|_2\|\Gamma\|_2 = k_0/2\gamma$. Therefore,

$$\dot{V} \leq -\alpha_3(\|x\|_2) + \frac{k_0^2}{4\gamma}$$

Since α_3 is class $\mathcal{K}_\infty$, it is always true that $\dot{V}$ is negative outside some ball. It follows from Theorem 5.1 that for any initial state $x(t_0)$, the solution of the closed-loop system is uniformly bounded. The Lyapunov redesign (13.19) is called *nonlinear damping*. We summarize our conclusion in the following lemma.

Lemma 13.1 *Consider the system* (13.18) *and let* $\psi(t, x)$ *be a stabilizing feedback control for the nominal system* (13.1) *with a Lyapunov function* $V(t, x)$ *that satisfies* (13.4)–(13.5) *for all* $t \geq 0$ *and all* $x \in R^n$, *with some class* $\mathcal{K}_\infty$ *functions* $\alpha_1(\cdot)$, $\alpha_2(\cdot)$, *and* $\alpha_3(\cdot)$. *Suppose the uncertain term* δ_0 *is uniformly bounded for* $(t, x, u) \in [0, \infty) \times R^n \times R^p$. *Let* v *be given by* (13.19) *and take* $u = \psi(t, x) + v$. *Then, for any* $x(t_0) \in R^n$ *the solution of the closed-loop system is uniformly bounded.* $\diamond$

Example 13.5 Consider the scalar system

$$\dot{x} = x^3 + u + x\delta_0(t)$$

where $\delta_0(t)$ is a bounded function of t. With the nominal stabilizing control $\psi(x) = -x^3 - x$, the Lyapunov function $V(x) = x^2$ satisfies (13.4)–(13.5) globally with $\alpha_1(r) = \alpha_2(r) = \alpha_3(r) = r^2$. The nonlinear damping component (13.19), with $\gamma = 1$, is given by $v = -2x^3$. The closed-loop system

$$\dot{x} = -x - 2x^3 + x\delta_0(t)$$

has a bounded solution no matter how large the bounded disturbance δ_0 is, thanks to the nonlinear damping term $-2x^3$. $\triangle$

13.2 Backstepping

We start with the special case of *integrator backstepping*. Consider the system

$$\dot{\eta} = f(\eta) + g(\eta)\xi \qquad (13.20)$$
$$\dot{\xi} = u \qquad (13.21)$$

where $[\eta^T, \xi]^T \in R^{n+1}$ is the state and $u \in R$ is the control input. The functions $f : D \to R^n$ and $g : D \to R^n$ are smooth[3] in a domain $D \subset R^n$ that contains $\eta = 0$ and $f(0) = 0$. We want to design a state feedback control to stabilize the origin ($\eta = 0$, $\xi = 0$). We assume that both f and g are known. This system can be viewed as a cascade connection of two components, as shown in Figure 13.1(a); the first component is (13.20), with ξ as input, and the second component is the integrator (13.21). Suppose the component (13.20) can be stabilized by a smooth state feedback control $\xi = \phi(\eta)$, with $\phi(0) = 0$; that is, the origin of

$$\dot{\eta} = f(\eta) + g(\eta)\phi(\eta)$$

is asymptotically stable. Suppose further that we know a (smooth, positive definite) Lyapunov function $V(\eta)$ that satisfies the inequality

$$\frac{\partial V}{\partial \eta}[f(\eta) + g(\eta)\phi(\eta)] \leq -W(\eta), \quad \forall \, \eta \in D \qquad (13.22)$$

where $W(\eta)$ is positive definite. By adding and subtracting $g(\eta)\phi(\eta)$ on the right-hand side of (13.20), we obtain the equivalent representation

$$\dot{\eta} = [f(\eta) + g(\eta)\phi(\eta)] + g(\eta)[\xi - \phi(\eta)]$$
$$\dot{\xi} = u$$

---

[3] We require smoothness of all functions for convenience. It will become clear, however, that in a particular problem we only need existence of derivatives up to a certain order.

which is shown in Figure 13.1(b). The change of variables

$$z = \xi - \phi(\eta)$$

results in the system

$$\dot{\eta} = [f(\eta) + g(\eta)\phi(\eta)] + g(\eta)z$$
$$\dot{z} = u - \dot{\phi}$$

that is shown in Figure 13.1(c). Going from Figure 13.1(b) to Figure 13.1(c) can be viewed as "backstepping" $-\phi(\eta)$ through the integrator. Since f, g, and ϕ are known, the derivative $\dot{\phi}$ can be computed using the expression

$$\dot{\phi} = \frac{\partial \phi}{\partial \eta}[f(\eta) + g(\eta)\xi]$$

Taking $v = u - \dot{\phi}$ reduces the system to the cascade connection

$$\dot{\eta} = [f(\eta) + g(\eta)\phi(\eta)] + g(\eta)z$$
$$\dot{z} = v$$

which is similar to the system we started from, except that now the first component has an asymptotically stable origin when the input is zero. This feature will be exploited in the design of v to stabilize the overall system. Using

$$V_a(\eta, \xi) = V(\eta) + \tfrac{1}{2}z^2$$

as a Lyapunov function candidate, we obtain

$$\dot{V}_a = \frac{\partial V}{\partial \eta}[f(\eta) + g(\eta)\phi(\eta)] + \frac{\partial V}{\partial \eta}g(\eta)z + zv \leq -W(\eta) + \frac{\partial V}{\partial \eta}g(\eta)z + zv$$

Choosing

$$v = -\frac{\partial V}{\partial \eta}g(\eta) - kz, \quad k > 0$$

yields

$$\dot{V}_a \leq -W(\eta) - kz^2$$

which shows that the origin ($\eta = 0$, $z = 0$) is asymptotically stable. Since $\phi(0) = 0$, we conclude that the origin ($\eta = 0$, $\xi = 0$) is asymptotically stable. Substituting for v, z, and $\dot{\phi}$, we obtain the state feedback control law

$$u = \frac{\partial \phi}{\partial \eta}[f(\eta) + g(\eta)\xi] - \frac{\partial V}{\partial \eta}g(\eta) - k[\xi - \phi(\eta)] \tag{13.23}$$

If all the assumptions hold globally and $V(\eta)$ is radially unbounded, we can conclude that the origin is globally asymptotically stable. We summarize our conclusions in the following lemma.

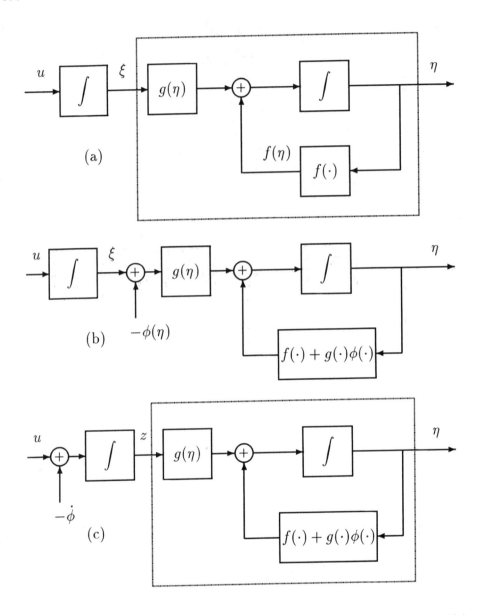

Figure 13.1: (a) The block diagram of system (13.20)–(13.21); (b) introducing $\phi(\eta)$; (c) "backstepping" $-\phi(\eta)$ through the integrator.

Lemma 13.2 *Consider the system* (13.20)–(13.21). *Let* $\phi(\eta)$ *be a stabilizing state feedback control for* (13.20) *with* $\phi(0) = 0$, *and* $V(\eta)$ *be a Lyapunov function that satisfies* (13.22) *with some positive definite function* $W(\eta)$. *Then, the state feedback control* (13.23) *stabilizes the origin of* (13.20)–(13.21), *with* $V(\eta) + \frac{1}{2}[\xi - \phi(\eta)]^2$ *as a Lyapunov function. Moreover, if all the assumptions hold globally and* $V(\eta)$ *is radially unbounded, the origin will be globally asymptotically stable.* ◇

Example 13.6 Consider the system

$$\dot{x}_1 = x_1^2 - x_1^3 + x_2$$
$$\dot{x}_2 = u$$

which takes the form (13.20)–(13.21) with $\eta = x_1$ and $\xi = x_2$. We start with the scalar system

$$\dot{x}_1 = x_1^2 - x_1^3 + x_2$$

where x_2 is viewed as input, and proceed to design feedback control $x_2 = \phi(x_1)$ to stabilize the origin $x_1 = 0$. With

$$x_2 = \phi(x_1) = -x_1^2 - x_1$$

we cancel the nonlinear term x_1^2 and obtain [4]

$$\dot{x}_1 = -x_1 - x_1^3$$

and $V(x_1) = \frac{1}{2}x_1^2$ satisfies

$$\dot{V} = -x_1^2 - x_1^4 \leq -x_1^2, \quad \forall \ x_1 \in R$$

Hence, by Lemma 13.2, the control

$$
\begin{aligned}
u &= \frac{\partial \phi}{\partial x_1}(x_1^2 - x_1^3 + x_2) - \frac{\partial V}{\partial x_1} - [x_2 - \phi(x_1)] \\
&= -(2x_1 + 1)(x_1^2 - x_1^3 + x_2) - x_1 - (x_2 + x_1^2 + x_1)
\end{aligned}
$$

stabilizes the origin $x = 0$ globally, and a composite Lyapunov function is given by

$$V_a(x) = \frac{1}{2}x_1^2 + \frac{1}{2}(x_2 + x_1^2 + x_1)^2$$

△

Application of integrator backstepping in the preceding example is straightforward due to the simplicity of scalar designs. For higher-order systems, we may retain this simplicity via recursive application of integrator backstepping, as illustrated by the following example.

---

[4] We do not cancel $-x_1^3$ because it provides nonlinear damping. See Example 12.16.

Example 13.7 The third-order system

$$
\begin{aligned}
\dot{x}_1 &= x_1^2 - x_1^3 + x_2 \\
\dot{x}_2 &= x_3 \\
\dot{x}_3 &= u
\end{aligned}
$$

is composed of the second-order system of the previous example with an additional integrator at the input side. We proceed to apply integrator backstepping as in the previous example. After one step of backstepping, we know that the second-order system

$$
\begin{aligned}
\dot{x}_1 &= x_1^2 - x_1^3 + x_2 \\
\dot{x}_2 &= x_3
\end{aligned}
$$

with x_3 as input, can be globally stabilized by the control

$$
x_3 = -(2x_1 + 1)(x_1^2 - x_1^3 + x_2) - x_1 - (x_2 + x_1^2 + x_1) \stackrel{\text{def}}{=} \phi(x_1, x_2)
$$

and

$$
V(x_1, x_2) = \tfrac{1}{2}x_1^2 + \tfrac{1}{2}(x_2 + x_1^2 + x_1)^2
$$

is the corresponding Lyapunov function. By viewing the third-order system as a special case of (13.20)–(13.21) with

$$
\eta = \begin{bmatrix} x_1 \\ x_2 \end{bmatrix}; \quad \xi = x_3; \quad f = \begin{bmatrix} x_1^2 - x_1^3 + x_2 \\ 0 \end{bmatrix}; \quad g = \begin{bmatrix} 0 \\ 1 \end{bmatrix}
$$

we can apply Lemma 13.2 to obtain the globally stabilizing state feedback control

$$
u = \frac{\partial \phi}{\partial x_1}(x_1^2 - x_1^3 + x_2) + \frac{\partial \phi}{\partial x_2}x_3 - \frac{\partial V}{\partial x_2} - [x_3 - \phi(x_1, x_2)]
$$

and the corresponding Lyapunov function

$$
V_a(x) = V(x_1, x_2) + \tfrac{1}{2}[x_3 - \phi(x_1, x_2)]^2
$$

$$\triangle$$

Let us move now from (13.20)–(13.21) to the more general system

$$
\begin{aligned}
\dot{\eta} &= f(\eta) + g(\eta)\xi & (13.24) \\
\dot{\xi} &= f_a(\eta, \xi) + g_a(\eta, \xi)u & (13.25)
\end{aligned}
$$

where f_a and g_a are smooth. If $g_a(\eta, \xi) \neq 0$ over the domain of interest, then the input transformation

$$
u = \frac{1}{g_a(\eta, \xi)}[u_a - f_a(\eta, \xi)] \tag{13.26}
$$

reduces (13.25) to the integrator form $\dot{\xi} = u_a$. Therefore, if there are a stabilizing state feedback control $\phi(\eta)$ and a Lyapunov function $V(\eta)$ that satisfy the conditions of Lemma 13.2 for (13.24), then the lemma, together with (13.26), yields

$$u = \phi_a(\eta, \xi) = \frac{1}{g_a(\eta, \xi)} \left\{ \frac{\partial \phi}{\partial \eta}[f(\eta) + g(\eta)\xi] - \frac{\partial V}{\partial \eta}g(\eta) - k[\xi - \phi(\eta)] - f_a(\eta, \xi) \right\}$$

(13.27)

for some $k > 0$, and

$$V_a(\eta, \xi) = V(\eta) + \tfrac{1}{2}[\xi - \phi(\eta)]^2 \tag{13.28}$$

as a stabilizing state feedback control and a Lyapunov function, respectively, for the overall system (13.24)–(13.25). By recursive application of backstepping, we can stabilize *strict-feedback* systems of the form

$$
\begin{aligned}
\dot{x} &= f_0(x) + g_0(x)z_1 \\
\dot{z}_1 &= f_1(x, z_1) + g_1(x, z_1)z_2 \\
\dot{z}_2 &= f_2(x, z_1, z_2) + g_2(x, z_1, z_2)z_3 \\
&\vdots \\
\dot{z}_{k-1} &= f_{k-1}(x, z_1, \ldots, z_{k-1}) + g_{k-1}(x, z_1, \ldots, z_{k-1})z_k \\
\dot{z}_k &= f_k(x, z_1, \ldots, z_k) + g_k(x, z_1, \ldots, z_k)u
\end{aligned}
$$

where $x \in R^n$, z_1 to z_k are scalars, and f_0 to f_k vanish at the origin. The reason for referring to such systems as "strict feedback" is that the nonlinearities f_i and g_i in the $\dot{z}_i$-equation $(i = 1, \ldots, k)$ depend only on $x, z_1, \ldots, z_i$; that is, on the state variables that are "fed back." We assume that

$$g_i(x, z_1, \ldots, z_i) \neq 0, \quad \text{for } 1 \leq i \leq k$$

over the domain of interest. The recursive procedure starts with the system

$$\dot{x} = f_0(x) + g_0(x)z_1$$

where z_1 is viewed as the control input. We assume that it is possible to determine a stabilizing state feedback control $z_1 = \phi_0(x)$ with $\phi_0(0) = 0$, and a Lyapunov function $V_0(x)$ such that

$$\frac{\partial V_0}{\partial x}[f_0(x) + g_0(x)\phi_0(x)] \leq -W(x)$$

over the domain of interest for some positive definite function $W(x)$. In many applications of backstepping, the variable x is scalar, which simplifies this stabilization problem. With $\phi_0(x)$ and $V_0(x)$ in hand, we proceed to apply backstepping in a systematic way. First, we consider the system

$$
\begin{aligned}
\dot{x} &= f_0(x) + g_0(x)z_1 \\
\dot{z}_1 &= f_1(x, z_1) + g_1(x, z_1)z_2
\end{aligned}
$$

as a special case of (13.24)–(13.25) with

$$
\eta = x, \quad \xi = z_1, \quad u = z_2, \quad f = f_0, \quad g = g_0, \quad f_a = f_1, \quad g_a = g_1
$$

We use (13.27) and (13.28) to obtain the stabilizing state feedback control and the Lyapunov function as

$$
\phi_1(x, z_1) = \frac{1}{g_1}\left[\frac{\partial\phi_0}{\partial x}(f_0 + g_0 z_1) - \frac{\partial V_0}{\partial x}g_0 - k_1(z_1 - \phi) - f_1\right], \quad k_1 > 0
$$

$$
V_1(x, z_1) = V_0(x) + \tfrac{1}{2}[z_1 - \phi(x)]^2
$$

Next, we consider the system

$$
\begin{aligned}
\dot{x} &= f_0(x) + g_0(x)z_1 \\
\dot{z}_1 &= f_1(x, z_1) + g_1(x, z_1)z_2 \\
\dot{z}_2 &= f_2(x, z_1, z_2) + g_2(x, z_1, z_2)z_3
\end{aligned}
$$

as a special case of (13.24)–(13.25) with

$$
\eta = \begin{bmatrix} x \\ z_1 \end{bmatrix}, \quad \xi = z_2, \quad u = z_3, \quad f = \begin{bmatrix} f_0 + g_0 z_1 \\ f_1 \end{bmatrix}, \quad g = \begin{bmatrix} 0 \\ g_1 \end{bmatrix}, \quad f_a = f_2, \quad g_a = g_2
$$

Using (13.27) and (13.28), we obtain the stabilizing state feedback control and the Lyapunov function as

$$
\phi_2(x, z_1, z_2) = \frac{1}{g_2}\left[\frac{\partial\phi_1}{\partial x}(f_0 + g_0 z_1) + \frac{\partial\phi_1}{\partial z_1}(f_1 + g_1 z_2) - \frac{\partial V_1}{\partial z_1}g_1 - k_2(z_2 - \phi_1) - f_2\right]
$$

for some $k_2 > 0$, and

$$
V_2(x, z_1, z_2) = V_1(x, z_1) + \tfrac{1}{2}[z_2 - \phi_2(x, z_1)]^2
$$

This process is repeated k times to obtain the overall stabilizing state feedback control $u = \phi_k(x, z_1, \ldots, z_k)$ and the Lyapunov function $V_k(x, z_1, \ldots, z_k)$.

Example 13.8 Consider a single-input–single output system in the special normal form

$$\dot{x} = f_0(x) + g_0(x)z_1$$
$$\dot{z}_1 = z_2$$
$$\vdots$$
$$\dot{z}_{r-1} = z_r$$
$$\dot{z}_r = [u - \alpha(x, z)]/\beta(x, z)$$
$$y = z_1$$

where $x \in R^{n-r}$, z_1 to z_r are scalars, and $\beta(x, z) \neq 0$ for all (x, z). This representation is a special case of the normal form (12.25)–(12.27) because the $\dot{x}$-equation takes the form $f_0(x) + g_0(x)z_1$ instead of the more general form $f_0(x, z)$. The system is in the strict feedback form. The origin can be globally stabilized by recursive backstepping if we can find a smooth function $\phi_0(x)$ and a smooth radially unbounded Lyapunov function $V_0(x)$ such that

$$\frac{\partial V_0}{\partial x}[f_0(x) + g_0(x)\phi_0(x)] \leq -W(x), \quad \forall \; x \in R^n$$

for some positive definite function $W(x)$. If the system is minimum phase, the origin of $\dot{x} = f_0(x)$ is globally asymptotically stable, and we know a Lyapunov function $V_0(x)$ that satisfies

$$\frac{\partial V_0}{\partial x}f_0(x) \leq -W(x), \quad \forall \; x \in R^n$$

for some positive definite function $W(x)$, then we can simply take $\phi_0(x) = 0$. Otherwise, we have to search for $\phi_0(x)$ and $V_0(x)$. This shows that backstepping allows us to stabilize nonminimum phase systems, provided the stabilization problem for the zero dynamics is solvable.   △

Example 13.9 The second-order system

$$\dot{x} = -x + x^2 z$$
$$\dot{z} = u$$

was considered in Example 12.14, where it was shown that $u = -\gamma z$, with sufficiently large $\gamma > 0$, can achieve semiglobal stabilization. In this example, we achieve global stabilization via backstepping. Starting with the system

$$\dot{x} = -x + x^2 z$$

it is clear that $\phi_0(x) = 0$ and $V_0(x) = \frac{1}{2}x^2$ result in

$$\frac{\partial V_0}{\partial x}[-x + x^2\phi_0(x)] = \frac{\partial V_0}{\partial x}(-x) = -x^2, \quad \forall \; x \in R$$

Using (13.27) and (13.28), we obtain

$$u = -(x^3 + kz), \quad k > 0$$

and

$$V(x, z) = \tfrac{1}{2}(x^2 + z^2)$$

The derivative of V along the trajectories of the closed-loop system is given by

$$\dot{V} = -x^2 - kz^2$$

which shows that the origin is globally exponentially stable.   △

Example 13.10 As a variation from the previous example, consider the system [5]

$$\begin{aligned} \dot{x} &= x^2 - xz \\ \dot{z} &= u \end{aligned}$$

This time,

$$\dot{x} = x^2 - xz$$

cannot be stabilized by $z = \phi_0(x) = 0$. It is easy, however, to see that $\phi_0(x) = x + x^2$
and $V_0(x) = \tfrac{1}{2}x^2$ result in

$$\frac{\partial V_0}{\partial x}[x^2 - x\phi_0(x)] = -x^4, \quad \forall\, x \in R$$

Using (13.27) and (13.28), we obtain

$$u = (1 + 2x)(x^2 - xz) + x^2 - k(z - x - x^2), \quad k > 0$$

and

$$V(x, z) = \tfrac{1}{2}[x^2 + (z - x - x^2)^2]$$

The derivative of V along the trajectories of the closed-loop system is given by

$$\dot{V} = -x^4 - k(z - x - x^2)^2$$

which shows that the origin is globally asymptotically stable.   △

In the previous section, we saw how to use Lyapunov redesign to robustly stabilize an uncertain system when the uncertainty satisfies the matching condition.

---

[5] With the output $y = z$, the system in nonminimum phase since the zero dynamics equation
$\dot{x} = x^2$ has unstable origin.

Backstepping can be used to relax the matching condition assumption. To illustrate the idea, let us consider the single-input system

$$\dot{\eta} = f(\eta) + g(\eta)\xi + \delta_\eta(\eta, \xi) \tag{13.29}$$

$$\dot{\xi} = f_a(\eta, \xi) + g_a(\eta, \xi)u + \delta_\xi(\eta, \xi) \tag{13.30}$$

defined on a domain $D \subset R^{n+1}$ that contains the origin ($\eta = 0$, $\xi = 0$). Suppose $g_a(\eta, \xi) \neq 0$ and all functions are smooth for all $(\eta, \xi) \in D$. Let f, g, f_a, and g_a be known and δ_η and δ_ξ be uncertain terms. We assume that f and f_a vanish at the origin, and the uncertain terms satisfy the inequalities

$$\|\delta_\eta(\eta, \xi)\|_2 \leq \alpha_1 \|\eta\|_2 \tag{13.31}$$

$$|\delta_\xi(\eta, \xi)| \leq \alpha_2 \|\eta\|_2 + \alpha_3 |\xi| \tag{13.32}$$

for all $(\eta, \xi) \in D$. Inequality (13.31) restricts the class of uncertainties because it restricts the upper bound on $\delta_\eta(\eta, \xi)$ to depend only on η. Nevertheless, it is less restrictive than the matching condition which would have required $\delta_\eta = 0$. Starting with the system (13.29), suppose we can find a stabilizing state feedback control $\xi = \phi(\eta)$ with $\phi(0) = 0$, and a (smooth, positive definite) Lyapunov function $V(\eta)$ such that

$$\frac{\partial V}{\partial \eta}[f(\eta) + g(\eta)\phi(\eta) + \delta_\eta(\eta, \xi)] \leq -c\|\eta\|_2^2 \tag{13.33}$$

for all $(\eta, \xi) \in D$ for some positive constant c. Inequality (13.33) shows that $\eta = 0$ is an asymptotically stable equilibrium point of the system

$$\dot{\eta} = f(\eta) + g(\eta)\phi(\eta) + \delta_\eta(\eta, \xi)$$

Suppose further that $\phi(\eta)$ satisfies the inequalities

$$|\phi(\eta)| \leq \alpha_4 \|\eta\|_2, \quad \left\|\frac{\partial \phi}{\partial \eta}\right\|_2 \leq \alpha_5 \tag{13.34}$$

over D. Consider now the Lyapunov function candidate

$$V_a(\eta, \xi) = V(\eta) + \tfrac{1}{2}[\xi - \phi(\eta)]^2 \tag{13.35}$$

The derivative of V_a along the trajectories of (13.29)–(13.30) is given by

$$\begin{aligned}
\dot{V}_a &= \frac{\partial V}{\partial \eta}(f + g\phi + \delta_\eta) + \frac{\partial V}{\partial \eta}g\,(\xi - \phi) \\
&\quad + (\xi - \phi)\left[f_a + g_a u + \delta_\xi - \frac{\partial \phi}{\partial \eta}(f + g\xi + \delta_\eta)\right]
\end{aligned}$$

Taking

$$u = \frac{1}{g_a} \left[\frac{\partial \phi}{\partial \eta}(f + g\xi) - \frac{\partial V}{\partial \eta}g - f_a - k(\xi - \phi) \right], \quad k > 0 \qquad (13.36)$$

and using (13.33), we obtain

$$\dot{V}_a \leq -c\|\eta\|_2^2 + (\xi - \phi)\left[\delta_\xi - \frac{\partial \phi}{\partial \eta}\delta_\eta \right] - k(\xi - \phi)^2$$

Using (13.31), (13.32), and (13.34), it can be shown that

$$\begin{aligned}
\dot{V}_a &\leq -c\|\eta\|_2^2 + 2\alpha_6\|\eta\|_2|\xi - \phi| - (k - \alpha_3)(\xi - \phi)^2 \\
&= -\left[\begin{array}{c} \|\eta\|_2 \\ |\xi - \phi| \end{array} \right]^T \left[\begin{array}{cc} c & -\alpha_6 \\ -\alpha_6 & (k - \alpha_3) \end{array} \right] \left[\begin{array}{c} \|\eta\|_2 \\ |\xi - \phi| \end{array} \right]
\end{aligned}$$

for some $\alpha_6 \geq 0$. Choosing

$$k > \alpha_3 + \frac{\alpha_6^2}{c}$$

yields

$$\dot{V}_a \leq -\sigma[\|\eta\|_2^2 + |\xi - \phi|^2]$$

for some $\sigma > 0$. Thus, we have completed the proof of the following lemma.

Lemma 13.3 *Consider the system* (13.29)–(13.30), *where the uncertainty satisfies inequalities* (13.31) *and* (13.32). *Let* $\phi(\eta)$ *be a stabilizing state feedback control for* (13.29) *that satisfies* (13.35), *and* $V(\eta)$ *be a Lyapunov function that satisfies* (13.34). *Then, the state feedback control* (13.36), *with* k *sufficiently large, stabilizes the origin of* (13.29)–(13.30). *Moreover, if all the assumptions hold globally and* $V(\eta)$ *is radially unbounded, the origin will be globally asymptotically stable.* $\diamond$

As an application of backstepping, consider the robust stabilization of the system

$$\left. \begin{array}{rcl} \dot{x}_i &=& x_{i+1} + \delta_i(x), \quad 1 \leq i \leq n - 1 \\ \dot{x}_n &=& [u - \alpha(x)]/\beta(x) + \delta_n(x) \end{array} \right\} \qquad (13.37)$$

defined on a domain $D \subset R^n$ that contains the origin $x = 0$, where $x = [x_1, \ldots, x_n]^T$. Suppose $\beta(x) \neq 0$ and all functions are smooth for all $x \in D$. Let α and β be known, and δ_i for $1 \leq i \leq n$ be uncertain terms. The nominal system is input-state linearizable. We assume that the uncertain terms satisfy the inequalities

$$|\delta_i(x)| \leq \alpha_i \sum_{k=1}^{i} |x_k|, \quad \text{for } 1 \leq i \leq n \qquad (13.38)$$

for all $x \in D$, where the nonnegative constants α_1 to α_n are known. Inequalities (13.38) restrict the class of uncertainties for $1 \leq i \leq n - 1$ because they restrict the upper bound on $\delta_i(x)$ to depend only on x_1 to x_i; that is, the upper bounds appear in a strict feedback form. Nevertheless, they are less restrictive than the matching condition which would have required $\delta_i = 0$ for $1 \leq i \leq n - 1$. To apply the backstepping recursive design procedure, we start with the scalar system

$$\dot{x}_1 = x_2 + \delta_1(x)$$

where x_2 is viewed as the control input, and $\delta_1(x)$ satisfies the inequality $|\delta_1(x)| \leq \alpha_1|x_1|$. In this scalar system, the uncertainty satisfies the matching condition. The origin $x_1 = 0$ can be robustly stabilized by the high-gain feedback control $x_2 = -k_1 x_1$, where $k_1 > 0$ is chosen sufficiently large. In particular, let $V_1(x_1) = \frac{1}{2}x_1^2$ be a Lyapunov function candidate. Then,

$$\dot{V}_1 = x_1[-k_1 x_1 + \delta_1(x)] \leq -(k_1 - \alpha_1)x_1^2$$

and the origin is stabilized for all $k_1 > \alpha_1$. From this point on, backstepping and Lemma 13.3 are applied recursively to derive the stabilizing state feedback control. The procedure is illustrated by the following example.

Example 13.11 We want to design a state feedback control to stabilize the second-order system

$$\begin{aligned} \dot{x}_1 &= x_2 + \theta_1 x_1 \sin x_2 \\ \dot{x}_2 &= \theta_2 x_2^2 + x_1 + u \end{aligned}$$

where θ_1 and θ_2 are unknown parameters that satisfy

$$|\theta_1| \leq a, \quad |\theta_2| \leq b$$

for some known bounds a and b. The system takes the form (13.37) with $\delta_1 = \theta_1 x_1 \sin x_2$ and $\delta_2 = \theta_2 x_2^2$. The function δ_1 satisfies the inequality $|\delta_1| \leq a|x_1|$ globally. The function δ_2 satisfies the inequality $|\delta_2| \leq b\rho|x_2|$ on the set $|x_2| \leq \rho$. Starting with the system

$$\dot{x}_1 = x_2 + \theta_1 x_1 \sin x_2$$

we take $x_2 = \phi_1(x_1) = -k_1 x_1$ and $V_1(x_1) = \frac{1}{2}x_1^2$ to obtain

$$\dot{V}_1 = x_1 \phi_1(x_1) + \theta_1 x_1^2 \sin x_2 \leq -(k_1 - a)x_1^2$$

We choose $k_1 = 1 + a$. Using (13.35) and (13.36), we take the control u and the Lyapunov function V_a as

$$u = -2x_1 - k_1 x_2 - k(x_2 + k_1 x_1)$$

$$V_a = \frac{1}{2}[x_1^2 + (x_2 + k_1 x_1)^2]$$

The derivative of V_a is given by

$$
\begin{aligned}
\dot{V}_a &= x_1(-k_1 x_1 + \theta_1 x_1 \sin x_2) \\
&\quad + (x_2 + k_1 x_1)(k_1 \theta_1 x_1 \sin x_2 + \theta_2 x_2^2) - k(x_2 + k_1 x_1)^2 \\
&\leq -x_1^2 + k_1 a |x_1|\, |x_2 + k_1 x_1| + b x_2^2 |x_2 + k_1 x_1| - k(x_2 + k_1 x_1)^2
\end{aligned}
$$

On the set

$$
\Omega_c = \{x \in R^2 \mid V_a(x) \leq c\}
$$

we have $|x_2| \leq \rho$ for some ρ dependent on c.[6] Restricting our analysis to Ω_c, we obtain

$$
\begin{aligned}
\dot{V}_a &\leq -x_1^2 + k_1 a |x_1|\, |x_2 + k_1 x_1| \\
&\quad + b\rho|x_2 + k_1 x_1 - k_1 x_1|\, |x_2 + k_1 x_1| - k(x_2 + k_1 x_1)^2 \\
&\leq -x_1^2 + k_1(a + b\rho)|x_1|\, |x_2 + k_1 x_1| - (k - b\rho)(x_2 + k_1 x_1)^2
\end{aligned}
$$

Choosing

$$
k > b\rho + k_1^2(a + b\rho)^2
$$

ensures that the origin is exponentially stable,[7] with Ω_c contained in the region of attraction. Since the preceding inequality can be satisfied for any $c > 0$ by choosing k sufficiently large, the feedback control can achieve semiglobal stabilization. $\triangle$

We conclude the section by noting that backstepping can be applied to multi-input systems in what is known as *block backstepping*, provided certain nonsingularity conditions are satisfied. To illustrate the idea, consider the system

$$
\begin{aligned}
\dot{\eta} &= f(\eta) + G(\eta)\xi & (13.39) \\
\dot{\xi} &= f_a(\eta, \xi) + G_a(\eta, \xi)u & (13.40)
\end{aligned}
$$

where $\eta \in R^n$, $\xi \in R^m$, and $u \in R^m$ where m could be greater than one. Suppose f, f_a, G, and G_a are (known) smooth functions over the domain of interest, f and f_a vanish at the origin, and the $m \times m$ matrix G_a is nonsingular. Suppose further that the component (13.39) can be stabilized by a smooth state feedback control $\xi = \phi(\eta)$ with $\phi(0) = 0$, and we know a (smooth, positive definite) Lyapunov function $V(\eta)$ that satisfies the inequality

$$
\frac{\partial V}{\partial \eta}[f(\eta) + G(\eta)\phi(\eta)] \leq -W(\eta)
$$

---

[6]ρ can be estimated by $\sqrt{2c(1 + k_1^2)}$.

[7]Note that we conclude exponential stability rather than asymptotic stability as guaranteed by Lemma 13.3. Why?

for some positive definite function $W(\eta)$. Using

$$V_a = V(\eta) + \tfrac{1}{2}[\xi - \phi(\eta)]^T[\xi - \phi(\eta)]$$

as a Lyapunov function candidate for the overall system, we obtain

$$\dot{V}_a = \frac{\partial V}{\partial \eta}(f + G\phi) + \frac{\partial V}{\partial \eta}G\ (\xi - \phi) + [\xi - \phi]^T \left[f_a + G_a u - \frac{\partial \phi}{\partial \eta}(f + G\xi) \right]$$

Taking

$$u = G_a^{-1} \left[\frac{\partial \phi}{\partial \eta}(f + G\xi) - \left(\frac{\partial V}{\partial \eta}G \right)^T - f_a - k(\xi - \phi) \right], \quad k > 0$$

results in

$$\dot{V}_a = \frac{\partial V}{\partial \eta}(f + G\phi) - k[\xi - \phi(\eta)]^T[\xi - \phi(\eta)] \leq -W(\eta) - k[\xi - \phi(\eta)]^T[\xi - \phi(\eta)]$$

which shows that the origin ($\eta = 0$, $\xi = 0$) is asymptotically stable.

13.3   Sliding Mode Control

Consider the system

$$\dot{x} = f_0(x) + \delta_1(x) + G(x)[u + \delta_2(x, u)] \tag{13.41}$$

where $x \in R^n$ is the state, $u \in R^p$ is the control input, and f_0, G, δ_1, and δ_2 are sufficiently smooth functions defined for $(x, u) \in D_0 \times R^p$ where D_0 is a domain that contains the origin. We assume that f_0 and G are known, while δ_1 and δ_2 are uncertain terms. We consider both matched δ_2 and unmatched δ_1 uncertainties. Let $T : D_0 \to R^n$ be a diffeomorphism such that

$$\frac{\partial T}{\partial x}G(x) = \left[\begin{array}{c} 0 \\ G_2(x) \end{array} \right] \tag{13.42}$$

where $G_2(x)$ is a $p \times p$ matrix that is nonsingular for all $x \in D_0$.[8] The change of variables

$$\left[\begin{array}{c} \eta \\ \xi \end{array} \right] = T(x), \quad \eta \in R^{n-p}, \ \xi \in R^p$$

transforms the system into the form

$$\dot{\eta} = f(\eta, \xi) + \delta_\eta(\eta, \xi) \tag{13.43}$$

$$\dot{\xi} = f_a(\eta, \xi) + G_a(\eta, \xi)[u + \delta_\xi(\eta, \xi, u)] \tag{13.44}$$

---

[8] The existence of T is explored in Exercise 13.5.

where $G_a(\eta,\xi) = G_2\left(T^{-1}\left(\begin{bmatrix} \eta \\ \xi \end{bmatrix}\right)\right)$ is nonsingular for all $(\eta,\xi) \in D = T(D_0)$.
The form (13.43)–(13.44) is usually referred to as the *regular form*. We note that
the term δ_η in (13.43) is due to the term δ_1 in (13.41), and $\delta_\eta = 0$ when $\delta_1 = 0$.
Therefore, we shall refer to δ_η as the unmatched uncertainty and to δ_ξ as the
matched uncertainty. To introduce sliding mode control, let us consider the robust
stabilization problem of designing a state feedback control to stabilize the origin
($\eta = 0$, $\xi = 0$) in the presence of the uncertainties δ_η and δ_ξ. We assume that f,
f_a, and δ_η vanish at the origin, but we do not require δ_ξ to vanish at the origin.
We approach the design via a backstepping approach. We start with the system

$$\dot{\eta} = f(\eta,\xi) + \delta_\eta(\eta,\xi)$$

and view ξ as the control input. We seek a stabilizing state feedback control $\xi = \phi(\eta)$
with $\phi(0) = 0$, so that the origin of

$$\dot{\eta} = f(\eta,\phi(\eta)) + \delta_\eta(\eta,\phi(\eta)) \tag{13.45}$$

is asymptotically stable. We will not dwell on this stabilization problem. We simply
assume that we can find a stabilizing continuously differentiable function $\phi(\eta)$. Let
us note, however, that this stabilization problem is the typical stabilization problem
we have studied in the previous two chapters, as well as the first two sections of the
current chapter. The tools of linearization, exact feedback linearization, Lyapunov
redesign, backstepping, or a combination of them could be used toward the design
of ϕ. To proceed with the design of the sliding mode control, set

$$z = \xi - \phi(\eta)$$

It is clear that if $z = 0$, we have $\xi = \phi(\eta)$ and the variable η approaches the origin
asymptotically. In the backstepping approach of the previous section, the control u
is designed to stabilize z and η simultaneously through a Lyapunov function for the
overall system. In sliding mode control, we approach the problem differently. We
design u to bring z to zero in finite time and then maintain the condition $z = 0$ for
all future time; that is, we make $z = 0$ a positively invariant set of the closed-loop
system. Toward that end, let us write the $\dot{z}$-equation

$$\dot{z} = f_a(\eta,\xi) + G_a(\eta,\xi)[u + \delta_\xi(\eta,\xi,u)] - \frac{\partial\phi}{\partial\eta}[f(\eta,\xi) + \delta_\eta(\eta,\xi)] \tag{13.46}$$

Take the control u as

$$u = u_{eq} + G_a^{-1}(\eta,\xi)v \tag{13.47}$$

where v will be determined shortly and u_{eq} is chosen to cancel the known terms on
the right-hand side of (13.46); that is,

$$u_{eq} = G_a^{-1}(\eta,\xi)\left[-f_a(\eta,\xi) + \frac{\partial\phi}{\partial\eta}f(\eta,\xi)\right] \tag{13.48}$$

The control component u_{eq} is called the *equivalent control*. In the absence of uncertainty, taking $u = u_{eq}$ results in $\dot{z} = 0$, which ensures that the condition $z = 0$ can be maintained for all future time. Substitution of (13.47) into (13.46) yields

$$\dot{z} = v + \Delta(\eta, \xi, v) \tag{13.49}$$

where

$$\Delta(\eta, \xi, v) = G_a(\eta, \xi)\delta_\xi(\eta, \xi, u_{eq} + G_a^{-1}(\eta, \xi)v) - \frac{\partial \phi}{\partial \eta}\delta_\eta(\eta, \xi)$$

We assume that Δ satisfies the inequality

$$\|\Delta(\eta, \xi, v)\|_\infty \le \rho(\eta, \xi) + k\|v\|_\infty, \quad \forall\, (\eta, \xi, v) \in D \times R^p \tag{13.50}$$

where $\rho(\eta, \xi) \ge 0$ (a continuous function) and $k \in [0, 1)$ are known. Using the estimate (13.50), we proceed to design v to force z toward the manifold $z = 0$. We write (13.49) as p scalar equations

$$\dot{z}_i = v_i + \Delta_i(\eta, \xi, v), \quad 1 \le i \le p \tag{13.51}$$

Using $V_i = \frac{1}{2}z_i^2$ as a Lyapunov function candidate for (13.51), we obtain

$$\dot{V}_i = z_i\dot{z}_i = z_iv_i + z_i\Delta_i(\eta, \xi, v) \le z_iv_i + |z_i|[\rho(\eta, \xi) + k\|v\|_\infty]$$

Take[9]

$$v_i = -\frac{\beta(\eta, \xi)}{1 - k}\,\text{sgn}(z_i), \quad 1 \le i \le p \tag{13.52}$$

where

$$\beta(\eta, \xi) \ge \rho(\eta, \xi) + b, \quad \forall\, (\eta, \xi) \in D$$

for some $b > 0$, and $\text{sgn}(\cdot)$ is the signum nonlinearity. Then,

$$\begin{aligned} \dot{V}_i &\le -\frac{\beta(\eta, \xi)}{1 - k}|z_i| + \rho(\eta, \xi)|z_i| + k\frac{\beta(\eta, \xi)}{1 - k}|z_i| \\ &= -\beta(\eta, \xi)|z_i| + \rho(\eta, \xi)|z_i| \le -b|z_i| \end{aligned}$$

The inequality $\dot{V}_i \le -b|z_i|$ ensures that if the trajectory happens to be on the manifold $z_i = 0$ at some time, it will be confined to that manifold for all future time because leaving the manifold requires $\dot{V}_i$ to be positive, which is impossible in view of the inequality. The inequality ensures also that if $z_i(0) \ne 0$, then[10]

$$|z_i(t)| \le |z_i(0)| - bt, \quad \forall\, t \ge 0$$

---

[9] The control law (13.52) can be viewed as a Lyapunov redesign for the scalar system (13.49); see Exercise 13.7.

[10] This inequality follows by application of the comparison lemma. Using the fact that $|z_i| = \sqrt{2V_i}$, it can be shown that $D^+|z_i| \le -b$. Hence, $|z_i(t)| \le y_i(t)$ where $y_i(t)$ is the solution of the scalar differential equation $\dot{y}_i = -b$ with $y_i(0) = |z_i(0)|$.

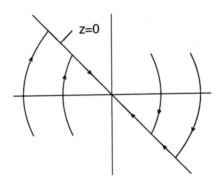

Figure 13.2: A typical phase portrait under sliding mode control.

which shows that trajectories starting off the manifold $z_i = 0$ must reach it in finite time. Thus, a typical motion under sliding mode control consists of a *reaching phase* during which trajectories starting off the manifold $z = 0$ move toward it and reach it in finite time, followed by a *sliding phase* during which the motion will be confined to the manifold $z = 0$ and the dynamics of the system will be represented by the reduced-order model (13.45). A sketch of this typical motion is shown in Figure 13.2. The manifold $z = \xi - \phi(\eta) = 0$ is called the *sliding manifold*.

The procedure for designing a sliding mode controller can be summarized by the following steps:

- Design the sliding manifold $\xi = \phi(\eta)$ to control the motion of the reduced-order system (13.45).

- Estimate $\rho(\eta, \xi)$ and k in (13.50).

- Take the control u as $u = u_{eq} + G_a^{-1}(\eta, \xi)v$, where u_{eq} is the equivalent control given by (13.48) and v is the switching (discontinuous) control given by (13.52).

This procedure exhibits model order reduction because the main design task is performed on the reduced-order system (13.45).

An important feature of sliding mode control is its robustness to uncertainties. To understand this feature, it is important to distinguish between the matched uncertainty δ_ξ and the unmatched uncertainty δ_η. We should also distinguish between the reaching and sliding phases. During the reaching phase, the tasks of forcing trajectories toward the sliding manifold and maintaining them on the manifold, once they are there, are achieved by the switching control (13.52). This task is affected by both the matched and unmatched uncertainties, and robustness is achieved by choosing $\beta(\eta, \xi) \geq \rho(\eta, \xi) + b$, where $\rho(\eta, \xi)$ is a measure of the size of uncertainty as determined by inequality (13.50). Since we do not require $\rho(\eta, \xi)$ to be small, the switching controller can handle fairly large uncertainties, limited only by practical constraints on the amplitude of control signals. During the sliding phase, the motion of the system, as determined by (13.45), is affected only by the unmatched uncertainty δ_η. Robustness to δ_η will have to be achieved through the design of $\phi(\eta)$. In general, robust stabilization at the reduced-order level will require restricting the size or structure of δ_η. In the special case when there is only matched uncertainty, that is, $\delta_\eta = 0$, sliding mode control will guarantee robustness for any nominal manifold design $\xi = \phi(\eta)$ that achieves the control objectives on the sliding manifold.

The sliding mode controller contains the discontinuous nonlinearity $\text{sgn}(z_i)$. As we noted in Section 13.1, discontinuous feedback control raises theoretical and practical issues. Theoretical issues like existence and uniqueness of solutions and validity of Lyapunov analysis will have to be examined in a framework that does not require the state equation to have locally Lipschitz right-hand side functions.[11] There is also the practical issue of chattering due to imperfections in switching devices and delays. The sketch of Figure 13.3 shows how delays can cause chattering. It depicts a trajectory in the region $z > 0$ heading toward the sliding manifold $z = 0$. It first hits the manifold at point a. In ideal sliding mode control, the trajectory should start sliding on the manifold from point a. In reality, there will be a delay between the time the sign of z changes and the time the control switches. During this delay period, the trajectory crosses the manifold into the region $z < 0$. When the control switches, the trajectory reverses its direction and heads again toward the manifold. Once again, it crosses the manifold, and repetition of this process creates the "zig-zag" motion (oscillation) shown in the sketch, which is known as chattering. Chattering results in low control accuracy, high heat losses in electrical power circuits, and high wear of moving mechanical parts. It may also excite unmodeled high-frequency dynamics, which degrades the performance of the system and may even lead to instability.

One approach to eliminate chattering is to use a continuous approximation of the discontinuous sliding mode controller.[12] By using continuous approximation, we also avoid the theoretical difficulties associated with discontinuous controllers.[13]

---

[11] To read about differential equations with a discontinuous right-hand side, consult [49], [176], [130], and [154].

[12] Other approaches to eliminate chattering include the use of observer [177] and extending the dynamics of the system using integrators [158]. It should be noted that the continuous approximation approach can not be used in applications where actuators have to be used in an on-off operation mode, like thyristors for example [177].

[13] While we do not pursue rigorous analysis of the discontinuous sliding mode controller, the reader is encouraged to use simulations to examine the performance of both the discontinuous controller and its continuous approximation.

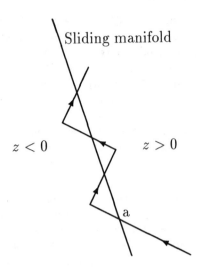

Figure 13.3: Chattering due to delay in control switching.

We approximate the signum nonlinearity by a saturation nonlinearity with high slope;[14] that is, the component v of the sliding mode control is taken as

$$v_i = -\frac{\beta(\eta, \xi)}{1 - k} \, \text{sat}\left(\frac{z_i}{\epsilon}\right), \quad 1 \leq i \leq p \tag{13.53}$$

where

$$\beta(\eta, \xi) \geq \rho(\eta, \xi) + b, \quad \forall \, (\eta, \xi) \in D$$

sat$(\cdot)$ is the saturation function defined by

$$\text{sat}(y) = \begin{cases} y, & \text{if } |y| \leq 1 \\ \text{sgn}(y) & \text{if } |y| > 1 \end{cases}$$

and ϵ is a positive constant. The signum nonlinearity and its approximation are shown in Figure 13.4. The slope of the linear portion of sat(z_i/ϵ) is $1/\epsilon$. A good approximation requires the use of small ϵ.[15] In the limit, as $\epsilon \to 0$, the saturation nonlinearity sat(z_i/ϵ) approaches the signum nonlinearity sgn(z_i). To analyze the performance of the "continuous" sliding mode controller, we examine the reaching

---

[14] Other approximations are discussed in Exercises 13.29 and 13.30.

[15] When ϵ is too small, the high-gain feedback in the linear portion of the saturation function may excite unmodeled high-frequency dynamics. Therefore, the choice of ϵ is a tradeoff between accuracy on one hand and robustness to unmodeled high-frequency dynamics on the other hand.

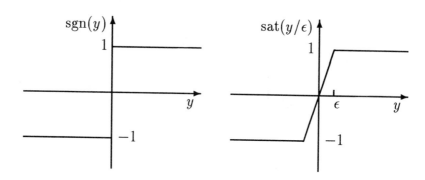

Figure 13.4: The signum nonlinearity and its saturation function approximation.

phase using the Lyapunov function $V_i = \frac{1}{2}z_i^2$. The derivative of V_i satisfies the inequality

$$\dot{V}_i \leq -\frac{\beta(\eta,\xi)}{1-k}z_i \, \text{sat}\left(\frac{z_i}{\epsilon}\right) + \rho(\eta,\xi)|z_i| + k\frac{\beta(\eta,\xi)}{1-k}|z_i|$$

In the region $|z_i| \geq \epsilon$, we have

$$\dot{V}_i \leq -\beta(\eta,\xi)|z_i| + \rho(\eta,\xi)|z_i| \leq -b|z_i|$$

which shows that whenever $|z_i(0)| > \epsilon$, $|z_i(t)|$ will be strictly decreasing until it reaches the set $\{|z_i| \leq \epsilon\}$ in finite time and remains inside thereafter. The set $\{|z_i| \leq \epsilon, \ 1 \leq i \leq p\}$ is called the *boundary layer*. To study the behavior of η, we assume that, together with the sliding manifold design $\xi = \phi(\eta)$, there is a (continuously differentiable) Lyapunov function $V(\eta)$ that satisfies the inequalities

$$\alpha_1(\|\eta\|) \leq V(\eta) \leq \alpha_2(\|\eta\|) \tag{13.54}$$

$$\frac{\partial V}{\partial \eta}[f(\eta,\phi(\eta)+z) + \delta_\eta(\eta,\phi(\eta)+z)] \leq -\alpha_3(\|\eta\|), \ \forall \ \|\eta\| \geq \gamma(\|z\|) \tag{13.55}$$

for all $(\eta,\xi) \in D$ where α_1, α_2, α_3, and γ are class $\mathcal{K}$ functions.[16] Noting that

$$|z_i| \leq c, \text{ for } 1 \leq i \leq p \ \Rightarrow \ \|z\| \leq k_1 c \Rightarrow \dot{V} \leq -\alpha_3(\|\eta\|), \text{ for } \|\eta\| \geq \gamma(k_1 c)$$

---

[16] Inequality (13.55) implies local input-to-state stability of the system

$$\dot{\eta} = f(\eta,\phi(\eta)+z) + \delta_\eta(\eta,\phi(\eta)+z)$$

when z is viewed as the input; see Theorem 5.2.

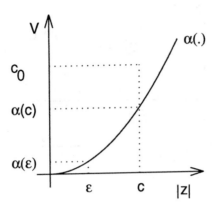

Figure 13.5: A representation of the set Ω for a scalar z. $\dot{V} < 0$ above the $\alpha(\cdot)$-curve.

for some positive constant k_1,[17] we define a class $\mathcal{K}$ function α by

$$\alpha(r) = \alpha_2(\gamma(k_1 r))$$

Then,

$$
\begin{aligned}
V(\eta) \geq \alpha(c) \quad &\Rightarrow \quad V(\eta) \geq \alpha_2(\gamma(k_1 c)) \;\Rightarrow\; \alpha_2(\|\eta\|) \geq \alpha_2(\gamma(k_1 c)) \\
&\Rightarrow \quad \|\eta\| \geq \gamma(k_1 c) \;\Rightarrow\; \dot{V} \leq -\alpha_3(\|\eta\|) \leq -\alpha_3(\gamma(k_1 c))
\end{aligned}
$$

which shows that the set $\{V(\eta) \leq c_0\}$ with $c_0 \geq \alpha(c)$ is positively invariant because $\dot{V}$ is negative on the boundary $V(\eta) = c_0$; see Figure 13.5. It follows that the set

$$\Omega = \{V(\eta) \leq c_0\} \times \{|z_i| \leq c,\ 1 \leq i \leq p\}, \quad \text{with } c_0 \geq \alpha(c) \tag{13.56}$$

is positively invariant whenever $c > \epsilon$ and $\Omega \subset D$. We assume that ϵ is chosen small enough that $c > \epsilon$ and $c_0 \geq \alpha(c)$ can be chosen such that $\Omega \subset D$. The compact set Ω serves as our estimate of the "region of attraction." For all initial states in Ω, the trajectories will be bounded for all $t \geq 0$. After some finite time, we have $|z_i(t)| \leq \epsilon$. It follows from the foregoing analysis that $\dot{V} \leq -\alpha_3(\gamma(k_1 \epsilon))$ for all $V(\eta) \geq \alpha(\epsilon)$. Hence, the trajectory will reach the positively invariant set

$$\Omega_\epsilon = \{V(\eta) \leq \alpha(\epsilon)\} \times \{|z_i| \leq \epsilon,\ 1 \leq i \leq p\} \tag{13.57}$$

in finite time. The set Ω_ϵ can be made arbitrarily small by choosing ϵ small enough. In the limit, as $\epsilon \to 0$, Ω_ϵ shrinks to the origin, which shows that the "continuous"

---

[17] The constant k_1 depends on the type of norm used in the analysis.

sliding mode controller recovers the performance of its discontinuous counterpart. Finally, we note that if all the assumptions hold globally, that is, $D = R^n$ and $V(\eta)$ is radially unbounded, we can choose Ω arbitrarily large to include any initial state.

Theorem 13.2 *Consider the system* (13.43)–(13.44). *Suppose there exist* $\phi(\eta)$, $V(\eta)$, $\rho(\eta,\xi)$, *and* k *which satisfy* (13.50), (13.54), *and* (13.55). *Let* $u = u_{eq} + G_a^{-1}(\eta,\xi)v$, *where* u_{eq} *and* v *are given by* (13.48) *and* (13.53), *respectively. Suppose* ϵ *is chosen small enough to ensure that* $c > \epsilon$ *and* c_0 *can be chosen such that the set* Ω, *defined by* (13.56), *is contained in* D. *Then, for all* $(\eta(0),\xi(0)) \in \Omega$, *the trajectory* $(\eta(t),\xi(t))$ *is bounded for all* $t \geq 0$ *and reaches the positively invariant set* Ω_ϵ, *defined by* (13.57), *in finite time. Moreover, if all the assumptions hold globally and* $V(\eta)$ *is radially unbounded, the foregoing conclusion holds for any initial state* $(\eta(0),\xi(0)) \in R^n$. $\diamond$

The theorem shows that the "continuous " sliding mode controller achieves ultimate boundedness with an ultimate bound that can be controlled by the design parameter ϵ. It also gives conditions for global ultimate boundedness. Since the uncertainty δ_ξ could be nonvanishing, ultimate boundedness is the best we can expect, in general. If, however, δ_ξ vanishes at the origin, then we may be able to show asymptotic stability of the origin. This issue is dealt with in Exercise 13.26.[18]

In introducing sliding mode control, we have dealt only with autonomous systems and considered only the stabilization problem. The foregoing discussion can be easily extended to the nonautonomous regular form

$$\dot{\eta} = f(t,\eta,\xi) + \delta_\eta(t,\eta,\xi) \tag{13.58}$$

$$\dot{\xi} = f_a(t,\eta,\xi) + G_a(t,\eta,\xi)[u + \delta_\xi(t,\eta,\xi,u)] \tag{13.59}$$

where $G_a(t,\eta,\xi)$ is nonsingular for all $t \geq 0$ and all (η,ξ) in the domain of interest. Such extension is pursued in Exercise 13.27. Control objectives other than stabilization, like tracking for example, can be handled by the same design procedure. The only difference comes at the design of the sliding surface $\xi = \phi(\eta)$, which should be designed to meet the control objective for the reduced-order system (13.45). The forthcoming Example 13.14 will illustrate this point.

Example 13.12 Consider the pendulum equation

$$\dot{x}_1 = x_2$$

$$\dot{x}_2 = -a\sin(x_1 + \delta_1) - bx_2 + cu$$

---

[18]The idea of showing asymptotic stability is similar to the one used in Corollary 13.1 with Lyapunov redesign. Basically, inside the boundary layer, the control $v_i = -\beta z_i/\epsilon(1-k)$ acts as a high-gain feedback controller for small ϵ. By choosing ϵ small enough, it can be shown that the high-gain feedback stabilizes the origin.

where $x_1 = \theta - \delta_1$, $x_2 = \dot{\theta}$, and u is the torque input. The goal is to stabilize the pendulum at the angle $\theta = \delta_1$. Equivalently, we want to design state feedback control to make $x = 0$ an asymptotically stable equilibrium point of the closed-loop system. Suppose the parameters a, b, c are uncertain, and let $\hat{a}$, $\hat{b}$, $\hat{c}$ be their known nominal values. The system can be represented in the form

$$\dot{x}_1 = x_2$$
$$\dot{x}_2 = -\hat{a}\sin(x_1 + \delta_1) - \hat{b}x_2 + \hat{c}[u + \delta(x, u)]$$

where

$$\delta(x, u) = \frac{1}{\hat{c}}[-(a - \hat{a})\sin(x_1 + \delta_1) - (b - \hat{b})x_2 + (c - \hat{c})u]$$

The system is already in the regular form (13.43)–(13.44) with $\eta = x_1$ and $\xi = x_2$. Note that the uncertain term $\delta(x, u)$ satisfies the matching condition. Note also that $\delta(x, u)$ vanishes at $(x = 0, \ u = 0)$ only if $\sin \delta_1 = 0$; that is, if the desired closed-loop equilibrium point is an open-loop equilibrium point. Otherwise, we are faced with a nonvanishing perturbation. We start with the system

$$\dot{x}_1 = x_2$$

where x_2 is viewed as the control input, and design $x_2 = \phi(x_1)$ to stabilize the origin $x_1 = 0$. The simple linear feedback control $x_2 = -\mu x_1$, where $\mu > 0$, will do the job. Thus, the sliding manifold is given by $z = x_2 + \mu x_1$. Next, we estimate the function $\rho(x)$ and the constant k of (13.50). We have

$$u_{eq} = \frac{1}{\hat{c}}[\hat{a}\sin(x_1 + \delta_1) + \hat{b}x_2 - \mu x_2]$$

and

$$\Delta(x, v) = -(a - \hat{a})\sin(x_1 + \delta_1) - (b - \hat{b})x_2 + (c - \hat{c})(u_{eq} + v/\hat{c})$$

It can be easily verified that $|\Delta|$ satisfies the inequality

$$|\Delta| \le k_1 + k_2|x_2| + k|v|$$

globally, where k, k_1, and k_2 are nonnegative constants chosen such that

$$\left| \frac{c - \hat{c}}{\hat{c}} \right| \le k, \quad \left| -a + \frac{c\hat{a}}{\hat{c}} \right| \le k_1, \quad \left| -b + \frac{c\hat{b}}{\hat{c}} - \mu\left(\frac{c - \hat{c}}{\hat{c}}\right) \right| \le k_2$$

We assume that $k < 1$. The function $\rho(x)$ can be taken as $\rho(x) = k_1 + k_2|x_2|$. Consequently, $\beta(x)$ is taken as $\beta(x) = k_1 + b_0 + k_2|x_2|$, for some $b_0 > 0$. The sliding mode controller is given by

$$u = u_{eq} - \frac{\beta(x)}{\hat{c}(1 - k)} \ \text{sgn}(x_2 + \mu x_1)$$

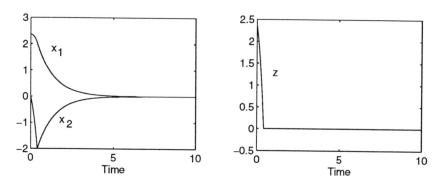

Figure 13.6: Simulation results for the sliding mode controller of Example 13.12 when $\delta_1 = \pi/4$.

and its continuous approximation is given by

$$u = u_{eq} - \frac{\beta(x)}{\hat{c}(1-k)} \, \text{sat} \left(\frac{x_2 + \mu x_1}{\epsilon} \right), \quad \epsilon > 0$$

We leave it as an exercise for the reader (Exercise 13.31) to verify that the conditions of Theorem 13.2 are satisfied globally. The "continuous" sliding mode controller achieves global ultimate boundedness, while the discontinuous controller achieves global stabilization of the origin. If $\delta_1 = 0$ or π, the uncertain term $\delta(x, u)$ vanishes at the origin and the "continuous" sliding mode controller achieves global stabilization of the origin, provided ϵ is chosen sufficiently small.[19] To run simulations, let us take the numerical values $\hat{a} = \hat{b} = \hat{c} = \mu = 1$, and assume

$$\tfrac{1}{2} \le a \le 2, \quad 0 \le b \le 3, \quad \tfrac{1}{2} \le c \le \tfrac{3}{2}$$

Then, the constants k, k_1, and k_2 can be taken as $k = \tfrac{1}{2}$, $k_1 = \tfrac{3}{2}$, and $k_2 = 2$. We take $b_0 = \tfrac{1}{2}$ so that $\beta(x) = 2(1 + |x_2|)$. Finally, the control u is given by

$$u = \sin(x_1 + \delta_1) - 4(1 + |x_2|) \, \text{sgn}(x_1 + x_2)$$

and $\text{sgn}(x_1 + x_2)$ is replaced by $\text{sat}((x_1 + x_2)/\epsilon)$ when the continuous approximation is implemented. The performance of the discontinuous sliding mode controller when $\delta_1 = \pi/4$, $a = \tfrac{3}{2}$, $b = \tfrac{3}{4}$, and $c = \tfrac{3}{4}$ is shown in Figures 13.6 and 13.7. The performance of the "continuous" sliding mode controller, with $\epsilon = 0.1$, is shown in Figure 13.8. The response is very close to that of the discontinuous controller, but actually the trajectory does not tend to the origin. It can be verified that the

---

[19] Asymptotic stability follows from Exercise 13.26.

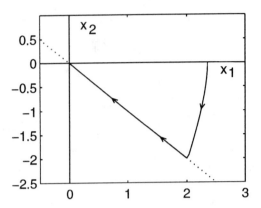

Figure 13.7: A state trajectory under the sliding mode controller of Example 13.12 when $\delta_1 = \pi/4$.

closed-loop system has an equilibrium point at $(\bar{x}_1, 0)$, where $\bar{x}_1$ is the root of

$$\bar{x}_1 = -\frac{\epsilon}{4}\sin(\bar{x}_1 + \delta_1)$$

For $\delta_1 = \pi/4$ and $\epsilon = 0.1$, the equilibrium point is $(-0.0174, 0)$. To exaggerate the effect of the continuous approximation, we show in Figure 13.9 the performance of the "continuous" sliding mode controller when $\delta_1 = \pi/4$ and $\epsilon = 1$. In this case, the equilibrium point is $(-0.1487, 0)$. The performance of the same controller, but with $\delta_1 = \pi$, is shown in Figure 13.10, where the origin is a globally asymptotically stable equilibrium point of the closed-loop system.

We have derived the sliding mode controller as the sum of the equivalent control and the switching control. It is always possible to implement the sliding mode controller using only the switching control. The idea is presented in Exercise 13.28 for the general case, and we illustrate it here using the pendulum example. Let us simply take $u_{eq} = 0$. Then, the derivative of z is given by

$$\dot{z} = \dot{x}_2 + \mu\dot{x}_1 = -a\sin(x_1 + \delta_1) - bx_2 + cu + \mu x_2 = v + \Delta(x, v)$$

where $u = v/\hat{c}$ and

$$\Delta(x, v) = -a\sin(x_1 + \delta_1) + (\mu - b)x_2 + \frac{c - \hat{c}}{\hat{c}}v$$

The function Δ satisfies the inequality

$$|\Delta| \le k_3 + k_4|x_2| + k|v|$$

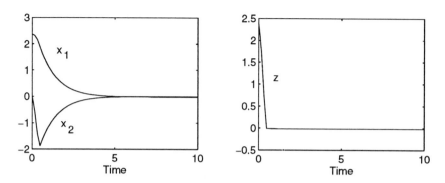

Figure 13.8: Simulation results for the "continuous" sliding mode controller of Example 13.12 when $\delta_1 = \pi/4$ and $\epsilon = 0.1$.

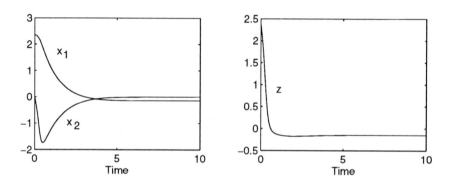

Figure 13.9: Simulation results for the "continuous" sliding mode controller of Example 13.12 when $\delta_1 = \pi/4$ and $\epsilon = 1$.

globally, where

$$\left| \frac{c - \hat{c}}{\hat{c}} \right| \leq k < 1, \quad |a| \leq k_3, \quad |\mu - b| \leq k_4$$

Thus, taking $\beta(x) = k_3 + b_0 + k_4|x_2|$, $b_0 > 0$, the control u is given by

$$u = -\frac{\beta(x)}{\hat{c}(1 - k)} \, \text{sgn}(x_2 + \mu x_1)$$

This control is easier to implement when compared with the version that includes the equivalent control. We can carry this simplification one step further by reducing the control to a simple switching function with constant gain. Over any bounded region of interest, like the set Ω of Theorem 13.2, the function $\beta(x)$ is bounded. Let

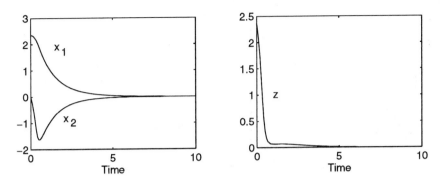

Figure 13.10: Simulation results for the "continuous" sliding mode controller of Example 13.12 when $\delta_1 = \pi$ and $\epsilon = 1$.

K be a positive constant such that

$$\frac{\beta(x)}{\hat{c}(1 - k)} \leq K$$

over the compact region of interest. Then, u can be taken as

$$u = -K \, \text{sgn}(x_2 + \mu x_1)$$

which is the simplest of the three versions we have seen. This version, however, does not achieve global stabilization like the other two versions. It can only achieve regional or semiglobal stabilization. $\triangle$

Example 13.13 Consider the second-order system

$$\begin{aligned}
\dot{x}_1 &= x_2 + \theta_1 x_1 \sin x_2 \\
\dot{x}_2 &= \theta_2 x_2^2 + x_1 + u
\end{aligned}$$

where θ_1 and θ_2 are unknown parameters that satisfy

$$|\theta_1| \leq a, \quad |\theta_2| \leq b$$

for some known bounds a and b. In Example 13.11, we designed a robust stabilizing state feedback controller using the backstepping method. Let us design a sliding mode controller to achieve the same objective. The starting point is similar to Example 13.11. We consider the system

$$\dot{x}_1 = x_2 + \theta_1 x_1 \sin x_2$$

and take $x_2 = -(1 + a)x_1$, which stabilizes the origin $x_1 = 0$ because

$$x_1 \dot{x}_1 = -(1 + a)x_1^2 + \theta x_1^2 \sin x_2 \leq -x_1^2$$

The sliding manifold is given by $z = x_2 + (1 + a)x_1 = 0$. The derivative $\dot{z}$ is given by

$$\dot{z} = \theta_2 x_2^2 + x_1 + u + (1 + a)(x_2 + \theta_1 x_1 \sin x_2)$$

To cancel the known term on the right-hand side, the equivalent control is taken as

$$u_{eq} = -x_1 - (1 + a)x_2$$

Taking $u = u_{eq} + v$ results in

$$\dot{z} = v + \Delta(x), \quad \text{where } \Delta(x) = \theta_2 x_2^2 + (1 + a)\theta_1 x_1 \sin x_2$$

It can be easily seen that

$$|\Delta(x)| \leq a(1 + a)|x_1| + b x_2^2$$

Therefore, we take

$$\beta(x) = a(1 + a)|x_1| + b x_2^2 + b_0, \quad b_0 > 0$$

and

$$u = u_{eq} - \beta(x) \, \text{sgn}(z)$$

This controller, or its continuous approximation with sufficiently small ϵ, globally stabilizes the origin.[20] $\triangle$

Example 13.14 Consider an input-output linearizable system represented in the normal form

$$
\begin{aligned}
\dot{\zeta} &= f_0(\zeta, x) \\
\dot{x}_1 &= x_2 \\
&\vdots \\
\dot{x}_{r-1} &= x_r \\
\dot{x}_r &= \frac{1}{\beta(\zeta, x)}[u - \alpha(\zeta, x)] + \delta(\zeta, x, u) \\
y &= x_1
\end{aligned}
$$

---

[20] Note that while the controller of Example 13.11 achieves only semiglobal stabilization, this is not due to the use of backstepping. It is due to the linear growth bound assumed in the controller design. It is possible to achieve global stabilization using the backstepping method; see Exercise 13.21.

where the input u and the output y are scalar variables, $\zeta \in R^{n-r}$, and $x = [x_1, \ldots, x_r]^T \in R^r$. We assume that α and β are known and $\beta(\zeta, x) \neq 0$ for all $(\zeta, x) \in R^n$. The term $\delta(\zeta, x, u)$ is an uncertain term. We want to design a state feedback control so that the output y asymptotically tracks a reference signal $y_R(t)$. We assume that $y_R(t)$ and its derivatives up to the rth derivative are continuous and bounded. Let us start by performing the change of variables

$$e_1 = x_1 - y_R(t), \ e_2 = x_2 - y_R^{(1)}(t), \ \ldots \ , \ e_r = x_r - y_R^{(r-1)}(t)$$

The state equations for e_1 to e_r are

$$\dot{e}_1 = e_2$$
$$\vdots \quad \vdots$$
$$\dot{e}_{r-1} = e_r$$
$$\dot{e}_r = \frac{1}{\beta(\zeta, x)}[u - \alpha(\zeta, x)] + \delta(\zeta, x, u) - y_R^{(r)}$$

Our objective now is to design state feedback control to ensure that $e = [e_1, \ldots, e_r]^T$ is bounded and $e(t) \to 0$ as $t \to \infty$. Boundedness of e will ensure boundedness of x, since y_R and its derivatives are bounded. We need also to ensure boundedness of ζ. This will follow from a minimum phase assumption. In particular, we assume that the system

$$\dot{\zeta} = f_0(\zeta, x)$$

is input-to-state stable. Hence, boundedness of x will imply boundedness of ζ. So, from this point on, we concentrate our attention on showing the boundedness and convergence of e. The $\dot{e}$-equation takes the regular form (13.43)-(13.44) with $\eta = [e_1, \ldots, e_{r-1}]^T$ and $\xi = e_r$. We start with the system

$$\dot{e}_1 = e_2$$
$$\vdots \quad \vdots$$
$$\dot{e}_{r-1} = e_r$$

where e_r is viewed as the control input. We want to design e_r to ensure the boundedness and convergence of $e_1, \ldots, e_{r-1}$. For this linear system (in a controllable canonical form), we can achieve this task by the linear control

$$e_r = -(k_1 e_1 + \cdots + k_{r-1} e_{r-1})$$

where k_1 to k_{r-1} are chosen such that the polynomial

$$s^{r-1} + k_{r-1} s^{r-2} + \cdots + k_1$$

is Hurwitz. Then, the sliding manifold is given by

$$z = e_r + (k_1 e_1 + \cdots + k_{r-1} e_{r-1}) = 0$$

and

$$\dot{z} = \frac{1}{\beta(\zeta, x)} [u - \alpha(\zeta, x)] + \delta(\zeta, x, u) - y_R^{(r)} + k_1 e_2 + \cdots + k_{r-1} e_r$$

To cancel the known terms on the right-hand side, the equivalent control is taken as

$$u_{eq} = \beta(\zeta, x)[-k_1 e_2 - \cdots - k_{r-1} e_r + y_R^{(r)}] + \alpha(\zeta, x)$$

With $u = u_{eq} + \beta(\zeta, x)v$, we obtain

$$\dot{z} = v + \delta(\zeta, x, u_{eq} + \beta(\zeta, x)v)$$

Suppose that

$$|\delta(\zeta, x, u_{eq} + \beta(\zeta, x)v)| \le \rho(\zeta, x, t) + k|v|, \quad 0 \le k < 1$$

for all $(\zeta, x) \in R^n$ and all $t \ge 0$, where ρ is a bounded function of t for every $(\zeta, x) \in R^n$; this is reasonable in view of the boundedness of y_R and its derivatives. Take

$$\beta_0(\zeta, x, t) \ge \rho(\zeta, x, t) + b, \quad b > 0$$

Then, the sliding mode controller is given by

$$u = u_{eq} - \frac{\beta_0(\zeta, x, t)\beta(\zeta, x)}{1 - k} \operatorname{sgn}(z)$$

and its continuous approximation is obtained by replacing $\operatorname{sgn}(z)$ by $\operatorname{sat}(z/\epsilon)$. We leave to the reader (Exercise 13.32) to show that with the "continuous" sliding mode controller, there exits a finite time T, possibly dependent on ϵ and the initial states, and a positive constant k, independent of ϵ and the initial states, such that $|y(t) - y_R(t)| \le k\epsilon$ for all $t \ge T$. $\triangle$

13.4 Adaptive Control

Consider a single-input–single-output plant described by the state-space model

$$\dot{x}_p = A_p x_p + b_p u \tag{13.60}$$
$$y_p = c_p^T x_p \tag{13.61}$$

or, equivalently, by the transfer function model

$$Y(s) = P(s)U(s) = k_p \frac{n_p(s)}{d_p(s)} U(s)$$

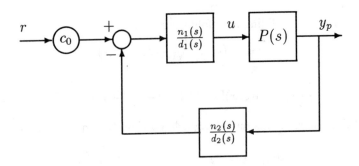

Figure 13.11: Controller structure.

where $U(s)$ and $Y(s)$ are the Laplace transforms of the control input $u(t)$ and the measured output $y(t)$, respectively. The polynomials $n_p(s)$ and $d_p(s)$ are monic[21] with degrees m and n, respectively. It is assumed that the relative degree $n - m > 0$. Our goal is to design a controller such that the closed-loop transfer function from the reference input r to the output y matches the reference model transfer function

$$M(s) = k_m \frac{n_m(s)}{d_m(s)}$$

where $n_m(s)$ and $d_m(s)$ are monic polynomials with

$$\deg d_m(s) = \deg d_p(s) = n$$

$$\deg n_m(s) = \deg n_p(s) = m$$

For this control system to make any sense we must choose $d_m(s)$ to be a Hurwitz polynomial; that is, all the roots of $d_m(s) = 0$ must have negative real parts.

13.4.1   Model Reference Controller

Consider the typical controller structure of Figure 13.11 with a cascade compensator $n_1(s)/d_1(s)$, a feedback compensator $n_2(s)/d_2(s)$, and a feedforward gain c_0. The transfer function from r to y_p is given by

$$\frac{Y_p(s)}{R(s)} = T(s) = \frac{c_0 k_p [n_p(s) n_1(s)/d_p(s) d_1(s)]}{1 + k_p [n_p(s) n_1(s) n_2(s)/d_p(s) d_1(s) d_2(s)]}$$

To simplify $T(s)$, let us choose $n_1(s) = d_2(s) = \lambda(s)$ so that the polynomials $n_1(s)$ and $d_2(s)$ cancel in the denominator of $T(s)$. This pole-zero cancellation necessitates

---

[21] A polynomial $a_{n+1}s^n + a_n s^{n-1} + \cdots + a_2 s + a_1$ is monic when $a_{n+1} = 1$.

that $\lambda(s)$ be chosen as a Hurwitz polynomial, so that the hidden modes would be stable. Without loss of generality, we take $\lambda(s)$ as a monic polynomial. To match $T(s)$ to $M(s)$, the numerator of $T(s)$ must contain the polynomial $n_m(s)$. Let $\lambda(s) = \lambda_0(s)n_m(s)$ and assume that both n_m and λ_0 are Hurwitz polynomials. Let $l = \deg \lambda_0$; then, $\deg \lambda = l + m$. With the given choices, the closed-loop transfer function is given by

$$T(s) = k_m \frac{n_m(s)}{d_m(s)} \left[\frac{c_0(k_p/k_m)n_p(s)\lambda_0(s)d_m(s)}{d_p(s)d_1(s) + k_p n_p(s)n_2(s)} \right]$$

To achieve the desired matching, the bracketed term must be identity. Divide the polynomial $\lambda_0(s)d_m(s)$ by $d_p(s)$:

$$\lambda_0(s)d_m(s) = d_p(s)q(s) + \rho(s)$$

The quotient $q(s)$ has degree l, and the remainder $\rho(s)$ has degree $n - 1$ (at most). Let

$$c_0^* = \frac{k_m}{k_p}; \ d_1^*(s) = n_p(s)q(s); \ n_2^*(s) = \frac{1}{k_p}\rho(s)$$

Then, with c_0, d_1, and n_2 equal to their starred values, we have

$$\begin{aligned} T(s) &= M(s) \left[\frac{(k_m/k_p)(k_p/k_m)n_p(s)\lambda_0(s)d_m(s)}{d_p(s)n_p(s)q(s) + n_p(s)\rho(s)} \right] \\ &= M(s) \left[\frac{n_p(s)\lambda_0(s)d_m(s)}{n_p(s)\lambda_0(s)d_m(s)} \right] = M(s) \end{aligned}$$

To arrive at the last step, we cancelled the polynomials λ_0, d_m, and n_p. We already know that λ_0 and d_m are Hurwitz polynomials. We need to assume that n_p is a Hurwitz polynomial. The feedback compensator is given by

$$\frac{n_2(s)}{d_2(s)} = \frac{n_2^*(s)}{\lambda(s)}$$

where $\deg n_2^* = \deg r = n - 1$. For this compensator to be realizable, we need to restrict its transfer function to be proper. This imposes the constraint $\deg \lambda \geq n-1$. Since we are usually interested in the simplest controller that will do the job, let us take $\deg \lambda = n - 1$. This implies that $l = \deg \lambda_0 = n - m - 1$. The cascade compensator is given by

$$\frac{n_1(s)}{d_1(s)} = \frac{\lambda(s)}{d_1^*(s)}$$

The denominator polynomial $d_1^*(s) = n_p(s)q(s)$ is not necessarily Hurwitz. Since it is easier to implement transfer functions with Hurwitz denominators, we will

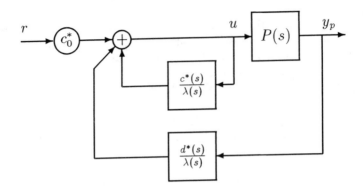

Figure 13.12: Controller structure; known plant.

rearrange the cascade compensator so that the transfer function to be implemented always has a Hurwitz denominator. Define

$$c^*(s) = \lambda(s) - d_1^*(s) = \lambda(s) - n_p(s)q(s)$$

Since λ and d_1^* are monic polynomials of degree $n - 1$, c^* has degree $n - 2$. Now

$$\frac{n_1(s)}{d_1(s)} = \frac{\lambda(s)}{\lambda(s) - c^*(s)} = \frac{1}{1 - [c^*(s)/\lambda(s)]}$$

which is the transfer function of a positive feedback connection with c^*/λ in the feedback loop. With this rearrangement, the transfer function to be implemented c^*/λ has a Hurwitz denominator. Thus, the final controller can be represented by the block diagram of Figure 13.12. Notice that the negative sign in the feedback loop of Figure 13.11 is now included in the feedback compensator

$$\frac{d^*(s)}{\lambda(s)} = \frac{-(1/k_p)\rho(s)}{\lambda(s)}$$

Note also that the denominator polynomial $\lambda(s)$ is independent of the plant parameters.

In preparation for adaptive control, let us find a state-space representation of the controller. Let

$$\Lambda = \begin{bmatrix} 0 & 1 & 0 & \cdots & 0 \\ 0 & 0 & 1 & \cdots & \vdots \\ \vdots & & \ddots & & 0 \\ \vdots & & & 0 & 1 \\ -\lambda_1 & -\lambda_2 & \cdots & \cdots & -\lambda_{n-1} \end{bmatrix}_{(n-1)\times(n-1)} \quad ; b_\lambda = \begin{bmatrix} 0 \\ 0 \\ \vdots \\ 0 \\ 1 \end{bmatrix}_{(n-1)\times 1}$$

where λ_i is the coefficient of s^{i-1} in the polynomial $\lambda(s)$. It can be easily verified that

$$(sI - \Lambda)^{-1}b_\lambda = \frac{1}{\lambda(s)} \begin{bmatrix} 1 \\ s \\ \vdots \\ \vdots \\ s^{n-2} \end{bmatrix}$$

Hence,

$$\frac{c^*(s)}{\lambda(s)} = (c^*)^T(sI - \Lambda)^{-1}b_\lambda$$

and

$$\frac{d^*(s)}{\lambda(s)} = d_0^* + (d^*)^T(sI - \Lambda)^{-1}b_\lambda$$

where $c^* \in R^{n-1}$ is determined from the coefficients of the polynomial $c^*(s)$ while $d^* \in R^{n-1}$ and $d_0^* \in R$ are determined from the coefficients of the polynomial $d^*(s)$. Let $w^{(1)}(t) \in R^{n-1}$ and $w^{(2)}(t) \in R^{n-1}$ be the state vectors of the two dynamic compensators in the feedback loop of Figure 13.12. The state equations for $w^{(1)}$ and $w^{(2)}$ are

$$\dot{w}^{(1)} = \Lambda w^{(1)} + b_\lambda u \tag{13.62}$$
$$\dot{w}^{(2)} = \Lambda w^{(2)} + b_\lambda y_p \tag{13.63}$$

and the control u is given by

$$u = c_0^* r + (c^*)^T w^{(1)} + d_0^* y_p + (d^*)^T w^{(2)}$$

Let

$$\theta^* = \begin{bmatrix} c_0^* \\ c^* \\ d_0^* \\ d^* \end{bmatrix}; \quad w = \begin{bmatrix} r \\ w^{(1)} \\ y_p \\ w^{(2)} \end{bmatrix}$$

and rewrite the control as

$$u = (\theta^*)^T w \tag{13.64}$$

We can now augment (13.62)–(13.64) with (13.60)–(13.61) to find a state-space model of the closed-loop system. For later use, we take $u = (\theta^*)^T w + v$. Substitution of this expression in (13.60) yields

$$\begin{aligned} \dot{x}_p &= A_p x_p + b_p (\theta^*)^T w + b_p v \\ &= A_p x_p + b_p \left[c_0^* r + (c^*)^T w^{(1)} + d_0^* y_p + (d^*)^T w^{(2)} \right] + b_p v \\ &= (A_p + b_p d_0^* c_p^T) x_p + b_p (c^*)^T w^{(1)} + b_p (d^*)^T w^{(2)} + b_p c_0^* r + b_p v \end{aligned}$$

Substitution of u in (13.62) yields

$$
\begin{aligned}
\dot{w}^{(1)} &= \Lambda w^{(1)} + b_\lambda \left[c_0^* r + (c^*)^T w^{(1)} + d_0^* y_p + (d^*)^T w^{(2)} \right] + b_\lambda v \\
&= b_\lambda d_0^* c_p^T x_p + \left[\Lambda + b_\lambda (c^*)^T \right] w^{(1)} + b_\lambda (d^*)^T w^{(2)} + b_\lambda c_0^* r + b_\lambda v
\end{aligned}
$$

Finally,
$$
\dot{w}^{(2)} = b_\lambda c_p^T x_p + \Lambda w^{(2)}
$$

Taking
$$
x_{pw} = \left[\begin{array}{c} x_p \\ w^{(1)} \\ w^{(2)} \end{array} \right]
$$

as a state vector for the closed-loop system, we arrive at the closed-loop state-space model

$$
\begin{aligned}
\dot{x}_{pw} &= A_m x_{pw} + b_m v + b_m c_0^* r \\
y_p &= c_m^T x_{pw}
\end{aligned}
$$

where

$$
A_m = \left[\begin{array}{ccc} A_p + b_p d_0^* c_p^T & b_p (c^*)^T & b_p (d^*)^T \\ b_\lambda d_0^* c_p^T & \Lambda + b_\lambda (c^*)^T & b_\lambda (d^*)^T \\ b_\lambda c_p^T & 0 & \Lambda \end{array} \right] ; \quad b_m = \left[\begin{array}{c} b_p \\ b_\lambda \\ 0 \end{array} \right]
$$

$$
c_m^T = \left[\begin{array}{ccc} c_p^T & 0 & 0 \end{array} \right]
$$

When $v = 0$, the transfer function from r to y_p is $M(s)$. Therefore,

$$
c_m^T (sI - A_m)^{-1} b_m = \frac{1}{c_0^*} M(s) \tag{13.65}
$$

The realization $\{A_m, b_m, c_m^T\}$ is a nonminimal realization of $(1/c_0^*)M(s)$ when $n > 1$ because the dimension of A_m is $3n - 2$ while the order of $M(s)$ is n. Since in the process of deriving the controller we have verified that all pole-zero cancellations are cancellations of Hurwitz polynomials, the realization $\{A_m, b_m, c_m^T\}$ is internally stable; that is, the matrix A_m is Hurwitz.

13.4.2   Model Reference Adaptive Control

Suppose now that the plant transfer function $P(s)$ is unknown, but we know the polynomial degrees n and m as well as the sign of the high frequency gain k_p.

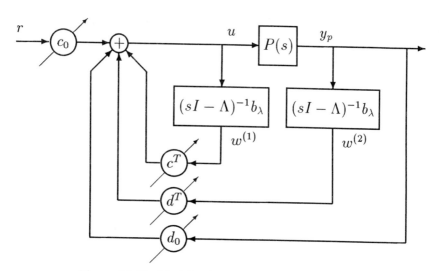

Figure 13.13: Model reference adaptive controller.

Without loss of generality, we take $k_p > 0$. When adaptive control is used, the constant parameters c_0^*, c^*, d_0^*, and d^* are replaced by the time-varying parameters c_0, c, d_0, and d, respectively. which are adjusted on line. A schematic block diagram representation of the adaptive controller is shown in Figure 13.13. With $\theta = \begin{bmatrix} c_0, & c^T, & d_0, & d^T \end{bmatrix}^T$, the control u can be expressed as $u = \theta^T w$. Adding and subtracting $(\theta^*)^T w$, we can write u as

$$u = (\theta^*)^T w + (\theta - \theta^*)^T w$$

Taking v in the previous derivation to be $v = (\theta - \theta^*)^T w$, we see that the closed-loop system under the adaptive controller is described by

$$\dot{x}_{pw} = A_m x_{pw} + b_m (\theta - \theta^*)^T w + b_m c_0^* r$$
$$y_p = c_m^T x_{pw}$$

To design the adaptive law for adjusting $\theta(t)$ on line, we need to derive an error model that describes how the output error $y_p - y_m$ depends on the parameter error $\theta - \theta^*$. In view of (13.65), we use

$$\dot{x}_m = A_m x_m + b_m c_0^* r \qquad (13.66)$$
$$y_m = c_m^T x_m$$

as a state-space representation for the reference model. Partition x_m compatibly with the partition of x_{pw} as

$$x_m = \begin{bmatrix} x_{pm} \\ w_m^{(1)} \\ w_m^{(2)} \end{bmatrix}$$

and define the state, output, and parameter errors

$$e = x_{pw} - x_m, \quad e_o = y_p - y_m, \quad \phi = \theta - \theta^*$$

Then, it can be easily seen that

$$\begin{aligned} \dot{e} &= A_m e + b_m \phi^T w \\ e_o &= c_m^T e \end{aligned}$$

To complete the description of the system, we express the dependence of w on e as

$$w = hr + Q x_{pw} = hr + Q(x_m + e) = w_m + Qe$$

where

$$h = \begin{bmatrix} 1 \\ 0 \\ 0 \\ 0 \end{bmatrix}, \qquad Q = \begin{bmatrix} 0 & 0 & 0 \\ 0 & I & 0 \\ c_p^T & 0 & 0 \\ 0 & 0 & I \end{bmatrix}$$

and

$$w_m(t) = hr(t) + Q x_m(t)$$

The signal $w_m(t)$ depends only on the reference input and the reference model. We assume that the reference input $r(t)$ is piecewise continuous and bounded for all $t \geq 0$. This implies that the signal $w_m(t)$ has the same properties because it is comprised of r and x_m, which is the solution of the state equation (13.66). Since A_m is Hurwitz, $x_m(t)$ will be bounded for any bounded input $r(t)$.

Relative Degree One Plants

We make the key assumption that $M(s)$ is strictly positive real (Definition 10.3). Note that a scalar transfer function $M(s)$ is strictly positive real only if $Re[M(j\omega)] > 0$ for all ω. This implies that the Nyquist plot of $M(j\omega)$ lies entirely in the closed right-half plane, which is true only when the relative degree is 1. Note also that

$k_p > 0$ implies that $c_0^* > 0$; hence, $(1/c_0^*)M(s)$ is strictly positive real. By the Kalman-Yakubovich-Popov lemma,[22] we know that there are matrices $P = P^T > 0$ and L and a positive constant ϵ such that

$$PA_m + A_m^T P = -L^T L - \epsilon P$$
$$Pb_m = c_m$$

Consider

$$V(e, \phi) = e^T Pe + \frac{1}{\gamma} \phi^T \phi$$

as a Lyapunov function candidate for the closed-loop system. The derivative of V along the trajectories of the system is given by

$$\dot{V} = \dot{e}^T Pe + e^T P\dot{e} + \frac{2}{\gamma} \phi^T \dot{\phi}$$

$$= e^T (PA_m + A_m^T P)e + 2e^T Pb_m \phi^T w + \frac{2}{\gamma} \phi^T \dot{\theta}$$

$$= -\epsilon e^T Pe - e^T L^T Le + \frac{2}{\gamma} \phi^T (\dot{\theta} + \gamma e_o w)$$

Choosing $\dot{\theta}$ as

$$\dot{\theta} = -\gamma e_o w \tag{13.67}$$

cancels the sign indefinite term and yields

$$\dot{V} = -\epsilon e^T Pe - e^T L^T Le \tag{13.68}$$

With this choice of the adaptive law, we complete the derivation of the model reference adaptive controller which is defined by the filters (13.62)–(13.63), the adaptive law (13.67), and the feedback control $u = \theta^T w$. The closed-loop system under this adaptive controller is described by

$$\dot{e} = A_m e + b_m \phi^T w \tag{13.69}$$
$$\dot{\phi} = -\gamma c_m^T ew \tag{13.70}$$

The right-hand side of (13.68) is only a negative semidefinite function of the state (e, ϕ). Therefore, we cannot conclude that the origin $(e, \phi) = (0, 0)$ is asymptotically stable. We can, however, conclude by Theorem 4.4 that, for all initial states $(e(t_0), \phi(t_0))$, the solution of the closed-loop system is bounded and

$$e^T(t)(\epsilon P + L^T L)e(t) \to 0, \text{ as } t \to \infty$$

---

[22] The Kalman-Yakubovich-Popov lemma was stated and proved in Lemma 10.2 for minimal realizations. An extension to nonminimal realizations (see [122, Section 2.6]) shows that there are P, L, and ϵ that satisfy the given equations.

Since P is positive definite, so is $\epsilon P + L^T L$; hence,

$$e(t) \to 0, \text{ as } t \to \infty$$

Notice that the bound on the state $(e(t), \phi(t))$ is dependent on the initial state $(e(t_0), \phi(t_0))$. This dependence can be seen as follows. The argument that the state is bounded is based on showing that the state will always belong to the positively invariant set $\{V \leq c\}$ whenever the initial state does so. The constant c must be chosen large enough so that the initial state of the system is included in this set. Thus, the constant c is dependent on the initial state, and so is the bound on the state. By showing that $e(t)$ and $\phi(t)$ are bounded, we see that all signals in the closed-loop system (13.69)–(13.70) are bounded since we have already seen that $w_m(t)$ is bounded.

By a simple Lyapunov argument and the application of Theorem 4.4 we have shown that the tracking error $e(t)$ approaches zero asymptotically. As for the parameter error $\phi(t)$, the only conclusion we have reached so far is that the signal $\phi(t)$ is bounded. Recall that the parameter error $\phi(t) = \theta(t) - \theta^*$, where $\theta(t)$ are the adjustable parameters of the controller and θ^* is a vector of nominal unknown controller parameters for which the closed-loop transfer function matches the reference model transfer function $M(s)$. Clearly, we are interested in achieving parameter convergence $\phi(t) \to 0$ so that, as the system reaches steady-state, the transfer function of the closed-loop system indeed matches that of the reference model. To show that $\phi(t) \to 0$ as $t \to \infty$, we need to show that the origin $(e, \phi) = (0, 0)$ of the closed-loop system is asymptotically stable. With the Lyapunov function V, we cannot do this by Theorem 4.4 but we may be able to do it by Theorem 4.5 if we can show that the derivative $\dot{V}$ satisfies the additional condition

$$\int_t^{t+\delta} \dot{V}(e(\tau)) \, d\tau \leq -\lambda V(e(t), \phi(t)), \quad 0 < \lambda < 1$$

for all $t \geq 0$ and some $\delta > 0$. In Example 4.14, we saw a special case of a linear system with a quadratic Lyapunov function for which the preceding condition is implied by uniform observability. Our adaptive control system can be cast into the special case of Example 4.14. Let

$$x = \begin{bmatrix} e \\ \phi \end{bmatrix}; \; A(t) = \begin{bmatrix} A_m & b_m w^T(t) \\ -\gamma w(t) c_m^T & 0 \end{bmatrix}; \; C = \begin{bmatrix} L & 0 \\ \sqrt{\epsilon P} & 0 \end{bmatrix}$$

where C has the same number of columns as A. We can represent the closed-loop system (13.69)–(13.70) as a linear time-varying system

$$\dot{x} = A(t)x \tag{13.71}$$

At first glance, it appears erroneous to represent a nonlinear system as a linear time-varying system. Notice, however, that we have already established that all signals in the closed-loop system are bounded. Therefore, $w(t)$ is bounded, which implies that the time-varying matrix $\mathcal{A}(t)$ is bounded for all $t \geq 0$. Hence, the linear system representation (13.71) is indeed well defined. The derivative of the Lyapunov function V is given by

$$\dot{V} = -x^T C^T C x \leq 0$$

Therefore, by Theorem 4.5 (as applied to Example 4.14) we see that the origin $x = 0$ is exponentially stable if the pair $(\mathcal{A}(t),\ \mathcal{C})$ is uniformly observable. To check observability, we need the state transition matrix of (13.71), which is difficult to find in a closed form. Fortunately, we can get around this difficulty. The pair $(\mathcal{A}(t),\ \mathcal{C})$ is uniformly observable if and only if the pair $(\mathcal{A}(t) - K(t)\mathcal{C},\ \mathcal{C})$, with $K(t)$ piecewise continuous and bounded, is uniformly observable.[23] This can be seen by writing the system

$$\dot{x}(t) = \mathcal{A}(t)x(t)$$
$$y(t) = \mathcal{C}x(t)$$

as

$$\dot{x}(t) = [\mathcal{A}(t) - K(t)\mathcal{C}]x(t) + K(t)y(t)$$
$$y(t) = \mathcal{C}x(t)$$

Since $K(t)y(t)$ is known, the two systems have the same observability properties. Let us choose $K(t)$ as

$$K(t) = \begin{bmatrix} 0 & A_m \left(\sqrt{\epsilon P}\right)^{-1} \\ 0 & -\gamma w(t) c_m^T \left(\sqrt{\epsilon P}\right)^{-1} \end{bmatrix}$$

Then,

$$\mathcal{A}(t) - K(t)\mathcal{C} = \begin{bmatrix} 0 & b_m w^T(t) \\ 0 & 0 \end{bmatrix}$$

has the state transition matrix

$$\Phi(\tau, t) = \begin{bmatrix} I & b_m \int_t^\tau w^T(\lambda)\, d\lambda \\ 0 & I \end{bmatrix}$$

---

[23] See [79, Lemma 4.8.1] for a rigorous proof of this claim.

Define

$$\sigma(\tau, t) = \int_t^\tau w(\lambda) \, d\lambda$$

and rewrite the state transition matrix as

$$\Phi(\tau, t) = \begin{bmatrix} I & b_m \sigma^T(\tau, t) \\ 0 & I \end{bmatrix}$$

The observability Gramian of the pair $(\mathcal{A}(t) - K(t)\mathcal{C}, \, \mathcal{C})$ is given by

$$
\begin{aligned}
W(t, t+\delta) &= \int_t^{t+\delta} \Phi^T(\tau, t) \mathcal{C}^T \mathcal{C} \Phi(\tau, t) \, d\tau \\
&= \int_t^{t+\delta} \begin{bmatrix} D & D b_m \sigma^T(\tau, t) \\ \sigma(\tau, t) b_m^T D & \sigma(\tau, t) b_m^T D b_m \sigma^T(\tau, t) \end{bmatrix} d\tau
\end{aligned}
$$

where $D = \epsilon P + L^T L$ is positive definite.

Lemma 13.4 *If $w(t)$ and $\dot{w}(t)$ are bounded for all $t \geq 0$, then there are positive constants δ and k such that $W(t, t+\delta) \geq kI$ for all $t \geq 0$ if there are positive constants δ_1 and k_1 such that*

$$\int_t^{t+\delta_1} w(\tau) w^T(\tau) \, d\tau \geq k_1 I, \quad \forall \, t \geq 0$$

$\diamond$

Proof: Appendix A.20

We have already seen that $w(t) = hr(t) + Qx_m(t) + Qe(t)$ is bounded. From (13.66) and (13.69), it is clear that $\dot{x}_m(t)$ and $\dot{e}(t)$ are bounded. Thus, if the reference input $r(t)$ has a bounded derivative for all $t \geq 0$, the signal $\dot{w}(t)$ will be bounded. It is reasonable to assume that the reference input has a bounded derivative.[24] This will enable us conclude that the pair $(\mathcal{A}(t), \, \mathcal{C})$ is uniformly observable if $\int_t^{t+\delta_1} w(\tau) w^T(\tau) \, d\tau$ is uniformly positive definite. The latter condition is a property of the signal $w(t)$. This property plays an important role in the analysis of identification and adaptive control systems. It is known as the *persistence of excitation condition*.

---

[24] It is sufficient to assume that $\dot{r}(t)$ is bounded almost everywhere, so that piecewise differentiable signals may be used; see [3].

Definition 13.1 *A vector signal $\nu(t)$ is said to be persistently exciting if there are positive constants α_1, α_2, and δ such that*

$$\alpha_2 I \geq \int_t^{t+\delta} \nu(\tau)\nu^T(\tau) \, d\tau \geq \alpha_1 I \tag{13.72}$$

Although the matrix $\nu(\tau)\nu^T(\tau)$ is singular for each τ, the persistence of excitation condition requires that its integral over an interval of length δ be uniformly positive definite. The upper bound in (13.72) is satisfied whenever $\nu(t)$ is bounded, which is the case with our signal $w(t)$.

In view of Lemma 13.4 and Definition 13.1, we can say that the origin of the closed-loop system (13.69)–(13.70) is exponentially stable if $w(t)$ is persistently exciting. For all initial states of the system, the state will approach the origin exponentially fast. However, we cannot say that the origin is globally exponentially stable. To claim that the origin is globally exponentially stable we must be able to show that the state of the system satisfies the inequality

$$\|x(t)\| \leq \|x(t_0)\| k \exp[-\alpha(t - t_0)] \tag{13.73}$$

for some positive constants k and α and for all initial states $x(t_0)$. This inequality is not just an exponentially decaying bound; it also requires that the bound depend linearly on the norm of the initial state. Although we arrive at our conclusions by studying the linear system $\dot{x} = \mathcal{A}(t)x$, which ordinarily would give global results, our linear system representation is valid only as long as $\mathcal{A}(t)$ is bounded, which is the case when $w(t)$ is bounded. The bound on $w(t)$ depends on the initial state $x(t_0)$. Therefore, we cannot conclude global exponential stability. What we can say is that given any constant $a > 0$, no matter how large it is, all solutions starting in the ball $\{\|x\| < a\}$ will satisfy (13.73). This is done by choosing c large enough to include the ball $\{\|x\| < a\}$ in the set $\{V \leq c\}$ (notice that V is radially unbounded). Then, all signals in the closed-loop system can be bounded by constants dependent on a but independent of the given initial state.

The persistence of excitation condition on $w(t)$ is not a verifiable condition since the signal $w(t)$ is generated inside the adaptive control system and is unknown *a priori*; so there is no way for us to check whether or not it is persistently exciting. Notice, however, that

$$w(t) = w_m(t) + Qe(t), \quad \text{where} \quad e(t) \to 0 \text{ as } t \to \infty$$

and $e(t)$ and $\dot{e}(t)$ are bounded. It is reasonable to conjecture that the persistence of excitation of $w(t)$ would follow from the persistence of excitation of $w_m(t)$, which is a model signal generated by a Hurwitz transfer function driven by $r(t)$. This is indeed the case, as we show next with the aid of the following lemma.

Lemma 13.5 *Let $\nu(t)$ and $u(t)$ be piecewise continuous. If ν is persistently exciting and $\int_0^\infty \|u(t)\|_2^2 \, dt \leq k < \infty$, then $\nu + u$ is persistently exciting.* ◇

Proof: Appendix A.21.

In our problem, $e(t)$ is a square-integrable function since $\dot{V} = -e^T De$, with a positive definite D, and the limit of V as $t \to \infty$ is finite. Hence, the persistence of excitation of $w(t)$ is implied by the persistence of excitation of $w_m(t)$. We can now summarize our conclusions in the following theorem.

Theorem 13.3 *Consider the adaptive control system* (13.69)–(13.70) *which corresponds to a plant of order n and relative degree 1. Assume that the reference model transfer function $M(s)$ is strictly positive real. If the input reference $r(t)$ is piecewise continuous and bounded, then, for any arbitrary initial state, all signals in the adaptive control system are bounded (with bounds dependent on the initial state) and the tracking error*

$$e(t) \to 0 \text{ as } t \to \infty$$

If, in addition, $\dot{r}(t)$ is bounded and $w_m(t)$ is persistently exciting, then the origin of (13.69)–(13.70) *is exponentially stable. In particular, for all initial states in a given ball of arbitrary radius, the state decays to the origin exponentially fast, satisfying* (13.73).   ◇

A significant question here is: What conditions on $r(t)$ would ensure the persistence of excitation of $w_m(t)$? Since $w_m(t)$ is the output of a linear system (with a Hurwitz transfer function) driven by $r(t)$, this question is addressed via linear analysis. We shall not pursue this issue here,[25] but let us note that the ultimate result of such investigation is that the input $r(t)$ should be "sufficiently rich" to guarantee that $w_m(t)$ is persistently exciting. For example, if $r(t)$ is the sum of distinct sinusoids, then it should have at least n distinct frequencies. The intuition behind the persistence of excitation condition is that the input should be rich enough to excite all modes of the system that is being probed.

Example 13.15 Consider the first-order plant $P(s) = k_p/(s - a_p)$ and let the reference model be $M(s) = 1/(s + 1)$. When $n = 1$, the filters (13.62)–(13.63) do not exist. The monic polynomial $\lambda(s)$ is taken simply as $\lambda(s) = 1$. Consequently, the constant c_0^* and the polynomials $c^*(s)$ and $d^*(s)$ of Figure 13.12 are given by $c_0^* = 1/k_p$, $c^*(s) = 0$, and $d^*(s) = -(a_p + 1)/k_p$. Hence,

$$\theta^* = \begin{bmatrix} 1/k_p \\ -(a_p + 1)/k_p \end{bmatrix}, \quad w = \begin{bmatrix} r \\ y_p \end{bmatrix}$$

The model reference adaptive controller is given by $u = \theta^T w$ and $\dot{\theta} = -\gamma e_o w$; that is,

$$u(t) = \theta_1(t)r(t) + \theta_2(t)y_p(t)$$

---

[25] For a frequency-domain approach, see [151, Chapter 2] while a time-domain approach can be found in [4, Chapter 2].

$$\dot{\theta}_1 = -\gamma[y_p(t) - y_m(t)]r(t)$$
$$\dot{\theta}_2 = -\gamma[y_p(t) - y_m(t)]y_p(t)$$

which is the same controller derived in Section 1.1.6. We have seen in Section 1.1.6 that the closed-loop system can be represented by

$$\dot{e}_o = -e_o + k_p\phi_1 r(t) + k_p\phi_2[e_o + y_m(t)]$$
$$\dot{\phi}_1 = -\gamma e_o r(t)$$
$$\dot{\phi}_2 = -\gamma e_o[e_o + y_m(t)]$$

where $e_o = y_p - y_m$, $\phi_1 = \theta_1 - \theta_1^*$, and $\phi_2 = \theta_2 - \theta_2^*$. In this example, the signal $w_m(t)$ is given by

$$w_m(t) = \left[\begin{array}{c} r(t) \\ y_m(t) \end{array} \right]$$

Supposing $r(t) \equiv 1$, the steady-state output of the model reference transfer function $1/(s + 1)$ is identically one. Take $y_m(t) \equiv 1$. The fact that we do not include transients in $y_m(t)$ can be justified in one of two ways. We can say that the initial condition of the model, which is at our disposal, has been chosen to cancel the transients. Alternatively, we can assume that $y_m(t)$ includes transients, but only the steady-state component of $y_m(t)$ is important when we study the asymptotic behavior of the system. For example, when we check the persistence of excitation condition (13.72) it is clear that transient components of the signal which decay to zero as $t \to \infty$ do not contribute to the positive definiteness which is required to hold uniformly for all $t \geq 0$. With $w_m(t) = [1 \ 1]^T$ for all t, we have

$$\int_t^{t+\delta} w_m(\tau)w_m^T(\tau) \, d\tau = \delta \left[\begin{array}{c} 1 \\ 1 \end{array} \right] [\, 1 \ \ 1 \,], \ \ \forall \, t \geq 0$$

which has rank one. Thus, $w_m(t)$ is not persistently exciting. All other assumptions of Theorem 13.3 are satisfied. Hence, we can conclude that $y_p(t) \to 1$ as $t \to \infty$, but we cannot conclude that the parameter errors ϕ_1 and ϕ_2 will converge to zero. With $r(t) \equiv y_m(t) \equiv 1$, the closed-loop system reduces to the autonomous system

$$\dot{e}_o = -e_o + k_p\phi_1 + k_p\phi_2(e_o + 1)$$
$$\dot{\phi}_1 = -\gamma e_o$$
$$\dot{\phi}_2 = -\gamma e_o(e_o + 1)$$

This system has an equilibrium set $\{e_o = 0, \ \phi_1 + \phi_2 = 0\}$, which shows clearly that the origin is not asymptotically stable because it is not even an isolated equilibrium point. Thus, in this example it is simple to see that the lack of persistence of excitation when $r(t) \equiv 1$ means that the parameter errors will not necessarily converge to zero.

Consider $r(t) = 1 + a \sin \omega t$. Again, we take $y_m(t)$ as the steady-state output of $1/(s+1)$, which is

$$y_m(t) = 1 + \frac{a}{\sqrt{1+\omega^2}} \sin(\omega t - \psi), \quad \tan \psi = \omega$$

To check the persistence of excitation of $w_m(t)$, consider the quadratic form

$$x^T w_m(\tau) w_m^T(\tau) x = \left\{ (1 + a \sin \omega \tau) x_1 + \left[1 + \frac{a}{\sqrt{1+\omega^2}} \sin(\omega \tau - \psi) \right] x_2 \right\}^2$$

To simplify the calculation while checking the persistence of excitation condition, we take $\delta = 2\pi/\omega$. Then,

$$\int_t^{t+\delta} x^T w_m(\tau) w_m^T(\tau) x \, d\tau = x^T N x$$

where

$$N = \frac{2\pi}{\omega} \begin{bmatrix} 1 + \frac{a^2}{2} & 1 + \frac{a^2}{2(1+\omega^2)} \\ 1 + \frac{a^2}{2(1+\omega^2)} & 1 + \frac{a^2}{2(1+\omega^2)} \end{bmatrix}$$

The determinant of N is given by

$$\det(N) = \frac{2\pi^2 a^2}{1+\omega^2} \left[1 + \frac{a^2}{2(1+\omega^2)} \right]$$

Hence, for all $\omega \neq 0$ and $a \neq 0$, the signal $w_m(t)$ is persistently exciting. It follows from Theorem 13.3 that with $r(t) = 1 + a \sin \omega t$, the origin of the closed-loop system is exponentially stable. Figures 13.14–13.17 show the response of the adaptive controller for the unstable plant $P(s) = 2/(s-1)$. In all cases, the initial conditions are taken as $y_p(0) = 0$, $y_m(0) = 0$, $\theta_1(0) = 1$, and $\theta_2(0) = -0.75$. Figure 13.14 shows the tracking and parameter errors when $r(t) = 1$ and $\gamma = 1$. The errors approach the equilibrium set $\{e_o = 0, \ \phi_1 + \phi_2 = 0\}$, but ϕ does not converge to zero. Figure 13.15 shows the same errors when $r(t) = 1 + \sin t$ and $\gamma = 1$. In this case, w_m is persistently exciting and ϕ converges to zero. In Figure 13.16, we lower the adaptation gain γ to $\gamma = 0.2$, which slows down the adaptation. Finally, we lower the amplitude of the sinusoidal term of $r(t) = 1 + a \sin t$ to $a = 0.7$. Figure 13.17 shows that decreasing a slows down the adaptation. A careful look at the figure show that the parameter errors move first, relatively fast, toward the set $\phi_1 + \phi_2 = 0$ and then converge slowly toward zero. This behavior is reasonable because a is a measure of the level of persistence of excitation of w_m. $\triangle$

Example 13.16 Consider the same plant and model as in the previous example, but this time assume that k_p is known and take it equal to one. In this case, only

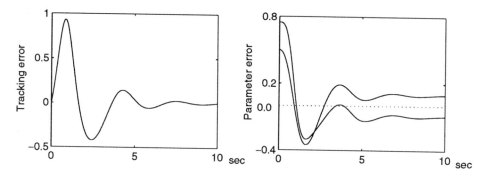

Figure 13.14: Response of the adaptive of controller of Example 13.15 when $r(t) = 1$ and $\gamma = 1$.

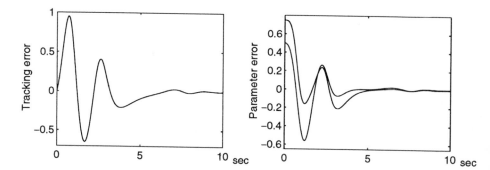

Figure 13.15: Response of the adaptive of controller of Example 13.15 when $r(t) = 1 + \sin t$ and $\gamma = 1$.

θ_2 needs to be adjusted on line. As we saw in Section 1.1.6, the closed-loop system is described by the second-order state equation

$$
\begin{aligned}
\dot{e}_o &= -e_o + k_p \phi(e_o + y_m(t)) \\
\dot{\phi} &= -\gamma e_o(e_o + y_m(t))
\end{aligned}
$$

where we dropped the subscript from ϕ_2. In this case, $r(t)$ does not appear in the equation; consequently, the signal $w_m(t)$ reduces to $w_m(t) = y_m(t)$. Consider a constant reference $r(t) \equiv r_0$. Then, $y_m(t) \equiv r_0$ and

$$
\int_t^{t+\delta} w_m(\tau) w_m^T(\tau) \, d\tau = r_0^2 \delta
$$

Thus, this time when there is only one parameter to adapt, a nonzero constant

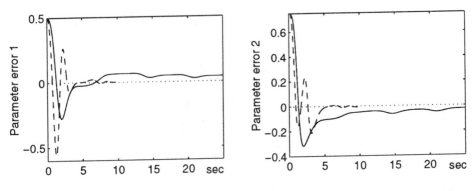

Figure 13.16: Response of the adaptive of controller of Example 13.15 with $r(t) = 1 + \sin t$ when $\gamma = 1$ (dashed) and $\gamma = 0.2$ (solid).

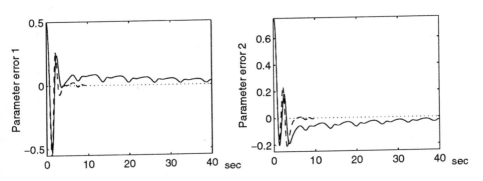

Figure 13.17: Response of the adaptive of controller of Example 13.15 with $\gamma = 1$ when $r(t) = 1 + \sin t$ (dashed) and $r(t) = 1 + 0.7 \sin t$ (solid).

reference is rich enough to guarantee persistence of excitation. From Theorem 13.3, we conclude that for any $r_0 \neq 0$, the origin will be exponentially stable. What about $r_0 = 0$? When $r_0 = 0$, the signal $w_m(t)$ is not persistently exciting, and we can only guarantee that $e_o(t) \to 0$ as $t \to \infty$. We saw in Example 3.10 that the set $\{e_o = 0\}$ is an equilibrium set; hence, the origin is not asymptotically stable. $\triangle$

A crucial question in studying the stability of any dynamical system is the robustness of the stability properties of the system to perturbations in the system's model. We have come across this question at several points, but we examined the question in some depth in Chapter 5. A fundamental fact that we gather from that chapter is that an exponentially stable equilibrium point is robust with respect to both vanishing and nonvanishing perturbations in the sense that the stability properties of the system will not be destroyed by arbitrarily small perturbations. Of course, with nonvanishing perturbations we cannot guarantee preservation of the exponential stability of the origin, but we can guarantee uniform ultimate boundedness. This fact is of fundamental importance when we analyze the adaptive control system under consideration. According to Theorem 13.3, if the persistence of excitation condition is satisfied, then the origin is exponentially stable. This statement carries with it a statement on the robustness of the adaptive control system. *We cannot ensure such robustness in the lack of persistence of excitation.*[26]

Example 13.17 Consider the adaptive control system of Example 13.15, but this time assume that the measured output is affected by measurement noise; that is, the plant is described by

$$\dot{x}_p = a_p x_p + k_p u$$
$$y_p = x_p + \nu(t)$$

where $\nu(t)$ is the measurement noise. To derive the equations of the closed-loop system, we have to repeat the derivation of Section 1.1.6. Doing so, it can be shown that the closed-loop system is described by

$$\dot{e} = -e + k_p \phi_1 r(t) + k_p \phi_2(e + y_m(t)) + k_p(\theta_2^* + \phi_2)\nu(t)$$
$$\dot{\phi}_1 = -\gamma e r(t) - \gamma r(t)\nu(t)$$
$$\dot{\phi}_2 = -\gamma e(e + y_m(t)) - \gamma(2e + y_m(t))\nu(t) - \gamma\nu^2(t)$$

where $e = x_p - y_m$. Denoting the state of the closed-loop system by $x = [e, \phi_1, \phi_2]^T$, we can rewrite the closed-loop state equation as

$$\dot{x} = f(t, x) + E(t)x + p(t)$$

where

$$f(t, x) = \begin{bmatrix} -x_1 + k_p x_2 r(t) + k_p x_3(x_1 + y_m(t)) \\ -\gamma x_1 r(t) \\ -\gamma x_1(x_1 + y_m(t)) \end{bmatrix},$$

$$E(t) = \begin{bmatrix} 0 & 0 & k_p \nu(t) \\ 0 & 0 & 0 \\ -2\gamma\nu(t) & 0 & 0 \end{bmatrix}, \text{ and } p(t) = \begin{bmatrix} k_p \theta_2^* \nu(t) \\ -\gamma r(t)\nu(t) \\ -\gamma y_m(t)\nu(t) - \gamma\nu^2(t) \end{bmatrix}$$

Assuming that $r(t)$ and $\dot{r}(t)$ are bounded and $w_m(t)$ is persistently exciting, the nominal system $\dot{x} = f(t, x)$ has an exponentially stable equilibrium point at the

---

[26] To ensure robustness in the lack of persistence of excitation, the adaptation law must be modified; see [122, Chapter 8], [151, Chapter 5], or [79, Chapter 8].

origin. The perturbed system has a vanishing perturbation $E(t)x$ and a nonvanishing perturbation $p(t)$; both are dependent on the measurement noise $\nu(t)$. Suppose that $|\nu(t)| \leq \delta_0 \leq 1$ for all $t \geq 0$. Since $r(t)$ and $y_m(t)$ are bounded, we have

$$\|E(t)\|_2 \leq k_1\delta_0; \quad \|p(t)\|_2 \leq k_2\delta_0$$

for some positive constants k_1 and k_2. To analyze the stability of the perturbed system, we start with a Lyapunov function of the nominal system. Since the origin of the nominal system is exponentially stable, (the converse Lyapunov) Theorem 3.12 guarantees the existence of a Lyapunov function $V(t, x)$ that satisfies the inequalities

$$c_1\|x\|_2^2 \leq V(t, x) \leq c_2\|x\|_2^2$$

$$\frac{\partial V}{\partial t} + \frac{\partial V}{\partial x}f(t, x) \leq -c_3\|x\|_2^2$$

$$\left\|\frac{\partial V}{\partial x}\right\|_2 \leq c_4\|x\|_2$$

over some domain $\{\|x\|_2 < r\}$ for some positive constants c_1, c_2, c_3, and c_4. Calculating the derivative of V along the trajectories of the perturbed system, we obtain

$$\dot{V} \leq -c_3\|x\|_2^2 + k_1c_4\delta_0\|x\|_2^2 + k_2c_4\delta_0\|x\|_2$$

Suppose $\delta_0 < c_3/2k_1c_4$. Then,

$$\dot{V} \leq -\tfrac{1}{2}c_3\|x\|_2^2 + k_2c_4\delta_0\|x\|_2$$

Suppose further that $\mu \overset{\text{def}}{=} (4k_2c_4\delta_0/c_3) < r\sqrt{c_1/c_2}$. Then,

$$\dot{V} \leq -\tfrac{1}{4}c_3\|x\|_2^2, \quad \forall \|x\|_2 \geq \mu$$

Hence, it follows from Theorem 5.1 (and Corollary 5.3) that for all $\|x(t_0)\|_2 < r\sqrt{c_1/c_2}$, the state of the perturbed system satisfies

$$\|x(t)\|_2 \leq k\|x(t_0)\|_2\exp[-\gamma_1(t - t_0)], \quad \forall t_0 \leq t < t_1$$

$$\|x(t)\|_2 \leq \mu\sqrt{c_2/c_1}, \quad \forall t \geq t_1$$

for some positive constants k and γ_1 and some finite time t_1. This shows that the state of the perturbed system will be bounded and will decay exponentially fast toward a neighborhood of the origin, where the norm of the state is guaranteed to be smaller than the uniform ultimate bound $\mu\sqrt{c_2/c_1}$. This ultimate bound is dependent on the size of the measurement noise as represented by δ_0. $\triangle$

We remarked before that the exponential stability statement of Theorem 13.3 is not global. This has an important implication when we study robustness of the adaptive control system. We know from Chapter 5 (or Chapter 6 on input-output stability) that if the origin of the nominal system is globally exponentially stable, then for any bounded disturbance, irrespective of its size, the state of the system will be bounded for all $t \geq 0$. In the lack of global exponential stability, we can guarantee boundedness of the state only if the size of the disturbance is limited as we have just done in the previous example. The next example shows that, if the size of the disturbance is not limited, a bounded disturbance may drive the state unbounded.

Example 13.18 Consider the adaptive control system of Example 13.16, and assume there is a disturbance $\nu(t)$ that affects the system at the plant input; that is, the plant is described by

$$\dot{y}_p = a_p y_p + u + \nu(t)$$

Take $y_m(t) \equiv r_0$. The closed-loop system is described by

$$\dot{e}_o = -e_o + \phi(e_o + r_0) + \nu(t)$$
$$\dot{\phi} = -\gamma e_o(e_o + r_0)$$

For $r_0 \neq 0$, the persistence of excitation condition is satisfied; hence, we can repeat the analysis of the previous example. We leave that to the reader (Exercise 13.44) and proceed to investigate the implication of having a large disturbance. Suppose $r_0 > 0$ and $\nu(t) = -\nu_0$ with $\nu_0 > r_0$. The closed-loop system is described by the second-order autonomous state equation

$$\dot{e}_o = -e_o + \phi(e_o + r_0) - \nu_0$$
$$\dot{\phi} = -\gamma e_o(e_o + r_0)$$

Let D denote the region of the state plane which lies to the left of the branch of the curve

$$\phi = \frac{e_o + \nu_0}{e_o + r_0} \tag{13.74}$$

on the left side of $e_o = -r_0$. The set D is a positively invariant set. This fact can be seen by noting that on the curve (13.74), $\dot{e}_o = 0$ and $\dot{\phi}$ is negative. Thus, the vector field on the curve (13.74) always points into the set D. It can be easily seen that there are no equilibrium points in D. Moreover, by an application of Bendixson's criterion,[27] it can be seen that there are no periodic orbits in D. Thus, for this second-order system, the only option left for the state is to escape to infinity. In fact, since $\dot{\phi}$ is always negative in D, it is clear that $\phi(t) \to -\infty$ as $t \to \infty$.

---

[27] See Theorem 7.2 and Exercise 7.5.

This confirms the point that the robustness of the system to bounded disturbance cannot be guaranteed if the disturbance is arbitrarily large. Notice that the size of the disturbance that drives the state unbounded is related to the size of the reference r_0. This can be given the following interpretation. For the persistence of excitation condition to hold, we must have $r_0 > 0$; the larger the value of r_0, the stronger the exponential stability of the nominal system and the more robust the system is. The analysis of this example has a serious implication in the case $r_0 = 0$. When $r_0 = 0$, the nominal system does not have exponential stability at the origin, although we were able to conclude that $e_o(t) \to 0$ as $t \to \infty$. The analysis we have just performed shows that with a constant disturbance $\nu(t) = -\nu_0$, the state of the system could go unbounded *even when ν_0 is arbitrarily small*. Thus, the property $e_o(t) \to 0$ of the nominal system is not robust because it cannot withstand arbitrarily small perturbations. Clearly, such an adaptive control system is not practical. $\triangle$

Relative Degree Two Plants

We describe briefly how to extend the foregoing results to plants with $n - m = 2$.[28] Let us start by considering the error model

$$\dot{e} = A_m e + b_m \phi^T w = A_m e + b_m[u - (\theta^*)^T w]$$
$$e_o = c_m^T e$$

or

$$E_o(s) = c_m^T(sI - A_m)^{-1}b_m[U(s) - (\theta^*)^T W(s)] = \frac{1}{c_0^*}M(s)[U(s) - (\theta^*)^T W(s)]$$

Since $M(s)$ has relative degree two, we cannot choose it to be strictly positive real. We can, however, choose $M(s)$ and $\alpha > 0$ such that $(s + \alpha)M(s)$ is strictly positive real. We rewrite the expression for $E_o(s)$ as

$$E_o(s) = \frac{1}{c_0^*}(s + \alpha)M(s)\left[\frac{1}{s + \alpha}U(s) - \frac{1}{s + \alpha}(\theta^*)^T W(s)\right]$$

Define $u_f(t)$ and $w_f(t)$ by

$$U_f(s) = \frac{1}{s + \alpha}U(s), \quad W_f(s) = \frac{1}{s + \alpha}W(s) \tag{13.75}$$

---

[28] Extension to plants of relative degree higher than two can be found in [79, Section 6.4.3]. The complexity of the controller increases considerably with the relative degree. For plants with high relative degree, other adaptive control approaches may be more appropriate.

Then, the error model can be written as

$$\dot{\bar{e}} = A_m \bar{e} + \bar{b}_m \phi^T w_f$$
$$e_o = c_m^T \bar{e}$$

where $\bar{b}_m$ is chosen such that[29]

$$c_m^T (sI - A_m)^{-1} \bar{b}_m = \frac{1}{c_0^*}(s + \alpha)M(s)$$

The new form of the error model resembles the form we encountered in the relative degree one case, with w replaced by w_f. The signal w_f can be obtained on line by filtering w using $1/(s + \alpha)$. By repeating the foregoing analysis, we arrive at the adaptive law

$$\dot{\theta} = -\gamma e_o w_f \qquad (13.76)$$

but now we need to choose u such that $u_f = \theta^T w_f$. From (13.75), it can be seen that

$$u = \theta^T w + \dot{\theta}^T w_f$$

results in

$$\frac{d}{dt}(u_f - \theta^T w_f) = -\alpha(u_f - \theta^T w_f)$$

Hence,

$$u_f = \theta^T w_f + e^{-\alpha t}[u_f(0) - \theta(0)w_f(0)]$$

The exponentially decaying term does not affect our earlier analysis,[30] and can be dropped for convenience. Substituting for $\dot{\theta}$ using the adaptive law (13.76), the control u can be implemented as

$$u = \theta^T w - \gamma e_o w_f^T w_f \qquad (13.77)$$

Equations (13.75) and (13.76), together with the filters (13.62), (13.63), and (13.70), describe the model reference adaptive controller. A result similar to Theorem 13.3 can be proved for relative degree two plants.[31]

---

[29] In particular, take $\bar{b}_m = A_m b_m + b_m \alpha$ and use the property $c_m^T b_m = 0$ which follows from the fact that $M(s)$ has relative degree two. Then,

$$\begin{aligned} c_m(sI - A_m)^{-1}(A_m b_m + b_m \alpha) &= c_m^T(sI - A_m)^{-1}[(-sI + A_m)b_m + b_m(s + \alpha)] \\ &= -c_m^T b_m + c_m^T(sI - A_m)^{-1}b_m(s + \alpha) \end{aligned}$$

[30] See Exercise 13.46.

[31] See [79, Theorem 6.4.2].

Example 13.19 Consider the plant and reference model transfer functions

$$P(s) = \frac{k_p}{s^2 + a_1 s + a_0}, \quad M(s) = \frac{\omega_n^2}{s^2 + 2\zeta\omega_n s + \omega_n^2}$$

where $\omega_n > 0$ and $0 < \zeta < 1$. The transfer function

$$(s + \alpha)M(s) = \frac{\omega_n^2(s + \alpha)}{s^2 + 2\zeta\omega_n s + \omega_n^2}$$

is strictly positive real for $0 < \alpha < 2\zeta\omega_n$.[32] Taking $\lambda(s) = s + \beta$ with $\beta > 0$, the model reference adaptive controller is given by

$$
\begin{aligned}
\dot{w}^{(1)} &= -\beta w^{(1)} + u \\
\dot{w}^{(2)} &= -\beta w^{(2)} + y_p \\
w &= [r, \ w^{(1)}, \ y_p, \ w^{(2)}]^T \\
\dot{w}_f &= -\alpha w_f + w \\
u &= \theta^T w - \gamma e_o w_f^T w_f \\
\dot{\theta} &= -\gamma e_o w_f
\end{aligned}
$$

$\triangle$

13.5 Exercises

Exercise 13.1 Consider the control law (13.10) in the case $n = 2$. Show that $w/\|w\|_2$ is a discontinuous function of w.

Exercise 13.2 Assume (13.4), (13.5), and (13.6) are satisfied with $\|\cdot\|_\infty$. Consider the following continuous approximation of the discontinuous control (13.11)

$$
v_i = \begin{cases}
-\frac{\eta(t,x)}{1-k}\,\text{sgn}(w_i), & \text{if } \eta(t,x)|w_i| \geq \epsilon \\[2mm]
-\frac{\eta^2(t,x)}{1-k} \cdot \frac{w_i}{\epsilon}, & \text{if } \eta(t,x)|w_i| < \epsilon
\end{cases}
$$

for $i = 1, 2, \ldots, p$, where $w^T = \frac{\partial V}{\partial x}(t, x)\, G(t, x)$.

(a) Show that

$$\dot{V} \leq -\alpha_3(\|x\|_\infty) + \sum_{i \in I}\left[\eta(t, x)|w_i| - \frac{\eta^2(t, x)|w_i|^2}{\epsilon}\right]\frac{1}{1-k}$$

where $i \in I$ if $\eta(t, x)|w_i| < \epsilon$.

---

[32] See Exercise 10.1.

(b) State and prove a theorem similar to Theorem 13.1 for the current controller.

Exercise 13.3 Consider the controller of Exercise 13.2. We want to prove a result similar to Corollary 13.1. Assume that $\alpha_3(\|x\|_\infty) \geq \phi^2(x)$, $\eta(t, x) \geq \eta_0 > 0$, and

$$|\delta_i| \leq \rho_1 \phi(x) + k|v_i|, \quad 0 \leq k < 1$$

for $i = 1, 2, \ldots, p$. These inequalities imply that (13.6) holds with $\|\cdot\|_\infty$ but they are more restrictive because the upper bound on $|\delta_i|$ depends only on $|v_i|$.

(a) Show that

$$\dot{V} \leq -\phi^2(x) + \sum_{i \in I} \left\{ -\eta_0^2 \frac{|w_i|^2}{\epsilon} + \rho_1 \phi(x)|w_i| \right\}$$

(b) State and prove a result similar to Corollary 13.1.

Exercise 13.4 Suppose inequality (13.6) takes the form

$$\|\delta(t, x, \psi(t, x) + v)\|_2 \leq \rho_0 + \rho_1 \phi(x) + k\|v\|_2, \quad 0 \leq k < 1$$

where $\phi(x) = \sqrt{\alpha_3(\|x\|_2)}$. Let $\eta(x) = \eta_0 + \eta_1 \phi(x)$, where $\eta_0 \geq \rho_0$ and $\eta_1 \geq \rho_1$. Consider the feedback control

$$v = \begin{cases} -\frac{\eta_0 + \eta_1 \phi(x)}{1-k} \cdot \frac{w}{\|w\|_2}, & \text{if } \|w\|_2 \geq \epsilon \\[2mm] -\frac{\eta_0 + \eta_1 \phi(x)}{1-k} \cdot \frac{w}{\epsilon}, & \text{if } \|w\|_2 < \epsilon \end{cases}$$

(a) Show that the derivative of V along the trajectories of the closed-loop system (13.9) satisfies

$$\dot{V} \leq -\phi^2(x) + \frac{\epsilon}{4}[\rho_0 + \rho_1 \phi(x)] \leq -\frac{1}{2}\alpha_3(\|x\|_2) + \frac{\epsilon^2 \rho_1^2}{32} + \frac{\epsilon \rho_0}{4}$$

(b) Apply Theorem 5.1 to arrive at a conclusion similar to Theorem 13.1.

(c) Compare this controller with (13.12).

Exercise 13.5 Consider the system

$$\dot{x}_1 = x_2$$
$$\dot{x}_2 = u + \delta(x)$$

where δ is unknown but we know an estimate ρ_1 such that $|\delta(x)| \leq \rho_1 \|x\|_2$. Let $u = \psi(x) = -x_1 - x_2$ be the nominal stabilizing control, and

$$v = \begin{cases} -\rho_1 \|x\|_2 \frac{w}{\|w\|_2}, & \text{if } \rho_1 \|x\|_2 \|w\|_2 \geq \epsilon \\[2mm] -\rho_1^2 \|x\|_2^2 \frac{w}{\epsilon}, & \text{if } \rho_1 \|x\|_2 \|w\|_2 < \epsilon \end{cases}$$

where $w^T = 2x^T PB$ and $V(x) = x^T Px$ is a Lyapunov function for the nominal closed-loop system. Apply the control $u = -x_1 - x_2 + v$.

(a) Verify that all the assumptions of Corollary 13.1 are satisfied except (13.16) which holds with $\eta_0 = 0$.

(b) Show that when $\delta(x) = 2(x_1 + x_2)$ and $\rho_1 = 2\sqrt{2}$, the origin is unstable.

Exercise 13.6 Consider the problem treated in Section 13.1.1 and assume that (13.6) is satisfied with 2-norm. Suppose further that (13.15) and (13.17) are satisfied. Show that the control law

$$u = \psi(t, x) - \gamma w; \quad w^T = \frac{\partial V}{\partial x} G(t, x)$$

with sufficiently large γ will stabilize the origin.

Exercise 13.7 Consider the problem treated in Section 13.1.1. Show that, instead of using the control law $u = \psi(t, x) + v$, we can simply use $u = v$, where $\rho(t, x)$ is taken from the inequality

$$\|\delta(t, x, u) - \psi(t, x)\|_2 \le \rho(t, x) + k\|u\|_2, \quad 0 \le k < 1$$

Exercise 13.8 Suppose that, in addition to the matched uncertainty δ, the system (13.1) has also unmatched uncertainty Δ; that is,

$$\dot{x} = f(t, x) + \Delta(t, x) + G(t, x)[u + \delta(t, x, u)]$$

Suppose that, over a domain $D \subset R^n$, all the assumptions of Theorem 13.1 are satisfied, inequalities (13.15)–(13.17) are satisfied, and the unmatched uncertainty satisfies $\|[\partial V/\partial x]\Delta(t, x)\|_2 \le \mu\phi^2(x)$, for some $\mu \ge 0$. Let $u = \psi(t, x) + v$, where v is determined by (13.12). Show that if $\mu < 1$, then the feedback control will stabilize the origin of the closed-loop system provided ϵ is chosen small enough to satisfy $\epsilon < 4(1 - \mu)\eta_0^2/\rho_1^2$.

Exercise 13.9 Consider the system $\dot{x} = f(x) + g(x)[u + \delta(x, u)]$ and suppose there are known smooth functions $\psi(x)$, $V(x)$, and $\rho(x)$, all vanishing at $x = 0$, and a known constant k such that

$$c_1\|x\|^2 \le V(x) \le c_2\|x\|^2, \quad \frac{\partial V}{\partial x}[f(x) + g(x)\psi(x)] \le -c_3\|x\|^2$$

$$\|\delta(x, \psi(x) + v)\| \le \rho(x) + k\|v\|, \quad 0 \le k < 1, \quad \forall\, x \in R^n,\ \forall\, v \in R^p$$

where c_1 to c_3 are positive constants.

(a) Show that it is possible to design a continuous state feedback control $u = \gamma(x)$ such that the origin of

$$\dot{x} = f(x) + g(x)[\gamma(x) + \delta(x, \gamma(x))]$$

is globally exponentially stable.

(b) Apply the result of part (a) to the system

$$\dot{x}_1 = x_2$$
$$\dot{x}_2 = (1 + a_1)(x_1^3 + x_2^3) + (1 + a_2)u$$

where a_1 and a_2 are unknown constants which satisfy $|a_1| \leq 1$ and $|a_2| \leq \frac{1}{2}$.

Exercise 13.10 Consider a single-input–single-output system in the normal form (12.25)–(12.27) and suppose that (12.25) is input-to-state stable. Let $\hat{\alpha}(x)$ and $\hat{\beta}(x)$ be nominal models of the functions $\alpha(x)$ and $\beta(x)$, respectively, and suppose that the modeling errors satisfy the inequalities

$$\left| \frac{\hat{\alpha}(x) - \alpha(x)}{\beta(x)} \right| \leq \rho_0(x), \quad \left| \frac{\hat{\beta}(x) - \beta(x)}{\beta(x)} \right| \leq k < 1$$

where the function $\rho_0(x)$ and the constant k are known. Let $y_R(t)$ be a reference signal and suppose y_R and its derivatives up to $y_R^{(r)}$ are continuous and bounded. Using the Lyapunov redesign technique, design a robust state feedback control (using x and ξ) such that the output y asymptotically tracks y_R with prespecified tolerance μ; that is, $|y(t) - y_R(t)| \leq \mu$ for all $t \geq T$ for some finite time T.

Exercise 13.11 Repeat the previous exercise for the case of constant reference using integral control; that is, augment $\int_0^t [y(\tau) - y_R] \, d\tau$ as an additional state variable and then design the robust state feedback control to ensure asymptotic regulation. Show that, due to the inclusion of integral control, the regulation error converges to zero.

Exercise 13.12 The tank system of Example 12.22 (with integrator included) is given by

$$\dot{y} = \frac{1}{\beta(y)}(u - c\sqrt{2y})$$
$$\dot{\sigma} = y - y_R$$

where y_R is a desired set point. Let $\hat{c}$ and $\hat{\beta}(y)$ be nominal models of c and $\beta(y)$, respectively, and suppose that you know nonnegative constants $\rho_1 < 1$ and ρ_2 such that

$$\left| \frac{\beta(y) - \hat{\beta}(y)}{\beta(y)} \right| \leq \rho_1, \quad |\hat{c} - c| \leq \rho_2$$

Use Lyapunov redesign to design a robust state feedback controller to ensure that all state variables are bounded and $|y(t) - y_R|$ converges to zero as $t \to \infty$.

Exercise 13.13 For each of the following scalar systems, use the nonlinear damping tool to design a state feedback control that guarantees boundedness of the state $x(t)$ as well as uniform ultimate boundedness by an ultimate bound μ. The function $\delta(t)$ is bounded for all $t \geq 0$ but we do not know an upper bound on $|\delta(t)|$.

$$(1) \quad \dot{x} = -x + x^2[u + \delta(t)], \quad (2) \quad \dot{x} = x^2[1 + \delta(t)] - xu$$

Exercise 13.14 Using backstepping, design a state feedback control law to globally stabilize the origin of the system

$$\dot{x}_1 = x_2 + a + (x_1 - a^{1/3})^3$$
$$\dot{x}_2 = x_1 + u$$

where a is a known constant.

Exercise 13.15 ([97]) Consider the system

$$\dot{x}_1 = x_1 x_2$$
$$\dot{x}_2 = x_1 + u$$

(a) Using backstepping, design a state feedback control law to globally stabilize the origin.

(b) Can you globally stabilize the origin via feedback linearization?

Exercise 13.16 ([97]) Consider the system

$$\dot{x}_1 = -x_2 - \tfrac{3}{2}x_1^2 - \tfrac{1}{2}x_1^3$$
$$\dot{x}_2 = u$$

(a) Using backstepping, design a **linear** state feedback control law to globally stabilize the origin.
   Hint: Avoid cancellation of nonlinear terms.

(b) Design a globally stabilizing state feedback control law using feedback linearization.

(c) Compare the two designs. Using computer simulations, compare their performance and the control effort used in each case.

Exercise 13.17 Consider the system

$$\dot{x}_1 = -x_1 + x_1^2[x_2 + \delta(t)]$$
$$\dot{x}_2 = u$$

where $\delta(t)$ is a bounded function of t for all $t \geq 0$, but we do not know an upper bound on $|\delta(t)|$. By combining backstepping and the nonlinear damping tools, design a state feedback control law that ensures global boundedness of the state x for all initial states $x(0) \in R^2$.

Exercise 13.18 Repeat Exercise 13.17 for the system

$$\dot{x}_1 = -x_1 x_2 + x_1^2[1 + \delta(t)]$$
$$\dot{x}_2 = u$$

Exercise 13.19 Consider the magnetic suspension system of Exercise 11.5 and suppose the ball is subject to a vertical force disturbance $d(t)$; that is,

$$m\ddot{y} = -k\dot{y} + mg + F(y, i) + d(t)$$

Suppose further that $|d(t)| \leq d_0$ for all $t \geq 0$, where the upper bound d_0 is known.

(a) Viewing the force F as the control input, use Lyapunov redesign to design a state feedback control $F = \gamma(y, \dot{y})$ such that $|y - y_R|$ is ultimately bounded by μ, where μ is a design parameter that can be chosen arbitrarily small. Design your control such that γ is a continuously differentiable function of its argument.

(b) Using backstepping, design a state feedback control law for the voltage input u that will ensure that $|y - y_R|$ is ultimately bounded by μ.

Exercise 13.20 Consider the system

$$\dot{x}_1 = x_2$$
$$\dot{x}_2 = x_1 - x_1^3 + u$$
$$y = x_1 - x_2$$

(a) Find a smooth state feedback $u = \psi(x)$ such that the origin is globally exponentially stable.

(b) Extend the dynamics of the system by connecting an integrator in series with the input, to obtain

$$\dot{x}_1 = x_2$$
$$\dot{x}_2 = x_1 - x_1^3 + z$$
$$\dot{z} = v$$

Using backstepping, find a smooth state feedback $v = \phi(x, z)$ such that the origin is globally exponentially stable.

Exercise 13.21 Using backstepping, redesign the state feedback control of Example 13.11 to make the origin globally asymptotically stable.
Hint: The first step is the same. In the second step, do not bound the term $\theta_2 x_2^2$ by a linear growth bound. Use Lyapunov redesign to cancel the destabilizing effect of this quadratic term.

Exercise 13.22 Repeat Exercise 13.12 using backstepping.

Exercise 13.23 Consider the system of Exercise 12.13.

(a) Starting with the $\dot{x}_1$-equation and stepping back to the the $\dot{x}_2$-equation, design a state feedback control $u = \psi(x)$ such that the origin $(x_1 = 0,\ x_2 = 0)$ of the first two equations is globally exponentially stable.

(b) Show that, under the feedback control of part (a), the origin $x = 0$ of the full system is globally exponentially stable.
Hint: Use input-to-state properties of the third equation.

(c) Compare this controller with the one designed in Exercise 12.13. Use computer simulation in your comparison and take $a = b = 1$.

Exercise 13.24 Consider the system

$$\begin{aligned}
\dot{x}_1 &= x_2 + \theta x_1^2 \\
\dot{x}_2 &= x_3 + u \\
\dot{x}_3 &= x_1 - x_3 \\
y &= x_1
\end{aligned}$$

where $\theta \in [0, 2]$. Using backstepping, design a state feedback control law such that $|y - a\sin t|$ is ultimately bounded by μ, where μ is a design parameter that can be chosen arbitrarily small. Assume that $|a| \le 1$ and $\|x(0)\|_\infty \le 1$.

Exercise 13.25 Consider the system (13.40).

(a) Let G be a constant matrix of rank p. Show that there is an $n \times n$ nonsingular matrix M such that $MG = [0,\ G_2^T]^T$, where G_2 is a $p \times p$ nonsingular matrix. Verify that $T(x) = Mx$ satisfies (13.42).

(b) Let $G(x)$ be a smooth function of x and assume that G has rank p for all x in a domain $D_1 \subset R^n$. Let $\Delta = \text{span}\{g_1, \ldots, g_p\}$, where $g_1, \ldots, g_p$ are the columns of G. Suppose Δ is involutive. Show that for every $x_0 \in D_1$ there exist smooth functions $\phi_1(x), \ldots, \phi_{n-p}(x)$, with linearly independent differentials $\frac{\partial \phi_1}{\partial x}, \ldots, \frac{\partial \phi_{n-p}}{\partial x}$ at x_0, such that $\frac{\partial \phi_i}{\partial x} G(x) = 0$ for $1 \leq i \leq n - p$. Show that we can find smooth functions $\phi_{n-p+1}(x), \ldots, \phi_n(x)$ such that $T(x) = [\phi_1(x), \ldots, \phi_n(x)]^T$ is a diffeomorphism in a neighborhood of x_0. Verify that $T(x)$ satisfies (13.42).

Hint: Use the differential geometric setup of Section 12.4 and apply Frobenius theorem.

Exercise 13.26 Suppose that, in addition to all the assumptions of Theorem 13.2 we have $\delta_\xi(0, 0, 0) = 0$ and the origin of $\dot\eta = f(\eta, \phi(\eta)) + \delta_\eta(\eta, \phi(\eta))$ is exponentially stable. Show that, for sufficiently small ϵ, the origin of the closed-loop system will be exponentially stable.

Exercise 13.27 Consider the nonautonomous regular form (13.58)–(13.59). Suppose there is a continuously differentiable function $\phi(t, \eta)$, with $\phi(t, 0) = 0$, such that the origin of $\dot\eta = f(t, \eta, \phi(t, y)) + \delta_\eta(t, \eta, \phi(t, y))$ is uniformly asymptotically stable. Let

$$u_{eq} = G_a^{-1}\left(-f_a + \frac{\partial \phi}{\partial t} + \frac{\partial \phi}{\partial \eta} f\right), \quad \text{and} \quad \Delta = G_a \delta_\xi - \frac{\partial \phi}{\partial \eta} \delta_\eta$$

Suppose Δ satisfies inequality (13.50) with $\rho = \rho(t, \eta, \xi)$. Taking $z = \xi - \phi(t, \eta)$ design a sliding mode controller to stabilize the origin. State and prove a theorem similar to Theorem 13.2.

Exercise 13.28 Consider the problem treated in Section 13.3. Show that, instead of using the control law $u = u_{eq} + G_a^{-1} v$, we can simply use $u = G_a^{-1} v$, where $\rho(\eta, \xi)$ is taken from the inequality

$$\left\| f_a + G_a \delta_\xi - \frac{\partial \phi}{\partial \eta}(f + \delta_\eta) \right\|_\infty \leq \rho(\eta, \xi) + k\|v\|_\infty, \quad 0 \leq k < 1$$

Exercise 13.29 Study the continuous approximation of the sliding mode controller (13.47), (13.52) when $\text{sgn}(z)$ is approximated by $z/(\epsilon + |z|)$, $\epsilon > 0$.

Exercise 13.30 Study the continuous approximation of the sliding mode controller (13.47), (13.52) when $\text{sgn}(z)$ is approximated by $\tanh(z/\epsilon)$, $\epsilon > 0$.

Exercise 13.31 With reference to Example 13.12, verify that the conditions of Theorem 13.2 are satisfied globally.

Exercise 13.32 With reference to Example 13.14, show that with the "continuous" sliding mode controller, there exists a finite time T, possibly dependent on ϵ and the initial states, and a positive constant k, independent of ϵ and the initial states, such that $|y(t) - y_R(t)| \leq k\epsilon$ for all $t \geq T$.

Exercise 13.33 Redesign the controller of Example 13.4 using sliding mode control.

Exercise 13.34 Repeat Exercise 13.12 using sliding mode control.

Exercise 13.35 Consider the system

$$\dot{x}_1 = x_2 + \sin x_1$$
$$\dot{x}_2 = \theta_1 x_1^2 + (1 + \theta_2)u$$
$$y = x_1$$

where $|\theta_1| \leq 2$ and $|\theta_2| \leq \frac{1}{2}$. Design a sliding-mode state feedback control so that the output $y(t)$ tracks a reference signal $y_R(t)$. Assume that y_R, $\dot{y}_R$, and $\ddot{y}_R$ are continuous and bounded.

Exercise 13.36 ([157]) Consider the controlled van der Pol equation

$$\dot{x}_1 = x_2$$
$$\dot{x}_2 = -\omega^2 x_1 - \epsilon\omega(1 - \mu^2 x_1^2)x_2 u$$

where ω, ϵ and μ are positive constants, and u is a control input.

(a) Show that for $u = 1$ there is a stable limit cycle outside the circle of radius $1/\mu$, and for $u = -1$ there is an unstable limit cycle outside the same circle.

(b) Let $\sigma = x_1^2 + x_2^2/\omega^2 - r^2$, with $r < 1/\mu$. Show that restricting the motion of the system to the surface $\sigma = 0$ (that is, $\sigma(t) \equiv 0$) results in a harmonic oscillator

$$\dot{x}_1 = x_2, \qquad \dot{x}_2 = -\omega^2 x_1$$

which produces a perfect sinusoidal oscillation of frequency ω and amplitude r.

(c) Design a state feedback sliding mode controller to drive all trajectories in the band $|x_1| < 1/\mu$ to the manifold $\sigma = 0$ and have it slide on that manifold.

(d) Simulate the response of this system for the ideal sliding mode controller and for a continuous approximation. Use $\omega = \mu = \epsilon = 1$.

Exercise 13.37 Use the Numerical data of Example 13.12 to simulate the Lyapunov redesign of Example 13.3. Choose the design parameters of the Lyapunov redesign to obtain the same level of control effort used in the sliding mode design. Compare the performance of the two controllers.

Exercise 13.38 Consider the pendulum equation $\ddot{\theta} + a\sin\theta + b\dot{\theta} = cu$, where $5 \le a \le 15$, $0 \le b \le 0.2$, and $6 \le c \le 12$. It is desired to stabilize the pendulum at the position $\theta = \delta_1$. It is also desired to use integral control, which can be achieved by augmenting the integral of $(\theta - \delta_1)$ with the given equation to obtain a third-order state model.

(a) Design a state feedback controller using Lyapunov redesign. Verify that the tracking error converges to zero.

(b) Design a state feedback controller using "continuous" sliding mode control. Verify that the tracking error converges to zero.

(c) Using simulation, compare the performance of the two controllers. In the simulation, you may use $a = 10$, $b = 0.1$, and $c = 10$.

Exercise 13.39 Design a model reference adaptive controller for

$$P(s) = \frac{k_p(s + b_0)}{s^2 + a_1 s + a_0} \quad \text{and} \quad M(s) = \frac{1}{s + 1}$$

Simulate the closed-loop system for $P(s) = (s+1)/(s-1)(s+2)$. In your simulation, use $r = 1$, $r = 1 + \sin t$, and $r = \sin t + \sin 1.5t$.

Exercise 13.40 Repeat Exercise 13.39 when k_p is known.

Exercise 13.41 Simulate the closed-loop system of Example 13.19 when $k_p = 1$, $a_1 = 1$, $a_0 = 0$, $\omega_n = 2$, and $\zeta = 0.5$. Try different choices of $r(t)$.

Exercise 13.42 Design a model reference adaptive controller for

$$P(s) = \frac{k_p(s + b_0)}{s^3 + a_2 s^2 + a_1 s + a_0} \quad \text{and} \quad M(s) = \frac{\omega_n^2}{s^2 + 2\zeta\omega_n s + \omega_n^2}, \ \omega_n > 0, \ 0 < \zeta < 1$$

Exercise 13.43 Consider the adaptive control system of Example 13.15 and assume that there is a disturbance $\nu(t)$ that affects the system at the plant input; that is,

$$\dot{y}_p = a_p y_p + k_p u + \nu(t)$$

Analyze the stability of the system when $r(t) = 1 + a\sin\omega t$.

Exercise 13.44 Consider Example 13.18 and assume $r_0 \ne 0$. Analyze the stability of the adaptive control system.

Exercise 13.45 Consider the adaptive control system of Example 13.16 and assume that there is measurement noise $\nu(t)$; that is,

$$\dot{x}_p = a_p x_p + k_p u, \quad y_p = x_p + \nu(t)$$

Analyze the stability of the system when $r(t) = r_0 \ne 0$.

Exercise 13.46 ([79]) Suppose equations (13.69)–(13.70) are modified to

$$\dot{e} = A_m e + b_m \phi^T w + \rho(t)$$
$$\dot{\phi} = -\gamma c_m^T e w$$

where $\rho(t)$ is an exponentially decaying function of time that can be modeled by

$$\dot{\eta} = A_0 \eta, \quad \rho = C_0 \eta$$

where A_0 is a Hurwitz matrix. Let P_0 be the unique positive definite solution of the Lyapunov equation $P_0 A_0 + A_0^T P_0 = -I$. Using a Lyapunov function candidate of the form $V_1 = V(e, \phi) + \gamma_1 \eta^T P_0 \eta$, where $\gamma_1 > 0$ and $V(e, \phi)$ is the same Lyapunov function used with (13.69)–(13.70), show that all the conclusions of Theorem 13.3 can be extended to the current problem.

Exercise 13.47 Consider the single-input–single-output nonlinear system

$$\dot{x}_i = x_{i+1}, \quad 1 \leq i \leq n-1$$
$$\dot{x}_n = f_0(x) + (\theta^*)^T f_1(x) + g_0(x)u$$

where f_0, f_1, and g_0 are known smooth functions of x, defined for all $x \in R^n$, while $\theta^* \in R^p$ is a vector of unknown constant parameters. The function $g_0(x)$ is bounded away from zero; that is, $|g_0(x)| \geq k_0 > 0$, for all $x \in R^n$. We assume that all state variables can be measured. It is desired to design a state feedback adaptive controller such that x_1 asymptotically tracks a desired reference signal $y_R(t)$, where y_R and its derivatives up to $y_R^{(n)}$ are continuous and bounded for all $t \geq 0$.

(a) Taking $e_i = x_i - y_R^{(i-1)}$ for $1 \leq i \leq n$ and $e = [e_1, \ldots, e_n]^T$, show that e satisfies the equation

$$\dot{e} = Ae + B[f_0(x) + (\theta^*)^T f_1(x) + g_0(x)u - y_R^{(n)}]$$

where (A, B) is a controllable pair.

(b) Design K such that $A + BK$ is Hurwitz and let P be the positive definite solution of the Lyapunov equation $P(A + BK) + (A + BK)^T P = -I$. Using the Lyapunov function candidate $V = e^T Pe + \phi^T \Gamma^{-1} \phi$, where $\phi = \theta - \theta^*$ and Γ is a symmetric positive definite matrix, show that the adaptive controller

$$u = \frac{1}{g_0(x)} \left[-f_0(x) - \theta^T f_1(x) + y_R^{(n)} + Ke \right]$$
$$\dot{\theta} = \Gamma f_1(x) e^T PB$$

ensures that all state variables are bounded and the tracking error converges to zero as $t \to \infty$.

(c) Let $w_m = f(\mathcal{Y}_d)$, where $\mathcal{Y}_d = [y_R, \ldots, y_R^{(n-1)}]^T$. Show that if w_m is persistently exciting, then the parameter error ϕ converges to zero as $t \to \infty$.

Appendix A

A.1 Proof of Lemma 2.5

The upper right-hand derivative $D^+ v(t)$ is defined by

$$D^+ v(t) = \lim_{h \to 0+} \sup \frac{v(t+h) - v(t)}{h}$$

where $\lim \sup_{n \to \infty}$ (the limit superior) of a sequence of real numbers $\{x_n\}$ is a real number y satisfying:

- for every $\epsilon > 0$ there exists an integer N such that $n > N$ implies $x_n < y + \epsilon$;

- given $\epsilon > 0$ and given $m > 0$, there exists an integer $n > m$ such that $x_n > y - \epsilon$.

The first statement means that ultimately all terms of the sequence are less than $y + \epsilon$, and the second one means that infinitely many terms are greater than $y - \epsilon$. One of the properties of $\lim \sup^1$ is that if $z_n \le x_n$ for each $n = 1, 2 \ldots$, then $\lim \sup_{n \to \infty} z_n \le \lim \sup_{n \to \infty} x_n$. From this property we see that if $|v(t+h) - v(t)|/h \le g(t, h)$, $\forall\, h \in (0, b]$ and $\lim_{h \to 0+} g(t, h) = g_0(t)$, then $D^+ v(t) \le g_0(t)$.

To prove lemma 2.5, consider the differential equation

$$\dot{z} = f(t, z) + \lambda, \qquad z(t_0) = u_0 \tag{A.1}$$

where λ is a positive constant. On any compact interval $[t_0, t_1]$, we conclude from Theorem 2.6 that for any $\epsilon > 0$ there is $\delta > 0$ such that if $\lambda < \delta$ then (A.1) has a unique solution $z(t, \lambda)$ defined on $[t_0, t_1]$ and

$$|z(t, \lambda) - u(t)| < \epsilon, \qquad \forall\, t \in [t_0, t_1] \tag{A.2}$$

---

[1] See [7, Theorem 12-4].

Claim 1: $v(t) \leq z(t, \lambda)$ for all $t \in [t_0, t_1]$.

This claim can be shown by contradiction, for if it was not true there would be times $a, b \in (t_0, t_1]$ such that $v(a) = z(a, \lambda)$ and $v(t) > z(t, \lambda)$ for $a < t \leq b$. Consequently,

$$v(t) - v(a) > z(t, \lambda) - z(a, \lambda), \quad \forall\, t \in (a, b]$$

which implies

$$D^+ v(a) \geq \dot{z}(a, \lambda) = f(a + z(a, \lambda)) + \lambda > f(a, v(a))$$

which contradicts the inequality $D^+ v(t) \leq f(t, v(t))$.

Claim 2: $v(t) \leq u(t)$ for all $t \in [t_0, t_1]$.

Again, this claim can be shown by contradiction, for if it was not true there would exist $a \in (t_0, t_1]$ such that $v(a) > u(a)$. Taking $\epsilon = [v(a) - u(a)]/2$ and using (A.2), we obtain

$$v(a) - z(a, \lambda) = v(a) - u(a) + u(a) - z(a, \lambda) \geq \epsilon$$

which contradicts the statement of Claim 1.

Thus we have shown that $v(t) \leq u(t)$ for all $t \in [t_0, t_1]$. Since this is true on every compact interval, we conclude that it holds for all $t \geq t_0$. If this was not the case, let $T < \infty$ be the first time the inequality is violated. We have $v(t) \leq u(t)$ for all $t \in [t_0, T)$, and by continuity $v(T) = u(T)$. Thus, we can extend the inequality to the interval $[T, T + \Delta]$ for some $\Delta > 0$, which contradicts the claim that T is the first time the inequality is violated.

A.2   Proof of Lemma 3.1

Since $x(t)$ is bounded, by the Bolzano-Weierstrass theorem,[2] it has an accumulation point as $t \to \infty$; hence the positive limit set L^+ is nonempty. For each $y \in L^+$, there is a sequence t_i with $t_i \to \infty$ as $i \to \infty$ such that $x(t_i) \to y$ as $i \to \infty$. Since $x(t_i)$ is bounded uniformly in i, the limit y is bounded; that is, L^+ is bounded. To show that L^+ is closed, let $\{y_i\} \in L^+$ be some sequence such that $y_i \to y$ as $i \to \infty$ and prove that $y \in L^+$. For every i, there is a sequence $\{t_{ij}\}$ with $t_{ij} \to \infty$ as $j \to \infty$ such that $x(t_{ij}) \to y_i$ as $j \to \infty$. We shall now construct a particular sequence $\{\tau_i\}$. Given the sequences t_{ij}, choose $\tau_2 > t_{12}$ such that $\|x(\tau_2) - y_2\| < \frac{1}{2}$; choose $\tau_3 > t_{13}$ such that $\|x(\tau_3) - y_3\| < \frac{1}{3}$; and so on for $i = 4, 5, \dots$. Of course, $\tau_i \to \infty$ as $i \to \infty$ and $\|x(\tau_i) - y_i\| < 1/i$ for every i. Now, given $\epsilon > 0$ there are positive integers N_1 and N_2 such that

$$\|x(\tau_i) - y_i\| < \frac{\epsilon}{2}, \quad \forall\, i > N_1 \quad \text{and} \quad \|y_i - y\| < \frac{\epsilon}{2}, \quad \forall\, i > N_2$$

---

[2] See [7].

The first inequality follows from $\|x(\tau_i) - y_i\| < 1/i$ and the second one from the limit $y_i \to y$. Thus,

$$\|x(\tau_i) - y\| < \epsilon, \quad \forall\, i > N = \max\{N_1, N_2\}$$

which shows that $x(\tau_i) \to y$ as $i \to \infty$. Hence, L^+ is closed. This proves that the set L^+ is compact because it is closed and bounded.

To show that L^+ is an invariant set, let $y \in L^+$ and $\phi(t, y)$ be the solution of (3.1) that passes through y at $t = 0$; that is, $\phi(0, y) = y$ and show that $\phi(t, y) \in L^+$, $\forall\, t \in R$. There is a sequence $\{t_i\}$ with $t_i \to \infty$ as $i \to \infty$ such that $x(t_i) \to y$ as $i \to \infty$. Write $x(t_i) = \phi(t_i, x_0)$, where x_0 is the initial state of $x(t)$ at $t = 0$. By uniqueness of the solution,

$$\phi(t + t_i, x_0) = \phi(t, \phi(t_i, x_0)) = \phi(t, x(t_i))$$

where, for sufficiently large i, $t + t_i \geq 0$. By continuity,

$$\lim_{i \to \infty} \phi(t + t_i, x_0) = \lim_{i \to \infty} \phi(t, x(t_i)) = \phi(t, y)$$

which shows that $\phi(t, y) \in L^+$.

Finally, to show that $x(t) \to L^+$ as $t \to \infty$, use a contradiction argument. Suppose this is not the case; then, there is an $\epsilon > 0$ and a sequence $\{t_i\}$ with $t_i \to \infty$ as $i \to \infty$ such that $\text{dist}(x(t_i), L^+) > \epsilon$. Since the sequence $x(t_i)$ is bounded, it contains a convergent subsequence $x(t_i') \to x^*$ as $i \to \infty$. The point x^* must belong to L^+ and at the same time be at a distance ϵ from L^+, which is a contradiction.

A.3   Proof of Lemma 3.3

Uniform Stability: Suppose there is a class $\mathcal{K}$ function α such that

$$\|x(t)\| \leq \alpha(\|x(t_0)\|), \quad \forall\, t \geq t_0 \geq 0, \ \forall\, \|x(t_0)\| < c$$

Given $\epsilon > 0$, let $\delta = \min\{c, \alpha^{-1}(\epsilon)\}$. Then, for $\|x(t_0)\| < \delta$, we have

$$\|x(t)\| \leq \alpha(\|x(t_0)\|) < \alpha(\delta) \leq \alpha(\alpha^{-1}(\epsilon)) = \epsilon$$

Now assume that, given $\epsilon > 0$, there is $\delta = \delta(\epsilon) > 0$ such that

$$\|x(t_0)\| < \delta \Rightarrow \|x(t)\| < \epsilon, \quad \forall\, t \geq t_0$$

For a fixed ϵ, let $\bar{\delta}(\epsilon)$ be the supremum of all applicable $\delta(\epsilon)$. Then,

$$\|x(t_0)\| < \bar{\delta}(\epsilon) \Rightarrow \|x(t)\| < \epsilon, \quad \forall\, t \geq t_0$$

and, if $\delta_1 > \bar{\delta}(\epsilon)$, then there is at least one initial state $\bar{x}(t_0)$ such that

$$\|\bar{x}(t_0)\| < \delta_1 \quad \text{and} \quad \sup_{t \geq t_0} \|x(t)\| \geq \epsilon$$

The function $\bar{\delta}(\epsilon)$ is positive and nondecreasing, but not necessarily continuous. We choose a class $\mathcal{K}$ function $\zeta(r)$ such that $\zeta(r) \leq k\bar{\delta}(r)$, with $0 < k < 1$. Let $\alpha(r) = \zeta^{-1}(r)$. Then, $\alpha(r)$ is class $\mathcal{K}$. Let $c = \lim_{r \to \infty} \zeta(r)$. Given $x(t_0)$, with $\|x(t_0)\| < c$, let $\epsilon = \alpha(\|x(t_0)\|)$. Then, $\|x(t_0)\| < \bar{\delta}(\epsilon)$ and

$$\|x(t)\| < \epsilon = \alpha(\|x(t_0)\|) \tag{A.3}$$

Uniform Asymptotic Stability: Suppose there is a class $\mathcal{KL}$ function $\beta(r, s)$ such that (3.17) is satisfied. Then,

$$\|x(t)\| \leq \beta(\|x(t_0)\|, 0)$$

which implies that $x = 0$ is uniformly stable. Moreover, for $\|x(t_0)\| < c$, the solution satisfies

$$\|x(t)\| \leq \beta(c, t - t_0)$$

which shows that $x(t) \to 0$ as $t \to \infty$, uniformly in t_0. Suppose now that $x = 0$ is uniformly stable and $x(t) \to 0$ as $t \to \infty$, uniformly in t_0, and show that there is a class $\mathcal{KL}$ function $\beta(r, s)$ for which (3.17) is satisfied. Due to uniform stability, there is a constant $c > 0$ and a class $\mathcal{K}$ function α such that for any $r \in (0, c]$, the solution $x(t)$ satisfies

$$\|x(t)\| \leq \alpha(\|x(t_0)\|) < \alpha(r), \quad \forall\, t \geq t_0,\ \forall\, \|x(t_0)\| < r \tag{A.4}$$

Moreover, given $\eta > 0$ there exists $T = T(\eta, r)$ (dependent on η and r, but independent of t_0) such that

$$\|x(t)\| < \eta, \quad \forall\, t \geq t_0 + T(\eta, r)$$

Let $\bar{T}(\eta, r)$ be the infimum of all applicable $T(\eta, r)$. Then,

$$\|x(t)\| < \eta, \quad \forall\, t \geq t_0 + \bar{T}(\eta, r)$$

and

$$\sup \|x(t)\| \geq \eta, \quad \text{for } t_0 \leq t < t_0 + \bar{T}(\eta, r)$$

The function $\bar{T}(\eta, r)$ is nonnegative and nonincreasing in η, nondecreasing in r, and $\bar{T}(\eta, r) = 0$ for all $\eta \geq \alpha(r)$. Let

$$W_r(\eta) = \frac{2}{\eta} \int_{\eta/2}^{\eta} \bar{T}(s, r)\, ds + \frac{r}{\eta} \geq \bar{T}(\eta, r) + \frac{r}{\eta}$$

The function $W_r(\eta)$ is positive and has the properties:

- for each fixed r, $W_r(\eta)$ is continuous, strictly decreasing, and $W_r(\eta) \to 0$ as $\eta \to \infty$;

- for each fixed η, $W_r(\eta)$ is strictly increasing in r.

Take $U_r = W_r^{-1}$. Then, U_r inherits the foregoing two properties of W_r and $\bar{T}(U_r(s), r) < W_r(U_r(s)) = s$. Therefore,

$$\|x(t)\| \le U_r(t - t_0), \quad \forall\, t \ge t_0,\ \forall\, \|x(t_0)\| < r \tag{A.5}$$

From (A.4) and (A.5), it is clear that

$$\|x(t)\| \le \sqrt{\alpha(\|x(t_0)\|)U_c(t - t_0)}, \quad \forall\, t \ge t_0,\ \forall\, \|x(t_0)\| < c$$

Thus, inequality (3.17) is satisfied with $\beta(r, s) = \sqrt{\alpha(r)U_c(s)}$.

Global Uniform Asymptotic Stability: If (3.17) holds for all $x(t_0) \in R^n$, then it can be easily seen, as in the previous case, that the origin is globally uniformly asymptotically stable. To prove the opposite direction, notice that in the current case the function $\bar{\delta}(\epsilon)$ has the additional property $\bar{\delta}(\epsilon) \to \infty$ as $\epsilon \to \infty$. Consequently, the class $\mathcal{K}$ function α can be chosen to belong to class $\mathcal{K}_\infty$ and inequality (A.3) holds for all $x(t_0) \in R^n$. Moreover, inequality (A.5) holds for any $r > 0$. Let

$$\psi(r, s) = \min\left\{\alpha(r),\ \inf_{\rho \in (r, \infty)} U_\rho(s)\right\}$$

Then,

$$\|x(t)\| \le \psi(\|x(t_0)\|, t - t_0), \quad \forall\, t \ge t_0,\ \forall\, x(t_0) \in R^n$$

If ψ would be class $\mathcal{KL}$, we would be done. This may not be the case, so we majorize ψ by the function

$$\phi(r, s) = \int_r^{r+1} \psi(\lambda, s)\, d\lambda + \frac{r}{(r+1)(s+1)}$$

The function ϕ is positive and has the properties:

- for each fixed $s \ge 0$, $\phi(r, s)$ is continuous and strictly increasing in r;

- for each fixed $r \ge 0$, $\phi(r, s)$ is strictly decreasing in s and tends to zero as $s \to \infty$;

- $\phi(r, s) \ge \psi(r, s)$.

Thus,

$$\|x(t)\| \le \phi(\|x(t_0)\|, t - t_0), \quad \forall\, t \ge t_0,\ \forall\, x(t_0) \in R^n \tag{A.6}$$

From (A.6) and the global version of (A.3), we see that

$$\|x(t)\| \le \sqrt{\alpha(\|x(t_0)\|)\phi(\|x(t_0)\|, t - t_0)}, \quad \forall\, t \ge t_0,\ \forall\, x(t_0) \in R^n$$

Thus, inequality (3.17) is satisfied globally with $\beta(r, s) = \sqrt{\alpha(r)\phi(r, s)}$.

A.4   Proof of Lemma 3.4

Since $\alpha(\cdot)$ is locally Lipschitz, the equation has a unique solution for every initial state $y_0 \geq 0$. Since $\dot{y}(t) < 0$ whenever $y(t) > 0$, the solution has the property that $y(t) \leq y_0$ for all $t \geq t_0$. Therefore, the solution is bounded and can be extended for all $t \geq t_0$. By integration, we have

$$-\int_{y_0}^{y} \frac{dx}{\alpha(x)} = \int_{t_0}^{t} d\tau$$

Let b be any positive number less than a, and define

$$\eta(y) = -\int_{b}^{y} \frac{dx}{\alpha(x)}$$

The function $\eta(y)$ is a strictly decreasing differentiable function on $(0, a)$. Moreover, $\lim_{y \to 0} \eta(y) = \infty$. This limit follows from two facts. First, the solution of the differential equation $y(t) \to 0$ as $t \to \infty$, since $\dot{y}(t) < 0$ whenever $y(t) > 0$. Second, the limit $y(t) \to 0$ can happen only asymptotically as $t \to \infty$; it cannot happen in finite time due to the uniqueness of the solution. Let $c = -\lim_{y \to a} \eta(y)$ (c may be ∞). The range of the function η is $(-c, \infty)$. Since η is strictly decreasing, its inverse η^{-1} is defined on $(-c, \infty)$. For any $y_0 > 0$, the solution $y(t)$ satisfies

$$\eta(y(t)) - \eta(y_0) = t - t_0$$

Hence,

$$y(t) = \eta^{-1}(\eta(y_0) + t - t_0)$$

On the other hand, if $y_0 = 0$, then $y(t) \equiv 0$ since $y = 0$ is an equilibrium point. Define a function $\sigma(r, s)$ by

$$\sigma(r, s) = \begin{cases} \eta^{-1}(\eta(r) + s), & r > 0 \\ 0, & r = 0 \end{cases}$$

Then, $y(t) = \sigma(y_0, t - t_0)$ for all $t \geq t_0$ and $y_0 \geq 0$. The function σ is continuous since both η and η^{-1} are continuous in their domains and $\lim_{x \to \infty} \eta^{-1}(x) = 0$. It is strictly increasing in r for each fixed s because

$$\frac{\partial}{\partial r} \sigma(r, s) = \frac{\alpha(\sigma(r, s))}{\alpha(r)} > 0$$

and strictly decreasing in s for each fixed r because

$$\frac{\partial}{\partial s} \sigma(r, s) = -\alpha(\sigma(r, s)) < 0$$

Furthermore, $\sigma(r, s) \to 0$ as $s \to \infty$. Thus, σ is a class $\mathcal{KL}$ function.

A.5   Proof of Lemma 3.5

Define $\psi(s)$ by

$$\psi(s) = \inf_{s \le \|x\| \le r} V(x) \quad \text{for } 0 \le s \le r$$

The function $\psi(\cdot)$ is continuous, positive definite, and increasing. Moreover, $V(x) \ge \psi(\|x\|)$ for $0 \le \|x\| \le r$. But $\psi(\cdot)$ is not necessarily strictly increasing. Let $\alpha_1(s)$ be a class $\mathcal{K}$ function such that $\alpha_1(s) \le k\psi(s)$ with $0 < k < 1$. Then,

$$V(x) \ge \psi(\|x\|) \ge \alpha_1(\|x\|) \quad \text{for } \|x\| \le r$$

On the other hand, define $\phi(s)$ by

$$\phi(s) = \sup_{\|x\| \le s} V(x) \quad \text{for } 0 \le s \le r$$

The function $\phi(\cdot)$ is continuous, positive definite, and increasing (not necessarily strictly increasing). Moreover, $V(x) \le \phi(\|x\|)$ for $\|x\| \le r$. Let $\alpha_2(s)$ be a class $\mathcal{K}$ function such that $\alpha_2(s) \ge k\phi(s)$ with $k > 1$. Then

$$V(x) \le \phi(\|x\|) \le \alpha_2(\|x\|) \quad \text{for } \|x\| \le r$$

If $V(x)$ is radially unbounded, then there are positive constants c and r_1 such that $V(x) \ge c$ for all $\|x\| > r_1$. The definitions of $\psi(s)$ and $\phi(s)$ are changed to

$$\psi(s) = \inf_{\|x\| \ge s} V(x), \quad \phi(s) = \sup_{\|x\| \le s} V(x), \quad \text{for } s \ge 0$$

The functions ψ and ϕ are continuous, positive definite, increasing, and tend to infinity as $s \to \infty$. Hence, we can choose the functions α_1 and α_2 to belong to class $\mathcal{K}_\infty$.

A.6   Proof of Theorem 3.14

The construction of a Lyapunov function is accomplished by using a lemma, known as Massera's lemma. We start by stating and proving Massera's lemma.

Lemma A.1 *Let $g : [0, \infty) \to R$ be a positive, continuous, strictly decreasing function with $g(t) \to 0$ as $t \to \infty$. Let $h : [0, \infty) \to R$ be a positive, continuous, nondecreasing function. Then, there exists a function $G(t)$ such that*

- *$G(t)$ and its derivative $G'(t)$ are class $\mathcal{K}$ functions defined for all $t \ge 0$.*

- *For any continuous function $u(t)$ which satisfies $0 \le u(t) \le g(t)$ for all $t \ge 0$, there exist positive constants k_1 and k_2, independent of u, such that*

$$\int_0^\infty G(u(t))\, dt \le k_1; \quad \int_0^\infty G'(u(t))h(t)\, dt \le k_2$$

$\triangle$

Proof of Lemma A.1:   Since $g(t)$ is strictly decreasing, we can choose a sequence t_n such that

$$g(t_n) \le \frac{1}{n+1}, \quad n = 1, 2, \ldots$$

We use this sequence to define a function $\eta(t)$ as follows:

(a) $\eta(t_n) = 1/n$.

(b) Between t_n and t_{n+1}, $\eta(t)$ is linear.

(c) In the interval $0 < t \le t_1$, $\eta(t) = (t_1/t)^p$ where p is a positive integer chosen so large that the derivative $\eta'(t)$ has a positive jump at t_1, $\eta'(t_1^-) < \eta'(t_1^+)$.

The function $\eta(t)$ is strictly decreasing and, for $t \ge t_1$, we have $g(t) < \eta(t)$. As $t \to 0^+$, $\eta(t)$ grows unbounded. The inverse of $\eta(t)$, denoted by $\eta^{-1}(s)$, is a strictly decreasing function which grows unbounded as $s \to 0^+$. Obviously,

$$\eta^{-1}(u(t)) \ge \eta^{-1}(g(t)) > \eta^{-1}(\eta(t)) = t, \quad \forall\, t \ge t_1$$

for any nonnegative function $u(t) \le g(t)$. Define

$$H(s) = \frac{\exp[-\eta^{-1}(s)]}{h(\eta^{-1}(s))}, \quad s \ge 0$$

Since η^{-1} is continuous and h is positive, $H(s)$ is continuous on $0 < s < \infty$ while $\eta^{-1}(s) \to \infty$ as $s \to 0^+$. Hence, $H(s)$ defines a class $\mathcal{K}$ function on $[0, \infty)$. It follows that the integral

$$G(r) = \int_0^r H(s)\, ds$$

exists and both $G(r)$ and $G'(r) = H(r)$ are class $\mathcal{K}$ functions on $[0, \infty)$. Now, let $u(t)$ be a continuous nonnegative function such that $u(t) \le g(t)$. We have

$$G'(u(t)) = \frac{\exp[-\eta^{-1}(u(t))]}{h(\eta^{-1}(u(t)))} \le \frac{e^{-t}}{h(t)}, \quad \forall\, t \ge t_1$$

Hence,

$$\int_{t_1}^\infty G'(u(t))h(t)\, dt \le \int_{t_1}^\infty e^{-t} \le 1$$

and

$$\int_0^\infty G'(u(t))h(t)\ dt \le \int_0^{t_1} G'(g(t))h(t)\ dt + 1 \le k_2$$

which shows that the second integral in the lemma is bounded. For the first integral, we have

$$\int_{t_1}^\infty G(u(t))\ dt = \int_{t_1}^\infty \int_0^{u(t)} \frac{\exp[-\eta^{-1}(s)]}{h(\eta^{-1}(s))}\ ds\ dt$$

$$\le \int_{t_1}^\infty \int_0^{\eta(t)} \frac{\exp[-\eta^{-1}(s)]}{h(0)}\ ds\ dt$$

For $0 \le s \le \eta(t)$, we have

$$-\eta^{-1}(s) \le -t$$

Hence,

$$\int_0^{\eta(t)} \frac{\exp[-\eta^{-1}(s)]}{h(0)}\ ds \le \int_0^{\eta(t)} \frac{e^{-t}}{h(0)}\ ds = \frac{e^{-t}}{h(0)}\eta(t) \le \frac{e^{-t}}{h(0)}$$

for $t \ge t_1$. Thus

$$\int_0^\infty G(u(t))\ dt \le \int_0^{t_1} G(g(t))\ dt + \int_{t_1}^\infty \frac{e^{-t}}{h(0)}\ dt \le k_1$$

Hence, the first integral of the lemma is bounded. The proof of the lemma is complete. □

To prove the theorem, let

$$V(t, x) = \int_t^\infty G(\|\phi(\tau, t, x)\|_2)\ d\tau$$

where $\phi(\tau, t, x)$ is the solution that starts at (t, x) and G is a class $\mathcal{K}$ function to be selected using Lemma A.1. To see how to choose G, let us start by checking the upper bound on $\partial V/\partial x$, which is given by

$$\frac{\partial V}{\partial x} = \int_t^\infty G'(\|\phi\|_2) \frac{\phi^T}{\|\phi\|_2} \phi_x\ d\tau$$

We saw in the proof of Theorem 3.12 that the assumption $\|\partial f/\partial x\|_2 \le L$, uniformly in t, implies that $\|\phi_x(\tau, t, x)\|_2 \le \exp[L(\tau - t)]$. Therefore,

$$\left\|\frac{\partial V}{\partial x}\right\|_2 \le \int_t^\infty G'(\|\phi(\tau, t, x)\|_2) \exp[L(\tau - t)]\ d\tau$$

$$\le \int_t^\infty G'(\beta(\|x\|_2, \tau - t)) \exp[L(\tau - t)]\ d\tau$$

$$\le \int_0^\infty G'(\beta(\|x\|_2, s)) \exp(Ls)\ ds$$

With $\beta(r_0, s)$ and $\exp(Ls)$ as the functions g and h of Lemma A.1, we take G as the class $\mathcal{K}$ function provided by the lemma. Hence, the integral

$$\int_0^\infty G'(\beta(\|x\|_2, s)) \exp(Ls)\, ds \stackrel{\text{def}}{=} \alpha_4(\|x\|_2)$$

is bounded for all $\|x\|_2 \leq r_0$, uniformly in x. Moreover, it is a continuous and strictly increasing function of $\|x\|_2$ because $\beta(\|x\|_2, s)$ is a class $\mathcal{K}$ function in $\|x\|_2$ for every fixed s. Thus, α_4 is a class $\mathcal{K}$ function, which proves the last inequality in the theorem statement. Consider now

$$\begin{aligned} V(t, x) &= \int_t^\infty G(\|\phi(\tau, t, x)\|_2)\, d\tau \\ &\leq \int_t^\infty G(\beta(\|x\|_2, \tau - t))\, d\tau = \int_0^\infty G(\beta(\|x\|_2, s))\, ds \stackrel{\text{def}}{=} \alpha_2(\|x\|_2) \end{aligned}$$

By Lemma A.1, the last integral is bounded for all $\|x\|_2 \leq r_0$. The function α_2 is a class $\mathcal{K}$ function. Recall from the proof of Theorem 3.12 that the assumption $\|\partial f/\partial x\|_2 \leq L$, uniformly in t, implies that $\|\phi(\tau, t, x)\|_2 \geq \|x\|_2 \exp[-L(\tau - t)]$. Therefore,

$$\begin{aligned} V(t, x) &\geq \int_t^\infty G(\|x\|_2 e^{-L(\tau - t)})\, d\tau = \int_0^\infty G(\|x\|_2 e^{-Ls})\, ds \\ &\geq \int_0^{(\ln 2)/L} G(\tfrac{1}{2}\|x\|_2)\, ds = \frac{\ln 2}{L} G(\tfrac{1}{2}\|x\|_2) \stackrel{\text{def}}{=} \alpha_1(\|x\|_2) \end{aligned}$$

Clearly, $\alpha_1(\|x\|_2)$ is a class $\mathcal{K}$ function. Hence, V satisfies the inequality

$$\alpha_1(\|x\|_2) \leq V(t, x) \leq \alpha_2(\|x\|_2)$$

for all $\|x\|_2 \leq r_0$. Finally, the derivative of V along the trajectories of the system is given by

$$\begin{aligned} \frac{\partial V}{\partial t} + \frac{\partial V}{\partial x} f(t, x) = \\ - G(\|x\|_2) + \int_t^\infty G'(\|\phi\|_2) \frac{\phi^T}{\|\phi\|_2} [\phi_t(\tau, t, x) + \phi_x(\tau, t, x) f(t, x)]\, d\tau \end{aligned}$$

Since

$$\phi_t(\tau, t, x) + \phi_x(\tau, t, x) f(t, x) \equiv 0, \quad \forall\, \tau \geq t$$

we have

$$\frac{\partial V}{\partial t} + \frac{\partial V}{\partial x} f(t, x) = -G(\|x\|_2)$$

Thus, the three inequalities of the theorem statement are satisfied for all $\|x\|_2 \le r_0$. Notice that, due to the equivalence of norms, we can state the inequalities in any $\|\cdot\|_p$ norm. If the system is autonomous, the solution depends only on $\tau - t$; that is, $\phi(\tau, t, x) = \psi(\tau - t, x)$. Therefore,

$$V = \int_t^\infty G(\|\psi(\tau - t, x)\|_2) \, d\tau = \int_0^\infty G(\|\psi(s, x)\|_2) \, ds$$

which is independent of t.

A.7 Proof of Theorems 4.1 and 4.3

The main element in the proofs of these two theorems is a contraction mapping argument, which is used almost identically in the two proofs. To avoid repetition, we shall state and prove a lemma which captures the needed contraction mapping argument and then use it to prove the two theorems. The statement of the lemma appears to be very similar to the statement of Theorem 4.1, but it has an additional claim that is needed in Theorem 4.3.

Lemma A.2 *Consider the system*

$$\dot{y} = Ay + f(y, z) \qquad (A.7)$$
$$\dot{z} = Bz + g(y, z) \qquad (A.8)$$

where $y \in R^k$, $z \in R^m$, the eigenvalues of A have zero real parts, the eigenvalues of B have negative real parts, and f and g are twice continuously differentiable functions which vanish together with their first derivatives at the origin. Then, there exist $\delta > 0$ and a continuously differentiable function $\eta(y)$, defined for all $\|y\| < \delta$, such that $z = \eta(y)$ is a center manifold for (A.7)–(A.8). Moreover, if $\|g(y, 0)\| \le k\|y\|^p$ for all $\|y\| \le r$ where $p > 1$ and $r > 0$, then there is $c > 0$ such that $\|\eta(y)\| \le c\|y\|^p$. $\diamond$

Proof: It is more convenient to show the existence of a center manifold when solutions in the manifold are defined for all $t \in R$. A center manifold for (A.7)–(A.8) may, in general, be only local; that is, a solution in the manifold may be defined only on an interval $[0, t_1) \subset R$. Therefore, the following idea is used in the proof. We consider a modified equation which is identical to (A.7)–(A.8) in the neighborhood of the origin, but has some desired global properties which ensure that a solution in a center manifold will be defined for all t. We prove the existence of a center manifold for the modified equation. Since the two equations agree in the neighborhood of the origin, this proves the existence of a (local) center manifold for the original equation.

Let $\psi : R^k \to [0,1]$ be a smooth (continuously differentiable infinitely many times) function[3] with $\psi(y) = 1$ when $\|y\| \leq 1$ and $\psi(y) = 0$ when $\|y\| \geq 2$. For $\epsilon > 0$, define F and G by

$$F(y,z) = f\left(y\psi\left(\frac{y}{\epsilon}\right),z\right); \quad G(y,z) = g\left(y\psi\left(\frac{y}{\epsilon}\right),z\right)$$

The functions F and G are twice continuously differentiable, and they, together with their first partial derivatives, are globally bounded in y; that is, whenever $\|z\| \leq k_1$, the function is bounded for all $y \in R^k$. Consider the modified system

$$\dot{y} = Ay + F(y,z) \tag{A.9}$$
$$\dot{z} = Bz + G(y,z) \tag{A.10}$$

We prove the existence of a center manifold for (A.9)–(A.10). Let X denote the set of all globally bounded, continuous functions $\eta : R^k \to R^m$. With $\sup_{y \in R^k} \|\eta(y)\|$ as a norm, X is a Banach space. For $c_1 > 0$ and $c_2 > 0$, let $S \subset X$ be the set of all continuously differential functions $\eta : R^k \to R^m$ such that

$$\eta(0) = 0, \quad \frac{\partial\eta}{\partial y}(0) = 0, \quad \|\eta(y)\| \leq c_1, \quad \left\|\frac{\partial\eta}{\partial y}(y)\right\| \leq c_2$$

for all $y \in R^k$. S is a closed subset of X. For a given $\eta \in S$, consider the system

$$\dot{y} = Ay + F(y,\eta(y)) \tag{A.11}$$
$$\dot{z} = Bz + G(y,\eta(y)) \tag{A.12}$$

Due to the boundedness of $\eta(y)$ and $[\partial\eta/\partial y]$, the right-hand side of (A.11) is globally Lipschitz in y. Therefore, for every initial state $y_0 \in R^k$, (A.11) has a unique solution defined for all t. We denote this solution by $y(t) = \pi(t,y_0,\eta)$, where $\pi(0,y_0,\eta) = y_0$; the solution is parameterized in the fixed function η. Equation (A.12) is linear in z; therefore, its solution is given by

$$z(t) = \exp[B(t-\tau)]z(\tau)$$
$$+ \int_{\tau}^{t} \exp[B(t-\lambda)]G(\pi(\lambda-\tau,y(\tau),\eta),\eta(\pi(\lambda-\tau,y(\tau),\eta)))\,d\lambda$$

---

[3] An example of such a function in the scalar ($k=1$) case is $\psi(y) = 1$ for $|y| \leq 1$, $\psi(y) = 0$ for $|y| \geq 2$, and

$$\psi(y) = 1 - \frac{1}{b}\int_1^{|y|} \exp\left(\frac{-1}{x-1}\right)\exp\left(\frac{-1}{2-x}\right)\,dx, \quad \text{for } 1 < |y| < 2$$

where

$$b = \int_1^2 \exp\left(\frac{-1}{x-1}\right)\exp\left(\frac{-1}{2-x}\right)\,dx$$

Multiply through by $\exp[-B(t-\tau)]$, move the integral term to the other side, and change the integration variable from λ to $s = \lambda - \tau$ to obtain

$$
\begin{aligned}
z(\tau) \;=\;& \exp[-B(t-\tau)]z(t) \\
&+ \int_{t-\tau}^{0} \exp(-Bs)G(\pi(s,y(\tau),\eta),\eta(\pi(s,y(\tau),\eta)))\; ds
\end{aligned}
$$

This expression is valid for any $t \in R$. Taking the limit $t \to -\infty$ results in

$$
z(\tau) = \int_{-\infty}^{0} \exp(-Bs)G(\pi(s,y(\tau),\eta),\eta(\pi(s,y(\tau),\eta)))\; ds \qquad (A.13)
$$

The expression on the right-hand side of (A.13) is dependent on $y(\tau)$ but independent of τ. Let us rewrite this expression with $y(\tau)$ replaced by y and denote it by $(P\eta)(y)$.

$$
(P\eta)(y) = \int_{-\infty}^{0} \exp(-Bs)G(\pi(s,y,\eta),\eta(\pi(s,y,\eta)))\; ds \qquad (A.14)
$$

With this definition, (A.13) says that $z(\tau) = (P\eta)(y(\tau))$ for all τ. Hence, $z = (P\eta)(y)$ defines an invariant manifold for (A.11)–(A.12) parameterized in η.

Consider (A.9)–(A.10). If $\eta(y)$ is a fixed point of the mapping $(P\eta)(y)$, that is,

$$
\eta(y) = (P\eta)(y)
$$

then $z = \eta(y)$ is a center manifold for (A.9)–(A.10). This fact can be seen as follows. First, using the properties of $\eta \in S$ and the fact that $y = 0$ is an equilibrium point of (A.11), it can be seen from (A.14) that

$$
(P\eta)(0) = 0; \quad \frac{\partial(P\eta)}{\partial y}(0) = 0
$$

Second, since $z = (P\eta)(y)$ is an invariant manifold for (A.11)–(A.12), $(P\eta)(y)$ satisfies the partial differential equation

$$
\frac{\partial}{\partial y}(P\eta)(y)[Ay + F(y,(P\eta)(y))] = B(P\eta)(y) + G(y,(P\eta)(y))
$$

If $\eta(y) = (P\eta)(y)$, then clearly $\eta(y)$ satisfies the same partial differential equation; hence, it is a center manifold for (A.9)–(A.10).

It remains now to show that the mapping $(P\eta)$ has a fixed point, which will be done by an application of the contraction mapping theorem. We want to show that the mapping $P\eta$ maps S into itself and is a contraction mapping on S. Using the definitions of F and G, there is a nonnegative continuous function $\rho(\epsilon)$ with $\rho(0) = 0$ such that

$$\|F(y,z)\| \le \epsilon\rho(\epsilon); \quad \left\|\frac{\partial F}{\partial y}(y,z)\right\| \le \rho(\epsilon); \quad \left\|\frac{\partial F}{\partial z}(y,z)\right\| \le \rho(\epsilon) \qquad (A.15)$$

$$\|G(y,z)\| \le \epsilon\rho(\epsilon); \quad \left\|\frac{\partial G}{\partial y}(y,z)\right\| \le \rho(\epsilon); \quad \left\|\frac{\partial G}{\partial z}(y,z)\right\| \le \rho(\epsilon) \qquad (A.16)$$

for all $y \in R^k$ and all $z \in R^m$ with $\|z\| < \epsilon$. Since the eigenvalues of B have negative real parts, there exist positive constants β and C such that for $s \le 0$,

$$\|\exp(-Bs)\| \le C\exp(\beta s) \qquad (A.17)$$

Since the eigenvalues of A have zero real parts, for each $\alpha > 0$ there is a positive constant $M(\alpha)$ (which may tend to ∞ as $\alpha \to 0$) such that for $s \in R$,

$$\|\exp(As)\| \le M(\alpha)\exp(\alpha|s|) \qquad (A.18)$$

To show that $P\eta$ maps S into itself, we need to show that there are positive constants c_1 and c_2 such that if $\eta(y)$ is continuously differentiable and satisfies

$$\|\eta(y)\| \le c_1; \quad \left\|\frac{\partial\eta}{\partial y}\right\| \le c_2$$

for all $y \in R^k$, then $(P\eta)(y)$ is continuously differentiable and satisfies the same inequalities for all $y \in R^k$. Continuous differentiability of $(P\eta)(y)$ is obvious from (A.14). To verify the inequalities, we need to use the estimates on F and G provided by (A.15)–(A.16); therefore, we take $c_1 < \epsilon$. Using (A.17) and the estimates on G and η, we have from (A.14),

$$
\begin{aligned}
\|(P\eta)(y)\| &\le \int_{-\infty}^{0} \|\exp(-Bs)\| \, \|G\| \, ds \\
&\le \int_{-\infty}^{0} C\exp(\beta s) \, \epsilon\rho(\epsilon) \, ds \;=\; \frac{C\epsilon\rho(\epsilon)}{\beta}
\end{aligned}
$$

The upper bound on $(P\eta)(y)$ will be less than c_1 for sufficiently small ϵ. Let π_y denote the Jacobian of $\pi(t, y, \eta)$ with respect to y. It satisfies the variational equation

$$\dot{\pi}_y = \left[A + \left(\frac{\partial F}{\partial y}\right) + \left(\frac{\partial F}{\partial z}\right)\left(\frac{\partial\eta}{\partial y}\right)\right]\pi_y, \quad \pi_y(0) = I$$

Hence, for $t \le 0$,

$$\pi_y(t) = \exp(At) - \int_{t}^{0} \exp[A(t-s)]\left[\left(\frac{\partial F}{\partial y}\right) + \left(\frac{\partial F}{\partial z}\right)\left(\frac{\partial\eta}{\partial y}\right)\right]\pi_y(s) \, ds$$

Using (A.18) and the estimates on F and η, we obtain

$$\|\pi_y(t)\| \leq M(\alpha)\exp(-\alpha t) + \int_t^0 M(\alpha)\exp[\alpha(s-t)](1+c_2)\rho(\epsilon)\|\pi_y(s)\| \, ds$$

Multiply through by $\exp(\alpha t)$ and apply the Gronwall-Bellman inequality to show that

$$\|\pi_y(t)\| \leq M(\alpha)\exp(-\gamma t)$$

where $\gamma = \alpha + M(\alpha)(1+c_2)\rho(\epsilon)$. Using this bound, as well as (A.17) and the estimates on G and η, we proceed to calculate a bound on the Jacobian $[\partial(P\eta)(y)/\partial y]$. From (A.14),

$$\frac{\partial(P\eta)(y)}{\partial y} = \int_{-\infty}^0 \exp(-Bs)\left[\left(\frac{\partial G}{\partial y}\right) + \left(\frac{\partial G}{\partial z}\right)\left(\frac{\partial \eta}{\partial y}\right)\right]\pi_y(s) \, ds$$

Hence,

$$\left\|\frac{\partial(P\eta)(y)}{\partial y}\right\| \leq \int_{-\infty}^0 C\exp(\beta s)(1+c_2)\rho(\epsilon)M(\alpha)\exp(-\gamma s) \, ds$$

$$\leq \frac{C(1+c_2)\rho(\epsilon)M(\alpha)}{\beta - \gamma}$$

provided ϵ and α are small enough that $\beta > \gamma$. This bound on the Jacobian of $(P\eta)(y)$ will be less than c_2 for sufficiently small ϵ. Thus, we have shown that, for sufficiently small c_1, the mapping $P\eta$ maps S into itself. To show that it is a contraction on S, let $\eta_1(y)$ and $\eta_2(y)$ be two functions in S. Let $\pi_1(t)$ and $\pi_2(t)$ be the corresponding solutions of (A.11) which start at y; that is,

$$\pi_i(t) = \pi(t, y, \eta_i), \quad i = 1, 2$$

Using the estimates (A.15)–(A.16), it can be shown that

$$\|F(\pi_2, \eta_2(\pi_2)) - F(\pi_1, \eta_1(\pi_1))\| \leq (1+c_2)\rho(\epsilon)\|\pi_1 - \pi_2\| + \rho(\epsilon)\sup_{y \in R^k}\|\eta_2 - \eta_1\|$$

$$\|G(\pi_2, \eta_2(\pi_2)) - G(\pi_1, \eta_1(\pi_1))\| \leq (1+c_2)\rho(\epsilon)\|\pi_1 - \pi_2\| + \rho(\epsilon)\sup_{y \in R^k}\|\eta_2 - \eta_1\|$$

From (A.11), $\|\pi_2 - \pi_1\|$ satisfies

$$\|\pi_2(t) - \pi_1(t)\| \leq \int_t^0 M(\alpha)\exp[\alpha(s-t)][\rho(\epsilon)\sup_{y \in R^k}\|\eta_2 - \eta_1\|$$
$$+ (1+c_2)\rho(\epsilon)\|\pi_1(s) - \pi_2(s)\|] \, ds$$

$$\leq \frac{1}{\alpha}M(\alpha)\rho(\epsilon)\sup_{y \in R^k}\|\eta_2 - \eta_1\|\exp(-\alpha t)$$
$$+ \int_t^0 (\gamma - \alpha)\exp[\alpha(s-t)]\|\pi_1(s) - \pi_2(s)\| \, ds$$

where we have used $\gamma = \alpha + M(\alpha)(1 + c_2)\rho(\epsilon)$ and the fact that $\pi_1(0) = \pi_2(0)$. Multiply through by $\exp(\alpha t)$ and apply the Gronwall-Bellman inequality to show that

$$\|\pi_2(t) - \pi_1(t)\| \leq \frac{1}{\alpha} M(\alpha)\rho(\epsilon) \sup_{y \in R^k} \|\eta_2 - \eta_1\| \exp(-\gamma t)$$

Using this inequality in

$$(P\eta_2)(y) - (P\eta_1)(y) = \int_{-\infty}^{0} \exp(-Bs) \left[G(\pi_2, \eta_2(\pi_2)) - G(\pi_1, \eta_1(\pi_1)) \right] \, ds$$

we obtain

$$
\begin{aligned}
\|(P\eta_2)(y) - (P\eta_1)(y)\| &\leq \int_{-\infty}^{0} C e^{\beta s} [(1 + c_2)\rho(\epsilon)\|\pi_2(s) - \pi_1(s)\| \\
&\quad + \rho(\epsilon) \sup_{y \in R^k} \|\eta_2 - \eta_1\| \,] \, ds \\
&\leq C\rho(\epsilon) \sup_{y \in R^k} \|\eta_2 - \eta_1\| \left[\frac{1}{\beta} \right. \\
&\quad \left. + \int_{-\infty}^{0} e^{\beta s}(1 + c_2)\frac{1}{\alpha} M(\alpha)\rho(\epsilon)e^{-\gamma s} \, ds \right] \\
&\leq b \sup_{y \in R^k} \|\eta_2 - \eta_1\|
\end{aligned}
$$

where

$$b = C\rho(\epsilon) \left[\frac{1}{\beta} + \frac{\gamma - \alpha}{\alpha(\beta - \gamma)} \right]$$

By choosing ϵ small enough, we can ensure that $b < 1$; hence, $P\eta$ is a contraction mapping on S. Thus, by the contraction mapping theorem, the mapping $P\eta$ has a fixed point in S.

Suppose now that $\|g(y, 0)\| \leq k\|y\|^p$. The function $G(y, 0)$ satisfies the same bound. Consider the closed subset

$$Y = \{\eta \in S \mid \|\eta(y)\| \leq c_3\|y\|^p\}$$

where c_3 is a positive constant to be chosen. To complete the proof of the lemma, we want to show that the fixed point of the mapping $P\eta$ is in Y, which will be the case if we can show that c_3 can be chosen such that $P\eta$ maps Y into itself. Using the estimate on G provided by (A.16), we have

$$\|G(y, \eta(y))\| \leq \|G(y, 0)\| + \|G(y, \eta(y)) - G(y, 0)\| \leq k\|y\|^p + \rho(\epsilon)\|\eta(y)\|$$

Since, in the set Y, $\|\eta(y)\| \leq c_3\|y\|^p$,

$$\|G(y, \eta(y))\| \leq [k + c_3\rho(\epsilon)]\|y\|^p$$

Using this estimate in (A.14) yields

$$\|(P\eta)(y)\| \le \int_{-\infty}^{0} C \exp(\beta s)\,[k + c_3\rho(\epsilon)]\|\pi(s, y, \eta)\|^p\, ds$$

Since

$$\pi_i(t, y, \eta) = \pi_i(t, 0, \eta) + \pi_{iy}(t, \zeta)y = y_i + \pi_{iy}(t, \zeta)y$$

for some ζ with $\|\zeta\| \le \|y\|$, we have

$$\|\pi(t, y, \eta)\| \le \|y\| + \|y\|M(\alpha)\exp(-\gamma t) \le [1 + M(\alpha)]\exp(-\gamma t)\|y\|$$

for $t \le 0$. Thus,

$$\|(P\eta)(y)\| \le \frac{C[k + c_3\rho(\epsilon)][1 + M(\alpha)]^p}{\beta - p\gamma}\|y\|^p \overset{\text{def}}{=} c_4\|y\|^p$$

provided ϵ and α are small enough so that $\beta - p\gamma > 0$. By choosing c_3 large enough and ϵ small enough, we have $c_4 < c_3$. Thus, $(P\eta)$ maps Y into itself, which completes the proof of the lemma. □

Proof of Theorem 4.1

It follows from Lemma A.2 with $A = A_1$, $B = A_2$, $f = g_1$, $g = g_2$. □

Proof of Theorem 4.3

Define $\mu(y) = h(y) - \phi(y)$. Using the fact that $\mathcal{N}(h(y)) = 0$ and $\mathcal{N}(\phi(y)) = O(\|y\|^p)$ where $\mathcal{N}(h(y))$ is defined by (4.11), we can show that $\mu(y)$ satisfies the partial differential equation

$$\frac{\partial\mu}{\partial y}(y)[A_1 + N(y, \mu(y))] - A_2\mu(y) - Q(y, \mu(y)) = 0 \tag{A.19}$$

where

$$N(y, z) = g_1(y, \phi(y) + z)$$

and

$$\begin{aligned} Q(y, z) =\ & g_2(y, \phi(y) + z) - g_2(y, \phi(y)) + \mathcal{N}(\phi(y)) \\ & - \frac{\partial\phi}{\partial y}(y)[g_1(y, \phi(y) + z) - g_1(y, \phi(y))] \end{aligned}$$

A function $\mu(y)$ satisfying (A.19) is a center manifold for an equation of the form (A.7)–(A.8) with $A = A_1$, $B = A_2$, $f = N$, and $g = Q$. Furthermore, in this case,

$$Q(y, 0) = \mathcal{N}(\phi(y)) = O(\|y\|^p)$$

Hence, by Lemma A.2, there exists a continuously differentiable function $\mu(y) = O(\|y\|^p)$ which satisfies (A.19). Therefore, $h(y) - \phi(y) = O(\|y\|^p)$. The reduced system is given by

$$\begin{aligned} \dot{y} &= A_1 y + g_1(y, h(y)) \\ &= A_1 y + g_1(y, \phi(y)) + g_1(y, h(y)) - g_1(y, \phi(y)) \end{aligned}$$

Since g_1 is twice continuously differentiable and its first partial derivatives vanish at the origin, we have

$$\left\| \frac{\partial g_1}{\partial z}(y, z) \right\| \leq k_1 \|y\| + k_2 \|z\|$$

in the neighborhood of the origin. By the mean value theorem,

$$g_{1i}(y, h(y)) - g_{1i}(y, \phi(y)) = \frac{\partial g_{1i}}{\partial z}(y, \zeta(y))[h(y) - \phi(y)]$$

where

$$\|\zeta(y)\| \leq \|\mu(y)\| + \|\phi(y)\| \leq k_3 \|y\|^p \leq k_3 \|y\|$$

for $\|y\| < 1$. Therefore,

$$\|g_1(y, h(y)) - g_1(y, \phi(y))\| \leq k_4 \|y\| \, \|\mu(y)\| = O(\|y\|^{p+1})$$

which completes the proof of the theorem.   □

A.8   Proof of Lemma 4.1

To show that R_A is invariant, we need to show that

$$x \in R_A \Rightarrow x(s) \overset{\text{def}}{=} \phi(s, x) \in R_A, \quad \forall \, s \in R$$

Since

$$\phi(t, \phi(s, x)) = \phi(t + s, x)$$

it is clear that $\lim_{t \to \infty} \phi(t, x(s)) = 0$ for all $s \in R$. Hence, R_A is invariant. To show that R_A is open, take any point $p \in R_A$ and show that every point in a neighborhood of p belongs to R_A. To that end, let $T > 0$ be large enough that $\|\phi(T, p)\| < a/2$, where a is chosen so small that the domain $\|x\| < a$ is contained in R_A. Consider the neighborhood $\|x - p\| < b$ of p. By continuous dependence of the solution on initial states, we can choose b small enough to ensure that for any point q in the neighborhood $\|x - p\| < b$, the solution at time T satisfies

$$\|\phi(T, p) - \phi(T, q)\| < \frac{a}{2}$$

Then,

$$\|\phi(T, q)\| \leq \|\phi(T, q) - \phi(T, p)\| + \|\phi(T, p)\| < a$$

This shows that the point $\phi(T, q)$ is inside R_A. Hence, the solution starting at q approaches the origin as $t \to \infty$. Thus, $q \in R_A$ and the set R_A is open. We leave it to the reader (Exercise 4.15) to show that R_A is connected. The statement about the boundary of R_A follows from the following lemma.

Lemma A.3 *The boundary of an open invariant set is an invariant set. Hence, it is formed of trajectories.* $\triangle$

Proof: Let M be an open invariant set and x be a boundary point of M. There exists a sequence $x_n \in M$ which converges to x. Since M is an invariant set, the solution $\phi(t, x_n) \in M$ for all $t \in R$. The sequence $\phi(t, x_n)$ converges to $\phi(t, x)$ for all $t \in R$. Hence, $\phi(t, x)$ is an accumulation point of M for all t. On the other hand, $\phi(t, x) \notin M$ since x is a boundary point of M. Therefore, the solution $\phi(t, x)$ belongs to the boundary of M for all t. $\square$

A.9   Proof of Theorem 6.4

We complete the proof of Theorem 6.4 by showing that the $\mathcal{L}_2$ gain is equal to $\sup_{\omega \in R} \|G(j\omega)\|_2$. Let c_1 be the $\mathcal{L}_2$ gain and $c_2 = \sup_{\omega \in R} \|G(j\omega)\|_2$. We know that $c_1 \leq c_2$. Suppose $c_1 < c_2$ and set $\epsilon = (c_2 - c_1)/3$. Then, for any $u \in \mathcal{L}_2$ with $\|u\|_{\mathcal{L}_2} \leq 1$ we have $\|y\|_{\mathcal{L}_2} \leq c_2 - 3\epsilon$. We will establish a contradiction by finding a signal u with $\|u\|_{\mathcal{L}_2} \leq 1$ such that $\|y\|_{\mathcal{L}_2} \geq c_2 - 2\epsilon$. It is easier to construct such signal if we define signals on the whole real line R. There is no loss of generality in doing so due to the fact (shown in Exercise 6.16) that the $\mathcal{L}_2$ gain is the same whether signals are defined on $[0, \infty)$ or on R. Now, select $\omega_0 \in R$ such that $\|G(j\omega_0)\|_2 \geq c_2 - \epsilon$. Let $v \in C^m$ be the normalized eigenvector ($v^* v = 1$) corresponding to the maximum eigenvalue of the Hermitian matrix $G^T(-j\omega_0)G(j\omega_0)$. Hence, $v^* G^T(-j\omega_0)G(j\omega_0)v = \|G(j\omega_0)\|_2^2$. Write v as

$$v = \left[\alpha_1 e^{j\theta_1}, \ \alpha_2 e^{j\theta_2}, \ \ldots, \ \alpha_m e^{j\theta_m} \right]^T$$

where $\alpha_i \in R$ is such that $\theta_i \in (-\pi, 0]$. Take $0 \leq \beta_i \leq \infty$ such that $\theta_i = -2\tan^{-1}(\omega_0/\beta_i)$, with $\beta_i = \infty$ if $\theta_i = 0$. Define the $m \times 1$ transfer function $H(s)$ by

$$H(s) = \left[\alpha_1 \frac{\beta_1 - s}{\beta_1 + s}, \ \alpha_2 \frac{\beta_2 - s}{\beta_2 + s}, \ \ldots, \ \alpha_m \frac{\beta_m - s}{\beta_m + s} \right]^T$$

with 1 replacing $(\beta_i - s)/(\beta_i + s)$ if $\theta_i = 0$. It can be easily seen that $H(j\omega_0) = v$ and $H^T(-j\omega)H(j\omega) = \sum_{i=1}^{m} \alpha_i^2 = v^*v = 1$ for all $\omega \in R$. Take $u_\sigma(t)$ as the output of $H(s)$ driven by the scalar function

$$z_\sigma(t) = \left(\frac{1}{1 + e^{-\omega_0^2 \sigma/2}}\right)^{1/2} \left(\frac{8}{\pi\sigma}\right)^{1/4} e^{-t^2/\sigma} \cos(\omega_0 t), \quad \sigma > 0, \ t \in R$$

It can be verified that $z_\sigma \in \mathcal{L}_2$ and $\|z_\sigma\|_{\mathcal{L}_2} = 1.^4$ Consequently, $u_\sigma \in \mathcal{L}_2$ and $\|u_\sigma\|_{\mathcal{L}_2} \leq 1$. The Fourier transform of $z_\sigma(t)$ is given by

$$Z_\sigma(j\omega) = \left(\frac{1}{1 + e^{-\omega_0^2 \sigma/2}}\right)^{1/2} \left(\frac{\pi\sigma}{2}\right)^{1/4} \left[e^{-(\omega-\omega_0)^2\sigma/4} + e^{-(\omega+\omega_0)^2\sigma/4}\right]$$

Let $y_\sigma(t)$ be the output of $G(s)$ when the input is $u_\sigma(t)$. The Fourier transform of $y_\sigma(t)$ is given by $Y_\sigma(j\omega) = G(j\omega)U_\sigma(j\omega) = G(j\omega)H(j\omega)Z_\sigma(j\omega)$. By Parseval's theorem,

$$
\begin{aligned}
\|y_\sigma\|_{\mathcal{L}_2}^2 &= \frac{1}{2\pi} \int_{-\infty}^{\infty} Z_\sigma^T(-j\omega)H^T(-j\omega)G^T(-j\omega)G(j\omega)H(j\omega)Z_\sigma(j\omega) \ d\omega \\
&= \frac{1}{2\pi} \int_{-\infty}^{\infty} H^T(-j\omega)G^T(-j\omega)G(j\omega)H(j\omega) \ |Z_\sigma(j\omega)|^2 \ d\omega
\end{aligned}
$$

Using

$$|Z_\sigma(j\omega)|^2 \geq \frac{1}{1 + e^{-\omega_0^2 \sigma/2}} \left(\frac{\pi\sigma}{2}\right)^{1/2} \left[e^{-(\omega-\omega_0)^2\sigma/2} + e^{-(\omega-\omega_0)^2\sigma/2}\right] \stackrel{\text{def}}{=} \psi_\sigma(\omega)$$

we obtain

$$\|y_\sigma\|_{\mathcal{L}_2}^2 \geq \frac{1}{2\pi} \int_{-\infty}^{\infty} H^T(-j\omega)G^T(-j\omega)G(j\omega)H(j\omega)\psi_\sigma(\omega) \ d\omega$$

By letting $\sigma \to \infty$, one can concentrate the frequency spectrum $\psi_\sigma(\omega)$ around the the two frequencies $\omega = \pm\omega_0.^5$ Hence, the right-hand side of the last inequality approaches

$$H^T(-j\omega_0)G^T(-j\omega_0)G(j\omega_0)H(j\omega_0) = \|G(j\omega_0)\|_2^2 = (c_2 - \epsilon)^2$$

as $\sigma \to \infty$. Therefore, we can choose a finite σ large enough such that $\|y_\sigma\|_{\mathcal{L}_2} \geq c_2 - 2\epsilon$. This, however, contradicts the inequality $\|y_\sigma\|_{\mathcal{L}_2} \leq c_2 - 3\epsilon$. The contradiction shows that $c_1 = c_2$.

---

[4] For the purpose of this discussion, the $\mathcal{L}_2$-norm is defined by $\|z\|_{\mathcal{L}_2}^2 = \int_{-\infty}^{\infty} z^T(t)z(t) \ dt$.

[5] $\psi_\sigma(\omega)$ approaches $\pi[\delta(\omega - \omega_0) + \delta(\omega + \omega_0)]$ as $\sigma \to \infty$, where $\delta(\cdot)$ is the impulse function.

A.10   Proof of Theorem 7.1

We prove the Poincaré-Bendixson theorem only for positive limit sets. The proof
for negative limit sets is similar. We start by introducing transversals with respect
to a vector field f. Consider the second-order equation

$$\dot{x} = f(x) \tag{A.20}$$

where $f : D \to R^2$ is locally Lipschitz over a domain $D \subset R^2$. A transversal with
respect to f is a closed line segment $L \in D$ such that no equilibrium points of
(A.20) lie on L and at every point $x \in L$, the vector field $f(x)$ is not parallel to the
direction of L. If L is a segment of a line whose equation is

$$g(x) = a^T x - c = 0$$

then L is a transversal of f if

$$a^T f(x) \neq 0, \quad \forall \, x \in L$$

If a trajectory of (A.20) meets a transversal L, it must cross L. Moreover, all such
crossings of L are in the same direction. In the next two lemmas, we state the
properties of the transversals which will be used in the proof.

Lemma A.4 *If y_0 is an interior point of a transversal L, then for any $\epsilon > 0$
there is $\delta > 0$ such that any trajectory passing through the ball $B_\delta(y_0) = \{x \in
R^2 \mid \|x - y_0\| < \delta\}$ at $t = 0$ must cross L at some time $t \in (-\epsilon, \epsilon)$. Moreover, if the
trajectory is bounded, then by choosing ϵ small enough the point where the trajectory
crosses L can be made arbitrarily close to y_0.* △

Proof: Let $\phi(t, x)$ denote the solution of (A.20) that starts at $\phi(0, x) = x$, and
define

$$G(t, x) = g(\phi(t, x)) = a^T \phi(t, x) - c$$

The trajectory of $\phi(t, x)$ crosses L if $G(t_1, x) = 0$ for some time t_1. For the function
$G(t, x)$ we have $G(0, y_0) = 0$ since $y_0 \in L$, and

$$\frac{\partial G}{\partial t}(0, y_0) = a^T f(\phi(t, y_0))\big|_{t=0} = a^T f(y_0) \neq 0$$

since L is a transversal. By the implicit function theorem, there is a continuously
differentiable function $\tau(x) : U \to R$ defined on a neighborhood U of y_0 such that
$\tau(y_0) = 0$ and $G(\tau(x), x) = 0$. By continuity of the map $\tau(x)$, given any $\epsilon >$ there
is $\delta > 0$ such that

$$\|x - y_0\| < \delta \Rightarrow |\tau(x)| < \epsilon$$

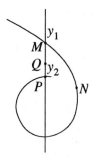

Figure A.1: Illustration of the proof of Lemma A.5.

Hence, for all x in the ball $B_\delta(y_0)$, the trajectory starting at x crosses L at $\tau(x) \in$ $(-\epsilon, \epsilon)$. If the trajectory $\phi(t, x)$ is bounded, then

$$
\begin{aligned}
\|\phi(\tau(x), x) - y_0\| &= \|\phi(\tau(x), x) - x + x - y_0\| \\
&\leq \|\phi(\tau(x), x) - \phi(0, x)\| + \|x - y_0\| \\
&\leq k|\tau(x)| + \delta \ < k\epsilon + \delta
\end{aligned}
$$

where k is a bound on $f(\phi(t, x))$. Since, without loss of generality, we can always choose $\delta < \epsilon$, the right-hand side of the last inequality can be made arbitrarily small by choosing ϵ small enough.   □

Lemma A.5 *If a trajectory γ, followed as t increases, crosses a transversal at three consecutive points y_1, y_2, and y_3, then y_2 lies between y_1 and y_3 on L.*   △

Proof: Consider the two consecutive points y_1 and y_2. Let C be a simple closed (Jordan) curve made up of the part of the trajectory between y_1 and y_2 (the arc MNP in Figure A.1) and the part of the transversal L between the same points (the segment PQM in Figure A.1). Let D be the closed bounded region enclosed by C. We assume that the trajectory of y_2 enters D; if it leaves, the argument is similar. By uniqueness of solutions, no trajectory can cross the arc MNP. Since L is a transversal, trajectories can cross L in only one direction. Hence, trajectories will enter D along the segment PQM of L. This means the set D is positively invariant. The trajectory of y_2 must remain in the interior of D which, for the case sketched in Figure A.1, implies that any further intersections with L must take place at a point below y_2.   □

The next lemma is concerned with the intersection of limit sets and transversals.

Lemma A.6 *Let y be a positive limit point of a bounded positive semiorbit γ^+. Then, the trajectory of y cannot cross a transversal at more than one point.*   △

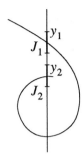

Figure A.2: Illustration of the proof of Lemma A.6.

Proof: Suppose that y_1 and y_2 are two distinct points on the trajectory of y and L is a transversal containing y_1 and y_2. Suppose $y \in \gamma^+(x_0)$ and let $x(t)$ denote the solution starting at x_0. Then, $y_k \in L^+(x_0)$ for $k = 1, 2$ because Lemma 3.1 proves the positive limit set L^+ is invariant. Let $J_k \subset L$ be an interval that contains y_k in its interior. Assume that J_1 and J_2 are disjoint (Figure A.2). By Lemma A.4, given any $\epsilon > 0$ there is $\delta > 0$ such that any trajectory that comes within a distance δ from y_1 will cross L. Since y_1 is a positive limit point of $x(t)$, there is a sequence t_n with $t_n \to \infty$ as $n \to \infty$ such that $x(t_n) \to y_1$ as $n \to \infty$. The solution $x(t)$ enters the ball $B_\delta(y_1)$ infinitely often; hence, it crosses L infinitely often. Let the sequence of crossing points be x_n, ordered as t increases. Since the trajectory is bounded, we can choose ϵ small enough to ensure that x_n is arbitrarily close to y_1. In particular, given any $\delta > 0$, there is $N > 0$ such that for all $n \geq N$, $\|x_n - y_1\| < \delta$. This shows that $x_n \to y_1$ as $n \to \infty$ and, for sufficiently large n, the sequence of crossing points x_n will lie within the interval J_1. By the same argument, the solution crosses J_2 infinitely often. Thus, there is a sequence of crossing points a_1, b_1, a_2, b_2, ..., taken as t increases, with $a_i \in J_1$ and $b_i \in J_2$. By Lemma A.5, the crossing points must be ordered on L in the same order a_1, b_1, a_2, b_2, ... However, this is impossible since J_1 and J_2 are disjoint. Therefore, the two crossing points y_1 and y_2 must be the same point. □

The last lemma states a property of bounded positive limit sets.

Lemma A.7 *Let L^+ be the positive limit set of a bounded trajectory. If L^+ contains a periodic orbit γ, then $L^+ = \gamma$.* △

Proof: Let $L^+ = L^+(x)$ for some point x. It is enough to show that

$$\lim_{t \to \infty} \text{dist}(\phi(t, x), \gamma) = 0$$

where $\text{dist}(\phi(t, x), \gamma)$ is the distance from the trajectory of x to γ. Let L be a transversal at $z \in \gamma$, so small that $L \cap \gamma = z$. By repeating the argument in the proof of the previous lemma, we know that there is a sequence t_n with $t_n \to \infty$ as $n \to \infty$ such that

$$x_n = \phi(t_n, x) \in L$$

$$x_n \to z \text{ as } n \to \infty$$

$$\phi(t, x) \notin L \text{ for } t_{n-1} < t < t_n, \quad n = 1, 2, \ldots$$

By Lemma A.5, $x_n \to z$ monotonically in L. Since γ is a periodic orbit, $\phi(\lambda, z) = z$ for some $\lambda > 0$. For n sufficiently large, $\phi(\lambda, x_n)$ will be within the ball $B_\delta(z)$ (as defined in Lemma A.4); hence, $\phi(t + \lambda, x_n) \in L$ for some $t \in (-\epsilon, \epsilon)$. Thus,

$$|t_{n+1} - t_n| < \lambda + \epsilon$$

which gives an upper bound for the set of positive numbers $t_{n+1} - t_n$. By continuous dependence of the solution on initial states, given any $\beta > 0$ there is $\delta > 0$ such that if $\|x_n - z\| < \delta$ and $|t| < \lambda + \epsilon$, then

$$\|\phi(t, x_n) - \phi(t, z)\| < \beta$$

Choose n_0 large enough that $\|x_n - z\| < \delta$ for all $n \geq n_0$. Then, the last inequality holds for $n \geq n_0$. Now, for all $n \geq n_0$ and $t \in [t_n, t_{n+1}]$ we have

$$\text{dist}(\phi(t, x), \gamma) \leq \|\phi(t, x) - \phi(t - t_n, z)\| = \|\phi(t - t_n, x_n) - \phi(t - t_n, z)\| < \beta$$

since $|t - t_n| < \lambda + \epsilon$.   □

We are now ready to complete the proof of the Poincaré-Bendixson theorem. Since γ^+ is a bounded positive semiorbit, by Lemma 3.1, its positive limit set L^+ is a nonempty, compact, invariant set. Let $y \in L^+$ and $z \in L^+(y) \subset L^+$. Define a transversal L at z; notice that z is not an equilibrium point because L^+ is free of equilibrium points. By Lemma A.6, the trajectory of y cannot cross L at more than one point. On the other hand, there is a sequence t_n with $t_n \to \infty$ as $n \to \infty$ such that $\phi(t_n, y) \to z$. Hence, the trajectory of y crosses L infinitely often. Since there can be only one crossing point, the sequence of crossing points must be a constant sequence. Therefore, we can find $r, s \in R$ such that $r > s$ and $\phi(r, y) = \phi(s, y)$. Since L^+ contains no equilibrium points, the trajectory of y is a periodic orbit. It follows from Lemma A.7 that L^+ is a periodic orbit, since it contains a periodic orbit.

A.11   Proof of Theorem 8.2

Without loss of generality, take $p^* = 0$ and $D = \{x \in R^n \mid \|x\| < r\}$. From (the converse Lyapunov) Theorem 3.12 and Lemma 5.2, we know that for sufficiently small $|\epsilon|$ and $\|\eta(0)\|$,

$$\|x(t, \epsilon)\| \leq k\|x(t_0, \epsilon)\|e^{-\gamma(t - t_0)} + c|\epsilon|$$

for some positive constants k, c, and γ. Hence, the solution of (8.1) is uniformly bounded. It is also uniformly ultimately bounded with an ultimate bound that is proportional to $|\epsilon|$. Note that for sufficiently small $\|\eta(0)\|$, the solution $x_0(t)$ of the nominal problem (8.7) will be uniformly bounded.

Consider the linear equations (8.8), which we solve to determine the Taylor series coefficients. We know from bounded input-bounded output stability (Theorem 6.1) that the solution of (8.8) will be uniformly bounded if the origin of $\dot{z} = A(t)z$ is exponentially stable and the input term g_k is bounded. The input g_k is a polynomial in $x_1, \ldots, x_{k-1}$ with coefficients depending on t and $x_0(t)$. The dependence on t comes through the partial derivatives of f which were assumed to be bounded for $\|x\| < r$. Since $x_0(t)$ is bounded, the polynomial coefficients are bounded for all $t \geq t_0$. Hence, boundedness of g_k will follow from boundedness of $x_1, \ldots, x_{k-1}$. The matrix $A(t)$ is given by

$$A(t) = \frac{\partial f}{\partial x}(t, x_0(t), 0)$$

where $x_0(t)$ is the solution of the nominal system (8.7). It turns out that exponential stability of the origin as an equilibrium for (8.7) ensures that the origin of $\dot{z} = A(t)z$ will be exponentially stable for every solution $x_0(t)$ that starts within the ball $\{\|x\| < r_0\}$ with sufficiently small r_0. To see this point, let

$$A_0(t) = \frac{\partial f}{\partial x}(t, 0, 0)$$

and write

$$A(t) = A_0(t) + [A(t) - A_0(t)] \overset{\text{def}}{=} A_0(t) + B(t)$$

so that the linear system $\dot{z} = A(t)z$ can be viewed as a linear perturbation of $\dot{y} = A_0(t)y$. We have

$$\|x_0(t)\| \leq k\|\eta(0)\|e^{-\gamma(t-t_0)}$$

Moreover, by uniform boundedness of the partial derivatives of f, we know that $[\partial f/\partial x](t, x, 0)$ satisfies a Lipschitz condition, uniformly in t. Hence,

$$\|B(t)\| = \left\| \frac{\partial f}{\partial x}(t, x_0(t), 0) - \frac{\partial f}{\partial x}(t, 0, 0) \right\| \leq L\|x_0(t)\| \leq Lk\|\eta(0)\|e^{-\gamma(t-t_0)}$$

On the other hand, by exponential stability of the origin of (8.7) and Theorem 3.13 we know that the origin of the linear system $\dot{y} = A_0(t)y$ is exponentially stable. Therefore, similar to Example 5.9, we conclude that the origin of the linear system $\dot{z} = A(t)z$ is exponentially stable. Since $\|x_0(t)\| < r$ and

$$g_1(t, x_0(t)) = \frac{\partial f}{\partial \epsilon}(t, x_0(t), 0)$$

we see that g_1 is bounded for all $t \geq t_0$. Hence, by the bounded input-bounded output stability theorem (Theorem 6.1), we conclude that $x_1(t)$ is bounded. By a simple induction argument, we can see that $x_2(t), \ldots, x_{k-1}(t)$ are bounded.

So far, we have verified that the exact solution $x(t, \epsilon)$ and the approximate solution $\sum_{k=0}^{N-1} x_k(t)\epsilon^k$ are uniformly bounded on $[t_0, \infty)$ for sufficiently small $|\epsilon|$. All that remains now is to analyze the approximation error $e = x - \sum_{k=0}^{N-1} x_k(t)\epsilon^k$. The error analysis is quite similar to what we have done in Section 8.1. The error satisfies (8.10), where ρ_1 and ρ_2 satisfy (8.11)–(8.13) for all $(t, e, \epsilon) \in [t_0, \infty) \times B_\lambda \times [-\epsilon_1, \epsilon_1]$ for sufficiently small λ and ϵ_1. The error equation (8.10) is viewed as a perturbation of $\dot{e} = A(t)e$, where the perturbation term satisfies

$$\|\rho_1(t, e, \epsilon) + \rho_2(t, \epsilon)\| \leq k_1(\|e\| + |\epsilon|)\|e\| + k_2|\epsilon|^N \leq k_1(\lambda + |\epsilon|)\|e\| + k_2|\epsilon|^N$$

Noting that $\|e(t_0, \epsilon)\| = O(\epsilon^N)$, we conclude from Lemma 5.7 that for sufficiently small λ and $|\epsilon|$, $\|e(t, \epsilon)\| = O(\epsilon^N)$ for all $t \geq t_0$.

A.12   Proof of Theorem 8.3

The change of time variable $s = \epsilon t$ transforms (8.23) into

$$\frac{dy}{ds} = f_{av}(y) + \epsilon q(s/\epsilon, y, \epsilon) \tag{A.21}$$

where $q(s/\epsilon, y, \epsilon)$ is ϵT-periodic in s and bounded on Γ for sufficiently small ϵ. By applying Theorems 2.5 and 2.6 on continuity of solutions with respect to initial states and parameters, we see that if, for a given initial state, the averaged system

$$\frac{dy}{ds} = f_{av}(y)$$

has a unique solution $\bar{y}(s)$ defined on $[0, b]$ and $\bar{y}(s) \in D$ for all $s \in [0, b]$, then there exists $\epsilon^* > 0$ such that for all $0 < \epsilon < \epsilon^*$ the perturbed system (A.21) will have a unique solution defined for all $s \in [0, b]$ and the two solutions will be $O(\epsilon)$ close, provided the initial states are $O(\epsilon)$ close. Since $t = s/\epsilon$ and $x - y = O(\epsilon)$ by (8.22), the solution of the averaged system (8.18) provides an $O(\epsilon)$ approximation for the solution of (8.17) over the time interval $[0, b/\epsilon]$ in the t time scale.

Suppose the averaged system (8.18) has an exponentially stable equilibrium point $x = p^* \in D$. Without loss of generality, take $p^* = 0$ and $D = \{x \in R^n \mid \|x\| < r\}$. By Theorem 5.3 on continuity of solutions on the infinite interval, we conclude that there is a domain $\|x\| < r_0$ such that for all initial states within that domain, the $O(\epsilon)$ approximation will be valid for all $s \geq 0$, that is, for all $t \geq 0$.

To prove the last part of the theorem, we write (8.23) as

$$\dot{y} = \epsilon A y + \epsilon \psi(t, y, \epsilon) \tag{A.22}$$

where

$$A = \frac{\partial f_{av}(y)}{\partial y}, \quad g(y) = f_{av}(y) - Ay, \quad \psi(t, y, \epsilon) = g(y) + \epsilon q(t, y, \epsilon)$$

Since the origin of the averaged system is exponentially stable, the matrix A is Hurwitz. The function $g(y)$ is twice continuously differentiable and

$$g(0) = 0, \quad \frac{\partial g}{\partial y}(0) = 0$$

The function $\psi(t, y, \epsilon)$ is T-periodic in t, continuously differentiable, and bounded on Γ for sufficiently small ϵ. If (8.23) has a T-periodic solution $y(t)$, then using (A.22) we can express this solution as

$$y(t) = e^{\epsilon A(t-t_0)}y(t_0) + \epsilon \int_{t_0}^{t} e^{\epsilon A(t-\sigma)}\psi(\sigma, y(\sigma), \epsilon) \, d\sigma, \quad \forall \, t \geq t_0$$

Using $y(t + T) = y(t)$ in the foregoing expression, we obtain

$$y(t) = e^{\epsilon T A}y(t) + \epsilon \int_{t}^{t+T} e^{\epsilon A(t+T-\sigma)}\psi(\sigma, y(\sigma), \epsilon) \, d\sigma$$

Changing the integration variable from σ to $s = \sigma - t$, we can express $y(t)$ as

$$y(t) = \epsilon M(\epsilon T) \int_0^T e^{\epsilon A(T-s)}\psi(t + s, y(t + s), \epsilon) \, ds \qquad (A.23)$$

where

$$M(\delta) = \left[I - e^{\delta A}\right]^{-1}, \quad \delta > 0$$

Since A is Hurwitz and $\delta > 0$, all the eigenvalues of $e^{\delta A}$ are strictly inside the unit circle.[6] Consequently, $\left[I - e^{\delta A}\right]$ is nonsingular. The fact

$$e^{\delta A} = I + \delta A + O(\delta^2)$$

shows that

$$M(\delta) = [I - I - \delta A - O(\delta^2)]^{-1} = -\frac{1}{\delta}A^{-1}[I + O(\delta)]$$

Thus, there exist positive constants k_1 and ϵ_1^* such that

$$\epsilon \|M(\epsilon T)\| \leq k_1, \quad \forall \, 0 < \epsilon < \epsilon_1^*$$

---

[6] This is a well-known fact in sampled-data control theory. It can be proven by transforming A into its Jordan form.

By the Hurwitz property of A, there is a positive constant k_2 such that

$$\|e^{aA}\| \le k_2, \quad \forall\, a \ge 0$$

We shall use the expression (A.23) to show that (A.22) has a T-periodic solution. Let $\mathcal{D}$ be the space of all T-periodic continuous functions $x(t)$, with the norm $\|x\|_{\mathcal{D}} = \max_{0 \le t \le T} \|x(t)\|$. It can be verified that $\mathcal{D}$ is a Banach space. Define a mapping P on $\mathcal{D}$ by the right-hand side of (A.23); that is,

$$(Px)(t) = \epsilon M(\epsilon T) \int_0^T e^{\epsilon A(T-s)} \psi(t+s, x(t+s), \epsilon)\, ds$$

Lemma A.8 *Equation* (A.22) *has a T-periodic solution if and only if the mapping P has a fixed point in $\mathcal{D}$.*   ◇

Proof: We have already shown the *only if* part. To show the *if* part, let x be a fixed point of Px in $\mathcal{D}$. Then,

$$
\begin{aligned}
x(t) &= \epsilon M(\epsilon T) \int_0^T e^{\epsilon A(T-s)} \psi(t+s, x(t+s), \epsilon)\, ds \\
     &= \epsilon M(\epsilon T) \int_t^{t+T} e^{\epsilon A(t+T-\sigma)} \psi(\sigma, x(\sigma), \epsilon)\, d\sigma
\end{aligned}
$$

and

$$
\begin{aligned}
\dot{x}(t) &= \epsilon A \epsilon M(\epsilon T) \int_t^{t+T} e^{\epsilon A(t+T-\sigma)} \psi(\sigma, x(\sigma), \epsilon)\, ds \\
           &\quad + \epsilon M(\epsilon T) \left[\psi(t+T, x(t+T), \epsilon) - e^{\epsilon T A} \psi(t, x(t), \epsilon) \right] \\
           &= \epsilon A x(t) + \epsilon M(\epsilon T) \left[I - e^{\epsilon T A} \right] \psi(t, x(t), \epsilon) \\
           &= \epsilon A x(t) + \epsilon \psi(t, x(t), \epsilon)
\end{aligned}
$$

where we have used the fact that $\psi(t, x(t), \epsilon)$ is T-periodic.   □

Lemma A.9 *There exist positive constants ρ and ϵ^* such that for all $0 < \epsilon < \epsilon^*$, the mapping P has a unique fixed point in $S = \{x \in \mathcal{D} \mid \|x\|_{\mathcal{D}} \le \epsilon\rho\}$.*   ◇

Proof: In the set $\{\|x\| \le \epsilon\rho\}$, we have

$$\|\psi(t, x, \epsilon)\| \le k_3 \epsilon^2 \rho^2 + \epsilon k_4$$

for some positive constants k_3, k_4 independent of ρ. Therefore,

$$\|(Px)(t)\| \le k_1 k_2 T (k_3 \epsilon^2 \rho^2 + \epsilon k_4), \quad \forall\, x \in S$$

Take $\rho = 2k_1k_2k_4T$ and $\epsilon_2^* = \min\{\epsilon_1^*, k_4/k_3\rho^2\}$. Then, for all $0 < \epsilon < \epsilon_2^*$, P maps S into S. Using the properties of $g(y)$, it can be shown that[7]

$$\|g(x_2) - g(x_1)\| \leq k_5\|x_2 - x_1\|^2$$

Hence, there exist positive constants k_6 and ϵ_3^* such that

$$\|\psi(t, x_2, \epsilon) - \psi(t, x_1, \epsilon)\| \leq \epsilon k_6\|x_2 - x_1\|$$

for all x_1, $x_2 \in \{\|x\| \leq \epsilon\rho\}$. Consequently, for all x_1, x_2 in S, we have

$$\|(Px_2) - (Px_1)\|_{\mathcal{D}} \leq \epsilon k_1 k_2 k_6 T \|x_2 - x_1\|_{\mathcal{D}}$$

By choosing ϵ^* small enough, we can ensure that P is a contraction mapping in S for all $0 < \epsilon < \epsilon^*$. Thus, by the contraction mapping theorem, we conclude that P has a unique fixed point in S. $\qquad\square$

Lemma A.10 *If $\bar{y}(t, \epsilon)$ is a T-periodic solution of (8.23) such that $\|\bar{y}(t, \epsilon)\| \leq k\epsilon$, then, for sufficiently small ϵ, $\bar{y}(t, \epsilon)$ is exponentially stable.* $\qquad\diamond$

Proof: A systematic procedure to study stability of $\bar{y}(t, \epsilon)$ is to apply the change of variables $z = y - \bar{y}(t, \epsilon)$ and study stability of the equilibrium point at $z = 0$. The new variable z satisfies the equation

$$\dot{z} \; = \; \epsilon\{f_{av}(z + \bar{y}(t, \epsilon)) - f_{av}(\bar{y}(t, \epsilon)) + \epsilon[q(t, z + \bar{y}(t, \epsilon), \epsilon) - q(t, \bar{y}(t, \epsilon), \epsilon)]\}$$
$$\stackrel{\text{def}}{=} \; \epsilon f(t, z)$$

The linearization about $z = 0$ is given by

$$\epsilon\left.\frac{\partial f}{\partial z}\right|_{z=0} \; = \; \epsilon\left(\left.\frac{\partial f_{av}}{\partial y}\right|_{z=0} + \epsilon\left.\frac{\partial q}{\partial y}\right|_{z=0}\right)$$
$$= \; \epsilon\left\{A + \left[\frac{\partial f_{av}}{\partial y}(\bar{y}(t, \epsilon)) - A\right] + \epsilon\frac{\partial q}{\partial y}(t, \bar{y}(t, \epsilon), \epsilon)\right\}$$

By continuity of $[\partial f_{av}/\partial y]$, we know that for any $\delta > 0$ there is $\epsilon^* > 0$ such that

$$\left\|\frac{\partial f_{av}}{\partial y}(\bar{y}(t, \epsilon)) - \frac{\partial f}{\partial x}(0)\right\| < \delta$$

for $\epsilon < \epsilon^*$. Since A is Hurwitz and $[\partial g/\partial x](t, \bar{x}, \epsilon)$ is $O(1)$, we conclude from Lemma 5.1 that, for sufficiently small ϵ, the linear system

$$\dot{x} = \epsilon\left\{A + \left[\frac{\partial f_{av}}{\partial y}(\bar{y}(t, \epsilon)) - A\right] + \epsilon\frac{\partial q}{\partial y}(t, \bar{y}(t, \epsilon), \epsilon)\right\}x$$

---

[7]This inequality follows from Taylor's theorem; see [7, Theorem 6-22].

has an exponentially stable equilibrium at $x = 0$. Therefore, by Theorem 3.11, $z = 0$ is an exponentially stable equilibrium point. $\square$

By Lemmas A.8, A.9, and A.10, we conclude that (8.23) has a unique, exponentially stable, T-periodic solution $\bar{y}(t, \epsilon)$ in an $O(\epsilon)$ neighborhood of p^*. By (8.22), we see that (8.17) has a T-periodic solution

$$\bar{x}(t, \epsilon) = \bar{y}(t, \epsilon) + \epsilon u(t, \bar{y}(t, \epsilon))$$

Since u is bounded, the periodic solution $\bar{x}(t, \epsilon)$ lies in an $O(\epsilon)$ neighborhood of p^*.

A.13   Proof of Theorem 8.4

By expressing (8.40) in the $s = \epsilon t$ time scale, applying Theorems 2.5 and 2.6, and using the change of variables (8.41), we can conclude that if the averaged system (8.41) has a unique solution defined for all $t \in [0, b/\epsilon]$, then there exists $\epsilon^* > 0$ such that for all $0 < \epsilon < \epsilon^*$, the original system (8.32) will have a unique solution defined for all $t \in [0, b/\epsilon]$ and the two solutions will be $O(\alpha(\epsilon))$ close, provided the initial states are $O(\alpha(\epsilon))$ close.

Suppose $x = 0$ is an equilibrium point of the original system (8.32); that is, $f(t, 0, \epsilon) = 0$ for all $(t, \epsilon) \in [0, \infty) \times [0, \epsilon_0]$. This assumption implies that $f_{av}(0) = 0$, so the origin is an equilibrium point of the averaged system. It also implies that $h(t, 0) = 0$ and $w(t, 0, \eta) = 0$. In view of the bound $\|\partial w/\partial x\| \leq \alpha(\eta)/\eta$, we see that the estimate on w can be revised to

$$\eta\|w(t, x, \eta)\| \leq \alpha(\eta)\|x\|$$

Furthermore, the assumption $f(t, 0, \epsilon) = 0$, together with differentiability of f with respect to ϵ, implies that $f(t, x, \epsilon)$ is Lipschitz in ϵ, linearly in x; that is,

$$\|f(t, x, \epsilon) - f(t, x, 0)\| \leq L_1 \epsilon \|x\|$$

Using these estimates, it can be verified that the function $q(t, y, \epsilon)$ in (8.40) satisfies the inequality

$$\|q(t, y, \epsilon)\| \leq L\|y\|$$

with some positive constant L for $(t, y, \epsilon) \in [0, \infty) \times D_0 \times [0, \epsilon_1]$, where $D_0 = \{\|y\| < r_0\}$ and r_0 and ϵ_1 are chosen small enough. Suppose now that the origin of the averaged system (8.41) is exponentially stable. By (the converse Lyapunov) Theorem 3.12 and Lemma 5.1, we conclude that, for sufficiently small ϵ, the origin will be an exponentially stable equilibrium point for the original system (8.32). Moreover, by Theorem 5.3 on continuity of solutions on the infinite interval, we conclude that the $O(\alpha(\epsilon))$ estimate on the approximation error will be valid for all $t \geq 0$.

A.14 Proof of Theorem 9.1

We shall work with the full problem (9.10)–(9.11) in the (x, y) variables. The error estimate for z will then follow from the change of variables (9.9). When we analyze (9.12) with the slowly varying variables t and x, we want to use the uniform exponential stability property of the boundary-layer model; in particular, inequality (9.15). This inequality is valid only when $x \in B_r$, so to use it we need to confirm that the slowly varying x will always be in B_r. We anticipate that this will be true because the solution $\bar{x}$ of the reduced problem (9.8) satisfies the bound $\|\bar{x}(t)\| \le r_1 < r$ by assumption, and we anticipate that the error $\|x(t, \epsilon) - \bar{x}(t)\|$ will be $O(\epsilon)$. Then, for sufficiently small ϵ, x will satisfy the bound $\|x(t, \epsilon)\| \le r$. However, the estimate $\|x(t, \epsilon) - \bar{x}(t)\| = O(\epsilon)$ has not been proven yet so we cannot start by using it. We use a special technique[8] to get around this difficulty. Let $\psi : R^n \to [0, 1]$ be a smooth (continuously differentiable infinitely many times) function with $\psi(x) = 1$ when $\|x\| \le \frac{1}{2}(r + r_1)$ and $\psi(x) = 0$ when $\|x\| \ge r$. We define F and G by

$$F(t, x, y, \epsilon) = f(t, x\psi(x), y + h(t, x\psi(x)), \epsilon) \qquad (A.24)$$

$$G(t, x, y, \epsilon) = g(t, x\psi(x), y + h(t, x\psi(x)), \epsilon) - \epsilon \frac{\partial h}{\partial t}(t, x\psi(x))$$

$$- \epsilon \frac{\partial h}{\partial x}(t, x\psi(x)) f(t, x\psi(x), y + h(t, x\psi(x)), \epsilon) \qquad (A.25)$$

It can be easily verified that for all $(t, x, y, \epsilon) \in [0, t_1] \times R^n \times B_\rho \times [0, \epsilon_0]$, we have

- F and G and their first partial derivatives with respect to ϵ are continuous and bounded.

- $F(t, x, y, 0)$ and $G(t, x, y, 0)$ have bounded first partial derivatives with respect to (x, y).

- The Jacobian $[\partial G(t, x, y, 0)/\partial y]$ has bounded first partial derivatives with respect to (t, x, y).

When $\|x\| \le \frac{1}{2}(r_1 + r)$, we have $x\psi(x) = x$; hence, the functions F and f are identical. The same is true for the functions G and $g - \epsilon[(\partial h/\partial t) + (\partial h/\partial x)f]$. Consider the modified singular perturbation problem

$$\dot{x} = F(t, x, y, \epsilon), \quad x(t_0) = \xi(\epsilon) \qquad (A.26)$$

$$\epsilon \dot{y} = G(t, x, y, \epsilon), \quad y(t_0) = \eta(\epsilon) - h(t_0, \xi(\epsilon)) \qquad (A.27)$$

The modified problem (A.26)–(A.27) is identical to the original problem (9.10)–(9.11) when $\|x\| \le \frac{1}{2}(r_1 + r)$. The constant $\frac{1}{2}(r_1 + r)$ has been chosen in anticipation

---

[8]The same technique has been used in the proof of the center manifold theorem; see Appendix A.7. The first footnote of that section gives a clue on how to choose the function ψ.

that the solution $x(t, \epsilon)$ will be confined to a ball of radius $\frac{1}{2}(r_1 + r)$, which is based on the fact that $\|\bar{x}(t)\| \leq r_1$. On the other hand, the function ψ has been chosen such that $\|x\psi(x)\| \leq r$ for all $x \in R^n$. This choice has been made to ensure that the equilibrium $y = 0$ of the boundary-layer model

$$\frac{dy}{d\tau} = G(t, x, y, 0) \tag{A.28}$$

will be exponentially stable uniformly in (t, x) for $t \in [0, t_1]$ and $x \in R^n$. Note that, since

$$G(t, x, y, 0) = g(t, x\psi(x), y + h(t, x\psi(x)), 0)$$

for any fixed $x \in R^n$, the boundary-layer model (A.28) can be represented as a boundary-layer model of the form (9.14) with $x\psi(x) \in B_r$ as the frozen parameter. Since inequality (9.15) holds uniformly in the frozen parameter, it is clear that the solutions of (A.28) satisfy the same inequality for all $x \in R^n$; that is,

$$\|y(\tau)\| \leq k\|y(0)\| \exp(-\gamma\tau), \ \forall \ \|y(0)\| < \rho_0, \ \forall \ (t, x) \in [0, t_1] \times R^n, \ \forall \ \tau \geq 0 \tag{A.29}$$

The reduced problem for (A.26)–(A.27) is

$$\dot{x} = F(t, x, 0, 0), \quad x(t_0) = \xi_0 \tag{A.30}$$

This problem is identical to the reduced problem (9.8) whenever $\|x\| \leq \frac{1}{2}(r + r_1)$. Since (9.8) has a unique solution $\bar{x}(t)$ defined for all $t \in [t_0, t_1]$ and $\|\bar{x}(t)\| \leq r_1$, it follows that $\bar{x}(t)$ is the unique solution of (A.30) and is defined for all $t \in [t_0, t_1]$. We proceed to prove the theorem for the modified singular perturbation problem (A.26)–(A.27). Upon completion of this task, we shall show that, for sufficiently small ϵ, the solution $x(t, \epsilon)$ of (A.26)–(A.27) satisfies the upper bound $\|x(t, \epsilon)\| \leq \frac{1}{2}(r_1 + r)$. This will establish that the original and modified problems have the same solution and proves the theorem for the original problem (9.10)–(9.11).

Consider the boundary-layer model (A.28). Since $[\partial G/\partial y]$ has bounded first partial derivatives with respect to (t, x) and $G(t, x, 0, 0) = 0$ for all (t, x), the Jacobian matrices $[\partial G/\partial t]$ and $[\partial G/\partial x]$ satisfy

$$\left\| \frac{\partial G}{\partial t} \right\| \leq L_1 \|y\|; \quad \left\| \frac{\partial G}{\partial x} \right\| \leq L_2 \|y\|$$

Using these estimates and inequality (A.29), we conclude from Lemma 5.11 that there is a Lyapunov function $V(t, x, y)$ which satisfies the inequalities:

$$c_1 \|y\|^2 \leq V(t, x, y) \leq c_2 \|y\|^2 \tag{A.31}$$

$$\frac{\partial V}{\partial y} G(t, x, y, 0) \leq -c_3 \|y\|^2 \tag{A.32}$$

$$\left\|\frac{\partial V}{\partial y}\right\| \le c_4 \|y\| \tag{A.33}$$

$$\left\|\frac{\partial V}{\partial t}\right\| \le c_5 \|y\|^2; \quad \left\|\frac{\partial V}{\partial x}\right\| \le c_6 \|y\|^2 \tag{A.34}$$

for all $y \in \{y \in R^m \mid \|y\| < \rho_0\}$ and all $(t, x) \in [0, t_1] \times R^n$. The derivative of V along the trajectories of the full system (A.26)–(A.27) is given by

$$\begin{aligned}
\dot{V} &= \frac{1}{\epsilon}\frac{\partial V}{\partial y}G(t, x, y, \epsilon) + \frac{\partial V}{\partial t} + \frac{\partial V}{\partial x}F(t, x, y, \epsilon) \\
&= \frac{1}{\epsilon}\frac{\partial V}{\partial y}G(t, x, y, 0) + \frac{1}{\epsilon}\frac{\partial V}{\partial y}[G(t, x, y, \epsilon) - G(t, x, y, 0)] \\
&\quad + \frac{\partial V}{\partial t} + \frac{\partial V}{\partial x}F(t, x, y, \epsilon)
\end{aligned}$$

Using inequalities (A.32)–(A.34) and the estimates

$$\|F(t, x, y, \epsilon)\| \le k_0; \quad \|G(t, x, y, \epsilon) - G(t, x, y, 0)\| \le \epsilon L_3$$

we obtain

$$\begin{aligned}
\dot{V} &\le -\frac{c_3}{\epsilon}\|y\|^2 + c_4 L_3 \|y\| + c_5 \|y\|^2 + c_6 k_0 \|y\|^2 \\
&\le -\frac{c_3}{2\epsilon}\|y\|^2 + c_4 L_3 \|y\|, \quad \text{for } \epsilon \le \frac{c_3}{2c_5 + 2c_6 k_0}
\end{aligned}$$

Thus, for all $\|y(t_0)\| < \rho_0\sqrt{c_1/c_2} \overset{\text{def}}{=} \mu$, the solution $y(t, \epsilon)$ of the full problem satisfies the exponentially decaying bound

$$\|y(t, \epsilon)\| \le k_1 \exp\left[\frac{-\alpha(t - t_0)}{\epsilon}\right] \|y(t_0)\| + \epsilon\delta, \quad \forall\, t \ge t_0 \tag{A.35}$$

where $k_1 = \sqrt{c_2/c_1}$, $\alpha = c_3/4c_2$ and $\delta = 2c_2 c_4 L_3/c_1 c_3$. Since $y(t_0) = \eta(\epsilon) - h(t_0, \xi(\epsilon))$, the condition $\|y(t_0)\| < \mu$ will follow, for sufficiently small ϵ, from $\|\eta(0) - h(t_0, \xi(0))\| < \mu$.

Consider (A.26). By rewriting the right-hand side as

$$F(t, x, y, \epsilon) = F(t, x, 0, 0) + [F(t, x, y, \epsilon) - F(t, x, 0, 0)]$$

we view (A.26) as a perturbation of the reduced system (A.30). The bracketed perturbation term satisfies

$$\begin{aligned}
\|F(t, x, y, \epsilon) - F(t, x, 0, 0)\| &\le \|F(t, x, y, \epsilon) - F(t, x, y, 0)\| \\
&\quad + \|F(t, x, y, 0) - F(t, x, 0, 0)\| \\
&\le L_4 \epsilon + L_5 \|y\| \\
&\le \theta_1 \epsilon + \theta_2 \exp\left[\frac{-\alpha(t - t_0)}{\epsilon}\right] \|y(t_0)\|
\end{aligned}$$

where $\theta_1 = L_4 + L_5\delta$ and $\theta_2 = L_5 k_1$. Define

$$u(t, \epsilon) = x(t, \epsilon) - \bar{x}(t)$$

Then,

$$
\begin{aligned}
u(t, \epsilon) \;=\;& \xi(\epsilon) - \xi(0) + \int_{t_0}^{t} [F(s, x(s, \epsilon), y(s, \epsilon), \epsilon) - F(s, \bar{x}(s), 0, 0)] \, ds \\
\;=\;& \xi(\epsilon) - \xi(0) + \int_{t_0}^{t} [F(s, x(s, \epsilon), y(s, \epsilon), \epsilon) - F(s, x(s, \epsilon), 0, 0)] \, ds \\
& + \int_{t_0}^{t} [F(s, x(s, \epsilon), 0, 0) - F(s, \bar{x}(s), 0, 0)] \, ds
\end{aligned}
$$

and

$$
\begin{aligned}
\|u(t, \epsilon)\| \;\leq\;& k_2 \epsilon + \int_{t_0}^{t} \left\{ \theta_1 \epsilon + \theta_2 \exp\left[\frac{-\alpha(s - t_0)}{\epsilon} \right] \|y(t_0)\| \right\} ds \\
& + \int_{t_0}^{t} L_6 \|u(s, \epsilon)\| \, ds \\
\;\leq\;& k_2 \epsilon + \left[\theta_1 \epsilon (t_1 - t_0) + \frac{\theta_2 \epsilon \mu}{\alpha} \right] + \int_{t_0}^{t} L_6 \|u(s, \epsilon)\| \, ds
\end{aligned}
$$

By the Gronwall-Bellman lemma, we arrive at the estimate

$$\|x(t, \epsilon) - \bar{x}(t)\| \leq \epsilon k_3 [1 + t_1 - t_0] \exp[L_6 (t_1 - t_0)] \qquad (A.36)$$

which proves the error estimate for x. We can also conclude that, for sufficiently small ϵ, the solution $x(t, \epsilon)$ is defined for all $t \in [t_0, t_1]$.

To prove the error estimate for y, consider (A.27) which, for convenience, is written in the τ time scale

$$\frac{dy}{d\tau} = G(t_0 + \epsilon\tau, x(t_0 + \epsilon\tau, \epsilon), y, \epsilon)$$

Let $\hat{y}(\tau)$ denote the solution of the boundary-layer model

$$\frac{dy}{d\tau} = G(t_0, \xi_0, y, 0), \quad y(0) = \eta(0) - h(t_0, \xi_0)$$

and set

$$v(\tau, \epsilon) = y(\tau, \epsilon) - \hat{y}(\tau)$$

By differentiating both sides with respect to τ and substituting for the derivatives of y and $\hat{y}$, we obtain

$$\frac{dv}{d\tau} = G(t_0 + \epsilon\tau, x(t_0 + \epsilon\tau, \epsilon), y(\tau, \epsilon), \epsilon) - G(t_0, \xi_0, \hat{y}(\tau), 0)$$

We add and subtract $G(t_0 + \epsilon\tau, x(t_0 + \epsilon\tau, \epsilon), v, 0)$ to obtain

$$\frac{dv}{d\tau} = G(t, x, v, 0) + \Delta G \tag{A.37}$$

where $t = t_0 + \epsilon\tau$, $x = x(t_0 + \epsilon\tau, \epsilon)$, $\Delta G = \Delta_1 + \Delta_2 + \Delta_3$, and

$$\begin{aligned}
\Delta_1 &= G(t, x, y, 0) - G(t, x, \hat{y}, 0) - G(t, x, v, 0) \\
\Delta_2 &= G(t, x, y, \epsilon) - G(t, x, y, 0) \\
\Delta_3 &= G(t, x, \hat{y}, 0) - G(t_0, \xi_0, \hat{y}, 0)
\end{aligned}$$

It can be verified that

$$\begin{aligned}
\|\Delta_1\| &\leq k_4\|v\|^2 + k_5\|v\|\,\|\hat{y}\|, \qquad \|\Delta_2\| \leq \epsilon L_3 \\
\|\Delta_3\| &\leq L_1|t - t_0|\,\|\hat{y}\| + L_2\|x - \xi_0\|\,\|\hat{y}\| \leq (L_1\epsilon\tau + L_2\epsilon a + L_2\epsilon\tau k_0)\|\hat{y}\|
\end{aligned}$$

for some nonnegative constants k_4, k_5, and a. Hence,

$$\begin{aligned}
\|\Delta G\| &\leq k_4\|v\|^2 + k_5 k\|v\|\,\|y(0)\|e^{-\gamma\tau} + \epsilon L_3 + \epsilon a_1 k\|y(0)\|(1 + \tau)e^{-\gamma\tau} \\
&\leq k_4\|v\|^2 + k_5 k\rho\|v\|e^{-\gamma\tau} + \epsilon a_2 \tag{A.38}
\end{aligned}$$

where $a_1 = \max\{L_2 a, L_1 + L_2 k_0\}$ and $a_2 = L_3 + a_1 k\rho \max\{1, 1/\gamma\}$. We have used the fact that $(1 + \tau)e^{-\gamma\tau} \leq \max\{1, 1/\gamma\}$. Equation (A.37) can be viewed as a perturbation of

$$\frac{dv}{d\tau} = G(t, x, v, 0) \tag{A.39}$$

which, by Lemma 5.11, has a Lyapunov function $V(t, x, v)$ that satisfies (A.31)–(A.34). Calculating the derivative of V along the trajectories of (A.37) and using the estimate (A.38), we obtain

$$\begin{aligned}
\dot{V} &= \frac{\partial V}{\partial t} + \frac{\partial V}{\partial x}F + \frac{1}{\epsilon}\frac{\partial V}{\partial v}[G(t, x, v, 0) + \Delta G] \\
&\leq c_5\|v\|^2 + c_6 k_0\|v\|^2 - \frac{c_3}{\epsilon}\|v\|^2 + \frac{c_4}{\epsilon}\|v\|\left(k_4\|v\|^2 + k_5 k\rho\|v\|e^{-\gamma\tau} + \epsilon a_2\right)
\end{aligned}$$

For $\|v\| \leq c_3/4c_4 k_4$ and $0 < \epsilon < c_3/4(c_5 + c_6 k_0)$, we have

$$\begin{aligned}
\dot{V} &\leq -\frac{c_3}{2\epsilon}\|v\|^2 + \frac{c_4 k_5 k\rho}{\epsilon}\|v\|^2 e^{-\gamma\tau} + c_4 a_2\|v\| \\
&\leq -\frac{2}{\epsilon}\left(k_a - k_b e^{-\gamma\tau}\right)V + 2k_c\sqrt{V}
\end{aligned}$$

where $k_a = c_3/4c_2$, $k_b = c_4 k_5 k\rho/2c_1$, and $k_c = c_4 a_2/2\sqrt{c_1}$. Taking $W = \sqrt{V}$ yields

$$\frac{dW}{d\tau} \leq -\left(k_a - k_b e^{-\gamma\tau}\right)W + \epsilon k_c$$

By the comparison principle (Lemma 2.5), we conclude that

$$W(\tau) \leq \phi(\tau, 0)W(0) + \epsilon \int_0^\tau \phi(\tau, \sigma)k_c \, d\sigma$$

where

$$\phi(\tau, \sigma) = \exp\left[-\int_\sigma^\tau \left(k_a - k_b e^{-\gamma\lambda}\right) \, d\lambda\right] \quad \text{and} \quad |\phi(\tau, \sigma)| \leq k_g e^{-\alpha_g(\tau-\sigma)}$$

for some $k_g, \alpha_g > 0$. Using the fact that $v(0) = O(\epsilon)$, we conclude that $v(\tau) = O(\epsilon)$ for all $\tau \geq 0$. This completes the proof that the solution of (A.26)–(A.27) satisfies

$$x(t, \epsilon) - \bar{x}(t) = O(\epsilon), \quad y(t, \epsilon) - \hat{y}\left(\frac{t}{\epsilon}\right) = O(\epsilon)$$

$\forall \ t \in [t_0, t_1]$ for sufficiently small ϵ. Since $\|\bar{x}(t)\| \leq r_1 < r$ there is $\epsilon_1^* > 0$ small enough such that $\|x(t, \epsilon)\| \leq \frac{1}{2}(r + r_1)$ for all $t \in [t_0, t_1]$ and all $\epsilon < \epsilon_1^*$. Hence, $x(t, \epsilon)$ and $y(t, \epsilon)$ are the solutions of (9.10)–(9.11). From (9.9), we have

$$z(t, \epsilon) - h(t, \bar{x}(t)) - \hat{y}\left(\frac{t}{\epsilon}\right) = y(t, \epsilon) - \hat{y}\left(\frac{t}{\epsilon}\right) + h(t, x(t, \epsilon)) - h(t, \bar{x}(t)) = O(\epsilon)$$

where we have used the fact that h is Lipschitz in x. Finally, since $\hat{y}(\tau)$ satisfies (A.29) and

$$\exp\left[\frac{-\gamma(t - t_0)}{\epsilon}\right] \leq \epsilon, \quad \forall \ \gamma(t - t_0) \geq \epsilon \ln\left(\frac{1}{\epsilon}\right)$$

the term $\hat{y}(t/\epsilon)$ will be $O(\epsilon)$ uniformly on $[t_b, t_1]$ if ϵ is small enough to satisfy

$$\epsilon \ln\left(\frac{1}{\epsilon}\right) \leq \gamma(t_b - t_0)$$

The proof of Theorem 9.1 is now complete.

A.15   Proof of Theorem 9.4

The proof of this theorem follows closely the proof of Theorem 9.1. We shall only point out the differences. Notice first that due to the extra assumptions, the Jacobian $[\partial F/\partial x](t, x, 0, 0)$ has bounded first partial derivatives with respect to x. Also, due to exponential stability of the origin of the reduced model, given $r_1 < r$ there exists $\mu_1 > 0$ such that for $\|\xi_0\| < \mu_1$ the solution $\bar{x}(t)$ of (9.8) is defined for all $t \geq t_0$ and satisfies $\|\bar{x}(t)\| \leq r_1$.

The main deviation from the proof of Theorem 9.1 comes when we view (A.26) as a perturbation of the reduced system (A.30) and define

$$u(t, \epsilon) = x(t, \epsilon) - \bar{x}(t)$$

Instead of using the Gronwall-Bellman lemma to derive an estimate of u, we employ a Lyapunov analysis that exploits the exponential stability of the origin of (A.30). The Lyapunov analysis is very similar to the one we used in the boundary-layer analysis of Theorem 9.1's proof. Therefore, we shall describe it only briefly. The error u satisfies the equation

$$\dot{u} = F(t, u, 0, 0) + \Delta F \tag{A.40}$$

where

$$\Delta F = [F(t, \bar{x} + u, 0, 0) - F(t, \bar{x}, 0, 0) - F(t, u, 0, 0)] + [F(t, x, y, \epsilon) - F(t, x, 0, 0)]$$

It can be verified that

$$\|\Delta F\| \leq \tilde{k}_4 \|u\|^2 + \tilde{k}_5 \|u\| \, \|\bar{x}\| + \tilde{k}_6 e^{-\alpha(t-t_0)/\epsilon} + \epsilon \tilde{k}_7$$

The system (A.40) is viewed as a perturbation of

$$\dot{u} = F(t, u, 0, 0) \tag{A.41}$$

Since the origin of (A.41) is exponentially stable, we can obtain a Lyapunov function $\tilde{V}(t, u)$ for it using Theorem 3.12. Using this function with (A.40), we obtain

$$
\begin{aligned}
\dot{\tilde{V}} &= \frac{\partial \tilde{V}}{\partial t} + \frac{\partial \tilde{V}}{\partial u} F(t, u, 0, 0) + \frac{\partial \tilde{V}}{\partial u} \Delta F \\
&\leq -\tilde{c}_3 \|u\|^2 + \tilde{c}_4 \|u\| \left[\tilde{k}_4 \|u\|^2 + \tilde{k}_5 \|u\| \, \|\bar{x}\| + \tilde{k}_6 e^{-\alpha(t-t_0)/\epsilon} + \epsilon \tilde{k}_7 \right]
\end{aligned}
$$

For $\|u\| \leq \tilde{c}_3/2\tilde{c}_4 \tilde{k}_4$, we have

$$\dot{\tilde{V}} \leq -2 \left[\tilde{k}_a - \tilde{k}_b e^{-\tilde{\alpha}(t-t_0)} \right] \tilde{V} + 2 \left[\epsilon \tilde{k}_c + \tilde{k}_d e^{-\alpha(t-t_0)/\epsilon} \right] \sqrt{\tilde{V}}$$

for some $\tilde{k}_a$, $\tilde{\alpha} > 0$ and $\tilde{k}_b$, $\tilde{k}_c$, $\tilde{k}_d \geq 0$. Application of the comparison principle yields

$$\tilde{W}(t) \leq \tilde{\phi}(t, t_0)\tilde{W}(0) + \int_{t_0}^{t} \tilde{\phi}(t, s) \left(\epsilon \tilde{k}_c + \tilde{k}_d e^{-\alpha(s-t_0)/\epsilon} \right) \, ds$$

where $\tilde{W} = \sqrt{\tilde{V}}$ and

$$|\tilde{\phi}(t, s)| \leq \tilde{k}_g e^{-\tilde{\sigma}(t-t_0)}, \quad \tilde{\sigma} > 0, \; \tilde{k}_g > 0$$

Since $u(0) = O(\epsilon)$ and

$$\int_{t_0}^{t} \exp[-\tilde{\sigma}(t-s)] \, \exp\left[\frac{-\alpha(s-t_0)}{\epsilon}\right] \, ds = O(\epsilon)$$

we can show that $\tilde{W}(t) = O(\epsilon)$ and, subsequently, $u(t, \epsilon) = O(\epsilon)$. The rest of the proof proceeds exactly as in Theorem 9.1. Notice that the boundary-layer analysis in that proof is valid for all $\tau \geq 0$.

A.16   Proof of Lemma 10.1

Sufficiency: Suppose the conditions of the lemma are satisfied. Let us note that in all three cases of the third condition, there exist positive constants ω_0 and σ_0 such that

$$\omega^2 \sigma_{\min}[Z(j\omega) + Z^T(-j\omega)] \geq \sigma_0, \quad \forall \, |\omega| \geq \omega_0 \tag{A.42}$$

Since $Z(s)$ is Hurwitz, there exist positive constants δ and μ^* such that poles of all the elements of $Z(s-\mu)$ have real parts less than $-\delta$, for all $\mu < \mu^*$. Let $\{\mathcal{A}, \mathcal{B}, \mathcal{C}, \mathcal{D}\}$ be a minimal realization of $Z(s)$. Then,

$$\begin{aligned}
Z(s-\mu) &= \mathcal{D} + \mathcal{C}(sI - \mu I - \mathcal{A})^{-1}\mathcal{B} \\
&= \mathcal{D} + \mathcal{C}(sI - \mathcal{A})^{-1}(sI - \mathcal{A})(sI - \mu I - \mathcal{A})^{-1}\mathcal{B} \\
&= Z(s) + \mu N(s)
\end{aligned} \tag{A.43}$$

where

$$N(s) = \mathcal{C}(sI - \mathcal{A})^{-1}(sI - \mu I - \mathcal{A})^{-1}\mathcal{B}$$

Since $\mathcal{A}$ and $(\mathcal{A} + \mu I)$ are Hurwitz, uniformly in μ, there is a positive constant k_1 such that

$$\sigma_{\text{mam}}\left[N(j\omega) + N^T(-j\omega)\right] \leq k_1, \quad \forall \, \omega \in R \tag{A.44}$$

Moreover, $\lim_{\omega \to \infty} \omega^2 N(j\omega)$ exists. Hence, there are positive constants k_0 and ω_1 such that

$$\omega^2 \|N(j\omega) + N^T(-j\omega)\| \leq k_0, \quad \forall \, |\omega| \geq \omega_1 \tag{A.45}$$

From (A.42), (A.43), and (A.45), we obtain

$$\sigma_{\min}[Z(j\omega - \mu) + Z^T(-j\omega - \mu)] \geq \frac{\sigma_0 - k_0\mu}{\omega^2}, \quad \forall \, |\omega| \geq \omega_2 \tag{A.46}$$

where $\omega_2 = \max\{\omega_0, \omega_1\}$. On the other hand, on the compact frequency interval $|\omega| \leq \omega_2$, we have

$$\sigma_{\min}[Z(j\omega - \mu) + Z^T(-j\omega - \mu)] \geq \sigma_1, \quad \forall \, |\omega| \leq \omega_2 \tag{A.47}$$

for some positive constant σ_1. From (A.43), (A.44), and (A.47), we obtain

$$\sigma_{\min}[Z(j\omega - \mu) + Z^T(-j\omega - \mu)] \geq \sigma_1 - k_1\mu, \quad \forall \; |\omega| \leq \omega_2 \qquad \text{(A.48)}$$

From (A.46) and (A.48), we see that by choosing $\mu \leq \min\{\sigma_0/k_0, \; \sigma_1/k_1\}$, we can make $Z(j\omega - \mu) + Z^T(-j\omega - \mu)$ positive semidefinite for all $\omega \in R$; thus showing that $Z(s - \mu)$ is positive real. Hence, $Z(s)$ is strictly positive real.

Necessity: Suppose $Z(s)$ is strictly positive real. There exists $\mu > 0$ such that $Z(s - \mu)$ is positive real. It follows that $Z(s)$ is Hurwitz and positive real. Consequently,

$$Z(j\omega) + Z^T(-j\omega) \geq 0, \quad \forall \; \omega \in R$$

Therefore,

$$Z(\infty) + Z^T(\infty) \geq 0 \qquad \text{(A.49)}$$

Let $\{\mathcal{A}, \mathcal{B}, \mathcal{C}, \mathcal{D}\}$ be a minimal realization of $Z(s)$. By Lemma 10.2, there exist P, L, W, and ϵ which satisfy (10.10)–(10.12). Let $\Phi(s) = (sI - \mathcal{A})^{-1}$. We have

$$Z(s) + Z^T(-s) = \mathcal{D} + \mathcal{D}^T + \mathcal{C}\Phi(s)\mathcal{B} + \mathcal{B}^T\Phi^T(-s)\mathcal{C}^T$$

Substitute for $\mathcal{C}$ using (10.11), and for $\mathcal{D} + \mathcal{D}^T$ using (10.12).

$$
\begin{aligned}
Z(s) + Z^T(-s) &= W^T W + (\mathcal{B}^T P + W^T L)\Phi(s)\mathcal{B} \\
&\quad + \mathcal{B}^T\Phi^T(-s)(P\mathcal{B} + L^T W) \\
&= W^T W + W^T L\Phi(s)\mathcal{B} + \mathcal{B}^T\Phi^T(-s)L^T W \\
&\quad + \mathcal{B}^T\Phi^T(-s)[-\mathcal{A}^T P - P\mathcal{A}]\Phi(s)\mathcal{B}
\end{aligned}
$$

Use of (10.10) yields

$$
\begin{aligned}
Z(s) + Z^T(-s) &= W^T W + W^T L\Phi(s)\mathcal{B} + \mathcal{B}^T\Phi^T(-s)L^T W \\
&\quad + \mathcal{B}^T\Phi^T(-s)L^T L\Phi(s)\mathcal{B} + \epsilon\mathcal{B}^T\Phi^T(-s)P\Phi(s)\mathcal{B} \\
&= [W^T + \mathcal{B}^T\Phi^T(-s)L^T] \; [W + L\Phi(s)\mathcal{B}] \\
&\quad + \epsilon\mathcal{B}^T\Phi^T(-s)P\Phi(s)\mathcal{B}
\end{aligned}
$$

From the last equation, it can be seen that $Z(j\omega) + Z^T(-j\omega)$ is positive definite for all $\omega \in R$, for if it was singular at some frequency ω, there would exist $x \in C^p$, $x \neq 0$ such that

$$x^*[Z(j\omega) + Z^T(-j\omega)]x = 0 \Rightarrow x^*\mathcal{B}^T\Phi^T(-j\omega)P\Phi(j\omega)\mathcal{B}x = 0 \Rightarrow \mathcal{B}x = 0$$

Also,

$$x^*[Z(j\omega) + Z^T(-j\omega)]x = 0 \Rightarrow x^*\left[W^T + \mathcal{B}^T\Phi^T(-j\omega)L^T\right] \; [W + L\Phi(j\omega)\mathcal{B}]\,x = 0$$

Since $Bx = 0$, the preceding equation implies $Wx = 0$. Hence,

$$x^*[Z(s) + Z^T(-s)]x \equiv 0, \quad \forall\, s$$

which contradicts the assumption that $\det[Z(s) + Z^T(-s)]$ is not identically zero. Now if $Z(\infty) + Z^T(\infty)$ is nonsingular, then in view of (A.49) it must be positive definite. If $Z(\infty) + Z^T(\infty) = 0$, then

$$\lim_{\omega \to \infty} \omega^2[Z(j\omega) + Z^T(-j\omega)] = \mathcal{B}^T(\epsilon P + L^T L)\mathcal{B}$$

In this case, the matrix $\mathcal{B}$ must have full column rank; otherwise, there is $x \neq 0$ such that $\mathcal{B}x = 0$. This implies

$$x^T[Z(j\omega) + Z^T(-j\omega)]x = 0, \quad \forall\, \omega \in R$$

which contradicts the positive definiteness of $Z(j\omega) + Z^T(-j\omega)$. Hence, $\mathcal{B}^T(\epsilon P + L^T L)\mathcal{B}$ is positive definite. Finally, if $Z(\infty) + Z^T(\infty)$ is singular but different than zero, let T be a nonsingular matrix such that $WT = [W_1\ 0]$, where W_1 has full column rank. Let $y \in C^p$ and partition $\mathcal{B}T$ and $x = T^{-1}y$, compatibly, as $\mathcal{B}T = [\mathcal{B}_1\ \mathcal{B}_2]$ and $x^* = [x_1^*\ x_2^*]$. Then,

$$\begin{aligned}
y^*[Z(s) + Z^T(-s)]y &= \{x_1^*[W_1 + L\Phi(-s)\mathcal{B}_1]^T + x_2^*[L\Phi(-s)\mathcal{B}_2]^T\} \\
&\quad \times \{[W_1 + L\Phi(s)\mathcal{B}_1]x_1 + L\Phi(s)\mathcal{B}_2x_2\} \\
&\quad + \epsilon[\mathcal{B}_1x_1 + \mathcal{B}_2x_2]^*\Phi^T(-s)P\Phi(s)[\mathcal{B}_1x_1 + \mathcal{B}_2x_2]
\end{aligned}$$

We note that B_2 must have full column rank; otherwise, there is $x_2 \neq 0$ such that $B_2x_2 = 0$. Taking $x_1 = 0$ yields

$$y^*[Z(j\omega) + Z^T(-j\omega)]y = 0, \quad \forall\, \omega \in R$$

which contradicts the positive definiteness of $Z(j\omega) + Z^T(-j\omega)$. Rewrite the expression for $y^*[Z(j\omega) + Z^T(-j\omega)]y$ as

$$y^*[Z(j\omega) + Z^T(-j\omega)]y = \alpha + \beta$$

where α contains the quadratic terms in x_1 as well as the cross product terms, while β contains the quadratic terms in x_2. It can be verified that

$$\lim_{\omega \to \infty} \alpha = x_1^* W_1^T W_1 x_1$$

and

$$\lim_{\omega \to \infty} \omega^2 \beta = x_2^* \mathcal{B}_2^T(\epsilon P + L^T L)\mathcal{B}_2 x_2$$

Since $W_1^T W_1$ and $\mathcal{B}_2^T(\epsilon P + L^T L)\mathcal{B}_2$ are positive definite matrices, there exists positive constants k_2, k_3, and k_4 such that

$$\omega^2 y^*[Z(j\omega) + Z^T(-j\omega)]y \geq k_2\omega^2 x_1^* x_1 + k_3 x_2^* x_2 \geq k_4 y^* y$$

for sufficiently large ω, which completes the proof of the lemma.

A.17   Proof of Lemma 10.2

Sufficiency: Suppose there exist P, L, W, and ϵ which satisfy (10.10)–(10.12); we need to show that $Z(s)$ is strictly positive real. We will do it by showing that $Z(s - \mu)$ is positive real for some $\mu > 0$. Take $\mu \in (0, \epsilon/2)$. From (10.10), we have

$$P(\mathcal{A} + \mu I) + (\mathcal{A} + \mu I)^T P = -L^T L - (\epsilon - 2\mu)P$$

which shows that $(\mathcal{A} + \mu I)$ is Hurwitz. Hence, $Z(s - \mu)$ is analytic in $Re[s] \geq 0$. Let $\Phi(s) = (sI - A)^{-1}$. We have

$$Z(s - \mu) + Z^T(-s - \mu) = \mathcal{D} + \mathcal{D}^T + \mathcal{C}\Phi(s - \mu)\mathcal{B} + \mathcal{B}^T\Phi^T(-s - \mu)\mathcal{C}^T$$

Substitute for $\mathcal{C}$ using (10.11), and for $\mathcal{D} + \mathcal{D}^T$ using (10.12).

$$
\begin{aligned}
Z(s - \mu) + Z^T(-s - \mu) &= W^T W + (\mathcal{B}^T P + W^T L)\Phi(s - \mu)\mathcal{B} \\
&\quad + \mathcal{B}^T\Phi^T(-s - \mu)(P\mathcal{B} + L^T W) \\
&= W^T W + W^T L\Phi(s - \mu)\mathcal{B} + \mathcal{B}^T\Phi^T(-s - \mu)L^T W \\
&\quad + \mathcal{B}^T\Phi^T(-s - \mu)[-2\mu P - A^T P - PA]\Phi(s - \mu)\mathcal{B}
\end{aligned}
$$

Use of (10.10) yields

$$
\begin{aligned}
Z(s - \mu) + Z^T(-s - \mu) &= W^T W + \mathcal{B}^T\Phi^T(-s - \mu)L^T L\Phi(s - \mu)\mathcal{B} \\
&\quad + W^T L\Phi(s - \mu)\mathcal{B} + \mathcal{B}^T\Phi^T(-s - \mu)L^T W \\
&\quad + (\epsilon - 2\mu)\mathcal{B}^T\Phi^T(-s - \mu)P\Phi(s - \mu)\mathcal{B} \\
&= [W^T + \mathcal{B}^T\Phi^T(-s - \mu)L^T][W + L\Phi(s - \mu)\mathcal{B}] \\
&\quad + (\epsilon - 2\mu)\mathcal{B}^T\Phi^T(-s - \mu)P\Phi(s - \mu)\mathcal{B}
\end{aligned}
$$

which shows clearly that $Z(j\omega - \mu) + Z^T(-j\omega - \mu)$ is positive semidefinite for all $\omega \in R$.

Necessity: The proof uses a spectral factorization result, which we quote without proof.

Lemma A.11 *Let the $p \times p$ proper rational transfer function matrix $U(s)$ be positive real and Hurwitz. Then, there exists an $r \times p$ proper rational Hurwitz transfer function matrix $V(s)$ such that*

$$U(s) + U^T(-s) = V^T(-s)V(s) \tag{A.50}$$

where r is the normal rank of $U(s) + U^T(-s)$, that is, the rank over the field of rational functions of s. Furthermore, rank $V(s) = r$ *for* $Re[s] > 0$. $\triangle$

Proof: See [190, Theorem 2].

Suppose now that $Z(s)$ is strictly positive real. Then, there exists $\delta > 0$ such that $Z(s - \delta)$ is positive real. This implies that $Z(s - \mu)$ is positive real for $0 \leq \mu \leq \delta$.[9] Since $\mathcal{A}$ is Hurwitz, we can choose ϵ small enough that $U(s) = Z(s - \epsilon/2)$ is positive real and $\mathcal{A}_\epsilon = (\epsilon/2)I + \mathcal{A}$ is Hurwitz. A minimal realization of $U(s)$ is given by

$$U(s) = \mathcal{C}(sI - \mathcal{A}_\epsilon)^{-1}\mathcal{B} + \mathcal{D}$$

We shall refer to this realization by the quadruple $\{\mathcal{A}_\epsilon, \mathcal{B}, \mathcal{C}, \mathcal{D}\}$. From Lemma A.11, there exists an $r \times p$ transfer function matrix $V(s)$ such that (A.50) is satisfied. Let $\{F, G, H, J\}$ be a minimal realization of $V(s)$. The matrix F is Hurwitz, since $V(s)$ is Hurwitz. It can be easily seen that $\{-F^T, H^T, -G^T, J^T\}$ is a minimal realization of $V^T(-s)$. Therefore,

$$\{\mathcal{A}_1, \mathcal{B}_1, \mathcal{C}_1, \mathcal{D}_1\} =$$
$$\left\{ \begin{bmatrix} F & 0 \\ H^T H & -F^T \end{bmatrix}, \begin{bmatrix} G \\ H^T J \end{bmatrix}, \begin{bmatrix} J^T H & -G^T \end{bmatrix}, J^T J \right\}$$

is a realization of the cascade connection $V^T(-s)V(s)$. By checking controllability and observability, and using the property that rank $V(s) = r$ for $Re[s] > 0$, it can be seen that this realization is minimal. Let us show the controllability test;[10] the observability test is similar. By writing

$$\begin{bmatrix} I & 0 \\ H(sI - F)^{-1} & I \end{bmatrix} \begin{bmatrix} sI - F & G \\ -H & J \end{bmatrix} = \begin{bmatrix} sI - F & G \\ 0 & H(sI - F)^{-1}G + J \end{bmatrix}$$
$$= \begin{bmatrix} sI - F & G \\ 0 & V(s) \end{bmatrix}$$

it can be seen that

$$\text{rank } V(s) = r, \ \forall \ Re[s] > 0 \ \Leftrightarrow \text{rank} \begin{bmatrix} sI - F & G \\ -H & J \end{bmatrix} = n_F + r, \ \forall \ Re[s] > 0$$

where n_F is the dimension of F. Now, show controllability of $(\mathcal{A}_1, \mathcal{B}_1)$ by contradiction. Suppose $(\mathcal{A}_1, \mathcal{B}_1)$ is not controllable. Then, there are a complex number λ

---

[9]Note that if $Z(s)$ is positive real, so is $Z(s + a)$ for any positive number a.

[10]The controllability argument uses the fact that $V^T(-s)$ has no transmission zeros at the poles of $V(s)$.

and a vector $w \in C^{n_F+r}$, partitioned into n_F and r subvectors, such that

$$w_1^* F + w_2^* H^T H = \lambda w_1^* \tag{A.51}$$

$$-w_2^* F^T = \lambda w_2^* \tag{A.52}$$

$$w_1^* G + w_2^* H^T J = 0 \tag{A.53}$$

Equation (A.52) shows that $Re[\lambda] > 0$, since F is Hurwitz, and (A.51) and (A.53) show that

$$\begin{bmatrix} w_1^* & w_2^* H^T \end{bmatrix} \begin{bmatrix} \lambda I - F & G \\ -H & J \end{bmatrix} = 0 \implies \text{rank } V(\lambda) < r$$

which contradicts the fact that rank $V(s) = r$ for $Re[s] > 0$. Thus, $(\mathcal{A}_1, \mathcal{B}_1)$ is controllable.

Consider the Lyapunov equation

$$KF + F^T K = -H^T H$$

Since the pair (F, H) is observable, there is a unique positive definite solution K. This fact is shown in Exercise 3.24. Using the similarity transformation

$$\begin{bmatrix} I & 0 \\ K & I \end{bmatrix}$$

we obtain the following alternative minimal realization of $V^T(-s)V(s)$:

$$\{\mathcal{A}_2, \mathcal{B}_2, \mathcal{C}_2, \mathcal{D}_2\} =$$
$$\left\{ \begin{bmatrix} F & 0 \\ 0 & -F^T \end{bmatrix}, \begin{bmatrix} G \\ KG + H^T J \end{bmatrix}, \begin{bmatrix} J^T H + G^T K & -G^T \end{bmatrix}, J^T J \right\}$$

On the other hand, $\{-\mathcal{A}_\epsilon^T, \mathcal{C}^T, -\mathcal{B}^T, \mathcal{D}^T\}$ is a minimal realization of $U^T(-s)$. Therefore,

$$\{\mathcal{A}_3, \mathcal{B}_3, \mathcal{C}_3, \mathcal{D}_3\} = \left\{ \begin{bmatrix} \mathcal{A}_\epsilon & 0 \\ 0 & -\mathcal{A}_\epsilon^T \end{bmatrix}, \begin{bmatrix} \mathcal{B} \\ \mathcal{C}^T \end{bmatrix}, \begin{bmatrix} \mathcal{C} & -\mathcal{B}^T \end{bmatrix}, \mathcal{D} + \mathcal{D}^T \right\}$$

is a realization of the parallel connection $U(s) + U^T(-s)$. Since the eigenvalues of $\mathcal{A}_\epsilon$ are in the open left-half plane while the eigenvalues of $-\mathcal{A}_\epsilon^T$ are in the open right-half plane, it can be easily seen that this realization is minimal. Thus, due to (A.50), $\{\mathcal{A}_2, \mathcal{B}_2, \mathcal{C}_2, \mathcal{D}_2\}$ and $\{\mathcal{A}_3, \mathcal{B}_3, \mathcal{C}_3, \mathcal{D}_3\}$ are equivalent minimal realizations of the same transfer function. Therefore, they have the same dimension and there is a nonsingular matrix T such that[11]

$$\mathcal{A}_2 = T\mathcal{A}_3 T^{-1}, \quad \mathcal{B}_2 = T\mathcal{B}_3, \quad \mathcal{C}_2 = \mathcal{C}_3 T^{-1}, \quad J^T J = \mathcal{D} + \mathcal{D}^T$$

---

[11] See [29, Theorem 5-20].

The matrix T must be a block-diagonal matrix. To see this point, partition T compatibly as

$$T = \begin{bmatrix} T_{11} & T_{12} \\ T_{21} & T_{22} \end{bmatrix}$$

Then, the matrix T_{12} satisfies the equation

$$FT_{12} + T_{12}\mathcal{A}_\epsilon^T = 0$$

Premultiplying by $\exp(Ft)$ and postmultiplying by $\exp(\mathcal{A}_\epsilon^T t)$, we obtain

$$\begin{aligned} 0 &= \exp(Ft)[FT_{12} + T_{12}\mathcal{A}_\epsilon^T]\exp(\mathcal{A}_\epsilon^T t) \\ &= \frac{d}{dt}[\exp(Ft)T_{12}\exp(\mathcal{A}_\epsilon^T t)] \end{aligned}$$

Hence, $\exp(Ft)T_{12}\exp(\mathcal{A}_\epsilon^T t)$ is constant for all $t \geq 0$. In particular, since $\exp(0) = I$,

$$T_{12} = \exp(Ft)T_{12}\exp(\mathcal{A}_\epsilon^T t) \to 0 \text{ as } t \to \infty$$

Therefore, $T_{12} = 0$. Similarly, we can show that $T_{21} = 0$. Consequently, the matrix T_{11} is nonsingular and

$$F = T_{11}\mathcal{A}_\epsilon T_{11}^{-1}, \quad G = T_{11}\mathcal{B}, \quad J^T H + G^T K = \mathcal{C}T_{11}^{-1}$$

Define

$$P = T_{11}^T K T_{11}, \quad L = HT_{11}, \quad W = J$$

It can be easily verified that P, L, and W satisfy the equations

$$\begin{aligned} P\mathcal{A}_\epsilon + \mathcal{A}_\epsilon^T P &= -L^T L \\ P\mathcal{B} &= \mathcal{C}^T - L^T W \\ W^T W &= \mathcal{D} + \mathcal{D}^T \end{aligned}$$

Since $\mathcal{A}_\epsilon = (\epsilon/2)I + \mathcal{A}$, the first equation can be rewritten as

$$P\mathcal{A} + \mathcal{A}^T P = -L^T L - \epsilon P$$

which completes the proof of the lemma.

A.18   Proof of Lemma 10.7

We start by writing a time-domain version of the infinite-dimensional equation (10.64). Consider the space $\mathcal{S}$ of all half-wave symmetric periodic signals of fundamental frequency ω, which have finite energy on any finite interval. A signal $y \in \mathcal{S}$ can be represented by its Fourier series

$$y(t) = \sum_{k \text{ odd}} a_k \exp(jk\omega t), \quad \sum_{k \text{ odd}} |a_k|^2 < \infty$$

Define a norm on $\mathcal{S}$ by

$$\|y\|^2 = \frac{\omega}{\pi} \int_0^{2\pi/\omega} y^2(t)\ dt = 2 \sum_{k\ \text{odd}} |a_k|^2$$

With this norm, $\mathcal{S}$ is a Banach space. Define $g_k(t - \tau)$ by

$$g_k(t - \tau) = \frac{\omega}{\pi}\{G(jk\omega)\exp[jk\omega(t - \tau)] + G(-jk\omega)\exp[-jk\omega(t - \tau)]\}$$

For odd integers m and $k > 0$, we have

$$\int_0^{\pi/\omega} g_k(t - \tau)\exp(jm\omega\tau)\ d\tau = \begin{cases} G(jk\omega)\exp(jk\omega t), & \text{if } m = k \\ G(-jk\omega)\exp(-jk\omega t), & \text{if } m = -k \quad \text{(A.54)} \\ 0, & \text{if } |m| \neq k \end{cases}$$

Define the linear mapping g and the nonlinear mapping $g\psi$ on $\mathcal{S}$ by

$$gy = \int_0^{\pi/\omega} \sum_{k\ \text{odd};\ k>0} g_k(t - \tau)y(\tau)\ d\tau$$

$$g\psi y = \int_0^{\pi/\omega} \sum_{k\ \text{odd};\ k>0} g_k(t - \tau)\psi(y(\tau))\ d\tau$$

where

$$y(t) = \sum_{k\ \text{odd}} a_k \exp(jk\omega t) \quad \text{and} \quad \psi(y(t)) = \sum_{k\ \text{odd}} c_k \exp(jk\omega t)$$

Using (A.54), it can be seen that

$$gy = \sum_{k\ \text{odd}} G(jk\omega)a_k \exp(jk\omega t)$$

$$g\psi y = \sum_{k\ \text{odd}} G(jk\omega)c_k \exp(jk\omega t)$$

With these definitions, the condition for existence of half-wave symmetric periodic oscillation can be written as

$$y = -g\psi y \tag{A.55}$$

Equation (A.55) is equivalent to (10.64). In order to separate the effect of the higher harmonics from that of the first harmonic, define a mapping P_1 by

$$P_1 y = y_1 = a_1 \exp(j\omega t) + \bar{a}_1 \exp(-j\omega t) = 2Re[a_1 \exp(j\omega t)]$$

and a mapping P_h by

$$P_h y = y_h = y - y_1 = \sum_{k \text{ odd}; \ |k| \neq 1} a_k \exp(jk\omega t)$$

Without loss of generality, we take $a_1 = a/2j$ so that $y_1(t) = a \sin \omega t$. Solving (A.55) is equivalent to solving both (A.56) and (A.57):

$$y_h = -P_h g \psi(y_1 + y_h) \tag{A.56}$$

$$y_1 = -P_1 g \psi(y_1 + y_h) \tag{A.57}$$

By evaluating the right-hand side of (A.57), it can be seen that (A.57) is equivalent to (10.79). The error term $\delta\Psi$, defined after (10.80), satisfies

$$|\delta\Psi| = \|P_1 g \psi y_1 - P_1 g \psi(y_1 + y_h)\| / \sqrt{2}a|G(j\omega)| \tag{A.58}$$

Thus, to obtain a bound on $\delta\Psi$ we need to find a bound on y_h, which we shall find using the contraction mapping theorem, without solving (A.56). By adding $[P_h g(\beta + \alpha)/2]y_h$ to both sides of (A.56), we can rewrite it as

$$\left(I + P_h g \frac{\beta + \alpha}{2}\right) y_h = -P_h g \left[\psi(y_1 + y_h) - \frac{\beta + \alpha}{2} y_h\right] \tag{A.59}$$

Let us consider the linear mapping $K = I + P_h g(\beta + \alpha)/2$, which appears on the left-hand side of (A.59). It maps $\mathcal{S}$ into $\mathcal{S}$. Given any $z \in \mathcal{S}$ defined by

$$z(t) = \sum_{k \text{ odd}} b_k \exp(jk\omega t)$$

we consider the linear equation $Kx = z$ and seek a solution x in $\mathcal{S}$. Representing x as

$$x(t) = \sum_{k \text{ odd}} d_k \exp(jk\omega t)$$

we have

$$\left(I + P_h g \frac{\beta + \alpha}{2}\right) x = x_1 + \sum_{k \text{ odd}; \ |k| \neq 1} \left[1 + \frac{\beta + \alpha}{2} G(jk\omega)\right] d_k \exp(jk\omega t)$$

Hence, the linear equation $Kx = z$ has a unique solution if

$$\inf_{k \text{ odd}; \ |k| \neq 1} \left|1 + \frac{\beta + \alpha}{2} G(jk\omega)\right| \neq 0 \tag{A.60}$$

In other words, condition (A.60) guarantees that the linear mapping K has an inverse. This condition is always satisfied if $\omega \in \Omega$ because the left-hand side of

(A.60) can vanish only if $\rho(\omega) = 0$. Denote the inverse of K by K^{-1} and rewrite (A.59) as

$$y_h = -K^{-1}P_h g\left[\psi(y_1 + y_h) - \frac{\beta + \alpha}{2}(y_1 + y_h)\right] \stackrel{\text{def}}{=} Ty_h$$

where we have used the fact that $P_h g y_1 = 0$. We would like to apply the contraction mapping theorem to the equation $y_h = Ty_h$. Clearly, T maps $\mathcal{S}$ into $\mathcal{S}$. We need to verify that T is a contraction on $\mathcal{S}$. To that end, consider

$$Ty^{(2)} - Ty^{(1)} = K^{-1}P_h g\left[\psi_T(y_1 + y^{(2)}) - \psi_T(y_1 + y^{(1)})\right]$$

where

$$\psi_T(y) = \psi(y) - \frac{\beta + \alpha}{2}y$$

Let

$$\psi_T(y_1 + y^{(2)}) - \psi_T(y_1 + y^{(1)}) = \sum_{k \text{ odd}; \; |k| \neq 1} e_k \exp(jk\omega t)$$

Then,

$$\left\|Ty^{(2)} - Ty^{(1)}\right\|^2 = 2 \sum_{k \text{ odd}; \; |k| \neq 1} \left|\frac{G(jk\omega)}{1 + [(\beta + \alpha)/2]G(jk\omega)}\right|^2 |e_k|^2$$

$$\leq \left\{\sup_{k \text{ odd}; \; |k| \neq 1} \left|\frac{G(jk\omega)}{1 + [(\beta + \alpha)/2]G(jk\omega)}\right|\right\}^2$$

$$\times \left\|\psi_T(y_1 + y^{(2)}) - \psi_T(y_1 + y^{(1)})\right\|$$

Due to the slope restriction on ψ, we have

$$\left|\psi_T(y_1 + y^{(2)}) - \psi_T(y_1 + y^{(1)})\right| \leq \left(\frac{\beta - \alpha}{2}\right)\left|y^{(2)} - y^{(1)}\right|$$

Moreover,

$$\sup_{k \text{ odd}; \; |k| \neq 1} \left|\frac{G(jk\omega)}{1 + [(\beta + \alpha)/2]G(jk\omega)}\right| \leq \frac{1}{\rho(\omega)}$$

where $\rho(\omega)$ is defined by (10.82). Hence,

$$\left\|Ty^{(2)} - Ty^{(1)}\right\| \leq \frac{1}{\rho(\omega)}\left(\frac{\beta - \alpha}{2}\right)\left\|y^{(2)} - y^{(1)}\right\|$$

Since

$$\frac{1}{\rho(\omega)}\left(\frac{\beta - \alpha}{2}\right) < 1, \quad \forall \, \omega \in \Omega$$

we conclude that as long as $\omega \in \Omega$, T is a contraction mapping. Thus, by the contraction mapping theorem, the equation $y_h = Ty_h$ has a unique solution. Noting that $T(-y_1) = 0$, we rewrite the equation $y_h = Ty_h$ as

$$y_h = Ty_h - T(-y_1)$$

and conclude that

$$\|y_h\| \leq \frac{1}{\rho(\omega)}\left(\frac{\beta - \alpha}{2}\right)\|y_h\| + \frac{1}{\rho(\omega)}\left(\frac{\beta - \alpha}{2}\right)a$$

Therefore,

$$\|y_h\| \leq \frac{a[(\beta - \alpha)/2]/\rho(\omega)}{1 - [(\beta - \alpha)/2]/\rho(\omega)} = \frac{a[(\beta - \alpha)/2]}{\rho(\omega) - [(\beta - \alpha)/2]} = \frac{2\sigma(\omega)a}{\beta - \alpha}$$

which proves (10.84). To prove (10.85), consider (A.57) rewritten as

$$y_1 + P_1 g \psi y_1 = P_1 g[\psi y_1 - \psi(y_1 + y_h)] \tag{A.61}$$

by adding $P_1 g \psi y_1$ to both sides. Taking norms on both sides of (A.61) gives

$$|1 + G(j\omega)\Psi(a)|a \leq |G(j\omega)|\left(\frac{\beta - \alpha}{2}\right)\|y_h\| \leq |G(j\omega)|\sigma(\omega)a$$

from which we can calculate the bound

$$|\delta\Psi| = \left|\frac{1}{G(j\omega)} + \Psi(a)\right| = \left|\frac{1 + G(j\omega)\Psi(a)}{G(j\omega)}\right| \leq \sigma(\omega)$$

which completes the proof of the lemma.

A.19   Proof of Theorem 10.9

If in the proof of Lemma 10.7 we had defined $P_1 = 0$ and $P_h = I$, then the mapping T would still be a contraction mapping if $\omega \in \tilde{\Omega}$; therefore $y \equiv y_h = 0$ would be the unique solution of $y_h = Ty_h$. This proves that there is no half-wave symmetric periodic solutions with fundamental frequency $\omega \in \tilde{\Omega}$. The necessity of the condition

$$\left|\frac{1}{G(j\omega)} + \Psi(a)\right| \leq \sigma(\omega)$$

shows that there would be no half-wave symmetric periodic solutions with fundamental frequency $\omega \in \Omega'$ if the corresponding error circle does not intersect the $-\Psi(a)$ locus. Thus, we are left with the third part of the theorem where we would like to show that for each complete intersection defining Γ, there is a half-wave symmetric periodic solutions with $(\omega, a) \in \bar{\Gamma}$. The proof of this part uses a result from degree theory, so let us explain that result first.

Suppose we are given a continuously differentiable function $\phi : D \rightarrow R^n$ where $D \subset R^n$ is open and bounded. Let $p \in R^n$ be a point such that $\phi(x) = p$ for some x inside D, but $\phi(x) \neq p$ on the boundary ∂D of D. We are interested in showing that $\tilde{\phi}(x) = p$ has a solution in D where $\tilde{\phi}(x)$ is a perturbation of $\phi(x)$. Degree theory achieves this by ensuring that no solution leaves D as ϕ is perturbed to $\tilde{\phi}$; that is why no solutions were allowed on the boundary ∂D. Assume that at every solution $x_i \in D$ of $\phi(x) = p$, the Jacobian matrix $[\partial \phi / \partial x]$ is nonsingular. Define the degree of ϕ at p relative to D by

$$d(\phi, D, p) = \sum_{x_i = \phi^{-1}(p)} \text{sgn} \left\{ \det \left[\frac{\partial \phi}{\partial x}(x_i) \right] \right\}$$

Notice that if $\phi(x) \neq p \ \forall \ x \in D$, the degree is zero. The two basic properties of the degree are:[12]

- If $d(\phi, D, p) \neq 0$ then $\phi(x) = p$ has at least one solution in D.

- If $\eta : \bar{D} \times [0, 1] \rightarrow R^n$ is continuous and $\eta(x, \mu) \neq 0$ for all $x \in \partial D$ and all $\mu \in [0, 1]$, then $d[\eta(\cdot, \mu), D, p]$ is the same for all $\mu \in [0, 1]$.

The second property is known as the homotopic invariance of d.

Let us get back to our problem and define

$$\phi(\omega, a) = \Psi(a) + \frac{1}{G(j\omega)}$$

on Γ; ϕ is a complex variable. By taking the real and imaginary parts of ϕ as components of a second-order vector, we can view ϕ as a mapping from Γ into R^2. By assumption, the equation $\phi(w, a) = 0$ has a unique solution (ω_s, a_s) in Γ. The Jacobian of ϕ with respect to (ω, a) at (ω_s, a_s) is given by

$$\begin{bmatrix} \frac{d}{da}\Psi(a)\big|_{a=a_s} & -\Psi^2(a_s) \left\{ \frac{d}{d\omega} Re[G(j\omega)] \right\}_{\omega=\omega_s} \\ 0 & -\Psi^2(a_s) \left\{ \frac{d}{d\omega} Im[G(j\omega)] \right\}_{\omega=\omega_s} \end{bmatrix}$$

The assumptions

$$\frac{d}{da}\Psi(a)\bigg|_{a=a_s} \neq 0; \quad \frac{d}{d\omega} Im[G(j\omega)]\bigg|_{\omega=\omega_s} \neq 0$$

---

[12] See [20] for the proof of these properties.

guarantee that the Jacobian is nonsingular. Thus,

$$d(\phi, \Gamma, 0) = \pm 1$$

We are interested in showing that

$$\tilde{\phi}(\omega, a) \stackrel{\text{def}}{=} \frac{1}{G(j\omega)} + \Psi(a) - \delta\Psi(\omega, a) = 0 \tag{A.62}$$

has a solution in $\bar{\Gamma}$. This will be the case if we can show that

$$d(\tilde{\phi}, \Gamma, 0) \neq 0$$

To that end, define

$$\eta(\omega, a, \mu) = (1 - \mu)\phi(\omega, a) + \mu\tilde{\phi}(\omega, a) = \phi(\omega, a) - \mu\delta\Psi(\omega, a)$$

for $\mu \in [0, 1]$, so that $\eta = \phi$ at $\mu = 0$ and $\eta = \tilde{\phi}$ at $\mu = 1$. It can be verified that

$$\left| \Psi(a) + \frac{1}{G(j\omega)} \right| \geq \sigma(\omega), \quad \forall\ (\omega, a) \in \partial\Gamma \tag{A.63}$$

For example, if we take the boundary $a = a_1$, then, with reference to Figure 10.23, the left-hand side of (A.63) is the length of the line connecting the point on the locus of $-\Psi(a)$ corresponding to $a = a_1$ to the a point on the locus of $1/G(j\omega)$ with $\omega_1 \leq \omega \leq \omega_2$. By construction, the former point is outside (or on) the error circle centered at the latter one. Therefore the length of the line connecting the two points must be greater than (or equal to) the radius of the error circle, that is, $\sigma(\omega)$. Using (A.63), we have

$$\begin{aligned} |\eta(\omega, a, \mu)| &\geq |\phi(\omega, a)| - \mu|\delta\Psi(\omega, a)| \\ &= \left| \Psi(a) + \frac{1}{G(j\omega)} \right| - \mu|\delta\Psi(\omega, a)| \geq \sigma(\omega) - \mu\sigma(\omega) \end{aligned}$$

where we have used the bound (10.85). Thus, for all $0 \leq \mu < 1$, the right-hand side of the last inequality is positive, which implies that $\eta(\omega, a, \mu) \neq 0$ on $\partial\Gamma$ when $\mu < 1$. There is no loss of generality in assuming that $\eta(\omega, a, 1) \neq 0$ on $\partial\Gamma$ because equality would imply we had found a solution as required. Thus, by the homotopic invariance property of d, we have

$$d(\tilde{\phi}, \Gamma, 0) = d(\phi, \Gamma, 0) \neq 0$$

Hence, (A.62) has a solution in $\bar{\Gamma}$ which concludes the proof of the theorem.

A.20 Proof of Lemma 13.4

To simplify the notation, let

$$\eta^T(t) = D^{1/2}b_m w^T(t)D^{1/2}, \quad \psi^T(\tau,t) = D^{1/2}b_m \sigma^T(\tau,t)D^{1/2}$$

$$U(t,t+\delta) = \int_t^{t+\delta} \begin{bmatrix} I & \psi^T(\tau,t) \\ \psi(\tau,t) & \psi(\tau,t)\psi^T(\tau,t) \end{bmatrix} d\tau$$

where $D^{1/2}$ is a square root of D. It can be easily seen that

$$\psi(\tau,t) = \int_t^\tau \eta(\lambda)\, d\lambda$$

$$W(t,t+\delta) = \begin{bmatrix} D^{1/2} & 0 \\ 0 & D^{-1/2} \end{bmatrix} U(t,t+\delta) \begin{bmatrix} D^{1/2} & 0 \\ 0 & D^{-1/2} \end{bmatrix}$$

and

$$\int_t^{t+\delta} \eta(\tau)\eta^T(\tau)\, d\tau = \left(b_m^T D b_m\right) D^{1/2} \int_t^{t+\delta} w(\tau)w^T(\tau)\, d\tau D^{1/2}$$

where $b_m^T D b_m$ is a positive scalar. It is also clear that $\eta(t)$ and $\dot{\eta}(t)$ are bounded. Thus, to prove the lemma it is sufficient to prove that uniform positive definiteness of $\int_t^{t+\delta_1} \eta(\tau)\eta^T(\tau)\, d\tau$ implies uniform positive definiteness of $U(t,t+\delta)$. We shall prove it in two steps:

Claim 1: If there are $\delta_1 > 0$ and $k_1 > 0$ such that

$$\int_t^{t+\delta_1} \eta(\tau)\eta^T(\tau)\, d\tau \geq k_1 I, \quad \forall\, t \geq 0 \tag{A.64}$$

then there are $\delta > 0$ and $k_2 > 0$ such that

$$\int_t^{t+\delta} \psi(\tau,t)\psi^T(\tau,t)\, d\tau \geq k_2 I, \quad \forall\, t \geq 0 \tag{A.65}$$

Claim 2: If there are $\delta > 0$ and $k_2 > 0$ such that (A.65) is satisfied, then there is $k > 0$ such that

$$U(t,t+\delta) \geq kI, \quad \forall\, t \geq 0 \tag{A.66}$$

Before we prove these claims, we state two general properties of signals which will be used in the proof. If, for a vector signal $f(t)$, we have $\|f(t)\| < a$ and $\|\ddot{f}(t)\| < b$ for all $t \geq 0$, then

$$\|\dot{f}(t)\| < 2\sqrt{2ab}, \quad \forall\, t \geq 0 \tag{A.67}$$

This can be seen as follows. From the mean value theorem,

$$f(t + T) = f(t) + T\dot{f}(t_1), \quad t < t_1 < t + T$$

Hence, $\|\dot{f}(t_1)\| < 2a/T$. On the other hand,

$$\|\dot{f}(t) - \dot{f}(t_1)\| = \left\| \int_t^{t_1} \ddot{f}(\tau)\, d\tau \right\| < Tb$$

Hence, $\|\dot{f}(t)\| < Tb + 2a/T$. Choosing $T = \sqrt{2a/b}$ proves (A.67). For a vector signal $g(t)$, we have

$$\frac{d}{dt} g^T(t)g(t) = 2g^T(t)\dot{g}(t)$$

Therefore,

$$|g^T(t)g(t) - g^T(t_0)g(t_0)| = 2\left| \int_{t_0}^t g^T(\tau)\dot{g}(\tau)\, d\tau \right|$$

By the Schwartz inequality (applied in the space of square integrable functions), we obtain

$$|g^T(t)g(t) - g^T(t_0)g(t_0)| \leq 2\sqrt{\int_{t_0}^t g^T(\tau)g(\tau)\, d\tau}\sqrt{\int_{t_0}^t \dot{g}^T(\tau)\dot{g}(\tau)\, d\tau} \tag{A.68}$$

Proof of Claim 1: We prove it by showing that failure of (A.65) implies failure of (A.64). Failure of (A.65) is equivalent to saying that for any $\mu > 0$ and $\epsilon > 0$, there is a time t and and vector x with $\|x\|_2 = 1$ such that

$$x^T \int_t^{t+\mu} \psi(\tau, t)\psi^T(\tau, t)\, d\tau\, x < \epsilon$$

Set $s(\tau, t) = \psi^T(\tau, t)x$. Using the estimate (A.68), we have

$$s^T(\tau, t)s(\tau, t) \leq 2\sqrt{\int_t^{t+\mu} s^T(\lambda, t)s(\lambda, t)\, d\lambda}\sqrt{\int_t^{t+\mu} x^T\eta(\lambda)\eta^T(\lambda)x\, d\lambda}$$

for $t \leq \tau \leq t + \mu$. Since $\eta(t)$ is bounded, we obtain

$$s^T(\tau, t)s(\tau, t) < \epsilon^{1/2}C_1$$

for some $C_1 > 0$. Using the estimate (A.67) with $s(\tau, t)$ as $f(\tau)$, we obtain

$$\|\eta^T(\tau)x\| = \left\|\frac{d}{d\tau}s(\tau, t)\right\| < \epsilon^{1/8}C_2$$

Hence,

$$x^T \int_t^{t+\mu} \eta(\tau)\eta^T(\tau) \, d\tau \; x < \epsilon^{1/4}\mu C_3$$

Since this inequality holds for arbitrary μ and ϵ, for any μ we can choose ϵ small enough so that $\epsilon^{1/4}\mu$ is arbitrarily small, which implies failure of (A.64). □

Proof of Claim 2: Consider the quadratic form

$$u(t, t+\delta) = x^T U(t, t+\delta)x = \int_t^{t+\delta} [x_1 + \psi^T(\tau, t)x_2]^T [x_1 + \psi^T(\tau, t)x_2] \, d\tau$$

where x is a vector with $\|x\|_2 = 1$, partitioned compatibly with the partition of U. Using the estimate (A.68), we have

$$\left| \|x_1\|_2^2 - \|x_1 + \psi^T(\tau, t)x_2\|_2^2 \right| \leq$$
$$2 \sqrt{\int_t^\tau \|x_1 + \psi^T(\lambda, t)x_2\|_2^2 \, d\lambda} \sqrt{\int_t^\tau \|\eta^T(\lambda)x_2\|_2^2 \, d\lambda}$$

Since $\eta(t)$ is bounded, we have

$$\|x_1\|_2^2 - \|x_1 + \psi^T(\tau, t)x_2\|_2^2 \leq C_4(\tau - t)^{1/2}\sqrt{u(t, t+\delta)}$$

for $t \leq \tau \leq t + \delta$. Integration of the last inequality over $[t, t+\delta]$ yields

$$\|x_1\|_2^2 \delta \leq u(t, t+\delta) + \delta^{3/2}C_5\sqrt{u(t, t+\delta)} \qquad (A.69)$$

On the other hand, from the triangle inequality

$$\|\psi^T(\tau, t)x_2\|_2^2 \leq [\|x_1\|_2 + \|x_1 + \psi^T(\tau, t)x_2\|_2]^2 \leq 2\|x_1\|_2^2 + 2\|x_1 + \psi^T(\tau, t)x_2\|_2^2$$

Integration over $[t, t+\delta]$ yields

$$\int_t^{t+\delta} \|\psi^T(\tau, t)x_2\|_2^2 \leq 2\|x_1\|_2^2 \delta + 2u(t, t+\delta)$$

Using (A.69), we arrive at

$$x_2^T \int_t^{t+\delta} \psi(\tau, t)\psi^T(\tau, t) \, d\tau \; x_2 \leq 4u(t, t+\delta) + 2\delta^{3/2}C_5\sqrt{u(t, t+\delta)} \qquad (A.70)$$

Inequalities (A.65), (A.69), and (A.70) imply (A.66). □

A.21   Proof of Lemma 13.5

Since $\nu(t)$ is persistently exciting, there are positive constants δ, α_1, and α_2 such that

$$\alpha_2 \geq \int_t^{t+\delta} [\nu^T(\tau)x]^2 \, d\tau \geq \alpha_1$$

where x is an arbitrary vector with $\|x\|_2 = 1$. By assumption

$$\int_0^\infty [u^T(\tau)x]^2 \, d\tau \leq \int_0^\infty \|u(\tau)\|_2^2 \, d\tau \leq k$$

Let $\delta_1 = \delta(1 + m)$, where $m \geq 2k/\alpha_1$ is an integer. Then

$$
\begin{aligned}
\int_t^{t+\delta_1} \left\{ [\nu^T(\tau) + u^T(\tau)]x \right\}^2 \, d\tau \; &\geq \; \tfrac{1}{2} \int_t^{t+\delta_1} [\nu^T(\tau)x]^2 \, d\tau - \int_t^{t+\delta_1} [u^T(\tau)x]^2 \, d\tau \\
&\geq \; \tfrac{1}{2}(1+m)\alpha_1 - k \; \geq \; \tfrac{1}{2}\alpha_1
\end{aligned}
$$

Similarly,

$$
\begin{aligned}
\int_t^{t+\delta_1} \left\{ [\nu^T(\tau) + u^T(\tau)]x \right\}^2 \, d\tau \; &\leq \; 2 \int_t^{t+\delta_1} [\nu^T(\tau)x]^2 \, d\tau + 2 \int_t^{t+\delta_1} [u^T(\tau)x]^2 \, d\tau \\
&\leq \; 2(1+m)\alpha_2 + 2k
\end{aligned}
$$

Thus,

$$\tfrac{1}{2}\alpha_1 \leq \int_t^{t+\delta_1} \left\{ [\nu^T(\tau) + u^T(\tau)]x \right\}^2 \, d\tau \leq 2(1+m)\alpha_2 + 2k$$

which shows that $\nu(t) + u(t)$ is persistently exciting.

Notes and References

The main references used in the preparation of this text are Hirsch and Smale [70], Hale [65], and Miller and Michel [118] for the theory of ordinary differential equations; Hahn [62], Krasovskii [96], and Rouche, Habets, and Laloy [137] for stability theory; and the texts by Vidyasagar [181](first edition) and Hsu and Meyer [76]. References for the specialized topics of Chapters 4 to 13 are listed under the respective chapters.

Chapter 1. The tunnel diode circuit and negative resistance oscillator are taken from Chua, Desoer, and Kuh [33]. The presentation of the mass-spring system is based on Mickens [117] and Southward [165]. The Hopfield neural network description is based on Hopfield [71] and Michel, Farrel and Porod [114]. The adaptive control example is based on Sastry and Bodson [151]. The classical material on second-order systems can be found in almost any text on nonlinear systems. Our presentation has followed closely the lucid presentation of Chua, Desoer, and Kuh [33]. Section 1.2.5 is based on Parker and Chua [132].

Chapter 2. The section on mathematical preliminaries has been patterned after similar sections in Bertsekas [21] and Luenberger [106]. For a complete coverage of topics in this section, the reader may consult any text on mathematical analysis. We have used Apostol [7]. Other standard texts are Rudin [140] and Royden [139]. The material in Sections 2.2–2.4 is standard and will be found in one form or the other in any graduate text on ordinary differential equations. Section 2.2 comes very close to the presentation in Vidyasagar [181], while Sections 2.3 and 2.4 are based on Hirsch and Smale [70] and Coppel [36]. The comparison principle is based on Hale [65], Miller and Michel [118] and Yoshizawa [189].

Chapter 3. Hahn [62], Krasovskii [96], and Rouche, Habets, and Laloy [137] are authoritative references on Lyapunov stability. The presentation style of Section 3.1 is influenced by Hirsch and Smale [70]. The proof of Theorem 3.1 is taken from Hirsch and Smale [70, Section 9.3], while that of Theorem 3.3 is based on Hahn [62, Section 25], Hale [65], and Miller and Michel [118]. The invariance principle presentation of Section 3.2 follows the original work by LaSalle [99]. The proof of Lemma 3.1 is based on Rouche, Habets, and Laloy [137, Appendix III]. The application of the invariance principle to neural networks is standard and can be found in Hopfield [71]. Our presentation is influenced by Salam [147]. See also Cohen and Grossberg [35] for a generalization of Example 3.11. The material on linear time-invariant systems in Section 3.3 is taken from Chen [29]. The proof of Theorem 3.6 is taken from Kailath [86]. The proof of Theorem 3.7 on linearization is guided by Rouche and Mawhin [138, Sections 1.6 and 1.7], which includes a careful treatment of the case when the linearization has at least one eigenvalue in the right-half plane with other eigenvalues on the imaginary axis. The proof of Lemma 3.3 is taken from Hahn [62, Section 35] (for the local part) and Lin, Sontag, and Wang [104] (for the global part). The statement and proof of Lemma 3.4 is guided by Hahn [62, Section 24E] and Sontag [162, Lemma 6.1]. The proof of Lemma 3.5 is taken from Hahn [62, Section 24A]. The proof of Theorem 3.8 combines ideas from Hahn [62, Section 25] and Rouche, Habets, and Laloy [137, Section I.6]. Section 3.5 is based on Vidyasagar [181]. The proof of Theorem 3.12 is taken from Krasovskii [96, Theorem 11.1]. The proof of Theorem 3.14 is based on Miller and Michel [118, Section 5.13] and Hahn [62, Section 49], with some insight from Hoppensteadt [72].

Chapter 4. Section 4.1 is based mostly on Carr [28], with help from Guckenheimer and Holmes [59]. The proof of Theorems 4.1 and 4.3 is taken from Carr [28, Chapter 2]. The proof of Theorem 4.2 using Lyapunov analysis is much simpler than the proof given in Carr [28]. We have dropped from the statement of the reduction theorem the case when the origin of the reduced system is stable. Dropping this case is justified by the fact that when we apply the theorem, we almost always use approximations of the center manifold. The only stability properties of the origin which can be concluded from an approximate model are asymptotic stability and instability since they are robust to perturbations. The version of the center manifold theorem quoted in Guckenheimer and Holmes [59] also drops the case of a stable origin. Corollary 4.1 deals with a stable origin in a special case. This corollary was suggested to the author by Miroslav Krstic. The proof of Lemma 4.1 is based on Hahn [62, Section 33]. Example 4.11 is taken from Willems [186]. The trajectory-reversing method is described in Genesio, Tartaglia, and Vicino [55]. Other methods for estimating the region of attraction are described or surveyed in [31, 55, 116, 127]. The proof of Lemma 4.2 is taken from Popov [135, page 211]. The proof of Theorem 4.5 is based on Sastry and Bodson [151, Theorem 1.5.2].

Chapter 5. The material of Sections 5.1 and 5.2 is based on a vast literature on robustness analysis in control theory. We can say, however, that the basic references are Hahn [62, Section 56] and Krasovskii [96, Sections 19 and 24]. Similar results are given in Coppel [36, Section III.3] for the case when the nominal system is linear, but the derivation is not based on Lyapunov theory; instead, it uses properties of fundamental matrices. The proof of Theorem 5.1 is guided by Miller and Michel [118, Theorem 9.14] and Corless and Leitmann [38]. The case of nonvanishing perturbations is referred to in Hahn [62] and Krasovskii [96] as the case of "persistent disturbance." We have decided not to use this term so that it would not be confused with the term "persistent excitation," which is common in the adaptive control literature. The results on nonvanishing perturbations are also related to the concept of total stability; see Hahn [62, Section 56]. The concept of input-to-state stability was introduced by Sontag [162], who proved a few basic results; see [163]. Our presentation benefited from a nice exposition in Krstic, Kanellakopolous, and Kokotovic [97]. The comparison method of Section 5.4 is based on sporadic use of the comparison lemma in the control literature. Theorem 5.3 on continuity of solutions on the infinite-time interval is novel in that continuity results are usually stated on compact time intervals, as we have them in Section 2.3. Inclusion of an infinite-time version is appropriate in view of the fact that in most engineering problems we are interested in solutions reaching steady-state. Technically, the theorem is a straightforward corollary of the results of Section 5.4. Section 5.6 on the stability of interconnected systems is based mostly on the tutorial paper Araki [8], with help from the research monographs Siljak [156] and Michel and Miller [115]. The neural network example is taken from Michel, Farrel, and Porod [114]. The treatment of slowly varying systems is based on Desoer and Vidyasagar [43, Section IV.8], Vidyasagar [181], Kokotovic, Khalil, and O'Reilly [95, Section 5.2], and Hoppensteadt [72]. Lemma 5.11 is a specialization of Lemma 2 of Hoppensteadt [72] to the case of exponential stability.

Chapter 6. The treatment of $\mathcal{L}$-stability in Sections 6.1 and 6.2 is based on Desoer and Vidyasagar [43] and Vidyasagar [181]. The concept of input-to-output stability is due to Sontag [162]. Section 6.4 on the $\mathcal{L}_2$-gain is based on van der Schaft [178], with help from the papers by Willems [185] and Hill and Moylan [67, 69] on dissipative systems. The proof of Theorem 6.4 is based on [43, Section 2.6], [180, Section 3.1.2] and [195, Section 4.3]

Chapter 7. The section on second-order systems is based on Hirsch and Smale [70, Chapters 10 and 11] and Guckenheimer and Holmes [59, Section 1.8]. The proof of Theorem 7.1 is based on Hirsch and Smale [70, Chapter 10] and Miller and Michel [118, Section 7.2]. The discussion on the stability of periodic solutions in Section 7.2 is based on Hahn [62, Section 81], Miller and Michel [118, Section 6.4], and Hale [65, Section VI.2]. The Poincaré map presentation is based on Hirsch and Smale [70, Chapter 13] and Guckenheimer and Holmes [59, Section 1.5].

Chapter 8. The perturbation method of Section 8.1 is classical and can be found in many references. A detailed treatment can be found in Kevorkian and Cole [89] and Nayfeh [124]. The asymptotic results of Theorems 8.1 and

8.2 are adapted from Hoppensteadt's work on singular perturbations; see Hoppensteadt [73]. The presentation of the averaging method in Section 8.3 is based on Sanders and Verhulst [150], Hale [65, Section V.3], Halanay [63, Section 3.5], and Guckenheimer and Holmes [59, Sections 4.1 and 4.2]. The vibrating pendulum example is taken from Tikhonov, Vasileva, and Volosov [174]. The application of averaging to weakly nonlinear oscillators in Section 8.4 is based on Hale [65, pp. 183–186]. The presentation of general averaging in Section 8.5 is based on Sanders and Verhulst [150], Hale [65, Section V.3], and Sastry and Bodson [151, Section 4.2].

Chapter 9. The presentation of the singular perturbation method follows very closely Kokotovic, Khalil, and O'Reilly [95]. The proofs of Theorems 9.1 and 9.4 adapt ideas from Hoppensteadt [72]. We have not included the construction of higher-order approximations. For that, the reader may consult Hoppensteadt [73], Butuzov, Vasileva, and Fedoryuk [24], or O'Malley[128]. The articles [72], [73], and [24] appear in Kokotovic and Khalil [94]. Example 9.8 is taken from Tikhonov, Vasileva, and Volosov [174]. For extensions of this example, see Grasman [58]. The proof of Theorems 9.2 and 9.3 is based on Saberi and Khalil [144].

Chapter 10. Absolute stability has a long history in the control theory literature. For a historical perspective, see Hsu and Meyer [76, Sections 5.9 and 9.5]. Our presentation of the circle and Popov criteria is based on Hsu and Meyer [76, Section 9.5 and Chapter 10], Vidyasagar [181], Siljak [155, Sections 8.6 to 8.9 and Appendix H], and Moore and Anderson [121]. A comprehensive coverage of absolute stability can be found in Narendra and Taylor [123]. The proof of Lemma 10.1 is based on Tao and Ioannou [171] and Wen [182]. The proof of Lemma 10.2 is based on Anderson [2]; see also Anderson and Vongpanitlerd [5] for an expanded treatment of the positive real lemma. Section 10.1.3 is based on Boyd and Yang [22] and Horisberger and Belanger [74]. The small-gain theorem presentation is based on Desoer and Vidyasagar [43] and the tutorial article by Teel, Georgiou, Praly, and Sontag [173]. Example 10.9 is taken from [173]. Section 10.3 on the passivity approach is based on several references, including [68, 26, 97, 173, 181]. Detailed treatment of the describing function method can be found in Atherton [13] and Hsu and Meyer [76, Chapters 6 and 7]. Our presentation in Section 10.4 follows Mees [111, Chapter 5]. The proof of Lemma 10.7 is taken from Mees and Bergen [113]. The proof of Theorem 7.2 is based on Mees and Bergen [113] and Bergen and Franks [19]. The presentation of the error analysis has gained a lot from a lucid presentation in Siljak [155, Appendix G], from which Example 10.24 is taken.

Chapter 11. The design via linearization approach of Section 11.2 is standard and can be found in almost any book on nonlinear control. The use of integral control is also standard, but the results given here can be traced back to Huang and Rugh [77] and Isidori and Byrnes [82]. The gain scheduling presentation of Section 11.3 is based Lawrence and Rugh [101] and Kaminer, Pascoal, Khargonekar, and Coleman [88], with help from Astrom and Wittenmark [11], Rugh [141], and Shamma and Athans [153].

Chapter 12. The exact feedback linearization chapter is based mainly on Isidori [80]. The lucid introduction of Section 12.1 is based on Spong and Vidyasagar [166, Chapter 10] and Isidori [80, Chapter 4]. The local stabilization and tracking results are based on Isidori [80, Chapter 4]. The global stabilization material is based on many references, as the global stabilization problem has caught the attention of several researchers in recent years. The papers [25, 39, 105, 109, 145, 169, 175] and the book [97] will guide the reader to the basic results on global stabilization, not only for the problem formulation presented here, but also for more general problem formulations, including problems treated in the backstepping Section 13.2. Examples 12.14, 12.15, and 12.16 are taken from [25], [169], and [52], respectively.

Chapter 13. The Lyapunov redesign section is guided by Corless and Leitmann [38], Barmish, Corless, and Leitmann [14], and Spong and Vidyasagar [166, Chapter 10]. See also Corless [37] for a survey of the use of Lyapunov redesign in the control of uncertain nonlinear systems. The robust stabilization design of Section 13.1.1 is also known as min-max control. The nonlinear damping of Section 13.1.2 is based on Krstic, Kanellakopolous, and Kokotovic [97, Section 2.5]. The backstepping section is based mainly on the same reference [97], with help from Qu [136] and Slotine and Hedrick [160]. The book [97] contains a comprehensive treatment of the backstepping design procedure, including its application to adaptive control of nonlinear systems. The sliding mode control section is based on the books by Utkin [176] and Slotine and Li [161], and the tutorial paper by DeCarlo, Zak, and Matthews [42]. Analysis of the continuous approximation of sliding mode controllers is based on [45]. The adaptive control section is based on Sastry and Bodson [151], Anderson and others [4], Narendra and Annaswamy [122], and Ioannou and Sun [79]. The proof of Lemma 13.4 is based on Anderson [3] and Anderson and others [4, Section 2.3]. The proof of Lemma 13.5 is based on Sastry and Bodson [151, Lemma 2.6.6].

Other references, which were consulted occasionally during the preparation of this text, are included in the bibliography.

Bibliography

[1] J. K. Aggarwal. *Notes on Nonlinear Systems.* Van Nostrand Reinhold, New York, 1972.

[2] B. D. O. Anderson. A system theory criterion for positive real matrices. *SIAM J. Control*, 5(2):171–182, 1967.

[3] B. D. O. Anderson. Exponential stability of linear equations arising in adaptive identification. *IEEE Trans. Automat. Contr.*, AC-22:83–88, February 1977.

[4] B. D. O. Anderson, R. R. Bitmead, C. R. Johnson, Jr., P. V. Kokotovic, R. L. Kosut, I. M. Y. Mareels, L. Praly, and B. D. Riedle. *Stability of Adaptive Systems.* MIT Press, Cambridge, Mass., 1986.

[5] B. D. O. Anderson and S. Vongpanitlerd. *Network Analysis and Synthesis: A Modern Systems Theory Approach.* Prentice Hall, Englewood Cliffs, New Jersey, 1973.

[6] A. M. Annaswamy. On the input-output behavior of a class of second-order nonlinear adaptive systems. In *American Control Conference*, pages 731–732, June 1989.

[7] T. M. Apostol. *Mathematical Analysis.* Addison-Wesley, Reading, Mass., 1957.

[8] M. Araki. Stability of large-scale nonlinear systems–quadratic-order theory of composite-system method using M-matrices. *IEEE Trans. Automat. Contr.*, AC-23(2):129–141, April 1978.

[9] D. K. Arrowsmith and C. M. Place. *Ordinary Differential Equations.* Chapman and Hall, London, 1982.

[10] R. B. Ash. *Real Analysis and Probability.* Academic Press, New York, 1972.

[11] K. J. Astrom and B. Wittenmark. *Adaptive Control*. Addison-Wesley, Reading, Mass., second edition, 1995.

[12] D. P. Atherton. *Stability of Nonlinear Systems*. John Wiley, New York, 1981.

[13] D. P. Atherton. *Nonlinear Control Engineering*. Van Nostrand Reinhold, London, student edition, 1982.

[14] B. R. Barmish, M. Corless, and G. Leitmann. A new class of stabilizing controllers for uncertain dynamical systems. *SIAM J. Control & Optimization*, 21(2):246–255, 1983.

[15] T. Basar and P. Bernhard. H_∞-*Optimal Control and Related Minimax Design Problems*. Birkhäuser, Boston, second edition, 1995.

[16] R. Bellman. *Introduction to Matrix Analysis*. McGraw-Hill, New York, second edition, 1970.

[17] R. E. Bellman, J. Bentsman, and S. M. Meerkov. Vibrational control of nonlinear systems: Vibrational controllability and transient behavior. *IEEE Trans. Automat. Contr.*, AC-31(8):717–724, August 1986.

[18] A. R. Bergen, L. O. Chua, A. I. Mees, and E. W. Szeto. Error bounds for general describing function problems. *IEEE Trans. Circuits Syst.*, CAS-29:345–354, June 1982.

[19] A. R. Bergen and R. L. Frank. Justification of the describing function method. *SIAM J. Control*, 9(4):568–589, November 1971.

[20] M. Berger and M. Berger. *Perspectives in Nonlinearity*. W. A. Benjamin, New York, 1968.

[21] D. P. Bertsekas. *Dynamic Programming*. Prentice Hall, Englewood Cliffs, N.J., 1987.

[22] S. Boyd and Q. Yang. Structured and simultaneous Lyapunov functions for system stability problems. *Int. J. Contr.*, 49(6):2215–2240, 1989.

[23] F. Brauer and J. A. Nohel. *The Qualitative Theory of Ordinary Differential Equations*. Dover, New York, 1989. Reprint; originally published by Benjamin, 1969.

[24] V. F. Butuzov, A. B. Vasileva, and M. V. Fedoryuk. Asymptotic methods in the theory of ordinary differential equations. In R. V. Gamkrelidze, editor, *Mathematical Analysis*, volume 8 of *Progress in Mathematics*, pages 1–82. Plenum Press, New York, 1970.

[25] C. I. Byrnes and A. Isidori. Asymptotic stabilization of minimum phase nonlinear systems. *IEEE Trans. Automat. Contr.*, 36(10):1122–1137, October 1991.

[26] C. I. Byrnes, A. Isidori, and J. C. Willems. Passivity, feedback equivalence, and the global stabilization of minimum phase nonlinear systems. *IEEE Trans. Automat. Contr.*, 36(11):1228–1240, November 1991.

[27] F. M. Callier and C. A. Desoer. *Multivariable Feedback Systems*. Springer-Verlag, New York, 1982.

[28] J. Carr. *Applications of Centre Manifold Theory*. Springer-Verlag, New York, 1981.

[29] C.-T. Chen. *Linear System Theory and Design*. Holt, Rinehart and Winston, New York, 1984.

[30] H.-D. Chiang, M. W. Hirsch, and F. F. Wu. Stability regions of nonlinear autonomous dynamical systems. *IEEE Trans. Automat. Contr.*, 33(1):16–27, January 1988.

[31] H.-D. Chiang and J. S. Thorp. Stability regions of nonlinear dynamical systems: a constructive methodology. *IEEE Trans. Automat. Contr.*, 34:1229–1241, December 1989.

[32] J. H. Chow, editor. *Time-Scale Modeling of Dynamic Networks with Applications to Power Systems*. Number 46 in Lecture Notes in Control and Information Sciences. Springer-Verlag, New York, 1982.

[33] L. O. Chua, C. A. Desoer, and E. S. Kuh. *Linear and Nonlinear Circuits*. McGraw-Hill, New York, 1987.

[34] L. O. Chua and Y.-S. Tang. Nonlinear oscillation via volterra series. *IEEE Trans. Circuits Syst.*, CAS-29(3):150–168, March 1982.

[35] M. A. Cohen and S. Grossberg. Absolute stability of global pattern formation and parallel memory storage by competitive neural networks. *IEEE Trans. Syst. Man, Cybern.*, 13(5):815–826, September 1983.

[36] W. A. Coppel. *Stability and Asymptotic Behavior of Differential Equations*. D. C. Heath, Boston, 1965.

[37] M. Corless. Control of uncertain nonlinear systems. *J. Dyn. Sys. Measurement and Control*, 115:362–372, 1993.

[38] M. Corless and G. Leitmann. Continuous state feedback guaranteeing uniform ultimate boundedness for uncertain dynamic systems. *IEEE Trans. Automat. Contr.*, AC-26(5):1139–1144, October 1981.

[39] J-M. Coron, L. Praly, and A. Teel. Feedback stabilization of nonlinear systems: sufficient conditions and Lyapunov and input-output techniques. In A. Isidori, editor, *Trends in Control*, pages 293–347. Springer-Verlag, New York, 1995.

[40] E. J. Davison. The robust control of a servomechanism problem for linear time-invariant multivariable systems. *IEEE Trans. Automat. Contr.*, AC-21(1):25–34, January 1976.

[41] R. A. Decarlo. *Linear Systems*. Prentice Hall, Englewood Cliffs, N.J., 1989.

[42] R. A. DeCarlo, S. H. Zak, and G. P. Matthews. Variable structure control of nonlinear multivariable systems: A tutorial. *Proc. of IEEE*, 76:212–232, 1988.

[43] C. A. Desoer and M. Vidyasagar. *Feedback Systems: Input-Output Properties.* Academic Press, New York, 1975.

[44] J. C. Doyle, K. Glover, P. P. Khargonekar, and B. A. Francis. State-space solutions to standard H_2 and H_∞ control problems. *IEEE Trans. Automat. Contr.*, 34(8):831–847, August 1989.

[45] F. Esfandiari and H. K. Khalil. Stability analysis of a continuous implementation of variable structure control. *IEEE Trans. Automat. Contr.*, 36(5):616–620, May 1991.

[46] F. Esfandiari and H.K. Khalil. Output feedback stabilization of fully linearizable systems. *Int. J. Contr.*, 56:1007–1037, 1992.

[47] M. Eslami. A unified approach to the stability robustness analysis of systems with multiple varying parameters. Technical Report UIC-EECS-90-3, University of Illinois-Chicago, April 1990.

[48] M. Fiedler and V. Ptak. On matrices with nonnegative off-diagonal elements and positive principal minors. *Czech. Math. J.*, 12:382–400, 1962.

[49] A. F. Filippov. Differential equations with discontinuous right-hand side. *Amer. Math. Soc. Translations*, 42(2):199–231, 1964.

[50] A. M. Fink. *Almost Periodic Differential Equations.* Number 377 in Lecture Notes in Mathematics. Springer-Verlag, New York, 1974.

[51] B. A. Francis. *A course in H_∞ control theory*, volume 88 of *Lect. Notes Contr. Inf Sci.* Springer-Verlag, New York, 1987.

[52] R. A. Freeman and P. V. Kokotovic. Optimal nonlinear controllers for feedback linearizable systems. In *Proc. American Control Conf.*, pages 2722–2726, Seattle, WA, June 1995.

[53] F. R. Gantmacher. *Theory of Matrices.* Chelsea Publ., Bronx, N.Y., 1959.

[54] F. M. Gardner. *Phaselock Techniques.* Wiley-Interscience, New York, 1979.

[55] R. Genesio, M. Tartaglia, and A. Vicino. On the estimation of asymptotic stability regions: State of the art and new proposals. *IEEE Trans. Automat. Contr.*, AC-30(8):747–755, August 1985.

[56] S. T. Glad. On the gain margin of nonlinear and optimal regulators. *IEEE Trans. Automat. Contr.*, AC-29(7):615–620, July 1984.

[57] G. H. Golub and C. F. Van Loan. *Matrix Computations.* The John Hopkins University Press, Baltimore, 1983.

[58] J. Grasman. *Asymptotic Methods for Relaxation Oscillations and Applications.* Number 63 in Applied Mathematical Sciences. Springer-Verlag, New York, 1987.

[59] J. Guckenheimer and P. Holmes. *Nonlinear Oscillations, Dynamical Systems, and Bifurcations of Vector Fields.* Springer-Verlag, New York, 1983.

[60] V. Guillemin and A. Pollack. *Differential Topology.* Prentice Hall, Englewood Cliffs, N.J., 1974.

[61] L. Guzzella and A. H. Glattfelder. Stability of multivariable systems with saturating power amplifiers. In *American Control Conference*, pages 1687–1692, June 1989.

[62] W. Hahn. *Stability of Motion.* Springer-Verlag, New York, 1967.

[63] A. Halanay. *Differential Equations: Stability, Oscillations, Time Lags*, volume 23 of *Mathematics in Science and Engineering.* Academic Press, New York, 1966.

[64] J. Hale and H. Kocak. *Dynamics and Bifurcations.* Springer-Verlag, New York, 1991.

[65] J. K. Hale. *Ordinary Differential Equations.* Wiley-Interscience, New York, 1969.

[66] P. Hartman. *Ordinary Differential Equations.* Wiley, New York, 1964.

[67] D. Hill and P. Moylan. The stability of nonlinear dissipative systems. *IEEE Trans. Automat. Contr.*, AC-21(5):708–711, October 1976.

[68] D. J. Hill and P. J. Moylan. Stability results for nonlinear feedback systems. *Automatica*, 13:377–382, 1977.

[69] D. J. Hill and P. J. Moylan. Dissipative dynamical systems: basic input-output and state properties. *J. of The Franklin Institute*, 309(5):327–357, May 1980.

[70] M. W. Hirsch and S. Smale. *Differential Equations, Dynamical Systems, and Linear Algebra.* Academic Press, New York, 1974.

[71] J. J. Hopfield. Neurons with graded response have collective computational properties like those of two-state neurons. *Proc. of the Natl. Acad. Sci. U.S.A.*, 81:3088–3092, May 1984.

[72] F. C. Hoppensteadt. Singular perturbations on the infinite interval. *Trans. Amer. Math. Soc.*, 123(2):521–535, 1966.

[73] F. C. Hoppensteadt. Properties of solutions of ordinary differential equations with small parameters. *Comm. Pure Appl. Math.*, 24:807–840, 1971.

[74] P. Horisberger and P. R. Belanger. Regulators of linear, time invariant plants with uncertain parameters. *IEEE Trans. Automat. Contr.*, AC-21(5):705–708, October 1976.

[75] D. Hrovat and M. Tran. Application of gain scheduling to design of active suspension. In *Proc. IEEE Conf. on Decision and Control*, pages 1030–1035, San Antonio, TX, December 1993.

[76] J. C. Hsu and A. U. Meyer. *Modern Control Principles and Applications.* McGraw-Hill, New York, 1968.

[77] J. Huang and W. J. Rugh. On a nonlinear multivariable servomechanism problem. *Automatica*, 26(6):963–972, 1990.

[78] R. A. Hyde and K. Glover. The application of scheduled H_∞ controllers to a VSTOL aircraft. *IEEE Trans. Automat. Contr.*, 38:1021–1039, 1993.

[79] P. A. Ioannou and J. Sun. *Robust Adaptive Control.* Prentice-Hall, Upper Saddle River, New Jersey, 1995.

[80] A. Isidori. *Nonlinear Control Systems.* Springer-Verlag, New York, third edition, 1995.

[81] A. Isidori and A. Astolfi. Disturbance attenuation and H_∞ control via measurement feedback in nonlinear systems. *IEEE Trans. Automat. Contr.*, 37(9):1283–1293, September 1992.

[82] A. Isidori and C. I. Byrnes. Output regulation of nonlinear systems. *IEEE Trans. Automat. Contr.*, 35(2):131–140, February 1990.

[83] A. Isidori, S. S. Sastry, P. V. Kokotovic, and C. I. Byrnes. Singularly perturbed zero dynamics of nonlinear systems. *IEEE Trans. Automat. Contr.*, 37(10):1625–1631, October 1992.

[84] J. Jiang. Optimal gain scheduling controllers for a diesel engine. *IEEE Contro Systems Magazine*, 14(4):42–48, 1994.

[85] Z. P. Jiang, A. R. Teel, and L. Praly. Small gain theorem for ISS systems and applications. *Mathematics of Control, Signals, and Systems*, 7(2):95–120 1995.

[86] T. Kailath. *Linear Systems*. Prentice Hall, Englewood Cliffs, N.J., 1980.

[87] R. E. Kalman and J. E. Bertram. Control system analysis and design via the "second method" of Lyapunov, parts I and II. *Journal of Basic Engineering* 82:371–400, June 1960.

[88] I. Kaminer, A. M. Pascoal, P. P. Khargonekar, and E. E. Coleman. A velocity algorithm for the implementation of gain scheduled controllers. *Automatica* 31:1185–1191, August 1995.

[89] J. Kevorkian and J. D. Cole. *Perturbation Methods in Applied Mathematics* Number 34 in Applied Mathematical Sciences. Springer-Verlag, New York 1981.

[90] H. K. Khalil. On the robustness of output feedback control methods to modeling errors. *IEEE Trans. Automat. Contr.*, AC-26(2):524–526, April 1981.

[91] H. K. Khalil. Stability analysis of nonlinear multiparameter singularly perturbed systems. *IEEE Trans. Automat. Contr.*, AC-32(3):260–263, March 1987.

[92] P. P. Khargonekar, I. R. Petersen, and M. A. Rotea. H_∞-optimal control with state feedback. *IEEE Trans. Automat. Contr.*, 33(8):786–788, August 1988.

[93] H. W. Knobloch and B. Aulbach. Singular perturbations and integral manifolds. *J. Math. Phys. Sci.*, 18:415–424, 1984.

[94] P. V. Kokotovic and H. K. Khalil, editors. *Singular Perturbations in System and Control*. IEEE Press, New York, 1986.

[95] P. V. Kokotovic, H. K. Khalil, and J. O'Reilly. *Singular Perturbations Methods in Control: Analysis and Design.* Academic Press, New York, 1986.

[96] N. N. Krasovskii. *Stability of Motion.* Stanford University Press, Stanford, 1963.

[97] M. Krstic, I. Kanellakopoulos, and P. Kokotovic. *Nonlinear and Adaptive Control Design.* Wiley-Interscience, New York, 1995.

[98] H. Kwakernaak and R. Sivan. *Linear Optimal Control Systems.* Wiley-Interscience, New York, 1972.

[99] J. P. LaSalle. Some extensions of Lyapunov's second method. *IRE Trans. Circuit Theory*, CT-7(4):520–527, December 1960.

[100] J. P. LaSalle. An invariance principle in the theory of stability. In J.K. Hale and J.P. LaSalle, editors, *Differential Equations and Dynamical Systems*, pages 277–286. Academic Press, New York, 1967.

[101] D. A. Lawrence and W. J. Rugh. Gain scheduling dynamic linear controllers for a nonlinear plant. *Automatica*, 31:381–390, mar 1995.

[102] S. Lefschetz. *Differential Equations: Geometric Theory.* Interscience, New York, 1963.

[103] S. Lefschetz. *Stability of Nonlinear Control Systems.* Academic Press, New York, 1965.

[104] Y. Lin, E. D. Sontag, and Y. Wang. A smooth converse Lyapunov theorem for robust stability. *SIAM J. Control & Optimization*, 34, January 1996.

[105] Z. Lin and A. Saberi. Robust semi-global stabilization of minimum-phase input-output linearizable systems via partial state and output feedback. *IEEE Trans. Automat. Contr.*, 40(6):1029–1041, 1995.

[106] D. G. Luenberger. *Optimization by Vector Space Methods.* Wiley, New York, 1969.

[107] D. G. Luenberger. *Introduction to Linear and Nonlinear Programming.* Addison-Wesley, Reading, Mass., 1983.

[108] I. M. Y. Mareels and D. J. Hill. Monotone stability of nonlinear feedback systems. *J. Mathematical Systems, Estimation and Control*, 2(3):275–291, 1992.

[109] R. Marino and P. Tomei. Dynamic output feedback linearization and global stabilization. *Systems Contr. Lett.*, 17:115–121, 1991.

[110] S. M. Meerkov. Principle of vibrational control: Theory and applications. *IEEE Trans. Automat. Contr.*, AC-25(4):755–762, August 1980.

[111] A. I. Mees. *Dynamics of Feedback Systems*. Wiley, New York, 1981.

[112] A. I. Mees. Describing functions: ten years on. *IMA J. Applied Mathematics*, 32:221–233, 1984.

[113] A. I. Mees and A. R. Bergen. Describing functions revisited. *IEEE Trans. Automat. Contr.*, AC-20(4):473–478, August 1975.

[114] A. N. Michel, J. A. Farrel, and W. Porod. Qualitative analysis of neural networks. *IEEE Trans. Circuits Syst.*, 36(2):229–243, February 1989.

[115] A. N. Michel and R. K. Miller. *Qualitative Analysis of Large Scale Dynamical Systems*. Academic Press, New York, 1977.

[116] A. N. Michel, N. R. Sarabudla, and R. K. Miller. Stability analysis of complex dynamical systems. *Circuits Systems Signal Process*, 1(2):171–202, 1982.

[117] R. E. Mickens. *Introduction to Nonlinear Oscillations*. Cambridge University Press, London, 1981.

[118] R. K. Miller and A. N. Michel. *Ordinary Differential Equations*. Academic Press, New York, 1982.

[119] R. K. Miller and A. N. Michel. An invariance theorem with applications to adaptive control. *IEEE Trans. Automat. Contr.*, 35(6):744–748, June 1990.

[120] N. Minorsky. *Nonlinear Oscillations*. Van Nostrand, Princeton, N.J., 1962.

[121] J. B. Moore and B. D. O. Anderson. Applications of the multivariable Popov criterion. *Int. J. Control*, 5(4):345–353, 1967.

[122] K. S. Narendra and A. M. Annaswamy. *Stable Adaptive Systems*. Prentice Hall, Englewood Cliffs, N.J., 1989.

[123] K. S. Narendra and J. Taylor. *Frequency Domain Methods for Absolute Stability*. Academic Press, New York, 1973.

[124] A. H. Nayfeh. *Introduction to Perturbation Techniques*. Wiley, New York, 1981.

[125] R. A. Nichols, R. T. Reichert, and W. J. Rugh. Gain scheduling for H_∞ controllers: a flight control example. *IEEE Trans. Contr. Syst. Tech.*, 1:69–75, 1993.

[126] H. Nijmeijer and A. J. van der Schaft. *Nonlinear Dynamic Control Systems.* Springer-Verlag, Berlin, 1990.

[127] E. Noldus and M. Loccufier. A new trajectory reversing method for the estimation of asymptotic stability regions. *Int. J. Contr.*, 61:917–932, 1995.

[128] R. E. O'Malley. *Singular Perturbation Methods for Ordinary Differential Equations.* Springer-Verlag, New York, 1991.

[129] R. Ortega and M. W. Spong. Adaptive motion control of rigid robots: a tutorial. *Automatica*, 25:877–888, 1989.

[130] B. E. Paden and S. S. Sastry. A calculus for computing Filippov's differential inclusion with application to the variable structure control of robot manipulators. *IEEE Trans. Circuits Syst.*, CAS-34:73–82, January 1987.

[131] M. A. Pai. *Power System Stability Analysis by the Direct Method of Lyapunov.* North-Holland, Amsterdam, 1981.

[132] T. S. Parker and L. O. Chua. *Practical Numerical Algorithms for Chaotic Systems.* Springer-Verlag, New York, 1989.

[133] R. V. Patel and M. Toda. Qualitative measures of robustness for multivariable systems. In *Joint Automatic Control Conference*, number TP8-A, 1980.

[134] W. R. Perkins and J. B. Cruz. *Engineering of Dynamic Systems.* John Wiley, New York, 1969.

[135] V. M. Popov. *Hyperstability of Control Systems.* Springer-Verlag, New York, 1973.

[136] Z. Qu. Robust control of nonlinear uncertain systems under generalized matching conditions. *Automatica*, 29(4):985–998, 1993.

[137] N. Rouche, P. Habets, and M. Laloy. *Stability Theory by Lyapunov's Direct Method.* Springer-Verlag, New York, 1977.

[138] N. Rouche and J. Mawhin. *Ordinary Differential Equations.* Pitman, Boston, 1973.

[139] H. L. Royden. *Real Analysis.* Macmillan, New York, 1963.

[140] W. Rudin. *Principles of Mathematical Analysis.* McGraw-Hill, New York, 3rd edition, 1976.

[141] W. J. Rugh. Analytical framework for gain scheduling. *IEEE Control Systems Magazine*, 11(1):79–84, January 1991.

[142] W. J. Rugh. *Linear System Theory*. Prentice Hall, Upper Saddle River, New Jersey, second edition, 1996.

[143] E. P. Ryan and N. J. Buckingham. On asymptotically stabilizing feedback control of bilinear systems. *IEEE Trans. Automat. Contr.*, AC-28(8):863–864, August 1983.

[144] A. Saberi and H. Khalil. Quadratic-type Lyapunov functions for singularly perturbed systems. *IEEE Trans. Automat. Contr.*, AC-29(6):542–550, June 1984.

[145] A. Saberi, P. V. Kokotovic, and H. J. Sussmann. Global stabilization of partially linear composite systems. *SIAM J. Control & Optimization*, 28(6):1491–1503, 1990.

[146] M. Safonov. *Stability and Robustness of Multivariable Feedback Systems*. MIT Press, Cambridge, Mass., 1980.

[147] F. M. A. Salam. A formulation for the design of neural processors. In *International Conference on Neural Networks*, pages I–173–I–180, July 1988.

[148] I. W. Sandberg. On the L_2-boundedness of solutions of nonlinear functional equations. *Bell Sys. Tech. J.*, 43:1581–1599, 1964.

[149] I. W. Sandberg. Global inverse function theorems. *IEEE Trans. Circuit Syst.*, CAS-27(11):998–1004, November 1980.

[150] J. A. Sanders and F. Verhulst. *Averaging Methods in Nonlinear Dynamical Systems*. Number 59 in Applied Mathematical Sciences. Springer-Verlag, New York, 1985.

[151] S. Sastry and M. Bodson. *Adaptive Control*. Prentice Hall, Englewood Cliffs, N.J., 1989.

[152] S. Sastry, J. Hauser, and P. Kokotovic. Zero dynamics of regularly perturbed systems are singularly perturbed. *Systems Contr. Lett.*, 13:299–314, 1989.

[153] J. S. Shamma and M. Athans. Gain scheduling: potential hazards and possible remedies. *IEEE Control Systems Magazine*, 12(3):101–107, June 1992.

[154] D. Shevitz and B. Paden. Lyapunov stability theory of nonsmooth systems. *IEEE Trans. Automat. Contr.*, 39:1910–1914, September 1994.

[155] D. D. Siljak. *Nonlinear Systems*. Wiley, New York, 1969.

[156] D. D. Siljak. *Large Scale Dynamic Systems: Stability and Structure*. North Holland, New York, 1978.

[157] H. Sira-Ramirez. Harmonic response of variable-structure-controlled van der Pol oscillators. *IEEE Trans. Circuits Syst.*, CAS-34:103–106, January 1987.

[158] H. Sira-Ramirez. A dynamical variable structure control strategy in asymptotic output tracking problem. *IEEE Trans. Automat. Contr.*, 38:615–620, April 1993.

[159] G. R. Slemon and A. Straughen. *Electric Machines.* Addison-Wesley, Reading, Mass., 1980.

[160] J-J. E. Slotine and J. K. Hedrick. Robust input-output feedback linearization. *Int. J. Contr.*, 57(5):1133–1139, 1993.

[161] J-J. E. Slotine and W. Li. *Applied Nonlinear Control.* Prentice-Hall, Englewood Cliffs, New Jersey, 1991.

[162] E. D. Sontag. Smooth stabilization implies coprime factorization. *IEEE Trans. Automat. Contr.*, 34(4):435–443, April 1989.

[163] E. D. Sontag. On the input-to-state stability property. *European J. Control*, 1, 1995.

[164] E. D. Sontag and Y. Wang. On characterizations of the input-to-state stability property. *Systems Contr. Lett.*, 24:351–359, 1995.

[165] S. C. Southward. *Modeling and Control of Mechanical Systems with Stick-Slip Friction.* PhD thesis, Michigan State University, East Lansing, 1990.

[166] M. W. Spong and M. Vidyasagar. *Robot Dynamics and Control.* Wiley, New York, 1989.

[167] G. Stein, G. Hartmann, and R. Hendrick. Adaptive control laws for F-8 flight test. *IEEE Trans. Automat. Contr.*, AC-22:758–767, 1977.

[168] B. L. Stevens and F. L. Lewis. *Aircraft Control and Simulation.* Wiley Interscience, New York, 1992.

[169] H. J. Sussmann and P. V. Kokotovic. The peaking phenomenon and the global stabilization of nonlinear systems. *IEEE Trans. Automat. Contr.*, 36(4):424–440, April 1991.

[170] F. L. Swern. Analysis of oscillations in systems with polynomial-type nonlinearities using describing functions. *IEEE Trans. Automat. Contr.*, AC-28:31–41, January 1983.

[171] G. Tao and P. A. Ioannou. Strictly positive real matrices and the Lefschetz-Kalman-Yakubovitch lemma. *IEEE Trans. Automat. Contr.*, 33(12):1183–1185, December 1988.

[172] A. Teel and L. Praly. Tools for semiglobal stabilization by partial state and output feedback. *SIAM J. Control & Optimization*, 33, 1995.

[173] A. R. Teel, T. T. Georgiou, L. Praly, and E. Sontag. Input-output stability In W. Levine, editor, *The Control Handbook*. CRC Press, 1995.

[174] A. N. Tikhonov, A. B. Vasileva, and V. M. Volosov. Ordinary differentia equations. In E. Roubine, editor, *Mathematics Applied to Physics*, pages 162–228. Springer-Verlag, New York, 1970.

[175] J. Tsinias. Partial-state global stabilization for general triangular systems *Systems Contr. Lett.*, 24:139–145, 1995.

[176] V. I. Utkin. *Sliding Modes in Optimization and Control*. Springer-Verlag New York, 1992.

[177] V. I. Utkin. Sliding mode control design principles. *IEEE Transactions o Industrial Electronics*, 40:23–36, February 1993.

[178] A. J. van der Schaft. L_2-gain analysis of nonlinear systems and nonlinea state feedback H_∞ control. *IEEE Trans. Automat. Contr.*, 37(6):770–784 June 1992.

[179] A. J. van der Schaft. Nonlinear state space H_∞ control theory. In H. L Trentelman and J. C. Willems, editors, *Essays on Control: Perspectives i the Theory and its Applications*, pages 153–190. Birkhäuser, Boston, 1993.

[180] M. Vidyasagar. *Large Scale Interconnected Systems*. Springer-Verlag, Berlir 1981.

[181] M. Vidyasagar. *Nonlinear Systems Analysis*. Prentice Hall, Englewood Cliff New Jersey, second edition, 1993.

[182] J. T. Wen. Time domain and frequency domain conditions for strict positiv realness. *IEEE Trans. Automat. Contr.*, 33(10):988–992, October 1988.

[183] S. Wiggins. *Introduction to Applied Nonlinear Dynamical Systems and Chao* Springer-Verlag, New York, 1990.

[184] J. C. Willems. *The Analysis of Feedback Systems*. MIT Press, Cambridg Mass., 1971.

[185] J. C. Willems. Dissipative dynamical systems, part I: general theory. *Arch. Rat. Mech. Anal.*, 45:321–351, 1972.

[186] J. L. Willems. The computation of finite stability regions by means of open Lyapunov surfaces. *Int. J. Control*, 10(5):537–544, 1969.

[187] H. H. Woodson and J. R. Melcher. *Electromechanical Dynamics, Part I: Discrete Systems.* John Wiley, New York, 1968.

[188] F. F. Wu and C. A. Desoer. Global inverse function theorem. *IEEE Trans. Circuit Theory*, CT-19:199–201, March 1972.

[189] T. Yoshizawa. *Stability Theory By Liapunov's Second Method.* The Mathematical Society of Japan, Tokyo, 1966.

[190] D. C. Youla. On the factorization of rational matrices. *IRE Trans. Information Theory*, IT-7:172–189, 1961.

[191] J. Zaborszky, G. Huang, B. Zheng, and T.-C. Leung. On the phase portrait of a class of large nonlinear dynamic systems such as the power system. *IEEE Trans. Automat. Contr.*, 33(1):4–15, January 1988.

[192] G. Zames. On the input–output stability of nonlinear time-varying feedback systems, part I. *IEEE Trans. Automat. Contr.*, AC-11(2):228–238, April 1966.

[193] G. Zames. On the input–output stability of nonlinear time-varying feedback systems, part II. *IEEE Trans. Automat. Contr.*, AC-11(3):465–477, July 1966.

[194] G. Zames. Feedback and optimal sensitivity: model reference transformations, multiplicative seminorms, and approximate inverses. *IEEE Trans. Automat. Contr.*, AC-26(2):301–320, April 1981.

[195] K. Zhou, J. C. Doyle, and K. Glover. *Robust and Optimal Control.* Prentice Hall, Upper Saddle River, New Jersey, 1996.

Symbols[1]

$\equiv$	identically equal
$\approx$	approximately equal
$\stackrel{\text{def}}{=}$	defined as
$\forall$	for all
$\in$	belongs to
$\subset$	subset of
$\rightarrow$	tends to
$\Rightarrow$	implies
$\sum$	summation
$\prod$	product
$|a|$	the absolute value of a scalar a
$\|x\|$	the norm of a vector x (58)
$\|x\|_p$	the p-norm of a vector x (58)
$\|A\|_p$	the induced p-norm of a matrix A (59)
max	maximum
min	minimum
sup	supremum, the least upper bound
inf	infimum, the greatest lower bound
R^n	the n-dimensional Euclidean space (58)
B_r	the ball $\{x \in R^n \mid \|x\| \le r\}$
$\overline{M}$	the closure of a set M
∂M	the boundary of a set M
$\text{dist}(p, M)$	the distance from a point p to a set M (114)
$f : S_1 \rightarrow S_2$	a function f mapping a set S_1 into a set S_2 (60)
$f_2 \circ f_1$	the composition of two functions (61)
$f^{-1}(\cdot)$	the inverse of a function f (61)
$f'(\cdot)$	the first derivative of a real-valued function f (61)

[1] The page where the symbol is defined is given in parentheses.

$D^+ f(\cdot)$	the upper right-hand derivative (651)
∇f	the gradient vector (62)
$\frac{\partial f}{\partial x}$	the Jacobian matrix (62)
$\text{diag}[a_1, \ldots, a_n]$	a diagonal matrix with diagonal elements a_1 to a_n
$\text{block diag}[A_1, \ldots, A_n]$	a block diagonal matrix with diagonal blocks A_1 to A_n
$A^T \ (x^T)$	the transpose of a matrix A (a vector x)
$\lambda_{\max}(P) \ (\lambda_{\min}(P))$	the maximum (minimum) eigenvalue of a symmetric matrix P
$P > 0$	a positive definite matrix P (103)
$P \geq 0$	a positive semidefinite matrix P (103)
$\text{Re}[z]$ or $\text{Re } z$	the real part of a complex variable z
$\text{Im}[z]$ or $\text{Im } z$	the imaginary part of a complex variable z
$\bar{z}$ or z^*	the conjugate of a complex variable z
Z^*	the conjugate transpose of a matrix Z
$\text{sat}(\cdot)$	the saturation function (73)
$\text{sgn}(\cdot)$	the signum function (454)
$O(\cdot)$	order of magnitude notation (315)
$\diamond$	designation of the end of theorems, lemmas and corollaries
$\triangle$	designation of the end of examples
$\square$	designation of the end of proofs

Index

Corollary Index[1]

Lemma Index

Theorem Index

---

[1] 3.1 (116) means Corollary 3.1 appears on page 116. The Lemma and Theorem Indexes are written similarly.